Epigenetics Territory and Cancer

Parvin Mehdipour
Editor

Epigenetics Territory and Cancer

Springer

Editor
Parvin Mehdipour
Department of Medical Genetics
School of Medicine, Tehran University
of Medical Sciences
Tehran
Iran

ISBN 978-94-017-9638-5 ISBN 978-94-017-9639-2 (eBook)
DOI 10.1007/978-94-017-9639-2

Library of Congress Control Number: 2014955196

Springer Dordrecht Heidelberg New York London

Printed on acid-free paper

Springer is part of Springer Science+Business Media (www.springer.com)

This Book is devoted to:
Cancer Care Taker, Cancer Activists.Cancer Globe.
The patients who deserve care.Those who create a harmonic, influential and multi-directional platforms for cancer management, flavored with LOVE and CARE.
This Book is also dedicated to those who:
Have contributed a discovery to the cancer world.Were cancer tracer.Globally, share their talent, ability and ideas with cancer media.Love all about cancer.

Preface

Cellular and molecular diversity are, partly, the roots of differences in different populations who are the trigger points of Genetic and Epigenetic alterations. However, it seems that cancer is a Cellular/Molecular-based disease.Cancer progression is partly due to the key role of different genes, methylation of tumor suppressor- and tumor related genes, and gene silencing which play crucial role at early or late stages of tumorigenic process. As the matter of fact methylation alteration may be affected by cellular miss-behavior or has influential capacity on tumorigenesis. However the main task in cancer management is to unmask the involved networks and the interactive statue of different genes to achieve the appropriate methylome based therapy.

The present book provides 17 chapters including 4 major sections as the fundamental aspects(Chaps. 1–5), brain (Chaps. 6–9), breast (Chaps. 10–12), sporadic section (Chaps. 13–16), and final mini-Chap. 17 entitles Essence of Cancer Epigenetic.

Chapter 1 is presentative of fundamental aspects of cancer epigenetic.

In this chapter different mechanisms at cellular and molecular levels are highlighted. In fact, this chapter reflects the multi-directional behaviour. The authors of chapter 1have explored the fundamental facts, required techniques and the essential instructions either in research or in diagnosis. In addition, they have linked constitutional chromosome aberrations to the epigenetic changes. Moreover, the epigenetic based therapeutic approaches are discussed.

Chapters 2–5 reflect the application aspect of epigenetic through which the translational insight may be immerged:

In **Chapter 2**, the circulating tumor cells (CTCs) and their characteristics through migration have been explored. Different aspects of Epigenetic alterations including silencing of tumor suppressor gene, metastatic suppressors genes, functional events and micro -environmental factors are presented. The aim of authors was to bridge between CTCs and personalized therapy by especial focusing on the role of miRNA in CTCs which may lead to therapeutic innovation.

Chapter 3 deals with Retrotransposons, especially LINE-1 elements in cell biology and cancer cell biology. However, the key points comprise diversity, variation and evolution.

Chapter 4 presents the essential role s of miRNA within biological territory and mechanisms in variety of cancers. The main aim of this chapter is introducing the ways towards innovation of epigenetic drugs.

Chapter 5 provides a platform through which soul and body interact according a programming system in individuals' life. Cancer is a cellular/molecular/psychosomatic disease which relies on diversity, variation, heterogeneity and evolution. The main aim of this chapter was bridging Schizophrenia (SCZ) to different cancers in human, at cell line level and in animal models.

In **chapter 6**, the Ataxia-telangiectasia mutated (ATM) gene as a master molecular target, its functional protein and its interaction with some key cellular/molecular targets are explored.

ATM as a magic- gene is present within divers territories in our body. However, ATM is, effectively, involved in both malignant and non-malignant diseases.

In **chapter 7,** the basic insights of MCPH1 gene, its protein and its role in cancer and selected non-cancerous diseases have been explored. Different molecular and functional aspects of MPCH1 in various neoplastic disorders including brain tumors have provided the peculiarities of this gene in diseases.

Chapter 8 presents *P53* gene as a remarkable Antitumorigenesis in different cancers. However, a cross talk between *p53* promoter methylation and its protein expression in brain tumors is required to be well defined. In this chapter the fundamental aspects of *p53* gene and the status of methylation are explored.

In **Chapters 9**, the predictive role of O^6-methylguanine DNA methyltransferase (MGMT), as a therapeutic tool in brain tumors, especially in malignant gliomas; and the methodological requirements are explored. The clinical managements are also discussed.

Chapter 10 deals with the Epigenetic alterations, diagnosis, tumor classification/ prognosis and treatment. Three remarkable receptors including estrogen receptor (ER), PR, and HER2/neu with diverse functions and impacts on breast cancer progression were discussed.

In **chapter 11,** Retinoic acid receptor gene beta 2 (RARβ2) as a hero in cancer, and manner of its cooperation have been explored.

Interactions of epigenetics with environmental factors, including nutrition, the role of chemopreventive agents in epigenetic, cancer stem cells, the main target receptors and genes, protein expression, miRNA and the therapeutic insight of *RAR*β are presented. Furthermore, detection of the methylated *RARβ2* in primary breast cancer is a key task to evaluate tumors with a positive responsiveness capacity to RA therapy.

Chapter 12 reflects the practicability of translational approach in Retinoic acid receptor-β at a glance. By considering the methylation status, the importance of genetic factors at a triangle level including DNA, RNA and protein; and the bridging system between functional and clinical insights are challenged. Hopefully RAR-β will be considered as an influential target in cancer prevention and therapy.

Chapter 13 deals with Methylation in the Colorectal Cancer. The interaction between hazard environmental factors and colorectal cancer (CRC) is highlighted. In this chapter the authors have focused on histhological progression and evolutionary pattern as well. The molecular characteristics and the mode of hypermethylation of involved genes are presented. The authors have emphasized on the key role of epigenetic on our life. The final aim in this chapter is the translational paradigm.

Chapter 14 is aimed on the epigenetic mechanism of tumorigenesis in Malignant Rhabdoid Tumor (MRT). In this chapter, the cellular senescence, apoptosis, mitotic control genes, pre-replication complex, chromatin behavior, cell proliferation and differentiation are discussed.

Chapter 15 presents Epigenetics of Thyroid Cancer in which the thyroid related genes have been explored. Besides the characteristics of the candidate genes in thyroid cancers have been discussed.

Chapter 16 presents an introduction to Bio-Energy and Bio-Resonance technology by linking this paradigm to Genetic and Epigenetic. This chapter is reflective of new insight on Bio-mechanical resonance within the cells, tissues and molecules. This chapter focuses on epigenetic modifications and transposans as well. In addition, an interesting link between environmental hazards, diverse electric oscillation in chromosome and epigenetic statue of organism has been also highlighted.

Chapter 17 presents an evolutionary insight in cancer epigenetic and was aimed to apply a bridging system between the provided chapters and also pave the way towards the more complementary manner for the future planning.

The interactive regulatory system characterizes and differentiates epigenetic as an extraordinary molecular biological territory. I do believe that, epigenetic is beyond DNA methylation, RNA interference and histone modification in cancer development, therefore, I aimed to provide three modeling schemes including (1) 'The map of cancer development'; (2) Epigenetic Programming; and (3) Diverse-methylation process, target and function at different ages.

The bridging system between Science and Medicine was the main design of this book which has been, mainly, edited on the basis of research paradigms. In this regard, I do hope that this book, would be useful as an educational and translational research package for Scientific and medical medias. I also appreciate receiving your comments.

Sincerely
Parvin Mehdipour

Contents

Chapter 1
Fundamental Aspects of Epigenetic in Cancer

Thomas Eggermann, Ulrike Gamerdinger and Gesa Schwanitz

Contents

G. Schwanitz (✉)
Institut für Humangenetik UKB, Sigmund-Freud-Str. 25, 53125 Bonn, Germany
e-mail: Gesa.Schwanitz@ukb.uni-bonn.de

T. Eggermann
Institut für Humangenetik RWTH, Aachen, Germany

U. Gamerdinger
Institut für Pathologie, Universität Gießen, Gießen, Germany

P. Mehdipour (ed.), *Epigenetics Territory and Cancer*,
DOI 10.1007/978-94-017-9639-2_1

Abstract The term epigenetics is defined as the interaction between genes and environment without an alteration of the DNA sequence. On molecular level epigenetics is based on the mechanisms of DNA methylation, RNA interference and histone modification (e.g. acetylation, methylation).

1.1 Introduction

The term epigenetics is defined as the interaction between genes and environment without an alteration of the DNA sequence. On molecular level epigenetics is based on the mechanisms of DNA methylation, RNA interference and histone modification (e.g. acetylation, methylation).

Epigenetic processes regulate gene expression and thereby cell differentiation in the process of embryogenesis, but also in the postnatal periods, and they cause abnormalities in higher age. These gene regulation processes can only proceed correctly if the transcription occurs at the right time in a defined gene locus. Thereby chromatin packing can show either an open or a closed configuration. The epigenetic gene regulation can reveal a DNA packing which is inherited over a high number of cell divisions in the different somatic cell systems.

Both somatic as well as germ cells can develop alterations of the epigenetic pattern and they can inherit these changes to their offspring. Tumor diseases show dysregulation of the epigenome as a relevant mechanism which leads to development and progression of the tumor. These epigenetic abnormalities can be reversible, in contrast to mutations in the tumorgenome which develop in parallel.

Depending on the type of aberrant DNA methylation and the altered histone modification the frequency of tumor development and the type of carcinoma can differ significantly. Recent investigations have analysed in which way genetic and epigenetic factors interact during tumor development. By the application of recently developed techniques an increasing number of relevant regions of the tumor methylome has been characterized and compared to type and extend of DNA mutations in the same region. These investigations are not restricted to basic research on the epigenome but it is also a relevant aim to gain informations which lead to improved possibilities of tumor therapy.

1.2 Historical Background

The phenomenon of a regulated gene expression was already known for a long time, before the term "Epigenetics" was introduced (Berger et al. 2009).

This special field of genetics covers the mechanisms which lead to regulatory processes of gene expression and maintain their degree of expression. Epigenetics furthermore includes the investigation how this condition is inherited in the course of cell divisions and finally how the interaction of the epigenetic pattern and the human genome works (Eggermann et al. 2013).

Epigenetic processes which lead to a change of genetic activity and inactivity have meanwhile been documented for all phyla of animals (Lyon 1974):

In the communities of social insects the different development of the animals (queen, worker, and soldier) is exogenously introduced by different feeding of the larvae and in single cases it is even reversible. In reptiles, as for example in crocodiles, sex determination is introduced by the temperature to which the fertilized egg is exposed.

Analyses of larvae of different insects showed that in the course of their ontogenetic development the regions of genetic activity change their position on the chromosomes in a characteristic order. This procedure becomes visible by the changing puff pattern of giant chromosomes (Chironomus, Drosophila).

The course and the intensity of gene activity are genetically determined, but they can be influenced exogenously by different physical and chemical factors. In mammals the first phenomenon being analysed in detail was the inactivation of the second X-chromosome in females. This inactivation is essential for dose compensation for genes localised on the X-chromosome and this regulation is also taking place in all cases of hyperploidy of the X-chromosome. It could be demonstrated that the second X-chromosome in females is inactivated at the beginning of ontogenesis and becomes facultative heterochromatic and late replicating in the cell cycle.

Furthermore it could be shown that from two normal X-chromosomes the paternal and the maternal one are inactivated at random and that all following cell divisions keep the pattern of their original cell. This ontogenetic peculiarity was first analysed from the coat patching in mice and it was defined as "Lyon hypothesis" according to its first description (Lyon 1961). In the following the phenomenon was also analysed more precisely in females with mutations in X-chromosomal recessive genes (i.e. Duchenne muscular dystrophy, anhydrotic ectodermal dysplasia, Martin-Bell syndrome, red-green blindness). If the second X-chromosome is structurally aberrant a screwed inactivation of the mutated X-chromosome takes place, and in X-autosomal translocations the normal X-chromosome is usually inactivated (Miller and Therman 2001; Gardner et al. 2012).

In interphases the inactivated X-chromosome shows a specific morphology and has been described as "Barr-body" or "drumstick" in different somatic cell systems. This X-chromatin was first described in the cat and subsequently in all mammals (Barr and Bertram 1949).

Experiments in mice and the investigation of uniparental human conceptus lead to the conclusion that the normal development of an embryo requires the combination of a maternal and paternal genome of which each develops a specific but different imprinting pattern (Mc Grath and Solter 1983).

Recent observations lead to the assumption that the differences between monozygotic twins can be delineated from differences in gene activation and deactivation which are caused by different conditions of intrauterine development (Galetzka et al. 2012; Engel and Antonarakis 2002). Furthermore, it has been documented that the normal imprinting pattern in early ontogenesis can be defective after assisted reproduction (for review: Amor and Halliday 2008).

In tumor cells the change of the epigenetic process can lead to cell transformation (Haaf 2006; Brena and Costello 2010). Epigenetic changes, their localization, the processes leading to their development, and the characterisation of their functional consequences are currently a relevant research topic in basic and applied science (Amor and Halliday 2008). In tumor genetics epigenetic aberration play an increasing role in the course of diagnostic investigations (Quante 2012).

It could be demonstrated that the best way of tumor characterisation includes both cytogenetic and molecular strategies. This combination can either take place as a cascade or the different techniques are applied parallel (Schwanitz and Raff 2005).

Therefore in the present publication the different aspects of relevant investigation methods will be described in detail.

1.3 Basic Mechanisms of Epigenetics

The cellular differentiation in multicellular organisms is based on the fine-tuned expression of thousands of genes. As nearly all cells in an organism carry the same genetic content but different cellular functions are displayed, cell specific regulation mechanisms of gene expression is needed. Epigenetic processes play a key role in these complex mechanisms, as they control packaging and function of chromatin and regulate gene expression without changing the DNA sequence itself. On the molecular level, epigenetic regulation includes specific patterns of DNA methylation, chromatin structure, and by post-translational histone modifications such as acetylation, ubiquitylation, phosphorylation and methylation, non-coding RNAs (ncRNAs) (for review: Delaval et al. 2006; Kacem and Feil 2009; Tollefsbol 2011).

Indeed, DNA methylation is one major form of epigenetic modification (Fig. 1.1). DNA methylation refers to the covalent addition of a methyl group to the C-5 atom of cytosine. This methylation is catalysed by several DNA methyltransferases which establish methylation marks in development and maintain it during later cell divisions. Cytosine methylation preponderantly occurs in so-called CpG islands (CGIs): These CpG-rich DNA sequences are often found at the transcription start site of genes (Cooper et al. 1983; Bird et al. 1985) where they coincide with promoter regions of 70 % of the human genes (Saxonov et al. 2006). The methylation of CGIs imposes transcriptional silencing which is transmitted by clonal inheritance in somatic cells. In general, the methylation of promoter-related CGIs causes gene silencing.

In mammals, DNA methylation stably alters the gene expression patterns in cells as the basis for a proper and orchestrated realisation of genetic information during development and cell differentiation. Furthermore, it contributes to genome stability, parent-of-origin specific expression of imprinted genes and X-chromosome inactivation in female organisms.

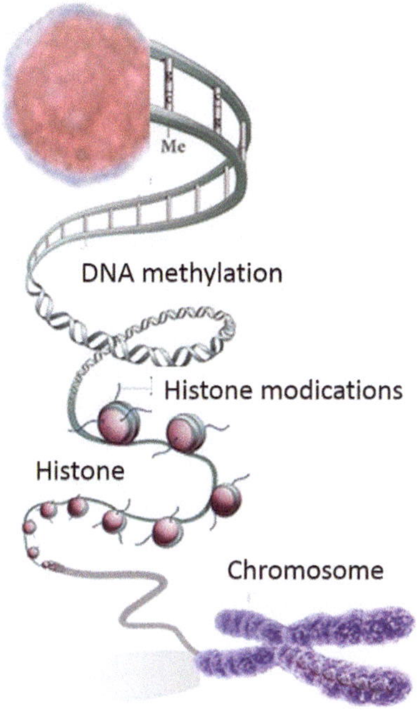

Fig. 1.1 Illustration of the different molecular processes involved in epigenetic mechanisms. (From Qiu 2006; with kind permission of the publisher).

As a result, studying the methylation status of CGIs in a mammalian genome is of major interest for deciphering the regulation mechanisms of gene expression. However, a major challenge is that there is only one genome in one individual, but hundreds of epigenomes as DNA methylation changes during development and is influenced by disease processes and environment (for review: Zhang et al. 2010).

Alterations of specific chromosome regions play, according to recent investigations, a relevant role in the origin and evolution of tumor cells.

In the aberrant tumor karyotype the high amount of intra- an interchromosomal rearrangements can lead to the formation of neocentromeres. This process is regulated by epigenetic mechanisms. The DNA remains unchanged by it has gained an alteration in function.

Furthermore, the centromere of one or more chromosomes might become unstable. This can be caused by hypomethylation of satellite DNA II and III.

Finally, aneuploidisation of tumor cells is a characteristic alteration in cancer. One of the processes which lead to this abnormality is a defect in the checkpoints which regulate the course of the cell cycle.

The Chromosomal Region 11p15 as an Example of an Epigenetically Regulated Segment The complex spatial and temporal expression of genes regulated by epigenetics can be illustrated by a more detailed description of the genomic region 11p15. The region belongs to the group of chromosomal loci underlying the so-called genomic imprinting, i.e. the parent-of-origin expression of gene-copies. Indeed, inborn imprinting markers are clustered, and approximately 60 imprinted

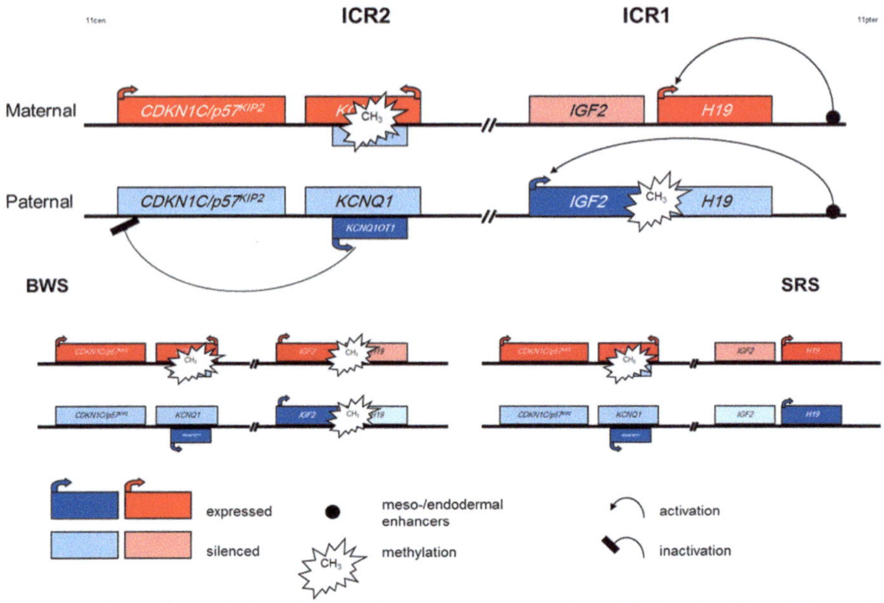

Fig. 1.2 Epigenetic regulation of the two imprinting centre regions (ICR) in 11p15 and illustration of the hyper-/hypomethylation of the ICR1 observed in BWS and SRS as an example for epigenetic regulation in Imprinting Disorders. (for a more detailed description of the complex interactions in these regions. (Choufani et al. 2013)

clusters/gene regions have been estimated in the human genome. Among these clusters, the region 11p15 is of central significance as it harbours genes encoding key factors in human development and growth.

The 11p15 region contains a number of imprinted genes whose expression is regulated by two different imprinting centres (ICR1 and ICR2) (for review: Choufani et al. 2013) (Fig. 1.2). The telomeric ICR1 confers a differential chromatin architecture to the two parental alleles leading to reciprocal expression of *H19* and *IGF2*. The two genes are coexpressed in entoderm- and mesoderm-derived tissues during embryonic development and compete for the same enhancers. The paternally expressed *IGF2* is a single chain polypeptide and is involved in fetal development and growth. The *H19* gene is realised from the maternal allele and encodes a 2.3 kb non-coding mRNA which is strongly expressed during embryogenesis. The physiological function of *H19* is unknown but a role as a primary micro RNA in post-transcriptional downregulation of specific mRNAs during vertebrate development has been postulated. The ICR1 contains seven CTCF target sites (CTCF1-CTCF7) in the DMR 2 kb upstream of H19 and shows allele specific methylation. The Zinc-finger binding factor CTCF binds to the maternal unmethylated ICR1 copy and thereby forms a chromatin boundary. This CTCF binding mechanism blocks *IGF2* and promotes *H19* transcription of the maternal 11p15 copy.

The centromeric ICR2 regulates the (reciprocal) expression of *CDKN1C*, *KCNQ1* and further genes and is methylated only on the maternal allele. Mutations in the paternally suppressed *CDKN1C* gene account for up to 40 % of familial Beckwith-Wiedemann syndrome (BWS) cases and 5–10 % of sporadic BWS patients. The gene encodes a cyclin dependent kinase inhibitor (p57KIP2) and is part of the p21CIP2Cdk inhibitor family. Its′ product is expressed in a tissue-specific manner with a 1.5 kb mRNA transcript in the placenta and at lower levels in other tissues. *CDKN1C* is a mainly maternally expressed growth inhibitor gene that negatively regulates cell proliferation. Overlapping with intron 9, exon 10 and intron 10 of the *KCNQ1* locus the untranslated *KCNQ1OT1* (*LIT1*) RNA is encoded. *KCNQ1OT1* is expressed by the paternal allele and probably represses realisation of the *CDKN1C* gene. Loss of methylation of the maternal ICR2 allele correlates with expression of *KCNQ1OT1*.

Disturbed Epigenetic Regulation *Genomic Imprinting—a Parent-of-Origin Specific Regulation of Gene Expression*

As briefly mentioned before, the term genomic imprinting describes an epigenetic marking of specific genes that allows expression from only one of the two paternal alleles (for review: Reik and Walter 2001). So far, more than 60 human genes are discussed to be imprinted by epigenetic mechanisms but probably there are much more (for review: Horsthemke 2010). The imprinting marks are inherited from the parental gametes and are then maintained in the somatic cells of an individual. Their programming is subject to a so-called "imprinting" cycle during life which leads to a reprogramming at each generation: During early development, methylation of the mammalian genome runs through dramatic changes and is linked to the rapid differentiation and formation of the various tissues and organs. The imprint marks are erased in the germ-line and re-established according to the sex of the contributing parent for the next generation.

As aforementioned, genes regulated by genomic imprinting mechanisms tend to cluster, thus the imprinting control is often not restricted to a single gene at an imprinted locus but affects the expression of several factors. Due to the numerous factors involved in these complex mechanisms, the balanced regulation of imprinted genes is prone to different disturbances, and indeed several disorders associated with altered genomic imprinting are known, belonging to the group of congenital imprinting disorders (IDs).

Currently eight different imprinting disorders have been reported, among them Angelman and Prader-Willi syndrome. Indeed, the clinical features are different in these diseases, but many clinical as well as molecular overlaps have been reported. Furthermore, similar molecular alterations can be observed in all imprinting disorders:

a. Aberrant methylation patterns at the differentially methylated regions
b. Chromosomal duplications or deletions
c. Point mutations in specific genes (e.g. CDKN1C in Beckwith-Wiedemann and UBE3A in Angelman syndrome)
d. Uniparental Disomy (UPD).

For the majority of imprinting disorders, patients with uniparental disomy were extremely valuable for the identification of regions harboring imprinted genes. Additionally, it has meanwhile become apparent that UPD can also occur in tumorigenesis.

Disturbed Epigenetic Regulation: 11p15 Alterations as an Example Causing Congenital Imprinting Disorders For the aforementioned imprinted region 11p15, two syndromes have been identified to be associated with molecular alterations in that region (Fig. 1.2): Beckwith-Wiedemann syndrome and Silver-Russell syndrome (BWS, SRS). Interestingly, the two syndromes show molecularly as well as clinically opposite features: Whereas BWS patients show overgrowth, the main clinical sign of SRS is growth retardation. On the molecular level, different types of mutations and epimutations affecting either the ICR1 or the ICR2 in 11p15 can be observed: In the ICR1, hypermethylation of the *H19* promotor and loss of imprinting of *IGF2* are detectable in 2–7 % of BWS patients while loss of methylation in the same region is a frequent finding (>40 %) in SRS. Nearly 50 % of BWS patients show hypomethylation in the ICR2. In 1–2 % of both syndromes, either paternal (BWS) or maternal duplications (SRS) are detectable. Paternal uniparental disomy of 11p15 accounts for 20 % of BWS patients, in SRS only single carriers of maternal uniparental disomy have been reported. In BWS, a genotype-phenotype correlation exists reflecting the functional consequences of the different molecular alterations, whereas in SRS it is too early to define such a correlation (for review: Engel and Antonarakis 2002).

Epigenetics in Cancer Aberrant methylation patterns are a well-established cause of cancer (Esteller 2008): hypermethylation of CGIs localized in tumor suppressor genes result in silencing of these genes, whereas hypomethylation leads to activation of oncogenes. Additionally, the well balanced expression of ncRNAs in the affected region is disturbed. Thus, epigenetic mechanisms control cell fate by maintaining a delicate balance between stability and susceptibility to developmental and environmental stimuli. These characteristics make them highly promising targets for molecular diagnostics and drug discovery: epigenetic biomarkers are highly compatible with clinical diagnostic procedures, and they are increasingly used for informing therapeutic decision-making as well as for suitable and personalized therapies for individual patients (Bock 2009). More details on epigenetic changes as cause of tumor development are given in the following chapter

Loss of Heterozygosity The aforementioned mechanism of unipaterenatal disomy (UPD) is one possible cause of Loss of Heterozygosity (LOH), a molecular finding which is a characteristic feature of cancer. LOH in general is defined by the functional loss of an allele of a gene of a cell (e.g. a tumor suppressor gene), in which the other allele has already been silenced. In tumor cells copy-neutral LOH can be biologically equivalent to the second hit in the Knudson hypothesis (Knudson 1971).

Further Examples of Epigenetic Disturbances For mutations responsible for the fragile X (FraX) syndrome it has been suggested that they interfere with the process of X-chromosome reactivation in oocytes, thus blocking the transcription of loci at or neighboring the fragile site (Xq27.3) and producing the clinical FraX phenotype. However, this hypothesis could not yet been confirmed.

Furthermore, a contribution of altered methylation patterns to the pathoetiology of chromosomal disturbances by gene silencing of proteins interfering with heterochromatin has been discussed (for review: Busson-Le Coniat et al. 1999).

1.3.1 Cell Cycle and Its Abnormalities in Cancer Cells

Changes in the course of the cell cycle are a characteristic feature of tumor cells and they cause,—depending on the type of disturbance—a broad spectrum of aberrations.

The course of the cell cycle is even in non-transformed meristematic soma cells not identical in all cells but shows tissue-specific peculiarities. But even one specific cells system as the T-lymphocytes of the peripheral blood show differences when comparing different probands caused by the individual genotype or the age. The somatic cell cycle is subdivided into two main subphases that is mitosis and interphase which are classified by their function. In mitosis these are prophase, metaphase, anaphase and telophase, and in interphase these are G1-, S-, and G2. Between the single phases the so-called "checkpoints" are interspersed, which regulate the progression from one phase to the next one.

Furthermore, the cell cycle is characterized by the duplication of the genome (in the S-phase), by the distribution of the chromosomes into two daughter cells in the telophase, and centriol-cycle which regulates the course of mitosis.

The G1-phase of the cell cycle is composed of one chromatid which is replicated during S-phase. This is the phase which leads to the majority of aberrations of tumor cells by the development and evolution of cells with aneuploid or polyploid karyotype.

After the end of the replication the transition into G2-phase can be blocked in tumor cells. In the normal course of the cell cycle this transition is induced by Cyclin and regulated by specific Cyclin-depending kinases. These proteins and their underlying genes are of general relevance in tumor development as they can function as oncogenes or they can inactivate tumorsuppressor genes.

If thus, the transition from S- to G2-phase is not performed correctly, S-phase is passed through a second or even several times, and the resulting karyotype will be polyploid by endoreduplication (4n to 8n).

The abnormal regulation of gene expression during the transition from S- to G2-phase is caused by the missing phosphorylation of Cyclin E and the following degradation and thus leads to an increased replication of the tumor cell.

Aneuploidisation of tumor cells can also be caused by other mistakes in DNA replication.

For example, the centromeric regions of one or more chromosomes can develop instability by hypomethylation of satellite DNA II and III. The consequence is the multiple passing through of the S-phase by the altered chromosomes.

Aneuploidisation of tumor cells can lead to the development of drug resistance in therapy in cases of specific mutations.

Finally, by replication error partial duplications of chromosome segments can originate. The aberrant chromosomes present as triradials. Starting regions of the doubled replication are among others certain fragile sites of the autosomes.

Another phase of the cell cycle causing an increased development of karyotype aberrations is the anaphase of mitosis. The basis of the origin of the different types of genome mutations are abnormalities of the spindle apparatus. It is either multipolar or shows degenerative alterations. If it is not developed at all, in anaphase the two chromatids of each chromosome separate but they are not distributed to the poles of the cell. In the succession either a tetraploid restitution of the nucleus develops or multiple micronuclei are formed.

The development of the structure of the micronuceli is regulated in interphase and mitosis by the centrosome. Its replication is controlled by the protein p53 and when it is missing this leads to the loss of the centriols or their amplification.

Beside the mutations in the gene T53 alterations in the gene STK15 are of relevance, as this gene regulates the centrosome associated kinase and thus plays a role in the aneuploidisation of tumor cells.

A further season for the development of tumor cells caused by an abnormal course of the cell cycle is the missing development of a new cell membrane after the end of mitosis between the two restituted nuclei. This leads to the development of dikaryotic and subsequently even to polykaryotic cells, of which the nuclei can fuse to a polyploidy genome.

A special type of polyploidisation of tumor cells is caused by endomitosis or amitosis, where the replication of the DNA shows a normal course in interphase but the checkpoint from G2 to mitosis is blocked (Cyclin B/CDK1). The membrane of the nucleus is not dissolved and the genome becomes tetraploid.

This pathologic course can be repeated several times and leads to highly polyploid cells.

A characteristic feature of tumor cells is the high instability of the genome. The aberrations which develop are inherited in different frequencies to their daughter cells, depending on their influence on the vitality and the specific length of the cell cycle.

Each pathologic cell can develop into an aberrant cell clone and leads to a “main line” or “side line” of the composite karyotype of a tumor.

Furthermore tumor cells are characterised by an increased number of secondary chromo-some aberrations such as gaps, breaks and exchanges. They are caused by defects in DNA repair. When developing in G2-phase, they present mainly as chromatid aberrations, in G1 phase as chromosome or isochromatid aberrations.

By secondary fusion of heterologous segments in the reactive breakpoints different types of derivative chromosomes can develop.

The development of tumor cells induced by mutations in oncogenes or tumor-suppressor genes can be caused by an increased or reduced production of gene products which then lead to a dysregulation of the cells

Often the cell cycle is significantly reduced in time and this can lead to dramatic increase of specific pathologic karyotypes.

As a rule it can be observed that the cell cycle of tumor cells is changed by different simul-taneously occurring structural and epigenetic mutations which lead to an aberrant course and explain the heterogeneity and the instability during the evolution of the tumor genome (Miller and Therman 2001).

1.4 Tumor Selection and Investigation Methods

Overview on techniques currently used in routine diagnostics in epigenetic disorders. As a huge number of techniques has been reported we can present only those procedures widely applied by many laboratories

1.4.1 Sample Collection and Transportation of Tumor Tissue

Special care is required when taking biopsies, on their transport and their storing, so that their vitality will not be reduced and the necessary investigations can be successfully performed without any restrictions. During the transport of the samples the right temperature (10–20 °C) and pH value must be guaranteed, a suitable buffer must be added to the medium as well as antibiotic and antimycotic solutions.

Before the first use of a new type of transport container they must be tested to exclude that cytotoxic substances are excreted from the plastic into the medium.

Biopsies of tumor probes always afford a maximum of sterile working.

Mailing of tumor samples is possible but the time period should not exceed 3 days. After their arrival at the investigation center they can additionally be kept for some days in the fridge.

If the sample is large enough part of it can be stored by cryoconservation.

Hematologic neoplasias are analysed from peripheral blood samples and bone marrow biopsies, these methods of investigation are well established (Wegner 1999) and usually chromosome analyses are combined with investigations in interphase nuclei by FISH.

1.4.2 Course and Type of Investigations

Vital Biopsies Depending on the amount of material of the solid tumor sample one or more investigation methods can be chosen.

If the tumor sample is very small (less than 5 mg) or if the tissue shows already degenerative alterations the direct preparation of the cells has to be preferred.

The tumor cells can be analysed by Interphase-FISH (see1.4.4) choosing DNA-probes relevant for the known changes of the tumor type investigated.

Further preferential techniques for the characterization of the tumor are microarray analyses (see 4.5), MLPA (see 4.6) and the different methods for documenting methylation defects (see 1.4.5, 1.4.6, 1.4.7).

If a higher amount of tumor material is available, chromosome analyses are also applied after direct preparation and long term cultivation of the cells. The best quality of the chromosomes for analysis is received after cell culture. Besides, special FISH and MS-MLPA investigations can be performed.

Cryconservation of Tumor Biopsies The temperature for conservation is − 80 or − 196 °C in the majority of cases. In the latter case the tissue can be stored for several years.

After defrosting the vitality of the cells usually is only slightly reduced and the desirable methods can be applied.

Paraffin Embedded Material After standard formalin fixation of the tumor tissue it is dehydrated and then embedded in paraffin. If the tumor is small and surrounded by normal somatic tissue the biopsies are cut, transferred on slides, and the tumor tissue is then prepared under microscopic control. Usually histologic investigations have been performed before and add a detailed characterisation of the tumor tissue. Then paraffin is removed and DNA is extracted.

1.4.3 Cell Preparation and DNA Extracts

In Situ Analyses Interphases and in vivo occuring mitoses are prepared after short time incubation and adding of colcemid as a block of mitoses.

In special problematic cases it might be useful to determine the spontaneous mitotic activity (mitotic index). The number of mitoses available after this procedure is usually small and they show a reduced structural resolution (about 200–250 bands per genome; Shaffer et al. 2013).

Cell Cultivation In Vitro The sample should not be older than 3 days when the cultivation is started. The tissue can be incubated after enzymatic digestion or it is cut into small pieces (2 × 2 mm).

The culture condition (type or medium, time of incubation and additives have to be chosen depending on the type of tumor).

Note:

While the number of mitoses after long term cell culture is usually high and their quality (bands per genome) good, it has to be taken into account that often, during cultivation, a selection of single cell lines of the tumor occurs or even that normal somatic cells overgrow the tumor cells.

Documentation of Growth Anomalies Tumor cells show, compared to not transformed cells, a characteristic change of behaviour. They do not grow in vitro two-dimensional in parallel alignment but irregular, partially overlapping each overlapping each other (criss-cross-growth) and, caused by a missing contact inhibition the cell growth becomes more and more three-dimensional (Fig. 1.3, 1.4).

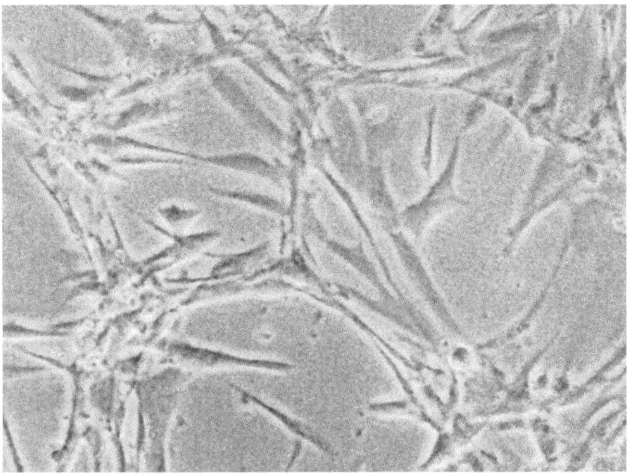

Fig. 1.3 Growth abnormalities in vitro after cultivation of tumor biopsies of a testicular tumor

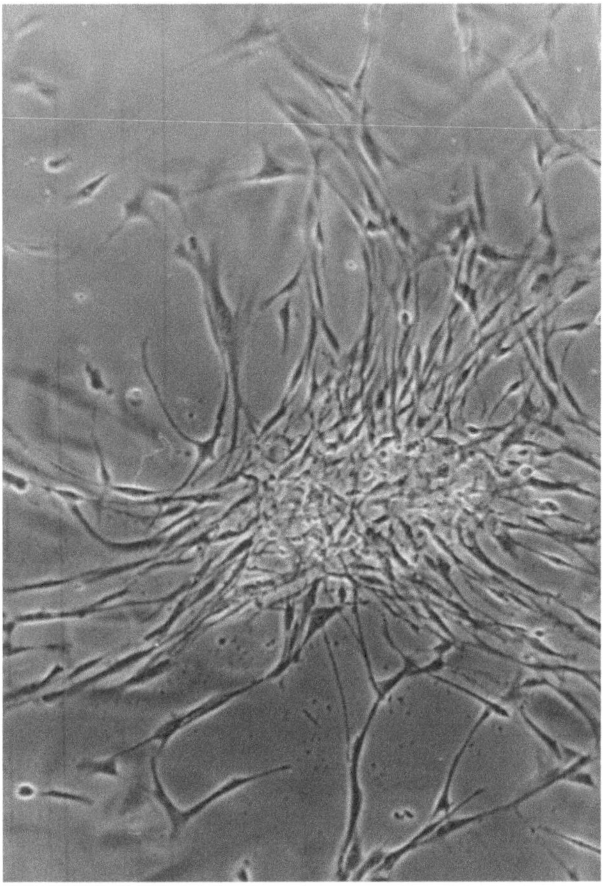

Fig. 1.4 Criss-cross growth of tumor cells by loss of two dimensional orientation of the cultured cells of a testicular tumor biopsy

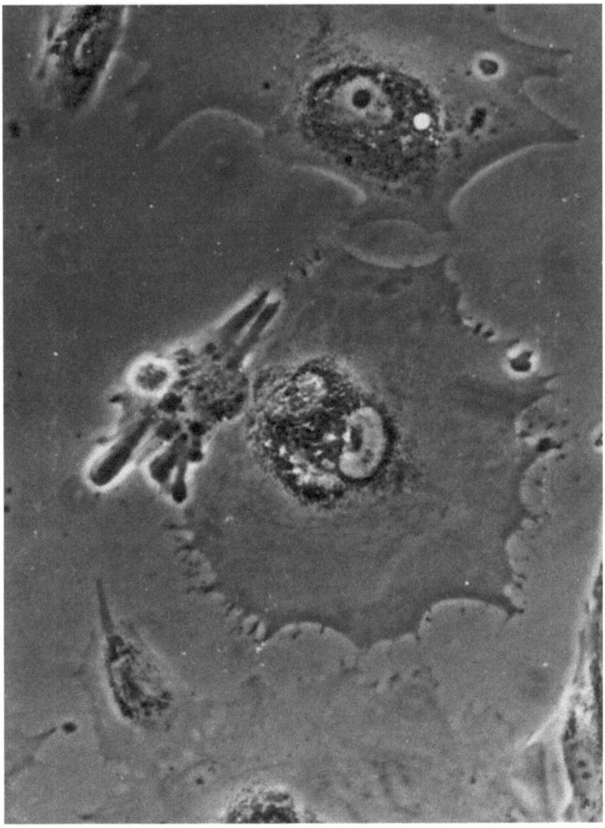

Fig. 1.5 Enlarged polykaryotic cell of kidney tumor with granular inclusions in the cytoplasm

The development and extend of these growth anomalies are a relevant parameter characterising tumor cells in vitro.

Characterization of an Abnormal Cell Morphology In vitro cultivation of tumor cells reveals a high variability of cell size, morphology and nucleus-cytoplasmic ratio. Often cells with long cytoplasmatic appendices are observed. There is an increased number of cytoplasmatic granulae espescially in the surrounding of the nucleus which are an indication for the abnormal metabolism of the cells.

Polykaryotic cells are frequent as well as cells with an abnormal morphology of the nucleus (formation of lobes, perforations, micronuclei). These different anomalies can be observed single or in different combinations and a quantitative documentation has proved to be useful (Fig. 1.5, 1.6).

Preparation of Interphases and Mitoses The preparation of cells in different stages of the cell cycle is done according to that of normal somatic cells (Wegner 1999).

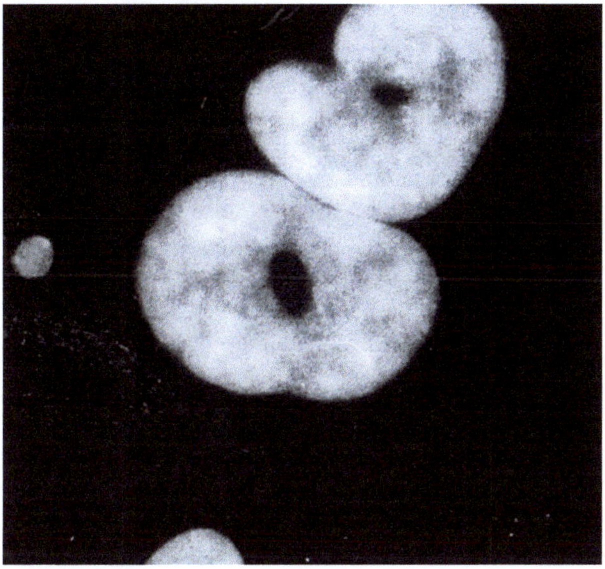

Fig. 1.6 Abnormalities of the nucleus morphology in a kidney tumor with pathologic structure and development of perforations

1.4.4 Analysis of Cells, Chromosomes and DNA

Structure and Function of the Interphase Nucleus in Normal Somatic and in Tumor Cells The combination of structure and function of the genome is defined as epigenome. This epigenome consists of different territories localised at defined areas of the interphase nucleus, each of them divided into domains the smallest entities of which are the genes (Haaf and Schmid 1991). The epigenome reveals different types of chromatin condensation and their special arrangements are of high relevance. In this context the celltype investigated as well as the age of the cell also have to be taken into consideration. In euchromatic regions the epigenome influences replication, transcription and repair mechanisms (Cremer 2010). In the course of the life cycle of a cell the epigenome passes through a number of alterations. As the structure of the nucleus is in tight connection with its function each change in structure has an influence on its function.

Similar to the euchromatin the constitutive heterochromatic regions show characteristic positions in the interphase nucleus. Usually these positions are in the periphery of the nucleus and have the tendency to form heterochromatin clusters. They are defined as chromocenters. As in the euchromatic regions the position and function of heterochromatin can differ in the various somatic cell types.

For example in the early pregnancy extra fetal cells are characterised by heterochromatin which is decondensed, hypomethylated and early replicating whereas in the differentiated placenta the same cell type is highly condensed, hypermethylated and late replicating. Furthermore, the centromeres show in the interphase nucleus a tendency to form clusters and besides a preferential position in the neighbourhood of the inner nucleus membrane. Epigenetic mutations in the centromeric regions influence the correct construction of the different kinetochore elements and thus the separation of the chromatids in the anaphase of the mitosis. Finally, even the telomeres show an increased association in interphases.

The different heterochromatic regions mentioned above have in many cases a relevant function in gene regulation and with reference to guaranteeing the stability of the genome. It could be demonstrated that the stability of the different genome regions is ensured in normal somatic cells by specific factors as the SNF2-family member Fun 30 (Neves-Costa et al. 2009).

In immortalised cells and in tumor cells the position of the territories in the nucleus is changed (Mai 2013; Wark et al. 2013). The genome shows a different grouping.

Subsequently, exemplary two types of heterochromatin mutations are given to show the relevance of these changes in the course of tumor evolution.

Changes in the length of the telomeres are known as a characteristic feature in the course of development of normal somatic cells into tumor cells. New techniques enabling 3-D-pictures of the tumor cells in interphase furthermore made it obvious, that telomeres play a role in gene regulations (Mai 2013; Wark et al. 2013). These dynamic regions show changes in their positions in the nucleus of solid tumors, and this change in position was partially reversible. It was the main topic of Mai and co-workers, to find characteristic changes of telomere position in the nucleus, which make it possible to delineate the development of normal somatic cells into precanceroses and finally to tumor cells.

Structural biomarkers can be investigated in vital tumor biopsies, in fixed tissue and in tissue sections.

The centromeres of the chromosomes are characterised by the histon H3 in the variant CenH3. In this chromosome region methylation is the condition for a correct synthesis of the kinetochores. If this reaction does not take place the consequence is a malsegregation of chromosomes (aneuploidisation) which is characteristic for a high number of tumors.

Chromosome Analyses Chromosome preparations are usually studied after GTG banding.

If peculiarities are observed in regions of constitutive heterochromatin special banding techniques are applied such as QFQ, CBG or DA/DAPI. The chromosomes are prepared in C-metaphase, this means that the contraction of the single chromosome has been increased and its position is no longer in the equatorial line but scattered over the whole volume of the cell (Wegner 1999).

Direct Preparations The chromosomes are highly contracted with a decreased pairing of the chromatids and an unequal spiralisation.

Partially the cells show a reduced reaction on the hypotonic treatment which leads to an increased overlapping of the chromosomes and thus makes the analyses even more difficult.

In general, the assessment of numerical aberrations is possible but that of structural anomalies is reduced.

Chromosome Investigations After Long Term Cell Culture Mitoses can be analysed according to the international quality standard (≥400 bands per genome; ISCN 2013). Thus, even small structural chromosome aberrations (balanced or unbalanced; Shaffer et al. 2013). can be registered, a "composite karyotype" can be delineated (summarising main and side lines).

A special type of structural chromosome aberration is the so-called complex chromosome rearrangement (CCR). By improvement of the investigation methods (FISH, array, next generation sequencing) it could be shown that the number of CCRs in tumor cells is much higher than originally suspected. They have an unequal inter- and intrachromosomal distribution, hotspots of aberration could be delineated and functional abnormalities of the cell were caused by the induction of changes in the methylation pattern.

The extent of the exchanged regions varies from kb to Mb with no preferential size. Investigations of interphase nuclei (see 1.4.1) showed that the nuclear architecture plays a relevant role in the frequency and types of interchromosomal exchanges. Furthermore, the new techniques enable the study of genomic rearrangements at high resolution. Thus hundreds of breakpoints could be characterised in detail.

Caused by these diagnostic improvements the phenomenon of chromothripsis could be defined (Molenaar et al. 2012). It shows in complex chromosome rearrangements changes even on the molecular level in gene expression caused by gene fusions, copy number changes, chromosome confirmation and even changes of the epigenetic program.

Note:

The frequencies of main and side lines of the karyotype can change during the course of cell cultivation or they can even arise de novo. Nontransformed soma cells cultivated by error can only partially be excluded from further investigation by their normal type of cell growth.

Fluorescence In Situ Hybridisation (FISH) Fluorescence in situ hybridisation can be applied to directly prepared samples and after cell cultivation and the methods are the same as for not transformed cells (Wegner 1999).

Three different types of DNA probes can be chosen for FISH: whole chromosome paints (wcp), centromer probes (cep), and locus specific probes (LSI) (Fig. 1.7; Gamerdinger et al. 2005).

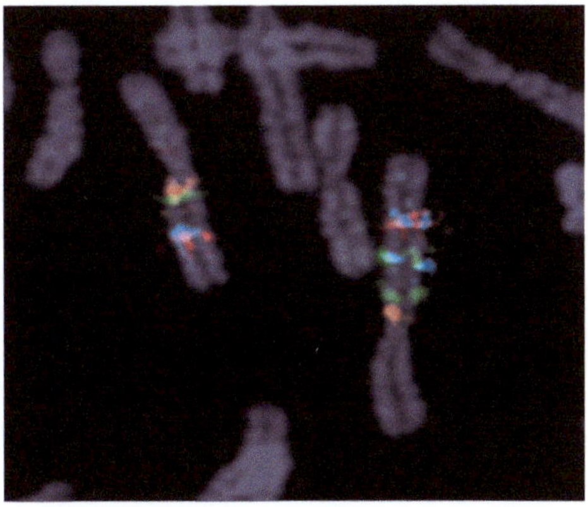

Fig. 1.7 Structural chromosome aberrations analysed by FISH (**a**) Duplication 3q12.3-q23 (YAC-probes: 949 C10, 967 F11, 766 D8, 800 G12, 925 B1) (**b**) Additional derivate of chromosome 14: wcp 14 in interphase nucleus showing two normal sized signals corresponding to two chromosomes 14 and one small signal representing the marker chromosome

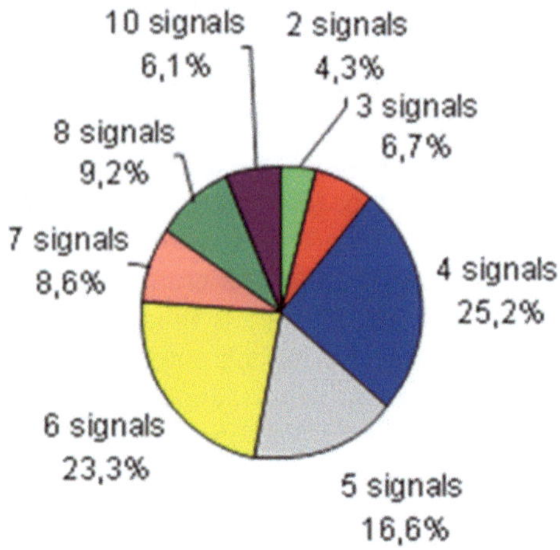

Fig. 1.8 Interphase FISH with DNA-probe cep 6: Quantitative analysis of gains and losses

Often they are applied in combinations or as so called break apart or double fusion probes for the detection of recurrent aberrations in some sophisticated probe. As in the majority of cases interphases are analysed, a high number of cells are available for investigation and therefore quantitative results of the occurring abnormalities can be achieved (Eggermann et al. 2005).

Note:

The efficiency of hybridisation is partially reduced in tumor tissue so that small side-lines of the karyotype or complex rearrangements cannot be detected in metaphases.

FISH with Selected DNA Probes Usually, in a first step of analyses, DNA probes are chosen (LSI, cep) which are known from literature (Wegner 1999) to cause typical changes in chromosome number (i.e. trisomy 8) or structure (i.e. del 13q14) in the tumor type investigated.

Additionally, by these analyses in cultivated tumor cells an asynchronous course of the cell cycle can be recognized or excluded and this peculiarity can be included in the characterisation of the tumor.

Interphase FISH is the only technique which allows to document the simultaneous occurrence of gains and losses of a single chromosome (monosomy, trisomy, polysomy, Fig. 1.8), and diagnostics of structural chromosome abnormalities by microscope requires an aberration size of minimal 5 Mb, while that recognised by FISH is less than 1 Mb.

It has to be taken into account that FISH is not a screening method, which means that not all aberrations occuring in the genome of the tumor can be recognized, but only those for which DNA-probes have been selected after the results of pre-investigations or according to the literature.

Partial improvement in the diagnostic pathway has been achieved by the application of special karyotyping (Wegner 1999).

Special application of wcp and LSI probes: In derivative chromosomes all chromosomes involved in the rearrangement have to be analysed and besides an exact characterisation of the breakpoints is necessary.

1.4.5 Microarray

The implementation of microarray-based high-resolution molecular karyotyping (Fig. 1.9) has significantly improved genetic diagnostics in children with congenital disabilities. The high resolution of this technique leads to an increase in the detection rate from 10% in conventional cytogenetic diagnostics to 20% for chromosomal imbalances (e.g. deletions) in patients with mental retardation and further abnormalities. Therefore, the application of molecular karyotyping has changed the diagnostic algorithms in cases of suspicious chromosomal abnormality, in that the new method is increasingly replacing conventional cytogenetics, at least if no known chromosomal syndrome, such as Down syndrome or Ullrich-Turner syndrome, is expected.

Although array-based molecular karyotyping leads to a higher detection rate of chromosomal imbalances, it is not suitable for the detection of balanced chromosomal rearrangements, pointing towards a potentially transmittable chromosome aberration, e.g. reciprocal translocation or inversion. For this purpose in these cases conventional cytogenetic analysis remains the method of choice.

In tumor genetics the composite karyotype of a tumor usually consists of a high number of pathologic karyotypes in different frequencies. If the same chromosome occurs monosomic as well as trisomic or tetrasomic it cannot be detected by microarray. As well, the relevant balanced rearrangements and the sidelines of the composite karyotype with a frequency of less than 10% cannot be detected.

Besides, it should also be mentioned that the currently available data do not allow to predict the clinical utility and validity of every detected chromosomal variant. Furthermore, guidelines how to deal with incidental findings are still not available. Given these challenges in interpretation and mediation of array results, the close case-related collaboration of clinicians, pathologists and human geneticists is substantially needed. This ensures an optimal support—within the frame of the genetic diagnosis act—for the patients and their families.

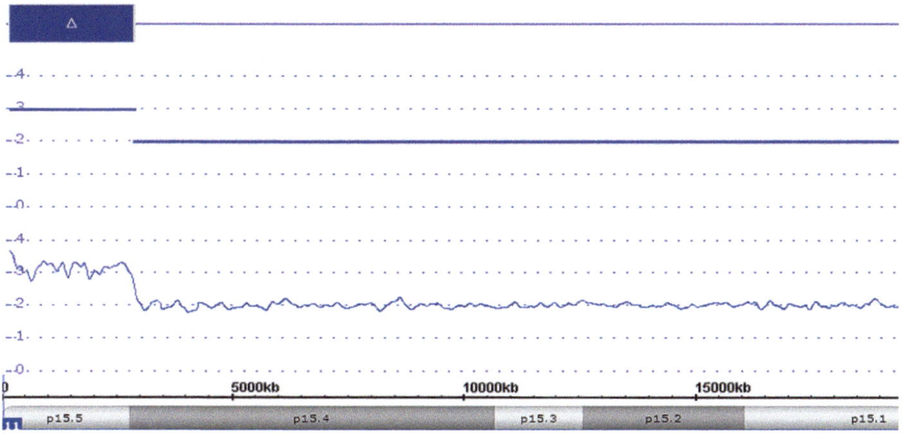

Fig. 1.9 Molecular karyotyping using SNP array illustrating the segmental deletion 11p15 in a patient with Beckwith Wiedemann syndrome

1.4.6 MLPA

In principle, DNA methylation can be analysed by the standard molecular methods, e.g. PCR, restriction digestion, MLPA (see www.mlpa.com) (multiplex ligation-dependent probe amplification) or Sanger sequencing. However, to obtain the DNA methylation information, methylation-specific (MS) modifications are required. These can either be performed by the use of MS sensitive restriction enzymes (e.g. in case of MS restriction digests or MS-MLPA), or by a bisulfite conversion of DNA. The latter conversion is based on the selective deamination but not 5-methylcytosine by sodium bisulfite treatment. This selective deamination leads to a conversion to uracil, whereas the methylated cytosine residues are not affected. In subsequent amplification reactions, the converted uracil will be amplified as thymine. In conclusion, by bisulfite treatment the methylated and the non-methylated cytosines can be distinguished according to sequence changes. After bisulfite conversion, the modified DNA can be analysed by DNA sequencing or methylation-specific PCR. In particular the sequencing of subcloned individual DNA has been used to determine the methylation status for every single CpG island (CGI) as it provides information in a qualitative and quantitative manner. However, these techniques are restricted to single CpGs or regions and do not allow the analysis of long CGI stretches. Furthermore, the sensitivity of bisulfite Sanger sequencing depends on the number of sequenced subclones and might therefore become laborious.

1.4.7 Sanger/Next Generation Sequencing

With the introduction of next generation sequencing (NGS) of bisulfite converted DNA, thousands of CGIs can be analyzed in parallel, and thousands to millions copies of the same fragment can be obtained ("ultra-deep sequencing") (Fig. 1.10).

This methodological improvement impressively decreases the sequencing costs per base and allows the generation of genome wide methylation data at single base resolution in a short time.

However, despite the considerable power of methylation NGS several issues have to be considered in the course of data generation and interpretation:

- The possibility of amplifying single converted DNA molecules which might give rise to numerous identical sequences biasing the result.
- The incomplete conversion of cytosine during the bisulfite conversion cannot be discriminated from methylation and thereby influences the result.

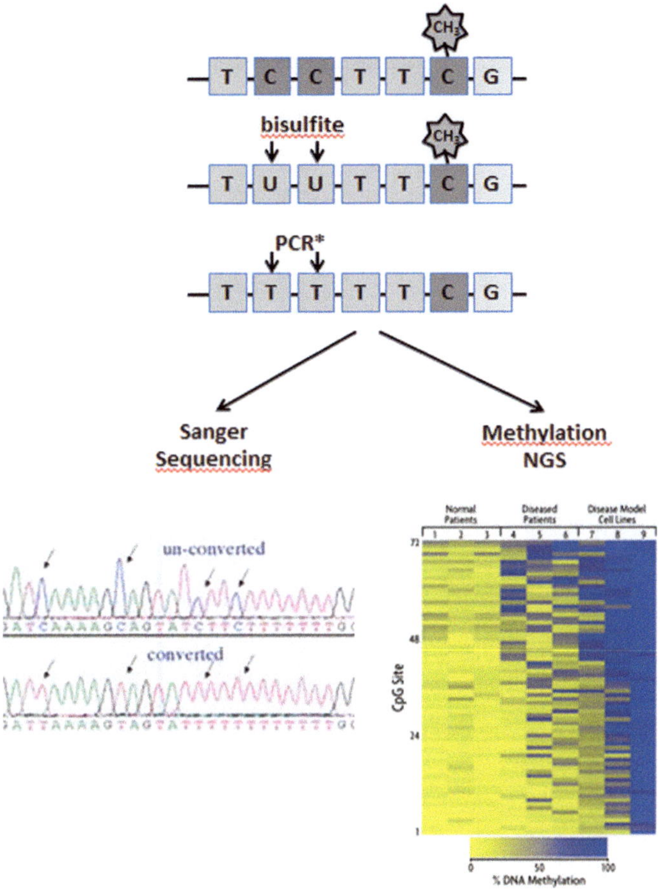

Fig. 1.10 Next generation sequencing

- The influence of the comprehensive statistical analysis significantly influences accuracy of the estimation of the genomic methylation level.
- The currently available software tools and algorithms still need to be optimized.

The advantages and disadvantages of the different cytogenetics and molecular methods are summarized in Table 1.1.

Table 1.1 Overview on techniques currently used in routine diagnostics in epigenetic disorders. As a huge number of techniques has been reported we can present only those procedures widely applied by many laboratories

Methods	Identification UPD	CNVs	Epimutation	Number of loci/test	Advantages	Disadvantages
Aiming at genomic imbalances, aberrant methylation is not detectable						
Microsatellite analysis (MSA)	Yes	Single loci	No	Single loci	Fast, cheap, quantification possible	No differentiation between UPD and CNV possible. A DNA sample of at least one parent is needed.
Conventional cytogenetics	No	>5 Mb	No	Whole genome	Genome-wide detection of large balanced/unbalanced rearrangements	Low resolution (>5 Mb), cell cultivation required
FISH	No	Single loci	No	Whole genome	Genome-wide detection of large (balanced) rearrangements	Microduplications are difficult to detect. Information on the affected region is necessary. Cell cultivation might be required.
Molecular karyotyping (arrayCGH/SNP array)	SNP array	Genome-wide	No	Whole genome	High resolution; genome-wide detection of unbalanced rearrangements	No detection of balanced rearrangements
Methylation-specific (MS) tests, aberrant methylation can be detected						
MS Southernblot	Yes	Single loci	Yes	Single loci	Quantitative	Large amount of DNA required; time-consuming; no differentiation between the different types of mutations/epimutations possible.
MS PCR	Yes	Single loci	Yes	Single loci	Fast, cheap, quantification possible	No differentiation between the different types of mutations/epimutations possible, only semi-quantitative.
MS pyrosequencing	Yes	Single loci	Yes	Single loci	Quantitative	No differentiation between the different types of mutations/epimutations possible.
QAMA real-time PCR based methylation assay	Yes	Single loci	Yes	Single loci	Quantitative	No differentiation between the different types of mutations/epimutations possible
Bisulfite sequencing	Yes	Single loci	Yes	Single loci	Quantitative	Time-consuming; cloning or next generation sequencing needed; no differentiation between the different types of mutations/epimutations possible
MS MLPA	Yes	Multiple loci	Yes	Up to 46 loci	Quantitative; differentiation between the different types of mutations/epimutations possible	Sensitive for DNA quality.
MS SnuPE	Yes	Multiple loci	Yes	Multiple loci	quantitative	No differentiation between the different types of mutations/epimutations possible

1.5 Guidelines for Quality Assessment in Genetic Tumor Diagnostics

The guidelines are delineated from the present international diagnostic standard relying on the methodical development which can be applied to characterise tumor cells. In future these guidelines must be continuously adapted to the possibilities of newly developed techniques. The guidelines are subdivided in essential and desirable standards of quality in diagnostic investigations. These demands must be regarded as a minimum of diagnostic quality. In its subgroups national regulations can differ from the general rules. The following parameters have to be taken into consideration:

Transport of the tumor material, investigation methods, type or tumor send for diagnostics (leukaemia, solid tumor), follow up analyses, detailed presentation of the results, and maximum of investigation time which is necessary.

The guidelines have been delineated from 38 relevant recent publications by the E.C.A. permanent working group for Cytogenetics and Society (Hastings et al. 2012).

1.6 Research Databases for Archiving Genome Data

In the last 10–15 years genome research has developed in an unexpected extent. By that the problem arose how to archive the genome data for a longer time: first of all to ensure their secure storage and secondly that the findings results would be available also for future research projects. In Germany the development of the required databases is supported by the Deutsche Forschungsgemeinschaft (DFG) and the German Council of Science and Humanities (Wissenschaftsrat). On the international part the European Commission and the National Institutes of Health (NIH) are involved. The scientists of the genome research projects are asked to document their data in a way so that they could be retrieved also after 10 years. At the same time the management of the research data has to take into consideration the social aspects as well as the scientific aspects. Different aspects have to be included as international standards of data representation, a reliable data backup by each chosen storage technology, the purpose specification of the backup and the secure protection of the data during the long-term backup.

The original data of research, the primary data, were saved as metadata. The metadata were upgraded by information of quality and by this way—following the general standards—these data could then show the completed primary data. The backup must exclude a loss of data by technical reasons and the unauthorized access.

The archiving of the genome data has to guarantee the patients' anonymization or at least their pseudonymization. Associated to this is the problem of the future structures of the legal requirements related to the archiving data.

Today it is already common to convert older formatted data into an actual data format. Programs exist to select relevant research data, which make a selection possible before the long-term archiving. For the long-term archiving different programs are available like the "Open Archival Information System" (OAIS) .

It is to the different genetic organizations to set up standards of the documentation of primary data and metadata in the near future. Therefore data centers have to be built up with a biomedical focus.

1.7 Predictive Genetic Diagnostics

By the genetic diagnostic it is possible to predict the occurrence of a genetic disease to a healthy human at a later age or at least the possibility to identify the genetic disease. There are certain diseases, in which there can be quite a few decades between the diagnostic setting and the onset of disease. The diagnostic setting is once possible prenatally investigating the embryo or the fetus or postnatally at any age. Indeed, in case of prenatal testing ethical issues as well as national laws have to be considered. In many cases the meaningfulness of the prediction is limited. Only in monogenic diseases with a high penetrance the evidence is relatively clear. If for example a mutation of the BRCA1 gene is ascertained to a patient, it can be assumed that he is carrier of the autosomal dominant mutant with risk for breast and ovarian cancer. Then there should be given recommendation to all the first degree relatives to perform a mutation testing and—if the mutation analysis is positive—early preventive steps and therapies should be recommended (Propping 2010).

Relatives of patients showing a mutation leading to a genetic tumor disease are usually recommended to have regular control investigations, sometimes over decades. By predictive genetic diagnostics it can be ascertained, if they are carrier of the familial mutation or not. If they are excluded as carriers of an increased genetic risk factor, regular control investigations and a high psychological stress can be avoided. In the other case of the increased risk for the disease the regular controls make it possible to diagnose a tumor at an early stage of development which improves the possibilities of the therapy significantly.

Besides family testing also populations with an increased risk of certain tumor types are recommended for predictive analyses.

In Germany the so called "Bundesärztekammer" has published guidelines for predictive diagnoses in cases of increased tumor risk. All diagnostic centers involved in these investigations are working according to these guidelines. When predictive diagnoses are planned the proband has to determine in advance which persons have to be informed about the results of the investigation (i.e. he, family, doctor).

However, there is no question that each person at risk can also refuse predictive analyses. Therefore any decision of the proband has to be preceded by a detailed genetic counseling which includes information about the relevance and the security of the investigation which is going to be performed and the general experiences about the relevance of predictive diagnostics in case of the tumor type in question. If children are concerned, the earliest known onset of the disease has to be taken into account, and older children (> 10 years) have to be informed in a way that is in relation to their age. In these cases positive results of such an early therapy must be known.

The counseling of probands, parents and family members is usually performed twice in Germany, first before the tests are performed and second after receiving the final results of investigations. In some cases there will be an interval of several months between the two data which will then make the second counseling more complicated. Only physicians with a certified qualification in Clinical Genetics can perform these discussions with the consultants.

In Germany the course, the content and the advices of the discussion should follow the guideline of the "Gendiagnostikgesetz" (2009) and it must be guarented that the counseling is non-directive but helps the consultant to the necessary informations to make independently a relevant decision for himself.

1.8 Epigenetics in Ontogenesis

Epigenetic imprints are stable marks in the genome, which control gene expression and which are established during ontogenesis during cell division and cell differentiation (Varrault et al. 2006).

The normal embryonic development requires the balanced maternal and paternal contribution to the genome of the zygote. In case of imprinted genes, either the maternal or the paternal copy is methylated or unmethylated, and as a result active or inactive. If one of these genes or chromosomal regions has been inherited only from one parent, a normal development of the embryo is not possible. In case of a total genome has been inherited from the mother, a rudimentary development of embryonic tissues (teratoma) can be observed. The pure paternal contribution leads to the exclusive development of extra-fetal tissues, a hydatidiform mole develops and subsequently the pregnant woman has an increased risk to develop an endometrial carcinoma. Usually a spontaneous abortion of the pathologic conceptus occurs in the first weeks of the 1st trimenon.

As mentioned before, the differentiation between maternal and paternal genome and their specific mode of gene activation is called genomic imprinting. The imprinting can be restricted to single genes or includes gene clusters.

The pattern of imprinting and epigenetic marks in general is set up in the early conceptus and changes during the ontogenesis. This "life cycle of imprinting" underlies complex sequence of methylation and demethylation enzymes. This balanced regulation is indeed prone to numerous deleterious influences, and a disturbance of the fine-tuned methylation results in pathological courses like imprinting disorders or tumor development.

Among others, the complex epigenetic machinery applies by the primary sex differentiation controlled by the gonosomes XX and XY. Of particular importance for the normal fetal development is the inactivation of the 2nd X-chromosome in the female (Lyonisation). If two structural normal X-chromosomes exist usually the paternal and maternal Xs are at randomly inactivated; as a result all female organisms are functional mosaics. The inactivation occurs in the 1000–2000 cell stage, however not in each cell system at the same time, but depends on the tissue and the state of development.

The inactivated X-chromosome forms a loop, the telomeres are at the nuclear membrane. During the cell cycle the inactivated X-chromosome is replicated as the last chromosome, and in pro- and metaphase of the mitosis it is often more condensed than the active early replicating X-chromosome.

During the S-phase of the cell cycle always after the synthesis of a new DNA region follows the respective methylation of inactivated genes and gene regions. If methylation errors during this process occur and lead to somatic mutations, this subsequently gives origin to the formation of mosaic.

The chronologically determined differentiation of totipotent cells of the conceptus into hundreds of different cell types by a changed gene imprinting pattern is of special ontogenetic importance. If one of the two parental genes remains cell physiologically active, it starts its replications earlier during S-phase of the cell cycle than the inactive gene of the other parent. This corresponds to the behaviour of the inactivated X-chromosome of females.

Induced by mutations the methylation of single genes can be significantly reduced in monogenic syndromes as well as in all types of tumors. An example of this type of mutation is the autosomal recessive inherited ICF-syndrome (Immunodeficiency, Centromeric region instability, Facial anomalies syndrome) which is characterised by a specific heterochromatin demethylation, especially of the satellite DNA 2.

The somatic hypomethylation of different genes can lead to a stronger gene expression, which subsequently produces an increased gene product. Such a demethylated gene can operate as an oncogene. The developing tumor can be benign or malign. On the other side, a tumor suppressor gene can be inactivated by hypermethylation, and its reduced expression results in tumor development, too. This process can already occur in the prenatal period (Retinoblastoma) or—in the majority of events and depending of the tumor type—at different ages. As many tumors are caused by an inborn constitutional mutation and a second somatic mutation occurring later in life, Knudson (1971) have postulated the so-called "two-hit modell" of carcinogenesis.

In summary different types of epigenetic mutations may be relevant a tumor genesis. They affect oncogenes, tumor suppressor genes and mutator genes. Mostly they occur in different combinations. According to the "two-hit model", a genetic predisposition exists in some cases. With increasing age the risk for tumorgenesis increases, because the number of the somatic mutations grows, while the effectiveness of the repair mechanism decreases.

1.9 Mosaic Formation by Specific Postzygotic Mechanisms

Genetic mosaicism is defined as the presence of different genomic constitutions in different somatic cell systems or the difference between germ and soma cells in the same individual. It is caused by gene mutations, by numeric or genome mutations, by structural or chromosome mutations or by epigenetic mutations during the ontogenetic development. The types of mutation can even show different combinations.

In more than 90 % of mosaic cases (own data) the carrier is composed of two different genomes. However, if the mutation is unstable (i.e. ring chromosomes), or if there is a strong selection against a pathologic genotype (i.e. trisomy 15, triploidy), because of subvitality/lethality of the aberrant cells the cell cycle shows an aberrant course in S-phase and anaphase followed by up to seven different genotypes in the daughter cells. Epigenetic mutations occur in different postnatal phases of life leading to mosaic formation with cells that kept their normal function besides the mutated ones with an aberrant function. With increasing age of the carrier the natural selection against mutated cells decreases. Thus age-correlated diseases will increase in number, and especially different types of tumor become more frequent. (Conlin et al. 2010).

Changes in the DNA methylation pattern can not only be diagnosed in meristematic cells, and here especially in G1- and S-phase of the cell cycle, but also in cells after differentiation. For the conservation of the normal methylation pattern a sufficient function of the DNA repair and methylation mechanisms are necessary and it is of special relevance for cells with long life expectancy. But with increasing age the repair efficiency decreases. Even the methylation maintenance of specific genes can change and as a result lead to an altered gene regulation of the cell and the stability of the genome. To analyse longitudinal changes of the epigenetic pattern, large epidemiological studies over a number of years are necessary. The first results of these investigations are promising, and they demonstrate that changes in the methylation pattern show significantly different frequencies in different families that means that genetic factors play a relevant role. In particular genes are affected which are involved in tumor development such as proto-oncogenes and tumor suppressor genes (Burell 2013).

A problem of the investigation of epigenetic changes and tumor development is the observation that many epigenetic mutations occur tissue specific. For some genes it could even be delineated that their epigenetic status is unstable and therefore they often change from methylation to demethylation.

Epidemiologic studies from the last years showed that the methylation pattern can also be changed by exogenous factors, such as mutagens. These results open new perspectives for preventive therapies in the future leading to a later onset of a disease like tumor development.

1.10 Peculiarities in Constitutional Chromosomal Aberrations

Balanced and unbalanced constitutional chromosome aberrations can lead to an increased tumor risk for the carrier.

Tumor genesis can already occur in prenatal life such as acute leukaemia in foetuses with Down's syndrome (trisomy 21) or the induction of precancerous lesions of the ovotestes in male fetuses with X0/XY mosaicism (Müller et al. 1999).

Unbalanced chromosome aberration with an increased tumor risk is well known in trisomy 21, where the risk of developing leukaemia is 20-fold increase compared to the normal population, but the majority of these cases shows spontaneous remission. Additionally, in males the frequency of germ cell tumors is increased (Satgé et al. 1998).

Constitutional chromosome mosaicism is caused by postzygotic mitotic abnormalities that can lead to numerical or structural aberrations. Best known are X0/XY mosaics with an increased risk for the development of gonadoblastoma in the dysgenetic gonad. It is observed in about 50 % of carriers of gonadal dysgenesis. The gonadoblastoma is not growing invasive but it is often associated with a malignant, invasive growing tumor, such as a dysgerminoma (Schüler and Schwanitz 2004).

Trisomy 8, an autosomal aberration, is only compatible with life when it occurs as mosaic and it is combined with a significantly increased tumor disposition. At the same time trisomy 8 mosaicism is known to be diagnosed as an acquired aberration in myeloid leukaemia and myelodysplastic diseases (Brady et al. 2000).

Besides, different types of balanced chromosome rearrangement are combined with an increased tumor risk.

For example, reciprocal translocations involving the region 3p14 are at risk of developing renal cancer. Here, the translocation, disrupting a tumor suppressor gene is the first step of tumor development. A second step is followed by a mitolic abnormality leading to the loss of the distal region 3p14 to pter. Thereafter the third step is a gene loss in the normal homolog of chromosome 3 (Valle et al. 2005).

Well known as a genetic disorder caused by chromosome as well as imprinting gene mutations is the imprinting disorder Beckwith-Wiedemann Syndrome characterised by mutations in the region 11p15.5. The critical region contains one gene that is imprinted in the male gametogenesis and another one in the female one. Absence of the paternal gene product leads to unregulated cell proliferation and as a consequence to the development of tumors of the genitourinary system, like nephroblastoma or Wilms' tumor. Besides the carrier of this syndrome shows phenotypical abnormalities such as gigantism, facial dysmorphisms, hemihypertrophia, d advanced bone age and others (for review: Choufani et al. 2013).

A further group of genetically caused syndromes with an increased tumor risk are the so called chromosome instability or breakage syndromes. They are caused by gene mutations which lead to a syndrome specific pattern of chromosome breaks, rearrangements and SCEs. The basic defect is a reduced repair of somatic mutations. The frequency of aberrations is influenced exogenously by mutagens like x-rays or different chemical substances, especially the alkylating agents.

Three of these syndromes are presented here in detail:

Fanconi Syndrome or Fanconi Anemia This syndrome shows an autosomal recessive inheritance. Eight subtypes are defined according to the position of the mutated genes and the following ones are defined exactly: A: 16q24.3, C: 9q22.3, D: 3p22–26, E: 6p21–22, G: 9p13. The rate of aberrant mitoses can exceed 30 %, the mitotic cycle is abnormal in S- and G2-phase. Beside the increased breakage rate the formation of quadrivalents caused by reciprocal translocations is a characteristic feature. The most frequent cancer is acute myeloid leukaemia.

The clinical features of the patients show variable expression. The most frequent symptoms are growth retardation, mental retardation, skeletal anomalies and facial dysmorphisms. Life expectancy is significantly reduced.

Bloom Syndrome The inheritance is autosomal recessive, a gene has been located in 15q26. Two subtypes could be defined by their different pattern of chromosome aberrations: subtype 1 shows besides an increased number of chromosome breaks a high frequency of reciprocal translocations between non homologous chromosomes while the exchanges occur preferentially between homologous in subtype 2.

The aberrations are mainly induced in the aberrant course of S to G2-phase of the cell cycle. The rate of sister chromatid exchanges (SCE rate) is increased in both subtypes by 10–20-times.

Frequent cancer types are acute myeloic leukaemia, lymphoma, adenocarcinoma, gastrointestinal and cervical tumors. Clinical features of the patients are dwarfism, facial dysmorphism and immunodeficiency. Life expectancy is reduced.

Ataxia Teleangiectasia The autosomal recessive disorder shows a gene localisation in 11q22.3. The number of aberrant cells is increased but less compared to Fanconi Anemia and Bloom syndrome. The defect in the mitotic cell cycle is an aberrant checkpoint between G1- and S-phase. Normally the repair of chromosome aberrations occuring in G1 is obligatory before the cell enters S-phase. This block of entry is missing in AT cells. Chromosome rearrangements often involve chromosome 14. A specific type of aberration is telomer fusion combined with telomer shortening. Characteristic tumor types are leukaemia, epithelial cancer, medulloblastoma, glioma and an increased breast cancer risk in heterozygotes. Phenotypic abnormalities include cerebellar ataxia, teleangiectasia of eyes and skin, increasing neurologic impairment, severe immunodeficiency premature and aging. Life expectancy is significantly reduced (Miller and Therman 2001).

The analysis of the combination of constitutional chromosome aberrations with epigenetic alterations is a new field of investigation in patients with primary chromosome abnormalities as well as in secondary changes observed in tumor genetics. It can be expected that this special field of applied and basic genetics will gain significant importance in the coming years.

1.11 Epigenetics and the Development of New Strategies in Tumor Therapy

Mechanisms leading to changes of the epigenetic pattern of an organism are an actual focus of investigation with concentration on constitutional genetic—monogenic and chromosomal—aberrations, but followed by analyses of characteristic changes in tumor diseases.

Induced changes in gene function were applied in first research projects on constitutional genetic changes. Jiang et al. (2013) performed in-vitro tests on cells with trisomy 21. They introduced the large X-inactivation gene (XIST) into the DYRK1A locus on chromosome 21. The recipient cells were pluripotent stem cells. The XIST non-coding RNA coated chromosome 21 and triggered stable chromatin modifications, leading to a chromosome-wide silencing of transcription.

This DNA methylation leads to a morphological change of the aberrant chromosome 21 by a higher condensation of the mutated chromosome.

Expression changes could be demonstrated and even defects in cell proliferation were shown to be reversible.

These inducible changes in gene function represent a first step approaching chromosome therapy.

The results of a second promising project have recently been published by Adorno et al. (2013). They investigated gene products of haematopoietic stem cells in two mouse models (Ts65Dn and Ts1Cje) with Down syndrome. These mice show phenotypic alterations which are also observed in humans with Down syndrome. An ubiquitereous relevant gene Usp16 could be proved to cause multiple developmental defects in the mouse model. Bone marrow transplantations with euploid cells showed the role of the third copy of Usp16 in pathologic cell development. These investigations on chimeric cell populations open new possibilities of silencing target genes.

Furthermore, completely artificial chromosomes of comparable length have been produced. They are stable during cell proliferation. Introduced genes or chromosome segments might become of therapeutic relevance by long-term correction of a pathologic gene function.

Genetic differences between normal and aberrant somatic cells in quantitative traits, in endophenotypes which present with risk factors for a tumor disease and an increased susceptibility to further somatic mutations can be the relevant factor leading to a changed gene expression.

Thus, candidates for a therapeutic interference might be selected in future on a specific and individual basis and will lead to an optimal choice of intervenience.

References

Adorno M, Sikandar S, Mitra SS, Kuo A, Nicolis Di Robilant B, Haro-Acosta V, Ouadah Y, Quarta M, Rodriguez J, Qian D, Reddy VM, Cheshier S, Garner CC, Clarke MF (2013) Usp16 contributes to somatic stem-cell defects in Down's syndrome. Nature 501(7467):380-384. doi: 10.1038/nature12530

Amor DJ, Halliday J (2008) A review of known imprinting syndroms and their association with assisted reproduction technologies. Hum Reprod 23:2826–2834

Barr ML, Bertram EG (1949) A morphological distinction between neurones of the male and female, and the behaviour of the nucleolar satellite during accelerated nucleoprotein synthesis. Nature 163:676–677

Berger SL, Kouzarides T, Shiekhattar R, Shilatifard A (2009) An operational definition of epigenetics. Genes Dev 23:781–783

Bird A, Taggart M, Frommer M, Miller OJ, Macleod D (1985) A fraction of the mouse genome that is derived from islands of nonmethylated, CpG-rich DNA. Cell 40:91–99

Bock C (2009) Epigenetic biomarker development. Epigenomics 1:99–110

Brady AF, Wataers CS, Pocha MJ, Brueton LA (2000) Chronic myelomonocytic leukaemia in a child with constitutional partial trisomy 8 mosaicism. Clin Genet 58:142–146

Brena RM, Costello JF (2010) The role of the epigenome in human cancers. In: Speicher MR et al (eds) Human genetics. Problems and approaches. Springer, Berlin

Burrell RA, Swanton C (2014) The evolution of the unstable cancer genome. Curr Opin Genet Dev 24:61–67.

Busson-Le Coniat M, Salomon-Nguyen, Mozziconacci MJ et al (1999) Fluorescence in situ hybridization analysis of chromosome 1 abnormalities in hematopoietic disorders: rearrangements of DNA satellite II and new recurrent translocations. Leukemia 13:1975–1981

Chandler LA, Jones PA (1988) Hypomethylation of DNA in the regulation of gene expression. Dev Biol 5:335-349

Choufani S, Shuman C, Weksberg R (2013) Molecular findings in Beckwith-Wiedemann syndrome. Am J Med Genet C Semin Med Genet 163C:131–140

Conlin LK, Thiel BD, Bonnemann CG, Medne L, Ernst LM, Zackai EH et al (2010) Mechanisms of mosaicism, chimerism and uniparental dissomy indentified by single nucleotide polymorphism array analysis. Hum Mol Genet 19:1263–1975

Cooper DN (1983) Eukaryotic DNA methylation. Hum Genet 64:315–333

Cooper DN, Taggart MH, Bird AP (1983) Unmethylated domains in vertebrate DNA. Nucleic Acids Res 11:647–658

Cremer T (2010) Von der Genetik zur Epigenetik und Epigenomforschung—Essay zur Geschichte der Vererbungsforschung und zur Zukunft der prädiktiven Medizin. In: Doerfler W, Ulrich HG, Böhm P (eds) Medicine at the interface between science and ethics, Nova Acta Leopoldina, Bd 98. Deutsche Akademie der Naturforscher Leopoldina, Halle pp 87–165

Delaval K, Wagschal A, Feil R (2006) Epigenetic deregulation of imprinting in congenital diseases of aberrant growth. Bioessays 28:453–459

Eggermann T, Gamerdinger U, Bosse K, Heidrich-Kaul C, Raff R, Meyer E et al (2005) Mosaic tetrasomy 14pter-q13due to a supernumerary isodicentric derivate of proximal chromosome 14q. Am J Med Genet 134A:305–308

Eggermann TB, Horsthemke B et al. (2013) Molekulargenetische Diagnostik von Imprintingerkrankungen. Med Genet 25:1–8

Engel E, Antonarakis SE (2002) Genomic imprinting and uniparentel disomy in medicine. Clinical and molecular aspects. Wiley-Liss, New York

Esteller M (2008) Epigenetics in cancer. N Engl J Med 358:1148–1159

Galetzka D, Hansmann T, El Hajj N, Weis E, Irmscher B, Ludwig M et al (2012) Monocygotic twins discordant for constitutive BRCA1 promotor, methylation, childhood cancer and secondary cancer. Epigenetics 7:47–54

Gamerdinger U, Bosse K, Eggermann T, Kalscheuer V, Schwanitz G, Engels H (2005) First report of a partial trisomy 3q12-q23 de novo-FISH breakpoint determination and phenotypic characterization. Eur J Med Genet 49:225–234

Gardner RJM, Sutherland GR, Shaffer LG (2012) Chromosome abnormalities and genetic counselling, 4th edn. Oxford University Press, Oxford

Gesetz über genetische Untersuchungen beim Menschen (Gendiagnostikgesetz GenDG) (2009). Bundesgesetzblatt 2009, Teil 1. Bonn. http://www.gesetze-im-internet.de/gendg/index.html

Gesetz über genetische Untersuchungen bei Menschen (Gendiagnostikgesetz - GenDG) Gendiagnostikgesetz vom 31. Juli 2009 (BGBl. I S. 2529, 3672), das durch Artikel 2 Absatz 31 u. Artikel 4 Absatz 18 des Gesetzes vom 7. August 2013 (BGBl. I S. 3154) geändert worden ist"

Haaf T (2006) Methylation dynamics in the early mammalian embryo: implications of genome reprogramming defects for development. Curr Top Microbiol Immunol 310:13–22

Haaf T, Schmid M (1991) Chromosome topology in mammalian interphase nuclei. Exp Cell Res 192:325–332

Hastings R (2012) A common European framework for quality assessment for constitutional, acquired and molecular cytogenetic investigations. Eur Cytogenet Assoc Newsl 30:11–19

Horsthemke B (2010) Mechanisms of imprint dysregulation. Am J Med Genet C Semin Med Genet 154C:321–328

Jacobs KB, Yeager M, Zhou W, Wacholder S, Wang Z, Rodriguez-Santiago B et al (2012) Detectable clonal mosaicism and its relationship to aging and cancer. Nat Genet 44:651–658

Jiang J, Jing Y, Cost GJ, Chiang JC, Kolpa HJ, Cotton AM, Carone DM, Carone BR, Shivak DA, Guschin DY, Pearl JR, Rebar EJ, Byron M, Gregory PD, Brown CJ, Urnov FD, Hall LL, Lawrence JB (2013) Translating dosage compensation to trisomy 21. Nature 500(7462):296–300

Kacem S, Feil R (2009) Chromatin mechanisms in genomic imprinting. Mamm Genome 20:544–556

Knudson AG Jr (1971) Mutation and cancer: statistical study of retinoblastoma. Proc Natl Acad Sci U S A 68:820–823

Lyon MF (1961) Geneaction in the X-chromosome of the mouse (*Mus musculus* L.). Nature 190:372–373

Lyon MF (1974) Mechanisms and evolutionary origins of variable X-chromosome activity in mammals. Proc R Soc Lond B 187:243–268

Mai S (2013) The 3D nuclear organization of the interphase nucleus in nirmal, immortalized and tumor cells. Chromosome Res 21:11

McGranahan N, Burrell RA, Endesfelder D, Novelli MR, Swanton C (2012) Cancer chromosomal instability: therapeutic and diagnostic Mallenges. EMBO Rep 13:528–538

Mc Grath J, Solter D (1983) Nuclear transplantation in mouse embryos. J Exp Zool 228:355–362

Miller OJ, Therman E (2001) Human chromosomes, 4th edn. Springer, Heidelberg

Molenaar JJ, Koster J, Zwijnenburg DA, van Sluis P, Valentijn LJ, van der Ploeg I et al (2012) Sequencing of neuroblastoma identifies chromothripsis and defects in neuritogenesis genes. Nature 483:589–593

Müller J, Ritzén EM, Ivarsson SA, Rajpert-De Meyts E, Norjavaara E, Skakkebaek NE (2009) Management of males with 46,X/46,XY gonadal dysgenesis. Horm Res 52:11–14

Neves-Costa A, Will WR, Vetter AT, Miller JR, Varga-Weisz P (2009) The SNF-2 family member Fun 30 promotes gene silencing in heterochromatic loci. PloS One 4(12):e8111

Propping P (2010) Prädiktive genetische Diagnostik: Möglichkeiten, Grenzen und Zukunft. Vortrag. Bundesministerium für Bildung und Forschung, Bonn

Qiu J (2006) Epigenetics: unfinished symphony. Nature 11:143–145

Quante T (2012) Mutant p53 is a transcriptional CO-factor that binds to G-rich regulatory regions of active genes and generates transcriptional plasticity. 43rd Annual conference of the German Genetics Society and GRK 143. September. Essen

Reik W, Walter J (2001) Genomic imprinting: parental influence on the genome. Nat Rev Genet 2:21–32

Richtlinien zur prädiktiven genetischen Diagnostik (2003) Bekanntmachungen: Richtlinien zur prädiktiven genetischen Diagnostik (verabschiedet vom Vorstand der Bundesärztekammer). Dtsch Ärztebl 100:1297–1305

Satgé D, Sommelet D, Geneix A, Nishi M, Malet P, Vekemans M (1998) A tumor profile in Down syndrome. Am J Med Genet 78:207–216

Saxonov S, Berg P, Brutlag DL (2006) A genome-wide analysis of CpG dinucleotides in the human genome distinguishes two distinct classes of promoters. Proc Natl Acad Sci U S A 103:1412–1417

Schüler HM, Schwanitz G (2004) Das 45,X/46,XY-Mosaik in der Pränataldiagnostik—Diagnostisches Vorgehen, Phänotyp-Analyse und Prognose. Med Genet 16:438–441

Schwanitz G, Raff R (2005) Application of specific cytologic, cytogenetic and molecular-cytogenetic techniques for the characterization of solid tumors. Rocz Akad Med Bialymst 50:91–96

Shaffer LG et al (eds) (2013) ISCN An international system for human cytogenetic nomenclature. Cytogenet Genome Res VI:1–40

Tollefsbol T (ed) (2011) Handbook of epigenetics. The new molecular and medical genetics. Elsevier, Amsterdam

Valle L, Cascón A, Melchor L, Otero I, Rodríguez-Perales S, Sánchez L et al (2005) About the origin and development of hereditary conventional renal cell carcinoma in a four-generation t(3;8)(p14.1, q24.23) family. Eur J Hum Genet 13:570–578

Varrault A, Gueydan C, Delalbre A, Bellmann A, Houssami S, Aknin C et al (2006) Zac1 regulates an imprinted gene network critically involved in the control of embryonic growth. Dev Cell 11:711–722

Wark L, Danescu A, Natarajan S, Zhu X, Cheng SY, Hombach-Klonisch S et al. (2013) Three dimensional (3D) telomere dynamics in follicular thyroid cancer. Thyroid 24(2):296–304 (PMID: 23819464)

Wegner RD (ed) (1999) Diagnostic cytogenetics. Springer, Berlin

Zhang TY, Hellstrom IC, Bagot RC, Wen X, Diorio J, Meaney MJ (2010) Maternal care and DNA methylation of a glutamic acid decarboxylase 1 promoter in rat hippocampus. J Neurosci 30:13130–13137

Chapter 2
Biodynamic Phenotypic and Epigenetics Changes of Circulating Tumor Cells: Their Application in Cancer Prognosis and Treatment

Ma José Serrano Fernández, Ma Jesús Alvarez-Cubero, Jose Luis García Puche, F Gabriel Ortega and Jose Antonio Lorente

Contents

Abstract This chapter focuses on a deep description on Circulating Tumor Cells (CTCs) and its main role in cancer progression and genetic changes related to metastasis. In solid tumors, like breast and lung cancer, is being more frequent to appear patients with resistance to chemo and radiation therapy, this event will lead to decreasing quality of life as well as less efficient medical treatment. As it is known, CTCs are tumor cells disseminated from primary and metastatic sites and they are current tumor biomarkers. Therefore, CTCs will allow a more efficient tumor characterization and offering a more personalized medicine and treatment to specific patients. In this chapter, we offer a deeper analysis in CTCs characterization in Epithelial Mesenchymal Transition (EMT) process, as well as epigenetic changes that are important for making a more specific characterization of CTCs. Epigenetic changes can lead to silence tumor suppressor and metastasis suppressors' genes, in addition to being important hallmarks giving clues of growth, proliferation, and invasiveness

J. A. Lorente (✉) · M. José Serrano Fernández · M. J. Alvarez-Cubero
J. L. G. Puche · F. G. Ortega
GENYO (Pfizer-University of Granada-Andalusian Government Centre for Genomics and Oncological Research), Granada, Spain
e-mail: jose.lorente@genyo.es

M. José Serrano Fernández · M. J. Alvarez-Cubero · J. A. Lorente
Laboratory of Genetic Identification, Legal Medicine and Toxicology Department, Facultad de Medicina, Universidad de Granada, Granada, Spain
e-mail: mjose.serrano@genyo.es

P. Mehdipour (ed.), *Epigenetics Territory and Cancer*,
DOI 10.1007/978-94-017-9639-2_2

of tumor cells. It is well known that microRNAs vary their concentration depending on the aggressiveness of the tumor as well as the epidermal characteristics of CTCs. Our main aim with this chapter is trying to give more clues on the genetic and phenotypic characterization of CTCs that will give important information in a personalized therapy, besides novel therapeutic targets and personalized medicine.

List of Abbreviations

CTCs Circulating Tumor Cells
EMT Epithelial Mesenchymal Transition
MET Mesenchymal-to-Epithelial Transition
MMP Matrix Metalloproteinases

2.1 Introduction

The paradigm metastatic define "Metastasis" as the process that involves the released of tumor cells from a primary organ to other target organs without anatomic direct relation with this primary site, where can growth and development new tumor focus (Valastyan and Weinberg 2011; Gupta and Massagué 2006). In addition, to develop of this process is necessary the circulation of tumor cells trough blood system or lymph system.

The development of genomic era supposed an extraordinary advance in the knowledge of tumor process. However, the most studies on metastasis have focused on measurement made at the end point of this process, this is: the establishment of micro or macrometastasis. Furthermore this best knowledge of complex metastasis process involved a change from a reductionist approach of metastasis process, focused only in the study of tumor mass, to study the interaction of these tumor cells with the microenvironment and the biology host.

The metastasis process involve multi-step, address the dynamic aspect of the metastatic process: cells exit the primary tumor, invade the local tissue, enter blood or lymphatic vessels (aprocess known as 'intravasation'), and are passively transported to the secondary site where the cells exit the vessels (a process known as 'extravasation') and enter the tissue (Details in Fig. 2.1). (Fidler 2003; Kim et al. 2009).

However this is a simplification of metastasis process, because these entire basic step have helped by multi epigenetic and genetic events which are in turn depends of different functional changes consequence of microenvironment (Cock-Rada and Weitzman 2013; Brabletz et al. 2013).

In addition, one fundamental step into of this metastasis cascade is the dissemination tumor cell phase. Through which, tumor cells acquire the capacity to circulate through bloodstream from primary tumor and finally colonize different organs to develop a metastasis (Pantel and Brakenhoff 2004).

Metastasis process presents inherent difficulties due to technological limitations in analyzing a complex in vivo process rich with heterotypic interactions. Invasion,

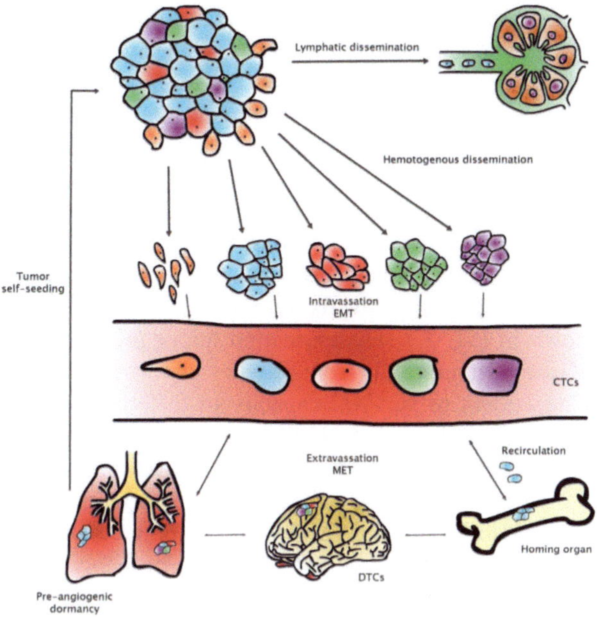

Fig. 2.1 Dissemination tumor process. Metastasis process involves the relapse of tumor cells from primary tumor to target organ. This metastasis process is dependent to survival capacity of different subpopulation of CTCs into blood system

survival in the circulation, and growth in distant organs are not amenable to methods that primarily use in vitro models. Despite technical challenges, elegant experiments that started in the 1950s were done with mouse xenograft models and resulted in an important descriptive understanding of the biology of metastasis. With the accumulation of knowledge from studying cancer cells in isolation, subsequent advances in metastasis built on the classic studies.

Unfortunately, metastasis remains responsible for the vast majority of cancer-related morbidity and mortality. Therefore, the understanding of this step into of metastasis process is a high priority.

In this chapter we will review our knowledge about the dissemination process, describing some important steps of this process in the metastasis of solid tumors.

2.2 Metastasis Process in Solid Tumor

The major cause of cancer-associated mortality is tumor metastasis. The metastasis process is known as the spread of cancer from the organ of origin (primary site) to distant tissues. Despite, our understanding about cell proliferation, cell death, genomic instability, and signal transduction pathways has rapidly progressed; our understanding of tumor metastasis is far from complete (Kovács et al. 2013; Lorusso and Rüegg 2012).

Cancer occurs after a cell is progressively genetically damaged and turns into a cell bearing a malignant phenotype. These cells are able to undergo uncontrolled abnormal mitosis, which leads to an increase of these cancerous cells at that location.

In absence of regular control mechanisms a heterogeneous population of cells is created and these cancerous cells together form the primary tumor. A tumor is considered benign if it lacks the ability to invade other tissue. When cells acquire the ability to penetrate and infiltrate surrounding normal tissues, the cancer is considered malignant and has the potential to metastasize (Yokota 2000; Chaffer and Weinberg 2011).

Before tumor cells can start to metastasize, they need to succeed in stimulating angiogenesis. In this way tumor cells gain direct access to the blood circulation. This leads to improved access to the nutrients and oxygen carried by the blood, but also an opportunity for the tumor cells to enter the blood stream. This process is shown in Fig. 2.1 (Dissemination tumor process. Metastasis process involves the relapse of tumor cells from primary tumor to target organ. This metastasis process is dependent to survival capacity of different subpopulation of CTCs into blood system). An alternative route for tumor cells to end up in the blood circulation is through the lymphatic system. The release of cancer cells from the primary tumor to the peripheral blood appears even in patients with small cancer primary tumors (Cho 2010; Nadal et al. 2013; Ross and Slodkowska 2009).

Circulating tumor cells (CTCs) are probably the origin of incurable metastatic disease and an active area of cancer research (Massard and Fizazi 2011). The first observation of tumor cells in blood was made by Thomas Ashworth in 1869 (Ashworth 1869). In subsequent reports CTCs where only observed in blood when present in high number. As technology advanced it became possible to detect the presence of CTCs in a much lower concentration (Panteleakou et al. 2009; Sleijfer et al. 2007).

Tumor cells circulating in the blood can reach in principle most sites of the body. However, despite apparent similarities in clinical and/or histological features, different cancer types do not exhibit the same proclivity to metastasize to the same organs, and the same cancer type can preferentially metastasize to different organs. For example breast cancer generally creates metastases in liver, lung and bone; while prostate cancer is often metastasizes to bone.

This tissue tropism has long been recognized and has intrigued clinicians and pathologists to seek an explanation. James Ewing and others argued that tissue tropism could be accounted for based on mechanical factors and circulatory patterns of the primary tumor. In contrast, Stephen Paget proposed his "seed and soil" hypothesis (Fidler 2003; Paget 1889). This stated that the propensity of different cancers to form metastases in specific organs was due to the dependence of the seed (the cancer) on the soil (the distant organ). The first is mechanical of nature, a large amount of CTC arrests in the first capillary bed they encounter. The second is more biological, the CTCs will form a metastasis in tissue only if they are able to extravasate out of the blood stream and the local environment is suitable for them to grow. Tumor cells thus have a preference for a certain site, and this opens an interesting research field to identify the cell surface molecules on the tumor cells and the endothelial cells align in the capillaries at the specific sites. Current understanding would suggest that both seed and soil factors and anatomic considerations contribute to metastatic tropism. A modern interpretation of the seed and soil hypothesis is an active area of investigation, with molecular definitions accumulating for both the cancer and the microenvironment (Langley and Fidler 2011).

Numerous sequential steps are needed for metastasis, multiple genetic changes are envisioned. A failure in any step would prevent metastasis altogether. Accordingly, tumor cells that can accumulate a full complement of needed alterations to endow them with metastatic ability should be rare. These ideas are supported by early experiments. Work by Fidler and colleagues showed that subpopulations of tumor cells that display significant variation in their metastatic ability and metastatic lesions likely arose from single progenitor cells. Recent studies confirmed that metastasis is an inefficient process. These studies revealed that less than 0.01 % of tumor cells gave rise to metastases (Zhe et al. 2011).

2.3 Epithelial Mesenchymal Transition Process: Contribution to Tumor Aggressiveness

One of the principal events related with initiation metastasis function is epithelial mesenchymal transition (EMT) (Yao et al. 2011). During development, the generation of many adult tissues and organs results from a series of EMT events and the reverse process, a mesenchymal-to-epithelial transition (MET) (Thiery 2002; Aokage et al. 2011). EMT is a multi-step morphogenetic process during which epithelial cells down-regulate their epithelial properties and up-regulate mesenchymal characteristic (Fig. 2.2). EMT is a process by which epithelial cell undergo remarkable morphological changes characterized by a transition from an epithelial phenotype to an elongated fibroblastic phenotype. This event is characterized by epithelial cells loosening their cell-cell adhesion, losing cell polarity, and gaining the ability to invade and migrate under controlled cues. Important regulators include Notch and Wnt/β-catenin pathways, TGF-β family members, and FGF proteins that serve to set up regulatory networks involving EMT transcription factors such as Snail and Twist (Wu and Zhou 2008). These networks do not necessarily regulate cell fate, but rather drive morphogenetic movements by repression of the cell-cell adhesion protein E-cadherin, promoting cytoskeletal rearrangement, and increasing MMP activity. After cells complete EMT-mediated morphogenetic migration, they can then transiently differentiate into epithelial structures.

Among different factors that involve EMT process, several transcription factors have been said to drive this process, including members of the Snail and basic Helix Loop Helix (bHLH) families, and two double zinc finger and homeodomain (ZEB) factors. Between most characterized are ZEB 1 and ZBE 2, snail, slug and Twist. Interesting all these factors are regulated by selective pressures as hypoxia, inflammation, apoptosis, senescence and need for proliferative, metabolic and self-renewal sufficiency (Peinado et al. 2007; Tan et al. 2012).

Interesting, hypoxia can induce Snail and Twist, a direct target of HIF-1α.34. Low oxygen enhances β-catenin activity by inhibiting the activity of glycogen synthase kinase-3β, which normally induces the destruction of β-catenin. Accordingly,

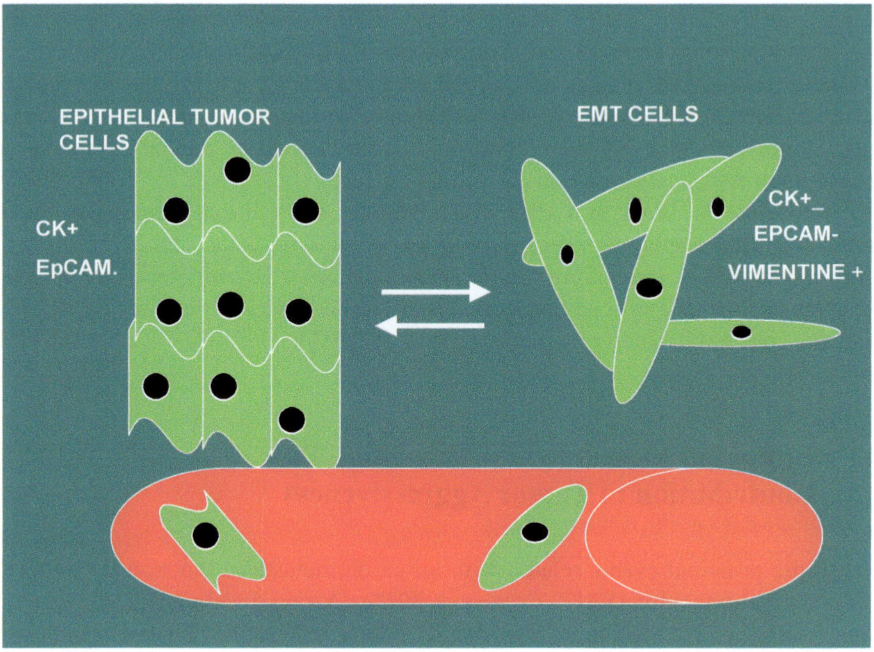

Fig. 2.2 EMT process. EMT process involves loss of epithelial marker and upregulation of mesenchymal markers. This event is necessary to initiation of metastasis

the presence of enhanced β-catenin signaling promotes Snail expression and subsequent EMT. Interestingly, the ability of hypoxia to liberate active β-catenin may set in place a feed-forward loop to help maintain EMT. Activation of Snail represses E-cadherin, which can then further enhance β-catenin and reinforce Snail expression (Tsai and Wu 2012; Lu and Kang 2010).

Similar to hypoxia, the inflammatory microenvironment can also promote EMT (Leibovich-Rivkin et al. 2013). It has recently been demonstrated that TNF-α, which is an inflammatory mediator secreted by TAMs, sets into motion a signaling cascade that funnels through NF-κB and glycogen synthase kinase-3β to stabilize Snail and β-catenin, and thus, enhances cancer cell migration. This ability of cancer cells to awaken an EMT program by co-opting an inflammatory microenvironment may be further reinforced by EMT itself. Snail-induced EMT is able to generate an environment of immunosuppressive T-regulatory cells and impaired dendritic cells partly through TGF-β and thrombospondin-1, helping to further perpetuate the inflammatory surroundings (Colotta et al. 2009).

Further hypoxia and inflammation, the need of CTCs to resist apoptosis and overcome senescence may be additional reasons to shoot EMT switch. Cells that have undergone EMT are associated with increased resistance to apoptosis (Savagner 2010). As mesenchymal cells do not necessarily have a direct contact to the basal lamina, they are not subject to anoikis and the EMT would therefore aid survival of the loosened cancer cells, possibly through prosurvival activity conferred by Snail and Twist. EMT can also help cancer overcome oncogene-induced senescence. Both Twist transcription factors and ZEB1 have been shown to suppress p21cip1 and/or p16ink4a, two p53-regulated cell cycle proteins that are critical in restraining oncogene-transformed cells via senescence (Smit and Peeper 2008). These findings suggest that the pressure to resist apoptosis and avoid senescence can result in activation of EMT transcriptional regulators. On the other hand, obtaining the ability to sustain long-term self-renewal would provide an advantage to cells in a growing tumor mass. Recent data suggest that EMT process is consistent with the acquisition of a "cancer stem–like" (CSC) phenotype (Pardal et al. 2003).

2.4 Circulating Tumor Cells and Cancer Stem Cells

The principal problems for the identification of CSC have result in problems to define CSCs in terms of origin and function. However it is well known that CSCs could be described like more or less differentiated cells; as well as, they are characterized by a peculiar phenotypic and genotypic plasticity (Tang 2012). Interestingly, we can detect these same characteristics into CTCs. In fact, similarly to CSCs, the molecular definition of CTCs presents the most significant difficulty to their clinical application. CTCs are a highly heterogeneous population of cancer cells, and their genotypic and phenotypic characterization not only to confirm their malignant origin but also to follow their immune-phenotypic changes with tumor progression and to identify diagnostically and therapeutically relevant targets for individual cancer therapies (Mavroudis 2010).

From the original observations of Bonnet and Dick on leukaemic stem cells (Bonnet and Dick 1997), many markers have been proposed for different CSCs (including CD133, CD44, CD24, and CD138; ABCG2, ABCB5, EpCAM, and ALDH1; CXCR4; various signaling pathways Hedgehog, Notch, Wnt/β-catenin, BM1, and PTEN; and different microRNAs) (Medema 2013).

On the other hand, recent studies have demonstrated the association among EMT process, metastasis development, recurrence with biology of CSCs. Recently it has been described the existence of stem cell-enriched subpopulations with a CD44 high/CD24 low cell surface marker profile, as well as having significantly higher levels of many EMT-related transcription factors, such as Snail and Twist, when compared with their CD44 low/CD24 high counterparts. Furthermore, these EMT transcription factors are able to directly increase the number of cells with stem-like characteristics. Although it remains to be determined whether EMT is occurring in cancer stem cells or whether EMT occurs in non–self-renewing cells that then give them stem-like properties, these data argue for a another compelling reason for cancer cells to acquire EMT properties (Tirino et al. 2013).

The connection between CSCs and CTC has been explained through the expression of EMT markers on CTCs suggesting that they are cancer stem cells. In fact, many CTCs do not express epithelial markers (EPCAM, Cytokeratin, etc) but express other markers like mesenchymal markers and stem markers (like Vimentine, Snail, N-cadherin, CD44, ALDH1).

The acquisition of these phenotypes involve cell-cell connection loss, while elasticity is gained by getting rid the rigid of cytoskeleton, facilitating passage of CTCs into the blood. More interesting, these EMT process is specially observed on single CTCs, while epithelial characteristics, like expression of EPCAM, are often observed in cell cluster (van de Stolpe et al. 2011). Thus, the gain of these capabilities to the extravasation and seeding in an organ tissue or evade the action of immune system are similar to normal mesenchymal stem cells. Finally, only the characterization of these CTCs like cancer stem cell could provide the assets to growth, like can be the reversion from EMT process to MET process (Mesenchymal Epithelial Transition) (Yao et al. 2011). Assuming this information is understandable that CTCs and CSC are the same population. Thus, CTSCs (circulating tumor stem cells) may hold to improve our known of metastasis process.

Recent results have shown that the tumor cells can adopt different characteristics, depending of type of microenvironment or niche (van de Stolpe 2013). Such that, some studies have demonstrated the expression of melanoma tumor marker in cells with endothelial mesenchymal, a phenomenon called "vascular mimicry" (Fan et al. 2013). Another example of this cell plasticity of CTCs, is their capacity to co-expression of specific tumor maker with the leucocyte marker CD45, symptomatic of the possibility the adaptation present in CTCs and suggesting the option that they represent a progeny of tumor stem cell (Conejo-Garcia et al. 2005).

In conclusion, the adaption capacity of CTCs, through the phenotypic and genetic plasticity present in these cells only can be explain because at least some CTCs presents feature of progenitor cancer cells. The presence of these specific characteristic involve the capacity of CTCs for adapting in the different microenvironments and opt for proliferation or enter in dormancy status.

2.5 Epigenetic Regulation of miRNA and Their Role in EMT Process

Cancer is a disease involving the failure of function of regulatory genes that control normal cellular homeostasis. Multiple mutations, almost all human cancers contain important epigenetic abnormalities that cooperate with genetic lesions to generate the cancer phenotype. Epigenetic aberrations arise early in carcinogenesis preceding gene mutations and therefore provide targets for early detection (Loeb et al. 2003).

MicroRNAs (miRNAs or miR) are small non-coding regulatory RNAs with sizes of 17–25 nucleotides (miRBase; http://microrna.sanger.ac.uk/). The first miRNA lin-4 and let-7 were identified by Ambros and Ruvkun in 1993 as a result of developmental studies on the worm *Caenorhabditis elegans* that identified important roles for noncoding RNAs (ncRNAs). In fact, in animal cells, post transcriptional regulation by miRNA requires an mRNA that is complementary to target sequence (position 2–7 of the mature miRNA). So, it is now appreciated that ncRNAs potentially influence cellular phenotypes by regulating of hundreds of target mRNAs (Gregory and Shiekhattar 2005).

The pattern of miRNA expression can be correlated with cancer type, stage, and other clinical variables. Thus, the analysis of miRNA's expression, suggests the main role in different aspects of cancer biology such as proliferation, apoptosis, invasion/metastasis, and angiogenesis. In fact, several microRNAs, such as miR-10b, miR-9, and members of the miR-200 family, play critical roles in EMT or EMT-related events. For example, miR-10b influences cell migration, invasion, and metastasis by repressing HOXD10, a known inhibitor of a RHOC-mediated promotility program (Zadran et al. 2013).

The EMT is a complex series of morphological changes that conclude in the loss of epithelial characteristics and the acquisition of a mesenchymal motile phenotype. In the context of cancer, this process facilitates the dissemination of cancer cells and survival of CTCs until to seeding a secondary site where this process is reversible, the reversal of EMT is known like Mesenchymal Epithelial Transition (MET) (Kalluri 2009). Due to the dynamic nature of this process, epigenetic alterations are involved in dissemination, EMT, MET and finally metastasis. Interesting, this epigenetic regulation impacts miRNA that regulate gene pathway involve EMT process (Taube et al. 2013). In fact, in the context of the epigenetic regulation of EMT, it was found that the CpG island near the miRNA-200c and miRNA-141 transcription start is unmethylated in miRNA-expressing tumor-normal cells and is closely methylated in miRNA-negative and invasive tumor cells (Bullock et al. 2012). miRNA expression is further facilitated by the enrichment of chromatin-permissive histone modifications (H3 acetylation and H3K4 trimethylation). Davalos et al. demonstrated that in epithelial cancer cell lines, the 5′-CpG islands of miRNA-200 family members are unmethylated, whereas the hypermethylation-mediated silencing of these miRNAs was found in transformed mesenchymal cells (Davalos et al. 2011). The reversibility of this methylation state mediates the shift between EMT and MET. Similar results were obtained in bladder cancer and breast cancer cell lines (Davalos et al. 2011). This results in increased histone acetylation and E-cadherin expression. Interestingly, the chemo- and radiosensitivity of these breast cancer cells was increased by enhanced p53-mediated apoptotic pathways (Mirzayans et al. 2012).

In this way, several miRNA has been involved in epigenetic EMT regulation. For example, miR-10b, miR-9, and members of the miR-200 family, play critical roles in EMT or EMT-related events (Samantarrai et al. 2013). The expression of miR-10b is under the control of the EMT transcription factor Twist. miR-10b influences cell migration, invasion, and metastasis by repressing HOXD10, a known inhibitor of a RHOC-mediated promotility program. The expression of miR-10b is under the control of the EMT transcription factor Twist. In addition, miR-10b is essential for Twist-induced EMT and for metastasis (Ma 2010). In contrast to miR-10b, which is specifically up-regulated in metastatic cancer cells, miR-9 expression is up-regulated in both metastatic and non-metastatic tumor cells relative to non-transformed cells. These observations, together with the fact that miR-9 expression can be increased by the actions of MYC and MYCN oncoproteins, cause us to propose that miR-9 expression is often induced at earlier stages of multi-step tumor progression (Nesbit et al. 1999). Thus miR-9 promotes metastasis by targeting E-cadherin, which results in β-catenin signaling and VEGF expression. Accordingly, motility, invasion, angiogenesis, and metastasis ensue. Consequently, miR-9 can be considered as a pro-metastatic miRNA and a negative regulator of the key metastasis suppressor, E-cadherin. Special importance has MiR-200 family members which also regulate EMT by inhibiting ZEB1 and ZEB2, through the suppression of E-cadherin (Gregory et al. 2008).

Other microRNAs such as miR-335, miR-126, miR-31, and let-7 also suppress distant spread. MiR-335 inhibits metastasis and invasion by targeting the LMS genes *SOX4* and *TNC*, while miR-125 suppresses overall tumor growth and proliferation.119 MiR-31 represses multiple steps in the metastatic cascade, an effect that is related to its influence on multiple metastasis genes (Shu et al. 2011).

Finally, must take into account miRNA like new option therapeutic in the context of metastasis process and especially in the context of EMT process. In fact, experiments in vivo and in vitro have led to development of new potential targeting miRNA. The ability of miRNAs to target genes that are implicated in the same pathway and/or in interacting pathways provides the justification for the use of miRNAs to achieve an orchestrated broad silencing of pro-tumoral pathways (Table 2.1). (Garzon et al. 2010).

As discussed above, miRNA activity can be depend on the cellular microenviorement and the same miRNA can be different targets in the same individual but in different cell types, and as a result opposite effects. Consequently the modulation on same miRNAs can have different effects on different cell types.

Table 2.1 miRNA and cancer

miRNA	Cancer involvement	Mechanism of action	Association with tumor progression
miR200	Breast cancer, lung cancer, prostate cancer, pancreatic cancer, ovarian cancer	miR-200 targets the E-cadherin transcriptional repressors ZEB1 and ZEB2	Promote the last step of metastasis in which migrating cancer cells undergo MET during their colonization at distant tissues
mir34	Colon cancer, lung cancer, breast cancer, Kidney cancer	miR-34a functions as a tumor suppressor, in part, through a SIRT1-p53 pathway	miR-34 inhibition of SIRT1 leads to an increase in acetylated p53 and expression of p21 and PUMA, transcriptional targets of p53 that regulate the cell cycle and apoptosis
mir 10b	Hepatocellular cancer (HCC), colon cancer	*mir-10b* gene is a promoter region containing a binding site for the twist transcription factor (Twist), leading to a reduced translation of the tumour suppressor HOXD10	Upregulation of miR-10 results in upregulation of RhoA/RhoC, Rho kinase activation and tumour cell invasion

Table 2.1 (continued)

miRNA	Cancer involvement	Mechanism of action	Association with tumor progression
miR17	Prostate cancer, lung cancer	Fibronectine and the fibronectin type-III domain containing 3A (FNDC3A) are two targets that have their expression repressed by miR-17	Down-regulation of miR-17-5p modulates androgen receptor transcriptional activity which is critical to the initiation and progression of prostate cancer.
		Suppress critical primary mitochondrial antioxidant enzymes, such as manganese superoxide dismutase, glutathione peroxidase-2 and thioredoxin reductase-2 (TrxR2)	Rbl2 (tumor suppressor) is identified as a target of miR-17 in lung cancer
		miR-17 is a target of c-Myc, which also inhibits E2F1 translation. The expression of E2F1 is negatively regulated by miR-17-5p by binding to target sites in its 3' untranslated region. Myc simultaneously activates E2F1 transcription and limits its translation, allowing a tightly controlled proliferative signal	
miR92	Breast cancer, HCC	miR-92 is regulated by oestrogen which is upstream of the ERβ1	–
		Biomarker for HCC	
miR21	Breast cancer, colon cancer, pancreatic cancer, prostate cancer	Oncomir: Oncomirs cause cancer by down-regulating genes by both translational repression and mRNA destabilization mechanisms	Increased oncomir activity: suppressing a tumor suppressor gene
		miR 21: down-regulate the tumor suppressor PDCD4	Underexpressed oncomirs, regulation is attenuated, allowing the cell to proliferate
MiR22	Prostate cancer, breast cancer	miR-22 can function as a tumour suppressor	Oncogenesis
		Principal targets: histone deacetylase 4 (HDAC4), which is known to have a critical role in cancer development MYC. regulator gene that codes for a transcription factor	

Table 2.1 (continued)

miRNA	Cancer involvement	Mechanism of action	Association with tumor progression
miR31	Breast cancer prostate cancer HCC	miR31 affect the levels of gene transcription factor p53, responsible for encoding the tumour suppressor protein p53	Oncogenesis
Let-7 family	Lung cancer, colon cancer, HCC, breast cancer	Regulator of *RAS* expression in human cells	Acts as a tumor suppressor
miR-9	Breast cancer, renal carcinoma	*miR-9* family microRNAs have been identified as a tumor suppressor miRNA in cancers by epigenetic modification	*miR-9* family microRNAs have been identified as a tumor suppressor miRNA in cancers
miR335	Breast cancer, ovarian cancer, lung cancer	Tumor suppressive miRNAs miR-335 suppresses metastasis and migration by targeting the progenitor cell transcription factor *SOX4* and the extracellular matrix component tenascin C (*TNC*)	Migration and mestastasis

In this context, the main cause of reducing expression of tumour suppressor miRNAs in human cancer are genetic deletion of the miRNA loci or epigenentic silencing via CpG island hypermethylation in the promoter of the miRNA genes. In this way, molecular approaching is focused in the reverting epigenetic silencing or enhances the biogenesis of miRNAs, thus restoring the molecular levels of miRNAs. However, nowadays strategies for block miRNA fuctions, oligonucleotide based are been explores. However, the concept of combination of miRNA cocktails with chemotherapy o biologic therapy might be benefactions to patients (Garzon et al. 2010).

Assuming that of miRNA are involved in the regulation of EMT and in the pathways gene regulation, the study of these molecules in relation with CTCs as essential precursors of metastasis is obvious. Thus, this miRNA could be responsible to the phenotype versatility present in these cells. Consequently, the identification of miRNA in CTCs could be identified new therapeutic targets, directly focused in CTCs. In other works, therapeutic targets that specifically take out the possibility to development a metastasis by remove of CTCs present in peripheral blood, as far as biologic characteristic present in these CTCs (Fig. 2.3).

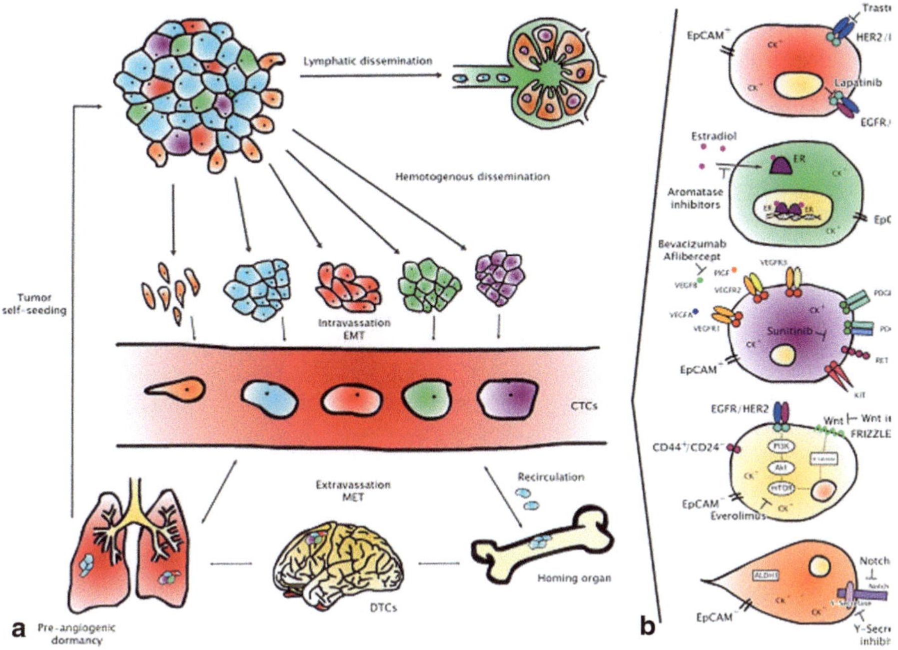

Fig. 2.3 Dissemination process in breast cancer model. **a** Cells tumor disseminate from primary tumor to target organ by blood system or lymph system. **b** Circulating Tumor Cells present high heterogeneity. This heterogeneity may affect to response to therapy

References

Aokage K, Ishii G, Ohtaki Y, Yamaguchi Y, Hishida T, Yoshida J et al (2011) Dynamic molecular changes associated with epithelial–mesenchymal transition and subsequent mesenchymal–epithelial transition in the early phase of metastatic tumor formation. Int J Cancer 128(7):1585–1595.

Ashworth T (1869) A case of cancer in which cells similar to those in the tumours were seen in the blood after death. Aust Med J 14(3):146–149

Bonnet D, Dick JE (1997) Human acute myeloid leukemia is organized as a hierarchy that originates from a primitive hematopoietic cell. Nat Med 3(7):30–737

Brabletz T, Lyden D, Steeg PS, Werb Z (2013) Roadblocks to translational advances on metastasis research. Nat Med 19(9):1104–1109

Bullock MD, Sayan AE, Packham GK, Mirnezami AH (2012) MicroRNAs: critical regulators of epithelial to mesenchymal (EMT) and mesenchymal to epithelial transition (MET) in cancer progression. Biol Cell 104(1):3–12

Chaffer CL, Weinberg RA (2011) A perspective on cancer cell metastasis. Science 331(6024):1559–1564

Cho W (2010) Recent progress in genetic variants associated with cancer and their implications in diagnostics development. Expert Rev Mol Diagn 10:699–703

Cock-rada A, Weitzman JB (2013) The methylation landscape of tumour metastasis. Biol Cell 105:73–90

Colotta F, Allavena P, Sica A, Garlanda C, Mantovani A (2009) Cancer-related inflammation, the seventh hallmark of cancer: links to genetic instability. Carcinogenesis 30(7):1073–1081
Conejo-Garcia JR, Buckanovich RJ, Benencia F, Courreges MC, Rubin SC, Carroll RG et al (2005) Vascular leukocytes contribute to tumor vascularization. Blood 105(2):679–681
Davalos V, Moutinho C, Villanueva A, Boque R, Silva P, Carneiro F et al (2011) Dynamic epigenetic regulation of the microRNA-200 family mediates epithelial and mesenchymal transitions in human tumorigenesis. Oncogene 31(16):2062–2074
Fan Y, Zheng M, Tang Y, Liang X (2013) A new perspective of vasculogenic mimicry: EMT and cancer stem cells (Review). Oncol Lett 6(5):1174–1180
Fidler IJ (2003) The pathogenesis of cancer metastasis: the 'seed and soil' hypothesis revisited. Nat Rev Cancer 3(6):453–458
Garzon R, Marcucci G, Croce CM (2010) Targeting microRNAs in cancer: rationale, strategies and challenges. Nat Rev Drug Discov 9(10):775–789
Gregory RI, Shiekhattar R (2005) MicroRNA biogenesis and cancer. Cancer Res 65(9):3509–3512
Gregory PA, Bert AG, Paterson EL, Barry SC, Tsykin A, Farshid G et al.(2008). The miR-200 family and miR-205 regulate epithelial to mesenchymal transition by targeting ZEB1 and SIP1. Nat Cell biol 10(5):593–601
Gupta GP, Massagué J (2006) Cancer metastasis: building a framework. Cell 127(4):679–695
Kalluri R (2009) EMT: when epithelial cells decide to become mesenchymal-like cells. J Clin Invest 119(6):1417
Kim M, Oskarsson T, Acharyya S, Nguyen DX, Zhang XH, Norton L et al (2009) Tumor self-seeding by circulating cancer cells. Cell 139(7):1315–1326
Kovács K, Hegedus B, Kenessey I, Tímár J (2013) Tumor type-specific and skin region-selective metastasis of human cancers: another example of the "seed and soil" hypothesis. Cancer Metastasis Rev 32(3–4):493–499
Langley RR, Fidler IJ (2011) The seed and soil hypothesis revisited—the role of tumor-stroma interactions in metastasis to different organs. Int J Cancer 128(11):2527–2535
Leibovich-Rivkin T, Liubomirski Y, Bernstein B, Meshel T, Ben-Baruch A (2013) Inflammatory factors of the tumor microenvironment induce plasticity in nontransformed breast epithelial cells: EMT, invasion, and collapse of normally organized breast textures. Neoplasia 15(12):1330-1346
Loeb LA, Loeb KR, Anderson JP (2003) Multiple mutations and cancer. Proc Natl Acad Sci U S A 100(3):776–781
Lorusso G, Rüegg C (2012) New insights into the mechanisms of organ-specific breast cancer metastasis. Semin Cancer Biol 22(3):226–233 (Elsevier)
Lu X, Kang Y (2010) Hypoxia and hypoxia-inducible factors: master regulators of metastasis. Clin Cancer Res 16(24):5928–5935
Ma L (2010) Role of miR-10b in breast cancer metastasis. Breast Cancer Res 12(5):210
Massard C, Fizazi K (2011) Targeting continued androgen receptor signaling in prostate cancer. Clin Cancer Res 17(12):3876–3883
Mavroudis D (2010) Circulating cancer cells. Annals Oncol 21(suppl 7):vii95–vii100
Medema JP (2013) Cancer stem cells: the challenges ahead. Nat Cell biol 15(4):338–344
Mirzayans R, Andrais B, Scott A, Murray D (2012) New insights into p53 signaling and cancer cell response to DNA damage: implications for cancer therapy. J Biomed Biotechnol 2012:170325
Nadal R, Lorente JA, Rosell R, Serrano MJ (2013) Relevance of molecular characterization of circulating tumor cells in breast cancer in the era of targeted therapies. Expert Rev Mol Diagn 13(3):295–307
Nesbit CE, Tersak JM, Prochownik EV (1999) MYC oncogenes and human neoplastic disease. Oncogene 18(19):3004–3016
Paget S (1889) The distribution of secondary growths in cancer of the breast. Lancet 133(3421):571–573
Pantel K, Brakenhoff RH (2004) Dissecting the metastatic cascade. Nat Rev Cancer 4(6):448–456
Panteleakou Z, Lembessis P, Sourla A, Pissimissis N, Polyzos A, Deliveliotis C et al (2009) Detection of circulating tumor cells in prostate cancer patients: methodological pitfalls and clinical relevance. Mol Med 15(3–4):101–114

Pardal R, Clarke MF, Morrison SJ (2003) Applying the principles of stem-cell biology to cancer. Nat Rev Cancer 3(12):895–902
Peinado H, Olmeda D, Cano A (2007) Snail, Zeb and bHLH factors in tumour progression: an alliance against the epithelial phenotype? Nat Rev Cancer 7(6):415–428
Ross JS, Slodkowska EA (2009) Circulating and disseminated tumor cells in the management of breast cancer. Am J Clin Pathol 132(2):237–245
Samantarrai D, Dash S, Chhetri B, Mallick B (2013) Genomic and epigenomic cross-talks in the regulatory landscape of miRNAs in breast cancer. Mol Cancer Res 11(4):315–328
Savagner P (2010) The epithelial–mesenchymal transition (EMT) phenomenon. Ann Oncol 21(suppl 7):vii89–vii92
Shu M, Zheng X, Wu S, Lu H, Leng T, Zhu W et al (2011) Targeting oncogenic miR-335 inhibits growth and invasion of malignant astrocytoma cells. Mol Cancer 10(1):59
Sleijfer S, Gratama J, Sieuwerts AM, Kraan J, Martens JW, Foekens JA (2007) Circulating tumour cell detection on its way to routine diagnostic implementation? Eur J Cancer 43(18):2645–2650
Smit MA, Peeper DS (2008) Deregulating EMT and senescence: double impact by a single twist. Cancer Cell 14(1):5–7
Tan E, Thuault S, Caja L, Carletti T, Heldin C, Moustakas A (2012) Regulation of transcription factor Twist expression by the DNA architectural protein high mobility group A2 during epithelial-to-mesenchymal transition. J Biol Chem 287(10):7134–7145
Tang DG (2012) Understanding cancer stem cell heterogeneity and plasticity. Cell Res 22(3):457–472
Taube JH, Malouf GG, Lu E, Sphyris N, Vijay V, Ramachandran PP et al (2013) Epigenetic silencing of microRNA-203 is required for EMT and cancer stem cell properties. Sci Rep 3:2687
Thiery JP (2002) Epithelial–mesenchymal transitions in tumour progression. Nat Rev Cancer 2(6):442–454
Tirino V, Desiderio V, Paino F, De Rosa A, Papaccio F, La Noce M et al (2013) Cancer stem cells in solid tumors: an overview and new approaches for their isolation and characterization. FASEB J 27(1):13–24
Tsai Y, Wu K (2012) Hypoxia-regulated target genes implicated in tumor metastasis. J Biomed Sci 19(1):102
Valastyan S, Weinberg RA (2011) Tumor metastasis: molecular insights and evolving paradigms. Cell 147(2):275–292
Van De Stolpe A (2013) On the origin and destination of cancer stem cells: a conceptual evaluation. Am J Cancer Res 3(1):107–116
Van De Stolpe A, Pantel K, Sleijfer S, Terstappen LW, Den Toonder JM (2011) Circulating tumor cell isolation and diagnostics: toward routine clinical use. Cancer Res 71(18):5955–5960
Wu Y, Zhou BP (2008) New insights of epithelial-mesenchymal transition in cancer metastasis. Acta Biochim Biophys Sin (Shanghai) 40(7):643–650
Yao D, Dai C, Peng S (2011) Mechanism of the mesenchymal–epithelial transition and its relationship with metastatic tumor formation. Mol Cancer Res 9(12):1608–1620
Yokota J (2000) Tumor progression and metastasis. Carcinogenesis 21(3):497–503
Zadran S, Remacle F, Levine R (2013) miRNA and mRNA cancer signatures determined by analysis of expression levels in large cohorts of patients. Proc Natl Acad Sci 110(47):19160–19165
Zhe X, Cher ML, Bonfil RD (2011) Circulating tumor cells: finding the needle in the haystack. Am J Cancer Res 1(6):740–751

Chapter 3
LINE-1 Retrotransposons and Their Role in Cancer

Raheleh Rahbari, Laleh Habibi, Jose L. Garcia-Puche, Richard M. Badge and Jose Garcia-Perez

Contents

R. Rahbari (✉)
The Wellcome Trust Sanger Institute, Wellcome Trust Genome Campus, Hinxton, UK
e-mail: rr11@sanger.ac.uk

L. Habibi
Faculty of Pharmacy, Nanotechnology Research Centre,
Tehran University of Medical Sciences, Tehran, Iran

J. L. Garcia-Puche · J. Garcia-Perez
Oncology Department, GENYO Centre for Genomics and Oncological Research,
Andalusian Regional Government, Granada, Spain

R. M. Badge
Department of Genetics, University of Leicester, Leicester, UK

P. Mehdipour (ed.), *Epigenetics Territory and Cancer*,
DOI 10.1007/978-94-017-9639-2_3

Abstract Retrotransposons comprise over 40% of the human genome and are a major contributor to genome diversity and evolution. They contribute to human genome variation through both germline and somatic retrotransposition. Over recent years, studies on the biology of cancer have revealed that somatic retrotransposition is a feature of many cancer genomes. The most recent comparison between 200 pairs of tumours and normal tissue, across 11 tumour types, has revealed frequent somatic retrotransposition in particular cancers; lung squamous cell, head and neck squamous cell, colorectal and endometrial carcinomas. Importantly some of these insertions occur in cancer-related genes underlining retrotransposition's role as a mutagen. It is now clear that retrotransposons contribute to genome instability during cancer progression. However, the exact role of retrotransposons in tumuorigenesis, tumour progression and prognosis still remains a subject of an active discussion in the field of cancer biology.

In this chapter, we have attempted to explain the biology of retrotransposons in the human genome, with the main focus on LINE-1 elements. We then have discussed how LINE-1 causes genome instability in the genome and the host defence mechanisms deployed to supress their retrotransposition. Next, we discuss the role of LINE-1 activity during tumourigenesis and consider the recent findings concerning their activity in different types of cancers. Finally, we explore how retrotransposons can be used as diagnostic tools in cancer.

3.1 Introduction

For many years geneticists assumed that a genome was an assembly of genes interrupted by their regulatory elements. However, it was soon recognised that the morphological complexity of an organism does not necessarily directly correlate with genome size (Thomas 1971; Gregory and Hebert 1999).

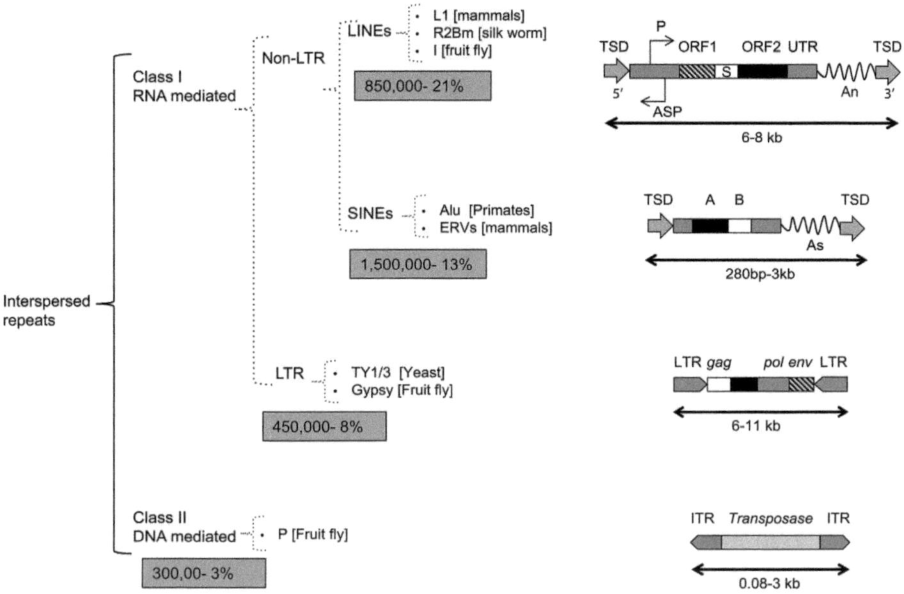

Fig. 3.1 Classes of interspersed repeats with their examples. (termini of the branches, and their taxa of origin in square brackets). For each class their copy number and contribution (%) to the human genome summarised in the grey boxes. *TSD*: Target Site Duplication, *P*: Promoter, *ORF1/2*: Open Reading Frames, *S*: Spacer, *An*: poly Adenosine tract, *A* and *B*: RNA polymerase III conserved regions, *LTR*: Long Terminal Repeat, *gag*: group specific antigen, *pol*: polymerase, *env*: envelope, *ITR*: Inverted Terminal Repeats, Adapted from lander et al. 2001.

Progressive developments in sequencing technologies and their large-scale application culminated in the first draft of the human genome (Lander et al. 2001). Although suspected for some time, the draft sequence confirmed that the human genome contained a very substantial amount of non-coding and repetitive sequences. The non-coding portion formed more than 95 % of the draft human genome sequence, ~50 % of which was clearly repetitive sequence. 10 years later, in 2011, Koning et al. revisited this analysis with new tools, and argued that up to 70 % of our genome has been generated by the activity of Transposable Elements (TEs) . These repetitive TE sequences have often described as "junk DNA" as there was not any evidence of beneficial function for the host (Ohno 1972; Pagel and Johnstone 1992). A comparatively small percentage of repetitive sequences are comprised of simple tandem repeats with short (microsatellites and telomeric repeats) or longer repeat periods (minisatellites, satellite DNAs and centromeric repeats), but the vast majority derives from TEs. Barbara McClintock first identified TEs in the late 1940s in maize, *Zea mays* (McClintock 1950). Today, many different kinds of mobile DNA have been identified in virtually all species, ranging from bacteria and yeast to plants and mammals, as illustrated in Fig. 3.1.

The question of why TEs have been so successful throughout evolution is the subject of ongoing discussion. TEs have been called "selfish genes" (Dawkins

1976) and "genomic parasites" (Yoder et al. 1997) in relation to their host genome, but evidence has accumulated over the last several decades demonstrating that, despite their disease-causing potential (reviewed in Kazazian 1998), TEs might have some overall beneficial effect. For example, TEs can increase genomic diversity and consequently drive genome evolution within a species (Boeke and Pickeral 1999; Nekrutenko and Li 2001; Seleme et al. 2006); they can play a role in the stress response of the host cell (Li and Schmid 2001); and in some lineages can take over vital cellular functions, such as telomere function (Pardue et al. 1996).

TEs can also have practical uses. For example, human specific mobile element insertions (mostly L1 and Alu) can be used for inferring human geographical origin, sex identification, DNA identification and quantification (Xing et al. 2007). However, while the contribution of mobile elements to host genomic architecture and fluidity is undeniable, relatively little is currently known about the evolutionary dynamics of their mobilisation in humans.

3.1.1 Human Transposable Elements

In *Homo sapiens*, TEs are responsible for the formation of at least 45 % of the genome (Lander et al. 2001). Figure 3.1 illustrates the different types of mobile elements that have been involved in mammalian and human genome expansion.

TEs can be classified into two groups based upon their genomic integration method (Pace and Feschotte 2007). Class I elements transpose via an RNA intermediate, utilising a reverse transcriptase activity, and include long and short interspersed elements (LINEs and SINEs), as well as long terminal repeat elements (LTR). The Class I transposition mechanism can be thought of as a 'copy and paste' method and as such is inherently replicative. Class II mobile elements integrate into the human genome, using a DNA intermediate, through a 'cut and paste' mechanism (Pace and Feschotte 2007; Kazazian et al. 2002).

3.1.1.1 DNA Transposons; Class II Transposable Elements

The mechanism of DNA transposition is a 'cut and paste' mechanism that is not inherently replicative. DNA transposons mobilise via a DNA intermediate, which is mediated by a transposase. Only about 3 % of the human genome is derived from DNA transposons (Fig. 3.1) (Lander et al. 2001).

The evolutionary history and genomic impact of transposons have been well studied in mammals. All ~300,000 DNA transposons identified in the human genome reference sequence are genomic fossils that have been inactive for at least 50 Myr (Lander et al. 2001; Pace and Feschotte 2007; Smit and Riggs 1996). Therefore any effects of transposition in contemporary human genomes must originate from a different class of transposable element. Indeed, the most active transposable elements in humans are L1 retrotransposons. Comparative genomic analysis between the human genome reference and the draft chimpanzee genome showed that 1174 human specific L1 insertions have accumulated in the 6–8 Myr since these species common ancestor (Mills et al. 2006). Due to their ongoing mobilization in humans, it is this group of retrotransposons that are the subject of this chapter.

3.1.1.2 Retrotransposons; Class I Mobile Elements

By far, the largest portion of human mobile DNA originates from retrotransposons. In contrast to DNA transposition, DNA retrotransposition is inherently replicative and functions via a 'copy-and-paste' mechanism, involving transcription of the complete element, reverse transcription of the RNA into cDNA, and integration of the cDNA into a new locus in the genome. Thus, one functional progenitor retrotransposon can generate multiple copies at new genomic locations. This circumstance, and the fact that there is at least one family of retrotransposons still active in humans (the L1Hs family), may account for the excess of retroelements in the human genome. Retrotransposons can be divided into two major classes that are phylogenetically and structurally unrelated (Craig et al. 2002). The long terminal repeat (LTR) retrotransposons account for 8% of the human genome, and are characterised by direct LTRs flanking the element's coding regions (Fig. 3.1). LTR and non-LTR retrotransposons do share some important structural characteristics. They each have a robust and functional promoter (Hata and Sakaki 1997), which is responsible for transcription of full-length RNA, and they each encode a reverse transcriptase enzyme in order to produce a cDNA copy of this RNA. However, there are also important differences: in the autonomous elements (LTR retrotransposons), the cDNA integrates into new genomic loci using its own unique protein machinery (Curcio and Derbyshire 2003) and the integration process is initiated by an element-encoded integrase (IN).

3.1.1.3 Long Terminal Repeat (LTR) Retrotransposons

LTR retrotransposons are also called 'retrovirus-like elements' or 'endogenous retroviruses' because their replication pathway is similar to that of retroviruses. They are thought to originate from retroviruses that have lost a functional *env*-gene, confining them to strictly intracellular replication (Esnault et al. 2008). Thus, endogenous retroviruses cannot infect other cells, and go through their replicative cycle within a single cellular lineage. With the possible exception of HERV-K, which is a putatively active human endogenous retrovirus, all known human LTR-retrotransposons are genomic fossils that have not been active for the last 40 Myr (Costas and Naveira 2000; Lander et al. 2001). However, there is currently no evidence for mobilization events in modern day humans, despite reports of LTR promoter reactivation in two cancers (Katoh and Kurata 2013).

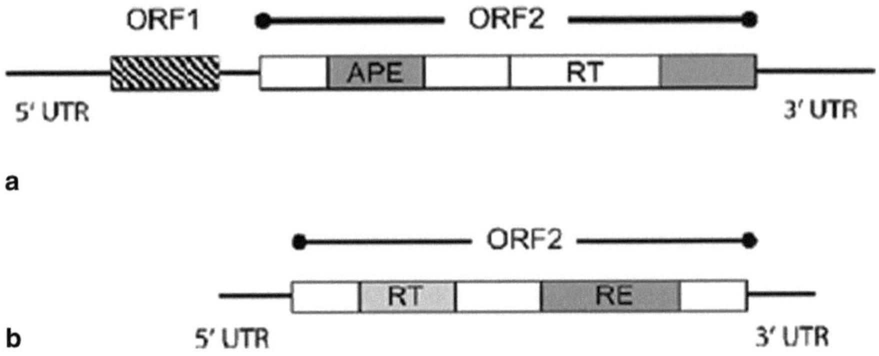

Fig. 3.2 Schematic diagrams of RE-type and APE-type non-LTR retrotransposons Illustrating their differences in structural organisation and in their coding capacity. **a.** APE-type non-LTR retrotransposons, **b.** RE-type non LTR retrotransposons, *UTR* untranslated region, *ORF* open reading frame, *APE* Apurinic/APyrimidinic Endonuclease, *RT* reverse transcriptase, *RE* restriction enzyme-like endonuclease (Craig et al. 2002)

3.1.1.4 Non-Long Terminal Repeat (non-LTR) Retrotransposons

Non-LTR retrotransposons are evolutionarily more ancient than LTR retrotransposons (Furano 2000). Sequence comparisons indicate that they share a common origin with RT-bearing group II introns of bacteria and mitochondria (Yang et al. 1999). Comprising more than one third of human DNA (32 %), non-LTR retrotransposons clearly have had a great impact.

Based on the structure of their coding regions, the autonomous non-LTR elements are further subdivided into the restriction enzyme (RE) type and the apurinic/apyrimidinic endonuclease (APE) type. The RE-type non-LTR retrotransposons are characterised by a single open reading frame (ORF) with a RE-like EN domain following the C-terminal end of the RT domain (Malik et al. 1999). RE-type elements represent the oldest lineage of non-LTR retrotransposons (Malik et al. 1999).

Most retrotransposons discovered so far are APE-type non-LTR retrotransposons. They are recognised by having one or two ORFs and the existence of an EN domain that is distantly related in sequence to the apurinic/apyrimidinic (AP) endonucleases (Martín et al. 1995; Feng et al. 1996). The EN domain is localised at the N-terminal end of ORF2p, upstream of the RT domain. Based on the elements' structures, and on phylogenetic analyses of their RT domains, we can currently distinguish four groups of APE-type non-LTR retrotransposons, and these can further be subdivided into a further 11 clades (Burke et al. 1999; Eickbush and Malik 1999; Lovsin et al. 2001). (Fig. 3.2).

3.1.2 Autonomous and Non-Autonomous Non-LTR Retrotransposons

The non-LTR retrotransposons can also be categorised as either autonomous or non-autonomous retrotransposons. Autonomous retrotransposons are able to encode the proteins required for their own retrotransposition. However, non-autonomous elements are unable to retrotranspose without appropriating the retrotransposition machinery of autonomous elements (Lander et al. 2001; Dewannieux et al. 2003).

3.2 Human Long Interspersed Elements (LINEs)

In this chapter we focus on human LINE1 elements, as they are the only active autonomous retrotransposons in our genome, and so are more likely to contribute to cancer. First, we will expand on what is known about the LINE1 family and their structure, the major roles of LINE1 in our genome and finally, we will discuss the role of LINE1 in different cancers and its potential to contribute to disease progression.

Long interspersed elements-1 (LINE-1s or L1s) are the only autonomous non-LTR retrotransposons in the human genome, *i.e.* they encode the proteins required for their own retrotransposition. LINE retrotransposons are further classified into three sub-groups in the human genome: LINE1 (L1), LINE2 (L2) and LINE3 (L3). LINE1 is the only active member of this family and it has a copy number of over 500,000, and makes up about 17 % of the genome. LINE2 and LINE3 are older lineages that together comprise less than 4 % of the genome. They have accumulated numerous mutations during the course of evolution, and so are unlikely to be still actively retrotransposing (Lander et al. 2001). In addition ~99 % of LINE1s are inactive due to a 5' truncation, internal rearrangements or deletions, but it has been estimated that in an average diploid human genome there are 80–100 full-length L1s with intact ORFs, which are likely to be competent for retrotransposition (RC-L1s) (Deininger et al. 2003; Brouha et al. 2003; Beck et al. 2010).

During their mobilization process, LINE-1 element proteins display strong *cis* preference, *i.e.* the proteins preferentially retrotranspose their encoding RNA, largely ensuring that only functional copies are propagated (Wei et al. 2001). This *cis* preference, from an evolutionary point of view, minimises the impact of the accumulation of mutated elements on active L1 retrotransposition. However, it is known that the LINE1 autonomous machinery can also act in *trans* to retrotranspose non-autonomous retrotransposons such as short Interspersed Elements (SINEs), SVA (SINE/VNTR/Alu) elements (Callinan et al. 2006) and other cellular RNAs (Esnault et al. 2000; Boeke 1997). In rare cases, the *cis* preference of LINEs is circumvented by spliced mRNAs of cellular genes. This results in an intronless and promoterless retropseudogene copy of the original gene transcript, followed by a poly-A tail flanked by target site duplications (Vanin 1985). Therefore processed retro-pseudogenes are also a direct result of LINE activity (Esnault et al. 2000). Indeed, recent studies have revealed that cancer genomes contain new processed pseudogenes absent from healthy tissues establishing that retrotransposition is ongoing in some cancers (Cook et al. 2014).

3.3 L1 Retrotransposon Structure and Retrotransposition

To date, the human LINE-1 element is the most thoroughly characterised mammalian APE-type non-LTR retrotransposon (Ostertag and Kazazian 2001a; Moran and Gilbert 2002). Human specific L1s are further divided into pre-Ta (Transcribed, subset a), Ta0, Ta1, Ta1nd, and Ta1d subfamilies based on lineage specific sequence variants.

The pre-Ta subfamily is characterised by an ACG diagnostic trinucleotide in its 3' UTR at nucleotide positions 5954–5956 (relative to the reference element L1.3, Accession: L19088, henceforth the basis for all element coordinates). Moreover, Salem et al. (2003) demonstrated that pre-Ta elements preferentially integrate into low GC content (36 %) genomic DNA. The majority of pre-Ta family elements are, 5' truncated but 29 full-length pre-Ta with intact ORFs have been reported. This fact and that a pre-Ta element insertion caused one case of human genetic disease (an integration into the factor VIII gene, resulting in haemophilia A) indicates the preTa family contains active members (Kazazian et al. 1988; Salem et al. 2003).

The Ta family (or Transcribed, subset a) is the youngest and most active L1 family, and has been found to cause ~ 100 identified clinical cases of various genetic disorders (Hancks and Kazazian 2012). Over 50 % of these elements show dimorphism (presence or absence) across human populations (Boissinot and Furano 2001). These families of L1 emerged after the divergence of humans from chimpanzees about 6 Myrs ago, and so are specific to humans. There are two main Ta subfamilies: L1 Ta0 and L1 Ta1 (Boissinot et al. 2000). ACA nucleotides at positions 5954–5956 of the 3' UTR are diagnostic for this family. Based on the nucleotides at positions 5557 and 5560 in ORF2 elements can be assigned to the distinct Ta1 and Ta0 classes. Ta1 elements have T and G nucleotides at these two positions and Ta0 have G and C respectively (Boissinot et al. 2000). The Ta0 subfamily is more similar in sequence to the non-Ta L1s, and therefore has been suggested to be an older family of elements. By contrast the Ta1 family are younger than the pre-Ta and Ta0 families, and so have accumulated fewer inactivating mutations. The Ta1 family still actively retrotransposes and is likely to be currently increasing in copy number in the human genome (Boissinot et al. 2000). It is estimated that the Ta1 family arose about 1.6 Myrs ago and can be further divided into two subfamilies: Ta1d and Ta1nd. The Ta1d group are recognised by a deletion at position 74 in the 5' UTR whilst the Ta1nd group lacks this deletion. There are around 90 full-length human L1s with intact ORFs in the human genome reference sequence, which are potentially RC-L1s (Brouha et al. 2003). However, cell culture retrotransposition assays demonstrated that only 6 of these elements account for 84 % of the total

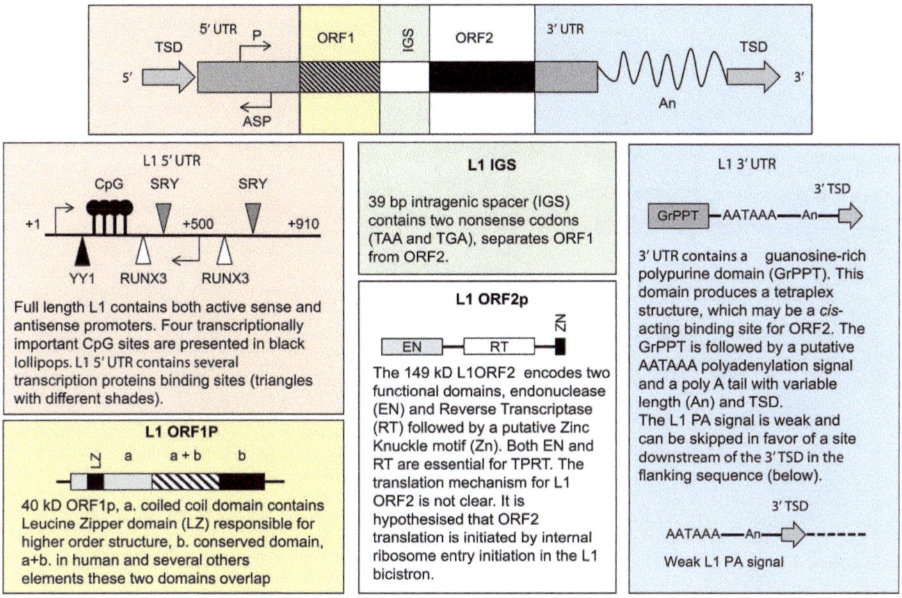

Fig. 3.3 L1 structure and the summaries of its components: L1 5' and 3' untranslated region (UTR), open reading frame 1 and 2 (ORF1/2) Intragenic spacer (IGS), poly A tail (An) and ta rget site duplication (TSD). L1 5' UTR structure is adapted from Badge, poster publication 2007, L1ORF1P adapted from Martin et al. 2000, L1ORF2p structure adapted from Goodier et al. 2004, and L1 3'UTR adapted from Craig et al. 2002

retrotransposition activity (Brouha et al. 2003). This data suggests that these very active elements dominate retrotransposition activity in the human genome. Four of the "hot" L1 elements characterised by Brouha et al. (2003) belong to the Ta1d family, with the other two elements belonging to the Ta1nd and Ta0 families (Brouha et al. 2003). Recent sequence-based studies have estimated the rate of L1 insertion into the human genome to be around 1 in 212 live births (Xing et al. 2009) and 1 in 140 (Ewing and Kazazian 2010). These estimates are much lower than was previously estimated (1 in 33 live births) for L1 insertions, based on the activity of disease-causing elements (Brouha et al. 2003; Beck et al. 2010). However unbiased capture of full-length elements and their retrotransposition activity suggests that presence/absence variation between individuals represents a substantial reservoir of active elements is segregating in human populations (Beck et al. 2010).

3.3.1 L1 Structure

A complete retrotransposition-competent (RC) L1 element is 6 kb in length and contains two non-overlapping open reading frames: ORF1 and ORF2 (Fig. 3.4). The 5' untranslated region (UTR) of a RC-L1 is approximately 900 bp in length.

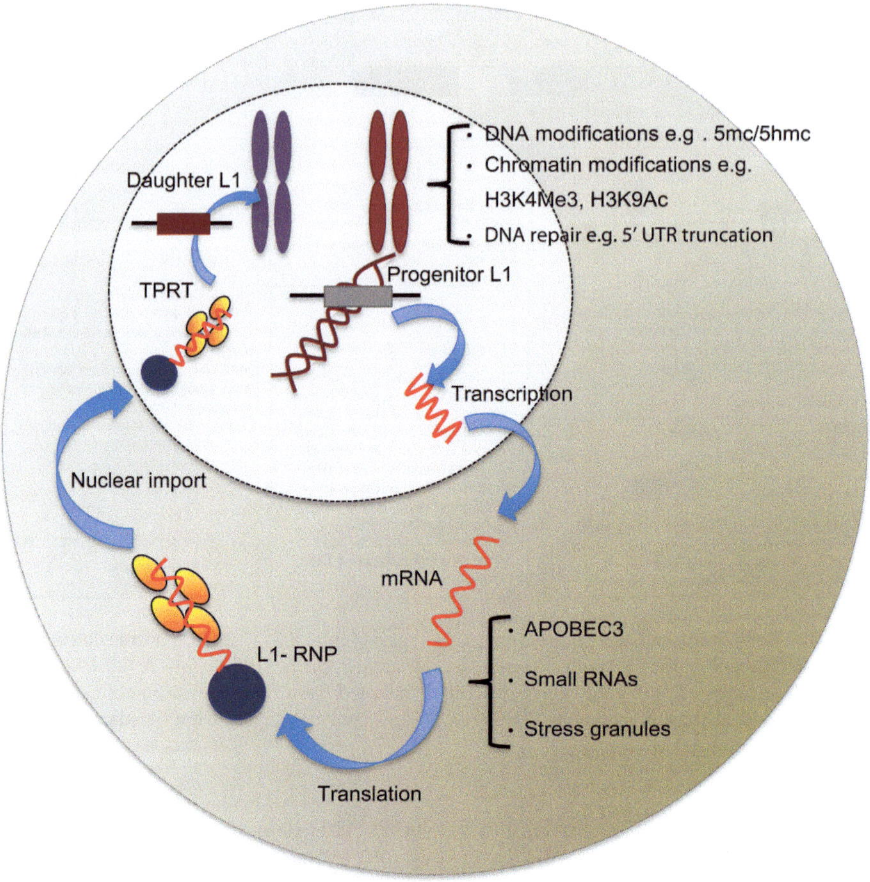

Fig. 3.4 Schematic diagram of host defence mechanisms deployed against endogenous L1 retrotransposition at different stages of L1 retrotransposition. Based on the location in the cell, the defence mechanisms against L1 retrotransposition can be divided into two categories: nuclear and cytoplasmic. Few of the defence mechanisms against endogenous L1 retrotransposition, and their timing are understood in detail. Many studies have contributed to this diagram, which are cited in Sect. 1.7

A major polymorphism of L1 elements occurs within this region: the presence or absence of a 131-bp sequence (Minakami et al. 1992). The L1 sense promoter is also located within the 5' UTR region and the first 155 bp have been demonstrated to be involved in L1 expression (Minakami et al. 1992; Athanikar et al. 2004). The structure of each L1 component and their role in L1 retrotransposition (where known) are discussed in more detail in the following sections. To illustrate these sections a schematic diagram of an intact L1 retrotransposon and its modules is presented in Fig. 3.3.

3.3.1.1 The L1 Promoter and Transcription of the L1 Element

The 5' UTR of the L1, which is about 900 bp in length, accommodates two internal promoters (+1 to +670). The region between +1 to +100 shows the highest promoter activity, although no TATA box is present (Swergold 1990). The L1 5' UTR contains a sense promoter (SP), which starts at +1 of the L1 sequence and an antisense promoter (ASP), positioned between +399 to +467 bp of the L1 sequence (Speek 2001). Both sense and antisense L1 promoter sites are highly conserved in human L1PA10-L1PA1 families, covering over 40 Myr of evolution. *In vitro*, luciferase reporter-based experiments have demonstrated that the L1PA6 elements have an active ASP (Macia et al. 2011), despite their antiquity. It has been suggested that over 1/3 of L1Hs contain highly active ASPs, which are capable of interfering with normal gene expression (Nigumann et al. 2002; Speek 2001) when located intragenically.

The L1 sense promoter possesses characteristics of both RNA polymerase II (Pol II) promoters, which control transcription of all protein-coding genes, and RNA polymerase III (Pol III) promoters that are responsible for synthesis of tRNA, 5S RNA and several small and non-coding RNAs (Kurose et al. 1995). The L1 transcript is about 6 kb long and it has two protein-coding regions and a polyadenylated extension at the 3' end of the transcript. These characteristics suggest L1's is a Pol II dependent promoter. However inhibition studies have shown the L1 promoter is less sensitive to α-amanitin, a Pol II inhibitor, and more sensitive to tagetitoxin, a Pol III inhibitor (Kurose et al. 1995). These data suggest that the L1 promoter is Pol III dependent, but produces transcripts more characteristic of Pol II. This unusual sensitivity may be explained by the importance of YY1 transcription factor in L1 transcription initiation (Athanikar et al. 2004), which is utilised at both Pol II and Pol III promoters.

The L1 sense promoter creates a long, protein encoding, polyadenylated transcript and the promoter acts as a Downstream Promoter Element (DPE), such that it initiates transcription at position +1 of the L1 sequence, but lacks features characteristic of PolII promoters such as upstream TATA and CAAT boxes (Kurose et al. 1995; Swergold 1990). The L1 5' UTR also contains several PolII transcription factor binding-sites, such as MECP2, SOX2 and RUNX3, which have been shown to be involved in the transcriptional regulation of L1s (Rosser and An 2012). The DNA methylation status of the L1 promoter has a great potential to impact its activity, especially during cancer progression. We discuss the methylation of the L1 promoter and its relation to cancer in later sections of this chapter.

YY1 Binding Site

The ubiquitous transcription factor YY1 (Yan Ying 1) binding site, which is a PolII and PolIII transcription, has been established as an important sequence in L1 transcription, and is located at +13 to +26 of the L1 5' UTR sequence (Becker et al. 1993; Kurose et al. 1995). Since YY1 is capable of both activating and repressing transcription, this protein may play a role in down-regulating L1 transcription in some cell types, while activating it in others (Becker et al. 1993). YY1 regulates L1 transcription by enhancing accurate transcription initiation rather than being required for initiation as even truncated L1s, which lack the YY1, site have functional promoters (Athanikar et al. 2004). It has been demonstrated that inhibition of the YY1 binding site in tissue culture assays has a minor effect on L1 transcription activation and retrotransposition (Athanikar et al. 2004). However, it has also been demonstrated that the deletion of the YY1 site in the first 20 bp significantly reduces (5 fold) L1 retrotransposition in cell culture assays (Singer et al. 2010).

Since deletion of the YY1 binding site does not abrogate L1 transcription, L1 elements must be able to be transcribed from upstream or downstream of this site. Transcription initiation from downstream of the YY1 binding site leads to 5' truncated progeny, which may not be retrotranspositionally competent, due truncation. It has been shown that most RC-L1s are transcribed from +1 or very nearby, such that their progeny are potentially able to retrotranspose autonomously (Athanikar et al. 2004).

Other L1 Transcription Factor Binding Sites

Previous studies have demonstrated that the L1 5' UTR contains four methyl-CP2-responsive elements at the following positions: +36, +101, +304 and +481 (Hata and Sakaki 1997). These C-methyl binding proteins bind to methylated DNA (Feng and Zhang 2001). Based on their recognition-binding site these proteins are divided into two types: the MBP group binds to methylated DNA, while the second group, MeCPs and MDBP, has no sequence specificity for methylated DNA (Feng and Zhang 2001). Among these, the MeCP2 are the most abundant methyl-Cytosine binding proteins and it has been demonstrated that MeCP2 binds to methylated DNA only in the context of chromatin contributing to long-term repression and nuclease-resistant methyl-CpGs (Meehan et al. 1992; Hata and Sakaki 1997).

Moreover, Tchenio et al. (2000) demonstrated that the human L1 promoter contains two functional sites for SRY (sex determining region Y) transcription factors. SRY transcription factors are members of the SOX protein family, and are expressed in the urogenital ridge of the embryo and in adult, testis, hypothalamus and midbrain (Lovell-Badge 2009). Cell culture studies have shown that ectopic overexpression of one of the SRY families, Sox11, results in 10 fold trans-activation of endogenous L1Hs (Tchenio et al. 2000). The two potential binding sites for SOX transcription factors are located in the first 670 nucleotides of the L1 promoter. It is possible that L1 activity in the brain is mediated by SOX2, as a decrease in SOX2 expression during the early stages of neuronal differentiation, when recapitulated in cell culture, is associated with increases in L1 transcription and retrotransposition (Muotri et al. 2005). The first site, SRYA, is located between nucleotides 427–477, and SRYB is located between 572–577. In addition *in vivo* experiments have demonstrated that SRY transcription factor binding at the L1 promoter can drive transcription in cell culture, and congruently mutations at the SOX binding site can inhibit L1 transcription (Tchenio et al. 2000). However, L1 transcription can be transiently stimulated by transcriptional binding switch from a SOX2/HDAC1 repressor complex to a wnt-mediated T-cell factor/lymphoid enhancer factor (TCF/LEF), which briefly activates L1 transcription in models of human and rodent neuronal differentiation (Muotri et al. 2010).

The RUNX3 family contains heterodimeric transcription factors, which can potentially bind to three regions in the L1 promoter: nucleotides +83 to +101 and +526–508 of the L1 5' UTR. These binding sites mean RUNX3 can potentially influence L1 transcription by regulating both sense and antisense promoters (Yang et al. 2003). Mutation analysis at each of the three sites has demonstrated that mutation of the first binding site reduces L1 transcription, while mutations at the other two binding sites do not have any significant effect (Yang et al. 2003). This may be due to the second and third binding sites being located outside the first 100 nucleotides of L1 5' UTR, which is important for transcription initiation (Yang et al. 2003). Moreover, a recent study of L1 5' UTR fragments driving luciferase reporter genes identified several novel transcription start sites at position +525 and +570. It is likely that these central sites are involved in the recruitment of transcription initiation complexes and it is possible that they can drive bi-directional L1 transcription (Alexandrova et al. 2012).

3.3.1.2 L1 ORF1 and ORF2 and Translation of the L1 Retrotransposition Machinery

Despite host genome defence mechanisms acting against L1 retrotransposition, these potentially mutagenic insertions occur in germline and somatic tissues, as shown by disease causing insertions. Because the L1 translational machinery has a strong *cis*-preference, functional protein crosstalk between individual elements is greatly reduced, and lack of competition from partially functional mutants may explain the longevity of L1 activity. However this hypothesis requires both ORFs to be expressed from the same transcript, so co-expression of the ORF encoded proteins is likely a marker of active L1 retrotransposition. Co-expression of the two L1-encoded proteins, ORF1p and ORF2p, has been detected by immunohistological analyses in pre-spermatogonia of human foetal testis and in germ cells of human adult testis (Ergün et al. 2004). Also, most disease-causing L1 insertions are apparently germline in origin (Kazazian 2004). These data and parallel observations of ORF1p expression in mouse pachytene spermatocytes (Martin and Bushman 2001) are consistent with the expectation that potentially mutagenic transposable elements confine their replication to germlines, where they can maximise their probability of transmission, without compromising host viability. As co-expression of both ORFs is required for retrotransposition their translation in quite different amounts from a bi-cistronic transcript is central to retrotransposition, but is far from clearly elucidated. However in the following section we review the current understanding of the structure, function and translation of each ORF in more detail.

Translation and Role of L1-ORF1 in L1 Retrotransposition

The first open reading frame of L1 (L1 ORF1) is 1017 bp in length and encodes a 338 amino acid cytoplasmic protein, also known as p40 (Hohjoh and Singer 1997). The centrally located leucine zipper (LZ) domain in human L1ORF1 is involved in the formation of higher order ORF1p multimers and it has been demonstrated that the LZ domain is required for RNP assembly and retrotransposition (Craig et al. 2002 and Doucet et al. 2010). The carboxyl domain of ORF1 is basic and contains several conserved amino acids that are likely to play a role in RNA binding. However, this carboxyl domain lacks common functional motifs, found in RNA binding proteins such as the RNP motif, and the Arginine-rich motif (Craig et al. 2002). The sequence of ORF1p is not related to any protein with known function and its role in the L1 retrotransposition cycle is incompletely understood (Basame et al. 2006). It has been demonstrated that efficient translation initiation of L1 5' UTR is strictly cap dependent, rather than as previously suggested via an internal ribosome entry site (IRES) mediated model (Dmitriev et al. 2007). Results of co-immunoprecipitation experiments demonstrate that ORF1p is a high affinity RNA binding protein with no sequence binding specificity (Kolosha and Martin 2003). It has also been demonstrated that the nucleic acid chaperone activity of ORF1p is important for successful L1 retrotransposition (Martin et al. 2005). Also, cell culture and in vivo experiments have each demonstrated that L1ORF1p exists in many copies in the cytoplasm (Hohjoh and Singer 1996). Moreover L1ORF1p contains non-canonical RNA recognition motifs (RRMs) that have RNA-binding properties, supporting its function as an unconventional RNA binding protein (Khazina and Weichenrieder 2009).

Several roles have been proposed for ORF1p in the L1 retrotransposition process. One concept is that the L1 RNA is unstable: ORF1p, with its RNA binding activity, is required to coat and protect the L1 RNA intermediate in the cytoplasm before its translocation to the nucleus where TPRT occurs. It is thought that *cis* preference acts to ensure that the L1 proteins associate with their functional encoding RNA (Moran and Gilbert 2002). Although ORF1p has only been definitively detected in the cytoplasm it could still be involved in the later stages of L1 retrotransposition, such as TPRT (Martin and Bushman 2001). It is hypothesised that the nucleic acid chaperone activity of ORF1p is involved in strand transfer, which allows the annealing of the DNA primer from the target site to the RNA primer during the process of reverse transcription (Martin and Bushman 2001). It is also possible that ORF1p facilitates the reverse transcription process by enabling movement of polymerase through RNA secondary structures formed during first cDNA synthesis (Martin and Bushman 2001).

Translation and Role of L1-ORF2 in L1 Retrotransposition

The L1 s open reading frame (ORF2) encodes a protein of ~ 150 kDa containing 1275 amino acids (Scott et al. 1987). The initiator methionine of ORF2 in the human L1 element is separated from ORF1 by a 66-bp in-frame spacer region containing three stop codons. It is not clear how the separate translation of both ORFs from the bi-cistronic RNA is accomplished; this problem is made even more intriguing by the fact that the spacer region is not conserved between L1 elements of different species (McMillan and Singer 1993). It was first suggested that translation of ORF2 must be accomplished either by reinitiating translation or by internal initiation via an internal ribosomal entry site (IRES) (McMillan and Singer 1993). However, using an engineered LINE1 retrotransposition assay, it was later demonstrated that L1-ORF2p is translated by an unconventional termination/re-initiation mechanism (Alisch et al. 2006).

The ORF2 protein has proven to be very hard to detect, largely due to the lack of robust and specific ORF2p antibodies (Wagstaff et al. 2011). Thus, indirect methods, such as measuring its enzymatic activity have been used to study the role of this protein in the L1 retrotransposition cycle. It seems that ORF2p has two major activities, each of which can be assigned to specific domains. The N-terminal contains a conserved endonuclease activity domain. Its sequence and crystal structure is similar to AP-like endonuclease APE1, which is involved in the base excision repair pathway (Ergun et al. 2004; Feng et al. 1996; Weichenrieder et al. 2004). Despite its conservation, it has been demonstrated that L1s lacking an EN domain are still able to retrotranspose, at a lower efficiency than wildtype, likely by using pre-existing nicked DNA sites for integration (Morrish et al. 2002). The central domain of ORF2p is responsible for the reverse transcriptase activity, and it contains a conserved Z-motif (Mathias et al. 1991). The L1 RT domain is related to those in other non-LTR elements (Malik et al. 1999) and also shows some sequence similarity to LTR retrotransposons and retroviruses (Xiong and Eickbush 1990). At the C-terminal end, there is a conserved "C-domain" containing a cysteine-rich region whose function is not clear. It has been suggested that this region has evolved in response to interactions with other L1 sequences or host factors (Wagstaff et al. 2011). Also, it has been shown that mutations in this region abolish the ability of ORF2p to interact with L1 RNA and ultimately block L1 retrotransposition in cultured cells (Feng et al. 1996; Moran et al. 1996; Doucet et al. 2010).

3.3.1.3 L1 3' UTR and Poly A Tail

The L1 3' UTR covers the terminal 205 bp of full-length elements, includes a polyadenylation (PA) signal, and terminates in a poly (A) tail. One of the characteristics of the L1 PA signal is the ability to transduce genomic DNA (up to 1.6 kb *in vitro)* downstream of its 3' UTR (Holmes et al. 1994). In the process of polyadenylation the poly-A tail is added to the putative AAUAAA polyadenylation specificity factor (CPSF1) binding site. However, the L1 PA signal lacks the conserved elements that normally reside downstream of the poly-A site in canonical RNA polymerase II transcripts. Hence it has been suggested that the L1 PA site is weak and can be bypassed by the transcription machinery in favour of a stronger PA site in the 3' flanking genomic sequence (Moran et al. 1999). L1's weak PA signal is suggested to be an evolutionary adaptation that allows L1 to reside within introns with minimum effect on gene expression through the induction of premature polyadenylation (Moran et al. 1999). Around a third of L1 elements carry a 3' transduction and they are estimated to have contributed ~33 Mb of DNA to the human genome (Moran et al. 1999; Pickeral et al. 2000; Goodier et al. 2000; Szak et al. 2003).

The L1 3' UTR also contains the sequence motif (CACAN$_5$GGGA) at position 5796–5884 nt, which has a high binding affinity for nuclear export factor 1 (NXF1) (Lindtner et al. 2002). The role of NXF1 is similar to that of constitutive transport elements (CTE), which facilitate the nuclear transport of viral intronless mRNA, such as simian type D retroviruses (Lindtner et al. 2002).

The 3' UTR of the L1 element is poorly conserved within and between species (Scott et al. 1987). Interruption of this region by additional nucleotides does not seem to have severe effects on retrotransposition, as illustrated by reporter assays, where L1 tolerates marker genes of up to 3500 bp in length in its 3' untranslated region (Moran et al. 1996; Ostertag et al. 2000; Gilbert et al. 2002; Symer et al. 2002).

All the classifications above apply to full-length copies of L1. However, only 5 % of endogenous human L1 elements are full length (6 kb). The remaining 95 % are 5' truncated, internally rearranged or deleted (Szak et al. 2002). Some of this damage to L1 structure may be the result of mutations and genomic rearrangements after integration of the retrotransposon. Indeed, Coufal et al. (2011) demonstrated that in ataxia telangiectasia mutated (ATM) deficient cells, there were either more or longer L1 retrotransposition events, compared to ATM wild type cells. This suggests that cellular proteins involved in the DNA damage response may modulate L1 retrotransposition. 5' truncation and inversion most probably occur during the retrotransposition process itself (Ostertag and Kazazian 2001a). In inverted L1 elements, the 5' truncated region is commonly orientated in an antisense direction to its 3' end. This structure is thought to be the consequence of a model of retrotransposition called 'twin priming', whereby second strand cDNA synthesis initiates before first strand cDNA is completed (Ostertag and Kazazian 2001b). Inversions can be detected in about 25 % of insertions in members of the Ta family (Ostertag and Kazazian 2001a; Skowronski et al. 1988).

L1 integrants are usually flanked by variable TSDs with lengths of up to 60 bp (Szak et al. 2002). These TSDs are generated during the process of L1 replication. Some TSDs are difficult to identify due to statistical uncertainties about the occurrence of short duplications; the presence of multiple mutations in TSDs of ancient integrants; the presence of blunt end nicking sites (Van Arsdell and Weiner 1984); or the presence of a staggered double strand break with a 5' overhang instead of a 3' overhang. The latter process causes a deletion of the target site instead of duplication (Gilbert et al. 2002; Symer et al. 2002). However the vast majority of L1 insertions have identifiable TSDs, suggesting they originate from an endonuclease dependent process.

3.3.2 Mechanism of L1 Retrotransposition

The mechanism of retrotransposition of non-LTR retrotransposons is not entirely understood. However, the first steps of integration of these classes of elements have been elucidated by biochemical experiments using the site-specific RE-type retrotransposon R2BM from the silkworm *Bombyx mori* (Luan et al. 1993). These studies led to the model of L1 retrotransposition called 'target primed reverse transcription' (TPRT) (Cost et al. 2002).

Although RE-type and APE-type elements belong to different families of non-LTR retrotransposons that share very few structural similarities, the basic mechanism of transposition initiation by TPRT is relatively conserved. This has been demonstrated by reconstitution of the initial steps of L1 element transposition *in vitro*, by providing only the complete L1 ORF2 protein, L1 RNA, and a target DNA (Cost et al. 2002). Also, further experiments have shown that the EN domains of the two types of retrotransposons (RE and APE) initiate the integration process by nicking the target DNA (Cost et al. 2002; Eickbush and Malik 2002). The resulting 3'-hydroxyl group serves as a primer for reverse transcription of the element's RNA. On the other hand, it has been demonstrated that L1 integration can also occur at pre-formed nicks and double strand breaks in the target DNA, known as endonuclease independent-TPRT (Morrish et al. 2002). However, this mode of insertion is prevalent only in cell lines with defects in DNA repair machinery. Therefore, endonuclease-independent insertion provides an alternative pathway for L1 retrotransposition in the human genome (Sen et al. 2007). As a result it is likely that nicking and reverse transcription are two independent steps in TPRT (Cost et al. 2002; Eickbush and Malik 2002). The EN domain, can also cleave the second strand of target DNA at a slower rate compared to the nicking of the first strand (Cost et al. 2002). Depending on the position of the second nicking site relative to the initial one, TPRT can generate a target site deletion, a simple 'blunt' integration, or a target site duplication (TSD) which flanks the inserted element (Cost et al 2002; Eickbush and Malik 2002).

A major unresolved issue regarding the mechanism of LINE retrotransposition is what occurs after second-strand cleavage. Despite extensive efforts, in vitro experiments with the R2 protein did not lead to the detection of intermediates expected for second-strand synthesis (Luan et al. 1993). In contrast, in vitro TPRT of L1 yielded 5' junctions between the L1 sequence and the target DNA. This result indicates that the RT is able to accept cDNA as a template for second-strand synthesis, probably by a second round of TPRT (Cost et al. 2002; Eickbush and Malik 2002).

However, this in vitro process is very inefficient and it does not necessarily reflect the natural mode of retrotransposon integration and still leaves open the major question of how the damaged genomic DNA is repaired. It is generally assumed that cellular DNA repair pathways are involved in these final steps of integration and that these activities generate the observed TSDs (Gilbert et al. 2005).

3.4 Genomic Distribution of Human L1s

Human LINEs are distributed across the genome, but not distributed evenly. There are some parts of the genome, which have very low repeat density. This could be because these regions cannot tolerate insertion of repeats due to essential *cis* regulatory architecture. An example of repeat poor regions is the homeobox (HOX) gene clusters, which contain the lowest reported density of interspersed repeats (Lander et al. 2001; Simons et al. 2006). In contrast to this, some parts of the genome are very rich in repeats, such as chromosome Xp11, which contains a 525 kb region comprised of 89 % repeats. Overall it is suggested that LINEs are more abundant in gene poor, and AT rich regions, which usually show low recombination rates (Lander et al. 2001). In comparison to Alu, LINEs have been reported to insert at a four fold higher density in GC poor regions, while Alus have a lower tendency (five fold lower) to insert in AT rich regions (Lander et al. 2001). One reason for this insertional bias of LINEs towards AT rich regions could be due to the consensus L1 endonuclease target site TT/AAAA, which is intrinsically more common in AT rich regions (Lander et al. 2001; Jurka 1997; Cost and Boeke 1998). However, Alu elements also use the L1 machinery in *trans* to integrate into the genome, but Alus have a high density in GC rich regions. Therefore, the biasing of L1 insertion in AT rich regions may not be only due to endonuclease site selection but also post-insertion selection. It has been suggested that L1 insertion occurs in AT and GC rich regions, but that insertions in GC-rich regions are lost through selection. It is clear that L1s inserted within genes can have a variety of negative effects on their host gene such as altered splicing, interference with gene regulation and level of expression, and premature polyadenylation (Cost and Boeke 1998; Lander et al. 2001).

3.5 Impact of L1 Integration on Human Genome Plasticity

Recently, efforts have been directed towards unveiling the molecular mechanisms by which L1 impacts gene expression and mammalian cell development, differentiation, and cancer. New L1 integrations have a great impact on host genome diversification and evolution. The ways that L1 retrotransposition can alter the host genome are discussed in detail below.

3.5.1 Increasing the Size of the Human Genome

An orthologous sequence comparison of the human and chimpanzee genomes suggested that the human genome continues to expand, either because of inherently more active insertional mutation processes or through being less efficient at deleting such events (Liu et al. 2003). Therefore, one of the greatest impacts of L1 on the human genome is their contribution to expanding genome size (Liu et al. 2003). Considering that L1 is also responsible for Alu retrotransposition in the genome, it has contributed about 750 Mb to the human genome (Lander et al. 2001). Moreover, the ongoing expansion of L1 has also created significant inter- and intra-individual variation by introducing L1 insertional polymorphisms (presence/absence) at orthologous loci.

3.5.2 Disease Causing L1 Retrotransposition

There are ~100 cases of human genetic diseases caused by L1 integration into genes (Hancks and Kazazian 2012). Based on L1 retrotransposition assays it has been suggested that about 10% of *de novo* L1 retrotransposition events occur in the introns of actively transcribed genes (Moran et al. 1999). In fact, it is likely that evolutionarily successful L1s (active L1s) preferentially insert into genes, which are transcriptionally active and therefore have an open chromatin configuration (Macia et al. 2011).

The first L1 disease-causing insertion was reported in two patients with haemophilia, where an L1 was integrated into exon 14 of the human factor-eight gene (Kazazian et al. 1988). Subsequently cases of L1 disruption of the dystrophin gene have been reported to cause muscular dystrophy and cardiomyopathy in four unrelated individuals (Holmes et al. 1994; Matsuo et al. 1991 and Yoshida et al. 1998). It has also been shown that a heritable full length L1 insertion into intron two of the ß-globin gene (L1ß-thal) is responsible for some cases of ß-thalassemia (Divoký et al. 1996; Kimberland et al. 1999). Additionally insertion of a full length L1 into an intron of the X-linked RP2 gene is responsible for progressive retinal degeneration and ultimately retinitis pigmentosa (XLRP) (Schwahn et al. 1998). Moreover, a case of colon cancer has reported to be caused by somatic insertion of a truncated L1 into the APC gene (Miki et al. 1992). More recently it has been reported that somatic *de novo* L1 retrotransposition events are detectable in lung cancer cells (Iskow et al. 2010). Also, up regulation of L1 RNA and ORF1p has been reported in several tumours including breast sarcomas and in 10% of tumours of germline origin, such as ovarian and testicular tumours (Asch et al. 1996; Bratthauer and Fanning 1993). The role of L1 in cancer will be covered in more detail in the following sections.

3.5.3 Genome Instability Caused by L1 Retrotransposition

In addition to mutagenic insertions, L1 retrotransposition can generate local genomic instability through several other mechanisms, which are explored in this section. All of these mechanisms are compatible with tumorigenic potential for these elements. DNA double strand breaks (DSBs) can be caused by the endonuclease activity endogenous L1ORF2p (Gasior et al. 2006). It is been shown that the number of

DNA DSBs generated by L1ORF2p is much higher than the number of actual L1 insertions (Gasior et al. 2006). However, the extent of genome instability induced by endogenous L1 retrotransposition is not clear due to a lack of sensitive antibodies to target ORF2p and also because the repair of L1-mediated DSBs may not leave any sign of L1ORF2p involvement. As a result, the attribution of L1ORF2p to genomic DSBs, which are highly mutagenic and prone to induce recombination, is likely underestimated (Cordaux and Batzer 2009). In addition to generating local genome instability, L1 can also cause genomic rearrangements through insertion-mediated deletions. Studies of L1 retrotransposition in cell culture have demonstrated that about 20 % of L1 insertions are associated with structural rearrangements, including flanking genomic deletions at the insertion site (Gilbert et al. 2002; Gilbert et al. 2005; Symer et al. 2002). Another study reported a lower frequency of deletion (2 %) than in cell culture assays, with endogenous L1 retrotransposition causing deletions with an average size of 800 bp in the human genome (Han et al. 2005). Since L1-mediated insertion deletions are generally grouped into two sizes classes (< 100 bp and > 1 kb), it has been suggested that each group is caused by a different mechanism. In general, small deletions may arise due to template switching with subsequent 5' to 3' exonuclease activity on both the exposed 5' ends. Larger deletions can be mediated by non-homologous end joining when the nascent cDNA invades a double strand break with a 3' overhang located upstream of the integration site. Subsequent gap repair will remove the cDNA and the adjacent segment to cause a large deletion (Han et al. 2005). A study by Chen et al., (2007) demonstrated a 46 kb full length L1 insertion-mediated deletion event that possibly occurred through the template jumping process. This deletion resulted in removal of seven exons of the pyruvate dehydrogenase complex, component X (PDHX) gene, which caused a case of pyruvate dehydrogenase complex deficiency (Chen et al. 2007).

3.5.4 Ectopic Recombination upon L1 Retrotransposition

Due to the high copy number of L1s in the human genome, they can also create structural variation at the post-integration stage, through non-allelic homologous recombination or ectopic recombination. Ectopic recombination events seem relatively rare and are usually mediated by truncated elements (Boissinot et al. 2000). Indeed there is no evidence of polymorphic L1 associated ectopic recombination in humans. This can be explained by the low activity of retrotransposition competent L1s in the modern human genome (Boissinot et al. 2000), or perhaps by the frequency with such mutations are deleterious (Wang et al. 2006). Ectopic recombination potentially causes various types of genomic rearrangements, including duplications, deletions, and inversions.

Segmentally duplicated regions can contain paralogous copies of genes, promoters and other regulatory components (Samonte and Eichler 2002). It is likely that segmentally duplicated regions are associated with the creation of novel genes and the formation of pseudogenes (Lynch and Conery, 2000). Alternatively, ectopic recombination can cause recombination-associated deletion events (RADs). Genome-wide comparisons of the human and chimpanzee genomes have identified 73 human specific L1RAD events that occurred following the divergence of humans from chimpanzees (Han et al. 2008). Although L1RAD events are not very common, it has been suggested that they are responsible for the deletion of about 450 kb of the human genome (Han et al. 2008). This event is most frequent in heterochromatic regions, which suggests that there may be negative selection against L1RADs in euchromatin (Graham et al. 2006).

As mentioned earlier, L1-mediated ectopic recombination is also involved in gene inversion events. It is suggested that L1 contributes to genomic inversion possibly through the formation of secondary structures or by providing a target site for double strand breaks (Lee et al. 2008). Among the characterised inversions mediated by L1 insertions, some loci include the exonic regions of known genes, which suggests that L1-mediated inversions can generate alterations in gene function (Lee et al. 2008; Cordaux and Batzer 2009). Therefore, although this type of recombination does not affect the size of the genome it can produce genomic variation.

3.5.5 L1-Mediated Sequence Transduction

In addition to duplicating themselves, L1s sometimes carry with them upstream or downstream flanking genomic sequences (termed 5' and 3' transduction, respectively), providing a novel mechanism for genome evolution. L1-mediated sequence transduction occurs when L1 transcripts extend upstream or downstream of the genomic flank and then transduce these sequences into new genomic locations through the L1 retrotransposition process. L1 5' sequence transduction is usually very short, ranging between 5–8 nt sequences and it is not a common process. Additionally, due to 5' truncation during L1 retrotransposition, there is a severe ascertainment bias to determine how often L1 mRNAs may contain 5' transduced sequences. This process occurs when L1 sequences are transcribed by a host promoter upstream of the L1 5' terminus, and subsequently mobilised during the L1 retrotransposition cycle (Pavlicek et al. 2002a; Pickeral et al. 2000; Szak et al. 2003). The 3' sequence transduction process is more common, and occurs when transcription of the L1 bypasses the weak polyadenylation (PA) signal in favour of a stronger canonical PA signal in the 3' genomic flank followed by mobilisation of the genomic flanking DNA to a new location. The sequence transduction process seems to be more common in active or recently active elements: it has been demonstrated, in cell culture assays that between 10–20% of recent active human insertions contains sequence transductions (Goodier et al. 2000). During the process of sequence transduction, exons, promoters and other regulatory sequences upstream and downstream of the L1 can be transduced into the new genomic location, causing exon shuffling and potentially altering the expression and or structure of the recipient gene (Moran et al. 1999). This process maintains genome plasticity and genome evolution (Goodier et al. 2000). Indeed if 5' truncation occurs during retrotransposition, removing the

L1 sequences, exon shuffling events are expected to be difficult to identify, and so their frequency may be greatly underestimated.

3.5.6 Regulation of Gene Expression

As mentioned above, L1s can affect the genome at the DNA level. In this section the effect of L1 at the RNA level are considered in more detail. It has been demonstrated that L1 can affect transcription in several distinct ways. They can generate alternative splice sites resulting in the exonization of L1 sequences, at least in rodents (Zemojtel et al. 2007; Huang et al. 2009). Also intronic L1s may sometimes interfere with transcriptional elongation and so produce different lengths of mRNA from a gene (Han et al. 2004). If the L1 inserts in the antisense orientation relative to the host gene, it can potentially produce truncated cellular transcripts by premature polyadenylation (Han et al. 2004). Moreover, L1 can produce novel transcripts through the activity of its antisense promoter (ASP). Nearly 1/3 of the L1Hs studied contain active ASPs (Speek et al. 2001). Therefore it is possible that some of the transcripts initiated from the L1 ASPs are competent for translation. On the other hand, it has been recently demonstrated that a large proportion of regulatory RNAs, termed long-non coding RNAs (lncRNAs), are derived mostly from TE sequences, and are frequently generated from TE-derived promoters (Kapusta et al. 2013). In addition, insertion of full length L1 sequences into intronic regions of a gene can potentially "break" a gene. "Gene breaking" occurs where an L1 inserted in the opposite orientation to a host gene can generate two novel partial transcripts: one from the endogenous promoter including exons upstream of the L1 insertion, and a second internal transcript driven by the L1 ASP. Indeed, bioinformatic analysis on the human genome has highlighted 15 genes and transcription units that have potentially been affected by L1 insertions in this way (Wheelan et al. 2005). Additionally a recent study of intragenic L1s in lung cancer cells has shown that L1 pre-mRNA binds to the Ago2 complex to suppress the transcription of cancer genes (Aporntewan et al. 2011). Therefore, with transcriptional interference from the endogenous L1 sense and antisense promoters, its polyadenylation signal, and alternative L1 transcripts, L1 exhibits a great potential to impact human transcriptome composition.

3.5.7 Epigenetic Regulatory Role of Human L1s

Because L1 elements are frequently found in or near genes, it is possible that heterochromatin formed at retrotransposons could spread and repress the transcription of nearby genes. It has been suggested that L1's principle epigenetic regulatory role is in X chromosome inactivation (XCi). XCi is a well-established mechanism of gene regulation that acts to achieve gene dosage compensation between male and female embryos (Heard and Disteche 2006). XCi initiates at the X inactivation centre

(XIC) (Rastan 1983), which contains several genes that produce non-coding RNAs (Chureau et al. 2002). Little is known about how inactivation spreads across the chromosome, although it has been proposed that L1s play a role in the *cis* spreading of X chromosome inactivation (Lyon 1998). L1s are enriched on the X chromosome compared to autosomes, and significantly so at Xq13 where the XIC is located. To support this idea, it has been demonstrated that genes on the X chromosome which escape X inactivation are generally located in L1 poor regions (Ross et al. 2005). For young L1s, the proposed involvement in X inactivation is also linked to methylation. Indeed, It has been shown that demethylation and activation of the L1 ASP can drive the transcription of neighboring genes: Weber et al. (2010) have shown that demethylation of the L1 ASP in colon cancer cell lines induces the expression of L1 and proto-oncogene cMet (L1-cMet) transcripts. This result demonstrated the involvement of L1 in gene regulation and a clear link to methylation. However, the formal demonstration of direct retrotransposon-mediated epigenetic control of neighboring genes in humans and the evaluation of the extent of this phenomenon at a genome-wide scale are active topics of investigation, and will be discussed more in following sections.

3.6 Host Defence Mechanisms Against L1 Retrotransposition

As well as the direct mutational effects of L1 insertion, various forms of genetic instability caused by L1 integration include the generation of L1 chimeras, intrachromosomal deletions (chromosomal deletions of > 11 kb), intrachromosomal duplications, and chromosomal inversions (approximately 120 kb in length) (Gilbert et al. 2002; Han et al. 2005; Symer et al. 2002). It was demonstrated by Gilbert et al. (2005) that the L1 reverse transcriptase can faithfully replicate its own transcript and has a base mis-incorporation rate of ~1 in 7000 bases. All these observations indicate that L1 retrotransposition can lead to a variety of genomic rearrangements suggesting that hosts should be under selection to restrict L1 activity, as integration of L1 and other retrotransposons poses a potential threat. As a result organisms have apparently evolved diverse mechanisms to combat retrotransposon activity. Indeed, the initial step in L1 retrotransposition was described as a host/parasite "battleground" that serves to limit the number of active L1s in the genome (Gilbert et al. 2005). Since L1 has been actively mutating mammalian genomes for millions of years, it is likely that the host has evolved multiple mechanisms to combat L1 mobility at discrete steps of the retrotransposition cycle. In the following sections the mechanistic strategies used by the host to restrict L1 retrotransposition are discussed in more detail.

3.7 Epigenetic Modifications Regulate L1 Retrotransposition

Different types of epigenetic regulation are suggested to keep L1 retrotransposition activity in check. Some of the well-studied epigenetic regulatory modes are outlined in the following subsections.

3.7.1 Cytosine Methylation in Host Defence and Genome Instability

A possible mechanism, by which the activity of many potentially active human L1s could be suppressed, is methylation of cytosine bases in their promoters, some of which are known to be critical for promoter activity (Hata and Sakaki 1997).

The majority of cytosine methylation in plants and mammals resides in repetitive elements and a large proportion of this lies in retrotransposons, which constitute more than 42% of the human genome (Goll and Bestor 2005). Transposons can only proliferate in genomes where the fitness of transposons is greater than that of the host. Therefore, host defence mechanisms are under selective pressure to suppress these elements (Bestor 2003); as judged by its distribution, DNA methylation is primarily a mechanism of transposon suppression. In somatic cells L1 promoters are generally hypermethylated, but in malignancy-derived cells, the global hypomethylation of CpG dinucleotides is correlated with L1 activity (Kitkumthorn and Mutirangura 2011). This correlation was supported by the recent identification of several *de novo* L1 insertions in a cohort of lung tumours (Iskow et al. 2010) with more frequent insertions being observed in tumours showing significant genomic hypomethylation.

As previously mentioned, a variety of studies have suggested that *de novo* L1 retrotransposition is more likely to occur in germ cells and/or during early embryonic development (Garcia-Perez et al. 2007b; Van den Hurk et al. 2007), where a pair of global de-methylation events occur at the genome reprogramming stages. Although it has been frequently suggested that methylation of CpG dinucleotides has a regulatory role, especially in suppressing repetitive elements, there is evidence against this hypothesis (Walsh and Bestor 1999), such as the somatic inheritance of genomic methylation patterns in mammals (Riggs 2002). Therefore, chromatin modifications such as DNA methylation could be a consequence of active transcription rather than a cause, and the causal relationship of these phenomena remains to be fully elucidated.

Studies on 5-methylcytosine residues in the L1 promoter, especially at the four transcriptionally important CpG sites, show that DNA methylation can repress L1 activity both in vivo and in vitro (Hata and Sakaki 1997). In contrast to the suppressive effect of DNA methylation on L1 promoters, it has been demonstrated that 5-hydroxylation of the methylcytosine moiety (hm5c) can be an activating factor. However, a study of hm5c protein interactions showed that it does not interact with the same proteins as the 5mc pathway, which suggests that hm5c must regulate the L1 promoter through other mechanisms (Williams et al. 2011). Indeed, Ficz et al. (2011) demonstrated that hm5c methylation modifications are enriched in euchromatic regions and show a positive correlation with L1 expression. Also, a recent study has demonstrated that the Tet protein can generate other cytosine modifications downstream of hm5c (Ficz et al. 2011). These modifications are 5-formylcytosine (5fc) and 5-carboxylcytosine (5ca5) (Ito et al. 2011). Whether these newly discovered DNA cytosine modifications have any direct and controlling effect on L1 promoters and L1 expression remains to be investigated, but their existence suggests that epigenetic DNA modification is more complex than suspected.

Many studies have shown that a variety of epigenetic modifications can regulate L1 activity, and these are not limited to DNA modifications. Chromatin modifications are also likely to have an important role in controlling L1 activity. For example, Teneng et al. (2011) have recently demonstrated the direct association of H3K4 and H3K9 modifications with L1 activity. In fact they have demonstrated that the exposure of HeLa cells to Benzo (a) pyrene (Bap) causes L1 reactivation in HeLa cells through induction of early enrichment of the transcriptionally active chromatin markers histone H3 trimethylation at lysine 4 (H3K4Me3) and histone H3 acetylation at lysine 9 (H3K9Ac), and also reduces the association of DNMT1 with the L1 promoter. These processes cause depletion in cellular DNMT1 expression, which subsequently reduces cytosine methylation within the L1 promoter CpG island (Teneng et al. 2011).

Other evidence for chromatin modifications regulating L1 activity was uncovered in hippocampus neural stem (HCN) cells. Muotri et al. (2005) showed that histone deacetylase 1 (HDAC1) and methylation of H3 at Lys9 (K9), which both associate with transcriptional silencing in undifferentiated HCN cells, was directly correlated with L1 reporter construct activity in transgenic mice. In contrast acetylation of H3K9 and methylation of H3K4 (associated with transcriptional activation) was associated with high levels of L1 transcripts in HCN differentiated cells. This data supports the idea that chromatin remodelling during the early stages of neuronal cell differentiation allows transient stimulation of L1 retrotransposition (Muotri et al. 2005). Additionally, recent studies of L1 expression in undifferentiated human embryonic stem cells have demonstrated that retrotransposition processes in pluripotent cells are subjected to strong epigenetic control (Macia et al. 2011; Munoz-Lopez et al. 2011).

3.7.2 Role of Small RNAs in Regulation of L1 Retrotransposition

Small RNAs inhibit retrotransposon proliferation in the host genome via two mechanisms, which are independently mediated by either small interfering RNAs (siRNAs) or PIWI-interacting RNAs (piRNAs) (Meister et al. 2004; Soifer and Rossi 2006). The mechanisms by which these small RNAs are generated and how they inhibit retrotransposon mRNAs are still not fully understood, but there is strong evidence for a connection. It has been reported that host siRNA can repress retrotransposition through the post-transcriptional disruption of L1mRNA (Yang and Kazazian 2006). It is suggested that L1 bidirectional transcripts can be processed into small interfering RNAs (siRNAs) that supress L1 retrotransposition by an RNA interference mechanism (Yang and Kazazian 2006). Multiple RNA silencing pathways might act as a defence mechanism against L1 retrotransposition. Consistently, very recently it has been demonstrated that Dicer and Ago2-dependant RNAi restricts L1 retrotransposition in undifferentiated mouse embryonic stem cells (Ciaudo et al. 2013).

Another independent mechanism that has been suggested to suppress retrotransposon mRNA are piRNAs, which are generated from genomic loci that encode long precursor RNAs containing the remnants of different families of TE elements (Malone et al. 2009). It is likely that small-RNA-based mechanisms may also play role in silencing the mammalian L1 elements. Indeed it has been demonstrated that an antisense promoter located within the human L1 5' UTR allows the production of an antisense RNA transcript (Speek et al. 2001) that, in principle, could base pair with sense-strand L1 mRNA to establish a dsRNA substrate for the Dicer protein (Levin et al. 2011). Furthermore, mouse mutants lacking the murine PIWI family proteins (MILI or MIWI2) exhibit a loss of methylation of L1 and IAP elements. This loss correlates with the elements transcriptional activation in male germ cells and suggests that MILI and MIWI2 play essential roles in establishing *de novo* DNA methylation of L1 retrotransposons in the fetal male germline (Kuramochi-Miyagawa et al. 2008). Recently it has been demonstrated that Drosha-DGCR8, components of the microprocessor machinery responsible for the generation of miRNAs, recognize and binds L1 RNA derived sequences; additionally, cultured cells lacking these proteins support elevated levels of L1 and Alu retrotransposition. Overall, these observations suggest that the microprocessor complex is involved in post-transcriptionally suppressing L1 and Alu retrotransposition (Heras et al. 2013).

3.7.3 RNA Editing Enzymes Modulating L1 Retrotransposition

Members of the apolipoprotein B mRNA editing complex polypeptide 1-like (APOBEC) family of enzymes exhibit modulatory activity against variants of exogenous and endogenous retrovirouses, including L1 retrotransposons. APOBEC3A, 3B and 3F suppress L1 retrotransposition in humans and IAP elements in mouse (Lovsin and Peterlin 2009). Recent knockdown study of APOBEC proteins and their effect on L1 retrotransposition in hESC and iPS cells has suggested that only knockdown of APOBEC3B enhances L1 retrotransposition in hESCs. Knockdown of other APOBEC3 family members has little effect on L1 retrotransposition (Wissing et al. 2011). Recently a study of chimpanzee and human iPSCs demonstrated that differences in APOBEC and PIWI expression might explain the reduced activity of L1s in humans, compared to chimpanzees (Marchetto et al. 2013).

Moreover, previous studies suggested that APOBEC3B and APOBEC3F repress the L1 retrotransposition process in a deamination-independent pathway. However, recent findings have demonstrated that indeed the L1 transcript is edited by APOBEC3A during retrotransposition (Richardson et al. 2014). Thus it is possible that APOBEC proteins may repress L1 retrotransposition by producing L1 integration barriers, and by inactivating the L1 transcript (Stenglein and Harris 2006). Recent studies on the activation-induced deaminase (AID)-like gene, which is the potential ancestral progenitor of the APOBEC lineages in mammals, demonstrated that AID could inhibit the retrotransposition of L1 through a DNA deamination-independent mechanism (MacDuff et al. 2009). This mechanism may manifest in the cytoplasmic compartment, co- or post-translationally, and suggests that APOBEC proteins might also exhibit similar inhibitory reactions in L1-mediated retrotransposition (MacDuff et al. 2009).

3.7.4 L1-Ribonucleoprotein Particles and Host Cellular Defence

Despite long study, the processes involved in the formation of L1 ribonucleoprotein (RNP) particles and their transportation to the nucleus remains unclear. Due to the suppression of L1 retrotransposon expression in most somatic cells and the association of L1 with many cellular mRNAs, it is difficult to detect and study endogenous L1RNPs. Goodier et al. (2008) have demonstrated the subcellular co-localisation of L1 RNA and proteins (ORF1p and ORF2p), in cytoplasmic RNP foci. One of the suggested host defence cellular mechanisms to repress the L1 retrotransposition process is the transport of L1 RNPs to stress granules. It had also previously been demonstrated that L1RNP foci also localise with nucleoli (Goodier et al. 2007).

Cytoplasmic RNA granules in somatic cells, stress granules, and processing bodies, have emerged as important players in post-transcriptional regulation of gene expression. Processing bodies and stress granules are related compartments that overlap, sharing some components depending upon the nature of the cellular stress. Goodier et al. (2007) demonstrated that ORF1p foci co-localise with cytoplasmic stress granules in both stress and unstressed conditions. However, in unstressed conditions fewer ORF1p foci engaged with stress granules. The discovery of L1ORF1p and L1 polyadenylated RNA in stress granules suggests a mechanism for host defence against the potential mutagenic effects of retrotransposition, by migrating L1 Ribonucleoprotein Particles (RNPs) to stress granules and subsequent degradation of L1mRNA in processing bodies. However, this does not rule out the possibility that the stress granules may be involved in the retrotransposition life cycle rather than their degradation, *i.e.* in stress conditions they may stop ORF1p translation and after the stress has passed they may redirect the L1RNPs to the polyribosomes for translation. Recently, co-immunoprecipitation experiments have identified many more proteins associated with L1 proteins and its RNP (Goodier et al. 2013). These include gene expression regulators and post-translational modifiers among a list of candidates (Doucet et al. 2010; and Goodier et al. 2013). Interestingly, some of the proteins, which co-localise with L1 RNPs, are strong inhibitors of HIV infection as well as other retroviruses (Goodier et al. 2013).

3.7.5 L1 Post-Translational Host Defence Mechanisms

As mentioned earlier, about 95% of L1 retrotransposons are 5' truncated in the human genome and therefore are not competent for retrotransposition. L1 5' truncation is perhaps a result of the low proccessivity of non-LTR endogenous reverse transcriptase, resulting in premature termination of reverse transcription. However non-LTR RTs are more processive than the reverse transcriptases encoded by retroviruses (Eickbush and Jamburuthugoda 2008). Hence it is more likely, that L1 5′ truncation is a result of a host defence mechanism acting post-translationally. This idea is also backed up by Coufal et al. (2011), in a study that reported the potential involvement of ATM in the process of L1 5' truncation, with mutant cells producing longer (or more) L1 insertions.

3.8 Ongoing L1 Retrotransposition in Different Tissues

Due to the disease-causing potential of L1 retrotransposition, the host is under selection to downregulate L1 activity in germline and somatic cells. However, since L1 can only propagate by vertical transmission, L1 expression and transposition must occur in cells contributing to the germline (e.g. germ cells or early embryonic cells) in order to proliferate (Ergün et al. 2004). Although it is estimated that up to 5% of newborns may contain a *de novo* L1-mediated retrotransposition event (Garcia-Perez et al. 2007b), relatively little is known about the developmental timing or cell types that accommodate endogenous LINE-1 retrotransposition in humans. In vivo studies using mouse models indicate that LINE-1 expression or retrotransposition can occur in male and female germ cells during early development, and also in select somatic tissues (Kidwell and Lisch 2000; Brouha et al. 2003). A study by Iskow et al. 2010, demonstrated that *de novo* L1 insertions can also occur in lung cancer. In addition, L1 retrotransposition events must occur in the germline or in early human embryogenesis before germline differentiation in order to be evolutionarily effective (Ergun et al. 2004; Van den Hurk et al. 2007). An *in vitro* retrotransposition assay has been used to demonstrate exogenous LINE-1 retrotransposition in a variety of human and rodent transformed cell lines (Ostertag et al. 2002; Ergun et al. 2004; Garcia-Perez et al. 2007b), in rat neuronal progenitor cells (Muotri et al. 2005), and at a low level in primary human fibroblasts (Bruke et al. 1998). Additionally, it has been shown that human embryonic stem cells can accommodate the retrotransposition of engineered LINE-1 elements in vitro (Garcia-Perez et al. 2007b, 2010). These data suggest that LINE-1 retrotransposition events may occur at early stages in human embryogenesis and that some individuals in the population may be genetic mosaics with respect to their LINE-1 content (Van den Hurk et al. 2007). In the next section four potential environments for *de novo* L1 retrotransposition are discussed in more detail.

3.8.1 L1 Retrotransposition in Neuronal Progenitor Cells

The human nervous system is complex, containing a diversity of neuronal cell types and connections that are influenced by complex and incompletely understood environmental and genetic factors (Tang et al. 2001). As mentioned earlier, L1s must retrotranspose in germ cells or during early embryogenesis to be evolutionarily successful, but the activity of these elements during this period and their effect on other somatic cells is not clear. A study on neuronal cells in transgenic mice reported that L1 constructs can retrotranspose, and that the activity of endogenous L1 promoter is strongly correlated with expression of the Sox2 gene (Muotri et al. 2005). Indeed, in transgenic mice the L1 promoter is repressed by the Sox2 gene in undifferentiated hippocampus neural cells (HCN cells) as well as in the early stages of HCN differentiation, and that depletion of Sox2 expression directly correlates with L1 transcript levels (Muotri et al. 2005). It is speculated that some of the genomic changes necessary for the uniqueness of individuals within a population, as defined by their neural circuitry, might be drivenpartly by the activities of mobile elements (Muotri et al. 2005). In addition, it has been demonstrated that neural progenitor cells isolated from human fetal brain and derived from human embryonic stem cells support the retrotransposition of engineered L1s (Coufal et al. 2009). Moreover, a moderate level of endogenous L1 transcripts have been detected in the hippocampus and several regions of the human brain, but few L1 transcripts were detected in other somatic cells, such as heart and liver, from the same individuals (Coufal et al. 2009). These data suggest that *de novo* L1 retrotransposition events may occur in the human brain and can contribute to brain somatic mosaicism (Coufal et al. 2009; Singer et al. 2010).

Finally it has been shown that the activity of L1 in human brain cells can vary due to environmental factors (Singer et al. 2010). Studies of neuronal progenitor cells derived from Rett syndrome (RTT) patients and human iPS cells have found that mutations in MeCP2 can influence the activity of L1 retrotransposition in human brain cells. As a result, if MeCP2 regulates L1 retrotransposition in a tissue-specific manner in human neuronal cells, this could add to the plasticity of human neuronal cells (Muotri et al. 2010). However, there are conflicting reports regarding the actual number of L1 insertions that may be found in the human brain. Direct high throughput sequence analysis of putative somatic L1 insertions in the hippocampus and caudate nucleus of three in three individuals has demonstrated in vivo L1 activity in the brain (Baillie et al. 2011). This suggests somatic genome mosaicism introduced by L1 retrotransposition can be feature of normal and abnormal neurological process. Despite the above observations, recent genome-wide L1 insertion profiling of 300 single neurons demonstrated only a low rate of (<0.6) somatic insertions per neuron derived from cerebral, cortex and caudate (Evrony et al. 2012). Analysing L1 diversity in 300 single neurones of 3 adults suggested that somatic L1 insertions are rare in adult human cortical pyramidal neurones and caudate neurones (Evrony et al. 2012). However this observation does not exclude the greater rate of L1Hs activity in other cell types or regions of the human brain. One explanation for the conflicting results may be inter-individual variation in the number of highly active L1s (Beck et al. 2010) that leads to variability in somatic retrotransposition rate among individuals. Based on single neuronal cell sequencing, the rate of L1 retrotransposition in human neuronal cells is estimated as 1/10,000 to 1/1000, which is consistent with the rates reported by Coufal et al. (2009). It is still not clear if L1 retrotransposons have any functional impact on neuronal cells and why neuronal cells might accommodate a high level of L1 retrotransposition, when compared to other somatic cells. Despite the recent report of low rate of somatic L1 retrotransposition in glial brain tumours and healthy adult cortex and caudate regions (Iskow et al. 2010; Lee et al. 2012; Evrony et al. 2012) it remains possible that neuronal L1 retrotransposition may occur at higher rates in other brain regions, such as the hippocampus, resulting in regional variation in L1 mutagenesis across the human brain.

3.8.2 L1 Retrotransposition in the Human Germline

As mentioned previously, for *de novo* L1 insertions to be evolutionary successful, they must occur in the germline or during early embryogenesis before germline differentiation (Levin and Moran 2011). To date, most of the discovered disease-causing insertions are thought to be germline in origin as deleterious embryonic mutations are likely to be lost during development (Freeman et al. 2011). Discovery of a *de novo* LRE3 element insertion in exon four of the CYBB gene of a chronic granulomatous disease (CGD) by Brouha et al. (2002) had suggested that the L1 insertion into the CYBB gene is most likely to be germline in origin and occurred during prophase of maternal meiosis II. This and other cases of L1 disease-causing insertions, suggest that L1 retrotransposition can occur early in female oogenesis and embryonic development. Although these findings suggest that L1s must actively retrotranspose in the female germline, direct study of the female germline is very limited due to the difficulty of obtaining oocytes (Freeman et al. 2011). Based on studies of L1 disease-causing insertions there is no direct evidence of *de novo* L1 retrotransposition in the male germline, but sperm provide an accessible resource for screening for *bona fide de novo* L1 insertions in the germline (Freeman et al. 2011). The sperm nucleus is a highly compact structure, and studies in mice have demonstrated that basic DNA associated proteins called protamines are important for post-meiotic chromatin condensation (Lee et al. 1995). Protamines are histone H1-derived, sperm-specific histone variants which associate with sperm DNA, thus permitting tight chromatin packaging (Lewis et al. 2004; Wouters-Tyrou et al. 1998). The dense packaging of DNA in sperm renders it transcriptionally inactive and so unlikely to be a good substrate for the L1 endonuclease, making it unlikely that L1 retrotransposition occurs in mature spermatozoa. By contrast, in oocytes there is no evidence of such tight chromatin packaging, and so they may be a preferential substrate for L1 retrotransposition. From this logic, retrotransposition must occur in the early stages of spermatogenesis rather than later. Indeed, immune-histochemical analysis has demonstrated the co-expression of ORF1 and ORF2 together in pre-spermatogonia, spermatocytes and immature spermatids but this is not detected in spermatogonia, suggesting L1 activity before and after meiosis (Ergun et al. 2004). Despite extensive studies on L1 retrotransposition in the male germline, there is no evidence of *de novo* L1 insertions in these cells, at least at the target loci studied, and so it is not clear at which stage of human spermatogenesis L1s are retrotranspositionally active (Freeman et al. 2011).

3.8.3 L1 Retrotransposition in Early Human Embryogenesis

Previously, L1 retrotransposition was thought to occur predominantly in the germline (Ostertag and Kazazian 2001; Bourchis and Bestor 2004). However, recent studies on transgenic mice have demonstrated that L1 retrotransposition in the germline is quite uncommon, and the bulk of engineered L1 retrotransposition occurs in early embryogenesis with only a fraction of these insertions partitioning into the germline and being transmitted to the progeny (Kano et al. 2009). Except in one reported case (Brouha et al. 2002), where L1 retrotransposition is more likely to have occurred during maternal meiosis I, many other disease causing *de novo* L1 insertions could, in principle, have occurred in early human embryogenesis, before germline partitioning (Kano et al. 2009). In support of this possibility Garcia-Perez et al. (2007b) showed that endogenous L1 elements are expressed during human embryogenesis. Study of isolated ribonucleoprotein particles (RNPs) from undifferentiated human embryonic stem cell lines revealed the presence of ORF1p and L1 mRNA, and subsequent L1 RT-PCR showed that RNAs belonged to both active and old (and largely inactive) L1 subfamilies (Garcia-Perez et al. 2007a). To investigate whether or not human embryonic stem cells (hESC) can support exogenous L1 retrotransposition, Garcia-Perez et al. (2007b) set up a tissue culture retrotransposition assay in which they transfected undifferentiated hESC cells with RC-L1s, driven by either the activity of their endogenous 5' promoter, or a cytomegalovirus immediate early promoter. Using this culture cell retrotransposition technique, Garcia Perez et al. (2007) found that human embryonic stem cells express endogenous L1 elements and can accommodate exogenous L1 retrotransposition in vitro.

Further supporting this thesis, Van den Hurk et al. (2007) studied a case of choroideremia, an X-linked progressive eye disease caused by an L1 insertion. This disease case resulted from the insertion of a full length L1, carrying two serial 3' transductions, into the CHM gene. By analyzing the sequence transduced by this L1 insertion ($L1_{CHM}$) the authors proposed the elements transposition path was from a precursor L1 on either Chromosome 10p15 or 18p11 that transposed to chromosome

6p21 and then to the CHM gene on Chromosome Xq21 (Van den Hurk et al. 2007). Using a PCR-based assay, the mutant CHM allele containing the $L1_{CHM}$ insertion was amplified from the patient's family using markers within the CHM gene, and this showed the presence of an $L1_{CHM}$ insertion in the mother. The results indicated that the mother was a previously undetected mosaic for the L1 insertion, and since the patient's mother showed both somatic and germline mosaicism for the L1 insertion into the CHM gene, the L1 retrotransposition event must have occurred during early embryogenesis, prior to germline segregation from the somatic lineages.

Based on this evidence, and the failure to find *de novo* L1 insertions in male germline cells (Freeman et al. 2011), it seems likely that *de novo* L1 retrotransposition occurs in early human embryogenesis, and this pathway may account for a majority of disease causing insertions.

3.8.4 L1 Retrotransposition in Malignant Derived Cells

As mentioned above, there are 97 known disease causing L1 insertions identified in humans (Hancks and Kazazian 2012; van der Klift HM 2012); of these insertions, 30 were reported in cases of cancer. Comparison between tumor genomes with their matched normal DNA revealed that certain cancers are particularly mutagenized by L1Hs, while no insertions were observed in normal tissues. Recently, a study by Lee et al. 2012 on five different types of cancer showed that somatic L1 insertions tend to occur in genes that are commonly mutated in cancer. As mentioned earlier, there are many mechanisms by which L1 insertional activity could have oncogenic effects, when inserted within oncogenes, or by disrupting tumour suppressor genes.

Many of these studies correlate genome-wide hypomethylation during cancer progression to L1 reactivation (Fig. 3.5) due to an increase in ORF1p and ORF2p expression in several malignancy-derived cell lines (Belancio et al. 2010). Indeed, the overexpression of ORF1p was observed in certain tumors (Harris et al. 2011). More specifically, Alves et al. (1996) demonstrated that sequences flanking the 5' ends of L1Hs are hypomethylated in the T-47D breast cancer cell line. Moreover, they compared hypomethylated L1 loci in cancer cells and germline cells, revealing that different subsets of L1Hs are hypomethylated in each of the cell types. Overall this suggests that L1 hypomethylation can be a feature of many tumors (Carreira et al. 2014; Shukla et al. 2013).

Recent studies using L1 specific high throughput sequencing of lung cancer have reported nine *de novo* L1 retrotransposition events in 6 lung tumors (Iskow et al. 2010). Interestingly, these *de novo* L1 insertions were detected in tumors bearing genome-wide hypomethylation, which is consistent with previous speculations that epigenome alterations can have an effect on L1 activity, and thus cancer biology. However, it is not clear whether L1 activation during cancer progression is a consequence of genome alteration during cancer cell growth or whether L1 has an active role in driving tumorigenesis (Rodić and Burns 2013). In the following sections, we discuss L1 expression and its epigenetic alteration during tumour progression and the load of ongoing L1 retrotransposition in various tumors.

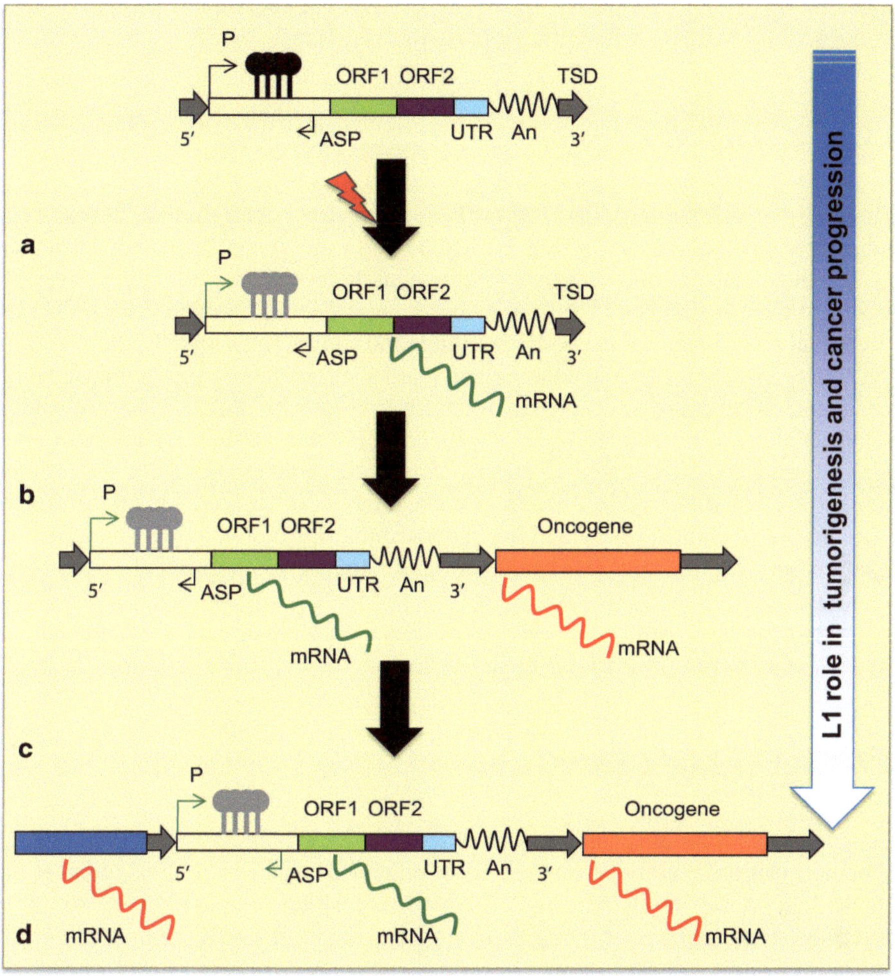

Fig. 3.5 Possible role of L1 in tumorigenesis and cancer progression. **a** Inactive L1: L1 promoters are hypermethylated in normal tissue and therefore transcriptionally inactive. **b** Mutagenic factors can activate L1s promoters through hypomethylation, resulting in L1 transcription. **c** Activated L1s retrotranspose upstream of an oncogene through TPRT and oncogene can transcriptionally be active through the L1 canonical promoter leading to tumorgenesis. **d** Eventually the activated L1 can progress cancer through its activated antisense promoter, which can transcribe nearby genes upstream of L1

3.9 LINE-1 Activity in Different Cancers (Epigenetically and Structurally)

Due to their density and activity in the human genome, the role of ongoing L1 retrotransposition in the origin or progression of cancer is an active area of research. As briefly discussed above, somatic L1 insertions tend to occur in genes that are commonly mutated in cancer, disrupting the expression of the target genes, and are biased toward regions of cancer-specific DNA hypomethylation, highlighting their potential impact in tumorgenesis (Lee et al. 2012). In this section, we will comment on the activity of L1 in different type of human tumours.

3.9.1 Germ Cell Tumours

Human germ cell tumours (GCTs) originate from germ cells, and are a heterogeneous group of neoplasms, which mainly occur in the gonad. Although testicular GCT is the most common cancer of young men, the genes controlling the development and differentiation of GCTs remain largely unknown (Neumann et al. 2011). These tumours can appear in neonates, infants and adults (Kesler and Einhorn 2009).

Gonocytes (fetal testis), secondary spermatocytes, immature spermatid (in adult testis) and GCTs express L1 ORF1p and ORF2p (Ergun et al. 2004; Su et al. 2007; Schulz 2006; Rosser and An 2012). Notably, GCTs that express L1-encoded proteins showed undifferentiated characters similar to carcinoma or yolk sac tumors (Rosser and An 2012). As the origin of these tumours often involved metastasis to other organs, Bratthauer and Fanning proposed that L1 proteins might function as oncoproteins in these cancers (Bratthauer and Fanning 1992; Bratthauer and Fanning 1993).

Despite these suggestive correlations, to date there is no direct information about the load of L1 retrotransposition in these tumours.

3.9.2 Colorectal Cancer

Genome instability, aberrant promoter CpG island hypomethylation and, global hypomethylation have been implicated in colorectal cancer (Chalitchagorn et al. 2004). Indeed, the involvement of L1 retrotransposition in colon cancer was first described from analysis of a *de novo* L1 insertion into the APC gene (Miki et al. 1992).

L1 promoter hypomethylation can be found in normal colonic mucosa in colon cancer patients but not in similar tissue in healthy people (Chalitchagorn et al. 2004). This phenomenon suggests that the effect of progressive demethylation of the L1 promoter may affect expression of nearby genes, which could eventually facilitate neoplastic progression event of normal cells (Suter et al. 2004). Indeed, the methylation level of L1 elements in normal colonic mucosa in colon cancer patients correlates significantly with common polymorphisms found in genes involved in DNA methylation processes (methylene tetrahydrofolate reductase and methylene tetrahydrofolate dehydrogenase) (Iacopetta et al. 2007). Additionally, L1 methylation levels are very diverse among patients (Baba et al. 2010), and may have a diagnostic value (Irahara et al. 2010).

Sequence analysis on 16 colorectal tumour and matched normal DNA by Solyom et al. (2012) have demonstrated a high rate of L1 retrotransposition in some, but not all, samples. Solyom et al. (2012) observed variable number of *de novo* L1 insertions in different tumours (up to 17 in some tumours) and the insertion number correlated with the age of patients. Additionally, this study demonstrated that L1 insertions targeted many genes with a known role in tumour origin/progression, including *ODZ3*, *ROBO2*, *PTPRM*, *PCM1*, and *CDH11*.

Interestingly, in stark contrast to germ line L1 insertions, the discovered *de novo* insertions in colorectal cancers are severely 5' truncated (Solyom et al. 2012). This suggests that in malignant cells, DNA repair pathways do not act as stringently as in normal cells to prevent the accumulation of *de novo* L1 insertions and the high rate of cell divisions in malignant cells may not allow sufficient time for TPRT to complete the integration of *de novo* insertions (Solyom et al. 2012). Alternatively while there is elevated L1 retrotransposition in colorectal cancer cells, the high 5' truncation rate may reflect alterations in the mode of action of the DNA repair system.

It is interesting to note that, in addition to L1 insertions, germ line mutations mediated by Alu insertions and recombination among TEs has been also detected in the APC gene (Halling et al. 1999; Su et al. 2007). However, analysis of secondary tumour sampling data suggests that the L1 insertions largely accumulated after the initiation of the tumour. So while the weight of evidence is that L1 insertions are bystander mutations, these studies demonstrate ongoing activity, so a role for L1 mutagenesis in tumour progression cannot be excluded. The analysis of a larger number of tumours or a deeper sequencing of retrotransposon insertions could shed more light on the role of L1 in this type of cancer (Solyom et al. 2012).

Notably, similar observations were made in studying the whole genome sequence data of five colorectal cancer patients; in one case, a high level of somatic L1 insertion along with microsatellite instability was observed, as well as a high frequency of non-silent SNVs, altered DNA repair pathways via MLH1 epigenetic silencing and a POLE missense mutation. This sample also showed the high CpG island methylator phenotype (CIMP-high) that can correlate with poor clinical outcomes (Lee et al. 2008). In contrast, the other colorectal tumours were microsatellite stable, had a low rate of SNVs and lacked aberration among DNA mismatch repair genes (Lee et al. 2008).

This observed variation among different tumours suggests the existence of different tumour subtypes with respect to L1 activity and L1 expression/activity could be potentially used to stratify different types of tumours. Yet, it is still unclear whether L1 retrotransposition is involved in colorectal cancer initiation or mainly contributes to cancer progression and a more aggressive phenotype.

3.9.3 Breast Cancer

Breast cancer is the most frequently diagnosed cancer and the leading cause of cancer death in females worldwide, accounting for 23% of total new cancer cases (Jemal et al. 2011).

Notably, L1 ORF1p and ORF2p are expressed in various types of breast cancers, including infiltrating ductal carcinoma, but not in non-malignant breast epithelial cell lines and normal breast tissue (Harris et al. 2011; Bratthauer et al. 1994; Chen et al. 2012). However it should be noted that ORF1p signals in positive tumours are heterogeneous so the correlation is not absolute (Asch et al. 1996). Significant association between the amount of ORF1p expression and clinical outcomes of breast cancer is not evident; however ORF1p nuclear localization might be correlated with poor prognosis in this type of cancer (Harris et al. 2011). Indeed, it has been suggested that L1 expression may have a prognostic value in this cancer (Chen et al. 2012). On the other hand, hypomethylation of L1 promoters in breast cancer cell lines result in expression of L1 chimeric transcript (cancer-specific) that are likely initiated from the L1 antisense promoter (Cruickshanks and Tufarelli 2009).

Finally, in a few cases the insertion of non-LTR retrotransposons (mainly Alu elements) in the BRCA1 and BRCA2 genes has been reported (Montagna et al. 1999; Teugels et al. 2005). However given the intense mutation screening directed at these genes it would be surprising if at least some retrotransposon insertion mutations were not detected, so the significance of these observations is not clear. Overall, further studies are required to fully elucidate the role of L1 retrotransposition in this type of cancer.

3.9.4 Hepatocellular Carcinoma

Liver cancer incidence rates are increasing world wide due to various reasons, such as the increasing rate of obesity and hepatitis C virus infection through intravenous drug use (Jemal et al. 2011). Among primary liver cancers, hepatocellular carcinoma (HCC) represents the major subtype (~80%) of liver cancer (Jemal et al. 2011). A recent study has shown that expression of L1 ORF1p in the HCC line HepG2 could result in cell proliferation and may also have effects on creating resistance to chemotherapeutic agents (Feng et al. 2013). The study also suggested that L1-ORF1p could be a good target to overcome chemotherapeutic resistance of hepatocellular carcinoma cells, but the connection between retrotransposon expression and cellular physiology is far from clearly established.

Similar to other tumour types, the hypomethylation of L1 promoters has been suggested to have diagnostic value for some malignant stages of this type of cancer. Indeed, L1 hypomethylation may result in increase retrotransposition and subsequent genomic instability (Takai et al. 2000). Recently, Shukla et al. (2013) reported L1 insertions in two tumour suppressor genes, named MCC (mutated in colorectal cancer) and ST18 (suppression of tumorigenicity 18), involved in the formation of hepatocellular carcinoma. It is thought that L1 insertions in the MCC gene may have resulted in the initiation of oncogenesis; moreover the normal function of ST18 seems to have been de-regulated by the *de novo* L1 insertion. Notably, this study demonstrated that the L1 promoter is hypomethylated in tumours compared to non-tumour and healthy tissue samples. Notably the authors demonstrated for the first time the causal role of a *de novo* L1 insertion in a human cancer (Shukla et al. 2013), but further studies are required to learn how often this phenomenon occurs in other cancer types.

3.9.5 Epithelial Cancers

Notably, the analysis of multiple cancer types has revealed that somatic L1 retrotransposition occurs mostly in cancers of epithelial origin (Lee et al. 2012). Analysing L1 insertions from NGS short read data has proven to be challenging and it is even more difficult to identify novel insertions in cancers due to the heterogeneity of their genome. Results from analysing whole genome sequencing data from tumour and blood samples from 43 colorectal, prostate, ovarian and multiple myeloma, glioblastoma revealed a high level of somatic L1 insertions (183 L1s, 10 Alus, and 1 ERV). Notably, it is demonstrated that in some of these tumours there were up to 29 insertions, indicating substantial heterogeneity (Lee et al. 2012). Comparing the presence/absence of insertions with matched blood samples revealed that the cancers with most insertions were (ovarian, prostate and colorectal) epithelial in origin, with colon tumours having the highest frequency of somatic L1 insertions (Lee et al. 2012).

A key issue in cancer biology is how a cell turns into a cancer stem cell. Cancer stem cells are remarkably similar to normal stem cells as they both have the ability to self renew, are multipotent, and express common surface markers (Zhu et al. 2009). However, it is not clear if cancer stem cells are the direct progeny of stem cells through an oncogenic mutation that yields a tumour rather than a normal differentiated cell (Zhu et al. 2009). On the other hand it is also possible that the cancer stem cell arises from reprogrammed differentiated cells, driven by an oncogenic mutation. Schwitalla et al. (2013) have demonstrated that NF $-_{k}$B can de-deferentiate epithelial cells in colon cancer into a cancer stem cell-like state.

Given the evidence that epithelial cells are able to de-deferentiate into cancer stem cells, and that epithelial cell-derived tumours are demonstrated to accommodate L1 retrotransposition, we might speculate that epithelial cancers likely originate from differentiated cells. Accordingly, perhaps L1 activity in these cells acts to generate driver oncogenic mutations that reprogram these cells into a tumour stem cell-like state (Carreira et al. 2014). Indeed, Wissing et al. (2011) have demonstrated that directed reprograming of epithelial cells into induced pluripotent stem cell (iPSC) types produces cell that can accommodate L1 retrotransposition and show restoration of L1 expression.

3.10 Role of L1 in Cancer

A clear correlation has been established between L1 mobilization and cancer. However, how frequently L1 activity generates mutations that are oncogenesis “drivers” rather than “passenger” mutations as a consequence of oncogenesis is an on going subject of debate.

Cancer is a complex disease attributed to the accumulation of multiple risk factors such as genetic predisposition and environmental exposure. Both L1 retrotransposition and tumourgenesis are affected by environmental factors such as diet, lifestyle and exposure to toxic compounds. It has been demonstrated that tumours often contain de novo L1 insertions, some of which interact with cancer genes. Moreover, inherited risk far exceeds the frequency of mutations already reported in cancer genes, suggesting that other contributing mechanisms or types of genetic alteration, such as rare genetic variants and retrotransposition events, may also substantially contribute to cancer development.

However, the specific pathways leading to L1 activation in cancer remain unknown. It is not known whether L1 activation in cancer is initiated by factors that contribute to cancer genome instability such as global hypomethylation, or other coincidental factors, such as mutations in the transcription factors that regulate L1 expression. Although direct evidence of L1 activity in cancer progenitor cells has not been demonstrated, knowledge derived from the study of pluripotent and developmentally plastic cells suggests that retrotransposons can become active upon alteration of cellular defence systems required for their suppression. This, combined with the recognition that L1 insertional mutagenesis occurs in tumours, in cancer cell lines, during development, and within the soma in both dividing and non-dividing cells, leads to a plausible model in which L1 activation is due to epigenetic or other perturbations of retrotransposon suppression within cancer stem cells. Given the substantial, predominantly deleterious, effects of intragenic L1 insertions upon host gene expression, L1 insertions may be more likely, on a per mutation basis, to have an impact on tumourigenesis than other genetic aberrations observed in cancer (Carreira et al. 2014).

On the other hand, if we take as given that retrotransposition is a stochastic process, and that most somatic cells have some basal L1 activity that has escaped host silencing defence mechanisms, it is plausible that environmental factors could increase the probability of a somatic L1 insertion affecting an oncogenic locus, thereby triggering neoplastic transformation.

3.11 L1 as a Diagnostic Tool for Cancer

In line with the observation that L1 can contribute to oncogenesis in some cases, they also can be used as biomarkers and diagnostic tools for malignancy and metastasis. Various studies have suggested that detectable levels of L1 mRNA and proteins are associated with poor cancer prognosis. Ogino et al. 2008 quantified L1 DNA methylation in 643 colon cancer patients as a measure of global DNA methylation level in colorectal carcinogenesis. They reported a linear association of L1 hypomethylation and increase in colon cancer mortality and therefore, suggested that tumour L1 hypomethylation is independently associated with shorter survival among colon cancer patients. Moreover, it is reported that the antisense promoter of L1 can direct transcription of adjacent unique genomic sequences resulting in the formation of chimeric RNAs, which can perturb transcription of neighboring genes. Indeed, Cruickshanks and colleagues have isolated novel chimeric transcripts that are unique to breast cancer cell lines, primary tumours and colon cancer cells. These findings indicate that the loss of L1 methylation in cancer cells is linked to the expression of L1-chimeric transcripts, which may therefore constitute biomarkers of malignancy (Cruickshanks et al. 2009). Although these findings may have considerable clinical implications, future studies are required to confirm the association of L1 genome-wide hypomethylation with cancer prognosis as well as examine the potential mechanisms by which genome-wide DNA hypomethylation affects tumor behavior.

From a clinical perspective, it is yet to be investigated whether all tumour cells or only a subset of cells from a neoplasm, show enhanced L1 activity and if this heterogeneity leads tumour cell evolution in response to chemotherapy or radiotherapy. There is growing evidence indicating a high level of endogenous RT activity associated with transformed/tumorigenic phenotypes in mammalian cells. Additionally inhibition of L1 RT through nuclear or non-nuclear inhibitors has been suggested as a promising approach in cancer therapy (Sbardella et al. 2011; Carlini et al. 2010; Jones et al. 2008). For example it has been demonstrated that ethyl-substituted derivatives 3a-h, belonging to the F2-DABOs class of non-nucleoside HIV-1 reverse transcriptase inhibitors, have an anti-proliferating role on A375 melanoma cells (Sbardella et al. 2011). In contrast, a study on the effect of different RT inhibitors against L1 RT activity and retrotransposition indicated that L1 RT is sensitive to nucleoside analog inhibitors (NRTIs), but non-nucleoside inhibitors (NNRTIs) inhibit L1 RT less efficiently. Also these authors demonstrated that Nevirapine, an RT inhibitor with reported anti-tumour function, has no effect on L1 RT activity (Dai et al. 2011). Therefore, in general it is yet unclear how or whether blocking L1 mobilization, for example using reverse transcriptase inhibitors, would in any way affect cancer progression or prognosis.

For future directions, it is important to elucidate the origins and importance of L1 retrotransposition in cancer development. This can be achieved by larger scale surveys of retrotransposition in human tumours. Finally, despite increasingly economical high throughput sequencing approaches, alternative in vivo studies using mouse models should not be neglected in providing a complementary picture of the role of L1 in cancer.

3.12 Acknowledgements

We apologize colleagues whose work has not been cited in this book chapter. We would like to thank Pierpaolo Maisano Delser, (Department of genetics, university of Leicetser, UK) for critically reviewing this chapter and for his constructive comments. J.L.G.P's lab is supported by CICE-FEDER-P09-CTS-4980, FIS-FEDER-PI11/01489, the European Research Council (ERC-Consolidator ERC-STG-2012-233764) and by an International Early Career Scientist grant from the Howard Hughes Medical Institute (IECS-55007420).

References

Alexandrova EA, Olovnikov IA, Malakhova GV, Zabolotneva AA, Suntsova MV, Dmitriev SE et al (2012) Sense transcripts originated from an internal part of the human retrotransposon LINE-1 5' UTR. Gene 511(1):46–53

Alves G, Tatro A, Fanning T (1996) Differential methylation of human LINE-1 retrotransposons in malignant cells. Gene 176(1–2):39–44

Aporntewan C, Phokaew C, Piriyapongsa J, Ngamphiw C, Ittiwut C, Tongsima S et al (2011) Hypomethylation of intragenic LINE-1 represses transcription in cancer cells through AGO2. PLoS One 6(3):e17934

Asch HL, Eliacin E, Fanning TG, Connolly JL, Bratthauer G, Asch BB (1996) Comparative expression of the LINE-1 p40 protein in human breast carcinomas and normal breast tissues. Oncol Res 8:239–247

Athanikar JN, Badge RM, Moran JV (2004) A YY1-binding site is required for accurate human LINE-1 transcription initiation. Nucl Acids Res 32(13):3846–3855

Baba Y, Huttenhower C, Nosho K, Tanaka N, Shima K, Hazra A (2010) Epigenomic diversity of colorectal cancer indicated by LINE-1 methylation in a database of 869 tumors. Mol Cancer 27(9):125

Baillie JK, Barnett MW, Upton KR, Gerhardt DJ, Richmond TA, De Sapio F et al (2011) Somatic retrotransposition alters the genetic landscape of the human brain. Nature 479(7374):534–537

Basame S, Wai-lun Li P, Howard G, Branciforte D, Keller D, Martin SL (2006) Spatial assembly and RNA binding stoichiometry of a LINE-1 protein essential for retrotransposition. Mol Biol 357(2):351–357

Beck CR, Collier P, Macfarlane C, Malig M, Kidd JM, Eichler EE et al (2010) LINE-1 retrotransposition activity in human genomes. Cell 141:1159–1170

Becker KG, Swergold GD, Ozato K, Thayer RE (1993) Binding of the ubiquitous nuclear transcription factor YY1 to a cis regulatory sequence in the human LINE-1 transposable element. Hum Mol Genet 2:1697–1702

Belancio VP, Roy-Engel AM, Deininger PL (2010) All y'all need to know 'bout retroelements in cancer. Semin Cancer Biol 20(4):200–210

Bestor TH (2003) Cytosine methylation mediates sexual conflict. Trends Genet 9(4):185–190

Boeke JD (1997) LINEs and Alus—the polyA connection. Nat Genet 16:6–7

Boeke JD, Pickeral OK (1999) Retroshuffling the genomic deck. Nature 398:108–111

Boissinot S, Furano AV (2001) Adaptive evolution in LINE-1 retrotransposons. Mol Biol Evol 18(12):2186–2194

Boissinot S, Chevret P, Furano AV (2000) L1 (LINE-1) retrotransposon evolution and amplification in recent human history. Mol Biol Evol 17:915–928

Borc'his D, Bestor TH (2004) Meiotic catastrophe and retrotransposon reactivation in male germ cells lacking Dnmt3 L. Nature 431:96–99

Bratthauer GL, Fanning TG (1992) Active LINE-1 retrotransposons in human testicular cancer. Oncogene 7(3):507–510

Bratthauer GL, Fanning TG (1993) LINE-1 retrotransposon expression in pediatric germ cell tumors. Cancer 71(7):2383–2386

Bratthauer GL, Cardiff RD, Fanning TG (1994) Expression of LINE-1 retrotransposons in human breast cancer. Cancer 73(9):2333–2336

Brouha B, Meischl C, Ostertag E, de Boer M, Zhang Y, Neijens H et al (2002) Evidence consistent with human L1 retrotransposition in maternal meiosis I. Am J Hum Genet 71(2):327–336

Brouha B, Schustak J, Badge RM, Lutz-Prigge S, Farley AH, Moran JV et al (2003) Hot L1s account for the bulk of retrotransposition in the human population. Proc Natl Acad Sci U S A 100:5280–5285

Burke WD, Malik HS, Jones JP, Eickbush TH (1999) The domain structure and retrotransposition mechanism of R2 elements are conserved throughout arthropods. Mol Biol Evol 16:502–511

Callinan PA, Batzer MA (2006) Retrotransposable elements and human disease. Genome Dyn 1:104–115

Carlini F, Ridolfi B, Molinari A, Parisi C, Bozzuto, G, Toccacieli L (2010) The reverse transcription inhibitor abacavir shows anticancer activity in prostate cancer cell lines. PLoS ONE 5(12):e14221

Carreira PE, Richardson SR, Faulkner GJ (2014) L1 retrotransposons, cancer stem cells and oncogenesis. FEBS J 281:63–73

Chalitchagorn K, Shuangshoti S, Hourpai N, Kongruttanachok N, Tangkijvanich P, Thong-ngam D et al (2004) Distinctive pattern of LINE-1 methylation level in normal tissues and the association with carcinogenesis. Oncogene 23(54):8841–8846

Chen JM, Férec C, Cooper DN (2007) Mechanism of Alu integration into the human genome. Genomic Med 1(1–2):9–17

Chen L, Dahlstrom JE, Chandra A, Board P, Rangasamy D (2012) Prognostic value of LINE-1 retrotransposon expression and its subcellular localization in breast cancer. Breast Cancer Res Treat 136(1):129–142

Chureau C, Prissette M, Bourdet A, Barbe V, Cattolico L, Jones L et al (2002) Comparative sequence analysis of the X-inactivation center region in mouse, human, and bovine. Genome Res 12(6):894–908

Ciaudo C, Jay F, Okamoto I, Chen CJ, Sarazin A, Servant N, et al. (2013) RNAi-dependent and independent control of LINE1 accumulation and mobility in mouse embryonic stem cells. PLoS Genet 9(11):e1003791

Cooke SL, Shlien A, Marshall J, Pipinikas CP, Martincorena I, Tubio JM (2014) Processed pseudogenes acquired somatically during cancer development. Nat Commun 9(5):3644

Cordaux R, Batzer MA (2009) The impact of retrotransposons on human genome evolution. Nat Rev Genet 10(10):691–703

Cost GJ, Boeke JD (1998) Targeting of human retrotransposon integration is directed by the specificity of the L1 endonuclease for regions of unusual DNA structure. Biochemistry 37:18081–18093

Cost GJ, Feng Q, Jacquier A, Boeke JD (2002) Human L1 element target-primed reverse transcription in vitro. EMBO J 2:5899–5910

Costas J, Naveira H (2000) Evolutionary history of the human endogenous retrovirus family ERV9. Mol Biol Evol 17(2):320–330

Coufal NG, Garcia-Perez JL, Peng GE, Yeo GW, Mu Y, Lovci MT et al (2009) L1 retrotransposition in human neural progenitor cells. Nature 460(7259):1127–1131

Coufal NG, Garcia-Perez JL, Peng GE, Marchetto MC, Muotri AR, Mu Y et al (2011) Ataxia telangiectasia mutated (ATM) modulates long interspersed element-1 (L1) retrotransposition in human neural stem cells. Proc Natl Acad Sci U S A 108(51):20382–20387

Craig NL, Craigie R, Gellert M, Lambowitz, AM (eds) (2002) Mobile DNA introduction. In Mobile DNA II. American Society for Microbiology, Washington, DC, pp 836–869

Cruickshanks HA, Tufarelli C (2009) Isolation of cancer-specific chimeric transcripts induced by hypomethylation of the LINE-1 antisense promoter. Genomic 94(6):397–406

Curcio MJ, Derbyshire KM (2003) The outs and ins of transposition: from mu to kangaroo. Nat Rev Mol Cell Biol 4:865–877

Dai L, Huang Q, Boeke JD (2011) Effect of reverse transcriptase inhibitors on LINE-1 and Ty1 reverse transcriptase activities and on LINE-1 retrotransposition. BMC Biochem 5(12):18

Dawkins R. (1976) The selfish gene. Oxford University Press, Oxford

Deininger PL et al (2003) Mobile elements and mammalian genome evolution. Curr Opin Genet Dev 13(6):651–658

Dewannieux M, Esnault C, Heidmann T (2003) LINE-mediated retrotransposition of marked Alu sequences. Nat Genet 35(1):41–48

Divoký V, Hammerová T, Sakalová A, Luhový M, Divoká M, Melichárková R, Indrák K (1996) Unstable Santa Ana hemoglobin or alpha 2 beta 2 88 (F4) Leu-Pro detected in a Slovak girl. Vnitr Lek 42(4):258–261

Dmitriev SE, Andreev DE, Terenin IM, Olovnikov IA, Prassolov VS, Merrick WC et al (2007) Efficient translation initiation directed by the 900-nucleotide-long and GC-rich 5' untranslated region of the human retrotransposon LINE-1 mRNA is strictly cap dependent rather than internal ribosome entry site mediated. Mol Cell Biol 27(13):4685–4697

Doucet AJ, Hulme AE, Sahinovic E, Kulpa DA, Moldovan JB, Kopera HC et al (2010) Characterization of LINE-1 ribonucleoprotein particles. PLoS Genet 6(10):e1001150

Eickbush TH, Jamburuthugoda VK (2008) The diversity of retrotransposons and the properties of their reverse transcriptases. Virus Res 134(1–2):221–234

Eickbush TH, Malik HS (2002) Origins and evolution of retrotransposons. Mobile DNA II Craig NL. In: Craigie R, Gellert M, Lambowitz AM (eds) American Society for Microbiology, Washington, DC, pp 1111–1144

Ergün S, Buschmann C, Heukeshoven J, Dammann K, Schnieders F, Lauke H et al (2004) Cell type- specific expression of LINE-1 ORF1 and ORF2 in fetal and adult human tissues. J Biol Chem 279(26):27753–27763

Esnault C, Maestre J, Heidmann T (2000) Human LINE retrotransposons generate processed pseudogenes. Nat Genet 24:363–367

Esnault C, Priet S, Ribet D, Heidmann O, Heidmann T (2008) Restriction by APOBEC3 proteins of endogenous retroviruses with an extracellular life cycle: ex vivo effects and in vivo "traces" on the murine IAPE and human HERV-K elements. Retrovirology 14(5):75

Evrony GD, Cai X, Lee E, Hills LB, Elhosary PC, Lehmann HS et al (2012) Single-neuron sequencing analysis of L1 retrotransposition and somatic mutation in the human brain. Cell 151(3):483–496

Ewing AD, Kazazian HH Jr (2010) High-throughput sequencing reveals extensive variation in human-specific L1 content in individual human genomes. Genome Research 20(9):1262–1270

Feng Q, Zhang Y (2001) The MeCP1 complex represses transcription through preferential binding, remodeling, and deacetylating methylated nucleosomes. Genes Dev 15:827–832

Feng Q, Moran J, Kazazian HH Jr, Boeke JD (1996) Human L1 retrotransposon encodes a conserved endonuclease required for retrotransposition. Cell 87:905–916

Feng F, Lu YY, Zhang F, Gao XD, Zhang CF, Meredith A et al (2013) Long interspersed nuclear element ORF-1 protein promotes proliferation and resistance to chemotherapy in hepatocellular carcinoma. World J Gastroenterol 19(7):1068–1078

Ficz G, Branco MR, Seisenberger S, Santos F, Krueger F, Hore TA et al (2011) Dynamic regulation of 5-hydroxymethylcytosine in mouse ES cells and during differentiation. Nature 473(7347):398–402

Freeman P, Macfarlane C, Collier P, Jeffreys AJ, Badge RM (2011) L1 hybridization enrichment: a method for directly accessing de novo L1 insertions in the human germline. Hum Mutat 32(8):978–988

Furano AV (2000) The biological properties and evolutionary dynamics of mammalian LINE-1 retrotransposons. Prog Nucleic Acid Res Mol Biol 64:255–294

Garcia-Perez JL et al (2007a) Distinct mechanisms for trans-mediated mobilization of cellular RNAs by the LINE-1 reverse transcriptase. Genome Res 17(5):602–611

Garcia-Perez JL et al (2007b) LINE-1 retrotransposition in human embryonic stem cells. Hum mol genet 16(13):1569–1577

Garcia-Perez JL, Morell M, Scheys JO, Kulpa DA, Morell S, Carter CC et al (2010) Epigenetic silencing of engineered L1 retrotransposition events in human embryonic carcinoma cells. Nature 466(7307):769–773

Gasior SL, Wakeman TP, Xu B, Deininger PL (2006) The human LINE-1 retrotransposon creates DNA double-strand breaks. J Mol Biol 357(5):1383–1393

Gilbert N, Lutz-Prigge S, Moran JV (2002) Genomic deletions created upon LINE-1 retrotransposition. Cell 110:315–325

Gilbert N, Lutz S, Morrish TA, Moran JV (2005) Multiple fates of L1 retrotransposition intermediates in cultured human cells. Mol Cell Biol 25:7780–7795

Goll MG, Bestor TH (2005) Eukaryotic cytosine methyltransferases. Annu Rev Biochem 74:481–514

Goodier JL, Kazazian HH Jr (2008) Retrotransposons revisited: the restraint and rehabilitation of parasites. Cell 135: 23–35

Goodier JL, Ostertag EM, Kazazian HH Jr (2000) Transduction of 3'-flanking sequences is common in L1 retrotransposition. Hum Mol Genet 9:653–657

Goodier JL, Ostertag EM, Engleka KA, Seleme MC, Kazazian HH Jr (2004) A potential role for the nucleolus in L1 retrotransposition. Hum Mol Genet 13(10):1041–1048

Goodier JL, Zhang L, Vetter MR, Kazazian HH Jr (2007) LINE-1 ORF1 protein localizes in stress granules with other RNA-binding proteins, including components of RNA interference RNA-induced silencing complex. Mol Cell Biol 27(18):6469–6483

Goodier JL, Cheung LE, Kazazian HH Jr (2013) Mapping the LINE1 ORF1 protein interactome reveals associated inhibitors of human retrotransposition. Nucleic Acids Res 41(15):7401–7419

Graham T, Boissinot S (2006) The genomic distribution of L1 elements: the role of insertion bias and natural selection. J Biomed Biotechnol 2006(1):75327

Gregory TR, and Hebert PD (1999) The modulation of DNA content: proximate causes and ultimate consequences. Genome Res 9:317–324

Halling KC, Lazzaro CR, Honchel R, Bufill JA, Powell SM, Arndt CA et al (1999) Hereditary desmoid disease in a family with a germline Alu I repeat mutation of the APC gene. Hum Hered 49(2):97–102

Han JS, Szak ST, Boeke JD (2004) Transcriptional disruption by the L1 retrotransposon and implications for mammalian transcriptomes. Nature 429(6989):268–274

Han K, Sen SK, Wang J, Callinan PA, Lee J, Cordaux R et al (2005) Genomic rearrangements by LINE-1 insertion-mediated deletion in the human and chimpanzee lineages. Nucl Acids Res 33:4040–4052

Han K, Lee J, Meyer TJ, Remedios P, Goodwin L, Batzer MA (2008) L1 recombination-associated deletions generate human genomic variation. Proc Natl Acad Sci U S A 105(49):19366–19371

Hancks DC, Kazazian HH Jr (2012) Active human retrotransposons: variation and disease. Curr Opin Genet Dev 22(3):191–203

Harris SL, Thorne LB, Seaman WT, Hayes DN, Couch ME, Kimple RJ (2011) Association of p16 (INK4a) overexpression with improved outcomes in young patients with squamous cell cancers of the oral tongue. Head Neck 33(11):1622–1627

Hata K, Sakaki Y (1997) Identification of critical CpG sites for repression of L1 transcription by DNA methylation. Gene 189(2):227–234

Heard E, Disteche CM (2006) Dosage compensation in mammals: fine-tuning the expression of the X chromosome. Genes Dev 20(14):1848–1867

Heras SR, Macias S, Plass M, Fernandez N, Cano D, Eyras E et al (2013) The Microprocessor controls the activity of mammalian retrotransposons. Nat Struct Mol Biol 20(10):1173–1181

Hohjoh H, Singer MF (1996) Cytoplasmic ribonucleoprotein complexes containing human LINE-1 protein and RNA. EMBO J 15:630–639

Hohjoh H, Singer MF (1997) Sequence-specific single-strand RNA binding protein encoded by the human LINE-1 retrotransposon. Embo J 16:6034–6043

Holmes SE, Dombroski BA, Krebs CM, Boehm CD, Kazazian HH Jr (1994) A new retrotransposable human L1 element from the LRE2 locus on chromosome 1q produces a chimaeric insertion. Nat Genet 7(2):143–148

Huang CJ, Lin WY, Chang CM, Choo KB (2009) Transcription of the rat testis-specific Rtdpoz-T1 and -T2 retrogenes during embryo development: co-transcription and frequent exonisation of transposable element sequences. BMC Mol Biol 10:74

Iacopetta B, Grieu F, Phillips M, Ruszkiewicz A, Moore J, Minamoto T et al (2007) Methylation levels of LINE-1 repeats and CpG island loci are inversely related in normal colonic mucosa. Cancer Sci 98(9):1454–1460

Irahara N, Baba Y, Nosho K, Shima K, Yan L, Dias-Santagata D et al (2010) NRAS mutations are rare in colorectal cancer. Diagn Mol Pathol 19(3):157–163

Iskow RC, McCabe M, Mills RE, Torene S, Pittard S, Neuwaid AF et al (2010) Natural Mutagenesis of Human Genomes by Endogenous Retrotransposons. Cell 141:1253–1261

Ito S, Shen L, Dai Q, Wu SC, Collins LB, Swenberg JA et al (2011) Tet proteins can convert 5-methylcytosine to 5-formylcytosine and 5-carboxylcytosine. Science 333(6047):1300–1303
Jemal A, Thun M, Yu XQ, Hartman AM, Cokkinides V, Center MM et al (2011) Changes in smoking prevalence among U.S. adults by state and region: estimates from the tobacco use supplement to the current population survey, 1992–2007. BMC Public Health 29(11):512
Jones S, Zhang X, Parsons DW, Lin JC, Leary RJ, Angenendt P (2008) Core signaling pathways in human pancreatic cancers revealed by global genomic analyses. Science 321(5897):1801–1806
Jurka J (1997) Sequence patterns indicate an enzymatic involvement in integration of mammalian retroposons. Proc Natl Acad Sci U S A 94:1872–1877
Kano H, Godoy I, Courtney C, Vetter MR, Gerton GL, Ostertag EM et al (2009) L1 retrotransposition occurs mainly in embryogen- esis and creates somatic mosaicism. Genes Dev 23:1303–1312
Kapusta A, Kronenberg Z, Lynch VJ, Zhuo X, Ramsay L, Bourque G et al (2013) Transposable elements are major contributors to the origin, diversification, and regulation of vertebrate long noncoding RNAs. PLoS Genet 9(4):e1003470
Katoh I, Kurata SI (2013) Association of endogenous retroviruses and long terminal repeats with human disorders. Front Oncol 3:234
Kazazian HH Jr (1998) Mobile elements and disease. Curr Opin Genet Dev 8:343–350
Kazazian HH Jr (2004) Mobile elements: drivers of genome evolution. Science 303(5664):1626–1632
Kazazian HH Jr, Goodier JL (2002) LINE drive. retrotransposition and genome instability. Cell 110(3):277–280
Kazazian HH Jr, Wong C, Youssoufian H, Scott AF, Phillip, DG, Antonarakis SE (1988) Haemophilia a resulting from de novo insertion of L1 sequences represents a novel mechanism for mutation in man. Nature 332:164–166
Kesler KA, Einhorn LH (2009) Multimodality treatment of germ cell tumors of the mediastinum. Thorac Surg Clin 19(1):63–69
Khazina E, and Weichenrieder O (2009) Non-LTR retrotransposons encode noncanonical RRM domains in their first open reading frame. Proc Natl Acad Sci U S A 106(3):731–736
Kidwell M, Lisch D (2000) Reply from M.G. Kidwell and D.R. Lisch. Trends Ecol Evol 15(7):288
Kimberland ML, Divoky V, Prchal J, Schwahn U, Berger W, Kazazian HH Jr (1999) Full-length human L1 insertions retain the capacity for high frequency retrotransposition in cultured cells. Hum Mol Genet 8(8):1557–1560
Kitkumthorn N, Mutirangura A (2011) Long interspersed nuclear element-1 hypomethylation in cancer: biology and clinical applications. Clin Epigenetics 2(2):315–330
Kolosha VO, Martin SL (2003) High-affinity, non-sequence-specific RNA binding by the open reading frame 1 (ORF1) protein from long interspersed nuclear element 1 (LINE-1). Biol Chem 278(10):8112–8117
Koning APJ, Gu W, Castoe TA, Batzer MA, Pollock DD (2011) Repetitive Elements May Comprise Over Two-Thirds of the Human Genome. PLoS Genet 7(12):e1002384
Kuramochi-Miyagawa S, Watanabe T, Gotoh K, Totoki Y, Toyoda A, Ikawa M et al (2008) DNA methylation of retrotransposon genes is regulated by Piwi family members MILI and MIWI2 in murine fetal testes. Genes Dev 22(7):908–917
Kurose K, Hata K, Hattori M, Sakaki Y (1995) RNA polymerase III dependence of the human L1 promoter and possible participation of the RNA polymerase II factor YY1 in the RNA polymerase III transcription system. Nucl Acids Res 23:3704–3709
Lander ES, Linton LM, Birren B, Nusbaum C, Zody MC et al (2001) Initial sequencing and analysis of the human genome. Nature 409:860–921
Lee K, Haugen HS, Clegg CH, Braun RE (1995) Premature translation of protamine 1 mRNA causes precocious nuclear condensation and arrests spermatid differentiation in mice. Proc Natl Acad Sci U S A 92(26):12451–12455
Lee S, Cho NY, Choi M, Yoo EJ, Kim JH, Kang GH (2008) Clinicopathological features of CpG island methylator phenotype-positive colorectal cancer and its adverse prognosis in relation to *KRAS/BRAF* mutation. Pathol Int 58:104–113

Lee E, Iskow R, Yang L, Gokcumen O, Haseley P, Luquette LJ 3rd, Lohr JG, Harris CC, Ding L, Wilson RK, Wheeler DA, Gibbs RA, Kucherlapati R, Lee C, Kharchenko PV, Park PJ, Cancer Genome Atlas Research Network (2012) Landscape of somatic retrotransposition in human cancers. Science 337:967–971

Levin HL, Moran JV (2011) Dynamic intractions between transposable elements and their hosts. Nature Rev Genet 12:615–627

Lewis SE, O'Connell M, Stevenson M, Thompson-Cree L, McClure N (2004) An algorithm to predict pregnancy in assisted reproduction. Hum Reprod 6:1385–1394

Li TH, Schmid CW (2001) Differential stress induction of individual Alu loci: implications for transcription and retrotransposition. Gene 276:135–141

Lindtner S, Felber BK, Kjems J (2002) An element in the 3' untranslated region of human LINE-1 retrotransposon mRNA binds NXF1 (TAP) and can function as a nuclear export element. RNA 8(3):345–356

Liu HT, Li B, Shang ZL, Li XZ, Mu RL, Sun DY et al (2003) Calmodulin is involved in heat shock signal transduction in wheat. Plant Physiol 132:1186–1195

Lovell-Badge R (2009) The early history of the Sox genes. Int J Biochem Cell Biol 42(3):378–380

Lovsin N, Peterlin BM (2009) APOBEC3 proteins inhibit LINE-1 retrotransposition in the absence of ORF1p binding. Ann N Y Acad Sci 1178:268–275

Lovsin E, Fazarinc G, Pogacnik A, Bavdek SV (2001) Growth dynamics of lipizzan horses and their comparison to other horse breeds. Pflugers Arch 442(6 Suppl 1):211–212

Luan DD, Korman MH, Jakubczak JL, Eickbush TH (1993) Reverse transcription of R2Bm RNA is primed by a nick at the chromosomal target site: a mechanism for non-LTR retrotransposition. Cell 72:595–605

Lyon MF (1998) X-chromosome inactivation spreads itself: effects in autosomes. Am J Hum Genet 63(1):17–19

Lynch M, Conery JS (2000) The evolutionary fate and consequences of duplicate genes. Science 290(5494):1151–1155

MacDuff DA, Demorest ZL, Harris RS (2009) AID can restrict L1 retrotransposition suggesting a dual role in innate and adaptive immunity. Nucl Acids Res 37(6):1854–1867

Macia A, Muñoz-Lopez M, Cortes JL, Hastings RK, Morell S, Lucena-Aguilar G et al (2011) Epigenetic control of retrotransposon expression in human embryonic stem cells. Mol Cell Biol 31(2):300–316

Malik HS, Burke WD, Eickbush TH (1999) The age and evolution of non-LTR retrotransposable elements. Mol Biol Evol 16(6):793–805

Malik HS, Eickbush TH (1999) Modular evolution of the integrase domain in the Ty3/Gypsy class of LTR retrotransposons. J Virol 73(6):5186–5190

Malone CD, Brennecke J, Dus M, Stark A, McCombie WR, Sachidanandam R et al (2009) Specialized piRNA pathways act in germline and somatic tissues of the Drosophila ovary. Cell 137(3):522–535

Marchetto MC, Narvaiza I, Denli AM, Benner C, Lazzarini TA, Nathanson JL et al (2013) Differential L1 regulation in pluripotent stem cells of humans and apes. Nature 503(7477):525–529

Martin SL, Li J, Weisz JA (2000) Deletion analysis defines distinct functional domains for protein-protein and nucleic acid interactions in the ORF1 protein of mouse LINE-1. J Mol Bio 304(1):11–20

Martin SL, Bushman FD (2001) Nucleic acid chaperone activity of the ORF1 protein from the mouse LINE-1 retrotransposon. Mol Cell Biol 21(2):467–475

Martín F, Maranon C, Olivares M, Alonso C, Lopez MC (1995) Characterization of a non-long terminal repeat retrotransposon cDNA (L1Tc) from Trypanosoma cruzi: homology of the first ORF with the ape family of DNA repair enzymes. J Mol Biol 247:49–59

Martin SL, Cruceanu M, Branciforte D, Wai-Lun Li P, Kwok SC, Hodges RS et al (2005) LINE-1 retrotransposition requires the nucleic acid chaperone activity of the ORF1 protein. Mol Biol 348(3):549–561

Mathias SL, Scott AF, Kazazian HH Jr, Boeke JD, Gabriel A (1991) Reverse transcriptase encoded by a human transposable element. Science 254(5039):1808–1810

Matsuo M, Masumura T, Nishio H, Nakajima T, Kitoh Y, Takumi T et al (1991) Exon skipping during splicing of dystrophin mRNA precursor due to an intraexon deletion in the dystrophin gene of Duchenne muscular dystrophy kobe. J Clin Invest 87(6):2127–2131

McClintock B (1950) The Origin and Behavior of Mutable Loci in Maize. Proc Natl Acad Sci U S A 36:344–355

McMillan JP, Singer MF (1993) Translation of the human LINE-1 element, L1Hs. Proc Natl Acad Sci U S A 90(24):11533–11537

Meehan RR, Lewis JD, Bird AP (1992) Characterization of MeCP2, a vertebrate DNA binding protein with affinity for methylated DNA. Nucl Acids Res 20(19):5085–5092

Meister G, Landthaler M, Patkaniowska A, Dorsett Y, Teng G, Tuschl T (2004) Human Argonaute2 mediates RNA cleavage targeted by miRNAs and siRNAs. Mol Cell 15(2):185–197

Miki Y, Nishisho I, Horii A, Miyoshi Y, Utsunomiya J, Kinzler KW et al (1992) Disruption of the APC gene by a retrotransposal insertion of L1 sequence in a colon cancer. Cancer Res 52(3):643–645

Mills RE, Walter K, Stewart C, Handaker RE, Chen K, Alkan C et al (2011) Mapping copy number variation by population-scale genome sequencing. Nature 470:59–65

Minakami R, Kurose K, Etoh K, Furuhata Y, Hattori M, Sakaki Y (1992) Identification of an internal cis-element essential for the human L1 transcription and a nuclear factor(s) binding to the element. Nucl Acids Res 20:3139–3145

Montagna M, Santacatterina M, Torri A, Menin C, Zullato D, Chieco-Bianchi L et al (1999) Identification of a 3 kb Alu-mediated BRCA1 gene rearrangement in two breast/ovarian cancer families. Oncogene 18(28):4160–4165

Moran JV, Gilbert N (2002) Mammalian LINE-1 retrotransposons and related elements. In: Craig NL, Craigie R, Gellert M, Lambowitz AM (eds) Mobile DNA II. American Society for Microbiology, Washington, D.C., pp 836–869

Moran JV, Holmes SE, Naas TP, DeBerardinis RJ, Boeke JD, Kazazian HH Jr (1996) High-frequency retrotransposition in cultured mammalian cells. Cell 87:917–927

Moran JV, DeBerardinis RJ, Kazazian HH Jr (1999) Exon shuffling by L1 retrotransposition. Science 283:1530–1534

Moran JV, Gilbert N (2002) Mammalian LINE-1 retrotransposons and related elements. In: Craig NL, Craigie R, Gellert M, Lambowitz AM (eds) Mobile DNA II. Washington, DC: American Society of Microbiology Press, pp 836–869

Morrish TA, Gilbert N, Myers JS, Vincent BJ, Stamato TD, Taccioli GE et al (2002) DNA repair mediated by endonuclease-independent LINE-1 retrotransposition. Nat Genet 31(2):159–165

Moutri AR, Marchetto MC, Deng W, Moran JV, Gage FH (2005) Somatic mosaicism in neuronal precursor cells mediated by L1 retrotranspo- sition. Nature 435:903–910

Muñoz-Lopez M, Macia A, Garcia-Cañadas M, Badge RM, Garcia-Perez JL (2011) An epi [c] genetic battle: LINE-1 retrotransposons and intragenomic conflict in humans. Mob Genet Elem 1(2):122–127

Muotri AR, Marchetto MC, Coufal NG, Oefner R, Yeo G, Nakashima K et al (2010) L1 retrotransposition in neurons is modulated by MeCP2. Nature 468(7322):443–446

Nekrutenko A, Li WH (2001) Transposable elements are found in a large number of human protein-coding genes. Trends Genet 17:619–621

Neumann JC, Chandler GL et al (2011) Mutation in the type IB bone morphogenetic protein receptor *alk6b* impairs germ-cell differentiation and causes germ-cell tumors in zebrafish. Proc Natl Acad Sci 108(32):13153–13158

Nigumann P, Redik K, Matlik K, Speek M (2002) Many human genes are transcribed from the antisense promoter of L1 retrotransposon. Genomics 79:628–634

Ogino S, Nosho K, Kirkner G, Kawasaki T, Chan AT, Schernhammer ES et al (2008) A cohort study of tumoral LINE-1 hypomethylation and prognosis in colon cancer. JNCI J Natl Cancer Inst 23:1734–1738

Ohno S (1972) So much "junk" DNA in our genome. In: Smith HH (ed) Evolution of genetic systems. Gordon and Breach, New York, pp 366–370

Ostertag EM, Kazazian HH Jr (2001a) Biology of mammalian L1 retrotransposons. Annu Rev Genet 35:501–538
Ostertag EM, Kazazian HH Jr (2001b) Twin priming: a proposed mechanism for the creation of inversions in L1 retrotransposition. Genome Res 11:2059–2065
Ostertag EM, Prak ET, DeBerardinis RJ, Moran JV, Kazazian HH Jr (2000) Determination of L1 retrotransposition kinetics in cultured cells. Nucl Acids Res 28(6):1418–23
Ostertag EM, DeBerardinis RJ, Goodier JL, Zhang Y, Yang N, Gerton GL et al (2002) A mouse model of human L1 retrotransposition. Nat Genet 32(4):655–660
Pace JK, Feschotte C (2007) The evolutionary history of human DNA transposons: evidence for intense activity in the primate lineage. Genome Res 17(4):422–432
Pagel M, Johnstone RA (1992) Variation across species in the size of the nuclear genome supports the junk-DNA explanation for the C-value paradox. Proc Biol Sci 249:119–124
Pardue ML, Danilevskaya ON, Lowenhaupt K, Slot F, Traverse KL (1996) Drosophila telomeres: new views on chromosome evolution. Trends Genet 12:48–52
Pavlicek A, Paces J, Elleder D, Hejnar J (2002a) Processed pseudogenes of human endogenous retroviruses generated by LINEs: Their integration, stability, and distribution. Genome Res 12:391–399
Pickeral OK, Makalowski W, Boguski MS, Boeke JD (2000) Frequent human genomic DNA transduction driven by LINE-1 retrotransposition. Genome Res 10:411–415
Rastan S (1983) Non-random X-chromosome inactivation in mouse X-autosome translocation embryos–location of the inactivation centre. J Embryol Exp Morphol 78:1–22
Richardson SR, Narvaiza I, Planegger RA, Weitzman MD, Moran JV (2014) APOBEC3A deaminates transiently exposed single-strand DNA during LINE-1 retrotransposition. Elife 24(3):e02008
Riggs AD (2002) X chromosome inactivation, differentiation, and DNA methylation revisited, with a tribute to Susumu Ohno. Cytogenet Genome Res 99:17–24
Rodić N, Burns KH (2013) Long Interspersed Element–1 (LINE-1): passenger or Driver in Human Neoplasms? PLoS Genet 9(3):e1003402
Ross MT, Grafham DV, Coffey AJ et al (2005) The DNA sequence of the human X chromosome. Nature 434(7031):325–337
Rosser JM, An W (2012) L1 expression and regulation in humans and rodents. Front Biosci (Elite Ed). 4:2203–2225
Salem AH, Myers JS, Otieno AC, Watkins WS, Jorde LB, Batzer MA (2003) LINE-1 preTa Elements in the Human Genome. Mol Biol 326:1127–1146
Samonte RV, Eichler EE (2002) Segmental duplications and the evolution of the primate genome. Nat Rev Genet 3(1):65–72
Sbardella G, Mai A, Bartolini S, Castellano S, Cirilli R, Rotili D et al (2011) Modulation of cell differentiation, proliferation, and tumor growth by dihydrobenzyloxopyrimidine non-nucleoside reverse transcriptase inhibitors. J Med Chem 54(16):5927–5936
Schulz WA (2006) L1 retrotransposons in human cancers. J Biomed Biotechnol 2006(1):83672
Schwahn U, Lenzner S, Dong J, Feil S, Hinzmann B, van Duijnhoven G et al (1998) Positional cloning of the gene for X-linked retinitis pigmentosa 2. Nat Genet 19(4):327–332
Schwitalla S, Fingerle AA, Cammareri P, Nebelsiek T, Göktuna SI, Ziegler PK et al (2013a) Intestinal tumorigenesis initiated by dedifferentiation and acquisition of stem-cell-like properties. Cell 152:25–38
Schwitalla S, Ziegler PK, Horst D, Becker V, Kerle I, Begus-Nahrmann Y et al (2013b) Loss of p53 in enterocytes generates an inflammatory microenvironment enabling invasion and lymph node metastasis of carcinogen-induced colorectal tumors. Cancer Cell 23(1):93–106
Scott AF, Schmeckpeper BJ, Abdelrazik M, Comey CT, O'Hara B, Rossiter JP et al (1987). Origin of the human L1 elements: proposed progenitor genes deduced from a consensus DNA sequence. Genomics 1:113–125
Seleme MC et al (2006) Extensive individual variation in L1 retrotransposition capability contributes to human genetic diversity. Proc Nat Acad Sci U S A 103(17):6611–6616

Sen SK, Huang CT, Han K, Batzer MA (2007) Endonuclease-independent insertion provides an alternative pathway for L1 retrotransposition in the human genome. Nucleic Acids Res 35(11):3741–3751

Shukla R, Upton KR, Muñoz-Lopez M, Gerhardt DJ, Fisher ME, Nguyen T et al (2013) Endogenous retrotransposition activates oncogenic pathways in hepatocellular carcinoma. Cell 153(1):101–111

Smit AF, Riggs AD (1996) Tiggers and DNA transposon fossils in the human genome. Proc Natl Acad Sci 93(4):1443–1448

Simons C, Pheasant M, Makunin IV, Mattick JS (2006) Transposon-free regions in mammalian genomes. Genome Res 16:164–172

Singer T, McConnell MJ, Marchetto MC, Coufal NG, Gage FH (2010) LINE-1 retrotransposons: mediators of somatic variation in neuronal genomes? Trends Neurosci 33(8):345–354

Skowronski J, Fanning TG Singer MF (1988) Unit-length LINE-1 transcripts in human teratocarcinoma cells. Mol Cell Biol 8:1385–1397

Soifer HS, Rossi JJ (2006) Small interfering RNAs to the rescue: blocking L1 retrotransposition. Nat Struct Mol Biol 13(9):758–759

Solyom S, Ewing AD, Rahrmann EP, Doucet T, Nelson HH, Burns MB et al (2012) Extensive somatic L1 retrotransposition in colorectal tumors. Genome Res 22(12):2328–2338

Speek M (2001) Antisense promoter of human L1 retrotransposon drives transcription of adjacent cellular genes. Mol Cell Biol 21(6):1973–1985

Stenglein MD, Harris RS (2006) APOBEC3B and APOBEC3F inhibit L1 retrotransposition by a DNA deamination-independent mechanism. J Biol Chem 281(25):16837–16841

Su Y, Davies S, Davis M, Lu H, Giller R, Krailo M et al (2007) Expression of LINE-1 p40 protein in pediatric malignant germ cell tumors and its association with clinicopathological parameters: a report from the Children's Oncology Group. Cancer Lett 247(2):204–212

Suter CM, Martin DI, Ward RL (2004) Hypomethylation of L1 retrotransposons in colorectal cancer and adjacent normal tissue. Int J Colorectal Dis 19(2):95–101

Swergold GD (1990) Identification, characterization, and cell specificity of a human LINE-1 promoter. Mol Cell Biol 10: 6718–6729

Symer DE, Connelly C, Szak ST, Caputo EM, Cost GJ, Parmigiani G et al (2002) Human L1 retrotransposition is associated with genetic instability in vivo. Cell 110:327–338

Szak ST, Pickeral OK, Makalowski W, Boguski MS, Landsman D, Boeke JD (2002) Molecular archeology of L1 insertions in the human genome. Genome Biol 19;3(10):research0052

Szak ST, Pickeral OK, Landsman D, Boeke JD (2003) Identifying related L1 retrotransposons by analyzing 3' transduced sequences. Genome Biol 4:R30

Takai D, Yagi Y, Habib N, Sugimura T, Ushijima T (2000) Hypomethylation of LINE1 retrotransposon in human hepatocellular carcinomas, but not in surrounding liver cirrhosis. Jpn J Clin Oncol 30(7):306–309

Tang Y, Nyengaard JR, De Groot DM, Gundersen HJ (2001) Total regional and global number of synapses in the human brain neocortex. Synapse 41(3):258–273

Tchenio T, Casella JF, Heidmann T (2000) Members of the SRY family regulate the human LINE retrotransposons. Nucl Acids Res 28:411–415

Teneng I, Montoya-Durango DE, Quertermous JL, Lacy ME, Ramos KS (2011) Reactivation of L1 retrotransposon by benzo(a)pyrene involves complex genetic and epigenetic regulation. Epigenetics 6(3):355–367

Teugels E, De Brakeleer S, Goelen G, Lissens W, Sermijn E, De Grève J (2005) De novo Alu element insertions targeted to a sequence common to the BRCA1 and BRCA2 genes. Hum Mutat 26(3):284

Thomas CA (1971) The genetic organization of chromosomes. Annu Rev Genet 5:237–256

Van Arsdell SW, Weiner AM (1984) Pseudogenes for human U2 small nuclear RNA do not have a fixed site of 3' truncation. Nucl Acids Res 12:1463–1471

Van den Hurk JA et al (2007) L1 retrotransposition can occur early in human embryonic development. Hum mol Genet 16(13):1587–1592 van der Klift HM, Tops CM, Hes FJ, Devilee P, Wijnen JT (2012) Insertion of an SVA element, a nonautonomous retrotransposon, in PMS2 intron 7 as a novel cause of Lynch syndrome. Hum Mutat 33(7):1051–1055

Vanin EF (1985) Processed pseudogenes: characteristics and evolution. Annu Rev Genet 19:253–272

Wagstaff BJ, Barnerssoi M, Roy-Engel AM (2011) Evolutionary conservation of the functional modularity of primate and murine LINE-1 elements. PLoS One 6(5):e19672

Walsh CP, Bestor TH (1999) Cytosine methylation and mammalian development. Genes Dev 13(1):26–34

Wang J, Song L, Grover D, Azrak S, Batzer MA, Liang P (2006) dbRIP: a highly integrated database of retrotransposon insertion polymorphisms in humans. Hum Mutat 27:323–329

Weber B, Kimhi S, Howard G, Eden A, Lyko F (2010) Demethylation of a LINE-1 antisense promoter in the cMet locus impairs Met signalling through induction of illegitimate transcription. Oncogene 29(43):5775–5784

Wei W, Gilbert N, Ooi SL, Lawler JF, Ostertag EM, Kazazian HH Jr et al (2001) Human L1 retrotransposition: cis preference versus trans complementation. Mol Cell Biol 21:1429–1439

Weichenrieder O, Repanas K, Perrakis A (2004) Crystal structure of the targeting endonuclease of the human LINE-1 retrotransposon. Structure 12(6):975–986

Wheelan SJ, Aizawa Y, Han JS, Boeke JD (2005) Gene-breaking: a new paradigm for human retrotransposon-mediated gene evolution. Genome Res 15(8):1073–1078

Williams K, Christensen J, Helin K (2011) DNA methylation: TET proteins-guardians of CpG islands? EMBO Rep 13(1):28–35

Wissing S, Montano M, Garcia-Perez JL, Moran JV, Greene WC (2011) Endogenous APOBEC3B restricts LINE-1 retrotransposition in transformed cells and human embryonic stem cells. J Biol Chem 286(42):36427–36437

Wouters-Tyrou D, Martinage A, Chevaillier P, Sautière P (1998) Nuclear basic proteins in spermiogenesis. Biochimie 80(2):117–128

Xing J et al (2007) Mobile DNA elements in primate and human evolution. Am J Phys Anthr (Suppl 45):2–19

Xing J, Zhang Y, Han K, Salem AH, Sen SK, Huff CD et al (2009) Mobile elements create structural variation: analysis of a complete human genome. Genome Res 19(9):1516–26

Xiong Y, Eickbush TH (1990) Origin and evolution of retroelements based upon their reverse transcriptase sequences. EMBO J 9(10):3353–3362

Yang N, Kazazian HH Jr (2006) L1 retrotransposition is suppressed by endogenously encoded small interfering RNAs in human cultured cells. Nat Struct Mol Biol 13(9):763–771

Yang J, Malik HS, Eickbush TH (1999) Identification of the endonuclease domain encoded by R2 and other site- specific, non-long terminal repeat retrotransposable elements. Proc Natl Acad Sci U S A 96:7847–7852

Yang N, Zhang L, Zhang Y, Kazazian HH Jr (2003) An important role for RUNX3 in human L1 transcription and retrotransposition. Nucl Acids Res 31:4929–4940

Yoder JA, Walsh CP, Bestor TH (1997) Cytosine methylation and the ecology of intragenomic parasites. Trends Genet 13:335–340

Yoshida K, Nakamura A, Yazaki M, Ikeda S, Takeda S (1998) Insertional mutation by transposable element, L1, in the DMD gene results in X-linked dilated cardiomyopathy. Hum Mol Genet 7(7):1129–1132

Zemojtel T, Penzkofer T, Schultz J, Dandekar T, Badge R, Vingron M (2007) Exonization of active mouse L1s: a driver of transcriptome evolution? BMC Genomics 8:392

Zhu L, Gibson P, Currle DS, Tong Y, Richardson RJ, Bayazitov IT et al (2009) Prominin 1 marks intial stem cells that are susceptible to neoplastic transformation. Nature 457:603–607

Chapter 4
Reciprocal Interconnection of miRNome-Epigenome in Cancer Pathogenesis and Its Therapeutic Potential

Seyed H. Ghaffari and Davood Bashash

Contents

Abstract MicroRNAs (miRNAs) are small noncoding RNAs that regulate gene expression post-transcriptionally. miRNAs are regarded both as targets of epigenetic changes and as regulators of the epigenetic machinery (epi-miRNAs). Studies over the past decade have demonstrated that deregulated cross-talk between miRNome-epigenome is functionally important in the pathogenesis of most human malignancies. While some miRNAs may be directly involved in cancer, others may

S. H. Ghaffari (✉)
Hematology, Oncology and Stem Cell Transplantation Research Center,
Tehran University of Medical Sciences, Tehran, Iran
e-mail: shghaffari200@yahoo.com

D. Bashash
Department of Hematology, Faculty of Allied Medicine,
Shahid Beheshti University of Medical Sciences, Tehran, Iran

P. Mehdipour (ed.), *Epigenetics Territory and Cancer*,
DOI 10.1007/978-94-017-9639-2_4

be involved by targeting the key players of carcinogenesis, including epigenetic machinery effectors, cancer oncogenes and/or tumor suppressors. Decoding the miRNome-epigenome interaction and comprehension of this reciprocal interconnection will open new avenues to better understanding of human cancerogenesis, leading to introduction and addition of novel promising drugs to the growing list of other new anti-cancer products. This chapter will explain the complicated network of reciprocal interconnections between miRNAs and epigenetics; also it will focus on those miRNAs which undergo epigenetic changes in some of the most common human malignancies. Further understanding of epigenetic mechanisms in miRNA regulation along with the effect of epigenetic drugs on specific miRNAs might help to reset the abnormal cancer epigenome.

4.1 Introduction

4.1.1 Epigenetic

Epigenetics is defined as all heritable changes in gene expression without concomitant association with alterations in the underlying DNA sequences (Egger et al. 2004). Classic genetics alone cannot elucidate the variety of phenotypes within a population, and it is not able to explain how monozygotic twins, in spite of the same genotypes can produce different phenotypes (Fraga et al. 2005). The concept of epigenetic, which has profound effects on various cellular phenomena, such as gene expression, cell signaling and carcinogenesis (Jones and Laird 1999; Jones and Baylin 2002), might offers a partial elucidation to this phenomenon. In the area of epigenetics, DNA cytosine methylation and histone modifications are regarded as the most important mechanisms; however, it is now widely accepted that RNA can also be regarded as an important epigenetic component that is involved in the formation of a repressive chromatin state (Mattick and Makunin 2006).

DNA methylation is a normal reversible process used by mammalian cells in order to maintain a normal expression pattern, and is achieved by the enzymatically addition of a single methyl group from S-adenosyl methionine to cytosine residue of CpG dinucleotide (Jones and Takai 2001). It is useful to note that the modification is generally repressive to transcription and is catalyzed by the activity of a family of DNA methyltransferase enzymes (DNMTs), namely DNMT1, DNMT3a and DNMT3b. DNMT1 (maintenance DNMT) is thought to preserve the methylation patterns by acting on hemi-methylated DNA throughout the cell division (Li et al. 1993); DNMT3a and DNMT3b (de novo DNMTs) are known to be responsible for establishing de novo methylation patterns for unmethylated DNA (Okano et al. 1999). Nearly all DNA methylation occurs on cytosine residues, located side by side to guanine nucleotides, forming cytosine-phosphate-guanine (CpG) dinucleotide. CpG sites are roughly 80% depleted in the genome, and are asymmetrically distributed into CpG poor and CpG dense regions (Takai and Jones 2003). The majority of the genome is rather CpG-poor due to the mutagenicity of a methylated cytosine which can undergo spontaneous deamination in the germline during evolution. On the other hand, "CpG islands" which refer to CpG dense regions, are often located in the promoter regions of nearly half of all the protein-coding genes, normally remain unmethylated, and their hypermethylated status prevents gene expression (Gardiner-Garden and Frommer 1987). DNA methylation is also the principal epigenetic factor governing allelic imprinting; the process by which only one allele of certain genes is expressed depending on parental origin (Kaneda et al. 2004).

A number of factors such as aging, diet and environment can influence the DNA methylation levels of a cell without requiring a change in genomic DNA sequence. With aging in certain tissues, there is a global genome hypomethylation whereas certain CpG islands become hypermethylated (Richardson 2003), a situation reminiscent of that is found in many cancer cells. Nutrition supplies methyl groups for DNA methylation via the folate and methionine pathway and some environmental agents such as arsenic and cadmium can have profound effects on DNA methylation (Okoji et al. 2002; Takiguchi et al. 2003).

Histones are the main protein components of chromatin, which is a highly ordered structure consisting of repeats of nucleosomes connected by linker DNA and functioning as the physiological template of all eukaryotic genetic information (Loizou et al. 2006). These dynamic structures provide not only a physical support to DNA, but also contribute to the transcriptional regulation, repair and replication. The histone N-terminal tails are crucial in helping to maintain chromatin stability and are subject to numerous modifications (Zhang and Dent 2005). Histone modifications, especially the posttranslational modifications (such as phosphorylation, methylation and acetylation) are important epigenetic mechanisms in controlling gene expression and can determine whether chromatin is in the accessible, decondensed and transcriptionally active form (called euchromatin), or in the inaccessible, densely compacted and transcriptionally inert form (called heterochromatin) (Santos-Rosa and Caldas 2005). Most modifications have some role to play in transcriptional regulation and so each has the potential to be oncogenic if deregulated deposition leads.The most well explained mechanisms are acetylation which decreases histones affinity for DNA, allow chromatin openness, and favoring gene transcription; the other one is histone methylation which may hinder gene transcription by the opposed mechanism, depending on which amino acids are methylated and DNA methylation itself (Jerónimo and Henrique 2011).

4.1.2 *miRNA Biogenesis and Its Mode of Action*

There are two major groups of non-coding RNAs (ncRNAs), the small and the long ncRNAs (Hassler and Egger 2012); small ncRNAs include transfer RNAs (tRNAs), ribosomal RNAs (rRNAs), piwi interacting RNAs (piRNAs), small nuclear RNAs (snoRNAs) and microRNAs (miRNAs). Long ncRNAs, with 200 bp to 100 kb, constitute a heterogeneous class of mRNA-like transcripts, yet non-coding (Hassler and Egger 2012).

miRNAs are a class of small, single-stranded, non-coding RNAs that post-transcriptionally control the translation and stability of mRNAs (Östling et al. 2011). They are transcribed within the nucleus by RNA polymerase II (Pol II) into a long primary miRNAs (pri-miRNAs) which contain both a 5'-cap structure as well as a 3'-end poly-A tail (Lee et al. 2004), and are then processed by the RNase III Drosha and DGCR8 (microprocessor complex) into the precursor miRNAs (pre-miRNAs) (Carthew and Sontheimer 2009). Following this nuclear processing, the pre-miRNAs are structured as imperfect stem loops, and are exported into the cytoplasm by Exportin-5 (Kim 2005). Here, pre-miRNAs are further processed by another RNase III enzyme Dicer into the final functional mature miRNAs. These mature miRNAs are ready to bind to their target mRNAs and interfere with their translation process (Kim 2005). This process requires an incorporation of the miRNA mature sequence into a complex called miRISC (miRNA-containing RNA-induced silencing complex), which contains AGO proteins and binds to target mRNA. miRNAs binding to their target mRNAs are usually restricted to the "seed" sequence at the 3' UTR of the target mRNA. When miRNA binds with complete complementarity, this can lead to the degradation of their target mRNAs, and when it binds with incomplete complementarity, this can lead to the translational suppression (Meltzer 2005). Each miRNA is predicted to control hundreds of target genes, and each mRNA may be regulated by more than one miRNA (Lewis et al 2003).

4.2 Reciprocal Interconnection Between Epigenome and miRNome

Participation of miRNAs in the epigenetic world represents a complicated regulatory loop interconnecting epigenome and miRNome. miRNAs are regarded both as targets of epigenetic changes and as regulators of the epigenetic machinery (epi-miRNAs). miRNAs are able to silence specific target molecules (including members of the epigenetic machinery) at the post-transcriptional level (Guo et al. 2010); on the other hand, they are also tightly regulated by epigenetic modifications. If epigenetic events such as DNA methylation or histones modifications have been shown to affect the miRNA expression, the other side of the coin is represented by the ability of this class of non-coding RNAs (epi-miRNAs) to control the epigenome through targeting its enzymatic components such as DNMTs, HDACs, and polycomb genes (Fabbri et al. 2007). This part will explain the complicated network of reciprocal interconnections between miRNAs and epigenetics; describing either how epigenetics can affect the miRNome, as well as how epi-miRNAs can control the epigenome.

4.2.1 Epigenetic Control of miRNA Expression

Although, evidence for the importance of miRNAs has been increased, the regulation of their expression is still poorly understood. It is now widely accepted that miRNAs undergo the same regulatory mechanisms as conventional protein-coding genes (PCG), including genetic regulations. Several studies have demonstrated that miRNA expression can be deregulated by several genetic alterations including mutations, defects in the miRNA biogenesis machinery, and altered activity of different transcription factors (Zhang et al. 2006). In this regard, some studies revealed a reduced expression of miR-15a and miR-16-1 due to inherited mutations in CLL (Calin et al. 2005); changes in microRNA levels consequent to altered Drosha or Dicer activity in different tumor types (Karube et al. 2005; Nakamura et al. 2007; Thomson et al. 2006), induction of miR-17-92 cluster, miR-34 family and miR-146a by transcription factors c-Myc, p53 and NF-kB, respectively (Chang et al. 2007; Mendell 2008; Raver-Shapira et al. 2007). In addition, it has been shown that the C/EBP alpha and NFI-A compete for binding to the miR-223 promoter, leading to an upregulated and downregulated expression of miR-223, respectively (Fazi et al. 2005). The finding that epigenetics can regulate the expression of many protein-coding genes, and that miRNAs are also generally transcribed by Pol II suggests that epigenetics can play essential roles in the regulation of the miRNA expression.

An extensive analysis of genomic sequences of microRNA genes shows that many miRNAs are located in the introns of protein-coding genes (Kim and Nam 2006), suggesting co-regulation of these miRNAs with their host genes (Ying and Lin 2005). However, it is probable that these miRNAs can have their own promoters; knowing that CpG islands within introns can act as promoters, it is reasonable to hypothesize that the transcription of CpG islands-containing intronic miRNAs could be subjected to epigenetic regulation such as DNA methylation. Several studies have shown that the epigenetic mechanisms can regulate the expressions of miRNAs and the deregulation of these mechanisms can contribute to the deregulated expression of these non-coding RNAs in cancer. In a study, Saito et al. (2006) found that miR-127, a putative tumor suppressor microRNA characterized by a CpG island promoter, was remarkably up-regulated in cancer cell lines after a treatment with DNMT inhibitor 5-Aza-2′-deoxycytidine (5-Aza-CdR), and HDAC inhibitor phenyl-butyric acid (PBA) that led to reduced DNA methylation levels, more open chromatin structures, and therefore re-expression of genes that had been silenced epigenetically, suggestive of miRNAs' regulation by epigenetic (Saito et al. 2006). The number of studies recording the epigenetic regulation of miRNAs has been increased dramatically; the list of the epigenetically regulated miRNAs is summarized in Table 4.1. However, much still needs to be discovered before a distinct regulatory role for epigenetic mechanisms in miRNAs expression can be established.

Table 4.1 Epigenetically regulated microRNAs

Epigenetically regulated miRNAs	Location	Cancer type	Target protein	References
miR-1–1 & -2	20q13.33 (miR-1-1) 18q11.2 (miR-1-2)	Hepatocellular carcinoma	FOXP1 MET HDAC4	(Datta et al. 2008)
Let-7i	12q14.1	Cholangiocytes	TLR4	(O'Hara et al. 2010)
miR-9	1q22	Leukemia Metastatic cancer Colorectal cancer		(Noonan et al. 2009)
let-7a-3	22q13.31	Lung cancer Ovarian cancer	IGF2	(Brueckner et al. 2007) (Lu et al. 2007)
miR-10a	21q21.32	ALL	HOXA3 HOXD10	(Roman-Gomez et al. 2009) (Han et al. 2007)
miR-10b	2q31.1	ALL		(Roman-Gomez et al. 2009)
miR-9-1	1q22	ALL Breast cancer	CDK6, FGFR1 NFKB1	(Rodriguez-Otero et al. 2011) (Lehmann et al. 2008)
miR-9-2	5q14.3	ALL Metastases	CDK6, FGFR1	(Rodriguez-Otero et al. 2011) (Lujambio et al. 2008)
miR-9-3	15q26.1	ALL Breast epithelial cells	CDK6, FGFR1	(Rodriguez-Otero et al. 2011) (Hsu et al 2009)
miR-21	17q23.2	Ovarian cancer Prostate cancer	PDCD4 TPM1 and MARCKS	(Iorio et al 2007) (Hulf et al 2011)
miR-29	7q32.3 1q32.2	Aggressive B-Cell Lymphomas CLL & AML	Mcl-1 DNMT3A DNMT3B SP1 Tcl-1 CDK6 and IGR1F	(Zhang et al. 2012) (Liu et al 2010) (Sampath et al. 2012)
miR-15a/miR-16	13q14	Chronic lymphocytic leukemia	BCL-2 and MCL-1	(Sampath et al. 2012)
miR-31	9p21.3	Melanoma Breast cancer	SRC, RAB27a, NIK, MET, RhoA and WAVE3	(Asangani et al. 2012) (Augoff et al. 2012)

Table 4.1 (continued)

Epigenetically regulated miRNAs	Location	Cancer type	Target protein	References
miR-34a	1p36.22	pancreas carcinoma cell, breast, colon, bladder, kidney, melanoma	CDK6	(Lodygin et al. 2008)
miR-34b/c	11q23.1	ALL Colorectal cancer Metastases	CDK6 c-MYC E2F3	(Roman-Gomez et al. 2009) (Vilas-Zornoza et al. 2011) (Toyota et al 2008) (Lujambio et al. 2008)
miR-107	10q23.31	Pancreatic cancer	CDK6	(Lee et al 2009)
miR-17-92 Cluster	13q31.3	Colorectal Cancer	PTEN BCL2L11 CDKN1A	(Humphreys et al. 2013)
miR-124a-1, -2 & -3	8p23.1 (miR-124a-1) 8q12.3 (miR-124a-2) 20q13.33 (miR-124a-3)	Colorectal cancer ALL Gastric cancer	CDK6 C/EBPa, VIM, SMYD3	(Lujambio et al. 2007) (Agirre et al. 2009) (Roman-Gomez et al. 2009) (Ando et al. 2009)
miR-125b	11q23 (b-1) 21q21 (b-2)	Hepatocellular carcinoma	PIGF	(Alpini et al. 2011)
miR-126	9q34.3	Bladder cancer Prostatic cancer	EGFL7 VEGFA PIK3R2	(Saito et al. 2009b) (Saito et al. 2009a)
miR-127	14q32.31	Bladder cancer clear cell renal cell carcinomas	BCL6	(Saito et al. 2009a) (Saito et al. 2006) (Wotschofsky et al. 2012)
miR-130b	22q11.21	Ovarian cancer		(Fabbri et al 2007)
miR-129-2	11p11.2	Gastric cancer		(Bandres et al. 2009)
miR-132	17p13.2	Prostate cancer	HB-EGF TALIN2	(Formosa et al. 2013) (Roman-Gomez et al. 2009)

Table 4.1 (continued)

Epigenetically regulated miRNAs	Location	Cancer type	Target protein	References
miR-137	1p21.3	colorectal cancer oral cancer	CDK6 E2F6 NCOA2	(Bandres et al. 2009) (Kozaki et al. 2008)
miR-143	5q32	ALL	MLL-AF4	(Dou et al. 2012)
miR-145	5q32–33	Prostate cancer and clear cell renal cell carcinomas	BNIP3 TNFSF10 PAK7	(Wotschofsky et al. 2012) (Zaman et al. 2010)
miR-148a	7p15.2	Metastases breast cancer, cervical cancer Cholangiocarcinoma	TGIF2 DNMT3b DNMT1	(Lujambio et al. 2008) (Lehmann et al. 2008) (Duursma et al. 2008) (Braconi et al. 2010)
miR-152	17q21.32	Bladder cancer Breast cancer	MLL, DNMT1	(Stumpel et al. 2011) (Benetti et al. 2008)
miR-155	21q21	Breast cancer		(Lujambio et al. 2008)
miR-181a/ b	9q33.3	Chronic lymphocytic leukemia	PLAG1	(Pallasch et al. 2009)
miR-181c	1q31.3	Gastric cancer	NOTCH4 KRAS	(Hashimoto et al. 2010)
miR-193a	17q11.2	oral cancer gastric cancer	E2F6 PTK2 MCL1	(Kozaki et al. 2008) (Ando et al. 2009)
miR-193b	16p13.12	Prostate cancer	ETS1 CCND1 PLAU	(Rauhala et al. 2010)
miR-196b	7p15.2	ALL prostate cancer	MYC	(Roman-Gomez et al. 2009) (Bhatia et al. 2010) (Hulf et al. 2011)
miR-200a/ b	1p36.33	ALL	ZEB1, ZEB2, E-cadherin	(Wiklund et al. 2011) (Stumpel et al. 2011)

Table 4.1 (continued)

Epigenetically regulated miRNAs	Location	Cancer type	Target protein	References
miR-203	14q32.11	ALL, AML hepatocellular carcinoma	ABL1 BCR-ABL1 Bmi-1	(Roman-Gomez et al. 2009) (Kozaki et al. 2008) (Furuta et al. 2010) (Bueno et al 2008)
miR-205	1q32.2	Prostate cancer bladder cancer	SIP1 and ZEP BCL-w	(Hulf et al. 2011) (Bhatnagar et al. 2010) (Wiklund et al. 2011)
miR-223	Xq12	AML	NFI-A MEF2C	(Fazi et al. 2007)
miR-224	Xq28	Hepatocellular carcinoma	API-5	(Wang et al. 2012)
miR-335	7q32	Hepatocellular carcinoma	SOX4 Rb1	(Dohi et al. 2013)
miR-200c/ miR-141	12p13.31	Breast cancer prostate cancer bladder cancer	ZEB2	(Vrba et al. 2010) (Wiklund et al. 2011)
miR-342	14q32.2	Colorectal cancer		(Grady et al. 2008)
miR-370	14q32.31	Cholangiocarcinoma Oral squamous cell carcinoma	MAP3K8 IRS-1	(Meng et al. 2008) (Chang et al. 2013)
miR-373	19q13.42	Hilar cholangiocarcinoma	MBD2	(Chen et al. 2011)
miR-449a/ b	5q11.2	Prostatic cancer Osteosarcoma Hepatocellular carcinoma	CDK6 CDC25A HDAC1 c-MET	(Noonan et al. 2009) (Yang et al. 2009) (Buurman et al. 2012)
miR-512-5p	19q13.41	Gastric cancer	Mcl-1	(Saito et al. 2009b)

4.2.2 miRNA Control of Epigenetic Mechanisms

To complicate the scenario connecting miRNAs and epigenetics, microRNAs themselves can regulate the expression of components of the epigenetic machinery, aberrant expression of these microRNAs called "epi-miRNAs" (Table 4.2). The epi-miRNAs not only are tightly regulated by epigenetic modifications, but they are also able to silence the expression of various epigenetic-modifying enzymes, representing a complicated regulatory feedback loop. An aberrant expression of epi-

Table 4.2 Epi-miRNAs

Epi-miRNAs	Location	Tissue type	Target protein	References
miR-1-1 & -2	20q13.33 (miR-1-1) 18q11.2 (miR-1-2)	Skeletal muscle tissue	HDAC4	(Chen et al. 2006)
miR-101-1 & -2	1p31.3 (miR-101-1) 9p24.1 (miR-101-2)	Prostatic cancer Bladder cancer	EZH2	(Varambally et al. 2008) (Friedman et al. 2009)
miR-140	8qD3	Mouse cartilage tissue	HDAC4	(Tuddenham et al. 2006)
miR-148a & b	7p15.2 (miR-148a) 12q13.13 (miR-148b)	Cervical cancer Cholangiocarcinoma	DNMT3b DNMT1	(Duursma et al. 2008) (Braconi et al. 2010) (Lujambio et al. 2007)
miR-152	17q21.32	Cholangiocarcinoma	DNMT1	(Braconi et al. 2010)
miR-290 cluster	7qA1	Dicer null cells, Pluripotent ES cells Mouse ES cells	DNMT1, -3a, -3b RBL2	(Scott et al. 2006) (Benetti et al. 2008) (Sinkkonen et al. 2008)
miR-29a/ b/ c	7q32.3 (miR-29a) 7q32.3 (miR-29b-1) 1q32.2 (miR-29b-2) 1q32.2 (miR-29c)	Lung cancer AML	DNMT-3a & -3b DNMT1, -3a & -3b & Sp1	(Fabbri et al. 2007) (Garzon et al. 2009)
miR-301	17q23.2	Cholangiocarcinoma	DNMT1	(Braconi et al. 2010)
miR-449a	5q11.2	Prostatic cancer	HDAC1	(Noonan et al. 2009)

miRNAs (those miRNAs which target, directly or indirectly, effectors of the epigenetic machinery) has been documented to be related to cancer pathogenesis. Study by Lujambio et al. (2008) on breast cancer cells shows that the hypermethylation of miR-148 led to its downregulation as a result of a reinforced overexpression of DNMTs which resulted in tumor growth and metastasis (Lujambio et al. 2008). Interestingly, upon the treatment of breast cancer cells with DNA demethylating agent, a reduced tumor growth and inhibition of metastasis were documented through the reactivation of miR-148 (Lujambio et al. 2008). As the first evidence of the existence of epi-miRNAs, Fabbri et al. (2007) reported that the enforced expression of miR-29 family directly induces disruption of de novo DNMT3a and DNMT3b, restores normal DNA methylation pattern and led to a global DNA hypomethylation of lung cancer cells (Fabbri et al. 2007). Moreover, the miR-29 family which has some interesting complementarity with the 3'UTR of DNMT3a and DNMT3b, also was shown to be able to induce the reactivation of silenced tumor suppressor genes and target the maintenance DNMT1 (Garzon et al 2009). In addition to DNA methylation, miRNAs may control the histone modification and chromatin structure by regulating key histone modifying enzymes such as HDACs; in this regard, Tuddenham et al. (2006) reported miR-140, which is a cartilage specific microRNA, targets histone modification through the regulation of HDAC-4 in mouse cells (Tuddenham et al 2006). Moreover, transfection of MiR-449a, which is a direct regulator of HDAC1, induces cell-cycle arrest, apoptosis and a senescent-like phenotype in prostate cancer cells (Yang et al. 2009). Also, it has been shown that upregulation of EZH2, a catalytic subunit of the polycomb repressive complex 2 (PRC2), by miR-101 resulted in an aberrantly tumor suppressor gene silencing via trimethylating histone H3 lysine 27 in bladder and prostate cancer (Friedman et al. 2009; Varambally et al. 2008).

4.3 miRNAs and Cancer Epigenetics

Studies over the past decade have demonstrated that deregulated cross-talks between miRNome-epigenome is functionally important in the pathogenesis of most human malignancies. Emerging evidence suggests the potential involvement of the deregulated miRNAs, which may be caused by various mechanisms such as epigenetic silencing, in cancer pathogenesis. While some miRNAs may be directly involved in cancer, the others may be involved by targeting the other key players of carcinogenesis, including epigenetic machinery effectors, cancer oncogenes and/or tumor suppressors. Decoding the miRNome-epigenome interaction and comprehension of this reciprocal interconnection will open new avenues to a better understanding of human cancerogenesis, therefore leading to introduction and addition of novel promising drugs to the growing list of the other new anti-cancer products. This part will focus on those miRNAs which undergo epigenetic changes in some of the most common human malignancies such as breast, prostate, lung and colorectal cancers as well as leukemias and melanoma.

4.3.1 Breast Cancer

The molecular mechanisms responsible for the initiation and progression of breast cancer are far from being understood. During the past decade, the somatic mutation theory of cancer, which refers to the genetic disorder of fatal acquisition of multiple mutations in key genes, has been revolutionized and became clear that the deregulation of epigenetic machinery and miRNAs play a role as equally essential as genetics in cancerogenesis. In this regard, Yu et al. (2007) demonstrated that depletion of let-7 is associated with enhanced tumorigenicity of breast cancer (Yu et al. 2007). Moreover, it has been shown that the overexpression of miR-21 in breast cancer confers increased invasion capacities and promotes tumor metastasis to the lung (Zhu et al. 2008). One of the first studies regarding epigenetic control of miRNA expression in breast cancer was conducted by Scott et al. (2006) in SkBr3 breast cancer cell line (Scott et al. 2006); In this study, they observed that upon treatment of SkBr3 cells with the HDAC inhibitor LAQ824, the expression levels of 5 miRNAs were up- and 22 miRNAs were down-regulated, indicative of epigenetic control of miRNAs in breast cancer development.

It is worthy to mention that miR-9, which is expressed from three genomic loci (miR-9-1, miR-9-2 and miR-9-3), is one of the most important miRNA involved in the pathogenesis of various malignancies including breast cancer (Bandres et al. 2009; Lehmann et al. 2008; Roman-Gomez et al. 2009). In this regard, Hsu et al. (2009) showed that xenoestrogen exposure may induce aberrant epigenetic repression of miR-9-3 in breast epithelial cells (Hsu et al. 2009). It has also been documented that in breast cancer, the miR-9-1 locus is highly methylated not only in invasive ductal carcinoma, but also in ductal carcinoma in situ and the intraductal component of invasive ductal carcinoma (Lehmann et al. 2008). Epigenetic silencing of miR-9 and miR-124a together with miR-148a, -152, and −663 was also reported by Lehmann et al. (2008) in breast cancer (Lehmann et al 2008); Interestingly, they found that treatment of breast cancer cell lines with 5-Aza-CdR, a DNA demethylating agent, reactivates miR-9-1, but not the other hypermethylated miRNAs. These findings suggest that epigenetic silencing of miR-9 loci constitutes an important event in breast carcinogenesis. In a study by Lujambio et al. (2008) on breast cancer cells, it has been shown that the hypermethylation of miR-148 led to its downregulation as a result of reinforced overexpression of DNMTs which resulted in tumor growth and metastasis (Lujambio et al 2008). Interestingly, upon treatment of the breast cancer cells with a DNA demethylating agent, reduced tumor growth and inhibition of metastasis were also documented through the reactivation of miR-148. In an study, Xu et al. (2013) found that DNMT1 expression, which is aberrantly upregulated in breast cancer and its overexpression is responsible for the hypermethylation of miR-148a and miR-152 promoters, and is inversely correlated with the expression levels of miR-148a/152 in breast cancer tissues; suggesting a negative feedback regulatory loop between miR-148a/152 and DNMT1 in breast cancer (Xu et al. 2013).

4.3.2 Prostate Cancer

Worldwide, prostate cancer is one of the three most common cancers among males (Siegel et al. 2012), is the second most commonly diagnosed neoplasia and the sixth leading cause of cancer death in males (Jemal et al 2011), despite all the recent improvements in diagnosis and treatment. Evolving data supports an important role for epigenetic processes in the development of prostate cancer in addition to the genetic mechanisms. Epigenetic events, including microRNAs (miRNAs) deregulation, have been recognized as critical players in prostate carcinogenesis (Shen and Abate-Shen 2010; Van der Poel 2007). In a study by Rauhala et al. (2010) it has been shown that miR-193b is an epigenetically silenced putative tumor suppressor in prostate cancer (Rauhala et al. 2010). They found an increased expression of 38 miRNAs upon treatment of prostate cancer cell lines with 5-Aza-CdR and trichostatin A; among these, a CpG island upstream of the miR-193b locus was methylated. They demonstrated that expressing miR-193b using pre-miR-193b oligonucleotides caused a significant growth reduction resulting from a decrease of cells in the S-phase of the cell cycle (Rauhala et al. 2010). MiR-145 is another example of epigenetically regulated microRNAs involved in various cancers including prostate. In seven cancer cell lines with miR-145 hypermethylation, 5-Aza-CdR treatment dramatically induced miR-145 expression. In a study by Suh et al. (2011) it has been reported that miR-145 is silenced in prostate cancer through DNA hypermethylation and p53 mutation (Suh et al. 2011). In prostate cancer, HDAC-1 is a direct target of miR-449a, and the downregulation of miR-449a causes an overexpression of HDAC-1; Thus, the aberrant expression of miR-449a may contribute to the abnormal epigenetic patterns which occurs in prostate cancer. Transfection of MiR-449a has been shown to induce cell-cycle arrest, apoptosis and a senescent-like phenotype in the prostate cancer cells (Yang et al 2009). Also, it has been shown that the upregulation of EZH2, a catalytic subunit of the polycomb repressive complex 2 (PRC2), by miR-101 results in an aberrant tumor suppressor gene silencing via trimethylating histone H3 lysine 27 in both bladder and prostate cancer (Friedman et al. 2009; Varambally et al. 2008). To screen for epigenetically silenced miRNAs in prostate cancer, Formosa et al. (2013) treated prostate normal epithelial and carcinoma cells with 5-Aza-CdR and subsequently examined for the expression changes of 650 miRNAs (Formosa et al. 2013). The results of this study point to miR-132 as a methylation-silenced miRNA with an antimetastatic role in prostate cancer controlling cellular adhesion.

Epigenetically regulated miRNAs not only are involved in the acquisition of prostate cancers invasive capabilities, but also they may contribute to a significant resistance to chemotherapy-induced apoptosis. Bhatnagar et al. (2010) found that miR-205 and miR-31 are significantly downregulated in WPE1-NB26 cell line, which is a highly malignant prostate cancer cell, as well as in other cell lines representing advanced-stage prostate cancers (Bhatnagar et al. 2010). They cloned the promoter region of the miR-205 gene and found this region to be hypermethylated in cell lines derived from advanced prostate cancers. Treatment with the DNA methylation inhibitor 5-Aza-CdR induced the expression of miR-205, downregulated Bcl-w, and sensitized prostate cancer cells to the chemotherapy-induced apoptosis; which indicates the key rule of miR-205 and miR-31 in the resistance to apoptosis in advanced prostate cancer (Bhatnagar et al. 2010).

4.3.3 Lung Cancer

In a study by Lujambio et al. (2007), the hypermethylation of miR-124a was reported in 13 of 27 (48 %) lung cancer specimens (Lujambio et al. 2007). Remarkably, immunostaining analyses of lung cancer specimens (n=27) showed that the hypermethylation of miR-124a was associated with strong CDK6 expression and Rb phosphorylation; indicating that the epigenetic silencing of miR-124a in cancer cells leads to the CDK6 up-regulation. Let-7a-3, an epigenetically regulated miRNA with an oncogenic function, belongs to the archetypal let-7 miRNA gene family and can be regulated by the DNMT1 and DNMT3B (Brueckner et al 2007). The gene was normally silenced by a promoter hypermethylation in normal human tissues but was hypomethylated in some lung adenocarcinomas. Brueckner et al. (2007) reported that an elevated expression of let-7a-3 in a human lung cancer cell line resulted in enhanced tumor phenotypes; which suggests epigenetic silencing of this oncogenic miRNA is a protective mechanism in lung cancer (Brueckner et al. 2007). Also, it has been demonstrated that miR-34a, which is a target of the tumor suppressor gene product p53, is silenced in seven of 24 (29.1 %) lung cancer specimens due to an aberrant CpG methylation of its promoter (Lodygin et al 2008). This miRNA is recognized as tumor suppressor microRNA and its epigenetic silencing was reported to be a mechanism responsible for lung cancer pathogenesis.

In invasive lung cancer cells, hypermethylation in the promoter region of miR-200C was found to be responsible for the loss of its expression as evaluated by 5-Aza-CdR treatment of two highly aggressive lung cancer cell lines, H1299 and Calu-1 (Ceppi et al. 2010). Furthermore, in the primary tumor specimens that were obtained from the non-small cell lung cancer (NSCLC) patients, a lower miR-200c expression level was found to be associated with a poor grade of differentiation and also with a higher propensity to lymph node metastases. Ceppi et al. (2010) found that the loss of miR-200c expression induces an aggressive phenotype in NSCLC; reintroduction of this miRNA into the highly invasive/aggressive NSCLC cells not only inhibits in vitro cell invasion, but also hinders in vivo metastasis formation as well (Ceppi et al. 2010). The MiR-29 family (miR-29a, b, and c) has been highlighted as a representative of epi-miRNA for targeting DNMT in various human cancers including lung cancer (Fabbri et al. 2007; Garzon et al. 2009; Nguyen et al. 2011); In this setting, an inverse correlation between the expression of miR-29s and DNMT-3A/-3B has been reported in lung cancer tissues. It has been shown that the elevated miR-29s can restore normal patterns of DNA methylation, leading to the release of overmethylated tumor suppressor genes, and inhibiting tumorigenicity in vitro and in vivo (Bartel 2009).

In addition to the DNA methylation, contribution of histone modifications was also reported in the epigenetic silencing of miRNAs in lung cancer. In this regard, Incoronato et al. (2010) identified histone modifications rather than DNA hypermethylation as epigenetic events that regulated miR-212 levels, which is strongly down-regulated in lung cancer (Incoronato et al. 2010). Moreover, this study showed that miR-212 silencing via histone modifications is correlated to the severity of the disease since it is significantly down-regulated in T3/T4 staging rather than in T1/T2 staging. It is worth mentioning that the epigenetic control of miRNA might be tissue specific, as none of the miRNAs showed a statistically significant change in the increased expression after treatments of A549 and NCI-H157 lung cancer cell lines with either demethylatin agent 5-azacytidine (5-aza-C) and/or HDAC inhibitor TSA (Yanaihara et al. 2006).

4.3.4 Colorectal Cancer (CRC)

To identify epigenetically silenced miRNAs in colorectal cancer (CRC), Toyota et al. (2008) screened for miRNAs induction in CRC cells by 5-Aza-CdR treatment (Toyota et al. 2008). They found that miR-34b and miR-34c are epigenetically silenced in CRC and that the 5-Aza-CdR treatment rapidly restored the expression of these miRNAs. Methylation of the miR-34b/c CpG island was frequently observed in 100 % (nine of nine) and in 90 % (101 of 111) of CRC cell lines and primary CRC tumors, respectively. Interestingly, transfection of precursor miR-34b or miR-34c into CRC cells induced dramatic changes in the gene expression profile, and there was a significant overlap between the genes that were down-regulated by miR-34b/c and those that were down-regulated by 5-Aza-CdR (Toyota et al. 2008). The relationship between miRNA and the cognate host gene epigenetic regulation was studied by Grady et al. (2008). Simultaneous epigenetic silencing of the intronic microRNA hsa-miR-342 and its host gene EVL (Ena/Vasp-like) was reported in 86 % of colorectal adenocarcinomas and in 67 % of adenomas, which indicates that aberrant methylation at this locus is an early common event in colorectal carcinogenesis. Grady et al. (2008) also showed that the combined treatment of 5-aza-C with an HDAC inhibitor restored simultaneous expression of EVL and miR-342. Furthermore, reconstitution of hsa-miR-342 in the colorectal cancer cell line HT-29 induced apoptosis, suggestive of a proapoptotic tumor suppressor function for this miRNA (Grady et al. 2008).

In order to analyze the epigenetic regulation of miRNA genes in colorectal cancer, Suzuki et al. (2011) conducted a genome-wide profiling of the histone modifications (H3K4me3, H3K27me3, and H3K79me2) (Suzuki et al. 2011). By comparing miRNA expression and histone modification before and after DNA demethylation, 47 miRNAs, including miR-1-1 which acts as a tumor suppressor, was found to be potential targets of epigenetic silencing in early and advanced CRCs (Suzuki et al. 2011). To identify tumor-supressor miRNAs that were silenced through aberrant epigenetic events in CRC, Bandres et al. (2009) identified 5 miRNAs located around/on a CpG island that were down-regulated in patient with colorectal cancer (Bandres et al. 2009). Combined treatment of 3 CRC cell lines with a DNA methyltransferase inhibitor and a HDAC inhibitor restored the expression of 3 of the 5 microRNAs (miR-9, miR-129 and miR-137); this suggests that the aberrant DNA methylation and the histone modifications work together to induce silencing of miRNAs in CRC (Bandres et al. 2009). In a study done by Balaguer et al. (2010), a contributing role was described for the epigenetic regulation of miR-137 in colorectal carcinogenesis (Balaguer et al. 2010). In this regard, methylation of the miR-137 CpG island was observed in virtually all CRC cell lines, 82 % of adenomas, and 82 % of CRCs, but only in 14 % of normal mucosae from the CRC patients and in 5 % of healthy subjects, which indicates a cancer-specific epigenetic event in CRC. Interestingly, using a systematic microarray and bioinformatics approaches, they identified LSD1, a histone demethylase, a target for miR-137 in the colon (Balaguer et al. 2010).

Using MBD-isolated Genome Sequencing (MiGS) to evaluate genome-wide DNA methylation patterns combined with a microarray analysis to determine miRNA expression levels, Yan et al. (2011) searched for candidates miRNAs that were regulated by DNA methylation in HCT-116 colorectal cancer cell and found that 64 miRNAs were robustly methylated (Yan et al. 2011). They also showed that miR-941, miR-1237 and miR-1247 were up-regulated by 5-Aza-CdR treatment and transcribed independent from their respective putative host genes (Yan et al. 2011). To address if the same epigenetic disruption can “hit” miRNAs in transformed cells, Lujambio et al. (2007) have used HCT-116 colon cancer cells and double knockout DNMT1 and DNMT3b (DKO) cells to compare the miRNA expression profile of DKO and HCT-116 wild-type cells (Lujambio et al. 2007). Among the dysregulated miRNAs, bisulfite genomic sequencing analyses of multiple clones of the original HCT-116 cells showed dense CpG island hypermethylation for miR-124a, miR-517c, and miR-373. Unlike miR-517c and miR-373 which were found to be densely methylated in normal colon tissues, the miR-124a–embedded CpG island was always unmethylated in normal counterparts. In the case of primary colorectal tumors, the miR-124a hypermethylation was observed in 75 % of patients. It is important to note that the presence of miR-124a hypermethylation was not a feature of this particular cell line, but analyzing a comprehensive collection of human cancer cell lines (n=22) and primary samples (n=171) from colon, breast, and lung carcinomas, leukemias, and lymphomas also showed a frequent presence of miR-124a hypermethylation (Lujambio et al. 2007).

4.3.5 Hepatocellular Carcinoma (HCC)

To identify any miRNA genes that are harboring CpG islands undergo a methylation-mediated silencing in hepatocellular carcinoma (HCC), Furuta et al. (2010) examined the methylation status of 43 loci containing CpG islands around 39 mature miRNA genes in a panel of HCC cell lines and non-cancerous liver tissues as controls (Furuta et al. 2010). Among 11 miRNA genes that were frequently methylated in HCC cell lines but not in non-cancerous liver tissues, miR-124, miR-203 and miR-375 were selected as silenced miRNAs through the CpG-island methylation. They also demonstrated that only miR-124 and miR-203 are silenced by CpG island methylation in the primary tumors of HCC (Furuta et al. 2010). In a similar study, Datta et al. (2008) analyzed the miRNA expression profile in HCC cell lines treated with 5-aza-C and/or trichostatin A and found that these epigenetic drugs differentially regulate expression of a few miRs, particularly miR-1-1 (Datta et al. 2008). The results of this study showed that the miR-1 expression is markedly reduced by an aberrant CpG island methylation in HCC compared with matching liver tissues. They found that the miR-1-1 gene was hypomethylated in DNMT1-null HCT-116 cells (but not in DNMT3B-null cells), this suggests an important role for the maintenance DNMT in the silencing of this particular miRNA. In addition, an ectopic expression of miR-1 in HCC cells caused an inhibition of cell growth and reduced replication potential (Datta et al 2008). All together, these findings indicate that specific miRNAs including miR-1, miR-124 and miR-203 are tumor-suppressor miRNAs that inhibit their target oncogenes and are epigenetically silenced during hepatocarcinogenesis.

To identify miRNAs which are involved in the regulation of the abnormal DNA methylation in HBV-related HCC, miR-152 was found to be frequently down-regulated in the HBV-related HCC tissues in comparison with adjacent noncancerous hepatic tissues (Huang et al. 2010). Huang et al. (2010) found that miR-152 was inversely correlated to the DNMT1 mRNA expression and may act as a tumor suppressor via suppression of this DNA methyltransferase. Interestingly, The forced expression of miR-152 in liver cell lines resulted in a marked reduction in the expression of DNMT1 at both mRNA and protein levels (Huang et al. 2010). In a recent study, Liu et al. (2013) reported that down-regulation of tumor suppressive miR-517a and miR-517c contribute to the HCC development; they found that ectopic expression of these miRNAs inhibits cell proliferation by blocking the G2/M transition, whereas down-regulation of miR-517a and miR-517c facilitates cell growth (Liu et al. 2013a). In addition to the DNA methylation, the histone acetylation has been shown to play important roles in the pathogenesis of the HCC, and aberrations in this important epigenetic mechanism have been frequently observed in this malignancy. In a study conducted by Yuan et al. (2011), they found that miR-200a and the level of histone H3 acetylation at its promoter were reduced in human HCC tissues as compared to adjacent noncancerous hepatic tissues (Yuan et al. 2011). They also found that a decreased expression of miR-200a and reduced histone acetylation level at the promoter region of this miRNA were induced through the activation of HDAC4 in a Sp1-dependent pathway. All together, the findings of this study suggest that the HDAC4/Sp1/miR-200a regulatory network induces the down-regulation of the miR-200a and the up-regulation of HDAC4 in HCC (Yuan et al. 2011).

miR-224 is one of the most commonly up-regulated microRNAs in HCC, it affects crucial cellular processes such as apoptosis and cell proliferation. In an effort to elucidate molecular mechanism that leads to the overexpression of miR-224 in

HCC, Wang et al. (2012) found that the overexpression of E1A binding protein p300 (EP300) may account, in part, for the up-regulation of this miRNA in patients with HCC (Wang et al. 2012). Also, in an effort to investigate the epigenetic mechanisms responsible for the increased expression of miR-191 in HCC, hypomethylation of this miRNA locus was reported as the causative reason of the aberrancy; leading to an increased cell invasion and to the transition of the HCC cells into mesenchymal-like cells. In this regard, treatment of normal liver cells with 5-aza-C also induced an up-regulation of miR-191 expression, which suggests miR-191 involvement in the HCC progression (He et al. 2011).

4.3.6 Leukemias

The role of aberrant epigenetic modifications, particularly DNA hypermethylation of gene promoters and miRNAs, is a frequent mechanism of gene silencing that contributes to the pathogenesis of acute leukemias including acute lymphoblastic leukemia (ALL) and acute myeloid leukemia (AML) (Alvarez et al. 2010; Davidsson et al. 2009; Figueroa et al. 2010; Lugthart et al. 2011; Martin-Subero et al. 2009; Milani et al. 2010; Román-Gómez et al. 2007; Stumpel et al. 2009). Overall, the methylation of miRNAs is found in a lower percentage of patients with AML in comparison with those with ALL. Using ALL cell lines, Roman-Gomez et al. (2009) demonstrated that 11 CpG islands that were embedded or closed to 13 miRNAs (miR-9-1, miR-9-2, miR-9-3, miR-10b, miR-34b, miR-34c, miR-124a1, miR-124a2, miR-124a3, miR-132, miR-196b, miR-203 and miR-212) disclosed in a closed chromatin conformation (decrease of 3mK4H3 and/or increase of 2mK9H3), which is associated with repressive gene expression (Roman-Gomez et al. 2009). Using bone marrow samples from 353 ALL patients at diagnosis, miRNAs methylation of at least one of the 13 miRNAs (methylated group) was also found in 65 % of the cases, supporting the role of the miRNA methylation in the early phases of lymphoid leukemogenesis. In addition, the downregulation of miRNAs expression reverted by a treatment with 5-Aza-CdR, suggesting that the expression of these miRNAs were regulated by epigenetic changes. They found that the patient-specific methylation profile provides important prognostic information in ALL and patients that belonged to the methylated group showed a significantly higher relapse and mortality rate (Roman-Gomez et al. 2009). Moreover, the methylation profile may be applied to redefine the prognosis in the selected ALL groups with well-established prognostic features; in this regard, the general poor outcome of BCR/ABL-positive or high-WBC-count ALL patients was improved in patients without miRNA hyper-methylation, whereas the general good outcome of the TEL/AML1-positive ALL patients was significantly worsened in those patients with the presence of miRNA methylation (Roman-Gomez et al. 2009). Findings in a study by Bueno et al. (2008) showed that genetic and epigenetic silencing of miR-203 enhanced ABL1 and BCR-ABL1 oncogene expression (Bueno et al. 2008); knowing that miR-203 is aberrantly methylated in ALL (Chim et al. 2011), it is reasonable to hypothesize that silencing of this miRNA may provide a proliferative advantage in the BCR-ABL1-positive leukemia. It has been identified that from 11 miRNAs which were downregulated in t(4;11)-positive infant ALL as a consequence of CpG hypermethylation, seven of which (miR-10a, miR-152, mir-200a, miR-220b, miR-429, miR-432 and miR-503) were re-activated after exposure to the DNA methyltransferase inhibitor Zebularine (Stumpel et al. 2011).

In an effort to explore the epigenetic alterations of miRNAs in ALL, Agirre et al. (2009) observed miR-124a hyermethylation in 59% of the 353 patients diagnosed with ALL and found that the epigenetic down-regulation of this miRNA contributed to the abnormal proliferation of ALL cells via CDK6-Rb pathway both in vitro and in vivo (Agirre et al. 2009). In the pathogenesis of ALL, miR-9 can be mentioned as another example of miRNAs which are epigenetically deregulated. In this regard, Rodriguez-Otero et al. (2011) analyzed the methylation status of the three members of the miR-9 family (miR-9-1, miR-9-2 and miR-9-3) in a uniformly treated cohort of 200 newly diagnosed ALLs and found miR-9 methylation in 54% of the patients. They found that the epigenetic downregulation of miR-9 induced the upregulation of FGFR1 and CDK6, while the treatment of ALL cells with FGFR1 and CDK6 inhibitors increased the rate of apoptosis in these cells (Rodriguez-Otero et al. 2011). Although, evidence for the deregulation of FGFR1 by miR-9 is limited, lines of evidence declared that regulation of the CDK6-Rb pathway is mediated by a number of miRNAs including miR-124a (Agirre et al. 2009) not only in leukemia, but also in other hematological malignancies such as myeloma and lymphoma (Wong et al 2011). It is worth to mention that CDK6 is also a target of both miR-34b and miR-34c, which are inappropriately methylated in 35% of ALL cases (Vilas–Zornoza et al. 2011). Investigating the role of miR-34a methylation, Chim et al. (2010) found the aberrant hypermethylation of miR-34a in various hematological malignancies such as non-Hodgkin's lymphoma (NHL), multiple myeloma (MM) and chronic lymphocytic leukemia (CLL), but not in samples from patients with ALL, AML and chronic myeloid leukemia (CML) (Chim et al. 2010).

CDK6 is also regulated by miR-124a in AML cells; the treatment of these cells with PD0332991 (CDK6 inhibitor) exerts a growth suppressive effect in AML cells (Vázquez et al. 2010). Aberrant methylation of miR-124a-1 and miR-124a-3 has been reported in cell lines and samples from the AML patients, independent from the cytogenetic subtype; Furthermore, it has been shown that the expression of the EVI1, a transcription factor implicated in the development and progression of high-risk AML, leads to the epigenetic silencing of miR-124a (Vázquez et al. 2010). Fazi et al. (2007) showed that transcription factors can recruit epigenetic effectors at miRNA promoter regions and also contributes to the regulation of their expression (Fazi et al. 2007). AML1/ETO, the most common AML-associated fusion protein, is the aberrant product of t(8;21) translocation and can bind to the pre-miR-223 region. This study identified miR-223 as a direct transcriptional target of the AML1/ETO fusion protein and showed that the expression of this oncoprotein induces epigenetic silencing of the myelopoiesis regulator miR-223 through the recruitment of chromatin remodeling enzymes (i.e., DNMTs, HDAC1, and MeCP2). An ectopic expression of miR-223, down-regulation of AML1/ETO protein levels via RNA interferance, or the use of demethylating agents has been shown to enhance miR-223 levels and to restore cell differentiation (Fazi et al. 2007). Finally in a study conducted by Gao et al. (2011) six miRNAs (miR-137 , miR-193a , miR-193b, miR-218, miR-221 and miR-222) were shown to bind to the 3'-UTR region and negatively regulate c-kit (Gao et al. 2011). Among these miRNAs, miR-193a was embedded in a CpG island and epigenetically repressed by promoter hypermethylation in both cell lines and primary samples from patients with AML, but not in the normal bone marrow cells. Importantly, miR-193a levels were inversely correlated with c-kit levels and restoring miR-193a expression, either by synthetic miR-193a transfection or by DNA hypomethylating agent 5-aza-C treatment, which resulted in a significant reduction in c-kit expression, reduced cell growth and induced differentiation of AML cells; this is suggestive of the role for methylation-repressed miRNA in myeloid leukemogenesis (Gao et al. 2011).

4.3.7 Melanoma

It has been reported that miR-34a expression is silenced in several types of cancers including breast, lung, colon, kidney, bladder and pancreatic carcinoma as well as in melanoma. The analyses of miR-34a CpG methylation revealed that 43.2 % and 62.5 % of melanoma cell lines and primary melanoma samples, respectively, were silenced due to the aberrant CpG methylation of its promoter (Lodygin et al. 2008). To identify epigenetically regulated miRNAs in melanoma, Mazar et al. (2011) treated a stage-III melanoma cell line (WM1552C) with 5AzadC and 4-PBA (Mazar et al. 2011). Among 15 hypermethylated miRNAs, miR-375 was highly methylated and involved in the human melanoma development. They showed that the ectopic expression of miR-375 inhibited melanoma cell proliferation, invasion, and cell motility, and also induced changes in cell shape, indicating an important function of this miRNA in the progression of human melanomas (Mazar et al. 2011). In a murine model of melanocyte malignant transformation, Molognoni et al. (2011) showed that the DNMT1 expression was increased through different stages of melanoma progression, and its maximum expression level was in malignant metastatic cells (Molognoni et al. 2011). In this study, they introduced epigenetic reprogramming as a key contributor to the melanocyte malignant transformation, since melanocytes treated with 5-Aza-CdR before each anchorage blockade, did not transform. Some epigenetic changes was shown to be responsible for the maintenance of the malignant phenotype, since melanoma cell lines treated with 5-Aza-CdR or Trichostatin A, showed a reduction in tumor growth in vivo (Molognoni et al. 2011).

Upon treatment of two highly metastatic human melanoma cell lines with 5-Aza-CdR and trichostatin A and then subsequent analyses of miRNA expression profile, it was found that miR-182 was significantly up-regulated in human melanoma cells. Methylation analysis showed that a prominent CpG island 8-10 kb upstream of the miR-182 locus was exclusively methylated in melanoma cells but not in human melanocytes, skin, or peripheral blood mononuclear cells; this is suggestive of an epigenetic mechanism involvement in modulating the expression level of miR-182 in melanoma (Liu et al. 2013b). Knowing that the genomic region in chromosome 9p21 that encompasses miR-31 is frequently deleted in solid cancers including melanoma, Asangani et al. (2012) queried the expression status and performed functional characterization of miR-31 in both melanoma tissues and cell lines (Asangani et al. 2012). They found that the down-regulation of miR-31, as a common event in melanoma tumors and cell lines, was a result of epigenetic silencing by DNA methylation and EZH2-mediated histone methylation. They showed that ectopic overexpression of miR-31 in various melanoma cell lines inhibited cell migration/invasion and resulted in down-regulation of EZH2, leading to de-repression of its target gene rap1GAP; suggesting a tumor suppressor role for miR-31 in melanoma (Asangani et al. 2012).

4.4 Deregulation of miRNAs by Epigenetic Drugs

There has been a significant amount of research aimed at developing cancer therapies that work by inhibiting methylation of tumor suppressor genes or by demethylating them and restoring their expression. A number of substances have been effective in demethylating tumor suppressor cells in cancers and restoring their functionality in in-vitro and in small animal experiments. Generally, five kinds of epigenetic drugs are known, including DNMT inhibitors, HDAC inhibitors, histone acetyl transferase (HAT) inhibitors, histone methyl transferase (HMT) inhibitors and histone demethylase (HDT) inhibitors (Boumber and Issa 2011); It is worth to mention that most of the research efforts are focused on the first two agent types. In this regard, two DNMT inhibitors, 5-Aza-C and 5-Aza-CdR, were approved by the FDA to treat MDS and AML (Rodriguez-Paredes and Esteller 2011). Furthermore, in 2006 the FDA approved HDAC inhibitor suberoyl anilide hydroxamic acid (SAHA) to treat cutaneous T-cell lymphoma (CTCL) as well (Marks and Breslow 2007).

In some cancer cells, an inappropriate methylation of miRNAs is associated with a decrease or complete lack of expression of the tumor suppressor miRNAs; demethylation and re-expression of miRNAs is associated with an inhibition of cancer cell proliferation and with an increase in apoptosis (Agirre et al. 2009; Rodriguez-Otero et al. 2011). The using of epigenetic drugs may help to control the expression of epigenetically regulated miRNAs for prevention or treatment of human cancers. Owing to the reversible nature of epigenetic alterations, therapeutic strategies targeting epigenetic machinery effectors might lead to introduction and addition of novel promising drugs to the growing list of other new anti-cancer products. As mentioned before, while some miRNAs may be directly involved in cancer, others may be involved by targeting the other key players of carcinogenesis including epigenetic machinery effectors. There is a growing list of miRNAs that their expressions are epigenetically regulated (Table 4.1); it is reasonable to hypothesize that part of the observed therapeutic effects of epigenetic drugs might be attributed to their effects on miRNAs. In a breast cancer cell line, Scott et al. (2006) was able to demonstrate that 27 miRNA expression levels were rapidly modified by a treatment with the HDAC inhibitor LAQ824, indicating that indeed epigenetic factors are involved in miRNA regulation (Scott et al. 2006). Similarly, treating bladder cancer cells with both a DNA demethylating agent (5-Aza-CdR) and an HDAC inhibitor (4-phenylbutyric acid), it has been shown that about 5 % of the human miRNAs increased their expression levels (Saito et al. 2006). Besides the two most commonly used inhibitors, DNMT and HDAC, arsenic trioxide (As_2O_3) is a novel demethylation agent that is able to reverse an aberrant methylation, possibly to regulate microRNA levels, and to restore for example the expression of tumor suppressor microRNAs.

A growing body of evidence indicates that arsenic is a potential demethylating agent and the effect induced by As_2O_3 is associated with extensive genomic-wide epigenetic changes with large-scale alterations in H3 acetylation and with a global DNA hypomethylation (Cui et al. 2006; Martens et al. 2010; Zhou et al. 2008). The mechanism of DNA hypomethylation after arsenic exposure is not clear. However, the unique arsenic metabolism may play an essential factor in the DNA demethylation of CpG islands in As_2O_3-treated cells. In a recent study using an acute promyelocytic leukemia (APL) cell line, we have shown that As_2O_3 by a biotransformation into the intracellular methylated metabolites through AS3MT catalyzes and by the inhibition of DNA methyltransferases (DNMT1, DNMT3a and DNMT3b) expression causes a depletion of the methyl donors (Khaleghian et al 2014). The depletion of methyl groups resulted in an inability to maintain methylated cytosine in DNA, resulting in hypomethylation (Fig. 4.1). Exposure to As2O3 has been shown to cause hypomethylation of several tumor suppressor genes leading to their re-expression in various cancer cells (Cui et al. 2006; Tong and Lin 2002).

As_2O_3 is a potential demethylating agent; the unique arsenic metabolism may play an essential factor in the DNA demethylation of CpG islands in As_2O_3-treated cells. The biotransformation of As_2O_3 in cell, involves a series of reduction and methylation. As_2O_3 is converted to methylated products by AS_3MT catalyzes. As_2O_3 exert demethylation effect by a direct inhibition of DNA methyltransferase (DNMT1, DNMT3a, and DNMT3b), and through depletion of the methyl donor. The depletion of methyl groups would result in inability to maintain methylated cytosine in DNA, resulting hypomethylation. Hypermethylated promoter regions of CpG island of the tumor suppressor genes including microRNAs are demthylated, leading to their re-expression in cancer cells.

In a recent study, we investigated the effect of As_2O_3 exposure on the expression profile of 88 cancer-related miRNAs in an APL cell line, NB4 (Ghaffari et al. 2012). Figure 4.2 shows the list of the 88 cancer-related miRNAs used in this experiment and their fold changes in the expression level after As_2O_3 exposure.

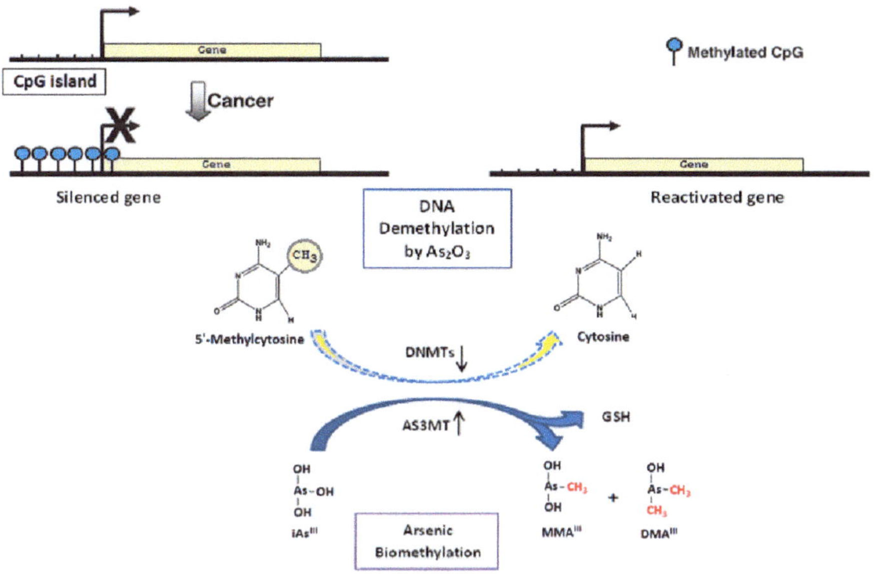

Fig. 4.1 Anticancer mechanism of As2O3 by epigenetic modulation. CpG islands can be found in the promoter regions of roughly half of the genes and normally remain unmethylated. When they become aberrantly hypermethylated, as can happen in many cancers, they lead to the silencing of downstream genes. During carcinogenesis, this change is much more dramatic, leading to a global hypermethylation of CpG islands. The results are silencing of some important tumor-suppressor genes and microRNAs

Among the 88 cancer-focused miRNAs, 80 miRNAs were found to be differentially up-regulated; statistical analysis of these differentially expressed miRNAs showed that 52 of these miRNAs were significantly dysregulated with p values of <0.01 and with more than two fold changes (Fig. 4.3). Among the up-regulated miRNAs, more than 45% of these were identified to be epigenetically regulated in different type of hematological malignancies as well as in solid tumors. Most of these miRNAs are known to function as a tumor and/or metastatic suppressor related to cell cycle regulation and apoptosis, as well as in inhibition of angiogenesis, cancer cell invasion and metastasis. A pathway analysis was conducted to determine the putative biological functions of the potential predicted genes for each relevant miRNA by the DIANA Pathway Software. Based on the target analysis, from 5331 putative target genes predicted by the software, 1168 genes were identified to be involved in 180 different biological processes. Table 4.3 summarized the number of predicted genes for each miRNA in some of the important biological processes, such as cell cycle, apoptosis, p53 signaling pathway, Wnt, MAPK and Jak-STAT signaling pathways and various cancers as well as in the regulation of actin cytoskeleton.

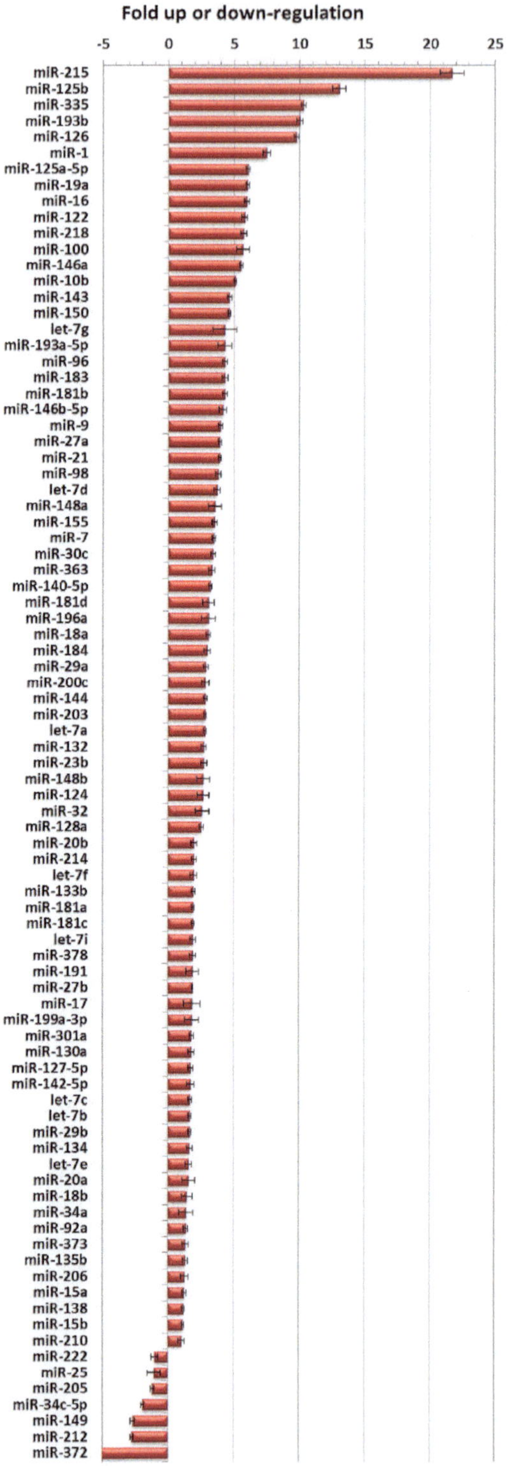

Fig. 4.2 Fold change in the expression level of 88 cancer-related miRNA after As2O3 exposure. The miRNA expression levels were quantified by the real-time PCR assays. Data are expressed in terms of fold change of miRNA levels detected in As2O3-treated (2 μM for 48 h) cells with respect to those found in the corresponding untreated controls. Each miRNA expression data is normalized to the average median of three housekeeping genes (U6, SNORD47 and SNORD48) in the array. Fold change of each miRNA was calculated from the expression levels between the As2O3 treated and untreated cells using 2 – ΔΔCt method. Values are given as mean ±S.D

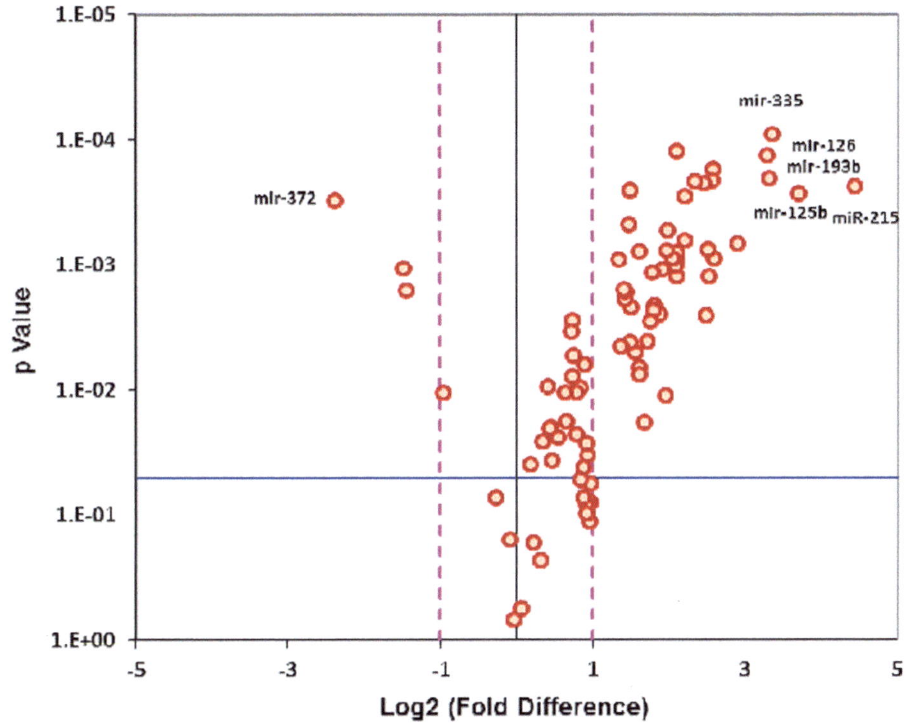

Fig. 4.3 Volcano plots of 88 cancer-related miRNAs expression in As2O3 treated versus untreated APL cell line (NB4). The volcano plot represents the distribution of 88 miRNAs in As2O3-treated (2 μM for 48 h) NB4 cells line as compared to the untreated control. X-axis indicates a difference in the expression level on a log2 scale, whereas the y-axis represents corresponding P values (Student's t-test) on a negative log scale. The *dashed-lines* (*vertical lines*) indicate a 2-fold change in miRNAs expression threshold, the blue line (*horizontal line*) indicate the significant level of *P*=0.05. Some of the miRNAs showing the highest fold change in their expression are labeled

Collectively, we hypothesize that the unique arsenic metabolism and its association with global DNA hypomethylation may contribute to the upregulation of a wide range of epigenetically regulated miRNAs in the As_2O_3-treated cancer cell. Most of these miRNAs are known to function as a tumor and/or metastatic suppressor in a wide range of biological functions including cell cycle regulation and apoptosis, as well as angiogenesis, invasion, and metastasis. Thus, it is reasonable to assume that at least part of the observed therapeutic effects of As_2O_3 might be attributed to its effects on the epigenetically regulated miRNAs.

Table 4.3 The number of predicted genes for each miRNA in some of the important biological processes by DIANA-mirPath.

miRNA Name	Total number of Genes	Number of Genes in Pathways										
		Total number of Genes in Pathways	Apoptosis	Cell Cycle	p53 signaling pathway	MAPK signaling pathway	Jak-STAT signaling pathway	Wnt signaling pathway	Toll-like receptor signaling pathway	VEGF signaling pathway	Regulation of actin cytoskeleton	Cancers[a]
miR-215	165	30	0	2	0	3	0	1	1	0	5	10
miR-125b	661	140	2	5	2	15	8	6	4	3	8	38
miR-335	308	62	1	2	1	7	0	4	3	0	7	12
miR-193b	308	65	0	5	3	8	2	5	1	2	6	51
miR-126	24	7	0	0	0	1	1	0	0	0	2	3
miR-1	562	147	1	5	6	13	6	13	3	5	11	87
miR-125a-5p	672	141	2	5	2	15	8	7	4	3	9	41
miR-16	978	212	6	12	9	19	12	18	7	10	18	121
miR-122	159	48	0	0	1	5	3	1	1	1	4	5
miR-218	810	161	2	3	1	8	11	7	1	2	8	41
miR-100	35	11	1	0	0	2	0	3	0	1	1	14
miR-146a	219	42	1	4	0	4	1	4	3	2	2	13
miR-143	456	94	1	1	4	8	4	3	1	2	10	37
let-7g	763	181	5	7	9	27	11	10	8	3	13	75
miR-193a-3p	300	62	0	5	3	8	2	5	1	2	6	52
miR-96	981	203	6	6	4	24	6	10	3	4	23	78
miR-183	391	91	2	2	2	14	2	10	1	2	11	21

Table 4.3 (continued)

miRNA Name	Total number of Genes	Number of Genes in Pathways										
		Total number of Genes in Pathways	Apoptosis	Cell Cycle	p53 signaling pathway	MAPK signaling pathway	Jak-STAT signaling pathway	Wnt signaling pathway	Toll-like receptor signaling pathway	VEGF signaling pathway	Regulation of actin cytoskeleton	Cancers[a]
miR-181b	990	206	6	5	2	22	14	10	6	6	16	100
miR-146b-5p	220	45	1	4	0	4	1	5	3	2	2	19
miR-372	1125	219	5	14	8	30	10	16	6	6	21	152
Union	5331	1168	23	39	30	118	51	68	36	27	97	436
-ln(p-value) (Union)		25.59	0.01	6.46	3.24	13.06	0.51	5.64	0.02	0.51	6.09	97.29

[a] Acute myeloid leukemia, chronic myeloid leukemia, glioma, colorectal cancer, prostate cancer, pancreatic cancer, renal cell carcinoma, melanogenesis, non-small cell lung cancer, thyroid cancer, bladder cancer, endometrial cancer, small cell lung cancer, basal cell carcinoma

-ln(p-value), the negative natural logarithm of the enrichment p-value calculated for the specific pathway

4.5 Conclusion

Epigenetics and miRNAs are two important subjects of study and the relationship between them is just beginning to be understood. microRNAs is considered part of a multilevel regulatory mechanism aimed to finely modulate specific target gene at a post-transcriptional level. miRNAs can be regulated by epigenetic events such as DNA methylation or histones modifications, similar to any other conventional protein-coding genes (PCG), and also can regulate effectors of the epigenetic machinery (epi-miRNAs). Thus, the participation of miRNAs in the epigenetic world represents a complicated regulatory loop interconnecting epigenome and miRNome that introduces new layers of complexity in gene regulation, and the relationship between them is just now beginning to be understood. The deregulated cross-talk between miRNome-epigenome is one of the mechanisms that lead to pathological conditions such as cancer. Owing to the reversible nature of epigenetic alterations, therapeutic strategies targeting epigenetic machinery effectors might provide promising tools for treatment of human cancers in the future.

References

Agirre X, Vilas-Zornoza A, Jiménez-Velasco A, Martin-Subero JI, Cordeu L, Gárate L et al (2009) Epigenetic silencing of the tumor suppressor microRNA Hsa-miR-124a regulates CDK6 expression and confers a poor prognosis in acute lymphoblastic leukemia. Cancer Res 69:4443–4453

Alpini G, Glaser SS, Zhang JP, Francis H, Han Y, Gong J et al (2011) Regulation of placenta growth factor by microRNA-125b in hepatocellular cancer. J Hepatol 55:1339–1345

Alvarez S, Suela J, Valencia A, Fernández A, Wunderlich M, Agirre X et al (2010) DNA methylation profiles and their relationship with cytogenetic status in adult acute myeloid leukemia. PloS One 5:e12197

Ando T, Yoshida T, Enomoto S, Asada K, Tatematsu M, Ichinose M et al (2009) DNA methylation of microRNA genes in gastric mucosae of gastric cancer patients: its possible involvement in the formation of epigenetic field defect. Int J Cancer 124:2367–2374

Asangani IA, Harms PW, Dodson L, Pandhi M, Kunju LP, Maher CA et al (2012) Genetic and epigenetic loss of microRNA-31 leads to feed-forward expression of EZH2 in melanoma. Oncotarget 3:1011

Augoff K, McCue B, Plow EF, Sossey-Alaoui K (2012) miR-31 and its host gene lncRNA LOC554202 are regulated by promoter hypermethylation in triple-negative breast cancer. Mol Cancer 11:5

Balaguer F, Link A, Lozano JJ, Cuatrecasas M, Nagasaka T, Boland CR et al (2010) Epigenetic silencing of miR-137 is an early event in colorectal carcinogenesis. Cancer Res 70:6609–6618

Bandres E, Agirre X, Bitarte N, Ramirez N, Zarate R, Roman-Gomez J et al (2009) Epigenetic regulation of microRNA expression in colorectal cancer. Int j cancer 125:2737–2743

Bartel DP (2009) MicroRNAs: target recognition and regulatory functions. Cell 136:215–233

Benetti R, Gonzalo S, Jaco I, Munoz P, Gonzalez S, Schoeftner S et al (2008) A mammalian microRNA cluster controls DNA methylation and telomere recombination via Rbl2-dependent regulation of DNA methyltransferases. Nat Struct Mol Biol 15:998

Bhatia S, Kaul D, Varma N (2010) Potential tumor suppressive function of miR-196b in B-cell lineage acute lymphoblastic leukemia. Mol Cell Biochem 340:97–106

Bhatnagar N, Li X, Padi S, Zhang Q, Tang M, Guo B (2010). Downregulation of miR-205 and miR-31 confers resistance to chemotherapy-induced apoptosis in prostate cancer cells. Cell Death Dis 1: e105
Boumber Y, Issa JP (2011) Epigenetics in cancer: what's the future? Int Soc Cell 25(220–226):228 (Williston Park)
Braconi C, Huang N, Patel T (2010) MicroRNA-dependent regulation of DNA methyltransferase-1 and tumor suppressor gene expression by interleukin-6 in human malignant cholangiocytes. Hepatology 51:881–890
Brueckner B, Stresemann C, Kuner R, Mund C, Musch T, Meister M et al (2007) The human let-7a-3 locus contains an epigenetically regulated microRNA gene with oncogenic function. Cancer Res 67:1419–1423
Bueno MJ, Pérez de Castro I, Gómez de Cedrón M, Santos J, Calin GA, Cigudosa JC et al (2008) Genetic and Epigenetic Silencing of MicroRNA-203 Enhances*ABL1*and*BCR-ABL1*Oncogene Expression. Cancer cell 13:496–506
Buurman R, Gurlevik E, Schaffer V, Eilers M, Sandbothe M, Kreipe H et al (2012) Histone deacetylases activate hepatocyte growth factor signaling by repressing microRNA-449 in hepatocellular carcinoma cells. Gastroenterology 143(811–820):e811–e815
Calin GA, Ferracin M, Cimmino A, Di Leva G, Shimizu M, Wojcik SE et al (2005) A microRNA signature associated with prognosis and progression in chronic lymphocytic leukemia. N Engl J Med 353:1793–1801
Carthew RW, Sontheimer EJ (2009) Origins and mechanisms of miRNAs and siRNAs. Cell 136:642–655
Ceppi P, Mudduluru G, Kumarswamy R, Rapa I, Scagliotti GV, Papotti M et al (2010) Loss of miR-200c expression induces an aggressive, invasive, and chemoresistant phenotype in non-small cell lung cancer. Mol Cancer Res 8:1207–1216
Chang T-C, Wentzel EA, Kent OA, Ramachandran K, Mullendore M, Lee KH et al (2007). Transactivation of miR-34a by p53 broadly influences gene expression and promotes apoptosis. Mol Cell 26:745–752
Chang KW, Chu TH, Gong NR, Chiang WF, Yang CC, Liu CJ et al (2013) miR-370 modulates insulin receptor substrate-1 expression and inhibits the tumor phenotypes of oral carcinoma. Oral Dis 19:611–619
Chen JF, Mandel EM, Thomson JM, Wu Q, Callis TE, Hammond SM et al (2006) The role of microRNA-1 and microRNA-133 in skeletal muscle proliferation and differentiation. Nat Genet 38:228–233
Chen Y, Gao W, Luo J, Tian R, Sun H, Zou S (2011) Methyl-CpG binding protein MBD2 is implicated in methylation-mediated suppression of miR-373 in hilar cholangiocarcinoma. Oncol Rep 25:443–451
Chim C, Wong K, Qi Y, Loong F, Lam W, Wong L et al (2010) Epigenetic inactivation of the miR-34a in hematological malignancies. Carcinogenesis 31:745–750
Chim CS, Wong KY, Leung CY, Chung LP, Hui PK, Chan SY et al (2011) Epigenetic inactivation of the hsa-miR-203 in haematological malignancies. J Cell Mol Med 15:2760–2767
Cui X, Wakai T, Shirai Y, Yokoyama N, Hatakeyama K, Hirano S (2006) Arsenic trioxide inhibits DNA methyltransferase and restores methylation-silenced genes in human liver cancer cells. Hum Pathol 37:298–311
Datta J, Kutay H, Nasser MW, Nuovo GJ, Wang B, Majumder S et al (2008) Methylation mediated silencing of MicroRNA-1 gene and its role in hepatocellular carcinogenesis. Cancer Res 68:5049–5058
Davidsson J, Lilljebjörn H, Andersson A, Veerla S, Heldrup J, Behrendtz M et al (2009). The DNA methylome of pediatric acute lymphoblastic leukemia. Hum Mol Genet 18:4054–4065
Dohi O, Yasui K, Gen Y, Takada H, Endo M, Tsuji K et al (2013) Epigenetic silencing of miR-335 and its host gene MEST in hepatocellular carcinoma. Int J Oncol 42:411–418
Dou L, Zheng D, Li J, Li Y, Gao L, Wang L et al (2012) Methylation-mediated repression of microRNA-143 enhances MLL-AF4 oncogene expression. Oncogene 31:507–517

Duursma AM, Kedde M, Schrier M, le Sage C, Agami R (2008) miR-148 targets human DNMT3b protein coding region. RNA 14:872–877
Egger G, Liang G, Aparicio A, Jones PA (2004) Epigenetics in human disease and prospects for epigenetic therapy. Nature 429:457–463
Fabbri M, Garzon R, Cimmino A, Liu Z, Zanesi N, Callegari E et al (2007) MicroRNA-29 family reverts aberrant methylation in lung cancer by targeting DNA methyltransferases 3A and 3B. Proc Natl Acad Sci U S A 104:15805–15810
Fazi F, Rosa A, Fatica A, Gelmetti V, De Marchis ML, Nervi C et al (2005) A minicircuitry comprised of microRNA-223 and transcription factors NFI-A and C/EBPα regulates human granulopoiesis. Cell 123:819–831
Fazi F, Racanicchi S, Zardo G, Starnes LM, Mancini M, Travaglini L et al (2007) Epigenetic silencing of the myelopoiesis regulator microRNA-223 by the AML1/ETO oncoprotein. Cancer cell 12:457–466
Figueroa ME, Lugthart S, Li Y, Erpelinck-Verschueren C, Deng X, Christos PJ et al (2010) DNA methylation signatures identify biologically distinct subtypes in acute myeloid leukemia. Cancer cell 17:13–27
Formosa A, Lena A, Markert E, Cortelli S, Miano R, Mauriello A et al (2013) DNA methylation silences miR-132 in prostate cancer. Oncogene 32:127–134
Fraga MF, Ballestar E, Paz MF, Ropero S, Setien F, Ballestar ML et al (2005) Epigenetic differences arise during the lifetime of monozygotic twins. Proc Natl Acad Sci U S A 102:10604–10609
Friedman JM, Liang G, Liu C-C, Wolff EM, Tsai YC, Ye W et al (2009) The putative tumor suppressor microRNA-101 modulates the cancer epigenome by repressing the polycomb group protein EZH2. Cancer Res 69:2623–2629
Furuta M, Kozaki K-i, Tanaka S, Arii S, Imoto I, Inazawa J (2010) miR-124 and miR-203 are epigenetically silenced tumor-suppressive microRNAs in hepatocellular carcinoma. Carcinogenesis 31:766–776
Gao X, Lin J, Li Y, Gao L, Wang X, Wang W et al (2011) MicroRNA-193a represses c-kit expression and functions as a methylation-silenced tumor suppressor in acute myeloid leukemia. Oncogene 30:3416–3428
Gardiner-Garden M, Frommer M (1987) CpG islands in vertebrate genomes. J Mol Biol 196:261–282
Garzon R, Liu S, Fabbri M, Liu Z, Heaphy CE, Callegari E et al (2009) MicroRNA-29b induces global DNA hypomethylation and tumor suppressor gene reexpression in acute myeloid leukemia by targeting directly DNMT3A and 3B and indirectly DNMT1. Blood 113:6411–6418
Ghaffari SH, Bashash D, Dizaji MZ, Ghavamzadeh A, Alimoghaddam K (2012) Alteration in miRNA gene expression pattern in acute promyelocytic leukemia cell induced by arsenic trioxide: a possible mechanism to explain arsenic multi-target action. Tumour Biol 33:157–172
Grady W, Parkin R, Mitchell P, Lee J, Kim Y, Tsuchiya K et al (2008) Epigenetic silencing of the intronic microRNA hsa-miR-342 and its host gene EVL in colorectal cancer. Oncogene 27:3880–3888
Guo H, Ingolia NT, Weissman JS, Bartel DP (2010) Mammalian microRNAs predominantly act to decrease target mRNA levels. Nature 466:835–840
Han L, Witmer PD, Casey E, Valle D, Sukumar S (2007) DNA methylation regulates MicroRNA expression. Cancer Biol Ther 6:1284–1288
Hashimoto Y, Akiyama Y, Otsubo T, Shimada S, Yuasa Y (2010) Involvement of epigenetically silenced microRNA-181c in gastric carcinogenesis. Carcinogenesis 31:777–784
Hassler MR, Egger G (2012) Epigenomics of cancer-emerging new concepts. Biochimie 94:2219–2230
He Y, Cui Y, Wang W, Gu J, Guo S, Ma K et al (2011) Hypomethylation of the hsa-miR-191 locus causes high expression of hsa-mir-191 and promotes the epithelial-to-mesenchymal transition in hepatocellular carcinoma. Neoplasia (New York, NY) 13:841
Hsu P-Y, Deatherage DE, Rodriguez BA, Liyanarachchi S, Weng Y-I, Zuo T et al (2009) Xenoestrogen-induced epigenetic repression of microRNA-9-3 in breast epithelial cells. Cancer Res 69:5936–5945

Huang J, Wang Y, Guo Y, Sun S (2010) Down-regulated microRNA-152 induces aberrant DNA methylation in hepatitis B virus-related hepatocellular carcinoma by targeting DNA methyltransferase 1. Hepatology 52:60–70
Hulf T, Sibbritt T, Wiklund ED, Bert S, Strbenac D, Statham AL et al (2011) Discovery pipeline for epigenetically deregulated miRNAs in cancer: integration of primary miRNA transcription. BMC Genomics 12:54
Humphreys KJ, Cobiac L, Le Leu RK, Van der Hoek MB, Michael MZ (2013) Histone deacetylase inhibition in colorectal cancer cells reveals competing roles for members of the oncogenic miR-17-92 cluster. Mol Carcinog 52:459–474
Incoronato M, Garofalo M, Urso L, Romano G, Quintavalle C, Zanca C et al (2010). miR-212 increases tumor necrosis factor-related apoptosis-inducing ligand sensitivity in non-small cell lung Cancer by targeting the antiapoptotic protein PED. Cancer Res 70:3638–3646
Iorio MV, Visone R, Di Leva G, Donati V, Petrocca F, Casalini P et al (2007) MicroRNA signatures in human ovarian cancer. Cancer Res 67:8699–8707
Jemal A, Bray F, Center MM, Ferlay J, Ward E, Forman D (2011) Global cancer statistics. CA Cancer J Clin 61:69–90
Jerónimo C, Henrique R (2011). Epigenetic biomarkers in urological tumors: a systematic review. Cancer Lett 342:264–274
Jones PA, Baylin SB (2002). The fundamental role of epigenetic events in cancer. Nat Rev Genet 3:415–428
Jones PA, Laird PW (1999) Cancer-epigenetics comes of age. Nat Genet 21:163–167
Jones PA, Takai D (2001) The role of DNA methylation in mammalian epigenetics. Science 293:1068–1070
Kaneda M, Okano M, Hata K, Sado T, Tsujimoto N, Li E et al (2004) Essential role for de novo DNA methyltransferase Dnmt3a in paternal and maternal imprinting. Nature 429:900–903
Karube Y, Tanaka H, Osada H, Tomida S, Tatematsu Y, Yanagisawa K et al (2005). Reduced expression of Dicer associated with poor prognosis in lung cancer patients. Cancer Sci 96:111–115
Khaleghian A, Ghaffari SH, Ahmadian S, Alimoghaddam K, Ghavamzadeh A (2014). Metabolism of arsenic trioxide in acute promyelocytic leukemia cells. J Cell Biochem 115(10):1729–1739
Kim VN (2005). MicroRNA biogenesis: coordinated cropping and dicing. Nat Rev Mol cell Biol 6:376–385
Kim VN, Nam J-W (2006) Genomics of microRNA. Trends Genet 22:165–173
Kozaki K, Imoto I, Mogi S, Omura K, Inazawa J (2008) Exploration of tumor-suppressive microRNAs silenced by DNA hypermethylation in oral cancer. Cancer Res 68:2094–2105
Lee Y, Kim M, Han J, Yeom KH, Lee S, Baek SH et al (2004). MicroRNA genes are transcribed by RNA polymerase II. EMBO J 23:4051–4060
Lee KH, Lotterman C, Karikari C, Omura N, Feldmann G, Habbe N et al (2009) Epigenetic silencing of MicroRNA miR-107 regulates cyclin-dependent kinase 6 expression in pancreatic cancer. Pancreatology 9:293–301
Lehmann U, Hasemeier B, Christgen M, Müller M, Römermann D, Länger F et al (2008) Epigenetic inactivation of microRNA gene hsa-mir-9-1 in human breast cancer. J Pathol 214:17–24
Lewis BP, Shih I-h, Jones-Rhoades MW, Bartel DP, Burge CB (2003) Prediction of mammalian microRNA targets. Cell 115:787–798
Li E, Beard C, Jaenisch R (1993) Role for DNA methylation in genomic imprinting. Nature 366:362–365
Liu S, Wu LC, Pang J, Santhanam R, Schwind S, Wu YZ et al (2010) Sp1/NFkappaB/HDAC/miR-29b regulatory network in KIT-driven myeloid leukemia. Cancer cell 17:333–347
Liu R-F, Xu X, Huang J, Fei Q-L, Chen F, Li Y-D et al (2013a). Down-regulation of miR-517a and miR-517c promotes proliferation of hepatocellular carcinoma cells via targeting Pyk2. Cancer Lett 329:164–173
Liu S, Howell PM, Riker AI (2013b) Up-regulation of miR-182 expression after epigenetic modulation of human melanoma cells. Ann Surg Oncol 20:1745–1752
Lodygin D, Tarasov V, Epanchintsev A, Berking C, Knyazeva T, Korner H et al (2008) Inactivation of miR-34a by aberrant CpG methylation in multiple types of cancer. Cell Cycle 7:2591–2600

Loizou JI, Murr R, Finkbeiner MG, Sawan C, Wang Z-Q, Herceg Z (2006) Epigenetic information in chromatin: the code of entry for DNA repair. Cell Cycle 5:696–701
Lu L, Katsaros D, de la LIA, Sochirca O, Yu H (2007) Hypermethylation of let-7a-3 in epithelial ovarian cancer is associated with low insulin-like growth factor-II expression and favorable prognosis. Cancer Res 67:10117–10122
Lugthart S, Figueroa ME, Bindels E, Skrabanek L, Valk PJ, Li Y et al (2011) Aberrant DNA hypermethylation signature in acute myeloid leukemia directed by EVI1. Blood 117:234–241
Lujambio A, Ropero S, Ballestar E, Fraga MF, Cerrato C, Setién F et al (2007). Genetic unmasking of an epigenetically silenced microRNA in human cancer cells. Cancer Res 67:1424–1429
Lujambio A, Calin GA, Villanueva A, Ropero S, Sánchez-Céspedes M, Blanco D et al (2008) A microRNA DNA methylation signature for human cancer metastasis. Proc Natl Acad Sci 105:13556–13561
Marks PA, Breslow R (2007) Dimethyl sulfoxide to vorinostat: development of this histone deacetylase inhibitor as an anticancer drug. Nat Biotechnol 25:84–90
Martens JH, Brinkman AB, Simmer F, Francoijs KJ, Nebbioso A, Ferrara F et al (2010) PML-RARalpha/RXR alters the epigenetic landscape in acute promyelocytic leukemia. Cancer Cell 17:173–185
Martin-Subero JI, Ammerpohl O, Bibikova M, Wickham-Garcia E, Agirre X, Alvarez S et al (2009) A comprehensive microarray-based DNA methylation study of 367 hematological neoplasms. PloS One 4:e6986
Mattick JS, Makunin IV (2006). Non-coding RNA. Hum Mol Genet 15:R17–R29
Mazar J, DeBlasio D, Govindarajan SS, Zhang S, Perera RJ (2011). Epigenetic regulation of microRNA-375 and its role in melanoma development in humans. FEBS Lett 585:2467–2476
Meltzer PS (2005) Cancer genomics: small RNAs with big impacts. Nature 435:745–746
Mendell JT (2008) miRiad roles for the miR-17-92 cluster in development and disease. Cell 133:217–222
Meng F, Wehbe-Janek H, Henson R, Smith H, Patel T (2008) Epigenetic regulation of microRNA-370 by interleukin-6 in malignant human cholangiocytes. Oncogene 27:378–386
Milani L, Lundmark A, Kiialainen A, Nordlund J, Flaegstad T, Forestier E et al (2010) DNA methylation for subtype classification and prediction of treatment outcome in patients with childhood acute lymphoblastic leukemia. Blood 115:1214–1225
Molognoni F, Cruz AT, Meliso FM, Morais AS, Souza CF, Xander P et al (2011) Epigenetic reprogramming as a key contributor to melanocyte malignant transformation. Epigenetics 6:451–465
Nakamura T, Canaani E, Croce CM (2007) Oncogenic All1 fusion proteins target Drosha-mediated microRNA processing. Proc Natl Acad Sci 104:10980–10985
Nguyen T, Kuo C, Nicholl MB, Sim M-S, Turner RR, Morton DL et al (2011) Downregulation of microRNA-29c is associated with hypermethylation of tumor-related genes and disease outcome in cutaneous melanoma. Epigenetics 6:388–394
Noonan EJ, Place RF, Pookot D, Basak S, Whitson JM, Hirata H et al (2009) miR-449a targets HDAC-1 and induces growth arrest in prostate cancer. Oncogene 28:1714–1724
O'Hara SP, Splinter PL, Gajdos GB, Trussoni CE, Fernandez-Zapico ME, Chen XM et al (2010) NFkappaB p50-CCAAT/enhancer-binding protein beta (C/EBPbeta)-mediated transcriptional repression of microRNA let-7i following microbial infection. J Biol Chem 285:216–225
Okano M, Bell DW, Haber DA, Li E (1999) DNA methyltransferases Dnmt3a and Dnmt3b are essential for de novo methylation and mammalian development. Cell 99:247–257
Okoji R, Yu R, Maronpot R, Froines J (2002) Sodium arsenite administration via drinking water increases genome-wide and Ha-ras DNA hypomethylation in methyl-deficient C57BL/6J mice. Carcinogenesis 23:777–785
Östling P, Leivonen S-K, Aakula A, Kohonen P, Mäkelä R, Hagman Z et al (2011). Systematic analysis of microRNAs targeting the androgen receptor in prostate cancer cells. Cancer Res 71:1956–1967
Pallasch CP, Patz M, Park YJ, Hagist S, Eggle D, Claus R et al (2009) miRNA deregulation by epigenetic silencing disrupts suppression of the oncogene PLAG1 in chronic lymphocytic leukemia. Blood 114:3255–3264

Rauhala HE, Jalava SE, Isotalo J, Bracken H, Lehmusvaara S, Tammela TL et al (2010). miR-193b is an epigenetically regulated putative tumor suppressor in prostate cancer. Int J cancer 127:1363–1372
Raver-Shapira N, Marciano E, Meiri E, Spector Y, Rosenfeld N, Moskovits N et al (2007). Transcriptional activation of miR-34a contributes to p53-mediated apoptosis. Mol Cell 26:731–743
Richardson B (2003) Impact of aging on DNA methylation. Ageing Res Rev 2:245–261
Rodriguez-Paredes M, Esteller M (2011) Cancer epigenetics reaches mainstream oncology. Nat Med 17:330–339
Rodriguez-Otero P, Román-Gómez J, Vilas-Zornoza A, José-Eneriz ES, Martín-Palanco V, Rifón J et al (2011) Deregulation of FGFR1 and CDK6 oncogenic pathways in acute lymphoblastic leukaemia harbouring epigenetic modifications of the MIR9 family. Br J Haematol 155:73–83
Román-Gómez J, Cordeu L, Agirre X, Jiménez-Velasco A, San José-Eneriz E, Garate L et al (2007) Epigenetic regulation of Wnt-signaling pathway in acute lymphoblastic leukemia. Blood 109:3462–3469
Roman-Gomez J, Agirre X, Jiménez-Velasco A, Arqueros V, Vilas-Zornoza A, Rodriguez-Otero P et al (2009) Epigenetic regulation of microRNAs in acute lymphoblastic leukemia. J Clin Oncol 27:1316–1322
Saito Y, Liang G, Egger G, Friedman JM, Chuang JC, Coetzee GA et al (2006) Specific activation of microRNA-127 with downregulation of the proto-oncogene BCL6 by chromatin-modifying drugs in human cancer cells. Cancer cell 9:435–443
Saito Y, Friedman JM, Chihara Y, Egger G, Chuang JC, Liang G (2009a) Epigenetic therapy up-regulates the tumor suppressor microRNA-126 and its host gene EGFL7 in human cancer cells. Biochem Biophys Res Commun 379:726–731
Saito Y, Suzuki H, Tsugawa H, Nakagawa I, Matsuzaki J, Kanai Y et al (2009b) Chromatin remodeling at Alu repeats by epigenetic treatment activates silenced microRNA-512-5p with downregulation of Mcl-1 in human gastric cancer cells. Oncogene 28:2738–2744
Sampath D, Liu C, Vasan K, Sulda M, Puduvalli VK, Wierda WG et al (2012) Histone deacetylases mediate the silencing of miR-15a, miR-16, and miR-29b in chronic lymphocytic leukemia. Blood 119:1162–1172
Santos-Rosa H, Caldas C (2005) Chromatin modifier enzymes, the histone code and cancer. Eur J Cancer 41:2381–2402
Scott GK, Mattie MD, Berger CE, Benz SC, Benz CC (2006). Rapid alteration of microRNA levels by histone deacetylase inhibition. Cancer Res 66:1277–1281
Shen MM, Abate-Shen C (2010). Molecular genetics of prostate cancer: new prospects for old challenges. Genes Dev 24:1967–2000
Siegel R, DeSantis C, Virgo K, Stein K, Mariotto A, Smith T et al (2012) Cancer treatment and survivorship statistics. CA Cancer J Clin 62:220–241
Sinkkonen L, Hugenschmidt T, Berninger P, Gaidatzis D, Mohn F, Artus-Revel CG et al (2008) MicroRNAs control de novo DNA methylation through regulation of transcriptional repressors in mouse embryonic stem cells. Nat Struct Mol Biol 15:259–267
Stumpel DJ, Schneider P, van Roon EH, Boer JM, de Lorenzo P, Valsecchi MG et al (2009) Specific promoter methylation identifies different subgroups of MLL-rearranged infant acute lymphoblastic leukemia, influences clinical outcome, and provides therapeutic options. Blood 114:5490–5498
Stumpel D, Schotte D, Lange-Turenhout E, Schneider P, Seslija L, De Menezes R et al (2011) Hypermethylation of specific microRNA genes in MLL-rearranged infant acute lymphoblastic leukemia: major matters at a micro scale. Leukemia 25:429–439
Suh SO, Chen Y, Zaman MS, Hirata H, Yamamura S, Shahryari V et al (2011) MicroRNA-145 is regulated by DNA methylation and p53 gene mutation in prostate cancer. Carcinogenesis 32:772–778
Suzuki H, Takatsuka S, Akashi H, Yamamoto E, Nojima M, Maruyama R et al (2011). Genome-wide profiling of chromatin signatures reveals epigenetic regulation of microRNA genes in colorectal cancer. Cancer Res 71:5646–5658
Takai D, Jones PA (2003). The CpG island searcher: a new WWW resource. Isilico Biol 3:235–240

Takiguchi M, Achanzar WE, Qu W, Li G, Waalkes MP (2003). Effects of cadmium on DNA-(Cytosine-5) methyltransferase activity and DNA methylation status during cadmium-induced cellular transformation. Exp Cell Res 286:355–365

Thomson JM, Newman M, Parker JS, Morin-Kensicki EM, Wright T, Hammond SM (2006). Extensive post-transcriptional regulation of microRNAs and its implications for cancer. Genes Dev 20:2202–2207

Tong H, Lin M (2002) [Arsenic trioxide induced p15INK4B gene expression in myelodysplastic syndrome cell line MUTZ-1]. Zhonghua Xue Ye Xue Za Zhi 23:638–641

Toyota M, Suzuki H, Sasaki Y, Maruyama R, Imai K, Shinomura Y et al (2008). Epigenetic silencing of microRNA-34b/c and B-cell translocation gene 4 is associated with CpG island methylation in colorectal cancer. Cancer Res 68:4123–4132

Tuddenham L, Wheeler G, Ntounia-Fousara S, Waters J, Hajihosseini MK, Clark I et al (2006). The cartilage specific microRNA-140 targets histone deacetylase 4 in mouse cells. FEBS Lett 580:4214–4217

Van der Poel H (2007). Molecular markers in the diagnosis of prostate cancer. Crit Rev Oncol Hematol 61:104–139

Varambally S, Cao Q, Mani R-S, Shankar S, Wang X, Ateeq B et al (2008) Genomic loss of microRNA-101 leads to overexpression of histone methyltransferase EZH2 in cancer. Science 322:1695–1699

Vázquez I, Maicas M, Marcotegui N, Conchillo A, Guruceaga E, Roman-Gomez J et al (2010) Silencing of hsa-miR-124 by EVI1 in cell lines and patients with acute myeloid leukemia. Proc Natl Acad Sci 107:E167–E168

Vilas-Zornoza A, Agirre X, Martín-Palanco V, Martín-Subero JI, San José-Eneriz E, Garate L et al (2011) Frequent and simultaneous epigenetic inactivation of TP53 pathway genes in acute lymphoblastic leukemia. PloS One 6:e17012

Vrba L, Jensen TJ, Garbe JC, Heimark RL, Cress AE, Dickinson S et al (2010) Role for DNA methylation in the regulation of miR-200c and miR-141 expression in normal and cancer cells. PloS One 5:e8697

Wang Y, Toh HC, Chow P, Chung AY, Meyers DJ, Cole PA et al (2012) MicroRNA-224 is up-regulated in hepatocellular carcinoma through epigenetic mechanisms. FASEB J 26:3032–3041

Wiklund ED, Bramsen JB, Hulf T, Dyrskjot L, Ramanathan R, Hansen TB et al (2011) Coordinated epigenetic repression of the miR-200 family and miR-205 in invasive bladder cancer. Int J Cancer 128:1327–1334

Wong KY, So CC, Loong F, Chung LP, Lam WWL, Liang R et al (2011) Epigenetic inactivation of the miR-124-1 in haematological malignancies. PloS One 6:e19027

Wotschofsky Z, Liep J, Meyer HA, Jung M, Wagner I, Disch AC et al (2012) Identification of metastamirs as metastasis-associated microRNAs in clear cell renal cell carcinomas. Int J Biol Sci 8:1363–1374

Xu Q, Jiang Y, Yin Y, Li Q, He J, Jing Y et al (2013). A regulatory circuit of miR-148a/152 and DNMT1 in modulating cell transformation and tumor angiogenesis through IGF-IR and IRS1. J Mol Cell Biol 5:3–13

Yan H, Choi A-j, Lee BH, Ting AH (2011) Identification and functional analysis of epigenetically silenced microRNAs in colorectal cancer cells. PloS One 6:e20628

Yanaihara N, Caplen N, Bowman E, Seike M, Kumamoto K, Yi M et al (2006) Unique microRNA molecular profiles in lung cancer diagnosis and prognosis. Cancer cell 9:189–198

Yang X, Feng M, Jiang X, Wu Z, Li Z, Aau M et al (2009). miR-449a and miR-449b are direct transcriptional targets of E2F1 and negatively regulate pRb-E2F1 activity through a feedback loop by targeting CDK6 and CDC25A. Genes Dev 23:2388–2393

Ying S-Y, Lin S-L (2005) Intronic microRNAs. Biochem Biophys Res Commun 326:515–520

Yu F, Yao H, Zhu P, Zhang X, Pan Q, Gong C et al (2007) let-7 regulates self renewal and tumorigenicity of breast cancer cells. Cell 131:1109–1123

Yuan Jh, Yang F, Chen B, Lu Z, Huo X, Zhou Wp et al (2011) The histone deacetylase 4/SP1/microrna-200a regulatory network contributes to aberrant histone acetylation in hepatocellular carcinoma. Hepatology 54:2025–2035

Zaman MS, Chen Y, Deng G, Shahryari V, Suh SO, Saini S et al (2010) The functional significance of microRNA-145 in prostate cancer. Br J Cancer 103:256–264

Zhang K, Dent SY (2005). Histone modifying enzymes and cancer: going beyond histones. J Cell Biochem 96:1137–1148

Zhang L, Huang J, Yang N, Greshock J, Megraw MS, Giannakakis A et al (2006) microRNAs exhibit high frequency genomic alterations in human cancer. Proc Natl Acad Sci 103:9136–9141

Zhang X, Zhao X, Fiskus W, Lin J, Lwin T, Rao R et al (2012) Coordinated silencing of MYC-mediated miR-29 by HDAC3 and EZH2 as a therapeutic target of histone modification in aggressive B-Cell lymphomas. Cancer cell 22:506–523

Zhou X, Sun H, Ellen TP, Chen H, Costa M (2008) Arsenite alters global histone H3 methylation. Carcinogenesis 29:1831–1836

Zhu S, Wu H, Wu F, Nie D, Sheng S, Mo Y-Y (2008). MicroRNA-21 targets tumor suppressor genes in invasion and metastasis. Cell Res 18:350–359

Chapter 5
Reduced Risk of Cancer in Schizophrenia, a Bridge Toward Etio-Pathology and Therapy of Both Diseases

Mohamad Reza Eskandari, Hamid Mostafavi Abdolmaleky, Jin-Rong Zhou and Sam Thiagalingam

Contents

M. R. Eskandari (✉) · H. M. Abdolmaleky · J.-R. Zhou
Department of Surgery, Laboratory of Nutrition and Metabolism at BIDMC,
Harvard Medical School, Boston, MA, USA
e-mail: eskandari@me.com

J.-R. Zhou
e-mail: jrzhou@bidmc.harvard.edu

M. R. Eskandari
Department of Psychiatry, Zanjan University of Medical Sciences, Zanjan, Iran

H. M. Abdolmaleky · S. Thiagalingam
Department of Medicine (Biomedical Genetics Section),
Boston University School of Medicine, Boston, MA, USA

S. Thiagalingam
e-mail: samthia@bu.edu

H. M. Abdolmaleky · S. Thiagalingam
Department of Genetics & Genomics, Boston University School of Medicine, Boston, MA, USA

P. Mehdipour (ed.), *Epigenetics Territory and Cancer*,
DOI 10.1007/978-94-017-9639-2_5

Abstract Schizophrenia (SCZ), a major mental disease, is associated with reduced risk for different types of cancers, with the exception of breast, endometrial and pancreatic cancers. The lower risk of cancer in first-degree relatives of SCZ patients rules out the drug effects as the cause of reduced cancer risk in SCZ. While genetic polymorphisms were attributed to the reduced cancer risk in SCZ, since hundreds of genes have been linked to SCZ and the effect size of each gene is small (<1%), the reduced cancer risk cannot be due to specific genetic polymorphisms. The same is true in cancer, as specific genetic changes are correlated to only 5–10% of all cancers.

In post-mortem brain studies we found abnormal DNA methylation of the regulatory regions of several genes such as RELN, MB-COMT, MAOA, 5-HTT and HTR2A associated with corresponding altered expression in SCZ patients. As a subset of these epigenetic alterations (e.g. MB-COMT, HTR2A, and 5-HTT) are retained in DNA derived from the saliva of drug naïve SCZ patients, these alterations are not limited to brain and are independent of drug use. In addition to RELN, MB-COMT and MAOA, dysfunctionality of TGFβ, immune system, VGEF, FOXO and specific miRNAs are linked to both SCZ and cancer. Interestingly, the changes in SCZ are often opposite in nature to those observed in cancer suggesting that they may predispose to SCZ, but reduce cancer risk. Therefore, knowledge from one disease may help to understand the molecular basis of the other disease and assist the implementation of therapeutic strategies.

Keywords Schizophrenia · Psychosis · Huntington's Disease · Huntingtin Associated Protein · Huntingtin Interacting Protein · Cancer · Maternal Immune Activation · Major Histocompatibility Complex · RELN · DRD1 · DRD2 · COMT · MB-COMT · MAOA · VMAT2 · 5-HTT · HTR2A · GABA · GAD1 · MIA · MHC · VEGF · VEGFR2 · FOXO · FOXO3 · FOXO3A · TGF-β · TGF-β1 · TGF-β2 · DNA Methylation · GABA-A receptor · Adiponectin · TNF-α · DNMT1 · VEGF · mir-126

Abbreviations

5-HTT	5-hydroxytryptamine (serotonin) transporter
BD	Bipolar Disorder
BMP	Bone Morphogenetic Protein
COMT	Catechol-O- Methyltransferase
DNMT1	DNA Methyltransferase 1
DRD1	Dopamine Receptor type 1
DRD2	Dopamine Receptor type 2
FOXO	O subclass of the Forkhead Family of Transcription Factors
GABA	Gamma Amino Butyric Acid
GAD1	Glutamate decarboxylase 1
HAP1	Huntingtin Associated Protein 1
HD	Huntington's Disease
HIP	Huntingtin Interacting Protein
HTR2A	Hydroxytriptamine (serotonin) type 2 receptor
MAOA	Mono Amine Oxidase A

MB-COMT	Membrane-Bound Catechol-O-Methyltransferase
MDD	Major Depressive Disorder
MHC	Major Histocompatibility Complex
MIA	Maternal Immune Activation
PD	Parkinson's Disease
RELN	Reelin
SCZ	Schizophrenia
SSRIs	Serotonin Specific Reuptake Inhibitors
TGF-β	Transforming growth factor beta
VEGF	Vascular Endothelial Growth Factor
VEGFR2	VEGF receptor 2

5.1 Introduction

Epidemiological studies have consistently demonstrated a reduced risk of cancer in patients with SCZ compared to the general population (Mortensen 1994; Dalton et al. 2005; Barak et al. 2005; Grinshpoon et al. 2005; Catts and Catts 2000; Cohen et al. 2002) particularly, in some specific organs such as lung (Dalton et al. 2005; Grinshpoon et al. 2005; Park 2004) and prostate (Mortensen 1989; Dalton et al. 2005; Torrey 2006), despite a higher frequency of smoking in SCZ patients (Vanable et al. 2003, de Leon and Diaz 2005). On the other hand, female SCZ patients may have an increased risk for some types of cancer, such as the breast (Mortensen 1989; Dalton et al. 2005; Grinshpoon et al. 2005) and pancreatic cancers (Mortensen 1989). However, the association between SCZ and breast cancer risk has been inconsistent. One study reported a reduced risk of breast cancer in female SCZ patients (Barak et al. 2005), but others reported an increased risk (Mortensen 1989; Dalton et al. 2005; Grinshpoon et al. 2005). Speculation as to the cause of this discrepancy in cancer rates has focused on genetic, pharmacological and environmental factors (Mortensen 1989). While genetic polymorphisms involved in SCZ pathogenesis were attributed to the reduced cancer risk in these patients, it is well established that functional genetic polymorphisms alone are insufficient by themselves to provide a molecular basis for the pathogenesis of SCZ (Abdolmaleky el al. 2011) as well as cancer. Although hundreds of genes have been implicated in SCZ, the effect size of each gene is very small (<1%). Thus, the reduced risk of cancer in SCZ patients cannot be due to specific genetic polymorphisms in recessive, dominant or even with a polygenic pattern. The same scenario is applicable for cancer based on findings that familial genetic factors are responsible for 5–10% of all cancers, including breast cancer (Coughlin and Piper 1999; Anand et al. 2008).

Based on the latest human studies, male SCZ patients and their first-degree relatives exhibited reduced cancer risk, but this cancer-protection trend may disappear with the use of antipsychotic drugs (Ji et al. 2013; Chen et al. 2013). On the contrary, female SCZ patients exhibit an increased risk for breast, cervical and endometrial cancers (Ji et al. 2013; Chen et al. 2013). Asian SCZ patients, regardless of their gender, reported to have more cancer risk than Caucasian patients (Catts et al. 2008; Chen et al. 2013). The increased cancer risk during the early deteriorating phase in male or female SCZ patients could be due to drug induced functional or epigenetic modifications in the immune and hormonal systems as well as change in nutritional habits (Pert et al. 1988; Khandaker et al. 2014; McAllister et al. 2014). Thus, the higher incidence of specific types of cancer in SCZ patients could be considered as "stage-specific" rather than a "lifetime risk" (Chen et al. 2013). In this chapter we discuss the findings that support the specific combinations of epigenetic and genetic aberrations involved in SCZ pathogenesis and influenced by the environment may facilitate apoptotic activity reducing the risk of cancer in SCZ patients and the reversal of these alterations may induce cancer phenotype in SCZ-free individuals. Therefore, we suggest that studies comparing genetic and epigenetic aberrations in cancer *versus* patients with SCZ may help to find targets for the prevention and treatment of both of these complex diseases.

5.2 Epigenetic Dysregulation of SCZ Related Genes in Cancer

5.2.1 The Link Between Epigenetic Down Regulation of Reelin (RELN) in SCZ and Cancer Risk

Although, the underlying mechanisms of increased risk of pancreatic and breast cancers in SCZ have been largely unknown, there is evidence that the same epigenetic dysregulation of specific genes in SCZ and particular types of cancer may be involved in higher risk of these cancers in SCZ patients. For example, RELN promoter DNA hypermethylation associated with reduced gene expression in SCZ (Abdolmaleky et al. 2005; Grayson et al. 2005) was also observed in pancreatic (Sato et al. 2006) as well as in breast cancers (Table 5.1) (Abdolmaleky and Thiagalingam, unpublished). Our studies uncovered that RELN gene promoter DNA is methylated in MCF7 (Abdolmaleky et al. 2005, 2008) and other breast cancer cell lines as well as in breast cancers tissues, but it is mostly unmethylated in MCF10A and HMEC, the normal breast cell lines and normal breast tissues (Abdolmaleky and Thiagalingam unpublished). Additionally, it has been shown that, while TGF-β1 suppresses RELN expression (mediated by snail), RELN is also a negative regulator of TGF-β1 induced cell migration at least in esophageal carcinoma cells (Fig. 5.1). Knockdown of RELN also results in increased expression of Fibronectin, vimentin and N-cadherin (but not E-cadheren), the markers of epithelial to mesenchymal transition suggesting that RELN is a cell migration suppressor (Yuan et al. 2012). These observations indicate that DNA hypermethylation of RELN in SCZ, associated with reduced expression, might be an underlying mechanism for high risk of pancreatic and breast cancer, and may also have a role in esophageal carcinoma progression towards invasion and metastasis. Interestingly, while SCZ patients exhibit a reduced risk of prostate cancer compared with the general population (Mortensen 1989; Dalton et al. 2005; Torrey 2006), it has been shown that, RELN gene/protein expression is increased in prostate carcinoma, presumably due to DNA hypomethylation of RELN gene promoter (Perrone et al. 2007). Therefore,

Table 5.1 The shared factors implicated in Psychiatry, with main focus on SCZ, and Cancer. The table summarizes the connected activities of factors in SCZ, BD, MDD and AD with Cancer. Symbols: (↑) indicates increase and (↓) indicates decrease

Factor	Activity in Psyc/Cancer	Mechanisms/activities	References
TGF-β	↑ in SCZ, ↓ by antipsychotics in SCZ[a];BD; ↓ in Cancer	A disease state marker in SCZ. Lithium inhibits Smad3/4-dependent TGFβ signaling. Induction of apoptosis involving SMAD or DAXX pathways. Regulation of immune system by FOXP3. Up-regulates miR-182 expression to promote gallbladder cancer metastasis by targeting CADM1. Acts as a tumor suppressor in normal epithelial cells, silenced by epigenetic mechanisms in some types of carcinomas	Borovcanin et al. 2012, 2013; Miller et al. 2011; Papageorgis et al. 2010; Qiu et al. 2014; Meulmeester and Ten Dijke 2011; Liang et al. 2008
TGFβ1	↑ in SCZ, ↓ in AD ↓ in Cancer (↓in early, ↑in late Cancer)	Cell growth, cell proliferation, cell differentiation and apoptosis. Inhibits interferon-γ, TNF-α. Suppresses RELN expression and RELN is a negative regulator of TGFβ1 induced cell migration in esophageal carcinoma cells. Promoter DNA hypermethylation in gastric cancer	Kim et al. 2004; Yuan et al. 2012; Kajdaniuk et al. 2013; Wang et al. 2013
TGFβ2	↑ in SCZ/BD ↓ in Cancer (↓in early, ↑in late Cancer)	Increased in SCZ and associated with decreased Wnt10A expression. Suppresses the effects of interleukin dependent T-cell tumors and disruption of the TGFβ/SMAD signaling is involved in diverse human cancers. Promoter DNA hypermethylation is linked to prostate cancer progression	Benes 2011; Humbert and Lebrun 2013; Liu et al. 2011b; Borovcanin et al. 2013
FOXP3	↑ in SCZ/ BD ↓↑ in Cancer	Potential tumor suppressor in gastric cancer. Down regulated in melanoma, breast, prostate, ovary and brain tumor cells but elevated level of expression reported in pancreatic adenocarcinoma, leukemia, hepatocellular carcinoma, bladder cancer, thyroid carcinoma and cervical cancer	Yang et al. 2013b; Hao et al. 2014; Drexhage et al. 2011; do Prado et al. 2013
FOXO	↑ in SCZ and BD ↓ in Cancer; Diabetes	Transcription factor; involved in cell metabolism, proliferation and apoptosis. Increased expression of FOXO3 in the brains of SCZ and BD patients. FOXO3A is linked to longevity. Loss of functions in several types of human cancers and diabetes. Some FOXO3A genotypes are linked to cancer, cardiovascular disease and deficit in cognitive functions. Metformin, Simvastatin, DRD2 blockers, Venlafaxine, SSRIs and ImipramineincreaseFOXO3a phosphorylation leading to its decreased nuclear localization	Monsalve and Olmos 2011; Eijkelenboom and Burgering 2013; Rodriguez et al. 2013; Jia et al. 2014; Pan et al. 2014; Polter et al. 2009; Weeks et al. 2010; Carbajo-Pescador et al. 2014; Di Bona et al. 2013; Takayama et al. 2014; Abdolmaleky and Thiagalingam, manuscript in preparation; You et al. 2006

Table 5.1 (continued)

Factor	Activity in Psyc/Cancer	Mechanisms/activities	References
TNF-α[b]	↑ in SCZ (in drug naïve and treated) and Depression; ↓ by Antidepressants; Cancer	Role in cancer treatment. Specific polymorphisms are associated with paranoid SCZ and others with cervical and gastric cancer. Mediates acute myeloid leukemia treatment response	Xia et al. 1996; Dowlati et al. 2010; Al-Asmari and Khan 2014; Song et al. 2014; Paul-Samojedny et al. 2013; Liu et al. 2012a; Zhu et al. 2014a; Dash et al. 2014
IL-1β[c]	Serum ↑ in SCZ (drug naïve and treated); ↓ by anti-depressants and antipsychotics; Cancer	Has pro-inflammatory property. IL-1β expression is induced by NF-κB after activation of immune cells. Through Wnt signaling, stimulates metastatic behavior and growth of colon and gastric cancer cells	Xia et al. 1996; Tourjman et al. 2013; Al-Asmari and Khan 2014; Song et al. 2014; Miller et al. 2011; Jedinak et al. 2010; Kaler et al. 2009; Li et al. 2014
INF-γ	CSF and serum ↑ in SCZ (drug naïve and treated); Reported ↓ in SCZ; ↓ by Anti-depressants; Cancer (anti-tumor effects)	Inhibits the growth of nasopharyngeal carcinoma and induces the cytotoxicity of daunorubicin against leukemic cells	Kim et al. 2004; Miller et al. 2011; Al-Asmari and Khan 2014; Dimitrov et al. 2013; Xia et al. 1996; Liu et al. 2012b; Zhang et al. 2014; Mohamed 2014
IL-2	↑ in SCZ (first episode psychosis and schizophrenia in relapse); Serum ↓ in stable SCZ; ↑ in plasma by antipsychotics; Cancer	Genotype TT and allele T is associated with paranoid SCZ. GG genotype and the G allele of the same polymorphism (rs2069762) are associated with higher risk of childhood lymphoma and gastric cardia cancer. IL2 is used in cancer immunotherapy	McAllister et al. 1995; Potvin et al. 2008; Tourjman et al. 2013; Paul-Samojedny et al. 2013; Song et al. 2012; Wu et al. 2009; Guma et al. 2014, Zhang et al. 2002
IL-6 Th2 prototype cytokine	Serum ↑ in SCZ (in drug naïve and treated) and Depression; ↓ by Anti-depressants in MDD and antipsychotics in SCZ; Cancer	Blocks apoptosis in cancer cells during inflammatory process. Induces tumor growth in colorectal cancer. Increases proliferation, migration, invasion and/or survival and chemoresistance in ovarian cancer cells. Th2 predominance hypotheses by meta-analysis in schizophrenia. Associated with paranoid SCZ	Avgustin et al. 2005; Potvin et al. 2008; Dowlati et al. 2010; Al-Asmari and Khan 2014; Song et al. 2014; Paul-Samojedny et al. 2013; Xia et al. 1996; Miller and Cole 2012; Zhang et al. 2002; Song et al. 2013a, b; Becker et al. 2005; Kumar and Ward 2014

Table 5.1 (continued)

Factor	Activity in Psyc/Cancer	Mechanisms/activities	References
IL-8	Serum ↑ in chronic SCZ; ↑ in Cancer	Increased in kidney and breast cancers. Elevated level of IL-8 during pregnancy increases likelihood of SCZ in offspring. High levels of IL-2 and IL-8 are indicators of poor treatment response in SCZ	Zhang et al. 2002; Miller et al. 2011 Li et al. 2013; Sheridan et al. 2006; Benoy et al. 2004; Liang-Kuan 2014; Brown et al. 2004; Zhang et al. 2004
IL-17	Serum ↓ in stable SCZ; Serum ↑ in first episode psychosis and schizophrenia in relapse; ↑ in Cancer	High levels of IL-17+FOXP3+CD4+T cells is associated with colon carcinoma. Increased level of IL-17+regulatory T cells is reported in blood and in ovarian and breast cancers, melanoma, and renal cell carcinoma. Prognostic factor in colorectal carcinoma	Dimitrov et al. 2013; Borovcanin et al. 2012; Dong 2006; Kryczek et al. 2011; Liu et al. 2011a
RELN	↓ in SCZ; ↑ in prostate cancer; ↓ in pancreatic and breast cancers	Neuronal migration, positioning and synaptic plasticity. RELN is a cell migration suppressor. RELN gene/protein expression is increased in prostate cancer. RELN promoter DNA hypermethylation in SCZ, pancreatic and breast cancers	Abdolmaleky et al. 2005, 2008; Grayson et al. 2005; Sato et al. 2006; Perrone et al. 2007
CXCR4	↑ DNA methylation in SCZ; ↑ expression in Cancer (23 types)	Chemokine receptor. Decreased expression and promoter DNA hypermethylation in SCZ. Marker of metastasis and cancer invasion. Promoter DNA hypermethylation in pancreatic cancer	Xu et al. 2013; Bousman et al. 2010; Toritsuka et al. 2013; Balkwill 2004; Aberg et al. 2014; Domanska et al. 2013
miRNA-126	↑ in SCZ (DLPFC#); ↓ in Diabetes; ↓ in Cancer	Angiogenesis control. Acts as a tumor suppressor and down regulated in various cancers: breast, gastric, prostate, colorectal, clear-cell renal and osteosarcoma. Down regulates CXCR4 and VEGF. Up-regulated in SCZ	Liu et al. 2014; Yang et al. 2013a; Zhou et al. 2013; Hansen et al. 2013; Meister and Schmidt 2010; Vergho et al. 2014; Watahiki et al. 2013
VEGF	↓ in SCZ (DLPFC); ↓ in AD; ↑ in Cancer	Involved in cancer progression and metastasis, Neuronal survival, neuroprotection, regeneration, growth and differentiation. Inhibited by miR-126	Fulzele and Pillai 2009; Rosenstein et al. 2010; Ye et al. 2013; Chen et al. 2014
GAD1	↓ in SCZ (PFC); ↑ in benign and malignant prostatic tissue; ↑ in oral squamous cell carcinoma, liver cancer	Key enzyme for GABA biosynthesis. Decreased expression and increased DNA methylation in SCZ. Overexpressed in oral squamous cell carcinoma and is associated with a higher degree of invasion/migration of cancer cells	Akbarian and Huang 2006; Huang and Akbarian 2007; Bharadwaj et al. 2013; Jaraj et al. 2011; Kimura et al. 2013

Table 5.1 (continued)

Factor	Activity in Psyc/Cancer	Mechanisms/activities	References
COMT/ MB-COMT	↑ in SCZ and BD patients ↓ in endometrial cancer	Monoamines, catecholamines and estrogen degradation. Hypomethylation of MB-COMT promoter DNA and increased gene expression in SCZ. Hypermethylation of promoter DNA and reduced expression in endometrial cancer. Increased expression in pancreatic cancer	Abdolmaleky et al. 2006; Nohesara et al. 2011; Sasaki et al. 2003; Lin et al. 2013b; He et al. 2012; Tian et al. 2014
MAOA	↑ DNA methylation in SCZ and cholangiocarcinoma; ↑ expression in advanced prostate cancer	Monoamines degradation. DNA Hypermethylation in male SCZ patients. MAOA inhibitors restrain prostate cancer	Flamand et al. 2010; Huang et al. 2012; Chen et al. 2012a

[a] Antipsychotics produce anti-inflammatory effects in SCZ (Tourjman et al. 2013)
[b] Pro-inflammatory cytokines in the innate immune system are IL-1β, TNF-α, IFN-α, IL-2 and IFN-γ (produced together with TNF-β by TH-1 cells). Anti-inflammatory cytokines (counterpart for TH-1 cells) are IL-4, IL-5, IL-6, IL-10 and IL-13 (Arolt and Ambrée 2013)
[c] IL-1 β, IL-6, and TGF- β reported as state markers for acute exacerbations in SCZ and IFN-γ and TNF-α reported as trait markers. (Miller et al. 2011)
[d] Dorso lateral Prefrontal Cortex

the reduced risk of prostate cancer in SCZ could be due to RELN hypermethylation in these patients. From a therapeutic point of view, considering the fact that RELN promoter DNA methylation is decreased by serotonin specific reuptake inhibitors (SSRIs) (Abdolmaleky et al. 2008) which are widely used as antidepressants, the potential use of SSRIs in breast and pancreatic cancers is of interest for further studies in the future. Of note, anti-psychotic drugs that attenuate the reduced expression of 5-HTT (due to its promoter DNA hypermethylation) observed in drug naive SCZ patients (Abdolmaleky et al. 2014) may increase these cancers via this mechanism.

In addition to RELN, the promoter region of CXCR4, a marker for cancer cell invasion is also hypermethylated (associated with reduced gene expression) in SCZ (Aberg et al. 2014) as well as in pancreatic cancer (Sato et al. 2005). CXCR4 is overexpressed in most cancers, including prostate and lung cancers (Sun et al. 2003; Singh et al. 2004; Lu et al. 2013; Spano et al. 2004), which is consistent to a reduced risk of these types of cancers in SCZ patients (Table 5.1). In prostate cancer the expression of CXCR4 (as well as BCL2) is decreased by ampelopsin, an anti-cancer flavonoid, associated with reduced cell proliferation of prostate cancer cell lines, but to a much lesser extent in the normal prostate cell line (PrEC). An inhibition of growth of PC-3 tumors and invasion to lymph node and metastasis was also observed with ampelopsin in animal studies using an orthotopic prostate cancer model in mice associated with increased apoptosis, inhibition of proliferation, reduced angiogenesis and reduced CXCR4 expression (Ni et al. 2012).

These observations along with the fact that the incidence of cancer is remarkably low in SCZ patients including lung cancer (Mortensen 1994; Dalton et al. 2005; Barak et al. 2005), despite a higher frequency of smoking (Vanable et al. 2003; de Leon and Diaz 2005) indicate that; epigenetic aberrations as key mediators for the pathogenesis of complex diseases, including cancer and SCZ may be responsible for the general reduced cancer risk and increased risk of certain types of cancer in SCZ patients.

5.2.2 Dysregulation of GABAergic Genes in Cancer and SCZ

Dysregulation of GABAergic system well known to be involved in SCZ pathogenesis is also linked to carcinogenesis. For instance, while epigenetic down-regulation of GAD1 is shown in SCZ patients (Bharadwaj et al. 2013), over-expression of GAD1 is observed in oral squamous cell carcinoma and it has been associated with a higher degree of invasion and migration of cancer cells (Kimura et al. 2013). In breast cancer, the metastatic cells entering into the brain exhibit GABAergic phenotype (higher levels of GABA-A receptors and RELN) in order to co-opt to the brain microenvironment (Neman et al. 2014). Notably, the expression of GABA-A receptor subunits is also increased in human liver cancer, and GABA could inhibit cell migration and invasion mediated by GABA-A receptors (Chen et al. 2012c). Considering that GABAB ligands also directly interact with the CXCR4 receptors (Guyon et al. 2013), the drugs known to modulate GABAergic system may be of use for treating specific cancers with higher levels of CXCR4 expression (Fig. 5.1).

The higher expression of GAD1 in benign and malignant prostatic tissue led to the conclusion that; GAD1 may be a prostate-specific tissue biomarker. However, GAD1 expression decreases as the Gleason score increases in prostate cancer (Jaraj et al. 2011) suggesting that in addition to RELN, the reduced risk of prostate cancer in SCZ patients may also be due to epigenetic down-regulation of GAD1 in these patients. These observations lend additional support for therapeutic utility of GABAergic drugs as well as epigenetic modifiers in specific types of cancers.

5.2.3 Involvement of Dopaminergic Genes in Cancer and SCZ

Catechol-O-methyltransferase (COMT) plays a major role in the metabolism of dopamine as well as the carcinogenic catechol estrogen. A recent meta-analysis concluded that the Val/Val polymorphism of COMT, the overactive genotype, is linked to a higher risk for endometrial cancer during the post-menopausal period in women (Lin et al. 2013b). However such an association was not found in an earlier meta-analysis (Qin et al. 2012), another meta-analysis confirmed that the Val158Met polymorphism of COMT is involved in breast cancer risk in Caucasians (He et al. 2012). An association with the Met/Met (AA) genotype for breast cancer risk has been confirmed by a different meta-analysis in Chinese population (Tian

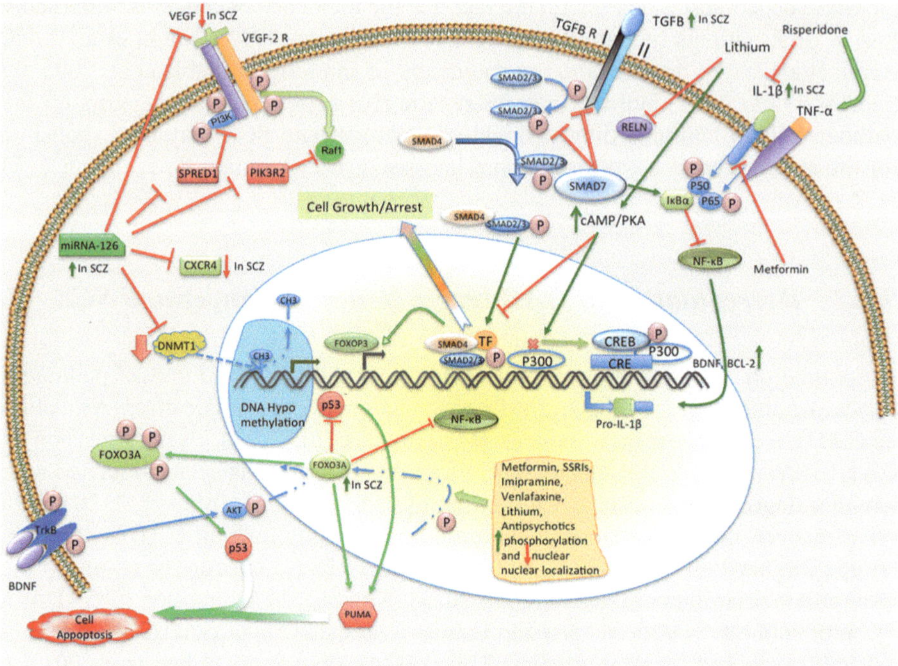

Fig. 5.1 TGF-β signaling prevents cancer development, but promote carcinogenesis and angiogenesis in advanced carcinomas. TGF-β regulates immune system by the mediation of FOXP3, which is over-expressed in tumor cells in pancreatic adenocarcinoma, melanoma, leukemia, hepatocellular carcinoma, bladder cancer, thyroid carcinoma and cervical cancer. TGF-β induces the expression of FOXP3, which is inversely regulated by smad7. Smad7 induces IκBα expression leading to NF-κB inhibition. Smad4 binds phosphorylated Smad2/Smad3 to create the Smad complex, which translocate into the nucleus to regulate target genes. Activation of PI3K and Akt leads to the phosphorylation of FOXO3A and departure of FOXO3A from the nucleus to the cytosol, and hence prevent its DNA binding and transcriptional activity. Smad7 inhibits the phosphorylated Smad2/3 complex formation with Smad4. Lithium inhibits Smad3/4-dependent transcription activity of TGF-β signaling through sequestration of transcriptional co-activator p300, which leads to increased CRE-mediated activation and expression of cell growth factors like BDNF and Bcl-2. Growth factors facilitate FOXO3A nuclear translocation. Lithium inhibits FOXO3A transcription activity. Phosphorylation of FOXO3A is increased by BDNF. Promoter region of CXCR4 is hypermethylated (associated with reduced gene expression) in SCZ as well as in pancreatic cancer and low levels of CXCR4 has been reported in >20 types of cancers. miRNA-126 has inhibitory effects on DNMT1 and negatively regulates VEGF signaling pathway through inhibition of SPRED1 and PIK3R2. Metformin, Venlafaxine, SSRIs, Imipramine, Lithium and Antipsychotic drugs increase FOXO3a phosphorylation leading to its decreased nuclear localization. An activated FOXO3a could regulate transcription or cytoplasmic accumulation of p53, modulating its apoptotic activity. *Green* lines and *arrows* indicate facilitating or activating roles and *red arrows* and symbols represent inhibitory roles

et al. 2014). The reduced activity of membrane-bound catechol-O-methyltransferase (MB-COMT) due to promoter DNA hypermethylation has also been observed in endometrial cancer cell lines as well as in endometrial cancer tissues (Table 5.1) (Sasaki et al. 2003).

We examined DNA methylation and the corresponding expression of MB-COMT in post-mortem brains as well as saliva samples from the SCZ and BD patients versus control subjects (Abdolmaleky et al. 2006; Nohesara et al. 2011). In contrast to cancer, the MB-COMT promoter DNA was significantly hypomethylated, and MB-COMT gene expression was significantly increased in SCZ and BD patients, compared to the control subjects (Abdolmaleky et al. 2006). Interestingly, while the rate of pancreatic cancer is higher in SCZ patients, an increased expression of COMT was also observed in pancreatic cancer cell lines as well as pancreatic ductal adenocarcinoma compared to the normal pancreatic tissues (Wu et al. 2012).

Notably, beside COMT, monoamine oxidase A (MAOA), another gene encoding an enzyme involved in the degradation of dopamine as well as serotonin was found to exhibit increased expression in advanced prostate cancer (Flamand et al. 2010). However, its expression is epigenetically silenced in cholangiocarcinoma due to promoter DNA hypermethylation (Huang et al. 2012). Hypermethylation of specific CpG sites of MAOA was also reported in male SCZ patients (Chen et al. 2012a). These findings, although diverse, support that epigenetic modifications of genes related to dopaminergic system might serve as targets for cancer therapy. This idea gains additional support from the observations that the inhibitory effects of Epigallocatechin 3-O-gallate (EGCG) on the growth of lung cancer cell lines is synergistically potentiated by COMT inhibitor drugs such as entacapone and tolcapone (Forester and Lambert 2014) and MAOA inhibitors exhibit antioncogenic effects in advanced prostate cancer (Zhao et al. 2009; Flamand et al. 2010).

5.3 Immune System Dysregulation in Cancer and SCZ

5.3.1 Maternal Infections and Dysregulation of Interleukins in Cancer and SCZ

While a defective immune system is a well-known causal factor in cancer development, over-activity of immune system is linked to higher rate of autoimmune diseases in SCZ as well as in relatives of patients with SCZ (Benros et al. 2011; Eaton et al. 2006; Chen et al. 2012b). Maternal infection as a factor in immune system activation also increases the risk of SCZ in the offspring. Hospital contacts resulting in viral or bacterial infections in childhood or adolescence can also increase the risk of SCZ (Nielsen et al. 2013). Maternal infections, as an environmental risk factor has been indicated as a contributor for the pathogenesis of at least one third of SCZ patients (Brown and Derkits 2010). It has been shown that maternal immune activation could induce a number of abnormalities in the offspring, such as alterations in the pattern of gene expression, neurochemistry and cortical connectivity as well as other neuropathologies similar to those observed in SCZ, including decreased cortical thickness and enlarged ventricles which are characteristic of SCZ pathology (Garbett et al. 2012; McAllister 2014). There is also strong evidence for a relationship between maternal immune system activation and the development of autism in offspring, both due to viral as well as bacterial infections (Abdallah et al. 2012a, b). One of the well-studied potential mechanisms is an increased level of maternal cytokines, especially IL-6, which may cause an "expanded adult forebrain neural precursor pool" following maternal infection perturbing the "olfactory neurogenesis in the offspring months after fetal exposure". This observation suggests that an acute and transient hyper-activation of IL-6 may have long-term impacts in the- IL6-dependent self-renewal pathway of neural stem cells altering the characteristics of neural precursors throughout the life of the offspring (Gallagher et al. 2013).

Since the genes encoding cytokines are among the main mediators of crosstalk between the immune system and brain, aberrations in a specific region of chromosome 6 which harbors immune genes, and specific haplotypes of immune genes especially those within the major histocompatibility complex (MHC) are highly relevant to cancer (Urayama et al. 2013) as well as SCZ pathogenesis (Shi et al. 2009; Stefansson et al. 2009; Purcell et al. 2009; Li et al. 2010; Jia et al. 2012). While deletion of this chromosomal region and down regulation of the associated genes are linked to cancer development (Feenstra et al. 1999), the expression of at least two out of five of the MHC class I genes, including Butyrophilin 2A2 (BTN2A2) and antigen HLA-B, which may have roles in synaptic development are increased in SCZ patients (Sinkus 2013).

The link between SCZ and the chromosome 6 region, containing MHC genes has also been confirmed by genetic and genome-wide association studies (Shi et al. 2009; Stefansson et al. 2009; Purcell et al. 2009; Li et al. 2010, Jia et al. 2012; Debnath et al. 2013). Interestingly, the TT genotype and the allele T of interleukin-2 (IL-2) are associated with paranoid type of SCZ (Paul-Samojedny et al. 2013). On the contrary, the GG genotype and the G allele of the same polymorphism of IL-2 (rs2069762) are associated with higher risk of childhood lymphoma (Song et al. 2012) as well as gastric cardia cancer (Wu et al. 2009). There are also several reports on cancer immunotherapy with IL-2 in diverse types of cancers (Antony and Dudek 2010; Kolitz et al. 2014; Guma et al. 2014).

It has been proposed that MHC I negatively regulates the synaptic densities and bidirectionally controls glutamatergic and GABAergic synaptic densities in the developing brains and modulates the region and age-specific gene expression patterns, and these alterations apparently dysregulate the immune response in SCZ patients. Furthermore, changes in MHC I expression in neurons of SCZ patients may also change channel properties of glutaminergic genes such as NMDA receptors, and may alter activity-dependent synaptic strength by limiting the NMDA-mediated AMPAR trafficking (McAllister 2014). These observations along with the fact that NMDA affects the growth of malignant glioma cells *in vitro* (Panchanathan et al. 2013) provide support that drugs acting on GABAergic system and/or NMDA receptors could have potential value in cancer therapy.

In addition to IL2, the serum levels of IL-6 and IL-8 are also elevated in chronic SCZ patients (Table 5.1) (Zhang et al. 2002; Song et al. 2013a, b). Although all studies do not support a role for IL-1β in SCZ susceptibility (Shibuya et al. 2014), higher serum levels of IL-1β as well as TNF-α and adiponectin have been reported in drug naïve SCZ patients (Song et al. 2013a, b) that might play a role in the maintenance of cancer stem cells. Interestingly, risperidone (an antipsychotic drug widely used in SCZ patient) decreases the serum levels of IL-1β (Fig. 5.1), however it may reach to the baseline level after 6 months. Notably, while the serum level of IL-6 is decreased during the first 3 months of risperidone treatment, it may also reach the baseline level in 6 months. Nevertheless, the serum level of TNF-α increases by risperidone treatment during this time period. Hence, risperidone treatment seems to be associated with an initial anti-inflammatory effect that is neutralized with long-term treatment (Song et al. 2014). A recent meta-analysis also concluded that antipsychotic drugs exhibit anti-inflammatory effects in SCZ through increasing the soluble IL-2 receptor and decreasing the IL-1β and interferon-γ plasma levels (Tourjman et al. 2013).

Beside expression changes of these cytokines in SCZ, there are reports indicating aberrant DNA methylation of genes coding for some of these cytokines in different kinds of cancer (Table 5.1). For instance, hypomethylation of IL-8 in human astrocytoma and clear cell renal cell carcinoma, aberrant DNA methylation of IL1β, IL6, and IL8 in non-small cell lung cancer, hypermethylation of TGF-β1 promoter DNA in gastric cancer have been reported (Venza et al. 2012; Tekpli et al. 2013; Wang et al. 2013; Yoo et al. 2013). Therefore, further investigations for the identification of these and other possible epigenetic mechanisms that could complete the missing part of this important line of evidence in SCZ and cancer is warranted.

5.3.2 *TGF-β Signaling in Cancer and SCZ*

The TGF-β super family of cytokines are involved in the regulation of cellular processes, including cell division, differentiation, motility, adhesion and death. TGF-βs and BMPs signal through binding to the TGF-β membrane receptors leading to transphosphorylation of R-Smads such as Smad1, Smad2, Smad3 and Smad5/8, which along with Smad4 (the Co-Smad) translocate to the nucleus and form transcriptional complexes with DNA binding factors and co-activators/co-repressors modulating the expression of many genes (Massague 2000; Papageorgis et al. 2015). In animal studies, high TGF-β1 and TGF-β3 expression were observed in cerebral cortex, hippocampus, central amygdaloid nucleus, substantia nigra and the brainstem reticular formation. In contrast, *TGF-β2* is reported to be highly expressed in deep cortical layers, dentate gyrus, cerebellum and areas of monoaminergic neurons (Vincze et al. 2010) known to be affected in SCZ and BD.

Hypo-activity of TGF-β signaling is known to be involved in the early stage of cancer development. However, high level of TGF-β in an already developed cancer may promote metastasis. For instance, in advanced breast cancer an activated TGF-β-Smad signaling silences the expression of several genes such as CDH1, CGN, CLDN4 and KLK10 by altering the binding capacity of DNMT1 to the CpGs in regulatory regions of these genes and play a role in epithelial to mesenchymal transition (Papageorgis et al. 2010; Papageorgis et al. 2015). Therefore, in the advanced breast cancer, the disruption of TGF-β signaling could decrease DNMT1 binding activity minimizing the malignant phenotype. On the other hand over-expression of SMAD7 which inhibits R-SMADs could reverse the malignant mesenchymal phenotype to epithelial-like phenotype (Papageorgis et al. 2010).

A recent study reported an increase in the production of TGF-β in first episode psychosis, psychotic patients and in relapsed SCZ patients suggesting that TGF-β could be a valuable marker for psychosis (Borovcanin et al. 2012). More importantly, a meta-analysis also concluded that TGF-β is a state marker in SCZ patients (Miller et al. 2011). These observations indicate that, while a reduced activity of TGF-β signaling is involved in early stage cancer development, an increased TGF-β expression in SCZ could help to reduce cancer risk. However, it may increase the risk of invasion and/or metastasis after the development of cancer in these patients. Although, the symptoms of SCZ are apparently linked to an accelerated gear for apoptosis of neuronal cells (Catts and Catts 2000; Jarskog et al. 2004; Glantz et al. 2006; Jia et al. 2010) and cortical atrophy (Francis et al. 2012), an increase in TGF-β signaling associated with a decrease in Wnt signaling may have a different effect and promote adult neuronal differentiation and migration, and inappropriate insertion into the neuronal network in SCZ patients (Kalkman et al. 2009). In support of this finding our recent gene expression profiling of post mortem brain samples uncovered an increased TGF-β2 expression associated with a decreased Wnt10A expression in the frontal lobe of SCZ patients (Abdolmaleky and Thiagalingam, manuscript in preparation).

Collectively, these findings support the hypothesis that an increased activity of TGF-β signaling which has a significant role in neuronal cell fate and apoptosis as well as SCZ pathogenesis might be the underlying mechanism for the reduced risk of cancer in these patients. However, antipsychotic drugs through modulating this pathway, combined with epigenetic silencing of RELN signaling may increase the risk of certain types of cancers such as the breast and pancreatic cancers. While these findings suggest that genes involved in SCZ may promote apoptosis reducing the risk of cancer, studies to compare genetic and epigenetic aberrations in cancer versus patients with SCZ may help to find targets for the prevention and treatment of both of these complex diseases. In this line, the use of metformin which is known to inhibit TGF-β signaling (Cufí et al. 2010) might be helpful in advanced cancer as well as SCZ treatment. Indeed, several recent *in vitro* and *in vivo* as well as clinical studies have shown anti-cancer activity of metformin in different types of cancers such as esophageal (Xu and Lu 2013), ovarian (Dilokthornsakul et al. 2013) breast (Zhu et al. 2014b; Hadad. et al. 2014) hepatic (Miyoshi et al. 2014; Lin et al. 2013a), bladder (Zhang et al. 2013), endometrial (Ko et al. 2014; Nevadunsky et al. 2014) and other cancers, in general (Yin et al. 2013; Beck and Scheen 2013). Lithium, a drug widely used for the treatment of BD and refractory SCZ, also inhibits Smad3/4-dependent TGF-β signaling in neurons (Fig. 5.1) through increasing the activity of cAMP/PKA signaling (Liang et al. 2008). It is likely that antipsychotic drugs that block DRD2 receptor and increase cAMP level may also have the same effect.

5.4 Adiponectin and Body Weight in Cancer and SCZ

Adiponectin is a protein hormone exclusively secreted from the adipose tissues and in contrast to SCZ, is found at reduced level in plasma of patients with several types of cancers as well as obesity linked to insulin resistance and type 2 diabetes (Kishida et al. 2014; Hebbard and Ranscht 2014). The use of adiponectin receptor agonists such as AdipoRon, which is an orally active small molecule and act on both AdipoR1 and AdipoR2 receptors, to activate AMPK and PPAR-α pathways, respectively, has been proposed for the treatment of obesity-related disorders such as type 2 diabetes and cancer (Okada-Iwabu et al. 2013). In fact, in a study that used human and mouse colon cancer cell lines, both adiponectin and metformin additively reduced the malignant potential of colon cancer. The major mechanism proposed for this effect is that; adiponectin and metformin inhibit the IL-1β signaling and decrease malignant potential through their effects on the expression of p53 (a tumor suppressor), p21, p27, and cyclin E2 (genes regulating cell cycle) involving AMPK/LKB1 pathways (Moon and Mantzoros 2013).

As mentioned above, in the first five years of SCZ diagnosis and before the use of antipsychotics, the rate of cancer is lower in SCZ compared to the general population. However, because of the poor nutritional state of SCZ patients and more importantly the use of atypical antipsychotic drugs with common side effects such as weight gain and metabolic imbalance, the innate lower risk of cancer is found to decline in this group of patients (Ji et al. 2013; Chen et al. 2013; Manzanares et al. 2014). It is important to note that, in SCZ patients, treatment with atypical antipsychotic drugs decreases the circulating adiponectin at levels comparable to patients with diabetes. Adiponectin is reported to exhibit anti-angiogenic and tumor growth-limiting properties during *in vitro* studies and its level is inversely correlated with several malignancies that occur later in life (Adachi et al. 2012; Tsai et al. 2011; Song et al. 2013a, b; Dalamaga et al. 2012). While the circulating levels of adiponectin with insulin-sensitizing, anti-inflammatory, proapoptotic, anti-proliferative properties and cancer protectiveness declines with the use of atypical antipsychotics in SCZ (and bipolar disorder patients), current data, considering the higher levels of TGF-β in SCZ patients indicates that; metformin may not only be useful in the treatment of metabolic syndrome and the increased cancer risk, but also psychotic symptoms of SCZ patients as an inhibitor of TGF-β signaling pathway. In fact, excess weight gain and/or obesity not only is an emerging worldwide heath problem in general population, it is also an important issue in psychiatric patients, especially in SCZ patients under atypical antipsychotic drug treatment which have well known weight gain as the side effect (Subramaniam et al. 2014). Excess body weight is also considered as a risk factor for postmenopausal breast cancer, endometrial and ovarian cancer, pancreatic cancer, renal cell cancer, esophageal adenocarcinoma, hematological malignancies, high-grade prostate cancer, colon, thyroid, and gallbladder cancers (Dalamaga et al. 2012). There is a tendency for a higher frequency of a number of these cancers in SCZ patients using antipsychotic drugs.

5.5 Vascular Endothelial Growth Factor (VEGF) and mir-126 in Cancer and SCZ

VEGF is a growth factor implicated in cancer progression and metastasis. A decreased level of VEGF mRNA has been reported in the dorsolateral prefrontal cortex of SCZ patients (Fulzele and Pillai 2009). Furthermore, a low level of serum VEGF in Alzheimer's disease (Mateo et al. 2007), and a significant increase in the VEGF serum level following the clinical improvement of drug resistant depressed patients treated by electroconvulsive therapy (ECT) support an important role for VEGF in neuropsychiatric disorders (Minelli et al. 2011). In an animal study, VEGF was found to mediate the anti-depressive effects of cAMP down-stream events in the adult hippocampus during the treatment by antidepressants (Lee et al. 2009). Chronic treatment with lithium could also attenuate the stress-induced decrease of VEGF expression in the hippocampus in stressed animals, thus the therapeutic efficacy of lithium as a mood stabilizer may be mediated by VEGF (Silva et al. 2007; Guo et al. 2009). VEGF is considered as a neurotrophic factor and has been implicated in neuronal survival, neuroprotection, regeneration, growth and differentiation (Rosenstein et al. 2010). As reported in a mouse model of diabetes, inhibition of the VEGF receptor 2 (VEGFR2) mediated signaling results in endothelial dysfunction and vascular problems, which could be potentially reversed with the use of antioxidants (Warren et al. 2014). It is noteworthy that VEGF is inhibited by miR-126 which its dysfunction is strongly associated with angiogenesis and is especially expressed in endothelial cells, and down-regulated under hypoxic condition as shown in both *in vitro* and *in vivo* studies (Ye et al. 2014). There is also supporting evidence that mir-126 is a tumor suppressor, and the reduced level of miR-126 is a significant predictor of poor survival in many cancers (Yang et al. 2013a; Yu et al. 2013; Sun et al. 2013). Mir-126 also suppresses DNMT1 (Zhao et al. 2011) as well as CXCR4 expression (Fig. 5.1), and its tumor suppressing potential is mediated by the AKT and ERK1/2 signaling pathways (Liu et al. 2014). Interestingly, along with the decreased level of VEGF, the expression level of mir-126 is up-regulated in the postmortem dorsolateral prefrontal cortex of the brain in SCZ patients (Beveridge and Cairns 2012). This signifies that multiple aspects of epigenetic alterations are inversely regulated in SCZ patients versus cancer and provide additional support for the idea that; the identification of disease pathogenesis in either illnesses could help the design of novel therapeutics for both of these diseases.

5.6 Cell Maintenance is Impaired in Both Cancer and Neuropsychiatric Diseases

FOXO genes, the O subclass of the forkhead family of transcription factors, mediate the effects of insulin and growth factors and are involved in cell metabolism, proliferation and apoptosis. The FOXO family members in humans are FOXO1,

FOXO3, FOXO4 and FOXO6. The shared nature of FOXO protein family members (with the exception of FOXO6, which is exclusively nuclear) is translocated out of the nucleus upon phosphorylation by Akt/PKB proteins of the PI3K signaling pathway (Brunet et al. 1999). The loss of FOXO functions, has been detected in several types of human cancers and diabetes, a known risk factor for cancer (Monsalve and Olmos 2011; Eijkelenboom and Burgering 2013). FOXO3 is widely distributed in the adult brain and exhibits an increased expression during the brain development (Barthel et al. 2005). Activation of PI3K and Akt lead to the phosphorylation of FOXO and translocation from the nucleus to the cytosol, to prevent its DNA binding and transcriptional activity (Fig. 5.1). However, dephosphorylated FOXO can return to the nucleus and induce expression of genes that are involved in cell cycle arrest, apoptosis and resistance to oxidative stress. Human FOXO3 (FOXO3A) has been associated with longevity and some of the FOXO3 genetic variations are also linked to cancer, cardiovascular disease and deficit in cognitive functions (Rodriguez et al. 2013; Jia et al. 2014; Pan et al. 2014; Carbajo-Pescador et al. 2014; Di Bona et al. 2013). Epigenetic modifications of FOXO3 such as methylation of the lysine 270 of histone protein can also inhibit DNA binding of FOXO3 and prevent the neuronal cell death induced by oxidative stress (Xie et al. 2012). The decreased expression of the FOXO3 is associated with poor prognosis of human breast cancer and primary gastric adenocarcinoma (Jiang et al. 2013; Yang et al. 2013b). Interestingly, in *C. elegans* all antipsychotic drugs activate AKT pathway and inhibit nuclear localization of DAF16, the homologous gene of the human FOXO3 (Weeks et al. 2010) suggesting this mechanism as one of the underlying causes of the increased breast cancer risk in medicated SCZ patients.

The phosphorylation of FOXO3 is also increased by brain derived neurotrophic factor (BDNF) that inhibits its transcriptional activity in differentiated human SH-SY5Y neuroblastoma cells. Treatment with lithium in mood disorders alongside the decrease in the levels of FOXO3, inhibits its transcriptional activity and alleviates the proposed BDNF deficiency in mood disorders based on an *in vivo* study (Mao et al. 2007). In other *in vivo* studies, an increased serotonergic activity was found to result in phosphorylation of FOXO1 and FOXO3 in several brain regions, and reduce the nuclear distribution of FOXO1 and FOXO3. Similarly, chronic treatment with imipramine, an antidepressant drug with serotonergic and noradrenergic properties, could also increase FOXO1 and FOXO3 phosphorylation in brain (Fig. 5.1). Additionally, metformin also increases FOXO3 phosphorylation leading to a decreased FOXO3 nuclear localization (Takayama et al. 2014). Of note, the selective deletion of FOXO1 from the brain reduces anxiety level, and the FOXO3a-deficient mice exhibit antidepressant-exposed behavior (Polter et al. 2009). In humans, while the loss of FOXO function, has been linked to cancer (Monsalve and Olmos 2011), we found a highly significant increase in the expression of FOXO3 during gene expression profiling of the post- mortem brains of SCZ and to lesser extent in BD patients (Abdolmaleky and Thiagalingam, manuscript in preparation). Therefore, similar to other genes/pathways mentioned above, FOXO also has opposite roles in cancer *versus* SCZ as well as BD proposing the indication/contraindication of specific psychiatric drugs in cancer therapy and *vice versa*.

5.7 Huntington's Disease, Cancer and Dopamine Related Drugs

Similar to SCZ, a decreased rate of cancer has been reported in patients with Huntington's disease (HD) in different countries (Sørensen and Fenger 1992; Ji et al. 2012). Interestingly, a large fraction of patients with HD exhibit SCZ-like symptoms before or after the appearance of HD symptoms. An increased CAG repeat expansion in the exon 1 of gene coding an expanded chain of glutamines in huntingtin (Htt) protein is responsible for HD as an inherited neurodegenerative disorder. Although the underlying mechanisms of the reduced cancer risk in patients with HD have not been fully explored, it has been attributed to an increased apoptotic capacity of the expanded polyglutamine repeat in an animal study using p53 deficient mice (Ryan and Scrable 2008).

The striatal neurons expressing dopamine receptors predominantly become degenerated in HD patients, and the degree of striatal neuronal loss is inversely linked to the age of death in HD patients. Thus the mutated Htt, which is predicted to exhibit prominent toxic activity in this brain region (Hadzi et al. 2012), may also be involved in the senescence of other cells. Interestingly, low doses of a selective dopamine type 1 (D1) receptor agonist such as 100 μM of dopamine which activate adenylate cyclase, accelerates the formation of mutant Htt nuclear aggregates and increases the rate of cell death in neuroblastoma cell lines (Robinson et al. 2008). Earlier studies also showed an accelerated formation of aggregates and cell death with 1 mM of dopamine in striatal primary cultures containing human HTT gene with expanded CAG repeats (Petersén et al. 2001). However, the dopamine D2 (dopamine type 2) receptor antagonists could prevent these effects of dopamine in primary cultures of striatal neurons transfected with GFP-tagged exon 1 of the mutant HTT. In this experiment, the D2 receptor agonist was also found to enhance the number of mutant Htt aggregates in the dendrites of neurons and increased cell death (Charvin et al. 2005). As dopamine system is involved in the striatal neuropathology, depletion of striatal dopamine by the 6-hydroxydopamine was shown to be neuroprotective in rodents of HD model via reduction of striatal glutamate (Stack et al. 2007). Considering these observations, tetrabenazine, as an inhibitor of vesicular monoamine transporter (VMAT2) was approved for use in the treatment of HD patients. Tetrabenazine acts by reducing dopaminergic input to the striatum and alleviates the behavioral deficits and neuronal death in the YAC128 mouse model of HD (Tang et al. 2007). In addition, treatment with haloperidol decanoate, a potent D2 antagonist, also protects neurons from expanded Htt-induced dysfunction in the rat striatum (Charvin et al. 2008).

These observations suggest that, manipulation of Htt function by the modulation of dopamine receptors activities may help to inhibit cancer progression. This hypothesis is supported by the fact that, other modulators of the normal Htt protein such as Huntingtin interacting protein (HIP) and huntingtin associated protein (HAP1) are also involved in cancerous cell fate. For instance, *in-vitro* studies on huntingtin-associated protein 1 (HAP1), the ligand of Htt that binds more tightly to Htt with an expanded glutamine repeat than to wild type Htt, showed reduced expression in human breast cancer tissues compared to the normal breast tissues. Interestingly, over-expression of HAP1 reduces cell growth in breast cancer cell lines (MDA-MB-231 and MCF-7) and suppresses the cell migration and invasion, and promotes apoptosis in these cell lines (Zhu et al. 2013). These lines of evidence suggest that in addition to Htt and the interlinked dopamine signaling pathway, other huntintin associated genes may also be involved in cancer development and progression, thus could be targeted for cancer therapy.

Conclusion Several lines of evidence provided here indicate that cancer and SCZ are inversely correlated in a disease stage specific manner and the molecular defects involved in SCZ pathogenesis might be protective against the development, especially the early stage of cancer. Certainly, follow up studies should reveal the key dysregulated genes up or downstream of affected genes/pathways such as RELN, dopamine, GABA and TGF-β2 and other genes involved in the pathogenesis of both SCZ and cancer to generate clues to deduce strategies for the prevention and to uncover novel molecular and epigenetic targets for therapeutic applications in SCZ as well as cancer.

References

Abdallah MW, Hougaard DM, Nørgaard-Pedersen B, Grove J, Bonefeld-Jørgensen EC, Mortensen EL (2012a) Infections during pregnancy and after birth, and the risk of autism spectrum disorders: a register-based study utilizing a Danish historic birth cohort. Turk Psikiyatri Derg 23(4, Winter):229–235

Abdallah MW, Larsen N, Grove J, Nørgaard-Pedersen B, Thorsen P, Mortensen EL et al (2012b) Amniotic fluid chemokines and autism spectrum disorders: an exploratory study utilizing a Danish Historic Birth Cohort. Brain Behav Immun 26(1):170–176

Abdolmaleky HM, Cheng K-H, Russo A, Smith CL, Faraone SV, Shafa R et al (2005) Hypermethylation of the Reelin (RELN) promoter in the brain of Schizophrenic patients: a preliminary report. Am J Med Genet B Neuropsychiatr Genet 134B:60–66

Abdolmaleky HM, Cheng KH, Faraone SV, Wilcox M, Glatt SJ, Gao F et al (2006) Hypomethylation of MB-COMT promoter is a major risk factor for schizophrenia and bipolar disorder. Hum Mol Genet 1;15(21):3132–3145

Abdolmaleky HM, Smith CL, Zhou JR, Thiagalingam S (2008) Epigenetic modulation of reelin function in schizophrenia and bipolar disorder. Fatemi SH (ed.), Reelin glycoprotein, Springer New York, pp. 365–384

Abdolmaleky HM, Yaqubi S, Papageorgis P, Lambert AW, Ozturk S, Sivaraman V et al (2011) Epigenetic dysregulation of HTR2A in the brain of patients with schizophrenia and bipolar disorder. Schizophr Res 129(2–3):183–190

Abdolmaleky HM, Nohesara S, Ghadirivasfi M, Lambert AW, Ahmadkhaniha H, Ozturk S et al (2014) DNA hypermethylation of serotonin transporter gene promoter in drug naïve patients with schizophrenia. Schizophr Res 152(2–3):373–380

Aberg KA, McClay JL, Nerella S, Clark S, Kumar G, Chen W et al (2014) Methylome-wide association study of schizophrenia: identifying blood biomarker signatures of environmental insults. JAMA Psychiatry 71(3):255–264

Adachi H, Yanai H, Hirowatari Y (2012) The underlying mechanisms for Olanzapine-induced hypertriglyceridemia. J Clin Med Res 4(3):206–208

Akbarian S, Huang HS (2006) Molecular and cellular mechanisms of altered GAD1/GAD67 expression in schizophrenia and related disorders. Brain Res Rev 52(2):293–304
Akbarian S, Kim JJ, Potkin SG, Hagman JO, Tafazzoli A, Bunney Jr et al (1995) Gene expression for glutamic acid decarboxylase is reduced without loss of neurons in prefrontal cortex of Schizophrenics. Arch Gen Psychiatry 52:258–266
Al-Asmari AK, Khan MW (2014) Inflammation and schizophrenia: alterations in cytokine levels and perturbation in antioxidative defense systems. Hum Exp Toxicol. 33(2):115–122
Anand P, Kunnumakkara AB, Sundaram C, Harikumar KB, Tharakan ST, Lai OS et al (2008) Cancer is a preventable disease that requires major lifestyle changes. Pharm Res. 25(9):2097–2116
Antony GK, Dudek AZ (2010) Interleukin 2 in cancer therapy. Curr Med Chem 17(29):3297–3302
Arolt V, Ambrée O (2013) The question of pro-inflammatory immune activity in schizophrenia and the potential importance of anti-inflammatory drugs. In: Halaris A, Leonard BE (eds) Inflammation in psychiatry, vol 28. Karger, Basel, pp 100–116 (Mod Trends Pharmacopsychiatry)
Atladóttir HO, Thorsen P, Østergaard L, Schendel DE, Lemcke S, Abdallah M et al (2010) Maternal infection requiring hospitalization during pregnancy and autism spectrum disorders. J Autism Dev Disord 40(12):1423–1430
Avgustin B, Wraber B, Tavcar R (2005) Increased Th1 and Th2 immune reactivity with relative Th2 dominance in patients with acute exacerbation of schizophrenia. Croat Med J 46(2):268–274
Balkwill F (2004) Cancer and the chemokine network. Nat Rev Cancer 4(7):540–550
Barak Y, Achiron A, Mandel M, Mirecki I, Aizenberg D (2005) Reduced cancer incidence among patients with schizophrenia. Cancer 104(12):2817–2821
Barthel A, Schmoll D, Unterman TG (2005) FoxO proteins in insulin action and metabolism. Trends Endocrinol Metab 16(4):183–189
Beck E, Scheen AJ (2013) Metformin, an antidiabetic molecule with anti-cancer properties. Rev Med Liege 68(9):444–449
Becker C, Fantini MC, Wirtz S, Nikolaev A, Lehr HA, Galle PR et al (2005) IL-6 signaling promotes tumor growth in colorectal cancer. Cell Cycle 4(2):217–20
Benes FM (2011) Regulation of cell cycle and DNA repair in post-mitotic GABA neurons in psychotic disorders. Neuropharmacology 60(7–8):1232–1242
Benoy IH, Salgado R, Van Dam P, Geboers K, Van Marck E, Scharpé S et al (2004) Increased serum interleukin-8 in patients with early and metastatic breast cancer correlates with early dissemination and survival. Clin Cancer Res 10(21):7157–7162
Benros ME, Nielsen PR, Nordentoft M, Eaton WW, Dalton SO, Mortensen PB (2011) Autoimmune diseases and severe infections as risk factors for schizophrenia: a 30-year population-based register study. Am J Psychiatry 168(12):1303–1310
Beveridge NJ, Cairns MJ (2012) MicroRNA dysregulation in schizophrenia. Neurobiol Dis. 46(2):263–271
Bharadwaj R, Jiang Y, Mao W, Jakovcevski M, Dincer A, Krueger W et al (2013) Conserved chromosome 2q31 conformations are associated with transcriptional regulation of GAD1 GABA synthesis enzyme and altered in prefrontal cortex of subjects with schizophrenia. J Neurosci 33(29):11839–11851
Borovcanin M, Jovanovic I, Radosavljevic G, Djukic Dejanovic S, Bankovic D, Arsenijevic N et al (2012) Elevated serum level of type-2 cytokine and low IL-17 in first episode psychosis and schizophrenia in relapse. J Psychiatr Res 46(11):1421–1426
Borovcanin M, Jovanovic I, Radosavljevic G, Djukic Dejanovic S, Stefanovic V, Arsenijevic N et al (2013) Antipsychotics can modulate the cytokine profile in schizophrenia: attenuation of the type-2 inflammatory response. Schizophr Res 147(1):103–109
Bousman CA, Chana G, Glatt SJ, Chandler SD, Lucero GR, Tatro E et al (2010) Preliminary evidence of ubiquitin proteasome system dysregulation in schizophrenia and bipolar disorder: convergent pathway analysis findings from two independent samples. Am J Med Genet B Neuropsychiatr Genet 153B(2):494–502
Brown AS, Derkits EJ (2010) Prenatal infection and schizophrenia: a review of epidemiologic and translational studies. Am J Psychiatry 167(3):261–280

Brown AS, Hooton J, Schaefer CA, Zhang H, Petkova E, Babulas V et al (2004) Elevated maternal interleukin-8 levels and risk of schizophrenia in adult offspring. Am J Psychiatry 161(5):889–895

Brunet A, Bonni A, Zigmond MJ, Lin MZ, Juo P, Hu LS et al (1999) Akt promotes cell survival by phosphorylating and inhibiting a Forkhead transcription factor. Cell 96(6):857–868

Carbajo-Pescador S, Mauriz JL, García-Palomo A, González-Gallego J (2014) FoxO proteins: regulation and molecular targets in liver cancer. Curr Med Chem 21(10):1231–1246

Catts VS, Catts SV (2000) Apoptosis and schizophrenia: is the tumour suppressor gene, p53, a candidate susceptibility gene? Schizophr Res 41(3):405–415

Catts VS, Catts SV, O'Toole BI, Frost AD (2008) Cancer incidence in patients with schizophrenia and their first-degree relatives—a meta-analysis. Acta Psychiatr Scand 117(5):323–336

Charvin D, Vanhoutte P, Pagès C, Borrelli E, Caboche J (2005) Unraveling a role for dopamine in Huntington's disease: the dual role of reactive oxygen species and D2 receptor stimulation. Proc Natl Acad Sci U S A 102(34):12218–12223

Charvin D, Roze E, Perrin V, Deyts C, Betuing S, Pagès C et al (2008) Haloperidol protects striatal neurons from dysfunction induced by mutated huntingtin in vivo. Neurobiol Dis 29(1):22–29

Chen Y, Zhang J, Zhang L, Shen Y, Xu Q (2012a) Effects of MAOA promoter methylation on susceptibility to paranoid schizophrenia. Hum Genet 131(7):1081–1087

Chen SJ, Chao YL, Chen CY, Chang CM, Wu EC, Wu CS et al (2012b) Prevalence of autoimmune diseases in in-patients with schizophrenia: nationwide population-based study. Br J Psychiatry 200(5):374–380

Chen ZA, Bao MY, Xu YF, Zha RP, Shi HB, Chen TY et al (2012c) Suppression of human liver cancer cell migration and invasion via the GABAA receptor. Cancer Biol Med 9(2):90–98

Chen YJ, Lin GM, Li YH (2013) Cancer risk before schizophrenia diagnosis in Taiwan, 1995–2009. Schizophr Bull 39(4):729–731

Chen J1, Tang D, Wang S, Li QG, Zhang JR, Li P et al (2014) High expressions of galectin-1 and VEGF are associated with poor prognosis in gastric cancer patients. Tumour Biol 35(3):2513–2519

Cohen M, Dembling B, Schorling J (2002) The association between schizophrenia and cancer: a population-based mortality study. Schizophr Res 57:139–146

Coughlin SS, Piper M (1999) Genetic polymorphisms and risk of breast cancer. Cancer Epidemiol Biomarkers Prev 8:1023–1032

Cufí S, Vazquez-Martin A, Oliveras-Ferraros C, Martin-Castillo B, Joven J, Menendez JA (2010) Metformin against TGFβ-induced epithelial-to-mesenchymal transition (EMT): from cancer stem cells to aging-associated fibrosis. Cell Cycle 9(22):4461–4468

Dalamaga M, Diakopoulos KN, Mantzoros CS (2012) The role of adiponectin in cancer: a review of current evidence. Endocr Rev 33(4):547–594

Dalton SO, Mellemkjaer L, Thomassen L, Mortensen PB, Johansen C (2005) Risk for cancer in a cohort of patients hospitalized for schizophrenia in Denmark, 1969–1993. Schizophr Res 75(2–3):315–324

Dash SK, Ghosh T, Roy S, Chattopadhyay S, Das D (2014) Zinc sulfide nanoparticles selectively induce cytotoxic and genotoxic effects on leukemic cells: involvement of reactive oxygen species and tumor necrosis factor alpha. J Appl Toxicol 34(11):1130–1144 (Epub ahead of print)

de Leon J, Diaz FJ (2005) A meta-analysis of worldwide studies demonstrates an association between schizophrenia and tobacco smoking behaviors. Schizophr Res 76(2–3):135–157

Debnath M, Cannon DM, Venkatasubramanian G. (2013) Variation in the major histocompatibility complex [MHC] gene family in schizophrenia: associations and functional implications. Prog Neuropsychopharmacol Biol Psychiatry. 42:49–62

Di Bona D, Accardi G, Virruso C Candore G, Caruso C (2013) Association between genetic variations in the insulin/insulin-like growth factor (igf-1) signaling pathway and longevity: a systematic review and meta-analysis. Curr Vasc Pharmacol 12(5):674–681 (Epub ahead of print)

Dilokthornsakul P, Chaiyakunapruk N, Termrungruanglert W, Pratoomsoot C, Saokeaw S, Sruamsiri R (2013) The effects of metformin on ovarian cancer: a systematic review. Int J Gynecol Cancer 23(9):1544–1551

Dimitrov DH, Lee S, Yantis J, Valdez C, Paredes RM, Braida N et al (2013) Differential correlations between inflammatory cytokines and psychopathology in veterans with schizophrenia: potential role for IL-17 pathway. Schizophr Res 151(1–3):29–35

do Prado CH, Rizzo LB, Wieck A, Lopes RP, Teixeira AL, Grassi-Oliveira R et al (2013) Reduced regulatory T cells are associated with higher levels of Th1/TH17 cytokines and activated MAPK in type 1 bipolar disorder. Psychoneuroendocrinology 38(5):667–676

Domanska UM, Kruizinga RC, Nagengast WB, Timmer-Bosscha H, Huls G, de Vries EG et al (2013) A review on CXCR4/CXCL12 axis in oncology: no place to hide. Eur J Cancer 49(1):219–230

Dong C (2006) Diversification of T-helper-cell lineages: finding the family root of IL-17-producing cells. Nat Rev Immunol 6(4):329–333

Dowlati Y, Herrmann N, Swardfager W, Liu H, Sham L, Reim EK et al (2010) A meta-analysis of cytokines in major depression. Biol Psychiatry 67(5):446–457

Drexhage RC, Hoogenboezem TH, Versnel MA, Berghout A, Nolen WA, Drexhage HA (2011) The activation of monocyte and T cell networks in patients with bipolar disorder. Brain Behav Immun 25(6):1206–1213

Eaton WW, Byrne M, Ewald H, Mors O, Chen CY, Agerbo E et al (2006) Association of schizophrenia and autoimmune diseases: linkage of Danish national registers. Am J Psychiatry 163(3):521–528

Eijkelenboom A, Burgering BM (2013) FOXOs: signalling integrators for homeostasis maintenance. Nat Rev Mol Cell Biol 14(2):83–97

Feenstra M, Veltkamp M, van Kuik J, Wiertsema S, Slootweg P, van den Tweel J et al (1999) HLA class I expression and chromosomal deletions at 6p and 15q in head and neck squamous cell carcinomas. Tissue Antigens 54(3):235–245

Fernandez-Egea E, Miller B, Bernardo M, Donner T, Kirkpatrick B (2008) Parental history of type 2 diabetes in patients with nonaffective psychosis. Schizophr Res 98(1–3):302–306

Flamand V, Zhao H, Peehl DM (2010) Targeting monoamine oxidase A in advanced prostate cancer. J Cancer Res Clin Oncol 136(11):1761–1771

Forester SC, Lambert JD (2014) Synergistic inhibition of lung cancer cell lines by (-)-epigallocatechin-3-gallate in combination with clinically used nitrocatechol inhibitors of catechol-O-methyltransferase. Carcinogenesis 35(2):365–372

Francis AN, Seidman LJ, Jabbar GA, Mesholam-Gately R, Thermenos HW, Juelich R et al (2012) Alterations in brain structures underlying language function in young adults at high familial risk for schizophrenia. Schizophr Res 141(1):65–71

Freude T, Braun KF, Haug A, Pscherer S, Stöckle U, Nussler AK et al (2012) Hyperinsulinemia reduces osteoblast activity in vitro via upregulation of TGF-β. J Mol Med (Berl) 90(11):1257–1266

Fulzele S, Pillai A (2009) Decreased VEGF mRNA expression in the dorsolateral prefrontal cortex of schizophrenia subjects. Schizophr Res 115(2–3):372–373

Gal G, Goral A, Murad H, Gross R, Pugachova I, Barchana M et al (2012) Cancer in parents of persons with schizophrenia: is there a genetic protection? Schizophr Res 139(1–3):189–193

Gallagher D, Norman AA, Woodard CL, Yang G, Gauthier-Fisher A, Fujitani M et al (2013) Transient maternal IL-6 mediates long-lasting changes in neural stem cell pools by deregulating an endogenous self-renewal pathway. Cell Stem Cell 13(5):564–576

Garbett KA, Hsiao EY, Kálmán S, Patterson PH, Mirnics K (2012) Effects of maternal immune activation on gene expression patterns in the fetal brain. Transl Psychiatry 2:e98

Ghadirivasfi M, Nohesara S, Ahmadkhaniha HR, Eskandari MR, Mostafavi S, Thiagalingam S et al (2011) Hypomethylation of the serotonin receptor type-2A Gene (HTR2A) at T102C polymorphic site in DNA derived from the saliva of patients with schizophrenia and bipolar disorder. Am J Med Genet B Neuropsychiatr Genet 156(5):536–545

Glantz LA, Gilmore JH, Lieberman JA, Jarskog LF (2006) Apoptotic mechanisms and the synaptic pathology of schizophrenia. Schizophr Res 81(1):47–63

Grayson DR, Jia X, Chen Y, Sharma RP, Mitchell CP, Guidotti A et al (2005) Reelin promoter hypermethylation in schizophrenia. Proc Natl Acad Sci U S A 28;102(26):9341–9346

Grinshpoon A, Barchana M, Ponizovsky A, Lipshitz I, Nahon D, Tal O et al (2005) Cancer in schizophrenia: is the risk higher or lower? Schizophr Res 73:333–341
Guma SR, Lee DA, Yu L, Gordon N, Hughes D, Stewart J et al (2014) Natural killer cell therapy and aerosol interleukin-2 for the treatment of osteosarcoma lung metastasis Pediatr Blood Cancer 61(4):618–626
Guo S, Arai K, Stins MF, Chuang DM, Lo EH (2009) Lithium upregulates vascular endothelial growth factor in brain endothelial cells and astrocytes. Stroke 40(2):652–655
Guyon A, Kussrow A, Olmsted IR, Sandoz G, Bornhop DJ, Nahon JL (2013) Baclofen and other GABAB receptor agents are allosteric modulators of the CXCL12 chemokine receptor CXCR4. J Neurosci 33(28):11643–11654
Hadad SM, Hardie DG, Appleyard V, Thompson AM (2014) Effects of metformin on breast cancer cell proliferation, the AMPK pathway and the cell cycle. Clin Transl Oncol 16(8):746–752
Hadzi TC, Hendricks AE, Latourelle JC, Lunetta KL, Cupples LA, Gillis T et al (2012) Assessment of cortical and striatal involvement in 523 Huntington disease brains. Neurology 79(16):1708–1715
Hansen TF, Christensen Rd, Andersen RF, Sørensen FB, Johnsson A, Jakobsen A (2013) MicroRNA-126 and epidermal growth factor-like domain 7-an angiogenic couple of importance in metastatic colorectal cancer. Results from the Nordic ACT trial. Br J Cancer 109(5):1243–1251
Hao Q, Zhang C, Gao Y, Wang S, Li J, Li M et al (2014) FOXP3 inhibits NF-κB activity and hence COX2 expression in gastric cancer cells. Cell Signal 26(3):564–569
He XF, Wei W, Li SX, Su J, Zhang Y, Ye XH et al (2012) Association between the COMT Val158Met polymorphism and breast cancer risk: a meta-analysis of 30,199 cases and 38,922 controls. Mol Biol Rep 39(6):6811–6823
Hebbard L, Ranscht B (2014) Multifaceted roles of adiponectin in cancer. Best Pract Res Clin Endocrinol Metab 28(1):59–69
Hsien-Sung H, Akbarian S (2007) GAD1 mRNA expression and DNA methylation in prefrontal cortex of subjects with schizophrenia. PLoS ONE 2(8):e809
Huang HS, Akbarian S (2007) GAD1 mRNA expression and DNA methylation in prefrontal cortex of subjects with schizophrenia. PLoS One. 29;2(8):e809
Huang L, Frampton G, Rao A, Zhang KS, Chen W, Lai JM et al (2012) Monoamine oxidase A expression is suppressed in human cholangiocarcinoma via coordinated epigenetic and IL-6-driven events. Lab Invest 92(10):1451–1460
Humbert L, Lebrun JJ (2013) TGF-beta inhibits human cutaneous melanoma cell migration and invasion through regulation of the plasminogen activator system. Cell Signal 25(2):490–500
International Schizophrenia Consortium, Purcell SM, Wray NR, Stone JL, Visscher PM, O'Donovan MC et al (2009) Common polygenic variation contributes to risk of schizophrenia and bipolar disorder. Nature 460(7256):748–752
Jakuš V, Sapák M, Kostolanská J (2012) Circulating TGF-β1, glycation, and oxidation in children with diabetes mellitus type 1. Exp Diabetes Res 2012:510902
Jaraj SJ, Augsten M, Häggarth L, Wester K, Pontén F, Ostman A et al (2011) GAD1 is a biomarker for benign and malignant prostatic tissue. Scand J Urol Nephrol 45(1):39–45
Jarskog LF, Selinger ES, Lieberman JA, Gilmore JH (2004) Apoptotic proteins in the temporal cortex in schizophrenia: high Bax/Bcl-2 ratio without caspase-3 activation. Am J Psychiatry 161(1):109–115
Jedinak A, Dudhgaonkar S, Sliva D (2010) Activated macrophages induce metastatic behavior of colon cancer cells. Immunobiology 215(3):242–249
Jeong JH, Kim HJ, Lee TJ, Kim MK, Park ES, Choi BS (2004) Epigallocatechin 3-gallate attenuates neuronal damage induced by 3-hydroxykynurenine. Toxicology 195(1):53–60
Ji J, Sundquist K, Sundquist J (2012) Cancer incidence in patients with polyglutamine diseases: a population-based study in Sweden. Lancet Oncol 13(6):642–648
Ji J, Sundquist K, Ning Y, Kendler KS, Sundquist J, Chen X (2013) Incidence of Cancer in patients with schizophrenia and their first-degree relatives: a population-based study in Sweden. Schizophr Bull 39: 527–536

Jia P, Wang L, Meltzer HY, Zhao Z (2010) Common variants conferring risk of schizophrenia: a pathway analysis of GWAS data. Schizophr Res 122(1–3):38–42

Jia P, Wang L, Fanous AH, Chen X, Kendler KS, International Schizophrenia Consortium et al (2012) A bias-reducing pathway enrichment analysis of genome-wide association data confirmed association of the MHC region with schizophrenia. J Med Genet 49(2):96–103

Jia Y, Mo SJ, Feng QQ, Zhan ML, Ouyang LS, Chen JC et al (2014) EPO-dependent activation of PI3K/Akt/FoxO3a signalling mediates neuroprotection in in vitro and in vivo models of Parkinson's Disease. J Mol Neurosci 53(1):117–124

Jiang Y, Zou L, Lu WQ, Zhang Y, Shen AG (2013) Foxo3a expression is a prognostic marker in breast cancer. PLoS ONE 8(8):e70746

Kajdaniuk D, Marek B, Borgiel-Marek H, Kos-Kudła B (2013) Transforming growth factor β1 (TGFβ1) in physiology and pathology. Endokrynol Pol 64(5):384–396

Kaler P, Augenlicht L, Klampfer L (2009) Macrophage-derived IL-1beta stimulates Wnt signaling and growth of colon cancer cells: a crosstalk interrupted by vitamin D3. Oncogene 28(44):3892–3902

Kalkman HO (2009) Altered growth factor signaling pathways as the basis of aberrant stem cell maturation in schizophrenia. Pharmacol Ther 121(1):115–122

Khandaker GM, Zammit S, Lewis G, Jones PB (2014) A population-based study of atopic disorders and inflammatory markers in childhood before psychotic experiences in adolescence. Schizophr Res 152(1):139–145

Kim YK, Myint AM, Lee BH, Han CS, Lee HJ, Kim DJ et al (2004) Th1, Th2 and Th3 cytokine alteration in schizophrenia. Prog Neuropsychopharmacol Biol Psychiatry 28(7):1129–1134

Kimura R, Kasamatsu A, Koyama T, Fukumoto C, Kouzu Y, Higo M et al (2013) Glutamate acid decarboxylase 1 promotes metastasis of human oral cancer by β-catenin translocation and MMP7 activation. BMC Cancer 13:555

Kishida K, Funahashi T, Shimomura I (2014) Adiponectin as a routine clinical biomarker. Best Pract Res Clin Endocrinol Metab 28(1):119–130

Ko EM, Walter P, Jackson A, Clark L, Franasiak J, Bolac C et al (2014) Metformin is associated with improved survival in endometrial cancer. Gynecol Oncol 132(2):438–442

Kolitz JE, George SL, Benson DM Jr, Maharry K, Marcucci G, Vij R et al (2014) Recombinant interleukin-2 in patients aged younger than 60 years with acute myeloid leukemia in first complete remission: results from cancer and leukemia group B 19808. Cancer 120(7):1010–1017

Kryczek I, Wu K, Zhao E, Wei S, Vatan L, Szeliga W et al (2011) IL-17+ regulatory T cells in the microenvironments of chronic inflammation and cancer. J Immunol 186(7):4388–4395

Kumar J, Ward AC (2014) Role of the interleukin 6 receptor family in epithelial ovarian cancer and its clinical implications. Biochim Biophys Acta 1845(2):117–125

Lee JS, Jang DJ, Lee N, Ko HG, Kim H, Kim YS et al (2009) Induction of neuronal vascular endothelial growth factor expression by cAMP in the dentate gyrus of the hippocampus is required for antidepressant-like behaviors. J Neurosci 29(26):8493–8505

Levav I, Lipshitz I, Novikov I, Pugachova I, Kohn R, Barchana M et al (2007) Cancer risk among parents and siblings of patients with schizophrenia. Br J Psychiatry 190:156–161

Li T, Li Z, Chen P, Zhao Q, Wang T, Huang K et al (2010) Common variants in major histocompatibility complex region and TCF4 gene are significantly associated with schizophrenia in Han Chinese. Biol Psychiatry 68(7):671–673

Li C, Guo S, Shi T (2013) Role of NF-κB activation in matrix metalloproteinase 9, vascular endothelial growth factor and interleukin 8 expression and secretion in human breast cancer cells. Cell Biochem Funct 31(3):263–268

Li S, Wang W, Zhang N, Ma T, Zhao C (2014) IL-1β mediates MCP-1 induction by Wnt5a in gastric cancer cells. BMC Cancer 14:480

Liang MH, Wendland JR, Chuang DM (2008) Lithium inhibits Smad3/4 transactivation via increased CREB activity induced by enhanced PKA and AKT signaling. Cell Signal 37(3):440–453

Liang-Kuan B, Nan Z, Cheng L, Fu-Ding L, Tian-Xin L, Xu-Jun X et al (2014) Kidney cancer cells secrete IL-8 to activate Akt and promote migration of mesenchymal stem cells. Urol Oncol 32(5):607–612

Lichtenstein P, Holm N, Verkasalo P, Iliadou A, Kaprio J, Koskenvuo M et al (2000) Environmental and heritable factors in the causation of Cancer. New Engl J Med 343:78–85

Lin HM, Lee JH, Yadav H, Kamaraju AK, Liu E, Zhigang D et al (2009) Transforming growth factor-beta/Smad3 signaling regulates insulin gene transcription and pancreatic islet beta-cell function. J Biol Chem 284(18):12246–12257

Lin F, Yan W, Wen T, Wu GY (2013a) Metformin induces apoptosis in hepatocellular carcinoma Huh-7 cells in vitro and its mechanism. Zhonghua Zhong Liu Za Zhi 35(10):742–746

Lin G, Zhao J, Wu J, Andreevich O R, Zhang WH, Zhang Y et al (2013b) Contribution of catechol-O-methyltransferase Val158Met polymorphism to endometrial cancer risk in postmenopausal women: a meta-analysis. Genet Mol Res 12(4):6442–6453

Liu J, Duan Y, Cheng X, Chen X, Xie W, Long H et al (2011a) IL-17 is associated with poor prognosis and promotes angiogenesis via stimulating VEGF production of cancer cells in colorectal carcinoma. Biochem Biophys Res Commun 407(2):348–354

Liu L, Kron KJ, Pethe VV, Demetrashvili N, Nesbitt ME, Trachtenberg J et al (2011b) Association of tissue promoter methylation levels of APC, TGFβ2, HOXD3 and RASSF1A with prostate cancer progression. Int J Cancer 129(10):2454–2462

Liu L, Yang X, Chen X, Kan T, Shen Y, Chen Z et al (2012a) Association between TNF-α polymorphisms and cervical cancer risk: a meta-analysis. Mol Biol Rep 39(3):2683–2688

Liu RY, Zhu YH, Zhou L, Zhao P, Li HL, Zhu LC et al (2012b) Adenovirus-mediated delivery of interferon-γ gene inhibits the growth of nasopharyngeal carcinoma. J Transl Med 10:256

Liu Y, Zhou Y, Feng X, An P, Quan X, Wang H et al (2014) MicroRNA-126 functions as a tumor suppressor in colorectal cancer cells by targeting CXCR4 via the AKT and ERK1/2 signaling pathways. Int J Oncol 44(1):203–210

Lu Z, Qi L, Bo XJ, Liu GD, Wang JM, Li G (2013) Expression of CD26 and CXCR4 in prostate carcinoma and its relationship with clinical parameters. J Res Med Sci 18(8):647–652

Manzanares N, Monseny R, Ortega L, Montalvo I, Franch J, Gutiérrez-Zotes A et al (2014) Unhealthy lifestyle in early psychoses: the role of life stress and the hypothalamic-pituitary-adrenal axis. Psychoneuroendocrinology 39:1–10

Mao Z, Liu L, Zhang R, Li X (2007) Lithium reduces FoxO3a transcriptional activity by decreasing its intracellular content. Biol Psychiatry 62(12):1423–1430

Massague, J (2000) How cells read TGF-beta signals. Nat Rev Mol Cell Biol 1(3):169–178

Mateo I, Llorca J, Infante J, Rodríguez-Rodríguez E, Fernández-Viadero C, Peña N et al (2007) Low serum VEGF levels are associated with Alzheimer's disease. Acta Neurol Scand 116(1):56–58

McAllister AK (2014) Major histocompatibility complex I in brain development and schizophrenia. Biol Psychiatry 75(4):262–268

McAllister CG, van Kammen DP, Rehn TJ, Miller AL, Gurklis J, Kelley ME et al (1995) Increases in CSF levels of interleukin-2 in schizophrenia: effects of recurrence of psychosis and medication status. Am J Psychiatry 152:1291–1297

Meister J, Schmidt MH (2010) miR-126 and miR-126*: new players in cancer. Scientific World Journal. 10:2090–100

Meulmeester E, Ten Dijke P (2011) The dynamic roles of TGF-β in cancer. J Pathol 223(2):205–218

Miller BJ, Buckley P, Seabolt W, Mellor A, Kirkpatrick B (2011) Meta-analysis of cytokine alterations in schizophrenia: clinical status and antipsychotic effects. Biol Psychiatry 70(7):663–671

Miller GE, Cole SW (2012) Clustering of depression and inflammation in adolescents previously exposed to childhood adversity. Biol Psychiatry. 1;72(1):34–40.

Minelli A, Zanardini R, Abate M, Bortolomasi M, Gennarelli M, Bocchio-Chiavetto L (2011) Vascular Endothelial Growth Factor (VEGF) serum concentration during electroconvulsive therapy (ECT) in treatment resistant depressed patients. Prog Neuropsychopharmacol Biol Psychiatry 35(5):1322–1325

Minzenberg MJ, Laird AR, Thelen S, Carter CS, Glahn DC (2009) Meta-analysis of 41 functional neuroimaging studies of executive function in schizophrenia. Arch Gen Psychiatry 66(8):811–822

Miyoshi H, Kato K, Iwama H, Maeda E, Sakamoto T, Fujita K et al (2014) Effect of the anti-diabetic drug metformin in hepatocellular carcinoma in vitro and in vivo. Int J Oncol 45(1):322–332

Mohamed MM, El-Ghonaimy EA, Nouh MA, Schneider RJ, Sloane BF, El-Shinawi M (2014) Cytokines secreted by macrophages isolated from tumor microenvironment of inflammatory breast cancer patients possess chemotactic properties. Int J Biochem Cell Biol 46:138–147

Monsalve M, Olmos Y (2011) The complex biology of FOXO. Curr Drug Targets 12(9):1322–1350

Moon HS, Mantzoros CS (2013) Adiponectin and metformin additively attenuate IL1β-induced malignant potential of colon cancer. Endocr Relat Cancer 20(6):849–859

Mortensen PB (1989) The incidence of cancer in schizophrenic patients. J Epidemiol Community Health 43(1):43–47

Mortensen PB (1994) The occurrence of cancer in first admitted schizophrenic patients. Schizophr Res 12:185–194

Neman J, Termini J, Wilczynski S, Vaidehi N, Choy C, Kowolik CM et al (2014) Human breast cancer metastases to the brain display GABAergic properties in the neural niche. Proc Natl Acad Sci U S A 111(3):984–989

Nevadunsky NS, Van Arsdale A, Strickler HD, Moadel A, Kaur G, Frimer M et al (2014) Metformin use and endometrial cancer survival. Gynecol Oncol 132(1):236–240

Ni F, Gong Y, Li L, Abdolmaleky HM, Zhou JR (2012) Flavonoid ampelopsin inhibits the growth and metastasis of prostate cancer in vitro and in mice. PLoS ONE. 7(6):e38802

Nielsen PR, Benros ME, Mortensen PB (2013) Hospital contacts with infection and risk of schizophrenia: a population-based cohort study with linkage of Danish national registers. Schizophr Bull 40(6):1526–1532 (Epub ahead of print)

Nohesara S, Ghadirivasfi M, Mostafavi S, Eskandari MR, Ahmadkhaniha H, Thiagalingam S et al (2011) DNA hypomethylation of MB-COMT promoter in the DNA derived from saliva in schizophrenia and bipolar disorder. J Psychiatr Res 45(11):1432–1438

Okada-Iwabu M, Yamauchi T, Iwabu M, Honma T, Hamagami K, Matsuda K et al (2013) A small-molecule AdipoR agonist for type 2 diabetes and short life in obesity. Nature 503(7477):493–499

Pan Q, Xie X, Guo Y, Wang H (2014) Simvastatin promotes cardiac microvascular endothelial cells proliferation, migration and survival by phosphorylation of p70 S6K and FoxO3a. Cell Biol Int 38(5):599–609

Panchanathan E, Ramanathan G, Lakkakula BV (2013) Effect of flupirtine on the growth and viability of U373 malignant glioma cells. Cancer Biol Med 10(3):142–147

Papageorgis P, Lambert AW, Ozturk S, Gao F, Pan H, Manne U et al (2010) Smad signaling is required to maintain epigenetic silencing during breast cancer progression. Cancer Res 70(3):968–978

Papageorgis P, Lambert AW, Ozturk S, Thiagalingam S (2015). TGF-β and BMP signaling in cancer, in System Biology of Cancer, S. Thiagalingam (ed). Cambridge University Press, UK., Chapter 14:204–221.

Park JK, Lee HJ, Kim JW, Park YH, Lee SS, Chang HI et al (2004) Differences in p53 gene polymorphisms between Korean schizophrenia and lung cancer patients. Schizophr Res 67(1):71–74

Patterson PH (2011) Maternal infection and immune involvement in autism. Trends Mol Med 17(7):389–394

Paul-Samojedny M, Owczarek A, Kowalczyk M, Suchanek R, Palacz M, Kucia K et al (2013) Association of interleukin 2 (IL-2), interleukin 6 (IL-6), and TNF-alpha (TNFα) gene polymorphisms with paranoid schizophrenia in a polish population. J Neuropsychiatry Clin Neurosci 25(1, Winter):72–82

Perrone G, Vincenzi B, Zagami M, Santini D, Panteri R, Flammia G et al (2007) Reelin expression in human prostate cancer: a marker of tumor aggressiveness based on correlation with grade. Mod Pathol 20(3):344–351

Pert CB, Knight JG, Laing P, Markwell MA (1988) Scenarios for a viral etiology of schizophrenia. Schizophr Bull 14(2):243–247
Petersén A, Larsen KE, Behr GG, Romero N, Przedborski S, Brundin P et al (2001) Expanded CAG repeats in exon 1 of the Huntington's disease gene stimulate dopamine-mediated striatal neuron autophagy and degeneration. Hum Mol Genet 10(12):1243–1254
Polter A, Yang S, Zmijewska AA, van Groen T, Paik JH, Depinho RA et al (2009) Forkhead box, class O transcription factors in brain: regulation and behavioral manifestation. Biol Psychiatry 65(2):150–159
Potvin S, Stip E, Sepehry AA, Gendron A, Bah R, Kouassi E (2008) Inflammatory cytokine alterations in schizophrenia: a systematic quantitative review. Biol Psychiatry 63(8):801–808
Qin X, Peng Q, Qin A, Chen Z, Lin L, Deng Y et al (2012) Association of COMT Val158Met polymorphism and breast cancer risk: an updated meta-analysis. Diagn Pathol 7:136
Qiu Y, Luo X, Kan T, Zhang Y, Yu W, Wei Y et al (2014) TGF-β upregulates miR-182 expression to promote gallbladder cancer metastasis by targeting CADM1. Mol Biosyst 10(3):679–685
Qu Y, Dang S, Hou P (2013) Gene methylation in gastric cancer. Clin Chim Acta 424:53–65
Ragland JD, Laird AR, Ranganath C, Blumenfeld RS, Gonzales SM, Glahn DC (2009) Prefrontal activation deficits during episodic memory in schizophrenia. Am J Psychiatry 166(8):863–874
Robinson P, Lebel M, Cyr M (2008) Dopamine D1 receptor-mediated aggregation of N-terminal fragments of mutant huntingtin and cell death in a neuroblastoma cell line. Neuroscience 153(3):762–772
Rodriguez M, Snoek LB, De Bono M, Kammenga JE (2013) Worms under stress: C. elegans stress response and its relevance to complex human disease and aging. Trends Genet 29(6):367–374
Rosenstein JM, Krum JM, Ruhrberg C (2010) VEGF in the nervous system. Organogenesis 6(2):107–114
Ryan A, Scrable H (2008) Mutant alleles of HD improve the life span of p53(-/-) mice. Mech Ageing Dev 129(4):238–241
Sasaki M, Kaneuchi M, Sakuragi N, Dahiya R (2003) Multiple promoters of catechol-O-methyltransferase gene are selectively inactivated by CpG hypermethylation in endometrial cancer. Cancer Res 63:3101–3106
Sato N, Matsubayashi H, Fukushima N, Goggins M (2005) The chemokine receptor CXCR4 is regulated by DNA methylation in pancreatic cancer. Cancer Biol Ther 4(1):70–76
Sato N, Fukushima N, Chang R, Matsubayashi H, Goggins M (2006) Differential and epigenetic gene expression profiling identifies frequent disruption of the RELN pathway in pancreatic cancers. Gastroenterology 130(2):548–565
Sernyak MJ, Leslie DL, Alarcon RD, Losonczy MF, Rosenheck R (2002) Association of diabetes mellitus with use of atypical neuroleptics in the treatment of schizophrenia. Am J Psychiatry 159(4):561–566
Sheridan C, Kishimoto H, Fuchs RK, Mehrotra S, Bhat-Nakshatri P, Turner CH et al (2006) CD44+/CD24- breast cancer cells exhibit enhanced invasive properties: an early step necessary for metastasis. Breast Cancer Res 8(5):R59
Shi J, Levinson DF, Duan J, Sanders AR, Zheng Y, Pe'er I et al (2009) Common variants on chromosome 6p22.1 are associated with schizophrenia. Nature 460(7256):753–757
Shibuya M, Watanabe Y, Nunokawa A, Egawa J, Kaneko N, Igeta H et al (2014) Interleukin 1 beta gene and risk of schizophrenia: detailed case-control and family-based studies and an updated meta-analysis. Hum Psychopharmacol 29(1):31–37
Silva R, Martins L, Longatto-Filho A, Almeida OF, Sousa N (2007) Lithium prevents stress-induced reduction of vascular endothelium growth factor levels. Neurosci Lett 429(1):33–38
Singh S, Singh UP, Grizzle WE, Lillard JW Jr (2004) CXCL12-CXCR4 interactions modulate prostate cancer cell migration, metalloproteinase expression and invasion. Lab Invest 84(12):1666–1676
Sinkus ML, Adams CE, Logel J, Freedman R, Leonard S. (2013) Expression of immune genes on chromosome 6p21.3-22.1 in schizophrenia. Brain Behav Immun. 32:51–62

Smith M, Hopkins D, Peveler RC, Holt RI, Woodward M, Ismail K (2008) First- v. second-generation antipsychotics and risk for diabetes in schizophrenia: systematic review and meta-analysis. Br J Psychiatry 192(6):406–411

Song N, Han S, Lee KM, Choi JY, Park SK, Jeon S et al (2012) Genetic variants in interleukin-2 and risk of lymphoma among children in Korea. Asian Pac J Cancer Prev 13(2):621–623

Song X, Fan X, Song X, Zhang J, Zhang W, Li X et al (2013a) Elevated levels of adiponectin and other cytokines in drug naïve, first episode schizophrenia patients with normal weight. Schizophr Res 150(1):269–273

Song XQ, Chen XM, Zhang W, Li X, Hei GR, Li YH et al (2013b) Study of adiponectin, IL-1β, IL-6 and TNF-α in first episode drug Naïve schizophrenia. Zhonghua Yi Xue Za Zhi 93(41):3256–3260

Song X, Fan X, Li X, Zhang W, Gao J, Zhao J et al (2014) Changes in pro-inflammatory cytokines and body weight during 6-month risperidone treatment in drug naïve, first-episode schizophrenia. Psychopharmacology (Berl) 231(2):319–325

Sørensen SA, Fenger K (1992) Causes of death in patients with Huntington's disease and in unaffected first degree relatives. J Med Genet 29(12):911–914

Spano JP, Andre F, Morat L, Sabatier L, Besse B, Combadiere C et al (2004) Chemokine receptor CXCR4 and early-stage non-small cell lung cancer: pattern of expression and correlation with outcome. Ann Oncol 15(4):613–617

Spelman LM, Walsh PI, Sharifi N, Collins P, Thakore JH (2007) Impaired glucose tolerance in first-episode drug-naive patients with schizophrenia. Diabet Med 24(5):481–485

Stack EC, Dedeoglu A, Smith KM, Cormier K, Kubilus JK, Bogdanov M et al (2007) Neuroprotective effects of synaptic modulation in Huntington's disease R6/2 mice. J Neurosci 27(47):12908–12915

Stefansson H, Ophoff RA, Steinberg S, Andreassen OA, Cichon S, Rujescu D et al (2009) Common variants conferring risk of schizophrenia. Nature 460(7256):744–747

Subramaniam M, Lam M, Guo ME, He VY, Lee J, Verma S et al (2014) Body mass index, obesity, and psychopathology in patients with schizophrenia. J Clin Psychopharmacol 34(1):40–46

Sun YX, Wang J, Shelburne CE, Lopatin DE, Chinnaiyan AM, Rubin MA et al (2003) Expression of CXCR4 and CXCL12 (SDF-1) in human prostate cancers (PCa) in vivo. J Cell Biochem 89(3):462–473

Sun X, Liu Z, Yang Z, Xiao L, Wang F, He Y et al (2013) Association of microRNA-126 expression with clinicopathological features and the risk of biochemical recurrence in prostate cancer patients undergoing radical prostatectomy. Diagn Pathol 8(1):208

Takayama H, Misu H, Iwama H, Chikamoto K, Saito Y, Murao K et al (2014) Metformin suppresses expression of the selenoprotein P Gene via an AMP-activated Kinase (AMPK)/FoxO3a pathway in H4IIEC3 Hepatocytes. J Biol Chem 289(1):335–345

Tan B, Anaka M, Deb S, Freyer C, Ebert LM, Chueh AC et al (2014) FOXP3 over-expression inhibits melanoma tumorigenesis via effects on proliferation and apoptosis. Oncotarget 5(1):264–276

Tang TS, Chen X, Liu J, Bezprozvanny I (2007) Dopaminergic signaling and striatal neurodegeneration in Huntington's disease. J Neurosci 27(30):7899–7910

Tekpli X, Landvik NE, Anmarkud KH, Skaug V, Haugen A, Zienolddiny S (2013) DNA methylation at promoter regions of interleukin 1B, interleukin 6, and interleukin 8 in non-small cell lung cancer. Cancer Immunol Immunother 62(2):337–345

Tian C, Liu L, Yang X, Wu H, Ouyang Q (2014) The Val158Met polymorphism in the COMT gene is associated with increased cancer risks in Chinese population. Tumour Biol 35(4):3003–3008 (Epub 2013 Dec 5)

Toritsuka M, Kimoto S, Muraki K, Landek-Salgado MA, Yoshida A, Yamamoto N et al (2013) Deficits in microRNA-mediated Cxcr4/Cxcl12 signaling in neurodevelopmental deficits in a 22q11 deletion syndrome mouse model. Proc Natl Acad Sci U S A 110(43):17552–17557

Tourjman V, Kouassi É, Koué MÈ, Rocchetti M, Fortin-Fournier S, Fusar-Poli P et al (2013) Antipsychotics' effects on blood levels of cytokines in schizophrenia: a meta-analysis. Schizophr Res 151(1–3):43–47

Tsai MC, Chang CM, Liu CY, Chang PY, Huang TL (2011) Association of serum levels of leptin, ghrelin, and adiponectin in schizophrenic patients and healthy controls. Int J Psychiatry Clin Pract 15(2):106–111
Urayama KY, Thompson PD, Taylor M, Trachtenberg EA, Chokkalingam AP (2013) Genetic variation in the extended major histocompatibility complex and susceptibility to childhood acute lymphoblastic leukemia: a review of the evidence. Front Oncol 3:300
Vanable PA, Carey MP, Carey KB, Maisto SA (2003) Smoking among psychiatric outpatients: relationship to substance use, diagnosis, and illness severity. Psychol Addict Behav. 17(4):259–265
Venza I, Visalli M, Fortunato C, Ruggeri M, Ratone S, Caffo M et al (2012) PGE2 induces interleukin-8 derepression in human astrocytoma through coordinated DNA demethylation and histone hyperacetylation. Epigenetics 7(11):1315–1330
Vergho D, Kneitz S, Rosenwald A, Scherer C, Spahn M, Burger M et al (2014) Combination of expression levels of miR-21 and miR-126 is associated with cancer-specific survival in clear-cell renal cell carcinoma. BMC Cancer 14(1):25
Vincze C, Pál G, Wappler EA, Szabó ER, Nagy ZG, Lovas G et al (2010) Distribution of mRNAs encoding transforming growth factors-beta1, -2, and -3 in the intact rat brain and after experimentally induced focal ischemia. J Comp Neurol 518(18):3752–3770
Wang YQ, Li YM, Li X, Liu T, Liu XK, Zhang JQ et al (2013) Hypermethylation of TGF-β1 gene promoter in gastric cancer. World J Gastroenterol 19(33):5557–5564
Warren CM, Ziyad S, Briot A, Der A, Iruela-Arispe ML (2014) A ligand-independent VEGFR2 signaling pathway limits angiogenic responses in diabetes. Sci Signal 7(307):ra1
Watahiki A, Macfarlane RJ, Gleave ME, Crea F, Wang Y et al (2013) Plasma miRNAs as biomarkers to identify patients with castration-resistant metastatic prostate cancer. Int J Mol Sci 14(4):7757–7770
Weeks KR, Dwyer DS, Aamodt EJ (2010) Antipsychotic drugs activate the C. elegans akt pathway via the DAF-2 insulin/IGF-1 receptor. ACS Chem Neurosci 1(6):463–473
Wu J, Lu Y, Ding YB, Ke Q, Hu ZB, Yan ZG et al (2009) Promoter polymorphisms of IL2, IL4, and risk of gastric cancer in a high-risk Chinese population. Mol Carcinog 48(7):626–632
Wu W, Zhang J, Zhou L, You L, Zhao Y, Li J (2012) Increased COMT expression in pancreatic cancer and correlation with clinicopathologic parameters. Sci China Life Sci 55(9):747–752
Xia Z, DePierre JW, Nässberger L (1996) Tricyclic antidepressants inhibit IL-6, IL-1 beta and TNF-alpha release in human blood monocytes and IL-2 and interferon-gamma in T cells. Immunopharmacology. 34(1):27–37
Xie Q, Hao Y, Tao L, Peng S, Rao C, Chen H et al (2012) Lysine methylation of FOXO3 regulates oxidative stress-induced neuronal cell death. EMBO Rep 13(4):371–377
Xu Y, Lu S (2013) Metformin Inhibits esophagus cancer proliferation through upregulation of USP7. Cell Physiol Biochem 32(5):1178–1186
Xu J, Sun J, Chen J, Wang L, Li A, Helm M et al (2013) RNA-Seq analysis implicates dysregulation of the immune system in schizophrenia. BMC Genomics 13(Suppl 8):S2 (Epub 2012 Dec 17)
Yadav H, Quijano C, Kamaraju AK, Gavrilova O, Malek R, Chen W et al (2011) Protection from obesity and diabetes by blockade of TGF-β/Smad3 signaling. Cell Metab 14(1):67–79
Yang C, Hou C, Zhang H, Wang D, Ma Y, Zhang Y et al (2013a) miR-126 Functions as a tumor suppressor in Osteosarcoma by targeting Sox2. Int J Mol Sci 15(1):423–437
Yang XB, Zhao JJ, Huang CY, Wang QJ, Pan K, Wang DD et al (2013b) Decreased expression of the FOXO3a gene is associated with poor prognosis in primary gastric adenocarcinoma patients. PLoS ONE 8(10):e78158
Ye P, Liu J, He F, Xu W, Yao K (2013) Hypoxia-Induced Deregulation of miR-126 and Its Regulative Effect on VEGF and MMP-9 Expression. Int J Med Sci 11(1):17–23
Yin M, Zhou J, Gorak EJ, Quddus F (2013) Metformin is associated with survival benefit in cancer patients with concurrent type 2 diabetes: a systematic review and meta-analysis. Oncologist 18(12):1248–1255
Yong-Qi W Yu-Min Li Xun Li Tao Liu Xiao-Kang Liu Jun-Qiang Z et al (2013) Hypermethylation of TGF-β1 gene promoter in gastric cancer. World J Gastroenterol 19(33):5557–5564

Yoo KH, Park YK, Chang SG (2013) DNA hypomethylation of interleukin 8 in clear cell renal cell carcinoma. Oncol Lett 5(1):39–42

You H, Yamamoto K, Mak TW (2006) Regulation of transactivation-independent proapoptotic activity of p53 by FOXO3a. Proc Natl Acad Sci U S A 103(24):9051–9056

Yu Q, Liu SL, Wang H, Shi G, Yang P, Chen XL (2013) miR-126 Suppresses the proliferation of cervical cancer cells and alters cell sensitivity to the chemotherapeutic drug Bleomycin. Asian Pac J Cancer Prev 14(11):6569–6572

Yuan Y, Chen H, Ma G, Cao X, Liu Z (2012) Reelin is involved in transforming growth factor-β1-induced cell migration in esophageal carcinoma cells. PLoS ONE 7(2):e31802

Zhang XY, Zhou DF, Zhang PY, Wu GY, Cao LY, Shen YC (2002) Elevated interleukin-2, interleukin-6 and interleukin-8 serum levels in neuroleptic-free schizophrenia: association with psychopathology. Schizophr Res 57(2–3):247–258

Zhang XY, Zhou DF, Cao LY, Zhang PY, Wu GY, Shen YC (2004) Changes in serum interleukin-2,-6, and -8 levels before and during treatment with risperidone and haloperidol: relationship to outcome in schizophrenia. J Clin Psychiatry 65(7):940–947

Zhang T, Guo P, Zhang Y, Xiong H, Yu X, Xu S et al (2013) The antidiabetic drug metformin inhibits the proliferation of bladder cancer cells in vitro and in vivo. Int J Mol Sci 14(12):24603–24618

Zhang J, Wang H, Wang L, Wang WD, Geng QR, Lu Y (2014) Adenovirus-mediated delivery of the human IFN-γ gene potentiates the cytotoxicity of daunorubicin against leukemic cells through downregulation of the α4β1 integrin/ILK/apoptosis pathway. Oncol Lett 7(2):361–368

Zhao S, Wang Y, Liang Y, Zhao M, Long H, Ding S et al (2011) MicroRNA-126 regulates DNA methylation in CD4+T cells and contributes to systemic lupus erythematosus by targeting DNA methyltransferase 1. Arthritis Rheum 63(5):1376–1386

Zhou Y, Feng X, Liu YL, Ye SC, Wang H, Tan WK (2013) Down-regulation of miR-126 is associated with colorectal cancer cells proliferation, migration and invasion by targeting IRS-1 via the AKT and ERK1/2 signaling pathways. PLoS ONE 8(11):e81203

Zhu L, Song X, Tang J, Wu J, Ma R, Cao H (2013) Huntingtin-associated protein 1: a potential biomarker of breast cancer. Oncol Rep 29(5):1881–1887

Zhu F, Zhao H, Tian X, Meng X (2014a) Association between tumor necrosis factor-α rs1800629 polymorphism and risk of gastric cancer: a meta-analysis. Tumour Biol 35(3):1799–1803 (Epub 2013 Oct 19)

Zhu P, Davis M, Blackwelder A, Bachman N, Liu B, Edgerton S et al (2014b) Metformin selectively targets tumor initiating cells in erbB-2 overexpressing breast cancer models. Cancer Prev Res (Phila) 7(2):199–210 (Epub 2013 Dec 9)

Chapter 6
Exploring ATM and Methylation in Cancer: Emphasizing on Brain Tumors

Parvin Mehdipour and Fatemeh Karami

Contents

P. Mehdipour (✉) · F. Karami
Department of Medical Genetics, School of Medicine, Tehran University of Medical Sciences, Poursina Street, Keshavarz Boulevard, P.O. Box: 14176-13151, Tehran, Iran
e-mail: mehdipor@tums.ac.ir

P. Mehdipour (ed.), *Epigenetics Territory and Cancer*,
DOI 10.1007/978-94-017-9639-2_6

Abstract Ataxia-telangiectasia mutated (ATM) molecule governs one of the major cellular DNA repair pathway and is found to be crucial to inhibit cancer progression. Herein, after introducing the basic description about the *ATM* gene, its protein structure and functions, then we will discuss about the role of *ATM* gene aberrations in various types of cancers. It will be explained that some overlaps between Ataxia Telangiectasia (AT) patient and some other metabolic disorders relying on the importance of *ATM* gene mutations in the pathogenesis of AT disease. Finally, the spectrum of *ATM* promoter methyltion in different types of cancers will be provided.

Abbreviations

AFP:	α-feto protein
ALL:	Acute lymphoblastic leukemia
AT:	Ataxia telangiectasia
ATM:	Ataxia-telangiectasia mutated
ATR:	ATM- and RAD3-related
ATRIP:	ATR interacting protein
BASC:	BRCA1 associated surveillance complex
B-CLL:	B-cell chronic lymphocytic leukemia
BNHL:	B-cell non-Hodgkin's lymphomas
Cdk5:	Cyclin-dependent kinase 5
CtIP:	C-terminal binding protein interacting protein
DDB:	DNA double strand break
DLBCL:	Diffuse large B-cell lymphoma
DNA-PKcs:	DNA-dependent protein kinase catalytic subunit
DNMT1:	DNA methyltransferase 1
4E-BP1:	eIF-4E-binding protein 1
EGC:	Early gastric cancer
ESR1:	Estrogen receptor 1
FAT:	FRAP-ATM-TRRAP
FATC:	FAT-C-terminal
FCL:	Follicular center cell lymphoma
GBM:	Glioblastoma multiform
G-CIMP:	Glioma-CpG island methylator phenotype
G6PD:	Glucose-6-phosphate dehydrogenase
HR:	Homologous recombination
HP1:	Heterochromatin protein 1
HIF:	Hypoxia induced factor
H_2O_2:	Hydrogen peroxide
HD:	Hodgkin's diseases
HG-IEN:	High-grade intraepithelial neoplasia
HNPCC:	Hereditary Non-Polyposis Colorectal Cancer
KAP-1:	KRAB associated protein 1
KD:	Kinase domains

LOH:	Loss of heterozygousity
3'UTR:	3' untranslated region
m-TOR:	Mammalian target of rapamycin
MASA:	Mutant allele-specific PCR amplification
MCL:	Mantle cell lymphoma
MPF:	Metaphase promoting factor
MRX:	Mre11–Rad50–Xrs2
MSI:	Microsatellite instability
mTORC1:	mTOR complex 1
NADPH:	Nicotinamide Adenine Dinucleotide Phosphate
NHEJ:	Non-homologous end-joining
NSCLC:	Non-small cell lung cancer
PIKKs:	Phosphatidylinositol 3-kinase-related kinases
PPP:	Pentose phosphate pathway
pRb:	Retinoblastoma protein
PRD:	PIKK-regulatory domain
Q-FISH:	Quantitative Fluorescence in situ hybridization
TDP1:	Tyrosyl phosphodiesterase 1
TL:	Telomere length
T-PLL:	T-cell prolymphocytic leukemia
TMPRSS2:	ERG: transmembrane protease/serine subfamily member 2: estrogen-regulated genes
Real time PCR:	Real time polymerization chain reaction
RNS:	Reactive nitrogen species
ROS:	Reactive oxygen species
SFN:	Stratifin

6.1 Introduction

6.1.1 Ataxia-Telangiectasia Disease

Ataxia-telangiectasia (A-T) is caused as the results of different genetic aberrations which occur within the gene with related name called *ATM*. AT disease has the global prevalence of approximately 1/40,000–300,000 and is inherited by usually two apparently normal but carrier parents in autosomal recessive mode. AT is characterized by the major clinical features including ataxia or uncoordinated movement and telangiectasia (McKinnon 2004). Ocular vascular dilatation leads to creation of dark red spots observable in foundoscopic eye examination. The same complication could be seen as the dark blemish areas on the skins of AT patients. AT usually begins at early childhood when the baby tries to walk around 1 year old but the progression of disease restricted him/her by walking on wheelchair. Complete ATM loss of function is also associated with varying neurological signs including choreoathetosis which is defined by abnormal ocular and skeletal muscles movements, progressive speech imperfectness called dysarthria and general neurodegeneration (Lavin et al. 2007).

Lack of normal ATM function results in decrease of serum levels of IgA, IgE and IgG2 leading to immunodeficiency. However, some of AT patients have milder neurological symptoms and intact immune system function due to detection of the type of *ATM* mutations. It was determined that missense or splice site alterations usually are associated with leaky function of ATM protein compared to deleterious and non-functional truncating mutations (Gilad et al. 1998; Toyoshima et al. 1998).

Although the liver function of AT patients is usually normal, increased α feto-protein (AFP) level is one of the most important biochemical marker found in the serum. They are prone to be affected with insulin- resistant diabetes and cancer and almost prematurely become old. Due to important role of ATM in DNA repair which will be discussed in detail in later parts of this chapter, AT patients are so sensitive to ionizing radiation. It was shown that γ-radiation caused reduced DNA synthesis and cell proliferation associated with increase in rate of chromosomal breaks. Cancer and chronic pulmonary diseases are the most two leading reasons of death in AT patients.

6.1.2 *ATM Gene*

The Ataxia Telangiectasia Mutated or *ATM* gene was mapped on 11q22-23 from base pair 108,093,558 to base pair 108,239,828 spanning 160 kbps of human genome (Fig. 6.1). Several non-sense and missense mutations had been found in *ATM* gene by positional cloning in cancer and AT patients. It has 66 exon and eight protein coding transcripts with different length (1.8, 2.6, 3.0, 4.7, and 5.7 kb) which are present in various levels in different cell types (Kapp 1992). Alternative splic-

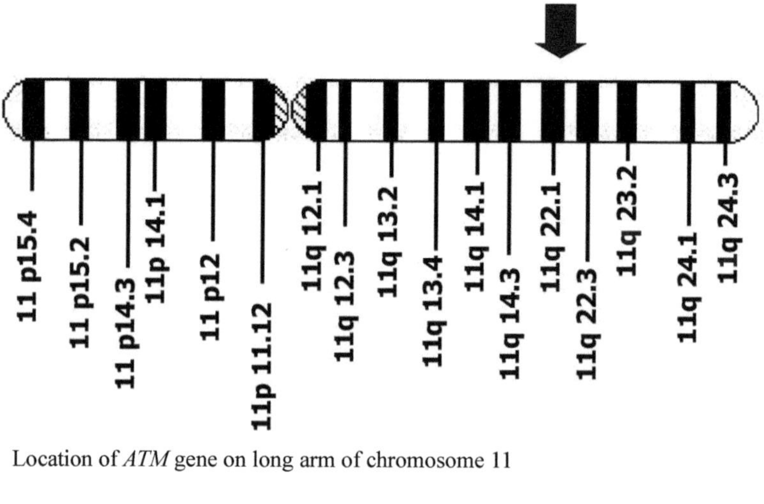

Fig 6.1 Location of *ATM* gene on long arm of chromosome 11

ing is based on the selection of either 1a or 1b as the first exon. *ATM* gene contains an internal start codon residing within the exon 4 and the 3' untranslated region (3'UTR) is located at approximately 4 kbps downstream of the last exon (Savitsky et al. 1995a).

ATM gene is, bidirectionally, transcribed and there is an intergenic region spanning 520 bps within this gene which is shared by the flanking gene named *NPAT*. The NPAT protein is induced by E-Cdk2 kinase to activate the expression of histone proteins during G1-S transition and early S phases (Gao et al. 2003).

There is a long list of *ATM* mutations containing approximately 400 alterations that most of them (70 %) lead to production of truncated protein and are 100 % penetrant (Concannon and Gatti 1997). Missense alterations have usually lower penetrance and are associated with milder phenotypes. Of note, AT patients are typically heterozygous for more than one mutation (compound heterozygous) and homozygous patients were hardly detected. No hotspot region was identified in *ATM* gene indicating that most of the ATM mutations are unique (Gatti et al. 1999), however this matter is diverse in different populations.

The *ATM* gene is predominantly expressed in adipose, blood, brain, adrenal, breast, colon and heart tissues.

6.1.3 ATM Protein

ATM protein encoded by ATM gene, harbors 3056 amino acids and weights 350 kDa. It is a member of a large serine tyrosine kinase family called Phosphatidylinositol 3-kinase-related kinases (PIKKs) . The name of this superfamily is due to the presence of a similar sequence in its members and phosphatidylinositol 3-kinases (PI3Ks). The other members of this family include ATM- and RAD3-related (ATR), DNA-dependent protein kinase catalytic subunit (DNA-PKcs) and mammalian target of Rapamycin (m-TOR) (Wyman 1998).

There are three known isoforms of ATM protein. The first isoform constitutes of 3056 amino acids and weights for 350.6 kDa which has shown high degree of homology with mammalian, Drosophila and yeast PI3 kinase. The second and the third isoforms have 1708 and 131 amino acids, respectively. The mouse ATM is more close to the isoform 1 which contains 3066 amino acids and is 84 % similar to human ATM in sequence (Pecker et al. 1996).

There are five major domains within the structure of ATM protein with different characteristic functions (Fig. 6.2). These domains include Huntington, Elongation factor 1A, protein phosphatase 2A A-subunit, TOR (HEAT) repeat domain. The HEAT domain lies at the N-terminal of ATM protein and it has several HEAT repeat of amino acids. HEAT stands for histidine (H), Glutamic acid (E), Argenine (A) and tyrosine (T) amino acids. It makes connection between ATM and NBS1, P53, c-Abl proteins to be involved in repair of DNA double strand break (DDB) (Perry and Kleckner 2003). The FRAP-ATM-TRRAP (FAT) and kinase (KD) domains follow the HEAT domain and interact together to stabilize the C-terminal of ATM protein.

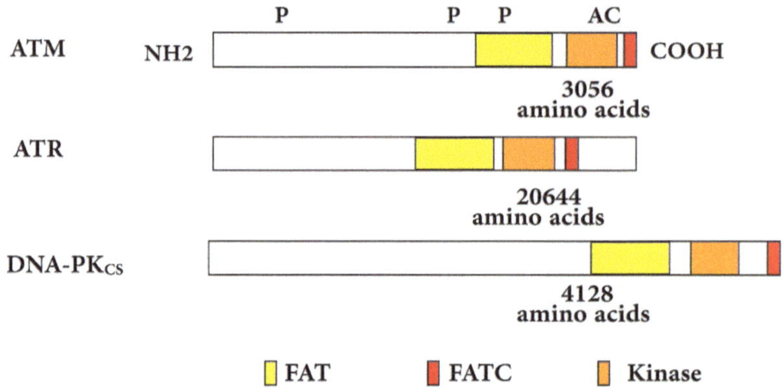

Fig. 6.2 Shared domains in PIKKs family members and the sites of post-translational modifications. (Modified from McKinnon 2012)

KD performs the kinase activity of ATM protein which is modulated by the adjacent domains of ATM protein including PIKK-regulatory domain (PRD) and the FAT-C-terminal (FATC) domain. The FATC domain may be the most conserved domain of ATM protein and is composed of 30 amino acids which form α-helix structure through disulfide bonding (Savitsky et al. 1995b). It was shown that FAT domain has also α-helix structure which in its tertiary structure provides a tunnel to enable the ATM to wrap around the damaged DNA. The importance of translational modification of FAT and FATC domains in regulation of ATM activity shows that they play pivotal role in modulating the conformational changes in ATM structure defining its activity. The acetylation of C2991 site within the FATC domain determines the phosphorylation status of S1981 site of FAT domain which is critical for gathering the non-functional dimers together (Fig. 6.2) (Rivera-Calzada et al. 2005).

Post-translational modifications of ATM almost occur in response to DNA damage and include phosphorylation at various sites of protein and acetylation. One of the most important of these modifications is phosphorylation at serine 1981 which was recently found to be necessary for ATM signaling and its activation (Bakkenist and Kastan 2003). It was stated that there was no effect on the overall process of DNA repair in mice whom were knocked out for other modifications, however, their critical roles in ATM signaling remains to be well elucidated (Daniel et al. 2008; So et al. 2009).In addition, the ATM protein status doesn't change in location and amount during all of the phases of cell cycle (Brown et al. 1997).

6.2 ATM Functions

ATM is well known for its critical role as a core of a triangle including DNA repair, cell cycle control and apoptosis. Therefore, the major functions of ATM can be descried for each apex of the mentioned triangle in details which are correlated to each other. The fate of ATM signaling depends on the determining factors including type of tissue and cell in addition to the stage of development wherein DNA damage has been occurred.

6.2.1 DNA Repair

ATM is well known for its critical activity in response to damage-specific DNA binding (DDB) protein to activate the main cellular check points avoiding the replication of damaged DNA and leading it to senescence or apoptosis. However, there are some evidences relying on the role of ATM in various other pathways induced by oxidative stress as a consequence of metabolic aberrations including autophagy (Zhang et al. 1997).

DDB is the most deleterious damage which could be occurred at DNA level. Various exogenous and endogenous factors are responsible for creating DDB. The first includes ionizing radiation not UV light and the latter comprises of toxic elements such as free radicals produced in oxidative metabolism. In addition, mitotic recombination during T cell and immunoglobulin genes rearrangement is also recognized as one of the most important cause of DDB (Wyman and Kanaar 2006).

It is interesting to note that ATM is actively involved in 10% of DDB which affects the heterochromatin DNA (Goodarzi et al. 2008). In this way, two proteins including 53 BP1 and KRAB associated protein 1 (KAP-1) are essential through which phosphorylation of the latter by ATM leads to alter the conformation of heterochromatin structure required for initiation of DNA repair. The 53BP1 enables ATM to damage site, activate KAP-1 protein and inducing repair in the late breaking DNA (Goodarzi et al. 2008; Ziv et al. 2006; Noon et al. 2010).

DDB can be repaired through two alternative pathways including homologous recombination (HR) and non-homologous end-joining (NHEJ). When DNA is targeted to be repaired by NHEJ pathway, the two ends of damaged DNA could be detected and occupied by highly conserved Ku70–Ku80 kinase heterodimer recruiting Artemis nuclease; these events are followed by an inaccurate cutting process of the broken DNA and leave the primary site for polymerization activities of μ and λ DNA polymerases. In yeast, when no homologous sequence being found to be as template of polymerization, NHEJ is driven through recognizing the broken ends by DNA–PKcs which interacts with Ku heterodimer. DNA–PKcs is another important member of PIKKs family that has a pivotal role in checkpoint signaling (Gottlieb and Jackson 1993; Dvir et al. 1992). Subsequently, the newly synthesized DNA segment being ligated by XRCC4 accompanying DNA ligase IV in a ligation complex without prior processing of the DNA ends. Of note, the homologous of DNA ligase IV, Cernunnos, and XLF factor encourage the initiation of ligation process in all animal species (Buck et al. 2006; Ahnesorg et al. 2006; Cavero et al. 2007). It is interesting to note that DNA–PKcs brings two ends of both strands of broken DNA together. It is alternatively carried out by Mre11–Rad50–Xrs2 (MRX) complex or other proteins in yeast (Manolis et al. 2001).

Although it is highly proficient, imprecise cutting and replacing the DNA nucleotides is the major reason for calling NHEJ as an error and mutation prone pathway of DNA repair (Pitcher et al. 2005; Lees-Miller and Meek 2003; Lieber 2008). NHEJ is the choice of DNA repair when the damage takes place in G0, G1 and in the initial times of S or replication phases.

Another choice of DDB repair is HR in which the broken part of DNA is replaced based on the sequence of sister chromatid as template of polymerization. This turn, the call for initiation of DNA repair is arisen when the ends of broken double strand DNA is filled with a triple protein complex known as MRN complex (MRE11–RAD50–NBS1/Xrs2). After cutting both flanking sites of broken part of DNA, replicated protein A or RPA binds to the created single strands and prevents rejoining of them until the polymerization of their corresponding opposite strands occur (Kadyk and Hartwell 1992; Sonoda et al. 2001). In yeast, RPA is dislodged from the single stranded DNA by cooperation between Rad51 and Rad52 proteins to provide the condition for initiation of polymerization. Rad51 actually carries out the task of search for the homologous sequence to be as template for replacing the missing cut nucleotides through wrapping around the nucleosome filaments (Sung 1994; Ogawa et al. 1993).

HR is the preferred choice of post replication DNA repair within the late S phase and throughout the G2 phase. The role of ATM in HR pathway may be tissue specific as testis and usually includes phosphorylation and activation of less critical elements of HR pathway (Barlow et al. 1997). In spite of the fact that HR is free of any mistake or mutation in replacing the missing broken DNA double strand sequences, it is intriguing to say that the cell prefers to use the error prone NHEJ pathway. Although the recent studies have refuted it, it may be due to the tight and complex infrastructure of DNA sequence wrapping around the histone proteins making homology searching more impossible and difficult (Ciccia and Elledge 2010; Bensimon et al. 2011; Kim et al. 1999; Matsuoka et al. 2007). However, the choice of DNA repair pathway is phase specific and it seems that it is the matter of authority, so cell determines and selects the appropriate one according to the cellular phase in which break occurs.

There are many reports on diseases including cancer and neurological disorders which are as consequences of defects in elements of either HR or NHEJ DNA repair pathways (Mu et al. 2007; Bensimon et al. 2010; Bennetzen et al. 2010; Marzano et al. 2012; McKinnon 2012; Schalch et al. 1970).

When DDB occurs, ATM is recruited to damage site through formation of MRN complex at the end of broken DNA (Shiloh 2003). Of note, Mre11 acts as an endonuclease, resects the broken DNA segment which should be replaced. Interaction of Nbs1 with C-terminal binding protein interacting protein (CtIP) is another fact indicating that MRN complex is actively contribute to DDB repair. In addition, attachment of MRN complex to damaged DNA leads to extension of coiled coil structure of Rad50 which makes the connection possible between two MRN complexes residing on both opposite DNA strands (Wiltzius et al. 2005; Hopfner et al. 2002).

It is assumed that recruitment of MRN complex to damage site is triggered by the presence of highly conserved identified motifs in Nbs1 whereas motifs in KU and ATR interacting protein (ATRIP) are required for motivating the recruitment of DNA-PKcs and ATR to damage site, respectively (Falck et al. 2005). ATM is tightened to the damage site through interaction of its C-terminus to Nbs1 protein. The same condition is held for ATRIP/ATR and KU/DNA-PKcs (Williams et al. 2009). However, some other proteins have been identified to be involved in activation of ATM after DDB including checkpoint protein, ring finger protein 8 and E3 ubiquitin ligases (Wu et al. 2011).

Dynamic ATM activation is dependent on the phosphorylation of S1981 within the FAT domain releasing the inactive ATM dimer to be bound to other molecules (Bakkenist and Kastan 2003). Phosphorylation and dephosphorylation of S1981 is performed by PP2A and WIP1 phosphatase, respectively (Shreeram et al. 2006). The same site of phosphorylation is targeted by ATM on tyrosyl phosphodiesterase 1(TDP1) which plays important role in DDB repair (Das et al. 2009). It was found that WIP1 regulates the methylation status of DNA sequences through interaction with ATM/BRCA1 to recruit heterochromatin protein 1 (HP1) and subsequently DNA methyltransferases (Filipponi et al. 2013). Moreover, there are other important serines as well as Ser367, Ser1893 and Ser2996 that their autophosphorylation have revealed major impacts on ATM signaling (Kozlov et al. 2011). Besides PP2A and WIP1, PP5 are another phosphatase which modulates the phosphorylation of ATM (Goodarzi et al. 2004).

Although the critical effect of ATM autophosphorylation on DDB repair have been confirmed in human cell line studies, animal and in vitro assays couldn't identify its importance. Activation of monomer ATM, in turn drives a cascade of phosphorylation and ubiquitinilation of several downstream molecules ending in DNA repair or apoptosis. There are some mediatory proteins which enhance and facilitate the relay of signals to target elements within the pathway (Li and Zou 2005; McGowan and Russell 2004).

The immediate downstream targets of ATM and ATR are Chk2 and Chk1 proteins, respectively. However, when both ATM and ATR become activated, formation of a complex is initially necessary for Chk1 and Chk2 stimulation. This complex comprises of four BRCT domain containing proteins including BRCA1, MDC1/NFBD1, MCPH1/BIRT1 that taken together recruit all the critical factors involved in cell cycle checkpoint and DNA repair. BRCA1 is phosphorylated by both ATM and Chk1 in response to DNA damage and resides on damage site. It has critical role in two intra-S and G2-M checkpoints (Deng 2006; Ouchi 2006).

The 53BP1 is required for p53 stabilization and is also involved in Chk2 phosphorylation and triggering intra-S and G2-M checkpoints when the DNA damage is occurred. There are approximately 15 sites within this protein which are targeted for phosphorylation by ATM and ATR in response to UV radiation (Wang et al. 2002).

Knock out study of MDC1 (mediator of DNA damage checkpoint protein 1) has led to disability of cell to induce its cycle checkpoints and apoptosis (Stewart et al. 2003).

In addition to recruitment of DNA repair complex, MCPH1 marks damaged DNA, which is wrapped in chromatin structure, open through interaction with SW1/SNF to be available for repair proteins (Peng et al. 2009). However, it will be discussed in further details in Chap. 7.

It is accepted that ATM and its counterpart molecule, ATR recognize and phosphorylate the serine and threonine within the S/T-Q motif of their substrates. It was determined that ATM and ATR recognize more than 900 S/T-Q motifs within about 700 target proteins (Shiloh 2003). ATM, initially, phosphorylates c-Abl which itself phosphorylate and activates Rad51, Rad52 and H2AX molecules through its special kinase function. Moreover, c-Abl phosphorylation is required for autophosphorylation of ATM at S1981 to help it to be maintained active. Phosphorylated H2AX or γ-H2AX (phosphorylation on SQE motif) marks the damaged DNA to be repaired through recruiting proteins responsible for the relevant pathway (Wang et al. 2011).

6.2.2 *Cell Cycle Control*

Most of the ATM functions in both cell cycle control and DNA repair pathways are overlapped and linked together. However, after the firing the phosphorylation cascade of DDB's repair, ATM switches its role from DNA repair to cell cycle control. It was shown that cells lacking functional ATM have aberrant G1-S, intra S and G2-M cell cycle checkpoints (Barlow et al. 1996; Borghesani et al. 2000; Elson et al. 1996; Herzog et al. 1998).

Study on postmitotic neurons has demonstrated that it is initiated through phosphorylation of ATM by cyclin-dependent kinase 5 (Cdk5) at S794. This phosphorylation is necessary for the next one on S1981 ensuring intact ATM signaling. The immediate targets of ATM phosphorylation are Chk1 and Chk2 which in turn phosphorylate p53 and Cdc25. The p53 phosphorylation stabilizes it to drive DNA damage response and when the Cdc25 become phosphorylated it will be no longer able to activate CycA- and CycE-Cdk2, leading to cell cycle stalling at G1/S transition (Lukas et al. 2003). As described above, H2AX is one of the major substrates of ATM which has shown that it shares common functions with ATM in DNA repair when damage has been occurred in G1 phase. However, it was found that H2AX can, independently, operate ATM in repairing DNA after its replication (Furuta et al. 2003).

ATM is also affects the cell cycle in S or DNA replication phase which is fulfilled through activation of BRCA1 associated surveillance complex (BASC) leading to induction of p21 followed by either apoptosis or cell cycle arrest (Fig. 6.3).

Activation of CycB-Cdk1 CycB-Cdk1 also named as metaphase promoting factor (MPF) by Cdc25C is necessary for transition of cell cycle from G2 to metaphase.

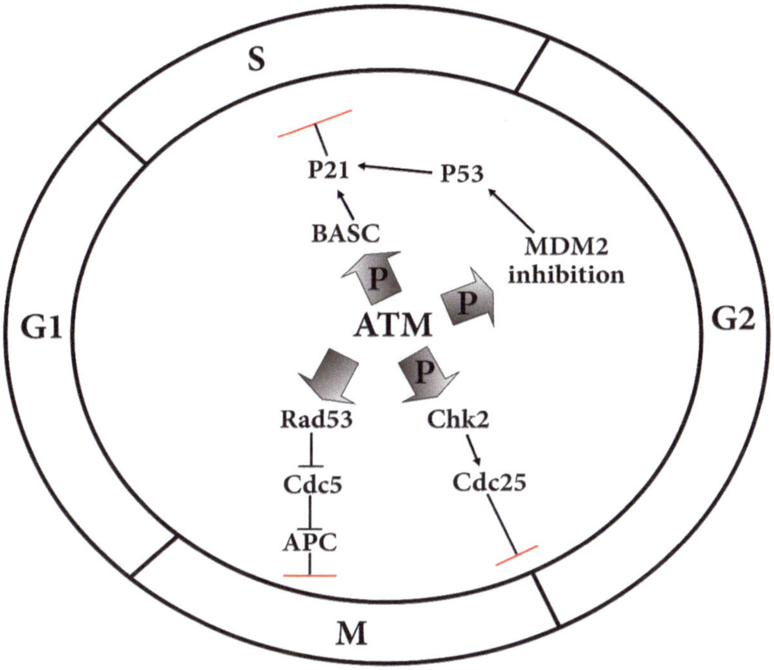

Fig. 6.3 Regulatory effects of ATM on the major cell cycle checkpoints

When DNA has been damaged, Wee1 protein provides a delay in S-G2 transition and by phosphorylation and inactivation of MPF arrests the cell cycle to enter the mitosis phase. In fission yeast, there is another kinase protein called Mik1 which function besides Wee1 to stall the cell cycle before getting entry to mitosis. In addition, phosphorylation of Cdc25 by ATM induced Chk2, would be another reason that the MPF being inactivated.

Finally, the other major contribution of ATM in cell cycle control is fulfilled through phosphorylation followed by inactivation of MDM2 setting p53 free to induce the expression of *p21* gene. The p21 protein is a putative suppressor of the entire cyclin-Cdk complex thought the cell cycle, so the cycle will be stopped to be repaired (Bartek and Lukas 2001).

Mitosis phase control of an ATM analogue known as Mec1 is performed by its activation after DNA damage leading to induction of Chk1 and Rad53 (Chk2 homologouse in yeast). Rad53 in turn, suppresses the Cdc5 which is the major factor of exiting from anaphase through induction of APC and destruction of Cdk1-Cyclin B. Therefore, Mec1 hinders the anaphase transition and controls the cell cycle at this main checkpoint when the DNA has been damaged (Zhang et al. 2009) (Fig. 6.3).

6.3 Other Functions of ATM Kinase

6.3.1 ATM and Telomere Length

It is worth to note that the role of ATM in telomere integrity and length is more important and critical than DNA repair in yeast. The similarity in structure of yeast proteins involved in telomere length (TL) including TEL1 and rad3 and their human homologous ATM could support the correlation between ATM and TL (Pandita and Dhar 2000). The presence of various chromosomal abnormalities as well as chromosomal breakage and short TL in AT patients could be the further evidences. It was shown that the putative contribution of ATM in DDB repair is not conserved in yeast (Peng et al. 2009).

Treatment of isolated cells either somatic or germ cells derived from AT patients have demonstrated that interaction between nuclear matrix and telomere required for telomere and chromosomal stability is governed by ATM (Pandita and Dhar 2000). *ATM* transfection of fibroblast cells improved the described interaction while decreased the overall rate of chromosomal breakage. Directed mutagenesis of *ATM* in neuroblast cells of Drosophila have shown that it is essential for maintaining the chromosomal stability and integrity and avoids of end to end chromosome fusion (Silva et al. 2004). Further studies are merited to focus on why only specific cells such as neurons are more sensitive to ATM deficiency producing neurologic symptoms in AT patients.

It was demonstrated that ATM and Mre11 are recruited to telomeric foci and detect it as a DSB when the human cells committed to enter the senescence phase. It was confirmed through knocking out of ATM which was associated with the gaining the cellular capability to continue the cell cycle. Strikingly, it is assumed that there is another switch in ATM functions from telomere protection to cell cycle arrest when the cell goes to be turned off. Recently, it was revealed that ATM differentially functions in repair of interstitial and telomeric DSBs. Any DSB near the telomere remains unrepaired in AT patients due to the lack of ATM function in repair machinery. ATM deficiency was associated with higher incidence of large deletions in telomeric regions whereas the rate of small deletions and NHEJ was greatly increased. These findings propose that ATM fulfill the telomere protection rather than telomere resection during DSB repair. Absence of ATM, therefore leads to induction of NHEJ alternative repair pathway whereas its presence impedes DNA repair through protection of telomere sequences (Muraki et al. 2013). Another study on cells lacking ATM demonstrated that ATM deficiency is responsible for telomere shortening in older mice increasing the risk of cancer at higher ages (Vaziri et al. 1997).

Contribution of ATM to TL integrity was further verified that ATM deficiency makes telomeric repeat sequences untraceable to be recognized by telomerase and other proteins involved in maintaining telomere integrity (Khanna 2000).

6.3.2 ATM in Cytoplasm

The presence of some clinical presentations in AT patients which seems to be unrelated to defect in DNA repair could be strong evidence on the role of ATM in other cellular functions rather than cell cycle and DNA repair control. Although it was shown that ATM is more present in nucleus versus cytoplasm in different cell types, the distribution of it is approximately equal between nucleus and cytoplasm in neural cells of cerebellum (Yang et al. 2011; Barlow et al. 1999; Boehrs et al. 2007). Finding ATM in cytoplasm is indicative of its special functions rather than nuclear activities (Li et al. 2009). It was demonstrated that ATM was associated with cytoplasm organelles including peroxisomes and endosomes (Watters et al. 1999).

The over-expression of *ATM* gene upon insulin treatment and the presence of insulin resistant type 2 diabetes in AT patients are complementary findings on the critical role of ATM in insulin metabolism pathway. Actually, ATM is contributed to insulin signaling through phosphorylation of eIF-4E-binding protein 1 (4E-BP1) at serine 111 and its separation from eIF-4E translation elongation factor leading to cap dependent translation (Yang and Kastan 2000). Insulin resistance in mouse models manipulated to have no functional ATM, was a further evidence of its role in insulin metabolism (Wu et al. 2010). It is conceivable that ATM plays pivotal role in insuline induced protein synthesis and glucose metabolism and may relay the surviving signals from IGF receptor within the neural cells (Yang and Kastan 2000).

Moreover, ATM is relocated to cytoplasm within the ATM-NEMO complex when its signaling has led to cell apoptosis. NEMO or NF-κB essential modulator is phosphorylated by ATM to be transported accompanying it to cytoplasm wherein NEMO and IκB kinase interaction begins the NF-κB induced apoptosis pathway (Wu et al. 2006).

Along these lines, ATM modulates the mTOR complex 1 (mTORC1) signaling in cytoplasm through stimulation of tuberous sclerosis complex 2, LKB1 and MAPK or phosphorylation of hypoxia induced factor (HIF) to prevent autophagy (Shiloh 2003; Zoncu et al. 2010). Inability of ATM deficient cells to up regulate the HIF level may be an important evidence on this fact that it is the major sensor of cellular oxygen level (Mongiardi et al. 2011). In addition, ATM indirectly controls the mitochondrial function by modulating HIF and its complex which is consisted of HIF-1α and HIF-1β (Cam et al. 2010). The ATM interaction with HIF factor would be a good reason to propose that ATM is an active member of cancer cell metastasis pathway through inducing angiogenesis.

Involvement of ATM in mitochondrial diseases is relying on its important roles in mitochondrial functions (Ambrose et al. 2007). Given the deregulation of mitochondrial homeostasis in structure, membrane potential and respiratory activity in cells lacking ATM and *ATM* gene alterations in various mitochondrial diseases, it may be the guardian of mitochondrial genome in addition to nuclear genome (Inomata et al. 2009). Moreover, ATM is involved in mitochondrial biogenesis through activation of AMPK in response to ATP deficiency and indirectly avoids lipid, carbohydrate and protein synthesis pathways (Reznick and Shulman 2006). In addition, it is assumed that ATM controls the number of mitochondria, although it has not been elucidated that this control is performed in negative or positive manner. For instance *In vivo*, shortage of ATM in thymocytes was associated with increased number of abnormal mitochondria that their degradation led to enhanced oxidative stress described below (Valentin-Vega et al. 2012).

Strikingly, most of the downstream molecules of ATM as well as p53 or CREB finally destined to mitochondria (Ditch and Paull 2012). The critical described functions in response to oxidative stress and its byproducts would be another linkage between ATM and mitochondria. Oxidative stress means exceeding of either reactive oxygen species (ROS) such as hydrogen peroxide (H_2O_2) or superoxide anion radical ($\bullet O^{-2}$) or nitrogen reactive species (RNS) over anti-oxidant reagents within the cell (Barzilai et al. 2002). It is the prominent feature of AT patients considered as a major cause of neurodegeneration in them (Ditch and Paull 2012). Of note, it was shown that the mitochondrial level of cytochrome C had been decreased in AT patients indicating the possible effect and interaction of ATM with the elements of respiratory circle (Patel et al. 2011). It was also demonstrated that ATM increases the antioxidant levels through controlling the mitochondrial pentose phosphate pathway (PPP). Extra free electrons are indirectly neutralized by ATM through induction of glucose-6-phosphate dehydrogenase (G6PD) activity and increase in cellular levels of Nicotinamide Adenine Dinucleotide Phosphate (NADPH). In addition, two ATM molecules form dimer with each other through disulfide bonding between their cysteine amino acids in C2991 site. This dimer formation is necessary to overcome the inhibitory effect of ROS on DSB repair (Guo et al. 2010).

6.4 ATM Defects in Human Malignant and Non-Malignant Diseases

6.4.1 ATM and Various Cancers Predisposition

There are many evidences relying on the importance of *ATM* gene aberrations in development of different cancers and support its critical role as a tumor suppressor gene. Firstly, it is based on the increased risk of being affected with leukemia and lymphoma (20%) in AT patients and secondly, decrease in *ATM* expression which was observed in various types of animal and human cancers (Gumy-Pause et al. 2004; Matei et al. 2006). According to the later evidence, even heterozygote carriers of ATM mutations especially amongst relatives of AT patients are prone to develop malignancies (Swift et al. 1987). It was stated that around 1% of general populations are heterozygote for *ATM* gene alterations, each *ATM* mutation would be the focus of more assessment (Matei et al. 2006).

6.4.2 *ATM* Aberrations in Breast Cancer

A key question is that how much does *ATM* alteration raise the risk of breast cancer in AT patients, their relatives and carriers of *ATM* mutations? Unfortunately the precise answer is still unclear among various populations. Swift was the first one who by studying on 110 AT families, suggested the enhanced probability of breast cancer in female and male relatives of AT patients which was revealed to be 3.1 and 2.3, respectively by 1987 (Swift et al. 1987). Review on the subsequent studies on the risk of breast cancer in AT patient's relative demonstrated that they had 3.9% risk compared to general population (Easton 1994). Taken together with other comprehensive carried out studies, the overall risk of breast cancer is almost 2.23 more times in relatives of AT patients versus general population which is doubled in individuals who are younger than 50 years (Thompson et al. 2005).

The initial study on comparing breast cancer patients with healthy controls could not support the hypothesis of higher frequency of breast cancer incidence in carriers of *ATM* mutations (FitzGerald et al. 1997). It's may be due to the different pattern of *ATM* mutations in AT patients and healthy carriers. Except of dominant negative and some of the splice site alterations, missense mutations usually have less effects on protein and are associated with residual function of protein versus truncating mutations influencing cancer susceptibility (Gatti et al. 1999). It was proposed that the role of *ATM* variants especially non-synonymous ones is not important in developing breast cancer in the carriers (Renwick et al. 2006). The *ATM* mutations causing AT was considered as low penetrant breast cancer alterations which are associated with two fold increase in risk of breast cancer in the carriers (Thompson et al. 2005). Moreover, the risk of breast cancer in carriers of *ATM* mutations was deduced to be similar to the risk of breast cancer in carriers of CHEK2*1100delC alteration (Nevanlinna and Bartek 2006).

The presence of premature stop codon mutations in *ATM* gene was associated with nine fold increase in risk of familial breast cancer amongst Dutch population (Broeks et al. 2000b). However, the chance of breast cancer development has shown to be two times more in British patients who were carrying ATM alterations (Inskip et al. 1999; Thompson et al. 2005). Although several studies couldn't show the contribution of AT in pathogenesis of breast cancer, Renwick and his colleagues have demonstrated that carriers of these mutations have two fold chance to be affected by breast cancer (Renwick et al. 2006; Teraoka et al. 2001; Sommer et al. 2003).

However, there are many reports relying on the significant characters while unknown roles are assumed to be due to some nucleotide substitutions of *ATM* gene in development of breast cancer. The c.7271T>G alteration enhances the risk of breast cancer in either homo- or heterozygous mutated *ATM* individuals (Stankovic et al. 1998).

The T7271G alteration is considered as another ATM gene alteration conferring special risk of breast cancer which was identified in three AT families and one case control study. It was identified in a family in which the heterozygous mother and daughters for T7271G were affected by breast cancer in 82, 44 and 50 years old, respectively. Breast cancer was also diagnosed in three paternal aunts of two AT brothers carrying T7271G accompanying a truncating mutation of *ATM* gene. Five breast cancer patients were also found to be carrier of T7271G substitution in an Australian family (Stewart et al. 2001; Stankovic et al. 1998). In another study, it was found that this mutation was associated with more than 15 fold increases in risk of breast cancer with 60 % penetrance (Chenevix-Trench et al. 2002). However, further studies are warranted to elucidate the role of this substitution in pathogenesis of breast cancer.

The significance of IVS 10-6T>G splice site mutation which is associated with the loss of exon 11 of ATM gene, in susceptibility, to breast cancer has been evaluated by three different studies. Broeks and colleagues have found that carriers of this alteration are at risk for development of bilateral and early onset breast cancer (Broeks et al. 2003). It was also demonstrated that this mutation was penetrant in 17.2 % by the age of 70 years old (Thompson et al. 2005).

It is stated that being carrier for *ATM* gene mutations didn't cause over-reaction to radiotherapy in breast cancer patients meriting further studies to clarify it (Broeks et al. 2000). However, whether it is safer to consider radiotherapy as an alternative option in treatment schedule of breast cancer patients carrying *ATM* alterations or not, that is a crucial challenge which varies between different populations.

Sporadic breast cancer is as a result of loss of heterozygousity (LOH) of various genes in 40 % of cases as the LOH of *ATM* is contributed in the earlier phases of breast cancer pathogenesis (Rio et al. 1998; Hampton et al. 1994). Although, LOH of *ATM* gene may be associated with higher risk of cancer, it is a further support for two-hit model of tumorgenesis. However, we have preliminary demonstrated the three hit hypothesis in astrocytoma which will be described in detail later in section of brain tumors.

The importance of D1853N in breast cancer was previously published. We have investigated 129 Iranian patients affected with breast cancer to determine the presence of D1853N polymorphism in patients compared to 248 healthy controls through Mutant allele-specific PCR amplification (MASA) assay confirmed by sequencing. This polymorphism was genotyped in 31.0, 26.9 and 12.5 % of the cases, internal- and external- controls, respectively. The relative risk was calculated to be 2.5. It was concluded that the significant difference between patient carriers and controls would be helpful in screening of cancer prone families (Mehdipour et al. 2011a).

6.4.3 *ATM* and Risk of Leukemia

ATM mutations are associated with both types of leukemia either in adult or childhood lymphoma. However, in T-cell prolymphocytic leukemia (T-PLL) patients whom are usually diagnosed in later ages, both of ATM alleles were non-functional.

Actually, the LOH of *ATM* gene was found to be completed with a missense mutation frequently occurs in PI-3K domain encoding part of other allele (Vorechovsky et al. 1997; Stilgenbauer et al. 1997; Stoppa-Lyonnet et al. 2000). Although the LOH of *ATM* gene was also seen in acute lymphoblastic leukemia (ALL), it seems that it is infrequent in this type of leukemia (Haidar et al. 2000; Gronbaek et al. 2002).

Follicular center cell lymphoma (FCL) is low grade B-cell non-Hodgkin's lymphomas (BNHL) with mild presentations (Cuneo et al. 2000) in which heterozygous deletion and missense point mutation of *ATM* gene were described in 11 and 8 % of all the FCL respectively.

The role of *ATM* gene alterations in pathogenesis of B-cell chronic lymphocytic leukemia (B-CLL) sounds to be more important as they were demonstrated in 26–40 % of this type of leukemia (Bullrich et al. 1999; Starostik et al. 1998; Stankovic et al. 2002; Schaffner et al. 1999). Missense mutations were found to be the leading type of *ATM* mutations in B-CLL patients which are usually associated with lower expression of ATM protein in homozygous status (Starostik et al. 1998). The B-CLL patients carrying *ATM* mutations had shown more severe phenotype than mutation in other genes that it may be due to their disability to take care of DNA against damages induced by treatment or every endogenous stresses (Stankovic et al. 2004). It was confirmed in another study on 57 CLL patients that, *ATM* gene was deleted or had point mutation in 25 % of patients (Guarini et al. 2012).

The t(11; 14) (q13;q32) translocation and 11q deletion are the frequent *ATM* alterations which were found in Mantle cell lymphoma (MCL). A microarray analysis on all types of lymphoma had revealed that MCL owned the highest frequency of *ATM* alterations, even though no association was found between *ATM* mutations and clinical presentations (Fang et al. 2003).

Diffuse large B-cell lymphoma (DLBCL) is the other type of lymphoma in which *ATM* mutations were determined in 13–20 % of these patients. However, no evidence was found relying on the hypermethylation of *ATM* promoter in DLBCL patients (Fang et al. 2003; Gronbaek et al. 2002).

Germ line *ATM* alterations were almost detected in Hodgkin's diseases (HD) that have increased the risk of breast cancer after radiotherapy (Broeks et al. 2000a; Offit et al. 2002).

Although there are a few reports on the role of *ATM* mutations in childhood leukemia, it was demonstrated that *ATM* alterations were found in 25 % of ALL patients which was significantly associated with the disease relapse (Gumy Pause et al. 2003).

6.4.4 ATM Role in Brain Tumors

Barlow and colleagues initially found that ATM protein has cytoplasmic localization in Purkinje cells as well as cerebellum. They demonstrated that the presence of ATM protein in cytoplasm is necessary and critical for balancing the number of cytoplasmic organelles, in particular lysosome. Accumulation of lysosomes in cytoplasm found in ATM−/− knocked out mice and was not associated with any neuronal degeneration (Barlow et al. 2000). Since then, ATM has been the focus of another study to determine the correlation between its expression and radioresistancy in glioblastoma multiform (GBM) in multiple GBM cell lines. Although there was inconsistent pattern in ATM expression among various types of cell, it was described that reducing the ATM expression may help to alleviate the mode of radio-resistance in GBM patients (Tribius et al. 2001).

Investigation of *ATM* gene mutations was followed in one GBM cell line (M059J) through yeast-based frameshift/stop codon assay. It was found that the absence of catalytic subunit of DNA-PKcs beside a truncating mutation in *ATM* were responsible for hypersensitivity of M059J to radiation (Tsuchida et al. 2002). However, in the attempts to find *ATM* mutations in medulloblastomas, no success was achieved in tumor samples, while the LOH of 11q region was identified in 25 % of the patients. Of note, the two putative D1853N and F858L polymorphisms of *ATM* gene were identified in approximately 20 % of patients (Liberzon et al. 2003).

We have described the D1853N polymorphism in a patient affected with astrocytoma and proposed the three hit hypothesis (Mehdipour et al. 2008). Based on this hypothesis, first alteration was inherited through germ line and the two novel splice sites including IVS 38- 63T→A and IVS38- 30 A→G polymorphisms constitute the second and third hits. Of note, D1853N polymorphism was found in one allele whereas the other two polymorphisms were found in another allele (Mehdipour et al. 2008). D1853N polymorphism was transmitted from proband's mother while the IVS 38- 63T→A polymorphism has been arisen in the early zygotic stage before differentiation of the peripheral blood. Although, the well-known two hits hypothesis proposed by knudson is a putative mechanism for tumor evolution, it was not shown in other cancer types rather than retinoblastoma.

Assessment of multiple loci LOH including *ATM* on brain metastasis in sporadic breast cancer patients demonstrated that multiple loci LOH have reduced the survival after brain metastasis (Hampl et al. 1998).

Regarding the *ATM* expression on the fresh brain tumor tissue samples, we have assayed the expression of mRNA of *Cyclin D2, P53, Rb* and *ATM* genes in 52 brain tumor patients by Real time polymerization chain reaction (Real time PCR) (Kheirollahi et al. 2011). It was demonstrated that *ATM, Rb, P53* and *Cyclin D2* had higher expression in astrocytoma compared to meningioma samples. However, *ATM* and *Cyclin D2* genes had higher expression in higher grades of astrocytoma while declined expression was detectable in Rb and *P53* genes. As a complementary insight and by considering the nature of protein expression, expression of ATM and p53 is reflective of low expression of ATM and p53 protein in tumor cells of patient affected with meningioma. In contrast both proteins revealed to have diverse pattern

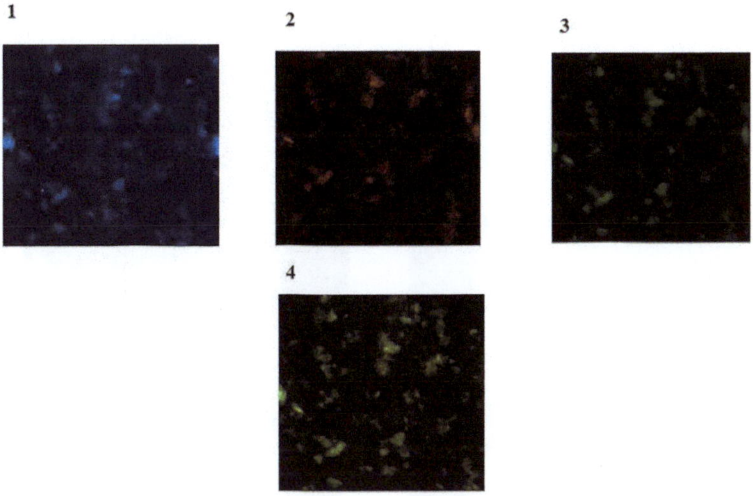

Fig. 6.4 Protein expressions of ATM and p53 in brain tumor cells. *1* Tumor cells of a patient with meningioma with dapi (*blue*); *2* Same cells conjugated with R-Pe (*orange*) representative of ATM protein characterized with low expression by IF. *3* Same cells conjugated with FITC (*green*), reflecting low expression of p53 protein; *4* Merged image of ATM and p53 presenting harmonized cooperation between these two proteins

as two clone of cells including one with low and another clone with high expression in patient with astrocytoma. As it is expected, both protein expression was high in control (Fig. 6.4).

However, the image of Cyclin D2/Cyclin E/C-Fos in a patient affected with astrocytoma is provided (Fig. 6.5). In this figure, there is more harmony between Cyclin D2 and c-Fos, but the status of protein expression in Cyclin E is rather low. Furthermore by considering the key role of cell cycle targets and transcription factor through tumor progression and to highlight the diverse pattern of protein expression in different pathological classifications of brain tumors, expression profile of glioblastoma multiform,

astrocytoma and meningioma has been provided (Fig. 6.6). In addition, expression profile of ATM/Rb/Ki67 in a meningioma patient is reflective of low expression of Ki 67, and more coordination between ATM and Rb protein (Fig. 6.7). To explore the crucial role of protein expression at cellular level, more complementary images and data is available in our recent publication (Mehdipour et al. 2014).

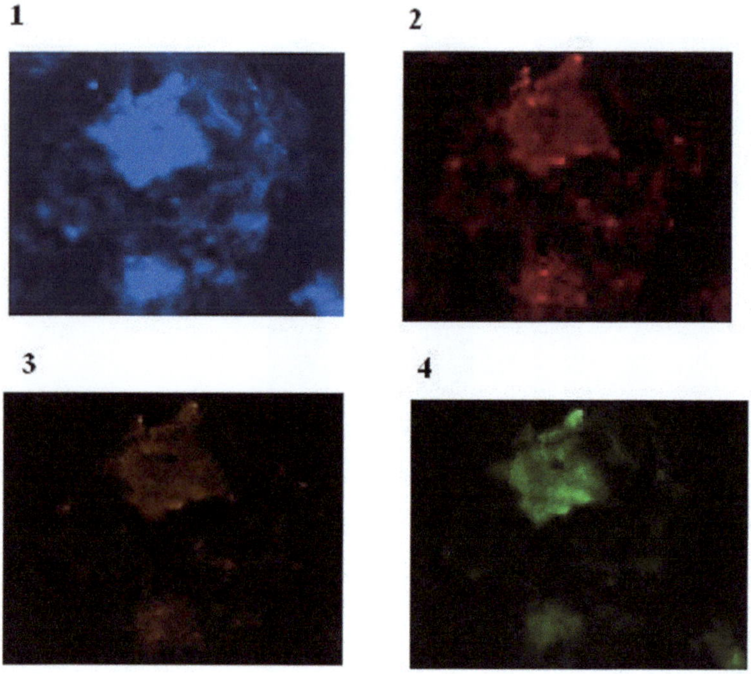

Fig. 6.5 Protein expression of Cyclin D2, Cyclin E and C-Fos in a patient affected with astrocytoma is provided . *1* Brain tissue cells of a patient affected with astrocytoma with dapi; *2* Same cells conjugated with Pe-cy5 reflecting diverse expression of Cyclin D2 including low and high; *3* The same cells conjugated with R-Pe illustrating very low expression of Cyclin E; and *4* Same cells with FITC presenting mixed expression mode including low, medium and high. In this figure, there is more harmony between Cyclin D2 and c-Fos, but the status of protein expression in Cyclin E is rather low. (From P.Mehdipour's archive)

6.4.5 ATM Gene Aberrations in Gastric and Pancreatic Carcinoma

It was initially demonstrated that the low level of *ATM* gene expression and protein phosphorylation was significantly associated with poor gastric tumor differentiation and patient survival. They also have found five polymorphisms within the *ATM* gene in 15 % of the advanced primary gastric cancer patients (Kang et al. 2008). In another study on ten human gastric cancer cell lines and fresh frozen tissues obtained from 604 gastric cancer patients, it was found that intronic *ATM* gene variations could be hot spots for driving microsatellite instability (MSI) in a specific group of gastric cancer cells (Kim et al. 2013). The recent study on larger sample size (321 gastric cancer cases) has verified the latter results and especially relied on the importance of MSI status and ATM expression in determination of the survival rate of gastric cancer patients after surgery (Kim et al. 2014).

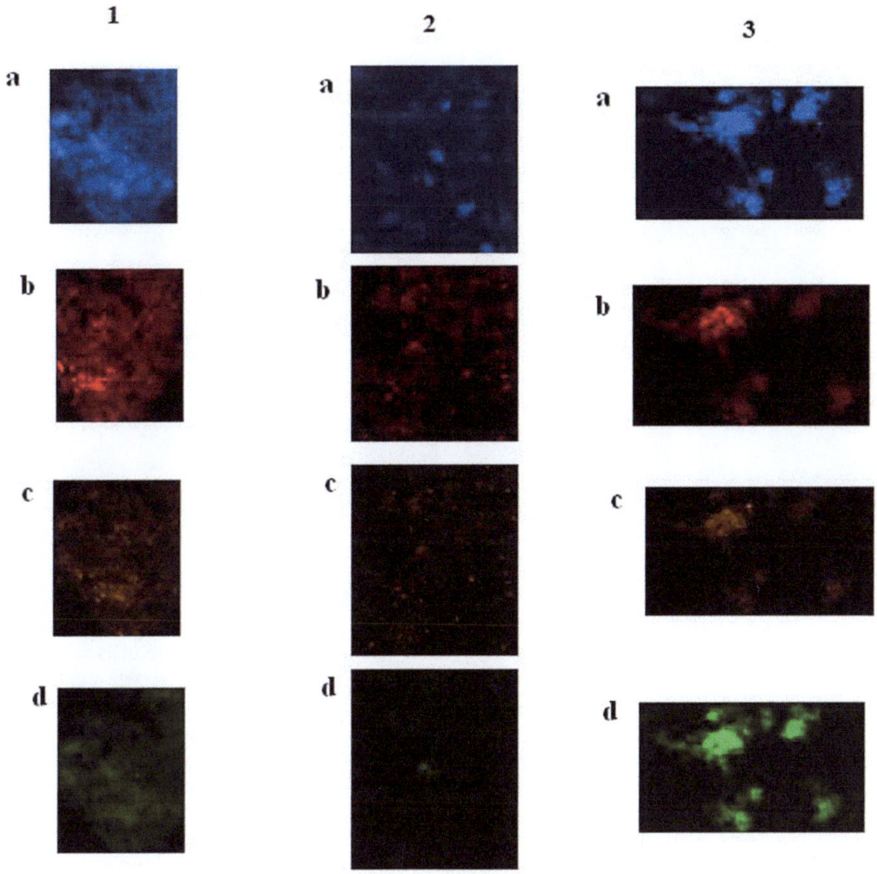

Fig. 6.6 Protein expressions of cyclin D2, CyclinE and c-Fos in brain tumor cells of patients affected with brain tumors. *GBM* Gleioblastoma multiform. *1* Tumor cells of brain tissue of a patient affected with GBM. **a** Tumor cells with dapi filter (*blue*), as counter stain; **b** cells conjugated with Pe-cy5 (*Texas red*) reflect overexpression of Cyclin D2; **c** same cells conjugated with, R-pe reflecting low expression of cyclin E protein; and **d** same cells conjugated with FITC (*green*), representative of c-Fos protein characterized with very low expression. *2* Tumor cells of brain tissue of a patient affected with astrocytoma. **a** Tumor cells with dapi filter (*blue*), as counter stain; **b** cells conjugated with Pe-cy5 (*Texas red*) reflect relativelyover-expression of Cyclin D2; **c** same cells conjugated with, R-pe reflecting low expression of cyclin E protein; and **d** same cells conjugated with FITC (*green*), representative of c-Fos protein characterized with very low expression. *3* Tumor cells of brain tissue of a patient affected with meningioma. **a** Tumor cells with dapi filter (*blue*), as counter stain; **b** cells conjugated with Pe-cy5 (*Texas red*) reflect both low and high expression of Cyclin D2; **c** same cells conjugated with, R-pe reflecting low expression of cyclin E protein; and **d** same cells conjugated with FITC (*green*), representative of c-Fos protein characterized with high expression. Magnification: x100. (Adopted from Mehdipour et al. 2014: P.Mehdipour's Archives)

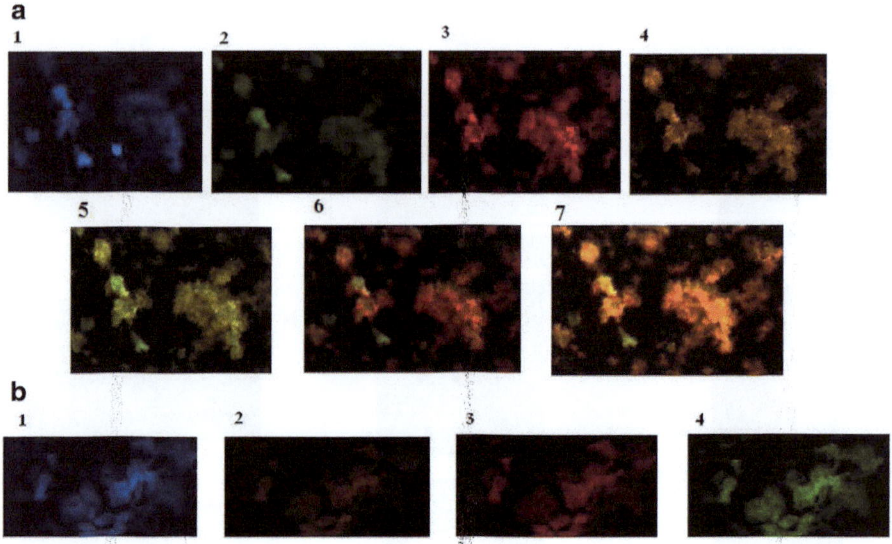

Fig. 6.7 Protein Expression profile of ATM, Rb and Ki67 in a patient affected with meningioma a. Tumor cells of brain tissue of a patient affected with meningioma: *1* Brain tissue cells with dapi; *2* Same cells conjugated with FITC reflecting low expression of Ki67; *3* Same cells conjugated with Pe-Cy5 showing diverse expression of Rb (*low, medium* and *high*); *4* Same cells with R-Pe indicating major cells with downregulated expression of ATM accompanied by very few cells with high expression; *5* co-expression of Ki67 and ATM in majority of cells accompanied by minor clone with prominent epression of Ki67; *6* Merged of Ki67 and Rb reflects diverse co-expression with prominent expression of Rb in majority of cells; *7* co-expression of Ki67, Rb, and ATM is indicative of a diverse interaction between these three targets and a noticeable harmonic expression between Rb and ATM as well. **b** Brain cells of a healthy diseased individual: *1* Brain tissue cells with dapi; *2* Same cells conjugated with FITC reflecting low expression of Ki67; *3* Same cells conjugated with Pe-Cy5 showing low expression; *4* Same cells with R-Pe indicating major cells with downregulated expression of ATM.Magnification: x200. (From P. Mehdipour's archive)

The pattern of *ATM* gene mutations has shown more than 50 % similarity in early gastric cancer (EGC) compared to high-grade intraepithelial neoplasia (HG-IEN). It may indicate that ATM alterations has no role in progression from early to intraepithelial gastric lesions (Fassan et al. 2013). Special mutation analysis of *ATM* gene in multiple pancreatic cancer families has revealed that *ATM* gene mutations has important role in families with more than three affected members (Roberts et al. 2012).

6.4.6 ATM Alterations in Other Cancers

In cervical cancer, there is a recent report in which no association was found between 5557G>A polymorphism of *ATM* gene and risk of cervical cancer (Paulikova et al. 2014).

Interestingly, inactivation of ATM gene led to formation of transmembrane protease/serine subfamily member 2: estrogen-regulated genes (TMPRSS2: ERG chromosomal rearrangement in non-malignant prostate epithelial cells (HPr-1AR) which has been frequently seen as a mechanism of chromosome instability induction (Chiu et al. 2012). Chromosomal rearrangement is expected in ATM deficiency as a consequence of genomic and chromosomal instability when the damaged DNA left unrepaired. Investigation of chromosomal rearrangement in other cancers lacking functional ATM may be helpful to clarify this correlation more obvious.

The significance of P1054R polymorphism was established in association with prostate cancer risk in two sequential studies that in the last one, it had shown two fold increases in risk of prostate cancer (Angele et al. 2004; Meyer et al. 2007).

The cancer protective effect of Gallic acid was found through increase in ATM induced CDC25A and CDC25C phosphorylation followed by cell cycle arrest (Agarwal et al. 2006).

6.4.7 ATM and Other Diseases

ATM shares its prominent involvement in clinic including diabetes mellitus, neurological symptoms and cancer predisposition through the genetic abnormalities in mitochondria and oxidative stress pathway (Schon and Manfredi 2003). The putative role of ATM in control of biogenesis and replication of mitochondria may support these similar clinical features. It was found that AT patients have malstructure in mitochondria with less membrane and cytochrome c oxidase activity (Patel et al. 2011). However, the role of ATM mutations in mitochondrial disorders and other diseases remains to be elucidated.

6.5 Methylation of *ATM* Promoter Gene in Various Cancers

6.5.1 ATM Methylation in Breast, Ovarian and Lung Cancers

In the initial mehtylation analysis on breast tumor tissues, 78 % methylation rate was found through methylation specific PCR of *ATM* promoter gene confirmed by bisulfite sequencing(Vo et al. 2004). Dense methylation of *ATM* promoter was significantly correlated with downregulation of its expression revealed in low mRNA abundance through Real-time PCR.

In contrast, the following study on more invasive tumor samples demonstrated that the *ATM* promoter was unmethylated in all of the analyzed samples. However, the ATM protein expression which was determined through immunohistochemistry has been decreased in around 39 % of the tumor samples and therefore no meaningful correlation was found between *ATM* promoter methylation and its protein expression (Treilleux et al. 2007). Seventeen genes including 5 known susceptible genes in familial breast cancer as well as *ATM* accompanying 12 genes which were found to be methylated in various sporadic breast cancer studies were selected to define their promoter methylation status in bilateral breast cancer patients. Microarray was implicated to determine the methylation status of 17 genes in peripheral blood in 14 patients compared to 14 healthy controls. The *ATM* promoter gene was not methylated in no one of the cases and controls while there was significant association between *ATM* gene body methylation and decreased expression (Flanagan et al. 2009).

In investigating the role of ATM in radioresistanc seen in Hereditary Non-Polyposis Colorectal Cancer (HNPCC), three cell lines derived from HNPCC patients were analyzed to determine the methylation status of *ATM* gene promoter (Kim et al. 2002). The HCT-116 cells demonstrated moderate radiosensitivity which was associated with decreased ATM expression and its proximal promoter hypermethylation. Treatment of HCT-116 cells with demethylating agent 5-azacytidine (AZA) re-established the normal *ATM* promoter methylation pattern and protein expression. These findings confirmed the previous hypothesis relying on the activation of some cellular responses which are common to those take place when the cell has been exposed to ionizing radiation.

The *ATM* promoter methylation status was examined in lymph nodes provided form non-small cell lung cancer (NSCLC) patients to determine the role of epigenetic aberrations in the high mortality rate of these patients (Safar et al. 2007). Methylation specific PCR analysis displayed that *ATM* promoter was methylated in 19% of all the resected nodes which had impact neither on patients' survival rate nor on tumor pathology and metastasis.

The same study, was carried out in epithelial ovarian cancer patients and ended in the same results (Flanagan et al. 2013). They have shown that except of stratifin (SFN) promoter methylation, epigenetic silencing of none of the studied genes including *ATM* had role in determining the survival rate of patients and risk of ovarian cancer. However, estrogen receptor 1 (ESR1) promoter methylation had meaningful association with level of CA125 tumor marker and neuropathy in the patient group to whom the paclitaxel have been prescribed. Their finding was in contrast to the previous study in the head and neck squamous cell carcinoma in which the *ATM* promoter methylation was found in 25% of the cultured cells. Moreover, the methylation rate was in strong correlation with early onset of cancer and poor prognosis and survival (Ai et al. 2004).

6.5.2 ATM Promoter Methylation in Brain Tumors

It was described that determination of the genome wide methylation status of brain tumor patients especially glioma, could help to identify the prognosis and survival of these patients. Stable hypermethylation of 263 genes, called as glioma-CpG island methylator phenotype (G-CIMP), was significantly associated with higher clinical outcomes in glioma patients (Noushmehr et al. 2010; Kloosterhof et al. 2013). Recently, 464 genes were identified to be hypermethylated in glioma patients that the top 10 of them include MTSS1, LDB3, HIPK2, PKD2, C11orf39, Ells1, C11orf2, FLJ36268, ZNF146 and GUP1 (Lai et al. 2014).

The first trial on seeking the *ATM* promoter methylation in brain tumors has performed in the study on three glioma cell lines including T98G, U118 and U87 to identify a possible role in radio-resistance of glioma patients (Roy et al. 2006). It was demonstrated that although both U87 and U118 cells have significant reduced ATM protein expression, but the *ATM* promoter was only methylated in U87 cell line. Radiosensitivity had two fold increases in U87 and U118 cell lines in comparison with T98G. Treatment of U87 cells with AZA which was associated with reversing the normal *ATM* promoter status and protein expression confirmed the results. This study revealed that ATM has heterogeneous promoter status and protein expression amongst different stages of brain tumors. Wang and his colleagues have shown decreased expression of ATM protein in glioma tissue samples compared to normal brain tissues. However, they have assayed only the methylation level of *Chk2* gene promoter which was in significant association with ATM expression, however the *ATM* methylation status in brain tumor tissues remains elusive (Wang et al. 2010).

In this regard, we performed a methylation specific PCR analysis on 30 available brain tumor tissues with various histopathology compared to two normal brain tissues to determine the methylation status of *ATM* gene promoter (Mehdipour et al. 2014). In addition, the ATM protein expression was assayed by immunofluorescence protein analysis using monoclonal mouse anti-human ATM. We have previously assayed the telomere length (TL) of brain tissues of these patients through southern blotting (Mehdipour et al. 2011b) which have been confirmed by Quantitative Fluorescence in situ hybridization (Q-FISH). By considering the molecular aspect, the *ATM* gene promoter was methylated in 73 % of the patients whereas it was unmethylated and semi-methylated in 3 and 5 patients, respectively (Fig. 6.8, 6.9).

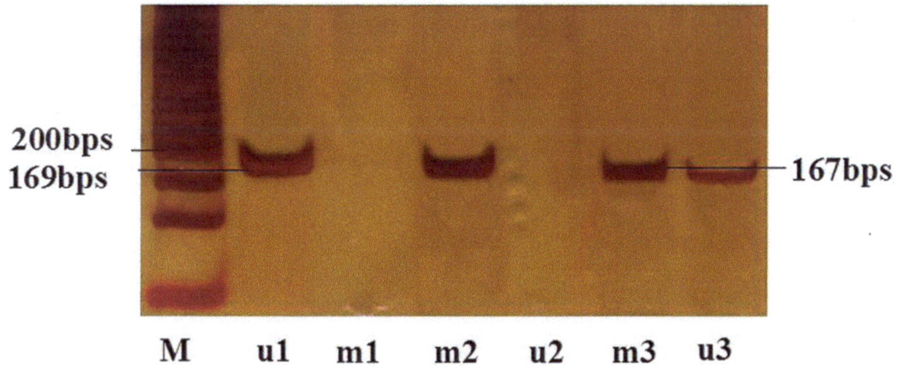

Fig. 6.8 Mehtylation specific PCR of ATM gene promoter. *u* PCR product of unmethylated template (169 bps), *m1* as unmethylated without methylated product; *m* PCR product of methylated template (167 bps); *m3* and *u3* are related to one of the semimethylated samples amplified by both methylated and unmethylated primer pairs. *M* marker

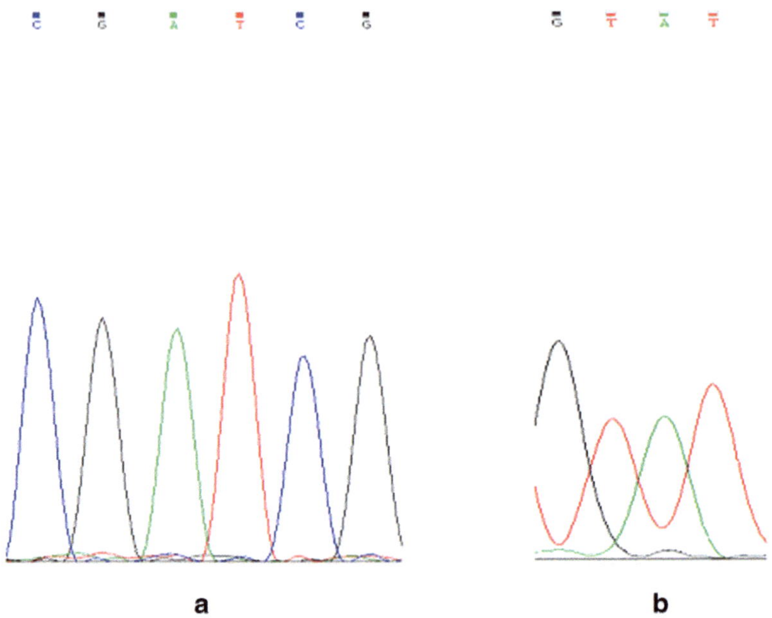

Fig. 6.9 Sequencing confirmation of mehtylated and unmethylated specific reaction for ATM gene. **a** Methylated. **b** Unmethylated

Among normal brain tissues, one was methylated and the other was unmethylated which were belonged to 82 and 71 years old individuals, correspondingly. ATM protein expression has shown diverse mixture pattern of expression (low, medium and high) in all the patients which was in strong correlation with *ATM* promoter methylation ($p<0.001$). However, *ATM* promoter methylation did not show any correlation with patient's age and tumor pathology including grade and stage. Moreover, the TL was not also meaningfully associated with *ATM* promoter methylation and protein expression while demonstrated significant correlation with grade and stage of brain tumors besides age of recruited patients ($p=0.01$). To our knowledge, we have primarily found the significant epigenetic silencing of *ATM* gene in various pathologies of brain tumors which was significantly correlated with its protein expression. Although, no association was found between TL and *ATM* promoter methylation, the crucial role of ATM in modulating TL would not be underestimated (Mehdipour et al. 2014). In this regard, it is required to highlight the pivotal interaction between ATM protein expression and TL in further studies. (Fig. 6.10). Figure 6.10 is indicating the impact of heterogenic signal mode at cellular level including two clones of cells characterized as cells lacking signals accompanied by very few cells with low telomeric signal intensity. Such characteristics is found to be in concordance with low ATM protein expression.

In addition, it seems that when the retinoblastoma protein (pRB) become inactivated, ATM accompanies the TIP60 protein which modulates the formation of a degradation complex to destruct DNA methyltransferase 1 (DNMT1)

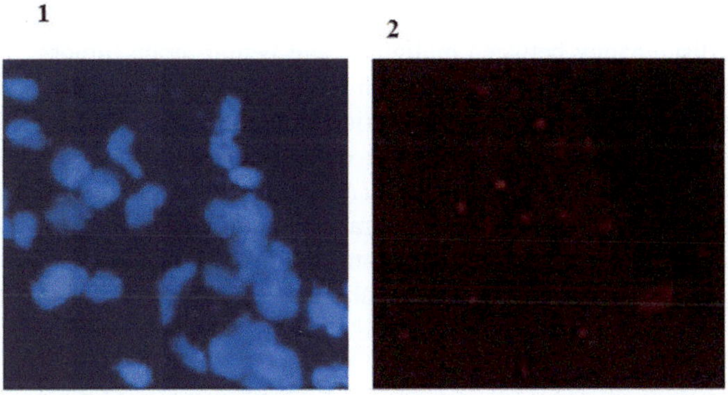

Fig. 6.10 Telomeric signal status of brain tumor from an astrocytoma patient by quantitative fluorescence in situ hybridization *1* Brain tumor cells from a patient affected with astrocytoma with dapi filter. *2* same cells conjugated with Cy3 characterized by lack of signals accompanied by very few cells with low intensity of telomeric signals. (Magnification: x400)

(Shamma et al. 2013). The reduced expression of RB protein in our brain tumor samples, in the presence of low expression and promoter methylation of *ATM*, may alter the methylation pattern of other genes which have to be silenced (Mehdipour et al. 2014).

Further investigations, therefore are warranted to complete our research in two paths; First, assessment of correlation between *ATM* promoter methylation, its protein expression and TL. Second, does *ATM* promoter methylation affect the expression of other genes which their expression had been restricted through promoter methylation? Such complementary research not only shed light on the new pathways of tumor progression and metastasis, but also by focusing on the controlling role of ATM on telomere length, we may clarify the determining factors in survival of cancer patients.

High rate of promoter methylation in ATM gene motivated us to search for the methylation of the promoter of MCPH1 and p53 genes interacting with it which will be further discussed in Chaps. 7 and 8 respectively.

6.6 Conclusion

The epigenetic modulation of ATM expression as the apex of the DNA repair cascade can affect the expression of the downstream targets indirectly. The *ATM* promoter methylation was confirmed in most of the methylation analysis focused on various types of cancer which was almost consistent with low protein expression. However, except of two studies which were carried out on in glioma patients (Noushmehr et al. 2010; Kloosterhof et al. 2013), no report of the other assays could find significant correlation between *ATM* promoter methylation and the grade and stage of tumors. In our study, the frequency of methylated *ATM* promoter had no overall discrepancy between malignant and benign brain tumors. It may highlight the earlier ATM suppression through promoter methylation which deviates the normal cell cycle toward hyperproliferation in the initial stages of tumor progression. Subsequent downstream molecular downregulations are capable to promote the development of malignant tumors. *ATM* promoter methylation, therefore, could be considered as an initial marker of alteration in normal molecular pattern of cells. Further confirmatory investigations are still required to include the *ATM* promoter methylation analysis in the first molecular analysis panel of different cancers.

References

Agarwal C, Tyagi A, Agarwal R (2006) Gallic acid causes inactivating phosphorylation of cdc25A/cdc25C-cdc2 via ATM-Chk2 activation, leading to cell cycle arrest, and induces apoptosis in human prostate carcinoma DU145 cells. Mol Cancer Ther 5:3294–3302

Ahnesorg P, Smith P, Jackson SP (2006) XLF interacts with the XRCC4-DNA ligase IV complex to promote DNA nonhomologous end-joining. Cell 124:301–313

Ai L, Vo QN, Zuo C, Li L, Ling W, Suen JY et al (2004) Ataxia-telangiectasia-mutated (ATM) gene in head and neck squamous cell carcinoma: promoter hypermethylation with clinical correlation in 100 cases. Cancer Epidemiol Biomarkers Prev 13:150–156

Ambrose M, Goldstine JV, Gatti RA (2007) Intrinsic mitochondrial dysfunction in ATM-deficient lymphoblastoid cells. Hum Mol Genet 16:2154–2164

Angele S, Falconer A, Edwards SM, Dork T, Bremer M, Moullan N et al (2004) ATM polymorphisms as risk factors for prostate cancer development. Br J Cancer 91:783–787

Bakkenist CJ, Kastan MB (2003) DNA damage activates ATM through intermolecular autophosphorylation and dimer dissociation. Nature 421:499–506

Barlow C, Hirotsune S, Paylor R, Liyanage M, Eckhaus M, Collins F et al (1996) Atm-deficient mice: a paradigm of ataxia telangiectasia. Cell 86:159–171

Barlow C, Liyanage M, Moens PB, Deng CX, Ried T, Wynshaw-Boris A (1997) Partial rescue of the prophase I defects of Atm-deficient mice by p53 and p21 null alleles. Nat Genet 17:462–466

Barlow C, Dennery PA, Shigenaga MK, Smith MA, Morrow JD, Roberts LJ et al (1999) Loss of the ataxia-telangiectasia gene product causes oxidative damage in target organs. Proc Natl Acad Sci U S A 96:9915–9919

Barlow C, Ribaut-Barassin C, Zwingman TA, Pope AJ, Brown KD, Owens JW et al (2000) ATM is a cytoplasmic protein in mouse brain required to prevent lysosomal accumulation. Proc Natl Acad Sci U S A 97:871–876

Bartek J, Lukas J (2001) Pathways governing G1/S transition and their response to DNA damage. FEBS Lett 490:117–122

Barzilai A, Rotman G, Shiloh Y (2002) ATM deficiency and oxidative stress: a new dimension of defective response to DNA damage. DNA Repair (Amst) 1:3–25

Bennetzen MV, Larsen DH, Bunkenborg J, Bartek J, Lukas J, Andersen JS (2010) Site-specific phosphorylation dynamics of the nuclear proteome during the DNA damage response. Mol Cell Proteomics 9:1314–1323

Bensimon A, Schmidt A, Ziv Y, Elkon R, Wang SY, Chen DJ et al (2010) ATM-dependent and -independent dynamics of the nuclear phosphoproteome after DNA damage. Sci Signal 3:rs3

Bensimon A, Aebersold R, Shiloh Y (2011) Beyond ATM: the protein kinase landscape of the DNA damage response. FEBS Lett 585:1625–1639

Boehrs JK, He J, Halaby MJ, Yang DQ (2007) Constitutive expression and cytoplasmic compartmentalization of ATM protein in differentiated human neuron-like SH-SY5Y cells. J Neurochem 100:337–345

Borghesani PR, Alt FW, Bottaro A, Davidson L, Aksoy S, Rathbun GA et al (2000) Abnormal development of Purkinje cells and lymphocytes in Atm mutant mice. Proc Natl Acad Sci U S A 97:3336–3341
Broeks A, Russell NS, Floore AN, Urbanus JH, Dahler EC, van TVMB et al (2000a) Increased risk of breast cancer following irradiation for Hodgkin's disease is not a result of ATM germline mutations. Int J Radiat Biol 76:693–698
Broeks A, Urbanus JH, Floore AN, Dahler EC, Klijn JG, Rutgers EJ (2000b) ATM-heterozygous germline mutations contribute to breast cancer-susceptibility. Am J Hum Genet 66:494–500
Broeks A, Urbanus JH, de Knijff P, Devilee P, Nicke M, Klopper K et al (2003) IVS10-6T>G, an ancient ATM germline mutation linked with breast cancer. Hum Mutat 21:521–528
Brown KD, Ziv Y, Sadanandan SN, Chessa L, Collins FS, Shiloh Y et al (1997) The ataxia-telangiectasia gene product, a constitutively expressed nuclear protein that is not up-regulated following genome damage. Proc Natl Acad Sci U S A 94:1840–1845
Buck D, Malivert L, de Chasseval R, Barraud A, Fondaneche MC, Sanal O (2006) Cernunnos, a novel nonhomologous end-joining factor, is mutated in human immunodeficiency with microcephaly. Cell 124:287–299
Bullrich F, Rasio D, Kitada S, Starostik P, Kipps T, Keating M et al (1999) ATM mutations in B-cell chronic lymphocytic leukemia. Cancer Res 59:24–27
Cam H, Easton JB, High A, Houghton PJ (2010) mTORC1 signaling under hypoxic conditions is controlled by ATM-dependent phosphorylation of HIF-1alpha. Mol Cell 40:509–520
Cavero S, Chahwan C, Russell P (2007) Xlf1 is required for DNA repair by nonhomologous end joining in *Schizosaccharomyces pombe*. Genetics 175:963–967
Chenevix-Trench G, Spurdle AB, Gatei M, Kelly H, Marsh A, Chen X et al (2002) Dominant negative ATM mutations in breast cancer families. J Natl Cancer Inst 94:205–215
Chiu YT, Liu J, Tang K, Wong YC, Khanna KK, Ling MT (2012) Inactivation of ATM/ATR DNA damage checkpoint promotes androgen induced chromosomal instability in prostate epithelial cells. PLoS ONE 7:e51108
Ciccia A, Elledge SJ (2010) The DNA damage response: making it safe to play with knives. Mol Cell 40:179–204
Concannon P, Gatti RA (1997) Diversity of *ATM* gene mutations detected in patients with ataxia-telangiectasia. Hum Mutat 10:100–107
Cuneo A, Bigoni R, Rigolin GM, Roberti MG, Milani R, Bardi A (2000) Acquired chromosome 11q deletion involving the ataxia teleangiectasia locus in B-cell non-Hodgkin's lymphoma: correlation with clinicobiologic features. J Clin Oncol 18:2607–2614
Daniel JA, Pellegrini M, Lee JH, Paull TT, Feigenbaum L, Nussenzweig A et al (2008) Multiple autophosphorylation sites are dispensable for murine ATM activation in vivo. J Cell Biol 183:777–783
Das BB, Antony S, Gupta S, Dexheimer TS, Redon CE, Garfield S et al (2009) Optimal function of the DNA repair enzyme TDP1 requires its phosphorylation by ATM and/or DNA-PK. EMBO J 28:3667–3680
Deng CX (2006) BRCA1: cell cycle checkpoint, genetic instability, DNA damage response and cancer evolution. Nucleic Acids Res 34:1416–1426
Ditch S, Paull TT (2012) The ATM protein kinase and cellular redox signaling: beyond the DNA damage response. Trends Biochem Sci 37:15–22
Dvir A, Peterson SR, Knuth MW, Lu H, Dynan WS (1992) Ku autoantigen is the regulatory component of a template-associated protein kinase that phosphorylates RNA polymerase II. Proc Natl Acad Sci U S A 89:11920–11924
Easton DF (1994) Cancer risks in A-T heterozygotes. Int J Radiat Biol 66:S177–182
Elson A, Wang Y, Daugherty CJ, Morton CC, Zhou F, Campos-Torres J et al (1996) Pleiotropic defects in ataxia-telangiectasia protein-deficient mice. Proc Natl Acad Sci U S A 93:13084–13089
Falck J, Coates J, Jackson SP (2005) Conserved modes of recruitment of ATM, ATR and DNA-PKcs to sites of DNA damage. Nature 434:605–611
Fang NY, Greiner TC, Weisenburger DD, Chan WC, Vose JM, Smith LM (2003) Oligonucleotide microarrays demonstrate the highest frequency of ATM mutations in the mantle cell subtype of lymphoma. Proc Natl Acad Sci U S A 100:5372–5377

Fassan M, Simbolo M, Bria E, Mafficini A, Pilotto S, Capelli P (2013) High-throughput mutation profiling identifies novel molecular dysregulation in high-grade intraepithelial neoplasia and early gastric cancers. Gastric Cancer 17:442–449

Filipponi D, Muller J, Emelyanov A, Bulavin DV (2013) Wip1 controls global heterochromatin silencing via ATM/BRCA1-dependent DNA methylation. Cancer Cell 24:528–541

FitzGerald MG, Bean JM, Hegde SR, Unsal H, MacDonald DJ, Harkin DP et al (1997) Heterozygous ATM mutations do not contribute to early onset of breast cancer. Nat Genet 15:307–310

Flanagan JM, Munoz-Alegre M, Henderson S, Tang T, Sun P, Johnson N et al (2009) Gene-body hypermethylation of ATM in peripheral blood DNA of bilateral breast cancer patients. Hum Mol Genet 18:1332–1342

Flanagan JM, Wilhelm-Benartzi CS, Metcalf M, Kaye SB, Brown R (2013) Association of somatic DNA methylation variability with progression-free survival and toxicity in ovarian cancer patients. Ann Oncol 24:2813–2818

Furuta T, Takemura H, Liao ZY, Aune GJ, Redon C, Sedelnikova OA et al (2003) Phosphorylation of histone H2AX and activation of Mre11, Rad50, and Nbs1 in response to replication-dependent DNA double-strand breaks induced by mammalian DNA topoisomerase I cleavage complexes. J Biol Chem 278:20303–20312

Gao G, Bracken AP, Burkard K, Pasini D, Classon M, Attwooll C et al (2003) NPAT expression is regulated by E2F and is essential for cell cycle progression. Mol Cell Biol 23:2821–2833

Gatti RA, Tward A, Concannon P (1999) Cancer risk in ATM heterozygotes: a model of phenotypic and mechanistic differences between missense and truncating mutations. Mol Genet Metab 68:419–423

Gilad S, Chessa L, Khosravi R, Russell P, Galanty Y, Piane M et al (1998) Genotype-phenotype relationships in ataxia-telangiectasia and variants. Am J Hum Genet 62:551–561

Goodarzi AA, Jonnalagadda JC, Douglas P, Young D, Ye R, Moorhead GB et al (2004) Autophosphorylation of ataxia-telangiectasia mutated is regulated by protein phosphatase 2A. EMBO J 23:4451–4461

Goodarzi AA, Noon AT, Deckbar D, Ziv Y, Shiloh Y, Lobrich M et al (2008) ATM signaling facilitates repair of DNA double-strand breaks associated with heterochromatin. Mol Cell 31:167–177

Gottlieb TM, Jackson SP (1993) The DNA-dependent protein kinase: requirement for DNA ends and association with Ku antigen. Cell 72:131–142

Gronbaek K, Worm J, Ralfkiaer E, Ahrenkiel V, Hokland P, Guldberg P (2002) ATM mutations are associated with inactivation of the *ARF-TP53* tumor suppressor pathway in diffuse large B-cell lymphoma. Blood 100:1430–1437

Guarini A, Marinelli M, Tavolaro S, Bellacchio E, Magliozzi M, Chiaretti S et al (2012) *ATM* gene alterations in chronic lymphocytic leukemia patients induce a distinct gene expression profile and predict disease progression. Haematologica 97:47–55

Gumy Pause F, Wacker P, Maillet P, Betts D, Sappino AP (2003) ATM gene alterations in childhood acute lymphoblastic leukemias. Hum Mutat 21:554

Gumy-Pause F, Wacker P, Sappino AP (2004) ATM gene and lymphoid malignancies. Leukemia 18:238–242

Guo Z, Deshpande R, Paull TT (2010) ATM activation in the presence of oxidative stress. Cell Cycle 9:4805–4811

Haidar MA, Kantarjian H, Manshouri T, Chang CY, O'Brien S, Freireich E (2000) ATM gene deletion in patients with adult acute lymphoblastic leukemia. Cancer 88:1057–1062

Hampl M, Hampl JA, Schwarz P, Frank S, Hahn M, Schackert G et al (1998) Accumulation of genetic alterations in brain metastases of sporadic breast carcinomas is associated with reduced survival after metastasis. Invasion Metastasis 18:81–95

Hampton GM, Mannermaa A, Winqvist R, Alavaikko M, Blanco G, Taskinen PJ et al (1994) Loss of heterozygosity in sporadic human breast carcinoma: a common region between 11q22 and 11q23.3. Cancer Res 54:4586–4589

Herzog KH, Chong MJ, Kapsetaki M, Morgan JI, McKinnon PJ (1998) Requirement for Atm in ionizing radiation-induced cell death in the developing central nervous system. Science 280:1089–1091

Hopfner KP, Craig L, Moncalian G, Zinkel RA, Usui T, Owen BA et al (2002) The Rad50 zinc-hook is a structure joining Mre11 complexes in DNA recombination and repair. Nature 418:562–566

Inomata K, Aoto T, Binh NT, Okamoto N, Tanimura S, Wakayama T et al (2009) Genotoxic stress abrogates renewal of melanocyte stem cells by triggering their differentiation. Cell 137:1088–1099

Inskip HM, Kinlen LJ, Taylor AM, Woods CG, Arlett CF (1999) Risk of breast cancer and other cancers in heterozygotes for ataxia-telangiectasia. Br J Cancer 79:1304–1307

Kadyk LC, Hartwell LH (1992) Sister chromatids are preferred over homologs as substrates for recombinational repair in Saccharomyces cerevisiae. Genetics 132:387–402

Kang B, Guo RF, Tan XH, Zhao M, Tang ZB, Lu YY (2008) Expression status of ataxia-telangiectasia-mutated gene correlated with prognosis in advanced gastric cancer. Mutat Res 638:17–25

Kapp LN, Painter RB, Yu L-C, van Loon N, Richard CW, James MR, Cox DR, Murnane JP (1992) Cloning of a candidate gene for ataxia-telangiectasia group D. Am J Hum Genet 51:45–54

Khanna KK (2000) Cancer risk and the ATM gene: a continuing debate. J Natl Cancer Inst 92:795–802

Kheirollahi M, Mehr-Azin M, Kamalian N, Mehdipour P (2011) Expression of cyclin D2, P53, Rb and ATM cell cycle genes in brain tumors. Med Oncol 28:7–14

Kim ST, Lim DS, Canman CE, Kastan MB (1999) Substrate specificities and identification of putative substrates of ATM kinase family members. J Biol Chem 274:37538–37543

Kim WJ, Vo QN, Shrivastav M, Lataxes TA, Brown KD (2002) Aberrant methylation of the ATM promoter correlates with increased radiosensitivity in a human colorectal tumor cell line. Oncogene 21:3864–3871

Kim JW, Im SA, Kim MA, Cho HJ, Lee DW, Lee KH (2013) Ataxia-telangiectasia-mutated protein expression with microsatellite instability in gastric cancer as prognostic marker. Int J Cancer 134:72–80

Kim JW, Im SA, Kim MA, Cho HJ, Lee DW, Lee KH et al (2014) Ataxia-telangiectasia-mutated protein expression with microsatellite instability in gastric cancer as prognostic marker. Int J Cancer 134:72–80

Kloosterhof NK, de Rooi JJ, Kros M, Eilers PH, Sillevis Smitt PA, van den Bent MJ et al (2013) Molecular subtypes of glioma identified by genome-wide methylation profiling. Genes Chromosomes Cancer 52:665–674

Kozlov SV, Graham ME, Jakob B, Tobias F, Kijas AW, Tanuji M (2011) Autophosphorylation and ATM activation: additional sites add to the complexity. J Biol Chem 286:9107–9119

Lai RK, Chen Y, Guan X, Nousome D, Sharma C, Canoll P (2014) Genome-wide methylation analyses in glioblastoma multiforme. PLoS One 9:e89376

Lavin MF, Gueven N, Bottle S, Gatti RA (2007) Current and potential therapeutic strategies for the treatment of ataxia-telangiectasia. Br Med Bull 81-82:129–147

Lees-Miller SP, Meek K (2003) Repair of DNA double strand breaks by non-homologous end joining. Biochimie 85:1161–1173

Li L, Zou L (2005) Sensing, signaling, and responding to DNA damage: organization of the checkpoint pathways in mammalian cells. J Cell Biochem 94:298–306

Li J, Han YR, Plummer MR, Herrup K (2009) Cytoplasmic ATM in neurons modulates synaptic function. Curr Biol 19:2091–2096

Liberzon E, Avigad S, Cohen IJ, Yaniv I, Michovitz S, Zaizov R (2003) ATM gene mutations are not involved in medulloblastoma in children. Cancer Genet Cytogenet 146:167–169

Lieber MR (2008) The mechanism of human nonhomologous DNA end joining. J Biol Chem 283:1–5

Lukas C, Falck J, Bartkova J, Bartek J, Lukas J (2003) Distinct spatiotemporal dynamics of mammalian checkpoint regulators induced by DNA damage. Nat Cell Biol 5:255–260

Manolis KG, Nimmo ER, Hartsuiker E, Carr AM, Jeggo PA, Allshire RC (2001) Novel functional requirements for non-homologous DNA end joining in Schizosaccharomyces pombe. EMBO J 20:210–221
Marzano V, Santini S, Rossi C, Zucchelli M, D'Alessandro A, Marchetti C et al (2012) Proteomic profiling of ATM kinase proficient and deficient cell lines upon blockage of proteasome activity. J Proteomics 75:4632–4646
Matei IR, Guidos CJ, Danska JS (2006) ATM-dependent DNA damage surveillance in T-cell development and leukemogenesis: the DSB connection. Immunol Rev 209:142–158
Matsuoka S, Ballif BA, Smogorzewska A, McDonald ER, Hurov KE, Luo J (2007) ATM and ATR substrate analysis reveals extensive protein networks responsive to DNA damage. Science 316:1160–1166
McGowan CH, Russell P (2004) The DNA damage response: sensing and signaling. Curr Opin Cell Biol 16:629–633
McKinnon PJ (2004) ATM and ataxia telangiectasia. EMBO Rep 5:772–776
McKinnon PJ (2012) ATM and the molecular pathogenesis of ataxia telangiectasia. Annu Rev Pathol 7:303–321
Mehdipour P, Habibi L, Mohammadi-Asl J, Kamalian N, Mehrazin M (2008) Three-hit hypothesis in astrocytoma: tracing the polymorphism D1853N in ATM gene through a pedigree of the proband affected with primary brain tumor. J Cancer Res Clin Oncol 134:1173–1180
Mehdipour P, Mohammadi-Asl J, Atri M (2011a) Importance of ATM gene as a susceptible trait: predisposition role of D1853N polymorphism in breast cancer. Med Oncol 3: 733–737
Mehdipour P, Kheirollahi M, Mehrazin M, Kamalian N, Atri M (2011b) Evolutionary hypothesis of telomere length in primary breast cancer and brain tumour patients: a tracer for genomic-tumour heterogeneity and instability. Cell Biol Int 35:915–925
Mehdipour P, Karami F, Javan F, Mehrazin M (2014) Linking ATM promoter methylation to cell cycle proteinexpression: cellular molecular triangle correlation in atm territory. Mol Neurobiol. doi:10.1007/s12035-014-8864-9
Meyer A, Wilhelm B, Dork T, Bremer M, Baumann R, Karstens JH et al (2007) ATM missense variant P1054R predisposes to prostate cancer. Radiother Oncol 83:283–288
Mongiardi MP, Stagni V, Natoli M, Giaccari D, D'Agnano I, Falchetti ML et al (2011) Oxygen sensing is impaired in ATM-defective cells. Cell Cycle 10:4311–4320
Mu JJ, Wang Y, Luo H, Leng M, Zhang J, Yang T et al (2007) A proteomic analysis of ataxia telangiectasia-mutated (ATM)/ATM-Rad3-related (ATR) substrates identifies the ubiquitin-proteasome system as a regulator for DNA damage checkpoints. J Biol Chem 282:17330–17334
Muraki K, Han L, Miller D, Murnane JP (2013) The role of ATM in the deficiency in nonhomologous end-joining near telomeres in a human cancer cell line. PLoS Genet 9:e1003386
Nevanlinna H, Bartek J (2006) The CHEK2 gene and inherited breast cancer susceptibility. Oncogene 25:5912–5919
Noon AT, Shibata A, Rief N, Lobrich M, Stewart GS, Jeggo PA et al (2010) 53BP1-dependent robust localized KAP-1 phosphorylation is essential for heterochromatic DNA double-strand break repair. Nat Cell Biol 12:177–184
Noushmehr H, Weisenberger DJ, Diefes K, Phillips HS, Pujara K, Berman BP et al (2010) Identification of a CpG island methylator phenotype that defines a distinct subgroup of glioma. Cancer Cell 17:510–522
Offit K, Gilad S, Paglin S, Kolachana P, Roisman LC, Nafa K et al (2002) Rare variants of ATM and risk for Hodgkin's disease and radiation-associated breast cancers. Clin Cancer Res 8:3813–3819
Ogawa T, Yu X, Shinohara A, Egelman EH (1993) Similarity of the yeast RAD51 filament to the bacterial RecA filament. Science 259:1896–1899
Ouchi T (2006) BRCA1 phosphorylation: biological consequences. Cancer Biol Ther 5:470–475
Pandita TK, Dhar S (2000) Influence of ATM function on interactions between telomeres and nuclear matrix. Radiat Res 154:133–139
Patel AY, McDonald TM, Spears LD, Ching JK, Fisher JS (2011) Ataxia telangiectasia mutated influences cytochrome c oxidase activity. Biochem Biophys Res Commun 405:599–603

Paulikova S, Petera J, Sirak I, Vosmik M, Drastikova M, Dusek L et al (2014) ATM and TGFB1 genes polymorphisms in prediction of late complications of chemoradiotherapy in patients with locally advanced cervical cancer. Neoplasma 61:70–76

Pecker I, Avraham KB, Gilbert DJ, Savitsky K, Rotman G, Harnik R (1996) Identification and chromosomal localization of Atm, the mouse homolog of the ataxia-telangiectasia gene. Genomics 35:39–45

Peng G, Yim EK, Dai H, Jackson AP, Burgt I, Pan MR et al (2009) BRIT1/MCPH1 links chromatin remodelling to DNA damage response. Nat Cell Biol 11:865–872

Perry J, Kleckner N (2003) The ATRs, ATMs, and TORs are giant HEAT repeat proteins. Cell 112:151–155

Pitcher RS, Wilson TE, Doherty AJ (2005) New insights into NHEJ repair processes in prokaryotes. Cell Cycle 4:675–678

Renwick A, Thompson D, Seal S, Kelly P, Chagtai T, Ahmed M et al (2006) ATM mutations that cause ataxia-telangiectasia are breast cancer susceptibility alleles. Nat Genet 38:873–875

Reznick RM, Shulman GI (2006) The role of AMP-activated protein kinase in mitochondrial biogenesis. J Physiol 574:33–39

Rio PG, Pernin D, Bay JO, Albuisson E, Kwiatkowski F, De Latour M, Bernard-Gallon et al (1998) Loss of heterozygosity of BRCA1, BRCA2 and ATM genes in sporadic invasive ductal breast carcinoma. Int J Oncol 13:849–853

Rivera-Calzada A, Maman JD, Spagnolo L, Pearl LH, Llorca O (2005) Three-dimensional structure and regulation of the DNA-dependent protein kinase catalytic subunit (DNA-PKcs). Structure 13:243–255

Roberts NJ, Jiao Y, Yu J, Kopelovich L, Petersen GM, Bondy ML et al (2012) ATM mutations in patients with hereditary pancreatic cancer. Cancer Discov 2:41–46

Roy K, Wang L, Makrigiorgos GM, Price BD (2006) Methylation of the ATM promoter in glioma cells alters ionizing radiation sensitivity. Biochem Biophys Res Commun 344:821–826

Safar AM, Spencer H, Su X, Cooney CA, Shwaiki A, Fan CY (2007) Promoter hypermethylation for molecular nodal staging in non-small cell lung cancer. Arch Pathol Lab Med 131:936–941

Savitsky K, Bar-Shira A, Gilad S, Rotman G, Ziv Y, Vanagaite L et al (1995a) A single ataxia telangiectasia gene with a product similar to PI-3 kinase. Science 268:1749–1753

Savitsky K, Sfez S, Tagle DA, Ziv Y, Sartiel A, Collins FS et al (1995b) The complete sequence of the coding region of the ATM gene reveals similarity to cell cycle regulators in different species. Hum Mol Genet 4:2025–2032

Schaffner C, Stilgenbauer S, Rappold GA, Dohner H, Lichter P (1999) Somatic ATM mutations indicate a pathogenic role of ATM in B-cell chronic lymphocytic leukemia. Blood 94:748–753

Schalch DS, McFarlin DE, Barlow MH (1970) An unusual form of diabetes mellitus in ataxia telangiectasia. N Engl J Med 282:1396–1402

Schon EA, Manfredi G (2003) Neuronal degeneration and mitochondrial dysfunction. J Clin Invest 111:303–312

Shamma A, Suzuki M, Hayashi N, Kobayashi M, Sasaki N, Nishiuchi T et al (2013) ATM mediates pRB function to control DNMT1 protein stability and DNA methylation. Mol Cell Biol 33:3113–3124

Shiloh Y (2003) ATM and related protein kinases: safeguarding genome integrity. Nat Rev Cancer 3:155–168

Shreeram S, Demidov ON, Hee WK, Yamaguchi H, Onishi N, Kek C et al (2006) Wip1 phosphatase modulates ATM-dependent signaling pathways. Mol Cell 23:757–764

Silva E, Tiong S, Pedersen M, Homola E, Royou A, Fasulo B et al (2004) ATM is required for telomere maintenance and chromosome stability during Drosophila development. Curr Biol 14:1341–1347

So S, Davis AJ, Chen DJ (2009) Autophosphorylation at serine 1981 stabilizes ATM at DNA damage sites. J Cell Biol 187:977–990

Sommer SS, Jiang Z, Feng J, Buzin CH, Zheng J, Longmate J et al (2003) ATM missense mutations are frequent in patients with breast cancer. Cancer Genet Cytogenet 145:115–120

Sonoda E, Takata M, Yamashita YM, Morrison C, Takeda S (2001) Homologous DNA recombination in vertebrate cells. Proc Natl Acad Sci U S A 98:8388–8394
Stankovic T, Kidd AM, Sutcliffe A, McGuire GM, Robinson P, Weber P (1998) ATM mutations and phenotypes in ataxia-telangiectasia families in the British Isles: expression of mutant ATM and the risk of leukemia, lymphoma, and breast cancer. Am J Hum Genet 62:334–345
Stankovic T, Stewart GS, Fegan C, Biggs P, Last J, Byrd PJ (2002) Ataxia telangiectasia mutated-deficient B-cell chronic lymphocytic leukemia occurs in pregerminal center cells and results in defective damage response and unrepaired chromosome damage. Blood 99:300–309
Stankovic T, Hubank M, Cronin D, Stewart GS, Fletcher D, Bignell CR et al (2004) Microarray analysis reveals that TP53- and ATM-mutant B-CLLs share a defect in activating proapoptotic responses after DNA damage but are distinguished by major differences in activating prosurvival responses. Blood 103:291–300
Starostik P, Manshouri T, O'Brien S, Freireich E, Kantarjian H, Haidar M, Lerner S (1998) Deficiency of the ATM protein expression defines an aggressive subgroup of B-cell chronic lymphocytic leukemia. Cancer Res 58:4552–4557
Stewart GS, Last JI, Stankovic T, Haites N, Kidd AM, Byrd PJ et al (2001) Residual ataxia telangiectasia mutated protein function in cells from ataxia telangiectasia patients, with 5762ins137 and 7271T→G mutations, showing a less severe phenotype. J Biol Chem 276:30133–30141
Stewart GS, Wang B, Bignell CR, Taylor AM, Elledge SJ (2003) MDC1 is a mediator of the mammalian DNA damage checkpoint. Nature 421:961–966
Stilgenbauer S, Schaffner C, Litterst A, Liebisch P, Gilad S, Bar-Shira A et al (1997) Biallelic mutations in the ATM gene in T-prolymphocytic leukemia. Nat Med 3:1155–1159
Stoppa-Lyonnet D, Lauge A, Sigaux F, Stern MH (2000) No germline ATM mutation in a series of 16T-cell prolymphocytic leukemias. Blood 96:374–376
Sung P (1994) Catalysis of ATP-dependent homologous DNA pairing and strand exchange by yeast RAD51 protein. Science 265:1241–1243
Swift M, Reitnauer PJ, Morrell D, Chase CL (1987) Breast and other cancers in families with ataxia-telangiectasia. N Engl J Med 316:1289–1294
Teraoka SN, Malone KE, Doody DR, Suter NM, Ostrander EA, Daling JR (2001) Increased frequency of ATM mutations in breast carcinoma patients with early onset disease and positive family history. Cancer 92:479–487
Thompson D, Duedal S, Kirner J, McGuffog L, Last J, Reiman A et al (2005) Cancer risks and mortality in heterozygous ATM mutation carriers. J Natl Cancer Inst 97:813–822
Toyoshima M, Hara T, Zhang H, Yamamoto T, Akaboshi S, Nanba E et al (1998) Ataxia-telangiectasia without immunodeficiency: novel point mutations within and adjacent to the phosphatidylinositol 3-kinase-like domain. Am J Med Genet 75:141–144
Treilleux I, Chapot B, Goddard S, Pisani P, Angele S, Hall J (2007) The molecular causes of low ATM protein expression in breast carcinoma; promoter methylation and levels of the catalytic subunit of DNA-dependent protein kinase. Histopathology 51:63–69
Tribius S, Pidel A, Casper D (2001) ATM protein expression correlates with radioresistance in primary glioblastoma cells in culture. Int J Radiat Oncol Biol Phys 50:511–523
Tsuchida R, Yamada T, Takagi M, Shimada A, Ishioka C, Katsuki Y et al (2002) Detection of ATM gene mutation in human glioma cell line M059J by a rapid frameshift/stop codon assay in yeast. Radiat Res 158:195–201
Valentin-Vega YA, Maclean KH, Tait-Mulder J, Milasta S, Steeves M, Dorsey FC (2012) Mitochondrial dysfunction in ataxia-telangiectasia. Blood 119:1490–1500
Vaziri H, West MD, Allsopp RC, Davison TS, Wu YS, Arrowsmith CH (1997) ATM-dependent telomere loss in aging human diploid fibroblasts and DNA damage lead to the post-translational activation of p53 protein involving poly(ADP-ribose) polymerase. EMBO J 16:6018–6033
Vo QN, Kim WJ, Cvitanovic L, Boudreau DA, Ginzinger DG, Brown KD (2004) The ATM gene is a target for epigenetic silencing in locally advanced breast cancer. Oncogene 23:9432–9437
Vorechovsky I, Luo L, Dyer MJ, Catovsky D, Amlot PL, Yaxley JC et al (1997) Clustering of missense mutations in the ataxia-telangiectasia gene in a sporadic T-cell leukaemia. Nat Genet 17:96–99

Wang B, Matsuoka S, Carpenter PB, Elledge SJ (2002) 53BP1, a mediator of the DNA damage checkpoint. Science 298:1435–1438

Wang H, Wang S, Shen L, Chen Y, Zhang X, Zhou J, Wang Z et al (2010) Chk2 down-regulation by promoter hypermethylation in human bulk gliomas. Life Sci 86:185–191

Wang X, Zeng L, Wang J, Chau JF, Lai KP, Jia D et al (2011) A positive role for c-Abl in Atm and Atr activation in DNA damage response. Cell Death Differ 18:5–15

Watters D, Kedar P, Spring K, Bjorkman J, Chen P, Gatei M et al (1999) Localization of a portion of extranuclear ATM to peroxisomes. J Biol Chem 274:34277–34282

Williams RS, Dodson GE, Limbo O, Yamada Y, Williams JS, Guenther G et al (2009) Nbs1 flexibly tethers Ctp1 and Mre11-Rad50 to coordinate DNA double-strand break processing and repair. Cell 139:87–99

Wiltzius JJ, Hohl M, Fleming JC, Petrini JH (2005) The Rad50 hook domain is a critical determinant of Mre11 complex functions. Nat Struct Mol Biol 12:403–407

Wu ZH, Shi Y, Tibbetts RS, Miyamoto S (2006) Molecular linkage between the kinase ATM and NF-kappaB signaling in response to genotoxic stimuli. Science 311:1141–1146

Wu ZH, Wong ET, Shi Y, Niu J, Chen Z, Miyamoto S (2010) ATM- and NEMO-dependent ELKS ubiquitination coordinates TAK1-mediated IKK activation in response to genotoxic stress. Mol Cell 40:75–86

Wu J, Chen Y, Lu LY, Wu Y, Paulsen MT, Ljungman M et al (2011) Chfr and RNF8 synergistically regulate ATM activation. Nat Struct Mol Biol 18:761–768

Wyman C, Kanaar R (2006) DNA double-strand break repair: all's well that ends well. Annu Rev Genet 40:363–383

Wyman MP, Pirola L (1998) Structure and function of phosphoinositide 3-kinases. Biochim Biophys Acta 1436:127–150

Yang DQ, Kastan MB (2000) Participation of ATM in insulin signalling through phosphorylation of eIF-4E-binding protein 1. Nat Cell Biol 2:893–898

Yang DQ, Halaby MJ, Li Y, Hibma JC, Burn P (2011) Cytoplasmic ATM protein kinase: an emerging therapeutic target for diabetes, cancer and neuronal degeneration. Drug Discov Today 16:332–338

Zhang N, Chen P, Khanna KK, Scott S, Gatei M, Kozlov S, Watters D (1997) Isolation of full-length ATM cDNA and correction of the ataxia-telangiectasia cellular phenotype. Proc Natl Acad Sci U S A 94:8021–8026

Zhang T, Nirantar S, Lim HH, Sinha I, Surana U (2009) DNA damage checkpoint maintains CDH1 in an active state to inhibit anaphase progression. Dev Cell 17:541–551

Ziv Y, Bielopolski D, Galanty Y, Lukas C, Taya Y, Schultz DC et al (2006) Chromatin relaxation in response to DNA double-strand breaks is modulated by a novel ATM- and KAP-1 dependent pathway. Nat Cell Biol 8:870–876

Zoncu R, Efeyan A, Sabatini DM (2010) mTOR: from growth signal integration to cancer, diabetes and ageing. Nat Rev Mol Cell Biol 12:21–35

Chapter 7
Molecular and Biological Aspects of *Microcephalin* Gene: Directions in Brain Tumor and Methylation

Fatemeh Karami and Parvin Mehdipour

Contents

Abstract *Microcephalin* gene (*MCPH1*) is located on chromosome 8p23.1 and initially was found to be involved in the pathogenesis of microcephaly. Its interaction with major cellular proteins especially cell cycle checkpoint molecules and DNA repair proteins demonstrated that it plays crucial roles in response to DNA damage and repair. In addition, some alterations have been detected in *MCPH1* gene which led to name it as a new tumor suppressor gene. On the top, promoter methylation of *MCPH1* gene and its direct effect on its protein expression has been proposed as a main mechanism of *MCPH1* gene inactivation in various cancers especially brain tumors. Moreover, there are some reports that relying on the effect of *MCPH1* gene mutations in development of non-cancerous diseases including neurocognitive disorders. In this chapter, at first the basics of MCPH1 gene and its protein will be

F. Karami (✉) · P. Mehdipour
Department of Medical Genetics, School of Medicine, Tehran University of Medical Sciences, Poursina Street, Keshavarz Boulevard, P.O. Box: 14176–13151, Tehran, Iran
e-mail: mehdipor@tums.ac.ir

P. Mehdipour (ed.), *Epigenetics Territory and Cancer*,
DOI 10.1007/978-94-017-9639-2_7

described and then we will provide a brief literature of the investigations carried out on the role of *MCPH1* gene alterations in various types of cancer and some other diseases. Moreover, the methylation data of our very recent research together with protein expression in brain tumors are provided.

Abbreviations

AD:	Alzheimer disease
APC:	Anaphase promoting complex
ASD:	Autism spectrum disorders
ASPM:	Abnormal spindle microcephaly
ATM:	Ataxia telangiectasia mutated
ATR:	ATM and Rad3-related
BC:	Breast cancer
BIRT1:	BRCT—repeat inhibitor of human telomerase reverse transcriptase expression 1
BMI:	Body mass index
BRCT:	BRCA1 carboxyl-terminal
CDK5RAP2:	Cyclin-dependent kinase5 regulatory subunit associated protein 2
CENPJ:	Centromere protein J
CML:	Chronic myeloid leukemia
COBRA:	Combined bisulfite restriction analysis
DSB:	Double strand DNA break
EOC:	Epithelial ovarian cancer
GBM:	Gleioblastoma multiform
HCC:	Hepatocellular carcinoma
IF:	Immunofluorescence
LOH:	Loss of heterozygosity
MCPH1:	Microcephalin
MMR:	Mismatch repair
MR:	Mental retardation
MSI:	Microsatellite instability
MSP:	Methylation specific PCR
MSRA:	Methylation specific restriction analysis
NPCs:	Neural progenitor cells
NSCLC:	Non-small cell lung cancer
PCC:	Premature chromosome condensation
p53BP:	p53 binding protein
Q-FISH:	Quantitative—fluorescent in situ hybridization
OR:	Odds ratio
OSCC:	Oral squamous cell carcinoma
SNPs:	Single nucleotide polymorphisms
SSCP:	Single strand conformation polymorphism

7.1 Introduction

7.1.1 Microcephalin Gene

Microcephalin1 (MCPH1) gene also known as *BIRT1* and *MCT* was mapped on chromosome 8p23.1 by Jackson AP team work in late 1990s it was located between 6251529 and 6493434 on plus strand (Fig. 7.1). It is the first gene, which reflectively, was proposed for microcephaly. It has 14 exons and 7 splice variants that three of those are protein coding and the remaining are only processed transcripts. It is expressed in most of the human tissues especially in brain, testes, pancreas, liver and fetal cerebral cortex (Jackson et al. 2002).

7.1.2 MCPH1 Protein

BIRT1 (BRCT–repeat inhibitor of human telomerase reverse transcriptase expression) was reported as a repressor of TERT transcription by 2003 (Lin and Elledge 2003). It encodes for a protein with 835 amino acids with 110 kDalton weight. It has homolouges in chimpanzee, dog, rat, mouse, zebrafish and drosophila. MCPH1 protein has demonstrated rapid evolution from simian ancestors to chimpanzee and then human, tending toward positive selection favoring 45 advantageous amino acids (Evans et al. 2004). This evolution may be related to enlargement of brain size in human. As a surprising tote, Rimol and his colleagues have shown that the brain size is controlled only by *MCPH1* gene in females whereas in men this critical task is ful-

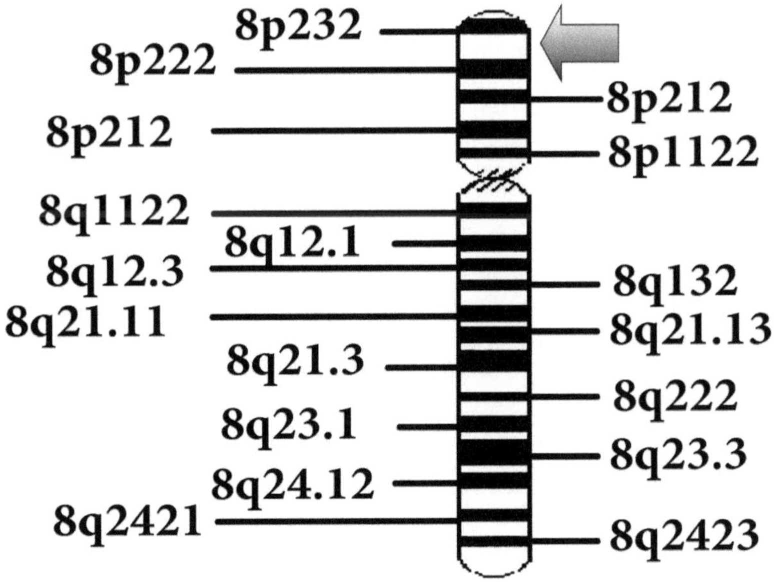

Fig. 7.1 Position of MCPH1 gene on chromosome 8p.

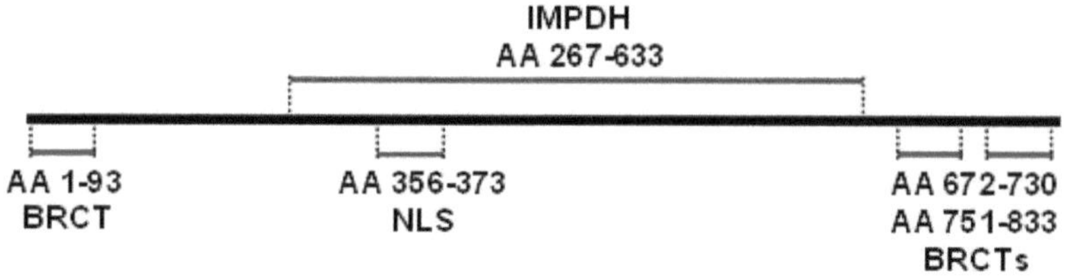

Fig. 7.2 Schematic representation of MCPH1 protein structure.

filled upon the expression of *CDK5RAP2* gene (Rimol et al. 2010). However, study on 2393 volunteers have revealed no correlation between the recent statement on evolution and normal discrepancies which are present in IQ (Mekel-Bobrov et al. 2007).

There are three BRCA1 carboxyl-terminal (BRCT) domains in the MCPH1 protein structure including one N-terminal BRCT (N-BRCT) and two C-terminal BRCT (C-BRCTs) domains. One large central IMPDH domain and nuclear localization signal motif constitute the other parts of MCPH1 protein localizing mostly in nucleus (Fig. 7.2).

N-BRCT domain is necessary for changing localization of MCPH1 protein toward centrosome within cells exposed to radiation and it is also important to prevent premature chromosome condensation (PCC) mediated by Condensin II (Yang et al. 2008). The C-BRCT domains are required for MCPH1 oligomer and foci formation after radiation which will be more discussed later. However, C-BRCTs are also attachment sites of phosphorylated proteins engaged for DNA damage response pathway similar to other BRCT containing proteins involved in DNA damage response (Yu et al. 2003; Wood et al. 2007) γH2A.X is one of the known direct target of MCPH1 which is phosphorylated upon the DNA damage (Wood et al. 2007). The function of IMPDH domain has been poorly studied, but it seems that it is indirectly involved in homologous recombination through binding to Condensin II (Fig. 7.2).

Although, *MCPH1* gene is one of the six genes that its mutation is associated with microcephaly, the role of this gene remains to be described in brain development and mental disabilities (Woods et al. 2006; Rushton et al. 2007). Both male and female knock out mouse bearing homozygote *MCPH1* gene deletion have demonstrated infertility, moderate hearing loss, chromosomal instability and cataract (Gerdin 2010). Although *MCPH1* is mutated in patients with microcephaly, no higher incidence of cancer and tumor has been reported in them. It may imply on this fact that *MCPH1* gene mutates in different domains or segments of protein lead to cancer and microcephaly (O'Driscoll and Jeggo 2006). For example, S25X mutation changes the start codon of translation and finally disrupts the N-BRCT domain which is necessary for the role of *MCPH1* in DNA repair (Leung et al. 2011).

7.2 Cellular MCPH1 Protein Functions

The MCPH1 protein has two main cellular contributions including cell cycle control and DNA damage response. It is involved in the latter by being involved in both major DNA damage response pathways, ataxia telangiectasia mutated (ATM) and

ATM and Rad3-related (ATR). All of the components of DNA damage response system with an exception of γ-H2AX are unable to reside on site of damage without MCPH1 owing to failure in phosphorylation of ATM and its downstream proteins. Phosphorylated H2AX stimulates accumulation of MCPH1 on the break site of double strand DNA break (DSB) to bind N-terminal of BRCA2 and Rad51 and thereby control their localization. Although, MCPH1 is not an obligatory element for localization of BRCA2 and Rad51, its absence leads to great diminished recruitment of them to damaged site (Wu et al. 2009). In ATR pathway, MCPH1 regulates cdk1 phosphorylation and thereby prevents the premature chromosome condensation and also marks the damaged DNA to be fragmented instead of running out from S to G2 phases (Alderton et al. 2006). It was shown that *MCPH1* mutation was associated with aberrant G2-M checkpoint and extra mitotic centrosomes leading to genomic instability and various chromosomal abnormalities (Rai et al. 2006). MCPH1 has pivotal role in regulation of cell cycle through modulating the expression of BRCA1 and ChK1 to impede premature start of mitosis after radiation exposure. It impels activation of p73 and renders it to make a complex with E2F1 to upregulate expression of genes which mediate E2F1 dependent apoptosis including Apafl, caspase 3 and caspase 7 (Yang et al. 2008; Urist et al. 2004). It was shown that under-expression of *MCPH1* gene was associated with decreased in success rate of chemotherapy due to impairment in activation of p73 to induce apoptosis (Yang et al. 2008).

Moreover, MCPH1 binds chromatin and changes its configuration through recruitment of MDC1 complex, p53 binding protein (p53BP) and NBS1 around the chromatin structure of damaged DNA (Wu et al. 2009).

MCPH1 by accompanying 101 other genes play important role in centrosome architecture and regulates its function. It was demonstrated that centrosome aberrations had great impact on progression and development of in situe and pre-invasive breast carcinoma. Centrosome duplication leads to inappropriate attachment of polar spindle fibers to kinetochores residing on centromere of each chromosome. Therefore, chromatids are no longer able to separate from each other and move on towards opposite poles in a proper manner resulting in numerical chromosomal abnormalities and aneuploidies which is one of the putative hallmarks of cancer. Recent studies have received great attentions to the role of proteins and other elements of mitotic checkpoints including centrosome in the etiology of aneuploidies, development and expansion of tumor cells. Centrosome controls progression of cell cycle from S-G2 and G2-M indicating that centrosome signaling may be the major governing key in cell division (Liang et al. 2010).

Given the importance of MCPH1 in cell cycle regulation, maintaining the genomic stability and centrosome structure and regulation in addition to its decreased level in various cancers and metastasis, it was introduced as a tumor suppressor gene (Rai et al. 2006).

In the following sections of current chapter, we were aimed to describe about the main functions of MCPH1 in details which have been reported in recent years. After that, we will discuss in detail regarding the role of aberrations of MCPH1 activity in development of brain tumors and some neurocognitive disorders.

7.2.1 *MCPH1: An Important Accessory Member of Cell Cycle Checkpoints*

Most of the neural cells are derived from neural progenitor cells (NPCs) lining the ventricles of human brain. NPCs follow two major particular patterns in their division including symmetric and asymmetric cell divisions (Brand and Rakic 1979). The first pattern results in two identical NPCs which are switched on to the latter type of division at specific time of brain development to generate one NPCs and one post mitotic neuron. It was found that every deviation at the proper time of switch between these two types of neural cells divisions can affect the number of neuron pools. It was also assumed that deregulations in asymmetric and symmetric cell divisions lead to micro- and macro-cephaly, respectively (Doe and Bowerman 2001). Functional studies on the orthologous of *MCPH1* gene in Drosophila have shown that it may have pivotal role in control of asymmetric cell division and also determining the right time of switching from symmetric to asymmetric cell divisions. Moreover, the important role of *MCPH1* gene in regulation of centrosome activity which itself plays crucial role in mitosis of neurons, would be a further evidence of determining activity of *MCPH1* in brain development (Yamashita et al. 2003). As mentioned above, centrosome governs two major cell cycle checkpoints before and after getting entry into the G2. There is an intra S checkpoint in which DNA synthesis is prohibited when the cell is exposed to radiation. This task is mediated by activation of ATM which induces CHK1 to stall the replication process through inactivation of Cdck25A and subsequently Cdk2-cyclin complex (Fig. 7.3) (Abraham 2001; Margolis and Kornbluth 2004). In this way, co working of MCPH1 seems to be necessary for the proper activity of the major cellular proteins including ATM and BRCA1 which guarantee the G2/M checkpoint to be intact. Cdc27

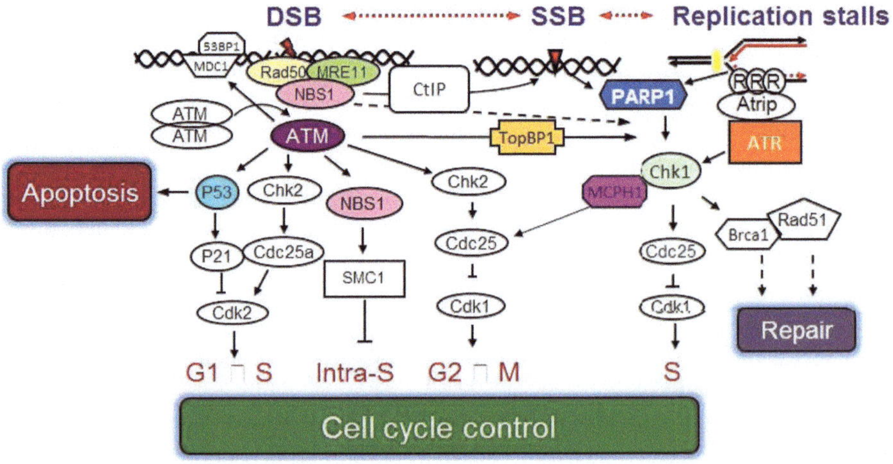

Fig. 7.3 The critical role of MCPH1 protein in cell cycle control in bridging between two main ATR and ATR DNA repair pathways.

and Cdc16 which are the elements of anaphase promoting complex (APC) has been shown to be phosphorylated with C-terminal BRCT domains of MCPH1 to drive the separation of chromosome during mitosis. MCPH1 can alternatively binds to another component of APC complex called Cdc26 when the Cdc27 has been mutated. Cdc26 differs from Cdc27 at the residue residing on + 1 amino acid relative to phosphoserine (Singh et al. 2012). These interactions of MCPH1 are necessary for APC complex stability and may shed light on this finding that loss of MCPH1 can mimics the loss of APC complex function.

Determination of *MCPH1* expression in seckel syndrome patients with severe microcephaly revealed that normal level of *MCPH1* transcription is required for active centrosome Chk1 expression and eventually induction of centrosome cyclin B-Cdk1 (Tibelius et al. 2009). It was demonstrated that directed mutagenesis of *MCPH1* gene was associated with arrest of manipulated cells in G2 phase and premature condensation of chromosomes (Alderton et al. 2006). Moreover, directed mutagenesis of *MCPH1* led to decrease in tyrosine phosphorylation of Cdk1 in S and G2 phases and consequently formation of PCC in manipulated cells. It also was associated with the aberrant G2-M checkpoint and nuclear fragmentation when DNA has been damaged (Alderton et al. 2006). It is obvious that MCPH1 has a major contribution in controlling the main cellular checkpoints throughout the cell cycle.

7.2.2 MCPH1 Activity in DNA Repair System

Presence of BRCT domains within the structure of MCPH1 protein would be a good predictor of its role in DNA repair system like other proteins carrying two carboxyl-terminal BRCT domains (PTCB). BRCT domain actually provides the suitable context for peptide and phosphopeptide binding activities (Rodriguez et al. 2008; Manke et al. 2003). Localization of MCPH1 in the DNA damage site and DNA repair complex could be a convincing evidence of its contribution in DNA repair pathway (Fig. 7.4) (Rai et al. 2006; Alderton et al. 2006; Lin et al. 2005). Loss of MCPH1 prevents activation of other DNA damage proteins including Rad51, BRCA2 and Chk1/Chk2 dependent pathways (Tibelius et al. 2009). It was shown that biallelic knock out of *MCPH1* gene in mice was associated with decreased length of life and genomic stability (Trimborn et al. 2010; Liang et al. 2010). It was described that the C-terminal domains of MCPH1 protein play pivotal role in DNA repair process (Yu et al. 2003). It is in line with the capability of N-terminus defective MCPH1 protein to avoid promotion of damaged cell to get entry in to the mitosis (Gavvovidis et al. 2010). Study on the role of *MCPH1* aberrations in endometrial cancer have revealed that deletion of one nucleotide within the repeated tracts of adenine in exons 4, 5 and 8 was compatible with mismatch repair (MMR) deficiency and microsatellite instability (MSI). It was proposed that this type of deletion is associated with non-functional C-terminal domains. The same inefficient MCPH1 protein had been reported prior that wherein deletion of 38 bps within the exon 10 was found in breast cancer (BC) (Alderton et al. 2006). Of note, there is a similar A tract in exon 10 of *ATR* gene as the major cooperators of *MCPH1* in DSB that any monoalleleic

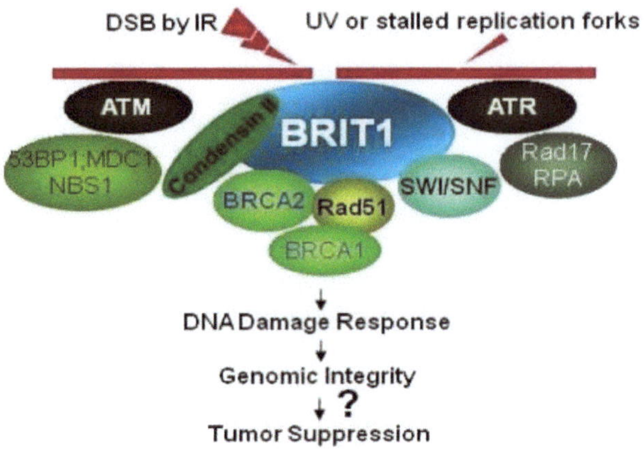

Fig. 7.4 Critical interactions of BIRT1/MCPH1 with multiple proteins involved in ATR and ATM dependent DNA repair pathways. Adopted from Lin et al. 2010 (Lin et al. 2010).

alteration of it was reflected by poor prognosis of endometrial cancer patients carrying MSI (Xu et al. 2004). It could be extrapolated that haplo-sufficiency of major components of DSB system is essential for maintaining the stability of genome and prevention of tumor progression in particular in endometrial cancer.

The better response to chemo- and radio-therapy in patients harboring mutations in genes encoding DSB repair components could be a strong reason to seek the harmful mutations within them especially in *MCPH1* gene (Li et al. 2004; Rodriguez et al. 2008).

7.3 MCPH1 Gene Aberrations in Various Cancer

It was demonstrated that the chromosome 8p was the fragile site and focal region for deletion in hepatocellular carcinoma (HCC) patients in particular, in larger tumors with poorer differentiation (Chan et al. 2002). This result motivated further group to search for the genes whose their deletion was associated with more aggressive tumors in two separate studies. They have found that the frequency of loss of heterozygosity (LOH) within the 8p23.1 was significantly more in metastatic (68 %) versus primary (19 %) tumors in HCC patients (Lu et al. 2007; Lu and Hano 2007).

Amongst 782 single nucleotide polymorphisms (SNPs) within the 101 centrosome related candidate genes which were sought in 798 BC patients and 843 controls, 40 SNPs demonstrated significant association with BC risk. Two SNPs of these including 3'UTR and non-synonymous variants (rs24433149 and rs1057091) were identified in *MCPH1* gene with equal odds ratio (OR) of 1.23 (Olson et al. 2011). In the next study of MCPH1 in BC patients, low level of MCPH1 expression was associated with T allele of rs2912010 in almost 50 % of BC patients, however the frequency of this variant was not significantly differs from healthy controls. In addition, T allele was meaningfully higher in patients with higher grades of tumor. Of note, MCPH1 protein was more localized in cytoplasm in higher grades of tumor (Jo et al. 2013).

Thirty four SNPs whose their association was previously found with BC risk, were screened in 1189 patients affected with pancreatic adenocarcinoma (Couch et al. 2010). They were undergone genotyping in comparison with 1126 healthy normal individuals through illumina next generation sequencing. The rs2433149 polymorphism within the *MCPH1* gene was found to be in significant association with pancreas cancer in patients with low body mass index (BMI). In addition, meaningful correlation was found between higher risk of pancreatic cancer and ever or former smokers carrying that polymorphism.

Expression assay of MCPH1 and ASPM proteins was extended on primary cultured cells derived from benign and epithelial ovarian cancer (EOC) tissues (Bruning-Richardson et al. 2011). In that study, cytoplasmic localization of MCPH1 and ASPM proteins and their expression level were determined through protein slot blotting and immunofluorescence. In contrast to the higher expression of ASPM protein in lower tumor's grades, the expression of mutant MCPH1 protein was correlated with higher grades of ovarian tumors and it was localized more in cytoplasm. Association between better survival of patients and nuclear localization of normal MCPH1 protein may indicate that mutant MCPH1 moves out from nucleus and unable to be participated in DNA repair and cell cycle checkpoint leading to higher tumor's grades. This finding is in line with the prior study which demonstrated that mutant MCPH1 was unable to reside the BRCA2 on DNA damage site. BRCA2 and MCPH1 proteins become connected together through peptide binding. Mutation in *MCPH1* gene prevents this connection and also recruitment of BRCA2 and Rad-51 repair proteins to the double strand break site (Wu et al. 2009). In summary, mutation in MCPH1 shifts the localization of it towards cytoplasm and keeps it away to be available for residing of BRCA2 and Rad-51 proteins on DNA damage site.

The expression level of *MCPH1* gene was determined in 55 non-small cell lung cancer (NSCLC) patients whom were undergone chemotherapy with EGFR inhibitor or erlotinib (Garcia-Campelo 2012). It was carried out due to the critical inhibitory effect of Erlotinib on DNA damage response proteins including BRCA2 and Rad51 which are localized at the damage site by MCPH1. The mRNA level of *MCPH1* gene was measured using a previously introduced NanoString nCounter gene expression system in which even one mRNA transcript is captured during detection procedure. The most and least median survivals were determined in patients who had the highest and lowest levels of *MCPH1* gene expression, respectively. It is worth to note that the efficacy of Erlotinib was improved when the expression of *MCPH1* was increased. It was proposed that it may be pertinent to the negative regulatory role of MCPH1 on 53BP and MDC1 proteins. It was also suggested that identification of *MCPH1* expression level could be a potential biomarker for clinical response to Erlotinib.

Study on patients affected with chronic myeloid leukemia (CML), and K562 cell line demonstrated under expression of *MCPH1* gene (Giallongo et al. 2011). No association was found between BCR/ABL fusion protein activity and the level of *MCPH1* expression. In addition, keeping on cell growth and proliferation in spite of treatment of cells with hydroxyurea is implying on the direct correlation between under expression of MCPH1 and aberrant G2/M checkpoint.

In the first *MCPH1* methylation study on BC, expression analysis of *MCPH1* and *ATM* genes and their proteins were performed through qRT-PCR and IHC on MCF7 cell line and 126 BC patients in comparison with 123 healthy volunteers. The promoter methylation statue of *MCPH1* and *ATM* genes were determined by means of methylation specific restriction analysis (MSRA) and methylation specific PCR (MSP). Moreover, single strand conformation polymorphism (SSCP) was implicated to screen the *MCPH1* gene of 30 randomly selected fractions of BC patients. They also assessed both *MCPH1/ATM* genes for the presence of deletions including intragenic micro deletions. Down-regulation in expression of either *ATM* or *MCPH1* gene was further replicated and frequent promoter methylation and deletions was found within the *MCPH1* gene. All the molecular alterations including methylation, deletion and expression of MCPH1 and ATM proteins and their genes were significantly more in ER/PR negative versus ER/PR positive patients. In addition, the lower expression and methylation of *MCPH1* promoter gene were in significant association with higher grades of BC. The strong expression of MCPH1 and ATM proteins in both nucleus and cytoplasm was detected in BC cells without any molecular defect. This is a further proof on that sufficient expression of intact *MCPH1* gene allows its protein product to be translocated from cytoplasm to nucleus wherein DNA repair process takes place.

Recently, a broad study was reported on 93 fresh oral squamous cell carcinoma (OSCC) tissue samples and three A549 (human lung adenocarcinoma), Hela (human cervical carcinoma) and KB (OSCC) cell lines (Venkatesh et al. 2013). The promoter methylation analysis of *MCPH1* gene was carried out in this study throughCombined Bisulfite Restriction Analysis (COBRA) and determination of the effect of 2'-deoxy-5-azacytidine (AZA) treatment on the expression of *MCPH1* gene. The *MCPH1* gene analysis was included loss of heterozygosity (LOH) assessment using three STR markers nearby the *MCPH1* gene locus, mutation screening and real time RT-PCR. The level of MCPH1 protein expression was determined through western blotting and immunohistochemistry (IHC). In addition, the tumor progression and apoptosis were tracked in an OSCC mouse model whom was transfected with an expression pcDNA vector to have high copy numbers of *MCPH1* gene.

MCPH1 promoter was methylated in 14/40 OSCC tumors and in neither of the corresponding normal tissues nor the examined cell lines except of SCC084. AZA treatment on the aforementioned cell lines induced increase in expression of *MCPH1* in SCC084 and SCC131 cell lines. Expression analysis at both the mRNA and protein levels demonstrated decrease in expression of *MCPH1* gene and protein within the 51 % of tumor tissues and OSCC cells, respectively. In mutation screening, c.151G >T and the out of frame alterations including c.321delA(p.Lys107fsX39) and c.1402delA(p.Thr468fsX32) were determined in oral tumor tissue samples and studied OSCC cell lines respectively. The *MCPH1* over-expression was associated with lower tumor growth in nude mice and it was led to decrease in cell proliferation besides induction of apoptosis in treated cell lines. Moreover, it was found that miR-27a modulates the expression of *MCPH1* in a negative manner. However, there was no meaningful association between *MCPH1* promoter

Table 7.1 Expression level of MCPH1 gene in different cancers

Cancer	Expression level	Considerations
Breast	Low	Associated with: T allele of rs2912010 polymorphism and promoter methylation of MCPH1 gene
Ovary	Low	Low expression level was associated with higher tumor's grades
Non small cell lung cancer (NSCLC)	Low	Low expression level was associated with poor response to erlotinib
Chronic myeloid leukemia (CML)	Low	No correlation between *MCPH1* expression level and bcr/abl activity
Oral squamous cell carcinoma (OSCC)	Low	Expression level was not associated with promoter methylation of *MCPH1* gene

methylation and expression and tumor's grade and stage in contrast to the study in BC. In general, according to the mentioned findings, the authors of the study have approved the *MCPH1* gene as a tumor suppressor gene.

However, expression analysis of *MCPH1* gene, solely, within the Hela cells previously determined that low expression level of this gene was associated with low binding to hTERT promoter and inhibition of telomerase activity (Shi et al. 2012).

The available data on the expression statue of *MCPH1* gene in various cancers is provided in Table 7.1.

7.3.1 How does MCPH1 Gene Behave in Brain Tumors?

The initial attempt on defining the role of *MCPH* gene family was carried out on primary cultured glioblastoma multiform (GBM) cells and four neural cell lines including U87, U251 and U373 and 30 brain tumor sections. The expression of both *MCPH1* and *Abnormal Spindle Microcephaly* (*ASPM)* genes was determined at either mRNA or protein levels. Although it had not any effect on patient's survival, *ASPM* gene expression was increased in higher grades of brain tumors whereas microcephalin expression level was not significantly different between various pathologies of tumors (Hagemann et al. 2008). The *ASPM* over-expression was associated with high proliferating tumor statue, provided it as a potential target of chemotherapy for higher grades of brain tumors. It was therefore, proposed that *MCPH1* may have no TSG activity in brain tumors and is not involved in their developmental process.

To clarify the exact role of *MCPH1* gene in brain tumors, we conducted a methylation based molecular study on tissue sections provided from various types of brain tumor patients and two healthy individuals (Karami et al. 2014). The methylation statue of *MCPH1* promoter was determined through MSP-PCR which its result was confirmed by full promoter sequencing (Fig. 7.5). The expression of MCPH1 protein besides CDC25A and cyclin E proteins as the cell cycle regulatory proteins were also analyzed by Immunofluorescence (IF) assay to be compared with the methylation status of *MCPH1* promoter gene (Fig. 7.6). The Quantitative fluorescent in situ hybrid-

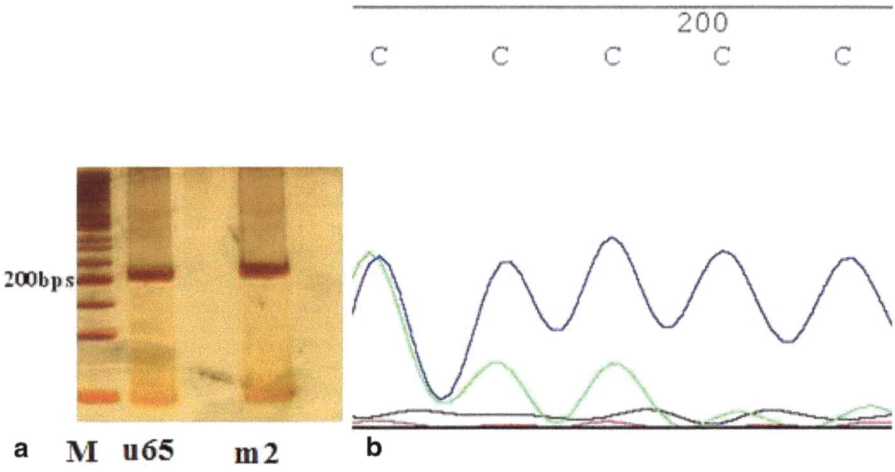

Fig. 7.5 MSP-PCR and sequencing analyses of brain tumor tissue samples (**a**) MSP-PCR analysis of brain tumor tissue samples; **M**: marker (*50 bp ladder*), **u65**: PCR using specific primers for unmethylated template 65 (*201 bps*), **m2**: PCR using specific primers for methylated template (*203 bps*); (**b**) Sequencing analysis of MSP-PCR of methylated template

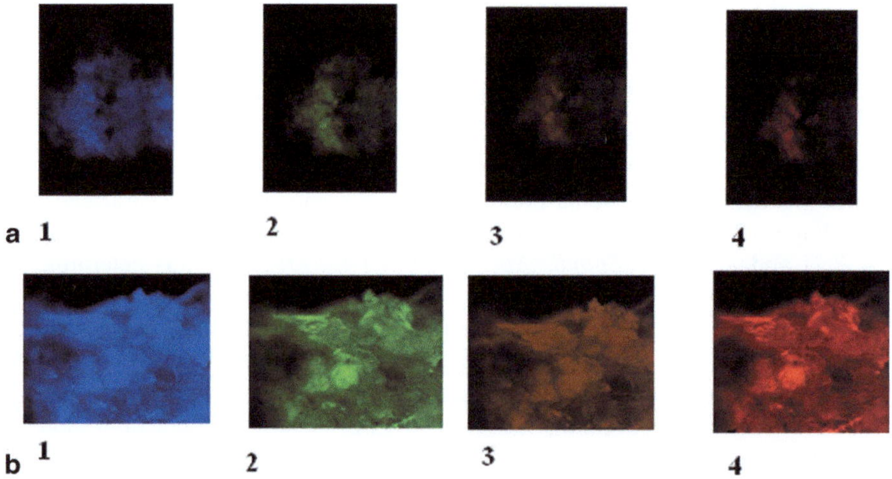

Fig. 7.6 MCPH1 protein expression of Cyclin E, CDC25A and MPCH1 in astrocytoma and meningioma tumors (a) Brain tumor cells (BTC) with astrocytoma: *1,* BTC with dapi filter; *2,* BTC conjugated with FITC presenting low expression of cyclin E; *3,* BTC conjugated with R-pep-resenting low expression of CDC25A; and *4,* BTC conjugated with Pe-cy5 reflecting low expression of MCPH1 *(x100)*. (b) brain tumor cells with meningioma reflecting *1,* BTC with dapi filter; *2,* BTC conjugated with FITC showing high expression of cyclin E; *3,* BTC conjugated with R-pe reflecting low expression of MCPH1; and *4,* BTC conjugated with Pe-cy5 showing high protein expression of MCPH1 *(x100)*

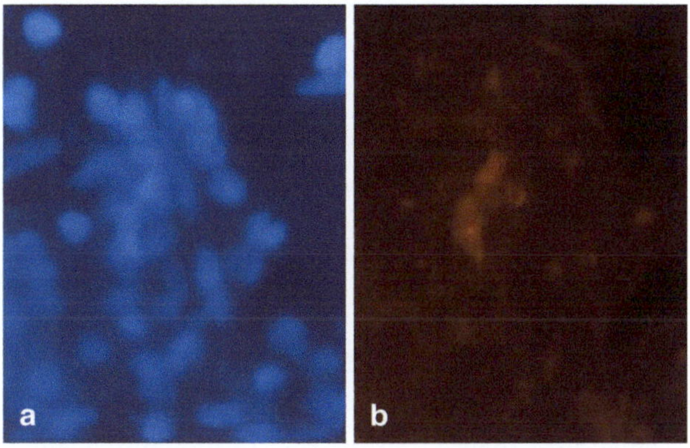

Fig. 7.7 Q-FISH illustrates status of telomere of an astrocytoma tumor in a patient affected with primary brain tumor (**a**) Dapi filter, (**b**) Cy3 showing tumor cell with low intensity and/or lacking signals. Magnification (*x400*)

ization (Q-FISH) was implicated to demonstrate the signal status of telomeres in all of the enrolled patients which was low and/or lacking any signals (Fig. 7.7).

More than 96% of the tissue samples were methylated in promoter of *MCPH1* gene (Fig. 7.5). One of the normal brain tissues was methylated and belonged to a woman who has died at 82 years old and a man with 71 years old whose brain tissue was unmethylated. MCPH1 and cyclin E proteins expression reflected a mixed expression mode including high and very low in patients affected with astrocytoma and meningioma, respectively. The methylation status of *MCPH1* gene was in direct negative correlation with their MCPH1 protein expression therein methylated promoter in astrocytoma patient has led to very low protein expression. However, the protein expression of CDC25A was detected to be low in both astrocytoma and meningioma patients. Strikingly, the expression of both positive cell cycle regulator proteins was shown to be greater in patients affected with meningioma than astrocytoma relying on the active proliferation stage of tumor progression in benign tumor (Fig. 7.6).

Given the critical role of MCPH1 in modulation of telomere repeat length, we analyzed the correlation between the *MCPH1* promoter methylation and telomere length of each patients through southern blotting (Kheirollahi et al. 2010, Mehdipour et al. 2011). Significant association was found between *MCPH1* promoter methylation and brain tumor's grade and stage. The clinicopathological features of all the brain tumor patients were in meaningful association with telomere length. The telomere length showed strong association with age of patients and the MCPH1 promoter methylation. The high intensity of signals obtained using Q-FISH was compatible with longer telomere length in higher stages and grades of tumors. This finding would be consistent with this fact that MCPH1 is no longer capable to inhibit the TERT and thereby prevents telomere reconstruction leading to higher telomere length in higher stages of tumor. To our knowledge, it was the first study investigated the methylation statue of MCPH1 gene in brain tumor tissue samples and its association with telomere length. Our study highlighted the significance of more methylation assays in brain tumors including larger samples and also study on the demethylation agents such as AZA in treatment of brain tumors especially in initial stages. In addition, it may lead to a further confirmation of the negative regulatory role of MCPH1 on modulation of telomere length which is a determining factor in survival and proliferation of tumor cells. In addition, according to the present data in brain tumors, this study may validate the tumor suppressing character of *MCPH1* gene.

7.4 MCPH1 Role in Other Diseases

Various studies have been focused on the role of polymorphisms within the *microcephalin* gene in development of microcephaly and mental retardation (MR). In this way, the frequency of c.940G > C was compared amongst Caucasian and African-American MR patients with or without microcephaly and healthy controls (Maghirang-Rodriguez et al. 2009). Although, G allele showed association with neither microcephaly nor MR, there was meaningful difference in frequencies of the alleles of this polymorphism between two studied populations.

Attempt to find correlation between *ASPM* and *MCPH1* gene variants (A44871G and G37995C, respectively) and either brain size or ratio of grey matter has been failed in genotyping of 118 healthy normal samples (Dobson-Stone et al. 2007). However, the critical role of *MCPH1* in brain size and positive selection in favoring of two haplotypes including two polymorphisms within *MCPH1* and *ASPM* genes (rs930557 and rs41310927) have motivated some investigators to search for the role of them in development of neurocognitive disorders like Alzheimer disease (AD) (Erten-Lyons et al. 2011). They also included variations in centromere protein J (CENPJ) and cyclin-dependent kinase 5 regulatory subunit associated protein 2 (CDK5RAP2) genes. Genotyping was carried out through PCR-sequencing for each polymorphism within all of the studied genes. None of polymorphisms in these four genes were associated with risk of cognitive disorders that warranted further study to reveal the effect of the variants of microcephaly genes on cognitive function and reserve.

Investigation on the role of *MCPH1* gene in other neurocognitive disorders was also included autism spectrum disorders (ASDs) (Ozgen et al. 2009). *MCPH1* gene was the most changed gene among ASDs patients harboring altered copy number or was lost during the inversion process and subsequently deletion of 8p locus. This finding may imply the importance of intact MCPH1 expression in specific features of brain development which are critical in avoiding ASDs.

7.5 Conclusion

The universal downregulated expression of *MCPH1* gene in multiple cancers would confirm that it acts as a tumor suppressor gene. However, further complementary researches are merited to define the exact role of *MCPH1* gene in various cancers when it is disturbed in different manners. It is important because of the primary reports stating that mutated MCPH1 gene in patients affected with microcephlay, doesn't increase the risk of cancer. Extra functional studies are required to determine that how various mutant form of MCPH1 protein acts in the backgrounds of different cancers. Finally, this chapter has emphasized on the crucial role of MCPH1 in brain tumors either malignant or benign types. Besides the mode of interaction between the nature of methylation with expression of MCPH1, cyclin E, CDC25A, and telomere length have opened an insight brain tumors.

References

Abraham RT (2001) Cell cycle checkpoint signaling through the ATM and ATR kinases. Genes Dev 15:2177–2196

Alderton GK, Galbiati L, Griffith E, Surinya KH, Neitzel H, Jackson AP et al (2006) Regulation of mitotic entry by microcephalin and its overlap with ATR signalling. Nat Cell Biol 8:725–733

Brand S, Rakic P (1979) Genesis of the primate neostriatum: [3H] thymidine autoradiographic analysis of the time of neuron origin in the rhesus monkey. Neuroscience 4:767–778

Bruning-Richardson A, Bond J, Alsiary R, Richardson J, Cairns DA, McCormack L et al (2011) ASPM and microcephalin expression in epithelial ovarian cancer correlates with tumour grade and survival. Br J Cancer 104:1602–1610

Chan KL, Lee JM, Guan XY, Fan ST, Ng IO (2002) High-density allelotyping of chromosome 8p in hepatocellular carcinoma and clinicopathologic correlation. Cancer 94:3179–3185

Couch FJ, Wang X, Bamlet WR, de Andrade M, Petersen GM, McWilliams RR (2010) Association of mitotic regulation pathway polymorphisms with pancreatic cancer risk and outcome. Cancer Epidemiol Biomarkers Prev 19:251–257

Dobson-Stone C, Gatt JM, Kuan SA, Grieve SM, Gordon E, Williams LM et al (2007) Investigation of MCPH1 G37995C and ASPM A44871G polymorphisms and brain size in a healthy cohort. Neuroimage 37:394–400

Doe CQ, Bowerman B (2001) Asymmetric cell division: fly neuroblast meets worm zygote. Curr Opin Cell Biol 13:68–75

Erten-Lyons D, Wilmot B, Anur P, McWeeney S, Westaway SK, Silbert L et al (2011) Microcephaly genes and risk of late-onset Alzheimer disease. Alzheimer Dis Assoc Disord 25:276–282

Evans PD, Anderson JR, Vallender EJ, Choi SS, Lahn BT (2004) Reconstructing the evolutionary history of microcephalin, a gene controlling human brain size. Hum Mol Genet 13:1139–1145

Garcia-Campelo MT, Benlloch S, Sanchez JJ, Costa C, Capitan AG, Bertran-Alamillo J et al (2012) MCPH1 (BRIT1) and outcome to erlotinib in non-small cell lung cancer (NSCLC) patients (p) harboring EGFR mutations. J Clin Oncol 30:e18131

Gavvovidis I, Pohlmann C, Marchal JA, Stumm M, Yamashita D, Hirano T et al (2010) MCPH1 patient cells exhibit delayed release from DNA damage-induced G2/M checkpoint arrest. Cell Cycle 9:4893–4899

Gerdin AK (2010) The Sanger Mouse genetics programme: high throughput characterisation of knockout mice. Acta Ophthalmol 88:0–0. doi:10.1111/j.1755-3768.2010.4142.x

Giallongo C, Tibullo D, La Cava P, Branca A, Parrinello N, Spina P et al (2011) BRIT1/MCPH1 expression in chronic myeloid leukemia and its regulation of the G2/M checkpoint. Acta Haematol 126:205–210

Hagemann C, Anacker J, Gerngras S, Kuhnel S, Said HM, Patel R et al (2008) Expression analysis of the autosomal recessive primary microcephaly genes MCPH1 (microcephalin) and MCPH5 (ASPM, abnormal spindle-like, microcephaly associated) in human malignant gliomas. Oncol Rep 20:301–308

Jackson A, Eastwood H, Bell SM, Adu J, Toomes C, Carr IM et al (2002) Identification of microcephalin, a protein implicated in determining the size of the human brain. Am J Hum Genet 71:136–142

Jo YH, Kim HO, Lee J, Lee SS, Cho CH, Kang IS et al (2013) MCPH1 protein expression and polymorphisms are associated with risk of breast cancer. Gene 517:184–190

Karami F, Javan F, Mehrazin M, Mehdipour P (2014) Key role of promoter methylation in inactivation of MCPH1 in brain tumors. J Neuro Res (under publication)

Kheirollahi M, Mehrazin M, Kamalian N, Mehdipour P (2010) Alterations of telomere length in human brain tumors. Med Oncol 28:864–870

Leung JW, Leitch A, Wood JL, Shaw-Smith C, Metcalfe K, Bicknell LS et al (2011) SET nuclear oncogene associates with microcephalin/MCPH1 and regulates chromosome condensation. J Biol Chem 286:21393–21400

Li HR, Shagisultanova EI, Yamashita K, Piao Z, Perucho M, Malkhosyan SR (2004) Hypersensitivity of tumor cell lines with microsatellite instability to DNA double strand break producing chemotherapeutic agent bleomycin. Cancer Res 64:4760–4767

Liang Y, Gao H, Lin SY, Peng G, Huang X, Zhang P et al (2010) BRIT1/MCPH1 is essential for mitotic and meiotic recombination DNA repair and maintaining genomic stability in mice. PLoS Genet 6:e1000826

Lin SY, Elledge SJ (2003) Multiple tumor suppressor pathways negatively regulate telomerase. Cell 113:881–889

Lin SY, Rai R, Li K, Xu ZX, Elledge SJ (2005) BRIT1/MCPH1 is a DNA damage responsive protein that regulates the Brca1-Chk1 pathway, implicating checkpoint dysfunction in microcephaly. Proc Natl Acad Sci U S A 102:15105–15109

Lin SY, Liang Y, Li K (2010) Multiple roles of BRIT1/MCPH1 in DNA damage response, DNA repair, and cancer suppression. Yonsei Med J 51:295–301

Lu T, Hano H (2007) Identification of minimal regions of deletion at 8p23.1-22 associated with metastasis of hepatocellular carcinoma. Liver Int 27:782–790

Lu T, Hano H, Meng C, Nagatsuma K, Chiba S, Ikegami M (2007) Frequent loss of heterozygosity in two distinct regions, 8p23.1 and 8p22, in hepatocellular carcinoma. World J Gastroenterol 13:1090–1097

Maghirang-Rodriguez R, Archie JG, Schwartz CE, Collins JS (2009) The c.940G variant of the Microcephalin (MCPH1) gene is not associated with microcephaly or mental retardation. Am J Med Genet A 149A:622–625

Manke I, Lowery DM, Nguyen A, Yaffe MB (2003) BRCT repeats as phosphopeptide-binding modules involved in protein targeting. Science 302:636–639

Margolis SS, Kornbluth S (2004) When the checkpoints have gone: insights into Cdc25 functional activation. Cell Cycle 3:425–428

Mehdipour P, Kheirollahi M, Mehrazin M, Kamalian N, Atri M (2011) Evolutionary hypothesis of telomere length in primary breast cancer and brain tumour patients: a tracer for genomic-tumour heterogeneity and instability. Cell Biol Int 35:915–925

Mekel-Bobrov NPD, Gilbert SL, Lind P, Gosso MF, Luciano M, Harris SE et al (2007) The ongoing adaptive evolution of ASPM and Microcephalin is not explained by increased intelligence. Hum Mol Genet 16:600–608

O'Driscoll M, Jeggo PA (2006) The role of double-strand break repair-insights from human genetics. Nat Rev Genet 7:45–54

Olson JE, Wang X, Pankratz VS, Fredericksen ZS, Vachon CM, Vierkant RA et al (2011) Centrosome-related genes, genetic variation, and risk of breast cancer. Breast Cancer Res Treat 125:221–228

Ozgen HM, van Daalen E, Bolton PF, Maloney VK, Huang S, Cresswell L et al (2009) Copy number changes of the microcephalin 1 gene (MCPH1) in patients with autism spectrum disorders. Clin Genet 76:348–356

Rai R, Dai H, Multani AS, Li K, Chin K, Gray J et al (2006) BRIT1 regulates early DNA damage response, chromosomal integrity, and cancer. Cancer Cell 10:145–157

Rimol LMAI, Djurovic S, Brown AA, Roddey JC, Kähler AK, Mattingsdal M et al (2010) Sex-dependent association of common variants of microcephaly genes with brain structure. Proc Natl Acad Sci U S A 107:384–388

Rodriguez R, Hansen LT, Phear G, Scorah J, Spang-Thomsen M, Cox A et al (2008) Thymidine selectively enhances growth suppressive effects of camptothecin/irinotecan in MSI+ cells and tumors containing a mutation of MRE11. Clin Cancer Res 14:5476–5483

Rushton JP, Vernon PA, Bons TA (2007) No evidence that polymorphisms of brain regulator genes Microcephalin and ASPM are associated with general mental ability, head circumference or altruism. Biol Lett 3:157–160

Shi L, Li M, Su B (2012) MCPH1/BRIT1 represses transcription of the human telomerase reverse transcriptase gene. Gene 495:1–9

Singh N, Wiltshire TD, Thompson JR, Mer G, Couch FJ (2012) Molecular basis for the association of microcephalin (MCPH1) protein with the cell division cycle protein 27 (Cdc27) subunit of the anaphase-promoting complex. J Biol Chem 287:2854–2862

Tibelius A, Marhold J, Zentgraf H, Heilig CE, Neitzel H, Ducommun B et al (2009) Microcephalin and pericentrin regulate mitotic entry via centrosome-associated Chk1. J Cell Biol 185:1149–1157

Trimborn M, Ghani M, Walther DJ, Dopatka M, Dutrannoy V, Busche A et al (2010) Establishment of a mouse model with misregulated chromosome condensation due to defective Mcph1 function. PLoS ONE 5:e9242

Urist M, Tanaka T, Poyurovsky MV, Prives C (2004) p73 induction after DNA damage is regulated by checkpoint kinases Chk1 and Chk2. Genes Dev 18:3041–3054

Venkatesh T, Nagashri MN, Swamy SS, Mohiyuddin SM, Gopinath KS, Kumar A (2013) Primary microcephaly gene MCPH1 shows signatures of tumor suppressors and is regulated by miR-27a in oral squamous cell carcinoma. PLoS ONE 8:e54643

Woods RP, Freimer NB, De Young JA, Fears SC, Sicotte NL, Service SK et al (2006) Normal variants of Microcephalin and ASPM do not account for brain size variability. Hum Mol Genet 15:2025–2029

Wood JL, Singh N, Mer G, Chen J (2007) MCPH1 functions in an H2AX-dependent but MDC1-independent pathway in response to DNA damage. J Biol Chem 282:35416–35423

Wu X, Mondal G, Wang X, Wu J, Yang L, Pankratz VS et al (2009) Microcephalin regulates BRCA2 and Rad51-associated DNA double-strand break repair. Cancer Res 69:5531–5536

Xu X, Lee J, Stern DF (2004) Microcephalin is a DNA damage response protein involved in regulation of CHK1 and BRCA1. J Biol Chem 279:34091–34094

Yamashita YM, Jones DL, Fuller MT (2003) Orientation of asymmetric stem cell division by the APC tumor suppressor and centrosome. Science 301:1547–1550

Yang SZ, Lin FT, Lin WC (2008) MCPH1/BRIT1 cooperates with E2F1 in the activation of checkpoint, DNA repair and apoptosis. EMBO Rep 9:907–915

Yu X, Chini CC, He M, Mer G, Chen J (2003) The BRCT domain is a phospho-protein binding domain. Science 302:639–642

Chapter 8
Sentinel Gene Within Cell Territory and Molecular Platforms in Cancer: Methylation Diversity of p53 Gene in Brain Tumors

Parvin Mehdipour and Fatemeh Karami

Contents

Abstract *TP53* is the most important tumor suppressor gene which plays critical functions to avoid tumor progression and development through inducing the DNA repair pathways or cell cycle arrest. High frequency of *p53* gene aberrations in various types of cancer has relied on its significant roles in prevention of arising tumors. However, the effect of promoter methylation of this gene on its protein

---

P. Mehdipour (✉) · F. Karami
Department of Medical Genetics, School of Medicine, Tehran University of Medical Sciences, Poursina Street, Keshavarz Boulevard, Tehran 14176-13151, Iran
e-mail: mehdipor@tums.ac.ir

P. Mehdipour (ed.), *Epigenetics Territory and Cancer*,
DOI 10.1007/978-94-017-9639-2_8

expression and function remains unclear and there is a strong controversy among various investigations especially those studies performed on brain tumors.

TP53 may be the most frequent gene which was studied in different aspects of cancer biology and therefore a considerable literature is available. In this chapter, we tried to provide a brief explanation about the basic knowledge which was found in investigations either at *p53* gene- or protein- level and then its function and aberrations in various cancers will be described. Finally, we will discuss about the methylation status of *p53* gene promoter in different types of cancers with emphasizing on brain tumors.

Abbreviations

AMPK	AMP-activated protein kinase
53BP1	53 binding protein 1
CPE	Core promoter element
CPT1	Carnitine palmitoyltransferase
COX	Cytochrome c oxidase
CRM1	Chromosomal region maintenance 1
C-terminal	Carboxy-terminal
DBD	DNA binding domain
ECM	Extracellular matrix
ER	Estrogen receptor
FAS	Fatty acid synthase
GAMT	Guanidinoacetate methyltransferase
GLS2	Glutaminase 2
HK2	Hexokinase 2
ISRE	Interferon stimulated response element
MPF	Maturation-promoting factor
mTORC1	Mammalian target of rapamycin1
NADPH	Nicotinamide adenine dinucleotide phosphate
NLS	Nuclear localization signals
N-terminal	NH3 terminal
PGM	Phosphoglycerate mutase
PKC δ	Protein kinase C δ
PP genotype	Proline proline genotype
PPP	Pentose phosphate pathway
PUMA	p53 upregulated modulator of apoptosis
PRD	Proline rich domain
PRMTs	Protein arginine methyltransferases
OD	Oligomerization domain
ROS	Reactive oxygen species
SAHFs	Senescence associated heterochromatin foci
SCO2	Cytochrome c oxidase

Sp1	Specificity protein 1
TSC2	Tuberous sclerosis complex 2
TIGAR	TP53-induced glycolysis regulator

8.1 Introduction

The p53 gene encodes for a protein amassed 53 kilo Dalton and due to its major functions in cell cycle regulation, maintaining the genome stability and suppressing tumor progression has been so called guardian of genome. In 1979, it was introduced by Arnold Levine, David Lane and William Old at Princeton University, Dundee University (UK) and Sloan-Kettering Memorial Hospital, respectively. Given the formation of complex between p53 and major oncoprotein of SV40 virus (large T-antigen), it had been previously assumed that p53 is an oncogene before 1989. Lane and Crawford then proposed that making this complex is a way for p53 to exert its neutralizing and inhibitory effect on T-antigen protein to control the proliferation rate of the cell (Lane and Crawford 1979). Following the various aspect of research on cancer cells containing mutant *p53* gene, the tumor suppressor character of *p53* gene was disclosed and after 4 years then, it was introduced as the molecule of the year in the Science magazine. Establishment of the idea of tumor suppressor nature of the *p53* was based on four main observations: (1) in the families with Li-Fraumenni's syndrome, in those, inheritance of mutated *p53* gene was associated with 100 % risk of being affecting with cancer with multiple and independent origins (Malkin et al. 1990); (2) Loss of function in *p53* mutations led to tumor development at younger age in knocked out mice (Lozano 2010; Donehower et al. 1992); (3) Biallelic mutations of *p53* gene observed in more than 50 % of the cancers (Robles and Harris 2010) and (4) Extremely abnormal high expression of *p53* in different tumors to prevent aberrant proliferation and cancer progression (Bartek et al. 1991). These reasons engaged extraordinary attentions toward characterizing the exact functions of p53 protein.

8.1.1 P53 Gene

The *p53* gene was mapped on the chromosome 17p13.1 and includes 19,198 nucleotides divided into 11 exons in which except of the first exon, the remaining 10 exons are coding domain (Fig. 8.1). It has an internal promoter within the fourth intron indicating that *p53* has a dual gene structure similar to its family members, i.e., *p63* and *p73*. It was demonstrated that this dual gene structure has been conserved during the human evolution relying on its importance in p53 protein functions (Chen et al. 2005; Chen et al. 2009). Alternative splicing of intron 9 of *p53* gene creates three p53 proteins with different carboxy-terminal (C-terminal) domains (Bourdon et al. 2005). Another alternative splicing of intron 2 results in p53

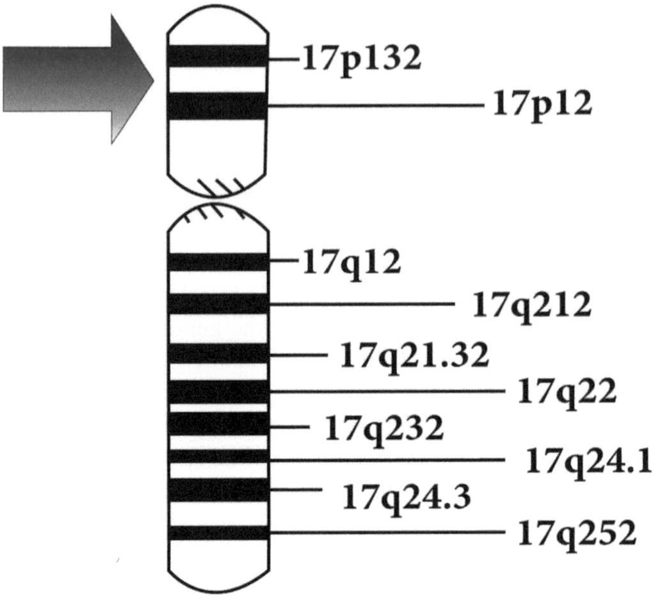

Fig. 8.1 Localization of p53 gene on chromosome 17p13.2

protein which hasn't the first 40 residues. All types of alternative splicing, promoter and translation start sites produce various types of p53 proteins including p53β, p53γ, Δ133p53, Δ133p53β, Δ133p53γ, Δ40p53, Δ40p53β and Δ40p53γ.

The origins of p53 gene's ancestors are initially seen in descendants of single cell choanoflagellates and early metazoan sea anemone. The ancestral *p53* gene is much related to both *p63* and *p73* genes. Interestingly, this *p63/p73* like genes are responsible for guarding the DNA of germ-line gametes against damage in sea anemone which is also conserved in insects, worms, clams, and vertebrates. It was proposed that *p53* gene is as a result of gene duplication of these ancestral genes to keep on their critical responsibility in both somatic and progenitor cells. The vertebral *p63* and *p73* genes are as consequences of second duplication which has been occurred in the ancestral genes. The functions of *p63/p73* genes have been complicated during the development of vertebrates and were extended to regulation of transcription within the skin and other organ cells (Yang et al. 2002; Yang and McKeon 2000; Hernandez-Acosta et al. 2011; Mills et al. 1999).

Transcription of *p53* expression is regulated through several transcription factor binding motifs which are distributed within the promoter sequence of *p53* gene and are evolutionary conserved (Neduva and Russell 2005; Neduva and Russell 2006) The Myc/Max, USF, YY1, NF1, AP-1 and NFκB were described to activate the expression of *p53* gene whereas binding the PAX2, PAX5, PAX8, and BCL6 transcription factors to their specific motifs negatively regulate the transcription of *p53* (Reisman et al. 1993; Kirch et al. 1999; Ronen et al. 1991; Roy et al. 1994). Moreover, HOXA5 was found as another activators of *p53* gene owing to the up-regulation of p53 following to its over-expression which was associated with apoptosis of breast cancer cells (Raman et al. 2000). Both the human and mouse have two interferon stimulated response element (ISRE) within their *p53* promoters that are induced by a complex including Stat1, Stat2 and IRF-9. These ISRE could be targeted for IFNα/β therapy in cancer in order to activate immune system against tumor cells (Takaoka et al. 2003; Pfeffer et al. 1998). There is an important regulatory element between the BCL6 binding motif and CpG Island of *p53* promoter which is called as CTCF motif that activates the expression of *p53* through blocking the negative effects of transcription silencers (Soto-Reyes and Recillas-Targa 2010; Su et al. 2009). Moreover, protein kinase C δ (PKC δ) is another transcription factor that by accompanying Btf, plays pivotal roles in induction of apoptosis in response to DNA damage through binding to its core promoter element (CPE) on *p53* promoter (Liu et al. 2007).

8.1.2 p53 Protein

In human, the full length of p53 protein is composed of 393 amino acids. It has a close homology in structure and function with its two famous family members including p63 and p73 (Fig. 8.2). Although it remains at low level in the cell by degradation effect of 26S proteasome, DNA damage or every other stress signals immediately stabilize and activate it. The first 42 amino acids build up the NH3 or amino terminal (N-terminal) of the p53 protein which its 22 and 23th residues forms the transactivation domain (Teufel et al. 2007). MDM-2, the major negative regulator of p53, exerts its inhibitory effects on transcription and ubiquitination of lysines residing in carboxyl terminal (C-terminal) domains (Rippin et al. 2002). The amino acids between the residues 43–63 construct the second transactivation domain of p53 protein probably designed to induce the transcription of other genes which are under the control of p53. Immediately after these transactivation domains, there is another domain spanning within the 61–94 residues which is proline rich and plays pivotal roles in apoptosis and protein-protein interactions through SH-3 signals (Walker and Levine 1996; Neduva et al. 2005a, b). DNA binding domain (DBD)

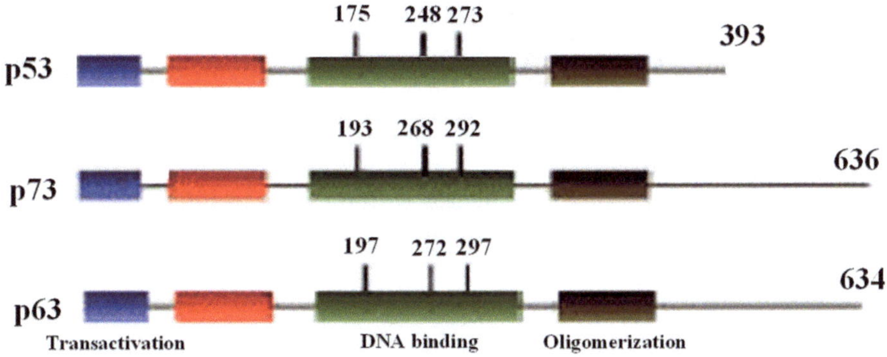

Fig. 8.2 Schematic structure of p53 and its two homologous proteins.

and oligomerization domain (OD) are the other domains of p53 protein which are composed of the 101–306 and 356–393 residues, respectively. Of note, there are three nuclear localization signals (NLS) in OD domain. The last domain at the C-terminal has regulatory function and is spanned within the 356–393 amino acids. Both C and N-terminals of p53 protein are actively involved in post-translational modifications including phosphorylation, ubiquitination, acetylation and methylation (Bode and Dong 2004).

8.1.2.1 p53 Protein Methylation

Protein methylation is almost defined as transferring a methyl group to amino acids especially lysines, arginines and histidines that is important in modulation of gene expression, RNA metabolism and also influences the function of protein (Chen et al. 1999; Shen et al. 1998). The p53 protein is one of the few proteins which its stabilization and transcription factor activity is regulated through methylation and demethylation of its residues.

Methylation of p53 protein is fulfilled through transferring methyl group to either lysines residues residing in the basic C-terminal domain or argenine amino acids within the DBD domain performing by histone lysine methyltransferases and arginine methyltransferases, respectively.

The histone lysine methyltransferases include KMT5 (Set9), KMT3C (Smyd2), and KMT5A (Set8) (Allis et al. 2007). KMT5 methylates the K372 and thereby enhances the stability and nuclear localization of p53 protein to be enough strong to activate the expression of its target genes as well as p21 and BAX and induce apoptosis through arresting cell cycle at G2/M checkpoint (Ivanov et al. 2007). The activity of KMT5 protein was shown to be increased in response to DNA damage without increasing its gene expression, possibly via post-translational modification with acetylation and phosphorylation. It was demonstrated that K372 has been methylated by KMT5 followed by acetylation of K373/382 when p53 induced the expression of p21. It is indicating that first methylation and then acetylation of p53 are required for inducing the expression of target genes. Interestingly, methylation of p53 by the two other methytransferases, KMT3C and KMT5A, leads to reducing the capability of p53 to induce the expression of p21, PUMA and MDM2 (Brown et al. 2006; Shi et al. 2007). Although, expression of KMT3C remains constant and is not changed by detecting the stress signals, but, methylation of K372 by KMT5 prevents methyltransferase activity of KMT3C which leads to p53 methylation (Huang et al. 2006).

Argenine methylation of p53 protein is performed by two protein groups belonged to class I of protein arginine methyltransferases (PRMTs) including PRMT1 and CARM1 through methylating the argenines within the transactivation and basic C-terminal domains, respectively. In contrast to lysine methyltransfrases, methylation of p53 by both PRMTs is associated with increase in its activity. They also regulate the p53 function through modulation of the methylation status of histone proteins wrapping around its target genes. p53 is able to induce the expression of

GADD45 in response to UV radiation when the CARM1 and PRMT have methylated the respective H3 and H4 histones which are enhanced via their prior acetylation by KAT3B (An et al. 2004).

Demethylation of lysine residues of p53 protein is carried out by lysine-specific demethylase KDM1 which has shown to pay critical roles in immediate p53 induction in response to cellular stresses. Knocking down the expression of KDM1 led to decrease in expression of p21 and MDM2 while its over-expression has changed neither the cellular proliferation rate nor the ability of p53 to activate its target genes (Scoumanne and Chen 2007). KDM1 can only demethylates p53 protein when it is dimethylated at K370 and K372 and thereby prevails the binding of it to its major activator, 53 binding protein 1 (53BP1) (Huang et al. 2007). In general, it is assumed that demethylase agents act as the repressors of p53 protein when the PRMTs increase its transactivation of target genes. Further studies are warranted to clarify the role of other demethylase enzymes in demethylation of p53 protein.

8.2 P53 Functions

8.2.1 Cell Cycle Control

There are four main cell cycle checkpoints ensuring that all parts of cell division have been perfectly occurred. When the cell encounters to a stress resulting in DNA damage, the division cycle should be stopped until the damage will be repaired or it immediately being deviated to be mortal. It was shown that p53 is involved in arresting the cell cycle at two major G1/S and G2/M checkpoints (Giono and Manfredi 2006). This function of p53 has been strongly investigated and various molecules have been determined in different pathways acting as upstream and downstream of it. The p21 is the most important downstream molecule of p53 which leads to G1/S arrest when it being activated. Given that inactivation of p21 was not as harmful as p53 loss, indicates that there are some other molecules that do this function as well (Brugarolas et al. 1995). The p53 induces G1/M arrest through repression of cyclin B1/cdc2 complex forming the maturation-promoting factor (MPF). After DNA damage, cell cycle arrest is carried out through two major mechanisms including inactivation of a phosphatase involved in promoting of mitosis known as cdc25c and, activation of 14-3-3σ (Hermeking et al. 1997; Clair St and Manfredi 2006). Increase in expression of 14-3-3σ is associated with interference with nuclear localization of cyclin B1/cdc2 complex following DNA damage. It was demonstrated that 14-3-3σ depletion in HCT116 cell line caused cell death after DNA damage (Chan et al. 1999).There are two final fates for cell cycle arrest: (1) Permanent arrest known as cell senescence; (2) Cell death (Fig. 8.3).

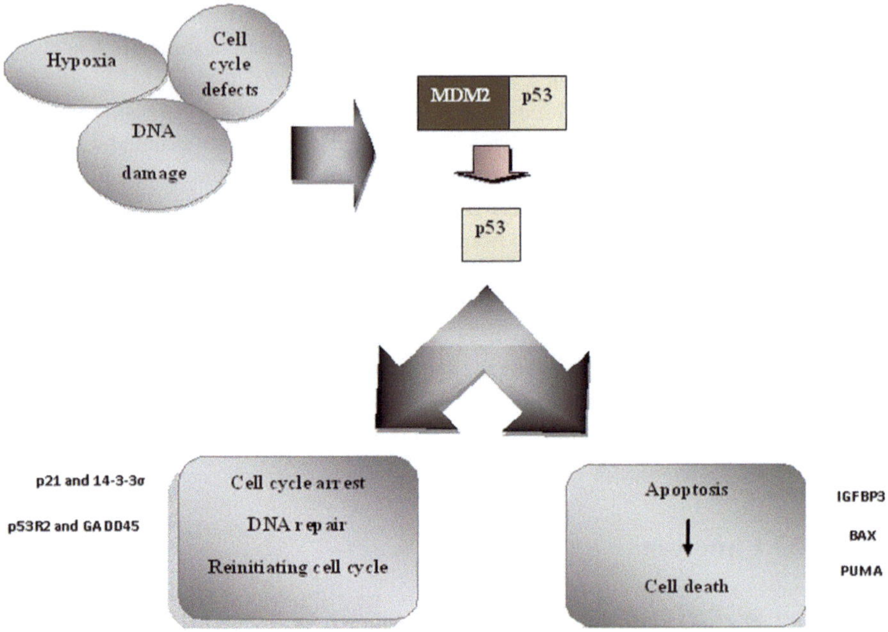

Fig. 8.3 p53 central role in modulation of DNA damage

8.2.2 Cell Senescence

Cellular senescence was initially proposed for normal fibroblast with human origin (Hayflick 1965). According to suggested hypothesis, cellular senescence can be harmful or beneficial to the involved cells. Based on the former hypothesis, cellular senescence could move the cell toward aging which is critical especially when the potential of regeneration is necessary. The later hypothesis reflects that cellular senescence would be an inhibitory mechanism against tumor's cells proliferation (Campisi and d'Adda di Fagagna 2007).

In the senescent cell, there is a remarkable but regular change in gene expression profile which is similar to the process of wound healing (Krizhanovsky et al. 2008). Although, there is no special marker to be assigned to a senescent cell, senescence associated β-galactosidase, increased in p16INK4a and p53INK4b expressions have been detected when the cell goes through senescence (Campisi and d'Adda di Fagagna 2007).

The diversity of cellular senescence are usually face some stresses including DNA damage, inactive telomere, abnormal chromatin structure and extra mitogenic signaling generated by oncogenes. It was demonstrated that viral oncoproteins to invade from senescence through inactivation of p53 (Stewart and Weinberg 2006). Many chemotherapeutic drugs are designed to induce senescence in tumor cells which can be inhibited in cells carrying mutant and inactive p53 (Roninson 2003).

Based on type of the involved cell and trigger molecule, both the p53-p21 and p16-RB pathways can drive the cellular senescence. p14ARF functions at the center of both mentioned pathways and links them together. The senescence signal is, firstly, received by p14ARF and then activates p53 followed by p21 which in turn suppresses the cyclin-dependent kinases upstream of RB (Martín-Caballero et al. 2001; Matheu et al. 2008). RB itself, induces subsequent activation of p14ARF and p53 by inhibition of E2F1 and its target genes through formation of senescence associated heterochromatin foci (SAHFs) around them (Narita et al. 2006). It was proposed that p53 exerts its tumor suppressive activity in induction of senescence through overexpression of extracellular matrix (ECM) degrading enzymes and inflammatory cytokines to target and kill the tumor cells by dynamic action of innate immune system (Xu 2008).

8.2.3 Apoptosis

The role of p53 in apoptosis was initially identified in mouse thymocytes after radiation (Clarke et al. 1993). There are many known targets for *p53* gene function in apoptosis indicating that such role of p53 is more indicative than its other functions. The most important targets of p53 include p53 upregulated modulator of apoptosis (PUMA) and Noxa which are the only proteins that have BH3 repeats, Bax, PIG3, CD95 or Fas, Killer/DR5, p53AIP1 and Perp (Riley et al. 2008). It seems that active involvement in apoptosis is shared by other family members of *p53* gene which is modulated with various external and internal signals and is also dependent on cell type (Fridman and Lowe 2003).

Induction of apoptosis is fulfilled through activation of cellular p53 depending on apoptosis pathway in mitochondria. Some apoptotic specific signals as well as radiation pull the p53 to be localized in mitochondria to induce secretion of pro-apoptotic factors from the intermediate space of mitochondria through increasing the permeability of outer membrane of mitochondria. The major effect of p53 in induction of apoptosis through mitochondria pathway could be observed through activation or repression of inhibitors of Bcl2 family including Bcl-X_L, Bak and Bax (Chipuk et al. 2005). When the p53 becomes activated in response to any stress, requiring cell to be undergone apoptosis, then it will immediately increase the level of PUMA to make the p53 free through binding with Bcl-XL.The TP53 then will be able to activate the Bax gene. Among all of the target genes of p53 in apoptosis, PUMA is as critical as p53 which its loss has shown the same defects as in p53 loss (Jeffers et al. 2003).

8.2.4 Response to Stress

The p53 protein protects cell against any type of stress including DNA damage, any temperature shock (heat or cold), spindle toxins, hypoxia, telomere shortening and even activation of oncogenes or inactivation of other tumor suppressor genes.

Activation of p53 in each of the mentioned cellular stresses is modulated through induction of its activators as well as HAUSP or inactivation of its negative regulators which include MDM-2, MDM-4, WIP-1 phosphatase (Perry 2010; Lu 2010; Meek and Anderson 2009). Some of these regulatory molecules, similar to WIP1- and MDM-2, are responsible for dimorphic pattern between two sexual identities of human. MDM-2 contains sequence recognized by estrogen receptor (ER) to be combined with the alleles of a SNP within the intron 1 which was associated with higher risk of breast cancer in ER positive premenopausal women (Hu et al. 2007). Of note, the cancer predisposition of this SNP and the age of onset of cancer incidence is different between male and females making sexual dimorphism (Grochola et al. 2010). *WIP-1* encodes a phosphatase and was found to be one of the estrogen regulated genes (Han et al. 2009).

Strikingly, p53 differently respond to stress in various types of cells even when the genetic and environmental conditions of the cells are the same (Lahav et al. 2004). It remained many questions to be answered especially about the roles of p53 regulators in these different patterns of response.

8.2.4.1 Metabolic Stress

The main factor that a cell may encounters to it, is metabolic stress. p53 plays critical roles in switching the metabolic pathways in cancers in order to disable the malignant cells to survive and proliferate under stress conditions (Gottlieb and Vousden 2010). Cell proliferation is induced with glucose through activation of AMP-activated protein kinase (AMPK) which phosphorylates and activates p53 (Hardie 2004). In metabolic stress, AMPK also induces the tuberous sclerosis complex 2 (TSC2) protein which in turn inhibits GTP-binding protein, Rheb, and then mammalian target of rapamycin1 (mTORC1) are noatable. In nutrient restriction, mTORC1 dephophorylates p53 on Ser15 (Inoki et al. 2003; Feng et al. 2006; Imamura et al. 2001) (Jones et al. 2005). The p53 targets which are activated one by one in mTORC1 and IGF-1/AKT pathways modulate their pathways in opposite directions. For instance, Sestrin 1 and 2 are induced by p53 in IGF-1/AKT pathway play role in response to oxidative stress through two critical functions: (1) Indirect activation of TSC1/2 by phosphorylation and activation of AMPK through interaction with its α-catalytic subunits which leads to suppression of m-TORC1 pathway; (2) Reducing the cellular level of reactive oxygen species (ROS) through deoxidizing the ROS products by oxidized peroxiredoxins (Budanov et al. 2002; Budanov et al. 2004; Velasco-Miguel et al. 1999).

In general, p53 is actively involved in maintaining the baseline metabolic status of cell and genomic integrity in response to stresses threatening them. These stresses are created through three major mechanisms which will be discussed in detail in the following sections including regulating the glycolysis, oxidative phosphorylation and metabolism of fatty acids.

8.2.4.2 Regulating Glycolysis

Glycolysis is a non-oxidative pathway of energy production mainly used in neoplasms including the malignant cells to supply their energy (Warburg 1956). In glycolysis pathway, there is a protein called TP53-induced glycolysis regulator (TIGAR) which under the induction of p53 shifts glycolysis toward pentose phosphate pathway (PPP) through fructose-2, 6-bisphosphate. TIGAR, actually by this switching, helps to keep the metabolic status of cell stable through reducing the generation of ROS and resistance against it through production of nicotinamide adenine dinucleotide phosphate (NADPH) in PPP pathway (Bensaad et al. 2006). Another attempt of p53 in stablizing the cell in metabolic stress is fulfilled through inhibition of phosphoglycerate mutase (PGM). Although, both the TIGAR and PGM are considered as intracellular antioxidants, in contrast to TIGAR, PGM increases the glycolysis to decrease the ROS generation by mitochondrial respiration (Brand and Hermfisse 1997). Nonetheless, enhancing of the glycolysis pathway in cancer cells would be dangerous, and PGM is suppressed by p53. Mutant p53 acts in reverse direction in cancerous cells in order to increase the glycolysis pathway through induction of PGM and hexokinase 2 (HK2) which is another driver of glycolysis oxidative pathway (Bustamante and Pedersen 1977; Mathupala et al. 1997).

8.2.4.3 Role of p53 in Oxidative Phosphorylation

Oxidative phosphorylation is the preferred and more proficient energy producing pathway in normal cells which is carried out in mithochondria. This pathway is aimed to oxidate the pyruvate, as the end product of glycolysis pathway, and generates ATP in a more efficient manner. P53 enhances the mithochondrial respiration and the expression of cytochrome c oxidase (SCO2) which is essential for formation of cytochrome c oxidase (COX) complex (Bensaad et al. 2006). Cancer cell with abnormal or missing p53 function had shown the pathway switching from oxidative phosphorylation to glycolysis (Ma et al. 2007). The p53 increases the mitochondrial respiration and ATP generation through glutaminase 2 (GLS2) induction which is required for converting the glutamine to glutamate and production of α-ketoglutarate (Hu et al. 2010).

8.2.4.4 How Does p53 Play Role in Metabolism of Fatty Acids

The p53 protein controls the metabolic status of the cell through regulating the functions of two important molecules in fatty acids metabolism pathways including guanidinoacetate methyltransferase (GAMT) and carnitine palmitoyltransferase (CPT1).

GAMT is involved in creatine pathway and changes the guanidineacetate to creatine to be used in ATP generation pathways (Ide et al. 2009). GAMT is related to fatty acid synthase (FAS) which has demonstrated high expression in various types of cancer cells to drive glucose independent energy generating pathways (Alo et al. 1996; Swinnen et al. 1997)

Another strategy of p53 in keeping the baseline status of cellular metabolism in a calm manner, is fulfilled by induction of β-oxidation of fatty acids through activation of CPT1 which attaches the carnitine to fatty acids to be imported in to the mitochondria (Buzzai et al. 2005; Vousden and Lu 2002).

8.2.5 Tumor Suppression

The successfulness of p53 in suppression of tumor is determined by three major factors including perfect detection of oncogenic signals, differentiating between healthy and cancerous signaling and efficient conquering on neoplastic cells (Tyner et al. 2002). Tumor suppressive activity of p53 can be overcome through five main mechanisms; (1) Oncogenic signaling below the threshold of p53 activation which is in line with first model described above. In these conditions, the cell containing damaged DNA can go through carcinogenesis prior to be repaired by ARF or p53 induced DNA repair pathways (Zilfou and Lowe 2009). Complementary studies on subordinating the threshold of p53 signal detection have shown that over-activity in tumor suppression role of p53 also increases its regenerative capacity which would be associated with premature aging (Cao et al. 2003; Lanni et al. 2008; Maier et al. 2004; Medrano et al. 2009; Campisi 2003). (2) Inactivating mutations in the upstream molecules of p53 which can be influenced by several factors including the type of mutation and gene, the extent of degeneracy of affected pathway and its contribution to tumor suppression. (3) Mutations in other contributory molecules rather than upstream which can abrogate the major functions of p53 in either apoptosis or cell senescence. Overexpression of BCL-2 or BCL-X_L, decreases the expression of apoptosis effectors as well as PUMA or NOXA and loss of INK4A and CDKN1A involved in cell senescence, are some examples. (4) Tumor may arise, when, the overall outcome of p53 activation is switching from cell senescence or apoptosis to temporary arrest or incomplete repair of damaged DNA in favor of cancer evolution. This mechanism makes the growth of tumor more plausible through unbalancing between statue of cell loss and cell gain. (5) Genetic or epigenetic mutations in *p53* gene which directly affect its function or expression and sometimes changes its tumor suppressor character to a new functionally oncogene discussed in detail in next section (Strano et al. 2007; Xu 2008; Lozano 2007).

8.3 Role of p53 Gene Mutations

P53 is the key gene which is altered in more than 50 % of different types of cancers. Li Fraumeni's syndrome (LFS) which is characterized with multiple generations susceptible to some types of cancers was the first evidence of the role of germline p53 mutations in risk of cancer (Li and Fraumeni 1969b; Li and Fraumeni 1969a). LFS is inherited with autosomal dominance pattern and affected families are prone to get breast cancer, sarcoma and different types of neoplasms (Malkin et al. 1990)

Single nucleotide non-synonymous mutations are the most common type of p53 alteration usually associated with replacement of amino acid with another one with new function. Mutations almost change the half life of p53 protein in order to be prolonged and this apparently gain of function led to assume that it is an oncoprotein. This special characteristics of p53 mutations, makes it unique among all the tumor suppressor proteins those alterations almost lead to insufficient expression with normal function (Levine et al. 1995). The second dominant aspect of p53 alteration could be described through this evidence that loss of heterozygosity (LOH) of an allele of *p53* gene is subsequently occurred after the episode of mutation in another allele (Brosh and Rotter 2009).

Similar to other genes, mutations can be seen in every nucleotides of p53 protein encoding gene, and some hot spots have been determined in this protein. DBD is the most common target of mutation in p53 protein wherein many various and famous alterations have been identified (Petitjean et al. 2007; Olivier et al. 2004).

8.3.1 p53 Gene Defects in Different Cancers

Involvement of the *p53* mutations in different cancers can open the promising windows toward improvements in treatment schedules and more precise determination of prognosis and survival for cancer patients (An et al. 2004). Interestingly, more than 80 % of the p53 gene mutations are missense alterations which usually ends in absolute intact and functional protein in contrast of non- or dysfunctional mutations in other tumor suppressor genes (Soussi and Beroud 2001). Loss of function of *p53* gene has shown controversial clinical effects on the outcome of chemotherapy which are usually cell specific (Soussi and Beroud 2003). Although *p53* gene mutations are the most frequent genetic alterations in various cancers, it seems that it may have no or scant role in some of cancers as well as testis cancer and melanoma. However, it was demonstrated that genetic mutations of *Apaf* gene, as one of the downstream molecules of p53, had influenced the success rate of chemotherapy in melanoma patients (Soussi and Beroud 2003). Furthermore, although the frequency of p53 gene mutations was reported to be low in inflammatory breast cancer and neuroblastoma, high amounts of normal p53 protein accumulated in cytoplasm of tumor cells may interfere with its immediate action after DNA damage (Zaika et al. 1999).

It is worth to note that more than 70 % of the *p53* mutations take place in the sequences that are not hot spot and around 4.4 % of them have been found once and their importance in the cancer progression merited to be determined (Soussi and Beroud 2003). Since most of the p53 mutation screening studies have focused on hot spot DNA binding domain; it has been famous to have the most frequent of alterations in p53 mutation database (Soussi and Beroud 2001). The most well-known hot spot regions of *p53* gene include the codons 175, 248 and 273, those alterations comprise 19% of all the *p53* mutations. Based on the effects of these alterations on structure and function of p53 protein, these could be classified into two major classes (Cao et al. 2006; Joerger et al. 2004). So mutations that change the amino acids which are involved in interaction of p53 with DNA sequence are categorized in the class I. These types of *p53* mutations as well as alterations in codon 248 comprise 7.6% of all the *p53* mutations recorded in *p53* gene mutations database (http://p53.free.fr). However, these mutations have no effect on overall p53 protein conformation determined by spatial monoclonal antibodies and low affinity of mutant p53 protein to be bound with chaperon molecules (Ory et al. 1994; Hinds et al. 1990; Vakifahmetoglu-Norberg et al. 2013). In contract, class II of mutations make mutant protein to be strongly recognized by the major chaperon, hsp70, due to their conformational changes (Ory et al. 1994). Alterations of codon 175 are the best examples of this type of p53 mutations. Class II mutations include 4.9% of the p53 mutations, however, most but not all of them have irreparable effect on protein conformation (Selivanova et al. 1997).

Mutations in hot spot regions have different frequency in various cancers and sometimes they have no crucial impact on the structure and function of p53 protein. An explanation for different frequency in various cancers may be due to the inactivation of p53 protein in special and diverse mutagenic pathways in each cancer (Soussi and Beroud 2003). For instance, G>T transversion is predominantly seen in lung cancer and head and neck cancer which have shown strong correlation with tobacco smoking. The G>A transversion is considered to be the most frequent mutation of codon 175 found in breast and colon cancers and, include the 5% of the total pool of *p53* mutations. This variant leads to substitution of an Arginine with one Histidine amino acid that accompanies excessive defects in normal biological and biochemical activities of p53 protein. It was shown that R175H mutation leads to increased cell proliferation and transformation in heterozygous mice carrying this alteration (Lang et al. 2004). Moreover, it was associated with a gain of function in p53 protein which was resistance to chemotherapeutic agents in cultured cells and breast cancer patients (Blandino et al. 1999; Aas et al. 1996). It was shown that p53 protein carrying R175H mutation is able to decrease the expression of Fas pro-apoptotic gene through binding with the sequence which is different from the site recognized by normal p53 protein to induce its expression (Zalcenstein et al. 2003). Although, it was demonstrated that the residues 62–69 of p53 protein play crucial roles in its apoptosis function, deletion of this region would be associated with decrease in induction of *BAX* and *PIG3* pro-apoptotic genes (Garcia and Attardi 2014; Venot et al. 1998). In lung cancer its importance had been attenuated by the alterations of codons 157 and 158 which are the hot spots of nucleotide change in exposing to benzo (α)pyrene adducts found in tobacco (Denissenko et al. 1996; Caron de Fromentel and Soussi 1992).

The GC>AT transition occurs in frequency of less than 5% in each of codons 175, 248 and 273 while comprise 51% of all the point mutations and 59% of all the CpG dinucleotide transitions within the *p53* gene. This high rate of transversion is due to the deamination of methylcytosine to thymidine nucleotide which could not be efficiently recognized by repair system and makes replacing of CpG with AT in the following rounds of replication (Denissenko et al. 1997). However, the position of CpG in the codons arrangement can affect the fate of transition. When the CpG lies in the CGN pattern of codon which is called the type I of CpG dinucleotide, it will always be associated with amino acid replacement. In the second and third types (NCG, or NNC, GNN), since there is degeneracy in genetic code including G nucleotide, mutation can only occurrs in the transition of modified C nucleotide. Given that the GC>AT transition usually provide no selectable growth advantage for infected cell, most of them are harmless and doesn't lead the cell to become cancerous (Smela et al. 2001; Soussi and Beroud 2003).

8.3.2 p53 Polymorphisms and Cancer

P72R is the first single nucleotide polymorphism identified in *p53* gene which has demonstrated heterogeneity in frequency of proline proline (PP) genotype among different populations from Scandinavia (16 %) to Africa (63 %) (Harris et al. 1986; Li and Fraumeni 1969b). The role of this polymorphism has been detected in open angle glaucoma, endometriosis, recurrent pregnancy loss, glioma, breast, cervix, and esophageal cancers (Hou et al. 2013; Jia et al. 2012; Zhao et al. 2013; Tang et al. 2011; Zhou et al. 2012; Shi et al. 2012; Guo et al. 2012). Codon 72 is located in the proline reach domain (PRD) of protein which is the major regulatory domain of p53 protein during apoptosis (Venot et al. 1998). It was shown that p53 protein containing proline has less potential proapoptotic activity relative to variant p53 proteins with Argenine that is may be due to stronger interaction of variant p53 protein with nuclear-export protein chromosomal region maintenance 1 (CRM1) (Bergamaschi et al. 2006). This would be associated with increased nuclear export of p53 protein leading to more entry of it into the mitochondria (Dumont et al. 2003). In addition, it was found that mutant R72 p53 protein provides a new malfunction which is greater affinity to p73 protein to inhibit its apoptotic activities. The later side effects of R72 mutant p53 would be a convincing explanation for poor prognosis and response to chemotherapy in head and neck cancer patients (Bergamaschi et al. 2003).

P47S is the second main exonic *p53* polymorphism that is associated with substitution of proline to serine in codon 47 and had demonstrated different frequency among various studies (Felley-Bosco et al. 1993; Kashima et al. 2007). Although the clinical importance of this variant remains elusive, Li and his coworkers have demonstrated that this amino acid conversion may interfere with the proper phosphorylation of adjacent S46 codon leading to defective apoptotic function of p53 protein (Li et al. 2005).

8.3.3 Other Genes Variants and Mutations Affecting p53 Signaling

Mutations and variants in other genes interacting with p53 can have some pivotal influences on its crucial functions that will describe the most important of them in two following sub-sections. Although it is established that the response element (RE) of p53 protein has a highly degenerate sequence, several studies have found some of SNPs in RE that can influence the p53 signaling. SNP309 in the *MDM2* gene is an example that include conversion of T to G nucleotide in intron 1 which is close to the p53 RE sequence. It causes the MDM2 mRNA and protein would be increased in response to enhanced recognition of DNA binding site with specifity protein 1 (Sp1) (Bond et al. 2004).

High level of MDM2 expression will be associated with decreased p53 expression and its defective apoptotic response to DNA damage (Bond et al. 2004).The GG genotype of T309G has a frequency of around 1 % and the G allele was found to be associated with earlier onset of cancer in family members of Li-Fraumeni syndrome. Interestingly, since the promoter of *MDM2* gene is recognized by hormonal signals, it was shown that penetrance of this variant was more in women (Bond et al. 2006). Another variant (SNP354) was found in the exon 12 of *MDM2* gene which has demonstrated strong association with risk of breast cancer, even though its biological effects remains undetermined (Boersma et al. 2006). *AKT* gene is negatively regulated by p53 through activation of Phosphatase and tensin homolog (PTEN) phosphatase and has a suppressive function on p53 activity by phosphorylating and stabilizing MDM2 (Zhou et al. 2001). Several SNPs have been identified in *AKT* gene that have shown association with overexpression of it and poor apoptotic response to irradiation (Harris et al. 2005). *Fms-like tyrosine kinase 1* (*FLT-1*) is another gene with a polymorphism in its p53 response element which can affect the strength of *FLT-1* gene expression induced by p53. FLT-1 is indirectly involved in angiogenesis through encoding a receptor which coupled with VEGF (Menendez et al. 2006).

Among the genes which their mutations could affect the p53 activity, BRCA1 seems to be the most important. Coexistence of *BRCA1* and *p53* mutations is a relatively common genetic alteration especially in breast cancer families. It was shown that *p53* mutations have risen in a mutated BRCA1 background in these families almost with different pattern more residing in non hot spot p53 gene codons (Crook et al. 1997). Although they were susceptible to get cancer, rescue of mice carrying *Brca1*Δ11 (deletion of carboxyl terminal of BRCA1 protein) in *p53+/−* background is relying on that *p53* mutations act as the passenger alterations (Cao et al. 2006). However, co-selection of *BRCA1* and *p53* mutations has shown to be associated with normal p53 protein functions. It may be due to that the lack of impact of these mutations on DNA repair activities of p53 owing to its close cooperation with BRCA1 or may alter some of the unknown p53 functions (Bourdon et al. 2005; Mihara et al. 2003).

Interestingly, the spectrum of p53 has shown to be low in colorectal cancers with positive status of microsatellite instability (MSI+). The tendency of MSI+ tumors to driving mutations in genes including *BAX*, *TGFβR* and *IGFR* which contain repeated polynucleotide tract provides the optimal condition for tumor development (Konishi et al. 1996; Young et al. 2001).

In general, further studies merited to define other genetic variations in genes interacting with p53 protein and affect the strength of its crucial activities.

8.4 Methylation Defects in Cancers

The primary evidences relying on the effect of promoter hypermethylation on p53 expression were reported *in vitro* studies on rat and human cell lines (Schroeder and Mass 1997; Pogribny et al. 2000). However, there are a few reports indicating the role of promoter methylation in regulation of its mRNA and protein expression which will be described separately in each cancer based on the available publications.

8.4.1 Leukemia

Molecular assessment of bone marrow samples of patients affected with acute lymphoblastic leukemia (ALL) demonstrated that *p53* promoter was methylated in 31 % of the patients. *p53* promoter methylation was compatible with decreased expression of p53 protein versus normal high expression in unmethylated samples (Agirre et al. 2003). Of note, *p53* mutations have very low frequency (2–3 %) in ALL patients and therefore promoter methylation may plays more pivotal role in suppression of p53 functions in ALL and different types of leukemia and myeloma (Hurt et al. 2006).

8.4.2 Breast Cancer

Almost low frequency of *p53* mutations in breast cancer families and patients (20–25 %) encouraged methylation analysis of *p53* promoter in two separate investigations (Pharoah et al. 1999). In the first one, 4/16 CpG dinucleotides sites were found to be methylated in 19 % of the patients (Kang et al. 2001). The second study was extended to analyze the promoter methylation of *TP53* gene in four groups including *BRCA1*/*BRCA2* mutation carriers with or without breast cancer, sporadic breast cancer patients and healthy controls. Interestingly, no one of the studied group had shown methylated *p53* promoter (Kontorovich et al. 2009).

8.4.3 Other Cancers

Methylation study of *TP53* promoter gene was failed to find methylated promoter in 10 patients affected with sporadic adrenocortical carcinoma (ACC) and 5 healthy controls (Sidhu et al. 2005). Methylation analysis in Li-Fraumeni's and Li-Fraumeni like syndromes patients has revealed strong association with splice site mutations and exonic methylation of *p53* gene (Kouidou et al. 2009).

8.4.4 Brain Tumors

Although there are more frequent studies on the methylation of brain tumors than other human cancers, literature of the methylation status of *p53* gene promoter is relatively poor and is restricted to almost cell line studies and a few pathological classification of brain tumors. In the primary trial carried out on 67 astrocytic gliomas patients, p53 promoter methylation was found in only 2 % of glioblastomas and 8 % of astrocytoma patients (Gonzalez-Gomez et al. 2003a; Gonzalez-Gomez et al. 2003b). In the following study on 41 oligodendroglial patients, none of the analyzed patients had methylated *p53* promoter (Alonso et al. 2003). The methylation status of *p53* promoter was then examined through methylation specific polymerase chain reaction (MSP-PCR) using the same primer pairs employed in the two previous mentioned studies in three U87MG, LNT-229, T98G glioma cell lines. It was revealed that the *p53* promoter methylation was in direct correlation with mRNA and protein expression in U87MG and T98G cell lines. They then analyzed the *p53* promoter methylation in various types of low grade glioma's primary tissues. *P53* promoter was methylated in 60 % of low-grade astrocytoma and oligoastrocytoma and, 74 % of all the oligodendroglioma tumors (Amatya et al. 2005). Huang and his coworkers have found the same correlation in their study on human GBM cell line and verified it using 5-Aza-2'-deoxycytidine (Aza) as a major DNA methyltransferase inhibitor (Huang et al. 2009). However, in spite of restoring the p53 expression in T98G cell line in both studies, in contrast to the results contributed by Amatya and his co-worker, *p53* promoter was not methylated in the study performed by Soto-Reyes et al (Soto-Reyes and Recillas-Targa 2010).

In another investigation on primary tissues derived from benign and metastatic brain tumors, it was demonstrated that *p53* promoter was methylated in 37.5 % of meningiomas, 30 % of schwannomas and 52.6 % of metastatic brain tumors. The presence of Arg72Pro and Pro47Ser polymorphisms was also examined in these samples through PCR-restriction fragment length polymorphism (PCR-RFLP). Both of two polymorphisms were associated with increased risk of brain tumor (Almeida et al. 2009). In another methylation study which was performed on four brain tumor cell lines including U87MG, U-118MG, LN-18 and Daoy in addition to 100 tissues provided by brain cancer patients, although *p53* promoter was hypomethylated in all of the samples and cell lines, mRNA expression was very low (Avci et al. 2011). This study is in line with our results obtained from analyzing the *p53* promoter of various grades of primary tissues provided by 30 brain tumor patients

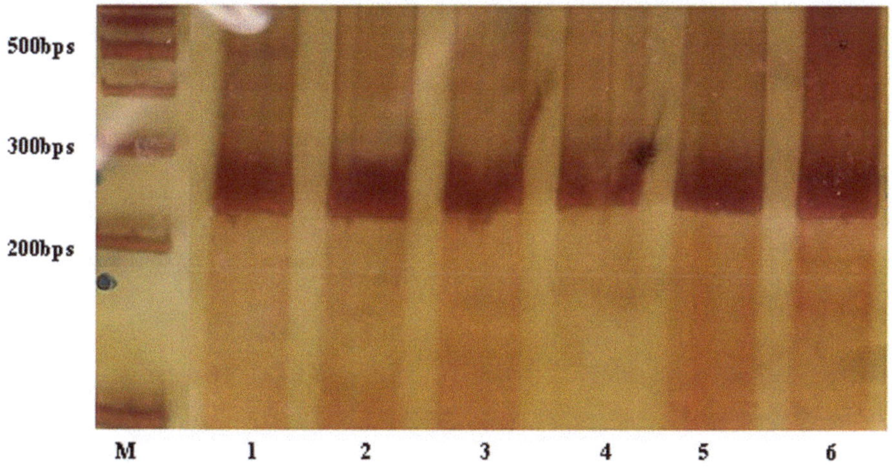

Fig. 8.4 PCR products of unmethylated templates (*M* Marker, *1–6* brain tumor samples)

compared to two healthy controls (unpublished data). The methylation status of *p53* gene was examined through MSP-PCR verified by sequencing and the expression level of its protein was determined using Immunofluorescence (IF) assay. The p53 expression was compared with its homologous protein, p63, and with its corresponding molecule as proliferation marker ki-67 (Johannessen et al. 2006; Scholzen and Gerdes 2000). Interestingly, all of our patients as well as one healthy control (71 yrs old) had unmethylated *p53* promoter (Fig. 8.4).

The expression level of p53 protein was low in 28 of patients and high in two others including oligodendroglioma and poleomorphic sarcoma patients (Fig. 8.5).

No significant correlation was found between methylation status of *p53* promoter gene and its protein expression. We couldn't find any meaningful correlation between *p53* promoter gene and its protein expression with RB/ATM proteins expression, *ATM* promoter methylation, D1853N polymorphism of *ATM* gene and telomere length (TL). Although it is commonly accepted that the activation of p53 is dependent on its phosphorylation by ATM (Shen et al. 2005). However in our results, the absence of correlation between protein expressions between both genes may rely on the p53 induction regardless of ATM expression (Fig. 8.5). It

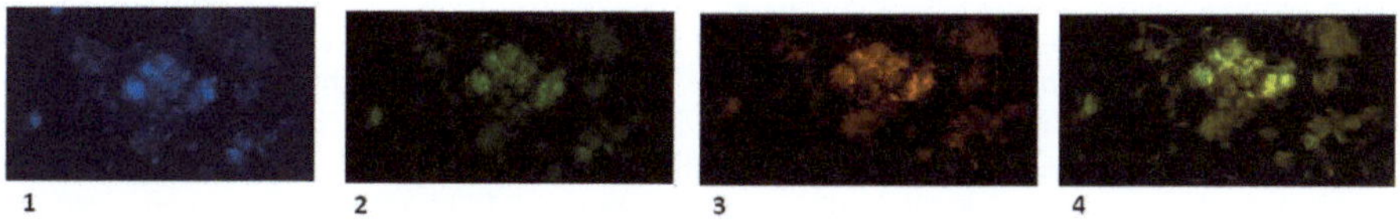

Fig. 8.5 Protein expressions of ATM and p53 in brain tumor cells of patient affected with Meningioma **a** Tumor cells with dapi filter (*blue*), as counter stain, **b** cells conjugated with FITC (*green*), representative of ATM protein characterized with two clones of cells with low and moderate expression by IF, **c** Cells conjugated with R-pe reflecting both low and high expression of p53 protein; **d** Merged image of ATM and p53. Magnification: x100 (From P.Mehdipour's archive)

was described that the cooperation between RB and p53 proteins are necessary for keeping the TL to be stable and thereby prevents changing the fate of cell toward tumorgenesis (Artandi and DePinho 2010; Garcia-Cao et al. 2002). In our study, it was revealed that the Rb and p53 proteins expression were in borderline association with each other ($p=0.05$). However, we couldn't find significant correlation between p53/RB proteins expression and TL. We have previously demonstrated that the expression of ATM/p53/Rb was higher at mRNA level in astrocytoma compared to meningioma patients (Kheirollahi et al. 2011). The expression of p63 protein has shown to be diverse among different grades and pathologies of our brain tumor patients (Fig. 8.6). The p63 expression has shown to be overexpressed in GBM

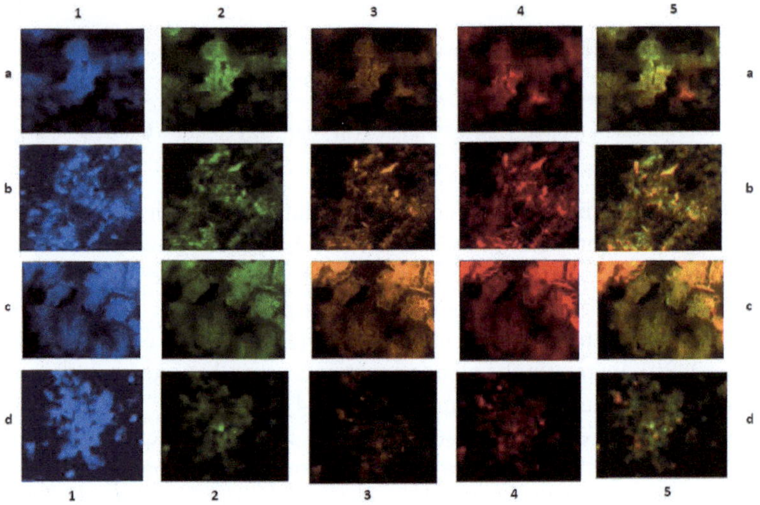

Fig. 8.6 Protein expression profile of p53, p63, and ki67 in brain tumor cells from a healthy deceased individual, and three patients affected with different brain tumors **a** Protein expression profile in a healthy individual as control **(1a)** Brain cells with dapi filter (*blue*). **(2a)** cells representative of p53 protein conjugated with FITC characterized with three clones of cells with low, moderate, and very few cells with high expression by IF. **(3a)** The same cells conjugated with R-Pe (*orange*) reflecting low and moderate expression of p63 protein. (4) The same cells conjugated with Pe-cy5 reflecting low, moderate protein expression accompanied by very few cells with high expression. (5) Merged image of p53/p63/ki67 with divese interaction between these proteins. **b** Protein expression profile in a patient affected with Gleioblastoma multiform, **(1b)** Tumor cells with dapi filter (blue), **(2b)** cells representative of p53 protein conjugated with FITC characterized with three clones of cells with moderate/low accompanied by few cells with high expression by IF. **(3b)** The same cells conjugated with R-Pe (*orange*) reflecting moderate/low/high expression of p63. **(4b)** The same cells conjugated with Pe-cy5 reflected of high/moderate/low protein expression. **(5b)** Merged image of p53/p63/ki67 with diverse interaction between these proteins. **c** Protein expression profile in a patient affected with astrocytoma **(1c)** Tumor cells with dapi filter (*blue*), **(2c)** cells representative of p53 protein conjugated with FITC characterized moderate/low/high by IF. **(3c)** The same cells conjugated with R-Pe (orange) reflecting moderate expression of p63 protein accompanied by few cells with higher expression. **(4c)** The same cells conjugated with Pe-cy5 reflected of low, moderate protein expression accompanied by a clone of cells with high expression. **(5c)** Merged image of p53/p63/ki67 with divese interaction between these proteins. d Protein expression profile in a patient affected with meningioma. Magnification: x200 (From P.Mehdipour's archive)

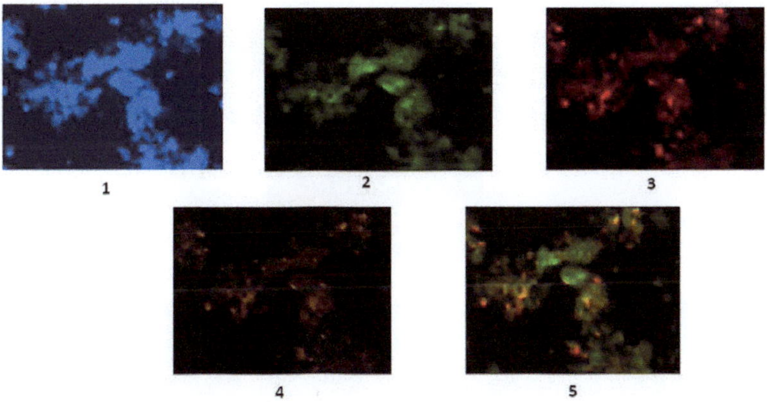

Fig. 8.7 Protein expressions of p53, Rb, and Ki-67 in brain tumor cells of patient affected with Meningioma **a** Tumor cells with dapi filter (*blue*). **b** The same cells representative of p53 protein conjugated with FITC (*green*) characterized with moderate and low expression of p53 protein accompanied by very few cells with high expression by IF. **c** The same cells conjugated with Pe-cy5 (*red*) are reflected of low, moderate an high protein expression of Rb. **d** The same cells conjugated with R-Pe (*orange*) reflecting low and moderate expression of Ki67 protein accompanied by very few cells with higher expression; and **e** Merged image of p53,Rb and Ki67 reflecting the diverse co-expression between these profile of proteins. Magnification: x200 (From P. Mehdipour's archive)

patients while a few clones of meningioma brain cells contained high expression of p63 protein (Fig. 8.6). Although it was expected that the expression of p53/p63 proteins expression would be high in healthy individuals, the expression pattern of these two proteins was high in a few clones of studied cells in deceased healthy controls (Fig. 8.6). The expression of Ki-67 protein was higher in GBM clone cells than astrocytoma clone cells whereas its expression pattern has shown to be lower in meningioma and healthy controls (Fig. 8.6). Further studies are required to clarify the exact correlation between p53 and other cell cycle target proteins especially RB/ATM with TL as well. Furthermore, the mode of protein expressions in p53, Rb, and Ki-67 in brain tumor cells of patient affected with Meningioma reveals to have the heterogenic pattern including high, moderate and low for p53/Rb and Ki-67 respectively, but with diverse mode of intensity (Fig. 8.7). However, these proteins have a diverse cooperationis. Interestingly p53 protein, apparently, has its own behavior within the tumor cells. In addition, p53 and Rb, relatively, cooperate. Besides, malignant brain tumor have different characteristics of protein expression mode of p53/Rb/Ki67 which demonstrate more interaction than in meningioma. However, the degree of intensity of p53 and ki67 is more upregulated in GBM than in astrocytoma, but both genes have high expression of *Rb* gene (Fig. 8.8).

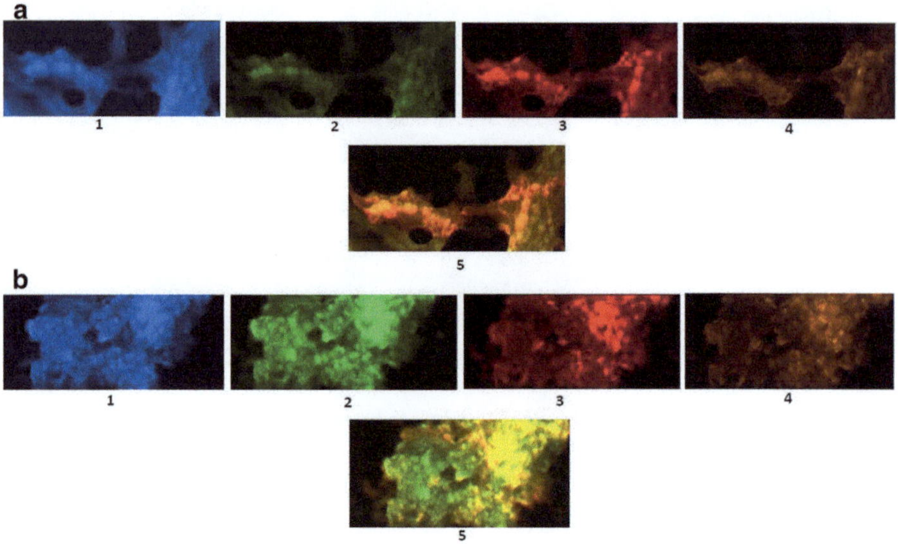

Fig. 8.8 Protein expressions of p53, Rb, and Ki-67 in brain tumor cells of patient affected with astrocytoma and Gleioblastoma multiform **a** Protein expression profile in a patient affected with astrocytoma (**a1**)Tumor cells from a patient affected, with dapi filter (*blue*), (**a2**) The same cells conjugated with FITC (*green*), representative of p53 protein, (**a3**) Same cells s indicative of Rb protein conjugated with Pe-cy5 (*red*) characterized with two clones of cells with low and high expression by IF, (**a4**) cells conjugated with R-Pe (*orange*) reflecting low expression of Ki67 protein accompanied by few cells with moderate and high expression.(**a 5**) Merged image of p53, Rb and Ki67. **b** Protein expression profile in a patient affected with Gleioblastoma multiform, (**b1**)Tumor cells from a patient affected, with dapi filter (*blue*), (**b2**) The same cells conjugated with FITC, representative of p53 protein, (**b3**) Same cells s indicative of Rb protein conjugated with Pe-cy5 characterized with two clones of cells with low and high expression by IF, (**b4**) cells conjugated with R-Pe (*orange*) reflecting low expression of Ki67 protein accompanied by few cells with moderate and high expression, (**b5**) Merged image of p53, Rb and Ki67 is indicative of cooperation between these genes. Magnification: x200 (From P. Mehdipour's archive).

8.5 Conclusions

By considering the occurrence of methylation in both *p53* gene and protein, it seems that methylation play pivotal role in regulation of p53 function. However, as described above, there is a strong controversy among various studies from unmethylation of all samples to methylation of the most either in cases or cell lines. This controversy is more significant in brain tumor investigations followed by breast cancer studies and may indicate that the pattern of p53 promoter methylation is influenced by the ethnicity of the analyzed population as well. In addition, there is no clear and ubiquitous pattern between *p53* promoter methylation status and its protein expression, as in our study, although the promoter was unmethylated in all the brain tumor patients, besides, the p53 protein expression is revealed to be low in 93.3 % of tumors. The question is 'what is the pattern of methylation at protein level?' Finally, due to the importance of p53 function in tumor suppression, by considering our aim and task within our ongoing project, more complementary studies are essential to shed light on the epigenetic changes of *p53* promoter in different stages of cancer development.

References

Aas T, Borresen AL, Geisler S, Smith-Sorensen B, Johnsen H, Varhaug JE et al (1996) Specific P53 mutations are associated with de novo resistance to doxorubicin in breast cancer patients. Nat Med 2:811–814

Agirre X, Vizmanos JL, Calasanz MJ, Garcia-Delgado M, Larrayoz MJ, Novo FJ (2003b) Methylation of CpG dinucleotides and/or CCWGG motifs at the promoter of TP53 correlates with decreased gene expression in a subset of acute lymphoblastic leukemia patients. Oncogene 22(7):1070–1072. doi:10.1038/sj.onc.1206236

Allis CD, Berger SL, Cote J, Dent S, Jenuwien T, Kouzarides T et al (2007) New nomenclature for chromatin-modifying enzymes. Cell 131:633–636

Almeida LO, Custodio AC, Pinto GR, Santos MJ, Almeida JR, Clara CA et al (2009) Polymorphisms and DNA methylation of gene TP53 associated with extra-axial brain tumors. Genet Mol Res 8(1):8–18

Alo PL, Visca P, Marci A, Mangoni A, Botti C, Di Tondo U (1996) Expression of fatty acid synthase (FAS) as a predictor of recurrence in stage I breast carcinoma patients. Cancer 77:474–482

Alonso ME, Bello MJ, Gonzalez-Gomez P, Arjona D, Lomas J, de Campos JM et al (2003) Aberrant promoter methylation of multiple genes in oligodendrogliomas and ependymomas. Cancer Genet Cytogenet 144(2):134–142

Amatya VJ, Naumann U, Weller M, Ohgaki H (2005) TP53 promoter methylation in human gliomas. Acta Neuropathol 110(2):178–184. doi:10.1007/s00401-005-1041-5

An W, Kim J, Roeder RG (2004) Ordered cooperative functions of PRMT1, p300, and CARM1 in transcriptional activation by p53. Cell 117:735–748

Artandi SE, DePinho RA (2010) Telomeres and telomerase in cancer. Carcinogenesis 31(1):9–18. doi:10.1093/carcin/bgp268

Avci CB, Susluer SY, Dodurga Y, Akalin T, Cogulu O, Dalbasti T, Oktar N, Gunduz C (2011) The emphasis of tumor suppressor genes and oncogenes in diagnosis and prognosis of anaplastic brain tumors. J Neurol Sci (Turkish) 28(4):563–580

Bartek J, Bartkova J, Vojtesek B, Staskova Z, Lukas J, Rejthar A et al (1991) Aberrant expression of the p53 oncoprotein is a common feature of a wide spectrum of human malignancies. Oncogene 6:1699–1703

Bensaad K, Tsuruta A, Selak MA, Vidal MN, Nakano K, Bartrons R et al (2006) TIGAR, a p53-inducible regulator of glycolysis and apoptosis. Cell 126:107–120

Bergamaschi D, Gasco M, Hiller L, Sullivan A, Syed N, Trigiante G et al (2003) p53 polymorphism influences response in cancer chemotherapy via modulation of p73-dependent apoptosis. Cancer Cell 3:387–402

Bergamaschi D, Samuels Y, Sullivan A, Zvelebil M, Breyssens H, Bisso A et al (2006) iASPP preferentially binds p53 proline-rich region and modulates apoptotic function of codon 72-polymorphic p53. Nat Genet 38:1133–1141

Blandino G, Levine AJ, Oren M (1999) Mutant p53 gain of function: differential effects of different p53 mutants on resistance of cultured cells to chemotherapy. Oncogene 18:477–485

Bode AM, Dong Z (2004) Post-translational modification of p53 in tumorigenesis. Nat Rev Cancer 4:793–805

Boersma BJ, Howe TM, Goodman JE, Yfantis HG, Lee DH, Chanock SJ et al (2006) Association of breast cancer outcome with status of p53 and MDM2 SNP309. J Natl Cancer Inst 98:911–919

Bond GL, Hu W, Bond EE, Robins H, Lutzker SG, Arva NC et al (2004) A single nucleotide polymorphism in the MDM2 promoter attenuates the p53 tumor suppressor pathway and accelerates tumor formation in humans. Cell 119:591–602
Bond GL, Hirshfield KM, Kirchhoff T, Alexe G, Bond EE, Robins H et al (2006) MDM2 SNP309 accelerates tumor formation in a gender-specific and hormone-dependent manner. Cancer Res 66:5104–5110
Bourdon JC, Fernandes K, Murray-Zmijewski F, Liu G, Diot A, Xirodimas DP et al (2005) p53 isoforms can regulate p53 transcriptional activity. Genes Dev 19:2122–2137
Brand KA, Hermfisse U (1997) Aerobic glycolysis by proliferating cells: a protective strategy against reactive oxygen species. FASEB J 11:388–395
Brosh R, Rotter V (2009) When mutants gain new powers: news from the mutant p53 field. Nat Rev Cancer 9:701–713
Brown MA, Sims RJ, 3rd, Gottlieb PD, Tucker PW (2006) Identification and characterization of Smyd2: a split SET/MYND domain-containing histone H3 lysine 36-specific methyltransferase that interacts with the Sin3 histone deacetylase complex. Mol Cancer 5:26
Brugarolas J, Chandrasekaran C, Gordon JI, Beach D, Jacks T, Hannon GJ (1995) Radiation-induced cell cycle arrest compromised by p21 deficiency. Nature 377:552–557
Budanov AV, Shoshani T, Faerman A, Zelin E, Kamer I, Kalinski H et al (2002) Identification of a novel stress-responsive gene Hi95 involved in regulation of cell viability. Oncogene 21:6017–6031
Budanov AV, Sablina AA, Feinstein E, Koonin EV, Chumakov PM (2004) Regeneration of peroxiredoxins by p53-regulated sestrins, homologs of bacterial AhpD. Science 304:596–600
Bustamante E, Pedersen PL (1977) High aerobic glycolysis of rat hepatoma cells in culture: role of mitochondrial hexokinase. Proc Natl Acad Sci U S A 74:3735–3739
Buzzai M, Bauer DE, Jones RG, Deberardinis RJ, Hatzivassiliou G, Elstrom RL et al (2005) The glucose dependence of Akt-transformed cells can be reversed by pharmacologic activation of fatty acid beta-oxidation. Oncogene 24:4165–4173
Campisi J (2003) Cancer and ageing: rival demons? Nat Rev Cancer 3:339–349
Campisi J, d'Adda di Fagagna F (2007) Cellular senescence: when bad things happen to good cells. Nat Rev Mol Cell Biol 8:729–740
Cao L, Wenmei L Kim S, Brodie SG, Deng CX (2003) Senescence, aging, and malignant transformation mediated by p53 in mice lacking the Brca1 full-length isoform. Genes Dev 17(2):201–213
Cao L, Kim S, Xiao C, Wang RH, Coumoul X, Wang X et al (2006) ATM-Chk2-p53 activation prevents tumorigenesis at an expense of organ homeostasis upon Brca1 deficiency. EMBO J 25:2167–2177
Caron de Fromentel C, Soussi T (1992) TP53 tumor suppressor gene: a model for investigating human mutagenesis. Genes Chromosomes Cancer 4:1–15
Chan TA, Hermeking H, Lengauer C, Kinzler KW, Vogelstein B (1999) 14-3-3Sigma is required to prevent mitotic catastrophe after DNA damage. Nature 401:616–620
Chen D, Ma H, Hong H, Koh SS, Huang SM, Schurter BT et al (1999) Regulation of transcription by a protein methyltransferase. Science 284:2174–2177
Chen J, Ruan H, Ng SM, Gao C, Soo HM, Wu W et al (2005) Loss of function of def selectively up-regulates Delta113p53 expression to arrest expansion growth of digestive organs in zebrafish. Genes Dev 19:2900–2911
Chen J, Ng SM, Chang C, Zhang Z, Bourdon JC, Lane DP et al (2009) p53 isoform delta113p53 is a p53 target gene that antagonizes p53 apoptotic activity via BclxL activation in zebrafish. Genes Dev 23:278–290
Chipuk JE, Bouchier-Hayes L, Kuwana T, Newmeyer DD, Green DR (2005) PUMA couples the nuclear and cytoplasmic proapoptotic function of p53. Science 309:1732–1735
Clair St, Manfredi JJ (2006) The dual specificity phosphatase Cdc25C is a direct target for transcriptional repression by the tumor suppressor p53. Cell Cycle 5:709–713
Clarke AR, Purdie CA, Harrison DJ, Morris RG, Bird CC, Hooper ML et al (1993) Thymocyte apoptosis induced by p53-dependent and independent pathways. Nature 362:849–852

Crook T, Crossland S, Crompton MR, Osin P, Gusterson BA (1997) p53 mutations in BRCA1-associated familial breast cancer. Lancet 350:638–639
Denissenko MF, Pao A, Tang M, Pfeifer GP (1996) Preferential formation of benzo[a]pyrene adducts at lung cancer mutational hotspots in P53. Science 274:430–432
Denissenko MF, Chen JX, Tang MS, Pfeifer GP (1997) Cytosine methylation determines hot spots of DNA damage in the human P53 gene. Proc Natl Acad Sci U S A 94:3893–3898
Donehower LA, Harvey M, Slagle BL, McArthur MJ, Montgomery CA Jr, Butel JS et al (1992) Mice deficient for p53 are developmentally normal but susceptible to spontaneous tumours. Nature 356(6366):215–221. doi:10.1038/356215a0
Dumont P, Leu JI, Della Pietra AC, 3rd, George DL, Murphy M (2003) The codon 72 polymorphic variants of p53 have markedly different apoptotic potential. Nat Genet 33:357–365
Felley-Bosco E, Weston A, Cawley HM, Bennett WP, Harris CC (1993) Functional studies of a germ-line polymorphism at codon 47 within the p53 gene. Am J Hum Genet 53:752–759
Feng Z, Jin S, Zupnick A, Hoh J, de Stanchina E, Lowe S et al (2006) p53 tumor suppressor protein regulates the levels of huntingtin gene expression. Oncogene 25:1–7
Fridman JS, Lowe SW (2003) Control of apoptosis by p53. Oncogene 22:9030–9040
Garcia PB, Attardi LD (2014) Illuminating p53 function in cancer with genetically engineered mouse models. Semin Cell Dev Biol 27c:74–85
Garcia-Cao M, Gonzalo S, Dean D, Blasco MA (2002) A role for the Rb family of proteins in controlling telomere length. Nat Genet 32(3):415–419. doi:10.1038/ng1011
Giono LE, Manfredi JJ (2006) The p53 tumor suppressor participates in multiple cell cycle checkpoints. J Cell Physiol 209:13–20
Gonzalez-Gomez P, Bello MJ, Arjona D, Lomas J, Alonso ME, De Campos JM et al (2003a) Promoter hypermethylation of multiple genes in astrocytic gliomas. Int J Oncol 22:601–608
Gonzalez-Gomez PB, Lomas MJ, Arjona D, Alonso ME, Amiñoso C, Lopez-Marin I, Anselmo NP, Sarasa JL, Gutierrez M, Casartelli C, Rey JA (2003b) Aberrant methylation of multiple genes in neuroblastic tumours. relationship with MYCN amplification and allelic status at 1p. Eur J Cancer 39(10):1478–1485
Gottlieb E, Vousden KH (2010) p53 regulation of metabolic pathways. Cold Spring Harb Perspect Biol 2:a001040
Grochola LF, Zeron-Medina J, Meriaux S, Bond GL (2010) Single-nucleotide polymorphisms in the p53 signaling pathway. Cold Spring Harb Perspect Biol 2:a001032
Guo Y, Zhang H, Chen X, Yang X, Cheng W, Zhao K (2012) Association of TP53 polymorphisms with primary open-angle glaucoma: a meta-analysis. Invest Ophthalmol Vis Sci 53:3756–3763
Han HS, Yu E, Song JY, Park JY, Jang SJ, Choi J (2009) The estrogen receptor alpha pathway induces oncogenic Wip1 phosphatase gene expression. Mol Cancer Res 7:713–723
Hardie DG (2004) The AMP-activated protein kinase pathway–new players upstream and downstream. J Cell Sci 117:5479–5487
Harris N, Brill E, Shohat O, Prokocimer M, Wolf D, Arai N et al (1986) Molecular basis for heterogeneity of the human p53 protein. Mol Cell Biol 6:4650–4656
Harris SL, Gil G, Robins H, Hu W, Hirshfield K, Bond E et al (2005) Detection of functional single-nucleotide polymorphisms that affect apoptosis. Proc Natl Acad Sci U S A 102:16297–16302
Hayflick L (1965) The limited *in vitro* lifetime of human diploid cell strains. Exp Cell Res 37:614–636
Hermeking H, Lengauer C, Polyak K, He TC, Zhang L, Thiagalingam S et al (1997) 14-3-3 sigma is a p53-regulated inhibitor of G2/M progression. Mol Cell 1:3–11
Hernandez-Acosta NC, Cabrera-Socorro A, Morlans MP, Delgado FJ, Suarez-Sola ML, Sottocornola R et al (2011) Dynamic expression of the p53 family members p63 and p73 in the mouse and human telencephalon during development and in adulthood. Brain Res 1372:29–40. doi:10.1016/j.brainres.2010.11.041
Hinds PW, Finlay CA, Quartin RS, Baker SJ, Fearon ER, Vogelstein B et al (1990) Mutant p53 DNA clones from human colon carcinomas cooperate with ras in transforming primary rat cells: a comparison of the "hot spot" mutant phenotypes. Cell Growth Differ 1:571–580

Hou J, Jiang Y, Tang W, Jia S (2013) p53 codon 72 polymorphism and breast cancer risk: a meta-analysis. Exp Ther Med 5:1397–1402

Hu W, Feng Z, Ma L, Wagner J, Rice JJ, Stolovitzky G et al (2007) A single nucleotide polymorphism in the MDM2 gene disrupts the oscillation of p53 and MDM2 levels in cells. Cancer Res 67:2757–2765

Hu W, Zhang C, Wu R, Sun Y, Levine A, Feng Z (2010) Glutaminase 2, a novel p53 target gene regulating energy metabolism and antioxidant function. Proc Natl Acad Sci U S A 107:7455–7460

Huang J, Perez-Burgos L, Placek BJ, Sengupta R, Richter M, Dorsey JA et al (2006) Repression of p53 activity by Smyd2-mediated methylation. Nature 444:629–632

Huang J, Sengupta R, Espejo AB, Lee MG, Dorsey JA, Richter M et al (2007) p53 is regulated by lysine demethylase LSD1. Nature 449:105–8

Huang J, Chen K, Huang J, Gong W, Dunlop NM, Howard OM et al (2009) Regulation of the leucocyte chemoattractant receptor FPR in glioblastoma cells by cell differentiation. Carcinogenesis 30(2):348–355. doi:10.1093/carcin/bgn266

Hurt EM, Thomas SB, Peng B, Farrar WL (2006) Reversal of p53 epigenetic silencing in multiple myeloma permits apoptosis by a p53 activator. Cancer Biol Ther 5(9):1154–1160

Ide T, Brown-Endres L, Chu K, Ongusaha PP, Ohtsuka T, El-Deiry WS et al (2009) GAMT, a p53-inducible modulator of apoptosis, is critical for the adaptive response to nutrient stress. Mol Cell 36:379–392

Imamura K, Ogura T, Kishimoto A, Kaminishi M, Esumi H (2001) Cell cycle regulation via p53 phosphorylation by a 5'-AMP activated protein kinase activator, 5-aminoimidazole- 4-carboxamide-1-beta-D-ribofuranoside, in a human hepatocellular carcinoma cell line. Biochem Biophys Res Commun 287:562–567

Inoki K, Zhu T, Guan KL (2003) TSC2 mediates cellular energy response to control cell growth and survival. Cell 115:577–590

Ivanov GS, Kurash J, Chuikov S, Gizatullin F, Herrera-Medina EM, Rauscher Fr et al (2007) Methylation-acetylation interplay activates p53 in response to DNA damage. Mol Cell Biol 27:6756–6769

Jeffers JR, Parganas E, Lee Y, Yang C, Wang J, Brennan J et al (2003) Puma is an essential mediator of p53-dependent and -independent apoptotic pathways. Cancer Cell 4:321–328

Jia S, Xu L, Chan Y, Wu X, Yang S, Yu H et al (2012) p53 codon 72 polymorphism and endometriosis: a meta-analysis. Arch Gynecol Obstet 285:1657–1661

Joerger AC, Allen MD, Fersht AR (2004) Crystal structure of a superstable mutant of human p53 core domain. Insights into the mechanism of rescuing oncogenic mutations. J Biol Chem 279:1291–1296

Johannessen AL, Torp SH (2006) The clinical value of Ki-67/MIB-1 labeling index in human astrocytomas. Pathol Oncol Res 12(3):143–147. doi:Paor.2006.12.3.0143

Jones RG, Plas DR, Kubek S, Buzzai M, Mu J, Xu Y et al (2005) AMP-activated protein kinase induces a p53-dependent metabolic checkpoint. Mol Cell 18:283–293

Kang JH, Kim SJ, Noh DY, Park IA, Choe KJ, Yoo OJ et al (2001) Methylation in the p53 promoter is a supplementary route to breast carcinogenesis: correlation between CpG methylation in the p53 promoter and the mutation of the p53 gene in the progression from ductal carcinoma in situ to invasive ductal carcinoma. Lab Invest 81(4):573–579

Kashima T, Makino K, Soemantri A, Ishida T (2007) TP53 codon 72 polymorphism in 12 populations of insular Southeast Asia and Oceania. J Hum Genet 52(8):694–697. doi:10.1007/s10038-007-0168-8

Kheirollahi M, Mehrazin M, Kamalian N, Mehdipour P (2010) Alterations of telomere length in human brain tumors. Med Oncol 28(3):864–870. doi:10.1007/s12032-010-9506-3

Kheirollahi M, Mehr-Azin M, Kamalian N, Mehdipour P (2011) Expression of cyclin D2, P53, Rb and ATM cell cycle genes in brain tumors. Med Oncol 28(1):7–14. doi:10.1007/s12032-009-9412-8

Kirch HC, Flaswinkel S, Rumpf H, Brockmann D, Esche H (1999) Expression of human p53 requires synergistic activation of transcription from the p53 promoter by AP-1, NF-kappaB and Myc/Max. Oncogene 18:2728–2738

Konishi M, Kikuchi-Yanoshita R, Tanaka K, Muraoka M, Onda A, Okumura Y et al (1996) Molecular nature of colon tumors in hereditary nonpolyposis colon cancer, familial polyposis, and sporadic colon cancer. Gastroenterology 111:307–317

Kontorovich T, Cohen Y, Nir U, Friedman E (2009) Promoter methylation patterns of ATM, ATR, BRCA1, BRCA2 and p53 as putative cancer risk modifiers in Jewish BRCA1/BRCA2 mutation carriers. Breast Cancer Res Treat 116:195–200

Kouidou S, Malousi A, Maglaveras N (2009) Li-Fraumeni and Li-Fraumeni-like syndrome mutations in p53 are associated with exonic methylation and splicing regulatory elements. Mol Carcinog 48:895–902

Krizhanovsky V, Xue W, Zender L, Yon M, Hernando E, Lowe SW (2008) Implications of cellular senescence in tissue damage response, tumor suppression, and stem cell biology. Cold Spring Harb Symp Quant Biol 73:513–522

Lahav G, Rosenfeld N, Sigal A, Geva-Zatorsky N, Levine AJ, Elowitz MB et al (2004) Dynamics of the p53-Mdm2 feedback loop in individual cells. Nat Genet 36:147–150

Lane DP, Crawford LV (1979) T antigen is bound to a host protein in SV40-transformed cells. Nature 278:261–263

Lang GA, Iwakuma T, Suh YA, Liu G, Rao VA, Parant JM et al (2004) Gain of function of a p53 hot spot mutation in a mouse model of Li-Fraumeni syndrome. Cell 119:861–872

Lanni C, Racchi M, Uberti D, Mazzini G, Stanga S, Sinforiani E et al (2008) Pharmacogenetics and pharmagenomics, trends in normal and pathological aging studies: focus on p53. Curr Pharm Des 14(26):2665–2671

Levine AJ, Wu MC, Chang A, Silver A, Attiyeh EF, Lin J et al (1995) The spectrum of mutations at the p53 locus. Evidence for tissue-specific mutagenesis, selection of mutant alleles, and a "gain of function" phenotype. Ann N Y Acad Sci 768:111–28

Li FP, Fraumeni JF Jr (1969a) Rhabdomyosarcoma in children: epidemiologic study and identification of a familial cancer syndrome. J Natl Cancer Inst 43:1365–1373

Li FP, Fraumeni JF Jr (1969b) Soft-tissue sarcomas, breast cancer, and other neoplasms. A familial syndrome? Ann Intern Med 71:747–752

Li X, Dumont P, Della Pietra A, Shetler C, Murphy ME (2005) The codon 47 polymorphism in p53 is functionally significant. J Biol Chem 280:24245–24251

Liu H, Lu ZG, Miki Y, Yoshida K (2007) Protein kinase C delta induces transcription of the TP53 tumor suppressor gene by controlling death-promoting factor Btf in the apoptotic response to DNA damage. Mol Cell Biol 27:8480–8491

Lozano G (2007) The oncogenic roles of p53 mutants in mouse models. Curr Opin Genet Dev 17:66–70

Lozano G (2010) Mouse models of p53 functions. Cold Spring Harb Perspect Biol 2:a001115

Lu X (2010) Tied up in loops: positive and negative autoregulation of p53. Cold Spring Harb Perspect Biol 2:a000984

Ma W, Sung HJ, Park JY, Matoba S, Hwang PM (2007) A pivotal role for p53: balancing aerobic respiration and glycolysis. J Bioenerg Biomembr 39:243–246

Maier B, Gluba W, Bernier B, Turner T, Mohammad K, Guise T et al (2004) Modulation of mammalian life span by the short isoform of p53. Genes Dev 18:306–319

Malkin D, Li FP, Strong LC, Fraumeni JF Jr, Nelson CE, Kim DH et al (1990) Germ line p53 mutations in a familial syndrome of breast cancer, sarcomas, and other neoplasms. Science 250:1233–1238

Martín-Caballero J, Flores JM, García-Palencia P, Serrano M (2001) Tumor susceptibility of p21Waf1/Cip1-deficient mice. Cancer Res 61(16):6234–6238

Matheu A, Maraver A, Serrano M (2008) The Arf/p53 pathway in cancer and aging. Cancer Res 68(15):6031–6034. doi:10.1158/0008-5472.can-07-6851

Mathupala SP, Heese C, Pedersen PL (1997) Glucose catabolism in cancer cells. The type II hexokinase promoter contains functionally active response elements for the tumor suppressor p53. J Biol Chem 272:22776–22780
Medrano S, Burns-Cusato M, Atienza MB, Rahimi D, Scrable H (2009) Regenerative capacity of neural precursors in the adult mammalian brain is under the control of p53. Neurobiol Aging 30:483–497
Meek DW, Anderson CW (2009) Posttranslational modification of p53: cooperative integrators of function. Cold Spring Harb Perspect Biol 1:a000950
Menendez D, Krysiak O, Inga A, Krysiak B, Resnick MA, Schonfelder G (2006) A SNP in the flt-1 promoter integrates the VEGF system into the p53 transcriptional network. Proc Natl Acad Sci U S A 103:1406–1411
Mihara M, Erster S, Zaika A, Petrenko O, Chittenden T, Pancoska P et al (2003) p53 has a direct apoptogenic role at the mitochondria. Mol Cell 11:577–590
Mills AA, Zheng B, Wang X-J, Vogel H, Roop DR, Bradley A (1999) p63 is a p53 homologue required for limb and epidermal morphogenesis. Nature 398(6729):708–713. doi:10.1038/19531
Narita M, Narita M, Krizhanovsky V, Nunez S, Chicas A, Hearn SA et al (2006) A novel role for high-mobility group a proteins in cellular senescence and heterochromatin formation. Cell 126:503–514
Neduva V, Russell RB (2005) Linear motifs: evolutionary interaction switches. FEBS Lett 579(15)3342–3345. doi:http://dx.doi.org/10.1016/j.febslet.2005.04.005
Neduva V, Russell RB (2006) DILIMOT: discovery of linear motifs in proteins. Nucleic Acids Res 34(Web Server issue):W350–355
Neduva VL, Su-Angrand R, Stark I, Masi A de, Gibson F, Lewis TJ, Serrano J, Russell L, R. B (2005a) Systematic discovery of new recognition peptides mediating protein interaction networks. PLoS Biol 3(12):e405
Neduva V, Linding R, Su-Angrand I, Stark A, de Masi F, Gibson TJ et al (2005b) Systematic discovery of new recognition peptides mediating protein interaction networks. PLoS Biol 3(12):e405. doi:10.1371/journal.pbio.0030405
Olivier M, Hussain SP, Caron de Fromentel C, Hainaut P, Harris CC (2004) TP53 mutation spectra and load: a tool for generating hypotheses on the etiology of cancer. IARC Sci Publ 157:247–270
Ory K, Legros Y, Auguin C, Soussi T (1994) Analysis of the most representative tumour-derived p53 mutants reveals that changes in protein conformation are not correlated with loss of transactivation or inhibition of cell proliferation. EMBO J 13:3496–3504
Perry ME (2010) The regulation of the p53-mediated stress response by MDM2 and MDM4. Cold Spring Harb Perspect Biol 2:a000968
Petitjean A, Achatz MI, Borresen-Dale AL, Hainaut P, Olivier M (2007) TP53 mutations in human cancers: functional selection and impact on cancer prognosis and outcomes. Oncogene 26:2157–2165
Pfeffer LM, Dinarello CA, Herberman RB, Williams BR, Borden EC, Bordens R et al (1998) Biological properties of recombinant alpha-interferons: 40th anniversary of the discovery of interferons. Cancer Res 58:2489–2499
Pharoah PD, Day NE, Caldas C (1999) Somatic mutations in the p53 gene and prognosis in breast cancer: a meta-analysis. Br J Cancer 80:1968–1973
Pogribny IP, Pogribna M, Christman JK, James SJ (2000) Single-site methylation within the p53 promoter region reduces gene expression in a reporter gene construct: possible in vivo relevance during tumorigenesis. Cancer Res 60:588–594
Raman V, Martensen SA, Reisman D, Evron E, Odenwald WF, Jaffee E et al (2000) Compromised HOXA5 function can limit p53 expression in human breast tumours. Nature 405:974–978
Reisman D, Elkind NB, Roy B, Beamon J, Rotter V (1993) c-Myc trans-activates the p53 promoter through a required downstream CACGTG motif. Cell Growth Differ 4:57–65
Riley T, Sontag E, Chen P, Levine A (2008) Transcriptional control of human p53-regulated genes. Nat Rev Mol Cell Biol 9:402–412

Rippin TM, Freund SM, Veprintsev DB, Fersht AR (2002) Recognition of DNA by p53 core domain and location of intermolecular contacts of cooperative binding. J Mol Biol 319:351–358
Robles AI, Harris CC (2010) Clinical outcomes and correlates of TP53 mutations and cancer. Cold Spring Harb Perspect Biol 2:a001016
Ronen D, Rotter V, Reisman D (1991) Expression from the murine p53 promoter is mediated by factor binding to a downstream helix-loop-helix recognition motif. Proc Natl Acad Sci U S A 88:4128–4132
Roninson IB (2003) Tumor cell senescence in cancer treatment. Cancer Res 63:2705–2715
Roy B, Beamon J, Balint E, Reisman D (1994) Transactivation of the human p53 tumor suppressor gene by c-Myc/Max contributes to elevated mutant p53 expression in some tumors. Mol Cell Biol 14:7805–7815
Scholzen T, Gerdes J (2000) The Ki-67 protein: from the known and the unknown. J Cell Physiol 182(3)311–322. doi:10.1002/(sici)1097-4652(200003)182:3<311::aid-jcp1>3.0.co;2-9
Schroeder M, Mass MJ (1997) CpG methylation inactivates the transcriptional activity of the promoter of the human p53 tumor suppressor gene. Biochem Biophys Res Commun 235:403–406
Scoumanne A, Chen X (2007) The lysine-specific demethylase 1 is required for cell proliferation in both p53-dependent and -independent manners. J Biol Chem 282:15471–15475
Selivanova G, Iotsova V, Okan I, Fritsche M, Strom M, Groner B et al (1997) Restoration of the growth suppression function of mutant p53 by a synthetic peptide derived from the p53 C-terminal domain. Nat Med 3:632–638
Shen EC, Henry MF, Weiss VH, Valentini SR, Silver PA, Lee MS (1998) Arginine methylation facilitates the nuclear export of hnRNP proteins. Genes Dev 12:679–691
Shen KC, Heng H, Wang Y, Lu S, Liu G, Deng CX et al (2005) ATM and p21 cooperate to suppress aneuploidy and subsequent tumor development. Cancer Res 65(19):8747–8753. doi:10.1158/0008-5472.can-05-1471
Shi X, Kachirskaia I, Yamaguchi H, West LE, Wen H, Wang EW et al (2007) Modulation of p53 function by SET8-mediated methylation at lysine 382. Mol Cell 27:636–646
Shi M, Huang R, Pei C, Jia X, Jiang C, Ren H (2012) TP53 codon 72 polymorphism and glioma risk: a meta-analysis. Oncol Lett 3:599–606
Sidhu S, Martin E, Gicquel C, Melki J, Clark SJ, Campbell P et al (2005) Mutation and methylation analysis of TP53 in adrenal carcinogenesis. Eur J Surg Oncol 31(5):549–554. doi:10.1016/j.ejso.2005.01.013
Smela ME, Currier SS, Bailey EA, Essigmann JM (2001) The chemistry and biology of aflatoxin B1: from mutational spectrometry to carcinogenesis. Carcinogenesis 22:535–545
Soto-Reyes E, Recillas-Targa F (2010) Epigenetic regulation of the human p53 gene promoter by the CTCF transcription factor in transformed cell lines. Oncogene 29(15):2217–2227. doi:10.1038/onc.2009.509
Soussi T, Beroud C (2001) Assessing TP53 status in human tumours to evaluate clinical outcome. Nat Rev Cancer 1:233–240
Soussi T, Beroud C (2003) Significance of TP53 mutations in human cancer: a critical analysis of mutations at CpG dinucleotides. Hum Mutat 21:192–200
Stewart SA, Weinberg RA (2006) Telomeres: cancer to human aging. Annu Rev Cell Dev Biol 22:531–557
Strano S, Dell'Orso S, Di Agostino S, Fontemaggi G, Sacchi A, Blandino G (2007) Mutant p53: an oncogenic transcription factor. Oncogene 26:2212–2219
Su CH, Shann YJ, Hsu MT (2009) p53 chromatin epigenetic domain organization and p53 transcription. Mol Cell Biol 29:93–103
Swinnen JV, Esquenet M, Goossens K, Heyns W, Verhoeven G (1997) Androgens stimulate fatty acid synthase in the human prostate cancer cell line LNCaP. Cancer Res 57:1086–1090
Takaoka A, Hayakawa S, Yanai H, Stoiber D, Negishi H, Kikuchi H et al (2003) Integration of interferon-alpha/beta signalling to p53 responses in tumour suppression and antiviral defence. Nature 424:516–523
Tang W, Zhou X, Chan Y, Wu X, Luo Y (2011) p53 codon 72 polymorphism and recurrent pregnancy loss: a meta-analysis. J Assist Reprod Genet 28:965–969

Teufel DP, Freund SM, Bycroft M, Fersht AR (2007) Four domains of p300 each bind tightly to a sequence spanning both transactivation subdomains of p53. Proc Natl Acad Sci U S A 104:7009–7014
Tyner SD, Venkatachalam S, Choi J, Jones S, Ghebranious N, Igelmann H et al (2002) p53 mutant mice that display early ageing-associated phenotypes. Nature 415:45–53
Vakifahmetoglu-Norberg H, Kim M, Xia HG, Iwanicki MP, Ofengeim D, Coloff JL et al (2013) Chaperone-mediated autophagy degrades mutant p53. Genes Dev 27:1718–1730
Velasco-Miguel S, Buckbinder L, Jean P, Gelbert L, Talbott R, Laidlaw J et al (1999) PA26, a novel target of the p53 tumor suppressor and member of the GADD family of DNA damage and growth arrest inducible genes. Oncogene 18:127–137
Venot C, Maratrat M, Dureuil C, Conseiller E, Bracco L, Debussche L (1998) The requirement for the p53 proline-rich functional domain for mediation of apoptosis is correlated with specific PIG3 gene transactivation and with transcriptional repression. EMBO J 17:4668–4679
Vousden KH, Lu X (2002) Live or let die: the cell's response to p53. Nat Rev Cancer 2(8):594–604. doi:10.1038/nrc864
Walker KK, Levine AJ (1996) Identification of a novel p53 functional domain that is necessary for efficient growth suppression. Proc Natl Acad Sci U S A 93:15335–15340
Warburg O (1956) On respiratory impairment in cancer cells. Science 124:269–270
Xu Y (2008) Induction of genetic instability by gain-of-function p53 cancer mutants. Oncogene 27:3501–3507
Yang A, McKeon F (2000) P63 and P73: P53 mimics, menaces and more. Nat Rev Mol Cell Biol 1(3):199–207. doi:10.1038/35043127
Yang A, Kaghad M, Caput D, McKeon F (2002) On the shoulders of giants: p63, p73 and the rise of p53. Trends Genet 18:90–95
Young J, Simms LA, Biden KG, Wynter C, Whitehall V, Karamatic R et al (2001) Features of colorectal cancers with high-level microsatellite instability occurring in familial and sporadic settings: parallel pathways of tumorigenesis. Am J Pathol 159:2107–2116
Zaika A, Marchenko N, Moll UM (1999) Cytoplasmically "sequestered" wild type p53 protein is resistant to Mdm2-mediated degradation. J Biol Chem 274:27474–27480
Zalcenstein A, Stambolsky P, Weisz L, Muller M, Wallach D, Goncharov TM et al (2003) Mutant p53 gain of function: repression of CD95(Fas/APO-1) gene expression by tumor-associated p53 mutants. Oncogene 22:5667–5676
Zhao L, Zhao X, Wu X, Tang W (2013) Association of p53 Arg72Pro polymorphism with esophageal cancer: a meta-analysis based on 14 case-control studies. Genet Test Mol Biomarkers 17:721–726
Zhou X, Gu Y, Zhang SL (2012) Association between p53 codon 72 polymorphism and cervical cancer risk among Asians: a HuGE review and meta-analysis. Asian Pac J Cancer Prev 13:4909–4914
Zhou BP, Liao Y, Xia W, Zou Y, Spohn B, Hung MC (2001) HER-2/neu induces p53 ubiquitination via Akt-mediated MDM2 phosphorylation. Nat Cell Biol 3:973–982
Zilfou JT, Lowe SW (2009) Tumor suppressive functions of p53. Cold Spring Harb Perspect Biol 1(5):a001883. doi:10.1101/cshperspect.a001883

Chapter 9
Predictive Role of O6-Methylguanine DNA Methyltransferase Status for the Treatment of Brain Tumors

Marina V. Matsko and Evgeny N. Imaynitov

Contents

Abstract Alkylating cytotoxic agents remain a backbone for the systemic therapy of brain tumors. However, the efficacy of temozolomide or other drugs of this class varies dramatically between patients, being negligible in some cases but critical for the disease outcome in others. Search for predictive markers led to discovery of the role of O^6-methylguanine DNA methyltransferase (MGMT), an enzyme involved in the removal of alkyl groups in N7 and O6 positions of guanine. Low expression of MGMT has been repeatedly shown to be associated with the pronounced tumor sensitivity to the systemic treatment. Nevertheless, methodological issues of MGMT status determination remain a subject of investigation. Furthermore, while grade IV malignant gliomas (glioblastomas) have been analyzed with sufficient level of comprehension, there are just a few studies dedicated to less aggressive varieties of CNS malignancies. This chapter presents an update on clinical studies on predictive role of MGMT status.

M. V. Matsko (✉)
Department of Surgery of the Brain and Spinal Cord, Polenov Research Institute of Neurosurgery, Mayakovskaya 12, Saint- Petersburg 191014, Russia
e-mail: marinamatsko@mail.ru

E. N. Imaynitov
Department of Tumor Growth Biology, N N Petrov Institute of Oncology, Saint- Petersburg 197758, Russia
e-mail: evgeny@imyanitov.spb.ru

P. Mehdipour (ed.), *Epigenetics Territory and Cancer*,
DOI 10.1007/978-94-017-9639-2_9

Abbreviations

ACNU	Nimustine
BCNU	Carmustine
BEV	Bevacizumab
CBDCA	Carboplatin
CCNU	Lomustine
CCRT-TMZ	Radiotherapy plus temozolomide
CRT	Chemoradiotherapy
CT	Chemotherapy
EP	Etoposide+carboplatin
GKSRS	Gamma Knife stereotactic radiosurgery
IFNβ	Human fibroblast interferon
Me	Methylated
MGMT	O^6-methylguanine-DNA-methyltransferase
mo	months
MSP	Methylation-specific polymerase chain reaction
OS	Overall survival
PAV	Procarbazine, nimustine, vincristine
PCV	Procarbazine, lomustine, vincristine
PFS	Progression free survival
RT	Radiotherapy
STB	Stereotactic biopsy
SR	Surgical resection
TMZ	Temozolomide
Um	Unmethylated
VCR	Vincristine
VP-16	Etoposide

9.1 Introduction

Cytotoxic drugs still remain the backbone of cancer therapy. First antitumor compounds have been developed from warfare poisons (Chabner and Roberts 2005). Most of cytotoxic agents are administered in the maximal tolerated dose, so the toxicity is considered as an inevitable component of the treatment. Unfortunately, only a fraction of cancer patients indeed benefit from the medical intervention, while virtually all of them suffer from various adverse effects. There are intensive efforts to develop so-called predictive markers, which will allow to chose the most effective drug for each particular patient, or, in the worst scenario, simply abstain from the therapy in case of expected tumor resistance.

Brain tumors are usually treated by alkylating cytotoxic agents. It has long been assumed that enzymes which interfere with alkylation of DNA residues may play a predictive role for this type of therapy. *O*^{6}-methylguanine DNA methyltransferase

(MGMT), an enzyme involved in the removal of alkyl groups in N7 and O6 positions of guanine, has been intensively studied in this respect, and so far remains the only established marker of drug sensitivity for CNS tumors. It has been repeatedly shown that low expression of MGMT renders tumor sensitivity to temozolomide or other alkylating agents, while high activity of MGMT counteracts with the cytotoxic effects of this class of the drugs (Table 9.1).

Here we systematically update the data on relationship between MGMT status and outcome of therapy of patients with CNS tumors.

9.2 Methodology of MGMT Status Determination

Immunohistochemical (IHC) analysis of protein expression remains the most accessible laboratory assay for the evaluation of the status of a given molecule. It utilizes binding of diagnostic antibody to an antigenic epitope of the studied protein, and uses conventional histological sections as a source of biological material. IHC allows to semi-quantitatively determine the amount the target as well as to control its intracellular localization (Cao et al. 2009; Sonoda et al. 2010). While the simplicity and availability constitute strong advantages of this method, it is important to acknowledge that IHC suffers from the lack of in-built internal control and poor interlaboratory reproducibility (Cao et al. 2009; Chinot et al. 2007; Lalezari et al. 2013; Lechapt-Zalcman et al. 2012; Metellus et al. 2009; Nakagawa et al. 2009; Quillien et al. 2012; Shah et al. 2011; Sonoda et al. 2010; Watanabe et al. 2011).

MGMT expression strongly depends on the methylation of MGMT gene promoter. Cytosine may exist both in unmethylated and methylated forms. Methylated cytosine is frequently called as the 5th DNA base: indeed, methylation status of cytosine is usually maintained through DNA replication. Methylation of cytosine is known to be a major mechanism of epigenetic regulation of gene activity. High content of methylation of promoter region is frequently accompanied by low level of expression of the corresponding gene, while cytosine demethylation is associated with activation of transcription (Esteller 2008; You and Jones 2012). Laboratory analysis of cytosine methylation usually utilizes bisulphite treatment of the target DNA; this procedure converts unmethylated cytosine to uracil, while methylated cytosines remain unaffected. Further determination of identity of the bases is performed by means of standard DNA analysis (i.e., sequencing, allele-specific PCR, etc.) (Umer and Herceg 2013).

Somewhat surprisingly, RNA measurement of MGMT level has been rarely utilized so far. Contrary to wide-sperad beliefs, RNA is decently well preserved in formalin-fixed paraffin-embedded tissue blocks. Therefore, despite well-known vulnerability of RNA, many studies of predictive markers routinely utilized the analysis of RNA expression in archival tumor material (Fairley et al. 2012; Imyanitov and Moiseyenko 2007). Furthermore, RNA-based assays are used in several commercial cancer diagnostic kits (e.g., Oncotype DX, http://www.oncotypedx.com/). RNA analysis may have significant advantages to the above methods, as it

Table 9.1 Predictive value of MGMT status for patients with brain tumors

Study	Number of patients	Treatment	Assay	MGMT status (frequency, %)	Survival		p	
					PFS	OS	PFS	OS
Newly diagnosed glioblastoma								
Jaeckle et al. 1998	40	SR-RT-BCNU	Protein expression: quantative immunofluores-cence	Low level: 33 % (≤60,000 molecules/ nucleus)		12 mo		0.0002
				High level: 67 % (>60,000 molecules/ nucleus)		7 mo		
Kamiryo et al. 2004	74	SR-RT-PAV/ SR-RT-PAV-IFNβ	DNA methylation	Me: 45 %	0.42 years	1.69 years	0.98	0.20
				Um: 55 %	0.43 years	1.34 years		
Hegi et al. 2005	206	SR-CCRT-TMZ	DNA methylation	Me: 45 %	10.3 mo	21.7 mo		<0.001
				Um: 55 %	5.3 mo	12.7 mo		
		SR-RT-TMZ		Me	5.9 mo	15.3 mo		
				Um	4.4 mo	11.8 mo		
Watanabe et al. 2005	29	SR-RT-ACNU-IFNβ	DNA methylation	Me: 34 %	14 mo	38 mo	0.009	0.008
				Um: 66 %	4 mo	7 mo		
Herrlinger et al. 2006	19	SR-RT-CCNU-TMZ	DNA methylation	Me: 42 %	19 mo	Not reached	0.014	0.037
				Um: 58 %	6 mo	12.5 mo		
Criniere et al. 2007	219	SR-RT	DNA methylation	Me: 58 %		10.2 mo		0.407
				Um: 42 %		15.1 mo		

Table 9.1 (continued)

Study	Number of patients	Treatment	Assay	MGMT status (frequency, %)	Survival		p	
					PFS	OS	PFS	OS
		SR-RT-CT (nitrosourea-based)		Me		17.1 mo		0.049
				Um		13.1 mo		
		SR-CRT (with nitrosourea)/CT (nitrosourea-based)		Me		19.9 mo		0.0004
				Um		13.5 mo		
Chinot et al. 2007	25 (inoperable glioblastoma)	STB-TMZ/RT-TMZ	Protein expression	Immunonegative: 44 % (< 35 % cells)	5.5 mo	16 mo	0.009	0.003
				Immunopositive: 56 % (≥ 35 % cells)	1.9 mo	5 mo		
Brandes et al. 2008	103	SR-CCRT-TMZ	DNA methylation	Me: 35 %	21.9 mo	43.6 mo	< 0.0001	< 0.0001
				Um: 65 %	9.2 mo	16.8 mo		
Capper et al. 2008	73	SR-RT-CT	Protein expression	Immunonegative: 58 % (< 15 % cells)		15.4 mo		0.0002
				Immunopositive: 42 % (> 15 % cells)		8 mo		
Smith et al. 2008	22	SR-BCNU (wafer implantation)- GKSRS	DNA methylation	Me: 41 %		103 weeks		0.0009
				Um: 59 %		45 weeks		
Brandes et al. 2009a	58 (elderly patients: ≥ 65 years)	SR-CCRT-TMZ	DNA methylation	Me: 43 %	22.9 mo	Not reached	< 0.01	0.05
				Um: 57 %	9.5 mo	13.7 mo		

Table 9.1 (continued)

Study	Number of patients	Treatment	Assay	MGMT status (frequency, %)	Survival		p	
					PFS	OS	PFS	OS
Cao et al. 2009	76	SR-CCRT-TMZ/SR-TMZ after recurrence	DNA methylation	Me: 61 %		19.87 mo		0.014
				Um: 39 %		12.3 mo		
	80		Protein expression	Immunonegative: 60 % (< 5 % cells)		17.43 mo		0.197
				Immunopositive: 27.5 % (< 30 % cells)		15.83 mo		
				Immunopositive: 12.5 % (≥ 30 % cells)		10.67 mo		
Clarke et al. 2009	48	SR-RT-dose-dense TMZ/ SR-RT-metronomic TMZ	DNA methylation	Me: 19 %	4.2 mo	28.1 mo		
				Um: 81 %	5 mo	14.9 mo		
Dunn et al. 2009	109	SR-CCRT-TMZ	DNA methylation	Me: 53 %	11.8 mo	16.8 mo	0.0001	0.00001
				Um: 47 %	8.3 mo	11.1 mo		
Felsberg et al. 2009	66	SR-RT-TMZ	DNA methylation	Me: 39 %	245 days	692 days	0.0003	0.004
				Um: 61 %	100 days	474 days		
		SR (near-complete)-RT-TMZ		Me	268 days	681 days		
				Um	163 days	507 days		
		SR (non-complete)-RT-TMZ		Me	166 days	499 days		
				Um	83 days	406 days		

Table 9.1 (continued)

Study	Number of patients	Treatment	Assay	MGMT status (frequency, %)	Survival		p	
					PFS	OS	PFS	OS
Gerstner et al. 2009	40 (elderly patients: ≥70 years)	SR-TMZ/SR-BCNU	DNA methylation	Me: 60 %	405 days	489 days	0.27	0.0021
				Um: 40 %	246 days	263 days		
Glas et al. 2009	23	SR-RT-TMZ/SR-CCNU	DNA methylation	Me: 48 %	19 mo	34.3 mo	0.0064	0.0009
				Um: 52 %	7 mo	12.5 mo		
Grossman et al. 2009	24	SR-CCRT-TMZ/talam-panel (6 mo)	DNA methylation	Me: 29 %		29.1 mo		
				Um: 71 %		16.3 mo		
Nakagawa et al. 2009	23	SR-RT-ACNU-CBDCA-VP-16-IFN-b	Protein expression	Immunonegative: 70 % (< 10 % cells)		15.3 mo		0.042
				Immunopositive: 30 % (≥ 10 % cells)		10 mo		
Prados et al. 2009	44	SR-CCRT-TMZ-erlotinib	DNA methylation	Me: 36 %		25.5 mo		0.006
				Um: 64 %		14.6 mo		
Park et al. 2009	48	SR-RT-ACNU-cisplatin	DNA methylation	Me: 54 %		17 mo		0.56
				Um: 46 %		17 mo		
Stupp et al. 2009	206	SR-RT	DNA methylation	Me: 45 %		15.3 mo		
				Um: 55 %		11.8 mo		
		SR-CCRT-TMZ		Me		23.4 mo	< 0.0001	
				Um		12.6 mo		

Table 9.1 (continued)

Study	Number of patients	Treatment	Assay	MGMT status (frequency, %)	Survival		p	
					PFS	OS	PFS	OS
Weller et al. 2009	295	SR-RT/SR-RT-TMZ	DNA methylation	Me: 45 %	7.5 mo	18.9 mo	< 0.001 (RT-TMZ)	< 0.001 (RT-TMZ)
				Um: 55 %	6.3 mo	11.1 mo		
Zawlik et al. 2009	371	SR/SR-RT	DNA methylation	Me: 44 %		10.1 mo		
				Um: 56 %		10 mo		
Ang et al. 2010	105	SR-CRT (with TMZ)/ SR-RT	DNA methylation	Me: 54 %	30 weeks	61 weeks	0.01	0.02
				Um: 46 %	24 weeks	42 weeks		
	78	SR-CRT (with TMZ)		Me		71 weeks		
				Um		42 weeks		
	25	SR-RT		Me		14 weeks		
				Um		Not reached		
Costa et al. 2010	80	SR-CCRT-TMZ	DNA methylation	Me: 48 %	9 mo	16 mo	0.77	0.58
				Um: 52 %	10 mo	13 mo		
Etcheverry et al. 2010	50	SR-CCRT-TMZ	DNA methylation	Me: 60 %		26.2 mo		< 0.01
				Um: 40 %		9.5 mo		
Karayan-Tapon et al. 2010	81	SR-CCRT-TMZ	DNA methylation	Me: 68 %		19 mo		0.005
				Um: 32 %		15 mo		
	81		DNA methylation	Me: 48 %		23 mo		< 0.0001
				Um: 52 %		14 mo		

Table 9.1 (continued)

Study	Number of patients	Treatment	Assay	MGMT status (frequency, %)	Survival		p	
					PFS	OS	PFS	OS
	79		DNA methylation	Me: 49 %		23 mo		0.0001
				Um: 51 %		14 mo		
	80		RNA expression	Low level: 50 %		20 mo		0.028
				High level: 50 %		14 mo		
	78		Protein expression	Immunonegative: 50 %		17 mo		0.595
				Immunopositive: 50 %		17 mo		
Minniti et al. 2010	87	SR-CCRT-TMZ	DNA methylation	Me: 36 %	5.6 mo	11.2 mo	0.0001	0.01
				Um: 64 %	9.8 mo	16.7 mo		
Morandi et al. 2010	159	SR-CCRT-TMZ/ SR-RT-TMZ	DNA methylation	Me: 44 %		36 mo		0.004
				Um: 56 %		20 mo		
Weiler et al. 2010	39	SR-TMZ (1 week on/1 week off) before RT-CCRT-TMZ (1 week on/1 week off) + Indometacin 25 mg twice daily	DNA methylation	Me: 41 %	15.8 mo	Not reached >21.5 mo	0.0002	0.10
				Um: 59 %	6.2 mo	15 mo		
Rivera et al. 2010	171	SR-RT	DNA methylation	Me: 32 %	31 weeks	63 weeks	0.009	0.019
				Um: 68 %	15 weeks	51 weeks		
Stupp et al. 2010	45	SR-CCRT-TMZ-cilen-gitide	DNA methylation	Me: 51 %	13.4 mo	23.2 mo	<0.001	0.022
				Um: 49 %	3.4 mo	13.1 mo		

Table 9.1 (continued)

Study	Number of patients	Treatment	Assay	MGMT status (frequency, %)	Survival		p	
					PFS	OS	PFS	OS
Sonoda et al. 2010	62	SR-RT-ACNU/ SR-RT-TMZ	DNA methylation	Me: 56 %	12 mo	26 mo	0.019	0.40
				Um: 44 %	5 mo	17 mo		
	73		Protein expression	Immunonegative: 36 % (<20 % cells)	13 mo	43 mo	0.045	0.30
				Immunopositive: 64 % (≥20 % cells)	7 mo	19 mo		
Balañá et al. 2011	70	SR-RT-BCNU/ SR-CCRT-TMZ	DNA methylation	Me: 39 %	34.2 weeks	73 weeks	0.03	0.04
				Um: 61 %	31.4 weeks	54.7 weeks		
Kreth et al. 2011	53	SR-CCRT-TMZ	RNA expression	Low level: 42 %	17.5 mo	21.6 mo		
				High level: 58 %	3.3 mo	10.4 mo		
Lai et al. 2011	70	SR-CCRT-TMZ-BEV	DNA methylation	Me: 41 %	17.5 mo	24.7 mo	<0.005	<0.005
				Um: 59 %	10.5 mo	15.9 mo		
Minniti et al. 2011	83 (elderly patients: ≥65 years)	SR-CCRT-TMZ	DNA methylation	Me: 51 %	10.5 mo	15.3 mo	0.0001	0.0001
				Um: 49 %	5.5 mo	10.2 mo		
Perez-Lar-raya et al. 2011	31 (elderly patients: ≥70 years)	SR-TMZ	DNA methylation	Me: 42 %	26 weeks	31 weeks	0.03	0.03
				Um: 58 %	11 weeks	19 weeks		
Shah et al. 2011	28	SR-CCRT-TMZ	DNA methylation	Me: 29 %	373 days		0.21	

Table 9.1 (continued)

Study	Number of patients	Treatment	Assay	MGMT status (frequency, %)	Survival		p	
					PFS	OS	PFS	OS
				Um: 71 %	224 days			
	28		DNA methylation	Me: 46 %	540 days		0.007	
				Um: 54 %	210 days			
	24		Protein expression	Immunonegative: 58 % (≤ 15 % cells)	540 days		< 0.0001	
				Immunopositive: 42 % (> 15 % cells)	197 days			
Thon et al. 2011	56 (inoperable glioblastoma)	STB-CCRT-TMZ	DNA methylation	Me: 54 %	56 weeks	104 weeks	< 0.0001	< 0.0001
				Um: 46 %	20 weeks	28 weeks		
Uno et al. 2011	29	SR-RT/SR-RT-BCNU	DNA methylation	Me: 41 %		27.4 mo		0.025
				Um: 59 %		12 mo		
	28	SR-RT/SR-RT-BCNU	DNA methylation	Me: 39 %		31.7 mo		0.004
				Um: 61 %		11.8 mo		
Zunarelli et al. 2011	77	SR-CCRT-TMZ/SR-RT	DNA methylation	Me: 31 %		17.8 mo		< 0.04
				Um: 69 %		11.1 mo		
	46	SR-CCRT-TMZ		Me: 30 %		20.1 mo		< 0.002
				Um: 70 %		12.9 mo		
	18	SR-RT		Me: 72 %		7.6 mo		
				Um: 28 %		5.3 mo		
Havik et al. 2012	86	SR-CCRT-TMZ	DNA methylation (MSP)	Me: 34 %		17.6 mo		
				Um: 66 %		12.2 mo		

Table 9.1 (continued)

Study	Number of patients	Treatment	Assay	MGMT status (frequency, %)	Survival		p	
					PFS	OS	PFS	OS
			DNA methylation (pyrosequencing)	Me: 51 %		16.1 mo		
				Um: 49 %		11.5 mo		
Kim et al. 2012	78	SR-CCRT-TMZ/SR-RT-TMZ/SR-RT-BCNU, CCNU, VCR	DNA methylation	Me: 45 %	18 mo	29 mo	0.017	0.002
				Um: 55 %	9 mo	20 mo		
	6 (age ≤50)	SR (gross-total or subtotal)-CCRT-TMZ/SR-RT-TMZ/SR-RT-BCNU, CCNU, VCR		Me		Not reached (38–77 mo)		
	13 (age ≤50)			Um		24 mo		
	5 (age ≤50)	SR (partial or biopsy)-CCRT-TMZ (6 cycles)/ SR-RT-TMZ/SR-RT-BCNU, CCNU, VCR		Me		38 mo		
	1 (age ≤50)			Um		15 mo		
	20 (age > 50)	SR (gross-total or subtotal)-CCRT-TMZ (6 cycles)/ SR-RT-TMZ/SR-RT-BCNU, CCNU, VCR		Me		21 mo		
	21 (age > 50)			Um		18 mo		
	4 (age > 50)	SR (partial or biopsy)-CCRT-TMZ (6 cycles)/ SR-RT-TMZ/SR-RT-BCNU, CCNU, VCR		Me		12 mo		
	8 (age > 50)			Um		13.2 mo		

Table 9.1 (continued)

Study	Number of patients	Treatment	Assay	MGMT status (frequency, %)	Survival		p	
					PFS	OS	PFS	OS
Lechapt-Zalcman et al. 2012	110	SR-BCNU (wafer implantation)-CCRT-TMZ	DNA methylation	Me: 57 %	10.7 mo	21.7 mo	0.155	0.025
				Um: 43 %	9.7 mo	15.1 mo		
	106		Protein expression	Immunonegative: 48 % (< 15 % cells)	11.5 mo	27 mo	0.049	0.021
				Immunopositive: 52 % (> 15 % cells)	9.6 mo	15.1 mo		
Malmstrom et al. 2012	203 (elderly patients: ≥60 years)	SR-TMZ	DNA methylation	Me: 45 %		9.7 mo		0.02
				Um: 55 %		6.8 mo		
		SR-RT		Me		8.2 mo		0.81
				Um		7 mo		
Niyazi et al. 2012	54	SR-CCRT-TMZ	DNA methylation	Me: 54 %	642 days		< 0.001	< 0.001
				Um: 46 %	231 days			
Quillien et al. 2012	99	SR-CCRT-TMZ	DNA methylation	Me: 60 %	12.2 mo	22.4 mo	0.0024	0.000081
				Um: 40 %	9.1 mo	14.9 mo		
			DNA methylation	Me: 33 %	13.9 mo	20.4 mo	0.0017	0.017
				Um: 67 %	9 mo	16.7 mo		
			DNA methylation	Me: 33 %	13.4 mo	20.4 mo	0.022	0.048
				Um: 67 %	9.1 mo	16.7 mo		
			DNA methylation	Me: 42 %	14.6 mo	26.2 mo	0.000012	0.000017
				Um: 58 %	9 mo	15.7 mo		

Table 9.1 (continued)

Study	Number of patients	Treatment	Assay	MGMT status (frequency, %)	Survival		p	
					PFS	OS	PFS	OS
			Protein expression	Immunonegative: 59 % (< 23 % cells)	12.2 mo	22.4 mo	0.036	0.00051
				Immunopositive: 41 % (> 23 % cells)	9 mo	14.6 mo		
Reifen-berger et al. 2012	233 (elderly patients: ≥ 70 years)	SR/biopsy	DNA methylation	Me: 58 %	1.8 mo	2.3 mo		0.388
				Um: 42 %	1.7 mo	2.0 mo		
		SR-RT		Me	4.5 mo	7.8 mo		
				Um	5.2 mo	8.8 mo		
		SR-TMZ/SR-Pro-carbacine + CCNU/ SR-Nitrosourea		Me	6.8 mo	7.2 mo		
				Um	0.5 mo	2.6 mo		
		SR-RT-TMZ/ SR-Nitrosourea		Me	7.3 mo	13.1 mo		
				Um	7.2 mo	10.4 mo		
Salvati et al. 2012	105	SR-CCRT-TMZ	DNA methylation	Me: 46 %		15.4 mo		< 0.0001
				Um: 54 %		10.1 mo		
	24	SR (gross-total)-CCRT-TMZ		Me		19.7 mo		< 0.006
	21			Um		14.8 mo		
	5	SR (subtotal)-CCRT-TMZ		Me		16.8 mo		< 0.009
	8			Um		11.3 mo		
	12	SR (partial)-CCRT-TMZ		Me		10.5 mo		< 0.001

Table 9.1 (continued)

Study	Number of patients	Treatment	Assay	MGMT status (frequency, %)	Survival		p	
					PFS	OS	PFS	OS
	10			Um		6.3 mo		
	7	Biopsy-CCRT-TMZ		Me		8 mo		<0.02
	18			Um		6.2 mo		
Stummer et al. 2012	79	SR-CCRT-TMZ	DNA methylation	Me: 37 %		26.4 mo		0.0005
				Um: 63 %		16.6 mo		
Abhinav et al. 2013	19 (elderly patients: ≥65 years)	SR-RT-TMZ	DNA methylation	Me: 47 %		242 days		0.29
				Um: 53 %		390 days		
	47	SR-RT/SR-RT-TMZ		Me: 60 %		167 days		0.068
				Um: 40 %		308 days		
Gilbert et al. 2013	762	SR-CCRT-TMZ	DNA methylation	Me: 32 %	8.8 mo	23.5 mo		
				Um: 68 %	7.1 mo	16.6 mo		
		SR-TMZ (dose-dense, 12 cycles)		Me	11.7 mo	21.9 mo		
				Um	8.2 mo	15.4 mo		
		SR-CCRT-TMZ (6–12 cycles)/SR- TMZ (dose-dense, 12 cycles)		Me	8.7 mo	21.2 mo	<0.001	<0.001
				Um	5.7 mo	14 mo		
Gutenberg et al. 2013	26	SR-CCRT-TMZ	DNA methylation	Me: 65 %	16 mo	18.1 mo		
				Um: 35 %	6.6 mo	13.8 mo		
	26	SR-RT-TMZ		Me	4.1 mo	11.9 mo		

Table 9.1 (continued)

Study	Number of patients	Treatment	Assay	MGMT status (frequency, %)	Survival		p	
					PFS	OS	PFS	OS
				Um	7.4 mo	17 mo		
Lalezari et al. 2013	312	SR-CCRT-TMZ/ SR-RT-TMZ	DNA methylation	Me: 49 %	11.5 mo	23.1 mo	< 0.0001	< 0.0001
				Um: 51 %	7.9 mo	15.6 mo		
	402		DNA methylation	Me: 37 %	13.3 mo	24.7 mo	< 0.0001	< 0.0001
				Um: 63 %	7.8 mo	16.2 mo		
	355		Protein expression	Immunonegative: 52 % (< 30 % cells)	10.9 mo	20.5 mo	< 0.0001	< 0.0001
				Immunopositive: 48 % (≥ 30 % cells)	7.8 mo	16.7 mo		
Park et al. 2013	75	SR-CCRT-TMZ	DNA methylation	Me: 28 %	24 mo	41 mo	0.02	0.02
				Um: 72 %	6 mo	18 mo		
		SR-RT-TMZ		Me	3 mo	17 mo	0.19	0.53
				Um	6 mo	17 mo		
Rapp et al. 2013	85	SR-CCRT-TMZ/ SR-RT-TMZ	DNA methylation	Me: 40 %	5 mo	22.1 mo	0.103	0.008
				Um: 60 %	6 mo	16.2 mo		
Sunwoo et al. 2013	26	SR-RT-TMZ	DNA methylation	Me: 19 %	14.5 mo		0.025	
				Um: 80 %	4 mo			
Capdevila et al. 2014	18	SR-TMZ-Cisplatin (2 cycles)/RT- TMZ-Cisplatin (until disease progression)	DNA methylation	Me	5.2 mo	3 mo	0.008	0.05
				Um	2.5 mo	9.1 mo		

Table 9.1 (continued)

Study	Number of patients	Treatment	Assay	MGMT status (frequency, %)	Survival		p	
					PFS	OS	PFS	OS
	16	SR-CCRT-TMZ		Me	10.5 mo	6.6 mo	0.09	0.02
				Um	7.4 mo	20.3 mo		
Quillien et al. 2014	89	SR-CCRT-TMZ	DNA methylation	Me: 44–46 %	14 mo	26.2 mo	0.00004	0.000006
				Um: 56–54 %	9 mo	15.9 mo		
Combined groups of patients with newly diagnosed glioblastoma or anaplastic astrocytoma								
Esteller et al. 2000	47	SR-Cisplatin-RT-BCNU	DNA methylation	Me: 40 %	21 mo		<0.001	<0.001
				Um: 60 %	8 mo			
Watanabe et al. 2005	45	SR-RT-ACNU-IFNβ	DNA methylation	Me: 38 %	22 mo	38 mo	0.14	0.07
				Um: 62 %	7 mo	13 mo		
Fabi et al. 2009	19 (recurrent)	SR-Fotemustine	DNA methylation	Me: 63 %	7 mo	45 mo	0.55	0.27
				Um: 37 %	6 mo	22 mo		
Liu et al. 2010	66	SR-Stereotactic radiosurgery-BCNU	DNA methylation	Me: 97 %		13.5 mo		<0.0001
				Um: 3 %		7 mo		
Kreth et al. 2011	63	SR-CCRT-TMZ	DNA methylation	Me: 51 %	18.3 mo	not reached (>22 mo)		<0.0001
				Um: 49 %	4.9 mo	9.6 mo		
			RNA expression	Low level: 51 %	17.5 mo	Not reached (>20 mo)		<0.0001
				High level: 49 %	5 mo	9.5 mo		

Table 9.1 (continued)

Study	Number of patients	Treatment	Assay	MGMT status (frequency, %)	Survival		p	
					PFS	OS	PFS	OS
Wick et al. 2012	209 (elderly patients: ≥65 years)	SR-TMZ	DNA methylation	Me: 35 %	8.4 mo			0.0001
				Um: 65 %	3.3 mo			
		SR-RT		Me	4.6 mo			
				Um	4.6 mo			
		SR-RT/SR-TMZ		Me		11.9 mo	0.014	
				Um		8.2 mo		
Capdevila et al. 2014	42	SR-TMZ-Cisplatin (2 cycles) -RT- TMZ-Cisplatin (until disease progression)/SR-CCRT-TMZ (6–12 cycles)	DNA methylation	Me: 38 %	5 mo	8.3 mo	0.53	0.78
				Um: 62 %	4.7 mo	11.2 mo		
Recurrent glioblastoma								
Brandes et al. 2006a	22	SR-RT-TMZ	DNA methylation	Me: 46 %	15.6 weeks	48.2 weeks	0.86	
				Um: 54 %	11.9 weeks	34.7 weeks		
Nagane et al. 2007	19	SR-RT-TMZ	Protein expression	Low level: 53 %	4.5 mo	10.3 mo	0.016	0.019
				High level: 47 %	1.6 mo	7.0 mo		
Wick et al. 2007	36	SR-CCRT-TMZ	DNA methylation	Me: 47 %	27 weeks		0.22	
				Um: 53 %	19 weeks			

Table 9.1 (continued)

Study	Number of patients	Treatment	Assay	MGMT status (frequency, %)	Survival		p	
					PFS	OS	PFS	OS
Metellus et al. 2009	19	SR-BCNU (wafer implantation)	DNA methylation	Me: 26 %	9 mo	18.8 mo	< 0.0001	0.006
				Um: 74 %	2.7 mo	8.4 mo		
	19		DNA methylation	Me: 32 %	8.9 mo	14.2 mo	0.003	0.03
				Um: 68 %	2.7 mo	9.2 mo		
	20		Protein expression	Immunonegative: 25 % (< 10 % cells)	9 mo	18.9 mo	0.002	0.013
				Immunopositive: 75 % (> 10 % cells)	3 mo	9.2 mo		
Watanabe et al. 2011	53	SR-CCRT-TMZ	Protein expression	Immunonegative: 47 % (< 10 % cells)	11.2 mo	26.7 mo	0.0015	0.038
				Immunopositive: 53 % (≥ 10 % cells)	6.8 mo	16.8 mo		
		SR-RT-EP		Immunonegative (< 10 % cells)	5.4 mo	9.8 mo	0.019	0.8
				Immunopositive (≥ 10 % cells)	2.3 mo	9.4 mo		
Norden et al. 2013	47	SR-CCRT-TMZ	DNA methylation	Me: 65 %	66 days	22.3 mo	0.11	0.01
				Um: 35 %	57 days	11.7 mo		
Newly diagnosed anaplastic astrocytoma								
Jaeckle et al. 1998	24	SR-RT-BCNU	Protein expression	Low level: 42 % (≤ 60,000 molecules/ nucleus)		62 mo		0.0002

Table 9.1 (continued)

Study	Number of patients	Treatment	Assay	MGMT status (frequency, %)	Survival		p	
					PFS	OS	PFS	OS
				High level: 58 % (>60,000 molecules/ nucleus)		14 mo		
Kamiryo et al. 2004	42	SR-RT-ACNU/ SR-RT-ACNU-INFβ	DNA methylation	Me: 45 %		6.66 years	0.018	0.013
				Um: 55 %	1.19 years	2.78 years		
Newly diagnosed anaplastic oligodendroglioma or oligoastrocytoma								
Brandes et al. 2006b	54	SR-RT-TMZ	DNA methylation	Me: 69 %	12 mo	40.9 mo	0.41	0.09
				Um: 31 %	13 mo	28.6 mo		
van den Bent et al. 2009	97	SR-RT	DNA methylation	Me: 84 %	17.9 mo	59.3 mo	0.01	0.002
				Um: 16 %	7.8 mo	12.3 mo		
		SR-RT-PCV		Me	49 mo	no reached	0.0011	0.0004
				Um	10.5 mo	19 mo		
Ducray et al. 2011	38 (elderly patients: ≥70 years)	SR-TMZ	DNA methylation	Me: 50 %	8.7 mo	16.1 mo	0.01	0.05
				Um: 50 %	5.7 mo	12.4 mo		
Recurrent anaplastic astrocytoma or oligoastrocytoma								
Sadones et al. 2009	11	SR-RT-TMZ	DNA methylation	Me: 36 %	36 weeks		0.52	
				Um: 64 %	20 weeks		0.028	

Table 9.1 (continued)

Study	Number of patients	Treatment	Assay	MGMT status (frequency, %)	Survival		p	
					PFS	OS	PFS	OS
Newly diagnosed grade II oligodendroglioma, oligoastrocytoma or astrocytoma								
Everhard et al. 2006	68	SR-TMZ (median 12 cycles)	DNA methylation	Me: 93 %	29.5 mo		< 0.00005	
				Um: 7 %	6 mo			
Capper et al. 2008	32	SR-RT-CT	Protein expression	Immunonegative: 81 % (< 35 % cells)		44 mo		0.005
				Immunopositive: 19 % (> 35 % cells)		10.4 mo		
Kesari et al. 2009	20 (newly diagnosed and recurrent)	SR-TMZ (75 mg/m2/ day—7 weeks on/4 weeks off) (for 6 cycles)	DNA methylation	Me: 60 %	47 mo	Not reached (> 72 mo)	0.15	0.008
				Um: 40 %	30 mo	29 mo		

ACNU nimustine, *BCNU* carmustine, *BEV* bevacizumab, *CBDCA* carboplatin, *CCNU* lomustine, *CCRT-TMZ* radiotherapy plus temozolomide, *CRT* chemoradiotherapy, *CT* chemotherapy, *EP* etoposide, carboplatin, *GKSRS* gamma knife stereotactic radiosurgery, *IFNβ* human fibroblast interferon, *Me* methylated, *MGMT* O^6 -methylguanine-DNA-methyltransferase, *mo* months, *MSP* methylation-specific polymerase chain reaction, *OS* overall survival, *PAV* procarbazine, nimustine, vincristine, *PCV* procarbazine, lomustine, vincristine, *PFS* progression free survival, *RT* radiotherapy, *STB* stereotactic biopsy, *SR* surgical resection, *TMZ* temozolomide, *Um* unmethylated, *VCR* vincristine, *VP-16* etoposide utilizes direct measurement of gene expression, appears perfectly quantitative and has in-built internal standard (gene-referee) (Karayan-Tapon et al. 2010; Kreth et al. 2011). Contrary to IHC, interlaboratory reproducibility of RNA measurement has not been assessed yet.

MGMT expression is a continuous variable, so the setting of cut-off between "MGMT-active" and "MGMT-inactive" tumors represents a significant problem. IHC studies usually rely on the proportion of MGMT antibody-stained cells, however the threshold varies from 5 to 35% in different studies (Cao et al. 2009; Lalezari et al. 2013; Lechapt-Zalcman et al. 2012; Metellus et al. 2009; Quillien et al. 2012). Methylation studies utilize the information regarding the number or proportion of methylated cytosines in the tested DNA fragment (Kim et al. 2012; Park et al. 2013; Quillien et al. 2014; Rapp et al. 2013). However, not all cytosine sites are functional in terms of regulation of transcription, so some of the methylation events may not be directly linked with the expression status of the gene. RNA studies are usually based on real-time reverse transcription PCR. Differences in RNA expression are determined by calculation of deltaCt values, but, as above, the choice of cut-off is more or less arbitrary (Karayan-Tapon et al. 2010; Kreth et al. 2011).

Overall, many interstudy differences or inconsistencies are likely to be attributed to insufficiency of the methodology of MGMT status detection. Significant efforts are being undertaken to provide researchers with better tools for the analysis of predictive markers.

9.3 Predictive Role of MGMT

According to WHO classification, malignant gliomas include glioblastomas (grade IV), as well as 2 categories of grade III tumors (anaplastic astrocytoma, anaplastic oligodendroglioma and anaplastic oligoastrocytoma). Whenever possible, these tumors are treated by complex therapy consisting of surgical intervention, radiotherapy and chemotherapy. Overall, MGMT status strongly influences the outcome of the patients, with the proression-free survival (PFS) being in the range of ~3–11 months for MGMT-active tumors vs. ~8–24 months for MGMT-inactive neoplasms (Table 9.1). The corresponding estimates for the overall survival (OS) are ~7–20 months vs. ~15–43 months (Table 9.1). As expected, the success of surgical debulking strongly influences outcomes both in MGMT-positive and in MGMT-negative cases (Felsberg et al. 2009; Salvati et al. 2012).

Importantly, those patients, who did not receive chemotherapy, show rather similar survival irrespectively of MGMT status (Criniere et al. 2007; Malmstrom et al. 2012; Stupp et al. 2009; Reifenberger et al. 2012; Wick et al. 2012; Zawlik et al. 2009; Zunarelli et al. 2011). These observations suggest that MGMT has limited if any independent prognostic role, and its impact on the results of complex treatment is purely attributed to the sensitivity of the tumor to cytotoxic agents. Furthermore, the predictive value of MGMT is usually maintained in patients who received chemotherapy only, but was omitted from the local intervention (Gerstner et al. 2009;

Gilbert et al. 2013; Malmstrom et al. 2012; Perez-Larraya et al. 2011; Reifenberger et al. 2012).

Many studies suggest that the addition of temozolamide to the radiotherapy should be limited to the subjects with inactive intratumoral MGMT. Indeed, MGMT-negative cases appear to benefit significantly from the combination of chemo- and radiotherapy as compared to radiotherapy only (PFS: 8.8–24 months vs. 3.0–14.5 months; OS: 15.3–43.6 months vs. 15.3–23 months; see Table 9.1). This benefit is significantly less pronounced for MGMT-positive tumors (PFS: 3.3–10 months vs. 3.3–6.0 months; OS: 9.5–18.0 months vs. 11.8–17.0 months; see Table 9.1).

The success of adjuvant use of temozolamide is also associated with MGMT status. Furthermore, patients with MGMT-inactive tumors benefit from extended temozolamide administration, while subjects with MGMT-active do not (Brandes et al. 2008, 2009b; Dunn et al. 2009; Gilbert et al. 2013; Gutenberg et al. 2013; Hegi et al. 2005; Stummer et al. 2012).

The results of the use of alkylating agents as a monotherapy for glioblastoma treatment remain rather poor, hence there are continuing attempts to supplement these schemes by additional compounds. Combination of standard alkylating therapy with other known cytotoxic drugs did not lead to improvement of outcomes, while the use of several targeted molecules (bevacizumab, cilengitide, erlotinib, talampanel) showed improved results as compared to historical controls (Grossman et al. 2009; Hegi et al. 2005; Lai et al. 2011; Prados et al. 2009; Stupp et al. 2010). Remarkably, predictive significance of MGMT status was maintained even upon the use combined treatment schemes (Gerstner et al. 2009; Glas et al. 2009; Kim et al. 2012; Sonoda et al. 2010).

As for all studies on predictive markers, MGMT data suffer from significant inconsistencies. Some patient series failed to demonstrate the predictive value of MGMT status (Abhinav et al. 2013; Capdevila et al. 2014; Minniti et al. 2010), or questioned its significance for certain patient categories (Gutenberg et al. 2013; Park et al. 2013). Nevertherless, most of clinical researchers accept the fact of low efficacy of alkylating therapy in MGMT-active tumors, therefore some trials are specifically focused on the search of alternative treatment options for presumably temozolamide-resistant glial tumors. For example, encouraging short-term results have been obtained with platinum based therapy (Tanaka et al. 2005, 2008) or with combination of irinotecan with bevacizumab (Herrlinger et al. 2013).

Data on predictive status on other tumors than malignant gliomas are rather limited. Nevertheless, there are some evidence that predictive significance of MGMT may be extended to e.g. melanomas, lymphomas etc. (Ma et al. 2003; Ohno et al. 2006; Pollack et al. 2006).

In conclusion, MGMT is one of the most intensively studied and most established markers for the use of cytotoxic therapy. Evaluation of MGMT status is instrumental for selection of the optimal treatment scheme in each given clinical case of malignant brain tumor, as well as for the meaningful selection of patients for experimental therapeutic trials. Lack of standardization of MGMT status evaluation presents a critical obstacle. Further studies are highly needed to improve reliability, reproducibility and accessibility of MGMT testing.

References

Abhinav K, Aquilina K, Gbejuade H, La M, Hopkins K, Iyer V (2013) A pilot study of glioblastoma multiforme in elderly patients: treatments, O-6-methylguanine-DNA methyltransferase (MGMT) methylation status and survival. Clin Neurol Neurosurg 115(8):1375–1378

Ang C, Guiot MC, Ramanakumar AV, Roberge D, Kavan P et al (2010) Clinical significance of molecular biomarkers in glioblastoma. Can J Neurol Sci 37(5):625–630

Balañá C, Carrato C, Ramírez JL, Cardona AF, Berdiel M, Sánchez JJ et al (2011) Tumour and serum MGMT promoter methylation and protein expression in glioblastoma patients. Clin Transl Oncol 13:677–685

Brandes AA, Tosoni A, Cavallo G, Bertorelle R, Gioia V, Franceschi E et al (2006a) Temozolomide 3 weeks on and 1 week off as first-line therapy for recurrent glioblastoma: phase II study from Gruppo Italiano Cooperativo di Neuro-Oncologia (GICNO). Br J Cancer 95:1155–1160

Brandes AA, Tosoni A, Cavallo G, Reni M, Franceschi E, Bonaldi L et al (2006b) Correlations between O6-methylguanine DNA methyltransferase promoter methylation status, 1p and 19q deletions, and response to temozolomide in anaplastic and recurrent oligodendroglioma: a prospective GICNO study. J Clin Oncol 24:4746–4753

Brandes AA, Franceschi E, Tosoni A, Blatt V, Pession A, Tallini G et al (2008) MGMT promoter methylation status can predict the incidence and outcome of pseudoprogression after concomitant radiochemotherapy in newly diagnosed glioblastoma patients. J Clin Oncol 26(13):2192–2197

Brandes AA, Franceschi E, Tosoni A, Benevento F, Scopece L, Mazzocchi V et al (2009a) Temozolomide concomitant and adjuvant to radiotherapy in elderly patients with glioblastoma. Cancer 115:3512–3158

Brandes AA, Tosoni A, Franceschi E, Sotti G, Frezza G, Amista`P et al (2009b) Recurrence pattern after temozolomide concomitant with and adjuvant to radiotherapy in newly diagnosed patients with glioblastoma: correlation with MGMT promoter methylation status. J Clin Oncol 27(8):1275–1279

Capdevila L, Cros S, Ramirez JL, Sanz C, Carrato C, Romeo M et al. (2014) Neoadjuvant cisplatin plus temozolomide versus standard treatment in patients with unresectable glioblastoma or anaplastic astrocytoma: a differential effect of MGMT methylation. J Neurooncol 117(1):77–84.

Capper D, Mittelbronn M, Meyermann R, Schittenhelm J (2008) Pitfalls in the assessment of MGMT expression and in its correlation with survival in diffuse astrocytomas: proposal of a feasible immunohistochemical approach. Acta Neuropathol (Berl) 115(2):249–259

Cao VT, Jung T-Y, Jung S, Jin S-G, Moon K-S, Kim I-Y et al (2009) The correlation and prognostic significance of MGMT promoter methylation and MGMT protein in glioblastomas. Neurosurgery 65(5):866–875

Chabner BA, Roberts TG Jr (2005) Timeline: chemotherapy and the war on cancer. Nat Rev Cancer 5(1):65–72

Chinot L, Barrié M, Fuentes S, Eudes N, Lancelot S, Metellus P et al (2007) Correlation between O^6-methylguanine-DNA methyltransferase and survival in inoperable newly diagnosed glioblastoma patients treated with neoadjuvant temozolomide. J Clin Oncol 25(12):1470–1475

Clarke JL, Iwamoto FM, Sul J, Panageas K, Lassman AB, DeAngelis LM et al (2009) Randomized phase II trial of chemoradiotherapy followed by either dose-dense or metronomic temozolomide for newly diagnosed glioblastoma. J Clin Oncol 27:3861–3867

Costa BM, Caeiro C, Guimaraes I, Martino O, Jaraquemada T, Augusto I et al (2010) Prognostic value of MGMT promoter methylation in glioblastoma patients treated with temozolomide-based chemoradiation: a Portuguese multicentre study. Oncol Rep 23:1655–1662

Criniere E, Kaloshi G, Laigle-Donadey F, Lejeune J, Auger N, Benouaich-Amiel A et al (2007) MGMT prognostic impact on glioblastoma is dependent on therapeutic modalities. J Neurooncol 83:173–179

Ducray F, Sierra del Rio M, Carpentier C, Psimaras D, Idbaih A, Dehais C et al (2011) Up-front temozolomide in elderly patients with anaplastic oligodendroglioma and oligoastrocytoma. J Neurooncol 101:457–462

Dunn J, Baborie A, Alam F, Joyce K, Moxham M, Sibson R et al (2009) Extent of MGMT promoter methylation correlates with outcome in glioblastomas given temozolomide and radiotherapy. Br J Cancer 101:124–131

Esteller M (2008) Epigenetics in cancer. N Engl J Med 358(11):1148–1159

Esteller M, Garcia-Foncillas J, Andion E, Goodman SN, Hidalgo OF, Vanaclocha V et al (2000) Inactivation of the DNA-repair gene MGMT and the clinical response of gliomas to alkylating agents. N Engl J Med 343:1350–1354

Etcheverry A, Aubry M, de Tayrac M, Vauleon E, Boniface R, Guenot F et al (2010) DNA methylation in glioblastoma: impact on gene expression and clinical outcome. BMC Genomics 11:701

Everhard S, Kaloshi G, Criniere E, Benouaich-Amiel A, Lejeune J, Marie Y et al (2006) MGMT methylation: a marker of temozolomid in low-grade gliomas. Ann Neurol 60:740–743

Fabi A, Metro G, Vidiri A, Lanzetta G, Carosi M, Telera S et al (2009) Low-dose fotemustine for recurrent malignant glioma: a multicenter phase II study. J Neurooncol 100(2):209–215

Fairley JA, Gilmour K, Walsh K (2012) Making the most of pathological specimens: molecular diagnosis in formalin-fixed, paraffin embedded tissue. Curr Drug Targets 13(12):1475–1487

Felsberg J, Rapp M, Loeser S, Fimmers R, Stummer W, Goeppert M et al (2009) Prognostic significance of molecular markers and extent of resection in primary glioblastoma patients. Clin Cancer Res 15:6683–6693

Gerstner ER, Yip S, Wang DL, Louis DN, Iafrate AJ, Batchelor TT (2009) MGMT methylation is a prognostic biomarker in elderly patients with newly diagnosed glioblastoma. Neurology 73:1509–1510

Gilbert MR, Wang M, Aldape KD, Stupp R, Hegi ME, Jaeckle KA et al (2013) Dose-dense temozolomide for newly diagnosed glioblastoma: a randomized phase III clinical trial. J Clin Oncol 31(32):4085–4091

Glas M, Happold C, Rieger J, Wiewrodt D, Bähr O, Steinbach JP et al (2009) Long-term survival of patients with glioblastoma treated with radiotherapy and lomustine plus temozolomide. J Clin Oncol 27:1257–1261

Grossman SA, Ye X, Chamberlain M, Mikkelsen T, Batchelor T, Desideri S et al (2009) Talampanel with standard radiation and temozolomide in patients with newly diagnosed glioblastoma:a multicenter phase II trial. J Clin Oncol 27:4155–4161

Gutenberg A, Bock HC, Reifenberger G, Brück W, Giese A (2013) Toxicity and survival in primary glioblastoma patients treated with concomitant plus adjuvant temozolomide versus adjuvant temozolomide: results of a single-institution, retrospective, matched-pair analysis. Acta Neurochir (Wien) 155(3):429–435

Havik AB, Brandal P, Honne H, Dahlback H-SS, Scheie D, Hektoen M et al (2012) MGMT promoter methylation in gliomasassessment by pyrosequencing and quantitative methylation-specific PCR. J Transl Med 10:36

Hegi ME, Diserens AC, Gorlia T, Hamou M-F, de Tribolet N, Weller M et al. (2005) Mgmt gene silencing and benefit from temozolomide in glioblastoma. N Engl J Med 352:997–1003

Herrlinger U, Rieger J, Koch D, Loeser S, Blaschke B, Kortmann R-D et al (2006) Phase II trial of lomustine plus temozolomide chemotherapy in addition to radiotherapy in newly diagnosed glioblastoma: UKT-03. J Clin Oncol 24:4412–4417

Herrlinger U, Schaefer N, Steinbach JP, Weyerbrock A, Hau P, Goldbrunner R et al. (2013) Bevacizumab, irinotecan, and radiotherapy versus standard temozolomide and radiotherapy in newly diagnosed, MGMT-nonmethylated glioblastoma patients: first results from the randomized multicenter GLARIUS trial. J Clin Oncol :31 (suppl; abstr LBA2000)

Imyanitov EN, Moiseyenko VM (2007) Molecular-based choice of cancer therapy: realities and expectations. Clin Chem Acta 379(1–2):1–13

Jaeckle KA, Eyre HJ, Townsend JJ, Schulman S, Knudson HM, Belanich M et al (1998) Correlation of tumor O6 methylguanine-DNA methyltransferase levels with survival of malignant astrocytoma patients treated with bis-chloroethyl nitrosourea: a southwest oncology group study. J Clin Oncol 16:3310–3315

Kamiryo T, Tada K, Shiraishi S, Shinojima N, Kochi M, Ushio Y (2004) Correlation between promoter hypermethylation of the O6-methylguanine-deoxyribonucleic acid methyltransferase gene and prognosis in patients with high-grade astrocytic tumors treated with surgery, radiotherapy, and 1-(4-amino-2-methyl-5-pyrimidinyl)methyl-3-(2-chloroethyl)-3-nitrosourea-based chemotherapy. Neurosurgery 54(2):349–357

Karayan-Tapon L, Quillien V, Guilhot J, Wager M, Fromont G, Saikali S et al (2010) Prognostic value of O6-methylguanine-DNA methyltransferase status in glioblastoma patients, assessed by five different methods. J Neurooncol 97(3):311–322

Kesari S, Schiff D, Drappatz J, LaFrankie D, Doherty L, Macklin EA et al (2009) Phase II study of protracted daily temozolomide for low-grade gliomas in adults. Clin Cancer Res 15:330–337

Kim YS, Kim SH, Cho J, Kim JW, Chang JH, Kim DS et al (2012) MGMT gene promoter methylation as a potent prognostic factor in glioblastoma treated with temozolomide-based chemoradiotherapy: a single-institution study. Int J Radiat Oncol Biol Phys 84(3):661–667

Kreth S, Thon N, Eigenbrod S, Lutz J, Ledderose C, Egensperger R et al (2011) O-methylguanine-DNA methyltransferase (MGMT) mRNA expression predicts outcome in malignant glioma independent of MGMT promoter methylation. PLoS One 6(2):e17156

Lai A, Tran A, Nghiemphu PL, Pope WB, Solis OE, Selch M et al (2011) Phase II study of bevacizumab plus temozolomide during and after radiation therapy for patients with newly diagnosed glioblastoma multiforme. J Clin Oncol 29:142–148

Lalezari S, Chou AP, Tran A, Solis OE, Khanlou N, Chen W et al (2013) Combined analysis of O6-methylguanine-DNA methyltransferase protein expression and promoter methylation provides optimized prognostication of glioblastoma outcome. Neuro Oncol 15(3):370–381

Lechapt-Zalcman E, Levallet G, Dugue AE, Vital A, Diebold M-D, Menei P et al. (2012) O6-Methylguanine-DNA Methyltransferase (MGMT) promoter methylation and low mgmt-encoded protein expression as prognostic markers in glioblastoma patients treated with biodegradable carmustine wafer implants after initial surgery followed by radiotherapy with concomitant and adjuvant temozolomide. Cancer 118:4545–4554

Liu B-L, Cheng J-X, Zhang W, Zhang X, Wang R, Lin H et al (2010) Quantitative detection of multiple gene promoter hypermethylation in tumor tissue, serum, and cerebrospinal fluid predicts prognosis of malignant gliomas. Neuro Oncol 12(6):540–548

Ma S, Egyhazi S, Ueno T, Lindholm C, Kreklau EL, Stierner U et al (2003) O6-methylguanine-DNA-methyltransferase expression and gene polymorphisms in relation to chemotherapeutic response in metastatic melanoma. Br J Cancer 89(8):1517–1523

Malmstrom A, Gronberg BH, Marosi C, Stupp R, Frappaz D, Schultz H et al (2012) Temozolomide versus standard 6-week radiotherapy versus hypofractionated radiotherapy in patients older than 60 years with glioblastoma: the Nordic randomised, phase 3 trial. Lancet Oncol 13:916–926

Metellus P, Coulibaly B, Nanni I, Fina F, Eudes N, Giorgi R et al (2009) Prognostic impact of O6-methylguanine-DNA methyltransferase silencing in patients with recurrent glioblastoma multiforme who undergo urgery and carmustine wafer implantation. Cancer 115:4783–4794

Minniti G, Amelio D, Amichetti M, Salvati M, Muni R, Bozzao A et al (2010) Patterns of failure and comparison of different target volume delineations in patients with glioblastoma treated with conformal radiotherapy plus concomitant and adjuvant temozolomide. Radiother Oncol 97(3):377–381

Minniti G, Salvati M, Arcella A, Buttarelli F, D'Elia A, Lanzetta G et al (2011) Correlation between O6-methylguanine-DNA methyltransferase and survival in elderly patients with glioblastoma treated with radiotherapy plus concomitant and adjuvant temozolomide. J Neurooncol 102:311–316

Morandi L, Franceschi E, de Biase D, Salvati M, Muni R, Bozzao A et al (2010) Promoter methylation analysis of O6-methylguanine-DNA methyltransferase in glioblastoma: etection by locked nucleic acid based quantitative PCR using an imprinted gene (SNURF) as a reference. BMC Cancer 10:48

Nagane M, Kobayashi K, Ohnishi A, Shimizu S, Shiokawa Y (2007) Prognostic significance of O6-methylguanine-DNA methyltransferase protein expression in patients with recurrent glioblastoma treated with temozolomide. Jpn J Clin Oncol 37:897–906

Nakagawa T, Ido K, Sakuma T, Takeuchi H, Sato K, Kubota T (2009) Prognostic significance of the immunohistochemical expression of O6-methylguanine-DNA methyltransferase, P-glycoprotein, and multidrug resistance protein-1 in glioblastomas. Neuropathology 29(4):379–388

Niyazi M, Schnell O, Suchorska B, Schwarz SB, Ganswind U, Geisler J et al (2012) FET-PET assessed recurrence pattern after radio-chemotherapy in newly diagnosed patients with glioblastoma is influenced by MGMT methylation status. Radiother Oncol 104:78–82

Norden AD, Lesser GJ, Drappatz J, Ligon KL, Hammond SN, Eudocia Q, Lee EQ et al (2013) Phase 2 study of dose-intense temozolomide in recurrent glioblastoma. Neuro Oncol 15(7):930–935

Ohno T, Hiraga J, Ohashi H, Sugisaki C, Li E et al (2006) Haruhiko Asano et al. Loss of O6-methylguanine-DNA methyltransferase protein expression is a favorable prognostic marker in diffuse large Bcell lymphoma. Int J Hematol 83(4):341–347

Park CK, Park SH, Lee SH, Kim C-Y, Kim D-W, Paek SH et al (2009) Methylation status of the MGMT gene promoter fails to predict the clinical outcome of glioblastoma patients treated with ACNU plus cisplatin. Neuropathology 29(4):443–449

Park CK, Lee SH, Kim TM, Choi SH, Park S-H, Heo DS et al (2013) The value of temozolomide in combination with radiotherapy during standard treatment for newly diagnosed glioblastoma. J Neurooncol 112(2):277–283

Perez-Larraya JG, Ducray F, Chinot O, Catry-Thomas I, Taillandier L, Guillamo J-S et al. (2011) Temozolomide in elderly patients with newly diagnosed glioblastoma and poor performance status: an ANOCEF phase II trial. J Clin Oncol 29:3050–3055

Pollack IF, Hamilton RL, Sobol RW, Burnham J, Yates AJ, Holmes EJ et al (2006) O6-methylguanine-DNA methyltransferase expression strongly correlates with outcome in childhood malignant gliomas: results from the CCG-945 Cohort. J Clin Oncol 24(21):3431–3437

Prados MD, Chang SM, Butowski N, DeBoer R, Parvataneni R, Carliner H et al (2009) Phase II study of erlotinib plus temozolomide during and after radiation therapy in patients with newly diagnosed glioblastoma multiforme or gliosarcoma. J Clin Oncol 27:579–584

Quillien V, Lavenu A, Karayan-Tapon L, Carpentier C, Labussie`re M, Lesimple T et al (2012) Comparative assessment of 5 methods (methylation-specific polymerase chain reaction, methylight, pyrosequencing, methylation-sensitive high-resolution melting, and immunohistochemistry) to analyze o6-methylguanine-dnamethyltranferase in a series of 100 glioblastoma patients. Cancer 118:4201–4211

Quillien V, Lavenu A, Sanson M, Legrain M, Dubus P, Karayan-Tapon L et al (2014) Outcome-based determination of optimal pyrosequencing assay for MGMT methylation detection in glioblastoma patients. J Neurooncol 116(3):487–496

Rapp M, Goeppert M, Felsberg J, Steiger HJ, Sabel M (2013) The impact of sequential vs. combined radiochemotherapy with temozolomide, resection and MGMT promoter hypermethylation on survival of patients with primary glioblastoma- a single centre retrospective study. Br J Neurosurg 27(4):430–435

Reifenberger G, Hentschel B, Felsberg J, Schackert G, Simon M, Schnell O et al (2012) Predictive impact of MGMT promoter methylation in glioblastoma of the elderly. Int J Cancer 131:1342–1350

Rivera AL, Pelloski CE, Gilbert MR, Colman H, De La Cruz C, Sulman EP et al (2010) MGMT promoter methylation is predictive of response to radiotherapy and prognostic in the absence of adjuvant alkylating chemotherapy for glioblastoma. Neuro Oncol 12(2):116–121

Sadones J, Michotte A, Veld P, Chaskis C, Sciot R, Menten J et al (2009) MGMT promoter hypermethylation correlates with a survival benefit from temozolomide in patients with recurrent anaplastic astrocytoma but not glioblastoma. Eur J Cancer 45:146–153

Salvati M, Pichierri A, Piccirilli M, Brunetto GMF, D'elia A, Artizzu S et al (2012) Extent of tumor removal and molecular markers in cerebral glioblastoma: a combined prognostic factors study in a surgical series of 105 patients. J Neurosurg 117:204–211

Shah N, Lin B, Sibenaller Z, Ryken T, Lee H, Yoon J-G et al. (2011) Comprehensive analysis of MGMT promoter methylation: correlation with MGMT expression and clinical response in GBM. PLoS One 6:E.16146

Smith KA, Ashby LS, Gonzalez F, Brachman DG, Thomas T, Coons SW et al (2008) Prospective trial of gross-total resection with Gliadel wafers followed by early postoperative gamma knife radiosurgery and conformal fractionated radiotherapy as the initial treatment for patients with radiographically suspected, newly diagnosed glioblastoma multiforme. J Neurosurg 109:106–117

Sonoda Y, Yokosawa M, Saito R, Kanamori M, Yamashita Y, Kumabe T et al (2010) O(6)-Methylguanine DNA methyltransferase determined by promoter hypermethylation and immunohistochemical expression is correlated with progression-free survival in patients with glioblastoma. Int J Clin Oncol 15:352–358

Stummer W, Meinel T, Ewelt C, Martus P, Jakobs O, Felsberg J et al (2012) Prospective cohort study of radiotherapy with concomitant and adjuvant temozolomide chemotherapy for glioblastoma patients with no or minimal residual enhancing tumor load after surgery. J Neurooncol 108:89–97

Stupp R, Hegi ME, Mason WP, van den Bent MJ, Taphoorn MJB, Janzer RC et al (2009) Effects of radiotherapy with concomitant and adjuvant temozolomide versus radiotherapy alone on survival in glioblastoma in a randomized phase III study: 5-year analysis of the EORTC-NCIC trial. Lancet Oncol 10:459–466

Stupp R, Hegi ME, Neyns B, Goldbrunner R, Schlegel U, Clement PMJ et al (2010) Phase I/IIa study of cilengitide and temozolomide with concomitant radiotherapy followed by cilengitide and temozolomide maintenance therapy in patients with newly diagnosed glioblastoma. J Clin Oncol 28(16):2712–2718

Sunwoo L, Choi SH, Park CK, Kim JW, Yi KS, Lee WJ et al. (2013) Correlation of apparent diffusion coefficient values measured by diffusion MRI and MGMT promoter methylation semiquantitatively analyzed with MS-MLPA in patients with glioblastoma multiforme. J Magn Reson Imaging 37:351–358

Tanaka S, Kobayashi I, Utsuki S, Oka H, Yasui Y, Fujii K (2005) Down-regulation of O6-methylguanine-DNA methyltransferase gene expression in gliomas by platinum compounds. Oncol Rep 14(5):1275–1280

Tanaka S, Akimoto J, Kobayashi I, Oka H, Ujiie H (2008) Individual adjuvant therapy for malignant gliomas based on O6- methylguanine-DNA methyltransferase messenger RNA quantitation by real-time reverse-transcription polymerase chain-reaction. Oncol Rep 20:165–171

Thon N, Eigenbrod S, Grasbon-Frodl EM, Lutz J, Kreth S, Popper G et al (2011) Predominant influence of MGMT methylation in non-resectable glioblastoma after radiotherapy plus temozolomide. J Neurol Neurosurg Psychiatry 82:441–446

Umer M, Herceg Z (2013) Deciphering the epigenetic code: an overview of DNA methylation analysis methods. Antioxid Redox Signal 18(15):1972–1986

Uno M, Oba-Shinjo SM, Camargo AA, Moura RP, de Aguiar PH, Cabrera HN et al (2011) Correlation of MGMT promoter methylation status with gene and protein expression levels in glioblastoma. Clinics (Sao Paulo) 66(10):1747–1755

van den Bent MJ, Dubbink HJ, Sanson M, van der Lee-Haarloo CR, Hegi M, Jeuken JW et al (2009) MGMT promoter methylation is prognostic but not predictive for outcome to adjuvant PCV chemotherapy in anaplastic oligodendroglial tumors: a report from EORTC brain tumor group study 26951. J Clin Oncol 27(35):5881–5886

Watanabe T, Katayama Y, Komine C, Yoshino A, Ogino A, Ohta T et al (2005) O6-methylguanine-DNA methyltransferase methylation and TP53 mutation in malignant astrocytomas and their relationships with clinical course. Int J Cancer 113(4):581–587

Watanabe R, Nakasu Y, Tashiro H, Mitsuya K, Ito I, Nakasu S et al (2011) O6-Methylguanine DNA methyltransferase expression in tumor cells predicts outcome of radiotherapy plus concomitant and adjuvant temozolomide therapy in patients with primary glioblastoma. Brain Tumor Pathol 28:127–135

Weiler M, Hartmann C, Wiewrodt D, Herrlinger U, Gorllia T, Bahr O et al (2010) Chemoradiotherapy of newly diagnosed glioblastoma with intensified temozolomide. Int J Radiat Oncol Biol Phys 77(3):670–676

Weller M, Felsberg J, Hartmann C, Berger H, Steinbach JP, Schramm J et al (2009) Molecular predictors of progression-free and overall survival in patients with newly diagnosed glioblastoma: a prospective translational study of the german glioma network. J Clin Oncol 27(34):5743–5750

Wick A, Felsberg J, Steinbach JP, Herrlinger U, Platten M, Blaschke B et al (2007) Efficacy and tolerability of temozolomide in an alternating weekly regimen in patients with recurrent glioma. J Clin Oncol 25(22):3357–3361

Wick W, Platten M, Meisner C, Felsberg J, Tabatabai G, Simon M et al (2012) NOA-08 study group of Neuro-oncology Working Group (NOA) of German cancer society. Temozolomide chemotherapy alone versus radiotherapy alone for malignant astrocytoma in the elderly: the NOA-08 randomised, phase 3 trial. Lancet Oncol 13(7):707–715

You JS, Jones PA (2012) Cancer genetics and epigenetics: two sides of the same coin? Cancer Cell 22(1):9–20

Zawlik I, Vaccarella S, Kita D, Mittelbronn M, Franceschi S, Ohgaki H (2009) Promoter methylation and polymorphisms of the MGMT gene in glioblastomas: a population-based study. Neuroepidemiology 32(1):21–29

Zunarelli E, Bigiani N, Sartori G, Migaldi M, Sgambato A, Maiorana A (2011) INI1 immunohistochemical expression in glioblastoma: correlation with MGMT gene promoter methylation status and patient survival. Pathology 43(1):17–23

Chapter 10
Epigenetics and Three Main Clinical Aspects of Breast Cancer Management

Pantea Izadi and Mehrdad Noruzinia

Contents

Abstract Breast cancer as a malignant disease is a common cancer in women worldwide. As any other malignancies there are three main aspects in the management of breast cancer: diagnosis (early detection is crucial), tumor classification/prognosis and treatment. This chapter focuses on the practical roles of epigenetic alterations (mainly DNA methylation) in these three clinical problems of breast cancer. DNA methylation signatures especially in cell free DNA in plasma or

---

P. Izadi (✉)
Department of Medical Genetics, Medical School, Tehran University of Medical Sciences, Tehran, Iran
e-mail: p-izadi@farabi.tums.ac.ir

M. Noruzinia
Department of Medical Genetics, School of Medical Sciences, Tarbiat Modares University, Tehran, Iran

P. Mehdipour (ed.), *Epigenetics Territory and Cancer*,
DOI 10.1007/978-94-017-9639-2_10

serum which originated from tumor cells, are promising tools for diagnosis and early detection of breast cancer. Also DNA methylation patterns in lymphocytes are a recent approach for breast cancer diagnosis. Epigenetic signatures in tumor tissues can classify tumors precisely and may provide new classification beyond conventional histopathological classifications. As epigenetic alterations such as DNA methylation and histone deacetylation are reversible, they are appropriate targets for epidrugs (DNMT inhibitors and HDACis) in breast cancer treatment. Finally micro RNAs as another epigenetic player in carcinogenesis will have a prominent role in the different clinical aspects of breast cancer diagnosis and prognostication.

Keywords Breast cancer · Epigenetic · DNA methylation

Abbreviations

APC	Adenomatous Polyposis Coli
5-aza	5-azacytidine
BRCA1 gene	Breast cancer gene 1, early onset
B-CIMP	Breast cancer CpG island methylator phenotype
cf DNA	Cell free DNA
DCIS	Ductal carcinoma in situ
DFS	Disease free survival
DNMTs	DNA methyl transferases
ECM	Extracellular matrix
ERα	Estrogen receptor α
EGCG	Epigallocatechin-3-gallate
GE	Genistein
HDAC	Histone deacetylases
HDACi	Histone deacetylase inhibitors
HER2	Human epidermal growth factor receptor 2
HME-1	Human mammary epithelial-specific marker
IDC	Invasive ductal carcinoma
LABC	Locally advanced breast cancer
MGMT	O_6-methylguanine-DNA methyltransferase
PRβ	Progesterone receptor β
PB	Peripheral blood
RAR-β	Retinoic acid receptor-β
RASSF1A	Ras-association domain family 1
TSA	Trichostatin A
WBC	White blood cells

10.1 Introduction

Breast cancer is the most common cancer and a leading cause of morbidity and mortality in women worldwide (www.cancer.org). Over the last few decades, the incidence rates of breast tumor have increased significantly, with more than one million new patients every year (Jemal et al. 2011). Detection of breast cancer at the early stage lead to a high survival rate (~98 %), while diagnosis at the advanced stage bring about a considerably lower survival rate (~27 %) (Radpour et al. 2011).

The molecular mechanisms involved in the development and progression process of breast tumors remain unclear. It is obvious that the accumulation of pathologically altered gene function is crucial for the triggering of both the induction of breast cancer and its transition towards the distinct breast cancer subtypes. Like in other cancers, almost any critical network, such as cell cycle checkpoints, detoxifying pathways, DNA repair, apoptosis, inflammation, cellular contact or migration is influenced by the increasing number of deregulated genes.

By the somatic mutation theory, cancer has long been considered by scientists as a genetic disorder of fatal acquisition of various mutations in important genes, which organize these vital networks. Such mutations can either lead to deactivation of tumor suppressor genes (e.g. TP53, BRCA1) or activation of proto-oncogenes (e.g. MYC); both of them provide the malignant condition of a transformed cell.

During the few last decades, the somatic mutation theory of cancer has been evolved as it became obvious that epigenetic modifications play a key role as equally pivotal as genetics in cancer progression. Although our findings on epigenetically modified genes in tumor is perpetually increasing, the fundamental mechanisms essential in both aberrant DNA methylation and the selection of genes that become methylated are merely partially understood (Veeck and Esteller 2010). In recent years, two epigenetic mechanisms including the DNA methylation and chemical histone tail modification have been shown as the most important players of transcriptional regulation. The detection of microRNAs, a class of tiny noncoding RNAs, accomplished as a third epigenetic mechanism. MiRNAs are 20–30 nucleotides in length with regulatory roles, that completely bind to the 3′untranslated regions of mRNAs, leading to its degradation or prohibition of mRNA translation (He and Hannon 2004). The function of the target mRNA determine the activity of miRNA either as a tumor suppressor (if perform opposed to the proto-oncogene transcripts) or oncogenic (if perform opposed to the tumor suppressor gene transcripts). The depletion of the let-7 family (containing at least 11 homologous miRNAs) as a significant members of miRNAs in breast, lung and colon cancer result in the enhanced tumorigenicity (Akao et al. 2006; Johnson et al. 2005). Other instance is miR-21, whose over expression in breast cancer leads to increased invasion and induces metastasis of the lung tumor (Iorio 2005; Zhu et al. 2008). The number of genes recognized to be mediated by miRNAs is rising quickly. The human miRNA disease database (**HMDD**) elucidate more than 617 miRNA genes, 438 diseases (Last update: Mar. 13, 2013) in the latest release, however, a large number of miRNAs are estimated to be described in the future.

In this chapter we will discuss about the role of epigenetic modifications (mainly focused on DNA methylation) in the diagnosis, classification & prognosis and treatment of the breast tumors.

10.2 Epigenetics Application in Breast Cancer Diagnosis

The crucial point to successful treatment and outcome of patients is the diagnosis of the cancer at an early stage (Hayes et al. 2001). The high mortality in cancer occurs with late onset of symptoms, the poor accessibility of the malignancy and the unpredictable course of the disease (Laird 2003). The physical examination, mammography and aspiration cytology are the common triple test for the diagnosis of breast cancer (Radpour et al. 2011). The improvement of efficient strategies in early detection and diagnosis of metastatic or recurrent conditions in preclinical or pre-symptomatic phases of the disease is beneficial. The investigation for more sensitive and specific prognostic markers, indicating the presence of tumor specific alterations in the blood specimens, is still in progress (Van_De_Voorde et al. 2012). Several findings have indicated that patients suffering from cancer have more than 90 % of the total free circulating DNA derived from tumor tissue (Hanash et al. 2011; Ignatiadis and Reinholz 2011; Kohler et al. 2011; Sharma et al. 2011).

Epigenetic silencing of tumor-associated genes owing to aberrant methylation of CpG islands plays a crucial role in breast tumorigenesis (Müller et al. 2003). DNA methylation is an early prevalent phenomenon in tumorigenesis and can be a pivotal predictor of cancer risk. The studies indicated that these epigenetic modifications is engaged in the whole procedure of oncogenic transformation (Jones and Baylin 2002).

Similarities of DNA methylation patterns in primary tumor tissue samples and in plasma show the potential application of a blood-based molecular characterization for breast cancer (Fabian et al. 2005; Jones and Liang 2009; Radpour et al. 2011; Wong et al. 2010; Yan et al. 2006; Zhu and Yao 2009).

10.2.1 Methylation in Serum DNA as a Biomarker for Breast Cancer Diagnosis

The presence of high concentration of cell free DNA (cf DNA) in serum/plasma of patients with breast cancer demonstrated that plasma DNA might be an appropriate target for the development of non-invasive diagnosis, prognosis and follow-up laboratory tests of cancer (Anker et al. 1999). Serum and plasma are more easily available samples and the collection of a specimen is patient-friendly and does not need a particular expertise (Dulaimi 2004; Qureshi et al. 2010). The reanalysis of data can be done at any time during the follow-up care planning. Based on the source of cf DNA, it is demonstrated that serum has a tendency to contain more DNA than plasma. There is some data that DNA is freed from the tumor as glyconucleoprotein structures that may keep it safe from degradation by nucleases (Hoque et al. 2009). Nevertheless partially contamination of serum by DNA released from leukocytes cannot be neglected (Radpour et al. 2011).

Tumor-specific DNA methylation patterns can be investigated in detached cancer cells in body fluid samples and biopsies, and they can be found in circulating DNA that is free from dead tumor cells (Sidransky 2002). DNA is a resistant and powerful analyte that can be amplified by PCR (Sidransky 1997). Even though, there are limited studies which investigated the DNA methylation patterns in tumor tissues and serum DNA in early and late stages of breast tumorigenesis.

Sensitive analysis methods are recommended for the detection of limited amounts of DNA released in breast fluids, plasma and serum (Suijkerbuijk et al. 2010). Some studies show that serum have a tendency to contain more DNA than plasma (Radpour et al. 2011). Despite the fact that sensitivity and specificity influence the efficiency of a biomarker analysis, there is no constant application of these bio statistical terms in the literature. The term of analytical sensitivity is categorized as an absolute or as a relative sensitivity, the first associated to the minimum level of the identification of pure, methylated target DNA, the latter refers to the minimal proportion of methylated DNA that the test could distinguish in the existence of an extra unmethylated DNA (Laird 2003). Nevertheless, the clinical sensitivity of a biomarker is influenced by the prevalence of marker in the tumor, the efficacy of transfer of the marker to the distant media being evaluated, and the analytical sensitivity of the investigation (Pepe et al. 2001).

Epigenetic markers of breast cancer can be detected using candidate gene approach or genome-wide identification studies.Recently, the Epigenomics Roadmap Program has been organized to provide epigenome reference maps (DNA methylation pattern and histone modification profile) of the breast tumor. Numerous important breast tumorigenesis-associated genes are detected as targets for epigenetic modifications. These genes control most of the cellular activity, including cell cycle, DNA repair, cell adherence and invasion. We explain below a number of significant DNA methylation biomarkers in serum/plasma of patients with breast cancer.

10.2.1.1 BRCA1

BRCA1 gene (breast cancer gene 1, early onset) is located on the q arm of chromosome 17 in human genome. This gene codes a crucial factor for the cellular DNA repair system and generates a protein involved in the cell cycle check point regulation, DNA repair, chromatin remodeling and protein ubiquitinylation. It is well known that germ line mutations in BRCA1 gene can increase the risk of breast and ovarian cancer (hereditary form). In addition, it seems that this tumor suppressor gene takes part in the sporadic forms of breast cancer. Hypermethylation of the CpG island promoter regions of BRCA1 gene is one of the important mechanisms for its functional deactivation, in addition most of these tumors accompany with the hypermethylatin of estrogen receptor α (ERα) and progesterone receptor β (PRβ) promoters, related to a more aggressive types of breast tumors (Mirza et al. 2007).

An investigation on 255 women with early onset breast cancer (under the age 40 years), showed that hypermethylation of BRCA1 promoter region of peripheral blood DNA was related to a 3.5-fold increased risk of early-onset of breast tumor (Wong et al. 2010). Also other researchers demonstrated a considerable difference between BRCA1 promoter methylation pattern in serum of sporadic breast cancer patients in comparison with the low BMI postmenopausal control women older than age 70 (Bosviel et al. 2012). In addition to breast cancer, aberrant CpG island methylation of BRCA1 gene has been observed in lung and ovarian cancers (Lee et al. 2007).

10.2.1.2 RASSF1A

The promoter region of the Ras-association domain family 1 (RASSF1 A) gene is one of the most common methylated statuses in primary breast cancer patients. Methylation of RASSF1 A results in the aggregation of Cyclin D1 and altering of the cell cycle control under the additional cell cycle pressure conditions. Hypermethylation of promoter RASSF1 A gene is not evident in serum DNA from control subjects and patients with inflammatory breast disease in comparison to breast cancer patients (Dulaimi 2004; Hoque et al. 2009; Parrella 2010; Shukla et al. 2006).

Based on recent studies, RASSF1 A is considered as an ideal surrogate marker for breast cancer in peripheral blood. First, we make assumption of its high specificity in breast tumor due to its rarely methylated state in normal tissue. Furthermore, it provides an ideal detection coverage, because of its high frequency of methylation in breast tumors (Dulaimi 2004; Hoque et al. 2009; Parrella 2010; Shukla et al. 2006).

Methylation of RASSF1 A promoter region has been identified in ovarian, bladder, lung and prostate cancers as well as breast cancer (Amin and Banerjee 2012; Dworkin et al. 2009; Ha et al. 2012; Kim et al. 2012)

10.2.1.3 APC

Adenomatous Polyposis Coli (APC) gene is a relatively large gene on the long arm of chromosome 5 in human (5q21-q22). It is well known as a gene which is involved in causing both sporadic and familial forms of colorectal cancer. APC is a tumor suppressor gene which its product acts as an antagonist of the Wnt signaling pathway. The protein encoded by the APC gene responsible for the pathways blocking metastasis: regulation of intercellular adhesion, controlling the cell cycle, and apoptosis. In addition to considerable difference in methylation pattern of APC between breast cancer patients and healthy controls, methylated APC is also remarkably different in metastatic and non-metastatic cancer cells (Matuschek et al. 2010). APC aberrant methylation has been identified in colorectal, lung and stomach cancers as well as breast tumors (Esteller et al. 2000; Virmani et al. 2001).

10.2.1.4 RAR-β

Retinoic acid receptor-β (RAR-β) is located on the 3p24.2 in human genome. This gene encodes receptor beta for retinoic acid (derived from vitamin A) which is a member of the thyroid-steroid hormone receptor superfamily of nuclear transcriptional regulators. This receptor binds retinoic acid and mediates cellular signaling in cell growth and differentiation.

Methylation of retinoic acid receptor-β (RAR-β) tumor suppressor gene in the promoter and first exon lead to decreased expression of protein and lack of response to retinoids. Methylation of this gene is an early and approximately common phenomena in benign intraductal papillomas and ductal carcinoma in situ (DCIS) (Widschwendter et al. 2000). Furthermore, the significantly higher level of methylation is evident in serum DNA of DCIS or invasive ductal carcinoma (IDC) than that in normal tissue (Kim et al. 2010). Aberrant methylation of CpG islands associated with decreased expression of RAR-β proteins may serve as a significant biomarker to indicate the risk of metastasis in breast tumor. In addition to breast tumors, RAR-β methylation has been identified in glioblastoma and lung cancer (Piperi et al. 2010; Su et al. 2012).

10.2.1.5 MGMT

O_6-methylguanine-DNA methyltransferase (MGMT) is located on the 10q26 in human genome. This gene encodes a protein which can eliminate mutagenic and cytotoxic adducts in O6 position of guanine base in DNA (Jacinto and Esteller 2007). DNA alkylation at the O6 position of guanine is a significant stage in inducing mutations in cancer, mainly in behalf of the tendency of O6 methylguanine to pair with thymine in replication, leading to the transition of guanine-cytosine to adenine-thymine base pairs in DNA. Furthermore, MGMT protects cells against the recombinogenic and cytotoxic lesions of the O6-methylguanine by transferring the alkyl group to a cysteine in its active site. Thus MGMT can directly reverse the damaged guanine base in DNA. Aberrant CpG islands methylation of MGMT mediated gene silencing is related to the loss of its protein expression. In a prospective study of 100 breast cancer patients with IDC (invasive ductal carcinoma), a significant correlation between MGMT methylation patterns in tumor tissue and serum DNA was detected (Sharma et al. 2010). Besides the breast cancer, MGMT methylation has been identified in gliomas (Hatziapostolou and Iliopoulos 2011).

10.2.1.6 14-3-3-sigma (stratifin)

14-3-3-sigma gene is located on 1p36.11 in human genome. The official name of this gene is stratifin (SFN).The name 14-3-3 for the protein refers to its elution and migration pattern on chromatography and electrophoresis. 14-3-3-sigma protein is a member of a group of 14-3-3 proteins. The 14-3-3 proteins are regulatory molecules involved in signal transduction, G2/M checkpoint and programmed cell death (apoptosis). Cdc2-cyclin B1 complex is segregated by in the cytoplasm, leading to G2 arrest (Yang et al. 2001b).

14-3-3-sigma is one of the most common and early hypermethylated genes in the development from normal cell to atypical ductal hyperplasia, to ductal carcinoma in situ (DCIS) and lastly to invasive breast carcinoma.

In many breast tumors, 14-3-3-sigma or human mammary epithelial-specific marker (HME-1) is deactivated by aberrant DNA methylation even in histological normal epithelial cell near the tumor area. Due to the contamination of hypermethylated alleles in the stroma and peripheral white blood samples, the application of this tumor marker is compromised partly (Bojesen et al. 2013; Pu et al. 2003; Umbricht et al. 2001). In addition to breast cancer, abberent methylation of 14-3-3-sigma has been identified in gastrointestinal tumors (Dim et al. 2011; Okumura et al. 2010; Pei et al. 2010).

10.2.1.7 E-cadherin

E-cadherin gene (CDH1) gene is located on chromosome 16 (16q22.1) in human genome and is a classical member from the cadherin superfamily. E-cadherin gene product is a glycoprotein which works as a calcium dependent cell-cell adhesion molecule.

E-cadherin gene methylation exerts a pivotal role in approximately 50 % of primary breast tumors, with aberrant methylation of the E-cadherin CpG island region in around 30 % of DCIS (Hoque et al. 2009; Yang et al. 2001b). Since this methylation is frequently observed in premalignant lesions like atypical ductal hyperplasia, it might be one of the prospective biomarkers for early diagnosis (Nass et al. 1999). Hypermethylation of the E-cadherin 5′ CpG island is shown in breast cancer cells and is said to be an important step in invasiveness and metastasis of poorly differentiated cancers (Graff et al. 1995; Hoque et al. 2009; Nass et al. 1999). Furthermore, the E-cadherin gene is rarely methylated in normal breast cells (Graff et al. 1995). Nass et al. (1999) indicated that co-occurring aberrant CpG island methylation of ER and E-cadherin gene occurs through tumor progression. CpG island methylator phenotype testing in serum of patients with sporadic breast cancer indicated a linked pattern of methylation for CDH1, RASSF1 A, BRCA1, and RAR-β2 as breast cancer markers. These methylation patterns provide a biomarker panel with high sensitivity and specificity value of 90 and 88 %, respectively (Jing et al. 2010). In addition to breast cancer, aberrant methylation of E-cadherin has been discovered in gastric cancer.

10.2.1.8 Slit

Three members (Slit1, Slit2 and Slit3) of Slit gene family encode both extracellular matrix (ECM) proteins and plasma membrane related glycoproteins that control axon growth, guidance, branching, and neuronal extension during brain development (Brose et al. 1999). In addition to its involvement in the nervous system, Slit2 is also transcribed in non-neuronal tissues like mammary tissue. Some investigations found that Slit2 seems to play a role as a novel tumor suppressor gene by inhibiting Sdfl/Cxcr4 in breast epithelial cells (Marlow et al 2008). Based on in vitro experiments, over expression of Slit2 gene lead to inhibition of breast cancer cell growth (Dallol et al. 2002; Dickinson et al. 2004). Since aberrant CpG island methylation of Slit2 gene has been shown in serum specimen of breast cancer patients, it is a good marker for epigenetic-based molecular detection in both DCIS and invasive breast cancer (Dallol et al. 2002; Kim et al. 2011). In addition to breast cancer, aberrant CpG island methylation of Slit2 gene has been demonstrated in ovarian, prostate and liver cancers (Bartholow et al. 2011; Dickinson et al. 2011; Jin et al. 2009).

10.2.1.9 Estrogen Receptor Alpha (ERα)

Three significant receptors present in breast tumor cells including estrogen receptor (ER), PR, and HER2/neu. Some investigations indicated that ERα has a more crucial function than ERβ in the proliferation and progression of breast tumor (Noruzinia et al. 2005). The first one is encoded by ESR1 gene and induces cell proliferation. ERβ is generated by ESR2 gene and plays a role in the development of tamoxifen resistance (Pujol et al. 2004). Aberrant CpG islands methylation of the ERα gene is concerned with the silencing of ER gene expression in breast cancers (ER negative status) (Izadi et al. 2012b).

Quantitative methylation analysis of ESR1 and 14-3-3-sigma in serum showed a significant difference between methylation pattern of these two genes in breast cancer patients and healthy women with sensitivity and specificity values of 81 and 88 %, respectively ($p < 0.0001$). Nevertheless, the specificity and sensitivity values acquired in this investigation were considered insufficient for their application in future screening studies (Martínez-Galán et al. 2008). In addition to breast cancer, ERα methylation has been identified in melanoma, lung and prostate cancers (Mori et al. 2006).

10.2.2 Breast Cancer Diagnosis Based on the Aberrant DNA Methylation in Peripheral Blood Lymphocytes Genome

The mechanism of gene inactivation through aberrant DNA methylation and its implication to cancer pathogenesis is well understood, with silencing of transcription in tumor suppressor genes, considered to be one of the most significant drivers of tumorigenesis. In the recent years, much attention is focused on the phenomenon of aberrant methylation of disease-associated genes in peripheral blood (PB) DNA and its engagement in the pathology of cancer (Iwamoto et al. 2011). The origins of this phenomenon are not understood. Nevertheless, it can be speculated that hypermethylation of genes in PB DNA specimens may be as a result of methylation alterations passed through germ line or somatic deviations that occurred before birth or during lifetime or under specific environmental exposures. Transmission of methylation alteration through germ line is still a dubious conception. Few studies show that methylation of particular genes e.g. MLH1 in some cases can be transmitted through germ line in non-Mendelian inheritance patterns (Hitchins et al. 2007). Two investigations indicated that paternal diet can have an effect on the promoter methylation status of the progeny (Carone et al. 2010; Ng et al. 2010). This further approves the importance of germ line transmission of methylation changes, while, these results have to be widely investigated in the future. Since the environmental conditions pressure on the promoter methylation status, the effect of various chemical substances on the methylation pattern of the somatic cell has been shown in the exposed animal models, and confirmed to be especially deleterious when the exposure happen in the stages of early growth (Jirtle and Skinner 2007). In humans there is consistent epidemiological evidence that chemical exposure can induce adult onset disorders. Yet, the mechanism of interactions between each organism and the environmental factors and their amount of influence on methylation alteration is not fully understood.

Overlooking the origin, the differences in intra individual methylation pattern of PB DNA specimens are being increasingly demonstrated in the investigations (Widschwendter et al. 2008). In addition, these alterations have been recommended to be a part of a disease predisposition mechanism, which could be based on the theory of constitutional methylation. At the beginning, constitutional gene methylation was determined as abnormal gene methylation evident in all body tissues (Dobrovic and Kristensen 2009). Constitutional methylation is probably influencing genes in a mono allelic inheritance pattern and if occurred during growth, it can be passed on through to all tissues in mosaic forms (Wong et al. 2010). Similar to somatic cell methylation in tumor, constitutional mono allelic methylation alterations are likely to put the carrying individual prone to neoplastic disorders development. Based on the Knudson's hypothesis of tumor suppressor inactivation, only one extra hit would be needed to silence the expression of the constitutionally mono allelic methylated gene and begin or cooperate to tumorigenesis. Additionally, allelic insufficiency might also be a disease-trigger factor.

Changes in DNA methylation status patterns, both at particular loci and in the whole genome, have been concerned with many various health outcomes. In cancer and other disorders, most of these alterations have been seen at somatic level. Findings on whether DNA methylation alterations in white blood cells (WBC) can be considered as a beneficial biomarker for various health outcomes are very restricted, but quickly emerging (Table 10.1). Evident proof for WBC methylation at particular loci and the risk of disease is more restricted, but important. Differences in WBC DNA methylation by particular risk factors such as demographic (age, gender, race), environmental factor (benzene, continuous organic pollutants, lead, arsenic, and air contaminations), and other risk exposure (cigarette smoke, alcohol drinking, body size, physical activity and diet) have been reported in epidemiologic investigations though the profiles are not consistent (Terry et al. 2011).

Table 10.1 Summary of DNA methylation studies on peripheral blood samples in breast cancer

Study	Year	Study population	Method	Main Result
1-(Snell et al. 2008)	2008	Familial breast cancer without BRCA1or BRCA2 mutations	MethyLight, MS-HRM (Methylation specific high resolution melting)	Hypermethylation of BRCA1 in tumors, low-level promoter methylation of BRCA1 in WBC
2-(Widschwendter et al. 2008)	2008	Post menopausal women	Methyl light assay ER-alpha target (ERT) genes, and polycomb group target (PCGT) genes	Factors like estrogens leave an imprint in the DNA of cells that are unrelated to the target organ and indicate the predisposition to develop a cancer
3-(Choi et al. 2009)	2009	Early stages breast cancer patients	Tandem mass spectrometry & quantitative bisulfite pyrosequencing	WBC DNA hypomethylation is independently associated with development of breast cancer
4-(Flanagan et al. 2009)	2009	Bilateral breast cancer	Enzyme based enrichment and microarray	ATM gene body hyper methylation (intragenic repetitive element) in cases versus controls
5-(Cho et al. 2010)	2010	Breast cancer patients	Methyl light (RASSF1A, APC, HIN1, BRCA1, CYCLIND2, RARbeta, CDH1 and TWIST1) and three repetitive elements (LINE1, Sat2 and Alu)	Significant correlations in the methylation of Sat2M1 between tumors, adjacent tissues and WBC DNA. A significant difference in methylation of Sat2M1 between cases and controls
6-(Wong et al. 2010)	2011	Breast cancer patients (<40 years) with no BRCA1 germ line mutations	Methylight & MS-HRM "for BRCA1	Peripheral blood methylation was associated with a 3.5-fold increased risk of having early onset breast cancer

Table 10.1 (continued)

Study	Year	Study population	Method	Main Result
7-(Wojdacz et al. 2011)	2011	Breast tumors with paired patients bloods	MS-HRM (RASSF1,APC, BRCA1)	Methylation of these genes in tumor and WB genome are not dependent
8-(Wu et al. 2011)	2011	Girls with family history of breast cancer/ girls without family history	MethyLight & pyrosequencing (LINE1, Sat2 and Alu)	Global WBC DNA was associated with family history of breast cancer
9-(Brennan et al. 2012)	2012	667 case769 control from three large cohort	Methyl light assay	WBC DNA methylation levels at ATM could be a marker for breast cancer risk
10-(Wu et al. 2012)	2012	Breast cancer patients &their unaffected sisters	Methylight & pyrosequencing (LINE1,sat2,Alu)	No association between breast cancer and LINE-1 and Alu methylation. Sat2 methylation was statistically significantly associated with breast cancer risk
11-(Xu et al. 2012)	2012	1055 Breast cancer patients & 1101 control	LUMA	Global promoter hypermethylation in WBC was associated with breast cancer risk
12-(Hansmann et al. 2012)	2012	Familial breast cancer without BRCA1or BRCA2 mutations	Bisulfite pyrosequencing	Constitutive epimutations in BRCA1 and RAD51C are relevant to ovarian and breast cancer pathogenesis
13-(Heyn et al. 2013)	2013	Monozygotic twin pairs discordant for breast cancer	High-resolution (450 K) DNA array	Hypermethylation of DOK7 occurs years before tumor diagnosis (a powerful epigenetic blood-based biomarker)

10.3 DNA Methylation Pattern for Breast Cancer Classification and Prognosis

Breast cancer is a heterogeneous disorder with many various outcomes and responses to therapy. It has basically been categorized by histopathological characteristics which are based on tumor size, number of involved lymph nodes and distant metastasis (TNM staging), and by immunochemical recognition of cell surface receptors, including estrogen receptor (ER), the progesterone receptor (PR) and the human epidermal growth factor receptor 2 (HER2). Yet, in many patients, staging breast tumors cannot predict prognosis or therapeutic outcome due to the heterogeneity of the disorder. In recent years, molecular procedures concentrating on the profiling of gene expression have been utilized for breast tumors. Alterations in gene expression pattern that bring up the program of a cell from a normal state to a malignant condition, include multiple genetic circuitries and creating a profile of gene expression that settle the cell's crystal clear identity. Such signatures have been reported to categorize subtypes of breast tumors. Categorizations according to the gene expression profiles have extended the exact classification of breast tumor by emerging the identity of cells, with certain persistence on existence of stem cells and the essence of the immune response to the tissue tumor. These current molecular-based categorizations are named 'intrinsic subtypes of breast tumors' since they define the molecular profile of the breast cancer cell instead of its stage. Many clear tumor types and normal breast-like intrinsic classification (including luminal A and B, Her-2, basal-like and normal breast like) of breast tumors were explained before (Perou et al. 2000).

These various subtypes are identified in all breast tumor stages, even in the initial steps, and utilize as predictor of initial prognosis and therapeutic responds. They have provided prognostic value in management of breast tumors and they are considered now as a guide in prediction of patient recurrence, survival and response to chemotherapy. Yet, there are already main challenges in exact initial prediction of breast tumor, prognostication, and therapeutic response prediction. There is critical space for developing our predictive and prognostic means, certainly in guidance of therapeutic choices.

Since DNA methylation is pivotal in programming gene expression, an alteration in methylation from a normal to diseased condition should be reflected in a DNA methylation pattern that includes multiple gene pathways. Recent investigations recommended that DNA methylation signatures will extend our ability to categorize breast cancer and predict outcome beyond what is now possible. DNA methylation is a robust biomarker, comprehensively more steady than RNA or proteins (which are needed in gene expression profiling in different levels from RNA to protein), and is so a promising target for the improvement of current methods for identification and prognosis of breast tumor and other disorders. There are two approaches in the investigations of DNA methylation markers for breast tumors classification and prognosis: candidate gene methylation studies and genomic approaches. Some researchers targeted candidate genes methylation in breast tumors and their association with different clinical outcomes such as response to therapy and disease free survival (DFS). In this approach candidate genes selection was based on our previous knowledge of candidate genes functions as tumor suppressor genes or genes with known prognostic values. While in genome wide methylation studies, researchers are seeking DNA methylation signatures which are associated with clinical outcomes and survival.

10.3.1 Candidate Gene Approaches

The basic concept driving study of DNA methylation changes in diseased conditions was that restricted sets of candidate genes were pivotal in initiation and progression of disease. Nevertheless, unbiased approaches could probably detect new genes and novel functional gene pathways that are concerned with a disease, while candidate approaches basically confirmed genes that are previously indicated to be involved. Initial investigations examined the association between aberrant methylation of particular CpGs islands in tumor suppressor genes and different stages of breast cancer (Dickinson et al. 2004). Some researchers are investigating the association between the methylation of genes with known prognostic values in breast cancer and different outcomes. For instance, we could indicate that hypermethylation of ERα is associated with poor prognosis subtypes breast tumors such as Her2 + and basal-like (Izadi et al. 2012a).

In other study methylation-specific PCR (MS-PCR) of six tumor suppressor genes was performed to provide a methylation signature of primary breast tumors, and the methylation status of various genes were shown to be significantly correlated with several prognostic factors (Shinozaki et al. 2005). However, our present information of the functional pathways involved in physiological and pathological processes recommend that it is highly improbable that investigation of a few particular CpG islands status will be adequate for staging and provide exact information about outcome of breast cancer.

10.3.2 Whole Genome Approaches

The expression profiles involve coordinated alterations in transcription of various genes creating a "signature" that characterizes the stage of breast cancer. Tumors with *BRCA1* and *BRCA2* mutations are discriminated with the Expression signatures (Hedenfalk et al. 2001), supporting the fact of unique molecular profiles for subtypes of breast tumor. So, it is probable that, similar to expression profiles, DNA methylation patterns involve several coordinate alterations in multiple genes and those particular signatures of DNA methylation over a wide spectrum of genes discriminate subtypes of breast tumors and their prognostic value with high accuracy.

Over the past decade, more universal procedures using differential methylation hybridization were developed to examine a considerable number of CpG islands in both cell lines and tumor specimen. This procedure utilized the restriction enzymes that sensitive to methylation states for enrichment of the methylated DNA fragments, after hybridization to CpG island arrays including 1000 CpG islands. The original concept of the informative DNA methylation status in cancer is aberrant methylation of CpG islands is remained. A pioneering investigation by the Huang group (Yan et al. 2000) used this method to find the signatures of DNA methylation by comparing 28 paired primary breast cancer and normal tissue specimens, and to evaluate whether patterns of particular CpG hypermethylation associated with pathological factors in the studied patients. The research show that decreased differentiation of the tumors correlated with the increasing the number of hypermethylated CpG islands. This was an initial evident of the potential of wide DNA methylation signatures for differentiating and classification of breast tumor. The main limit of this examination is its bias towards aberrant methylation of CpG islands.

Cell-culture-based approach is another way to obtain DNA methylation signatures that determine breast cancer subtypes and prognosis. Recently, The utility of distinctly phenotyped but highly associated breast tumor cell lines for defining patterns of DNA methylation that differentiate and classify breast cancers has been investigated (Andrews et al. 2010; Fang et al. 2011). Two MDA-MB-231 breast cancer cell lines including MDA-MB-468GFP and MDA-MB-468GFP-LN (the further derived from a lymphatic metastasis) compared in the survey. This investigation (Andrews et al. 2010) demonstrated wide changes in DNA methylation that involve both hypomethylation and hypermethylation, and evaluate their correlation with gene expression signature and copy number variation. The association between several hypomethylation and hypermethylation events with the copy number variations recommended a linkage of these two phenomena that requires further investigations. The modifications in DNA methylation was found in highly affected particular networks and functional pathways in a highly organized way. These data support the assumption that wide signatures determine variations in different metastatic state between closely associated breast cancer cells. Nevertheless, the main limitation of this investigation is the utility of breast cancer cell lines. It is not obvious what proportion of the DNA methylation signature detected *in vitro* will be correlated with primary breast tumors. If this is accurate in tumor tissues as well, such broad signatures could be useful in prognosis and have a great effect on therapeutic strategies of breast cancer.

Another approach is the use of genome-wide procedures to define stage-specific DNA methylation patterns for classification of primary breast cancers. Recently, several procedures have been developed to define a genome-wide signature of the DNA methylation, such as bisulfite conversion coupled to next-generation sequencing;methylated DNA immunoprecipitation (MeDIP) after hybridization to high-density illumina 27 and 450 K arrays or next-generation sequencing that evaluate the methylation changes in particular CpG islands in the genome wide scale. Although genome-wide sequencing is still exclusively expensive for larger sample sizes in the population studies, array procedure are being often used to define DNA methylation signatures of tumors in primary clinical specimen instead of cell lines. Recently, a whole genome-procedure was used to detected a group of genes that demonstrated a correlation with disease-free survival (Hill et al. 2011). Furthermore, Fang et al. (2011) utilize the 27 K array to define DNA methylation patterns that would classify breast tumors based on their metastatic state. The investigation first detected a “methylator” phenotype, a harmonic methylation of a group of CpG islands in some of tumors, that they described “breast cancer CpG island methylator phenotype” (B-CIMP).These observations correspond to the previously defined methylator phenotype in colorectal cancer. The methylator phenotype was related to the low risk of metastasis and improved outcome autonomously from different breast cancer prognostic factors, such as estrogen receptor status in tumors. This provides evident proof for the potency of DNA methylation signatures to differentiate breast tumors prognosis beyond common classifications. In the recent years, the application of genome-wide methods has enabled more study on the classification with prognostic value of DNA methylation signatures in breast tumors. Recent investigation recommends that DNA methylation signatures might provide knowledge both in the origin of tumor cells in a breast tumor and the microenvironment, especially the immune cell types, that are involved in the cancer (Dedeurwaerder et al. 2011a).

According to this evidence, a definite signature of T cell subtype gene expression could be identified in the tumor stroma samples (Kristensen et al. 2012). In this study they used a comprehensive method described ‘Pathway Recognition Algorithm using Data Integration on Genomic Models’ (PARADIGM), unified DNA methylation signature, expression profiling of mRNA with microRNA and DNA copy number variation. The analysis was performed on about 110 breast tumors and then the PARADIGM clusters of the analyzed specimen were evaluated in two other breast tumor cohorts. The researchers detected important tumor and stromal signatures in the tumor and stroma populations, indicating that it is likely to acquire stromal molecular signatures without dissection of the stromal cells. Furthermore, they obtained informative chronic inflammatory signature in all breast tumors as well as molecular signatures that categorize subtypes of breast tumor cell. The robust predictor of better prognosis was a high T-helper 1 (Th1)/cytotoxic T-lymphocyte signature. The PARADIGM clustering extends classification beyond traditional immunohistochemistry, since discrimination was detected between two clusters within luminal A and luminal B breast tumors.DNA methylation signature will become significant as a diagnostic and prognostic marker in breast cancer just if it provides the classification beyond commonly used methods such as immunohistochemistry and mRNA expression analysis. Recently, some investigators (Dedeurwaerder et al. 2011a)analyzed whole-genome DNA methylation pattern by using the illumina 27 K arrays and recommend that DNA methylation profiling could extend common classifications of breast tumor subtypes. The investigation of 248 breast tissue tumor specimens showed an immune ‘signature’ in mixed tumor-stromal samples. DNA methylation profiles displayed six classes, three of which determined new classifications that were not categorized by expression subtypes, and these might represent different cells of origin (Dedeurwaerder et al. 2011b). Actually, if DNA methylation profiling is just informative in tumor tissue specimens, this restricted the use of such markers for utility in common follow-up.Biopsies are invasive procedures; therefore, it is highly improbable that they will take apart in common screening methods. Furthermore, even in breast cancer patients, biopsies are not appropriate for common follow-up procedure following surgery and notably when there is no obvious growth of tumor. Noninvasive procedures are required for initial classification and follow-up of therapeutic effect after surgery. However, it is likely that free-circulating tumor cells reveal the profiles of DNA methylation that are representative of the status of methylation in the tissue tumor. Informative DNA methylation profiles in breast tissue tumor cells observed in blood samples would be highly significant in early screening, diagnosis, classification and follow-up of the therapy respond. An important area of epigenetic studies in breast tumors is DNA methylation profiling of free-floating tumor cells to determine DNA methylation patterns of breast tumors in these free-floating cells. The initial studies have been focused on hypermethylated genes that are feathers of numerous tumors. For instance, researchers have indicated that it is probable to detect DNA methylation alteration in serum of breast cancer patients. Furthermore, It demonstrated that a set of methylated genes could generat highly sensitive and specific markers for breast tumor as well as prognostic value, as CIMP + in patients' serum was correlated with a relative risk of relapse of 8.6 (Jing et al. 2010). However the prognostic and predictive values of DNA methylation based markers in breast cancer must be evaluated in future investigations. The pivotal challenge in this area is to obtain high-quality DNA methylation profiles that are validated in future investigations as specific and sensitive predictors for determination of prognosis and fallow up response to treatment. An additional research question is to ascertain whether DNA methylation profile would have advantages over common histopathological and immunochemical procedures.

10.4 Epigenetic Changes as Therapeutic Targets in Breast Cancer

Since the epigenetic modifications are potentially reversible processes, a large number of investigations have been mediated to provide information on this mechanism with the purpose of identifying effective treatments that target these modifications. The histone deacetylase inhibitors (HDACi) as well as demethylating drugs are under examination as single agents or in compositions with other systemic therapies. A large number of the preclinical studies have utilized the epigenetic treatments to re-express the silenced genes in cell lines. The high challenging issue has been the clinical utility of the laboratory findings and clinical effectiveness of them. For example, preclinical studies clearly showed that re-expression of the maspin gene can be achieved in breast cancer cell lines using histone deacetylase inhibitors (HDACi) and demethylating drugs, but the important inquiry is whether such a re-expression was obtained in a patient's tumor or in another word "*would it have any influence on clinical outcomes*?" In spite of these challenging features of the therapeutic application of epidrugs, knowledge of the critical role of such therapies has been increased in breast cancer. Currently, all of the epigenetic therapies are in research stages in cancers and not sill considered standard of care in the bedside. The information about the potential application of these agents in breast cancer is providing in combination with targeted therapies, chemotherapy and radiotherapy to overcome therapeutic resistance and improve respond to treatment. Another interesting and developing field is investigation the role of epigenetic modifications in breast cancer prevention (Lustberg and Ramaswamy 2011).

10.4.1 DNMT Inhibitor Therapy

DNA methyl transferases (DNMTs) are enzymes which can transfer methyl group to DNA. DNMT inhibitors can block the function of these enzymes. Nucleoside analogues 5-vazacytidine (5-aza) and 5-aza-2′-deoxycytidine (decitabine) are most frequently researched DNMT inhibitors. One of the significant challenges of breast cancer treatment is tumors with estrogen receptor negative status which are not responding to antiestrogen therapies such as tamoxifen. The role of epigenetic modifications (especially CpG island methylation) in the regulation of ER gene expression is well understood. Despite the fact that more than two thirds of breast tumors synthesize estrogen receptor (ER), both de novo and acquired drug resistance to tamoxifen therapy are major clinical problems. Response to endocrine therapy is associated with the level of ER expression, and silencing of ER gene expression during the course of breast cancer treatment is one of the significant mechanisms of endocrine therapy resistance in ER positive tumors. A deep knowledge of the epigenetic modulation of ER gene give possibility for planning effective epigenetic therapies to override endocrine resistance to tamoxifen and other endocrine therapies including the aromatase inhibitors. Additionally, this approach creates new possibility to treat ER-negative tumors using the combination of epigenetic agents and hormonal therapy.

Demethylation of the ER gene CpG island and activation of ER gene expression and synthesis of functional ER protein have been promoted in ER negative human breast tumor cell lines with 5-aza treatment (Yang et al. 2000). Inhibition of DNMT1 by antisense or siRNA may increase responsiveness to 5-aza in ER-negative breast cancer cells (Robert et al. 2003). Furthermore, the combination of HDACi and DNMT inhibitors has synergistic effects (Yang et al. 2001a). Primary evidences about the influence of epidrugs on ER gene re-expression achieved from treatment of ER negative cell lines with DNMT inhibitors and HDACi. For example to investigate the influence of epidrugs on ER reactivation, ER-negative breast cell line MDA-MB-231, treated with demethylating agent 5-aza-2'-deoxycytidine. Interestingly treated ER negative cells restored the expression of ER mRNA and ER protein and the growth of cells treated with tamoxifen were inhibited significantly (Wang et al. 2006).

In another experience on MDA-MB-435 as an ER negative breast cancer cell line, combinatory therapy by aza and TSA (trichostatin A) showed that the mRNA of estrogen and progestron receptors was re-expressed (Fan et al. 2008) The proliferation assay in the treated cells showed that their growth was suppressed. Growth suppression in the treated cells was further decreased by addition of tamoxifen to the culture medium. In contrast, the proliferation of cells treated only with tamoxifen showed no difference compared with the untreated cells. In animal model study xenograft volume of MDA-MB-435 cells treated with aza and TSA was smaller than that of the untreated control cells. Ovariectomy in these animals could suppress the growth of aza and TSA treated xenograft.

Currently, we have not enough data on the clinical function of 5-aza or decitabine as a single drug for breast cancer treatment. Recently, A phase I multi-center clinical trial of decitabine in patients with advanced breast cancer who had poor respond to standard treatment was finished (NCT00030615) (http://clinicaltrials.gov). But the results of this clinical trial not formally published. Other nucleoside analogues as DNMT inhibitors including zebularine and 5-fluoro-2′-deoxycitine are in clinical development process (Cheng et al. 2004). In addition, an anti-sense oligonucleotide known as MG98 that particularly targets DNMT1 is under clinical development process.

In another investigation in a phase I trial, the combination of decitabine and vorinostat given consecutively showed tumor stabilization in 7 of 22 (32 %) examined patients with advanced solid tumors. Bone marrow suppression and gastrointestinal toxicities were dose limiting toxicities (DLT) in the mentioned study (Stathis et al. 2009).

The influence of addition of hydralazine and magnesium valproate to neoadjuvant doxorubicin and cyclophosphamide treatment for locally advanced breast cancer (LABC) was investigated in a proof-of-principle phase I study (Arce et al. 2006). Starting day 7 through completion of four steps of chemotherapy, selected cases received oral hydralazine and valproate. On day 8 these patients were evaluated with core biopsies and significant decrease of the global methylation was detected. Also HDAC suppression was detected in peripheral blood specimens on day 8 after this combination epigenetic therapy. The combination of chemotherapy and epigenetic therapy was well tolerated in most patients and the incidence of drowsiness was related to the increase of valproate to the treatment regimen. Exclusively, one (6.6 %) patient had a complete pathologic response in the condition that most cases had a clinical response. Followed by this proof-of-concept trial, patients with advanced breast cancer were gradually improving to overcome chemotherapy resistance in a phase II trial of adding hydralazine and valproate to the same chemotherapy regime (Candelaria et al. 2007). On this study, three patients were treated, two patients who had continued on paclitaxel were progressed on the same in addition to hydralazine and valproate and both continued. Another patient was progressed on the same ineffective hormone therapy in addition to the combined epidrugs treatments and had disease stabilization for 4.5 months. While responses to epigenetic therapies were observed in a few patients with genitourinary cancers, this way has not been progressed considerably in breast cancer.

10.4.2 HDACi Therapy

Histone deacetylases (HDAC) are a class of chromatin changing enzymes which remove acetyl groups from a histone protein. This change causes a more tight interaction between DNA and histones and compact chromatin conformation. Genes which are located in such chromatin environment cannot transcribe and will be silent. HDACi (histone deacetylase inhibitors) are a group of compounds that can inhibit the enzymatic activity of Histone deacetylases in cell. These inhibitors can reverse the silencing effects of histone deacetylation on gene expression. Also HDACi possess the anti-proliferative and pro-apoptotic properties affect on different malignant cell types (Stearns et al. 2007). Currently, Vorinostat (suberoylanilide hydroxamic acid) is the only HDACi approved by the US Food and Drug Administration (FDA).

HDACi up regulates the transcription of proapoptotic genes and cyclin-dependent kinase inhibitors like p21 and p27. Furthermore, HDACi increases acetylation of the Hsp90 molecular chaperone system and leading to decreasing their stabilization (Isaacs et al. 2003). As a consequence, proteosome targeting and degradation of various target proteins, such as AKT, ER, HER2 and c-Raf, are enhanced. Then degradation of client protein influences on multiple downstream pathways, such as increasing of pro apoptotic proteins including Bak and Bim and decreasing of antiapoptic proteins Bcl-2 and Bcl-xL (Bali et al. 2005). In preclinical condition, estrogen receptor is one of the Hsp90 client proteins which are most sensitive to downstream inhibition of Hsp90 (Thomas and Munster 2009). HDACi sensitize breast tumor cells to endocrine, HER2-targeted, and cytotoxic therapies by attenuating these signaling pathways.

Vorinostat is a component of the hydroxamic acid family of HDACi. It can suppress the proliferation of ER-positive as well as ER-negative breast cancer cell lines (Munster et al. 2001). Vorinostat and other HDACi such as LAQ824, down regulates p-Akt, Akt, and c-Raf, sensibilising ER-positive breast tumor cells to endocrine therapies such as tamoxifen (Fiskus et al. 2007). Suppression of HDAC2 by siRNA silences of both ER and PR expression and increases the influence of tamoxifen in ER-positive breast cancer cells (Biçaku et al. 2008). In addition, the combinatory therapy with vorinostat and docetaxel or trastuzumab leads to attenuated levels of c-Raf and AKT with synergistic effect in breast tumor cells (Bali et al. 2005). Furthermore, breast tumor cells sensitize to topoisomerase inhibitors after prolonged exposure to vorinostat (Marchion et al. 2004). Regarding to the encouraging preclinical findings, there are many ongoing clinical trials in breast cancer evaluating the influence of vorinostat together with endocrine therapy and cytotoxic drugs.

The results of a phase I clinical trial in patients suffering from advanced solid tumor and hematologic malignancies revealed that intravenously administration of vorinostat well tolerate. In the phase II study performed in advanced breast cancer with single-drug oral vorinostat, 4 of 14 patients (29 %) acquired stabilization of their tumors (range, 4–14 months). However, no case showed an objective responses. The most prominent side effects involved nausea, diarrhea, fatigue and lymphopenia (Luu et al. 2008).

10.4.3 Combination of HDACi and Endocrine Therapy

Based on hopeful preclinical result of combining HDACi with endocrine therapy, a phase II trial study of vorinostat together with tamoxifen was performed in metastatic breast tumor patients who had disease progression on past lines of hormone therapy (Munster et al. 2011). Seven patients (21 %) had a limited response and one with exclusive bone disease had an objective tumor response by PET/CT scan, while another four (12 %) patients had stable disease of more than 6 months. Observing two patients (5 %) with pulmonary emboli, the toxicity profile of the combination was confirmed in the median response duration of 8 months. Associated investigation indicated acetylation of histone H3 and H4 at day 8, recommending sufficient amount of vorinostat in the most of the patients.

In another effort, selected postmenopausal patients who had disease progression in spite of aromatase inhibitor therapy continued on the same aromatase suppressor and entinostat at 5 mg weekly on a 28-day cycle was added to their treatment regime. One patient had an approved limited response and one patient had stable disease more than 6 months. Based on initial biomarker evaluation, there was expansive lysine acetylation in histones and apoptotic state in peripheral blood cells with the addition of HDACi therapy.

10.4.4 Epigenetic Therapy for Chemoprevention of Breast Cancer by Natural Ingredients

Epidemiological investigations have demonstrated that Asian women are less predisposed to breast tumor than the western counterparts because of their high soy food intake. Furthermore, Asian people generally utilized complementary/and or alternative medicines that are rich in bioactive components recognized to be chemopreventive against carcinogenesis. Instance of such compounds include dietary isothiocyanates from plant foods, resveratrol from grapes, epigallocatechin-3-gallate (EGCG) from green tea, sulforaphane from some vegetables (crucifer), genistein from soybean, curcumin from turmeric and polyphenols. These bioactive ingredients are able to modify the epigenetic alterations, and their epigenetic targets are correlated with prevention and treatment of breast cancer. This approach could promote the development of new drugs for breast cancer therapy (Khan et al. 2012).

Sulforaphane is an isothiocyanate compound which is found in broccoli sprouts. The anti-cancer effect of this compound has been demonstrated through histone acetylation, following the induction of P21 and Bax, and triggering of cell cycle arrest and apoptosis. Consumption of one cup of broccoli suppressed HDAC function in peripheral blood mononucleated cells 3–6 h after ingestion, leading to trigger of H3 and H4 acetylation, providing information to use this dietary component as a chemopreventive agent. Based on this study, a phase II placebo-controlled trial of broccoli sprout extract in patient with early detection of ductal carcinoma in situ and/or atypical ductal hyperplasia is in progress at present (http://clinicaltrials.gov).

Results from this investigation may determine a role for epigenetic therapy in breast tumor prevention.

It is well known that ER negative breast tumors are more aggressive and hormone refractory cancers. Hormone therapy by tamoxifen can not restrict the tumor growth in ER negative cancers. Recently it has been shown that soybean isoflavone, genistein (GE) can activate ERα gene expression in ERα-negative MDA-MB-231 breast cancer cells. The ER gene reactivation was synergistically increased when combined with TSA, a histone deacetylase inhibitor (Li et al. 2013).

Also GE treatment reactivated cellular responses to tamoxifen as an ER antagonist. Further studies showed that GE can contribute to the chromatin remodeling in the ERα promoter thereby leading to ERα gene expression. Consistently, in mouse model of breast cancer, dietary GE significantly suppressed cancer development and reduced the growth of ERα-negative tumors.

Another natural ingredient which has anticancer properties is green tea polyphenol, (-)-epigallocatechin-3-gallate (EGCG). It is believed that EGCG as an anticancer agent works by regulatory effect on cellular epigenetic.

In another investigation on ER-negative MDA-MB-231 breast cancer cells, researchers found that EGCG can reactivate ERa gene expression (Li et al. 2013).

Combination therapy using EGCG with the TSA as a histone deacetylase inhibitor was synergistically enhanced ER reactivation in ERa-negative breast cancer cells and re sensitized cells to tamoxifen. EGCG reactivated ER via changing in epigenetic status of the promoter by altering histone acetylation and methylation.

10.5 Micro RNAs Landscape as Another Epigenetic Player in Breast Cancer

MicroRNAs (miRNA (are other part of epigenetic machinery of the living cells. The role of miRNA in the cancer biology is confirmed by profound evidences of experimental investigations, that has regularly switched from signature studies, as the initial breast cancer profiling described in 2005 (Iorio et al. 2005) determining an aberrant microRNA profiling in various tumor types, to biological description of the causal role of these small fragments in the carcinogenic procedure, and the potential application in biomarkers or therapeutic approaches. These current investigations have generally shown that microRNAs can regulate both oncogenic and tumor suppressor pathways, as a result, their expression can be modulated by oncogenes or tumor suppressor genes.

Among the most important differentially expressed miRNAs, some were widely investigated as their initial detection and demonstrated a crucial role in the biology of breast cancer: miR-21, over expression in breast tumorogenesis, has been shown to induce cell survival straightly targeting the tumor suppressor genes including PTEN, PDCD4 and TPM1, and it has been correlated with high stages, lymph nodes involvement and poor outcome in patients (Qian et al. 2009) also in pregnancy associated breast cancer (Walter et al. 2011). Furthermore, MiR-21 has been discovered as circulating miRNA, which freely presents in the peripheral blood (Asaga et al. 2011). In the past recent, some investigations have interestingly shown that additional expression of miR-21 can be detected in bone marrow of breast cancer patients and the amount of this miRNA and PDCD4 (its target) have a prognostic effect in these patients (Wang et al. 2010).

Another important miRNA, Let-7, has a tumor suppressor function. Firstly it detected in *C. elegans*, where it triggers growth arrest and differentiation, has been shown as a novel regulator of self renewal and carcinogenicity of breast tumor cells (Ota et al. 2011). Additional expression of let-7 miRNA family can inhibit tumor progression in mouse models of both breast and lung tumors (Yu et al. 2007).

MiRNAs have been shown to have a pivotal role not only in mediating the tumor growth by modulating proliferation pathways and cell cycle control, but also to be decisive in regulating migration and invasion, mechanisms associated with the achievement of a more malignant phenotype and inducing the onset of the metastatic state. (Iorio et al. 2011).

Regarding to existing evidences in the miRNAs field and their contributions in breast cancer there is no doubt that they will be future epigenetic markers in three different aspects in management of breast cancer: diagnosis, prognosis and treatment.

References

Akao Y, Nakagawa Y, Naoe T (2006) Let-7 MicroRNA functions as a potential growth suppressor in human colon cancer cells. Biol Pharm Bull 29:903–906

Amin K, Banerjee P (2012) The cellular functions of RASSF1 A and its inactivation in prostate cancer. J Carcinog 11:3

Andrews J, Kennette W, Pilon J, Hodgson A, Tuck AB, Chambers AF et al (2010) Multi-platform whole-genome microarray analyses refine the epigenetic signature of breast cancer metastasis with gene expression and copy number. PLoS One 5:e8665

Anker P, Mulcahy H, Chen XQ, Stroun M (1999) Detection of circulating tumour DNA in the blood (plasma/serum) of cancer patients. Cancer Metastasis Rev 18:65–73

Arce C, Pérez-Plasencia C, González-Fierro A, de la Cruz-Hernández E, Revilla-Vázquez A, Chávez-Blanco A et al (2006) A proof-of-principle study of epigenetic therapy added to neoadjuvant doxorubicin cyclophosphamide for locally advanced breast cancer. PLoS One 1:e98

Asaga S, Kuo C, Nguyen T, Terpenning M, Giuliano AE, Hoon DSB (2011) Direct serum assay for microRNA-21 concentrations in early and advanced breast cancer. Clin Chem 57:84–91

Bali P, Pranpat M, Swaby R, Fiskus W, Yamaguchi H, Balasis M et al (2005) Activity of suberoylanilide hydroxamic Acid against human breast cancer cells with amplification of her-2. Clin Cancer Res 11:6382–6389

Bartholow TL, Becich MJ, Chandran UR, Parwani AV (2011) Immunohistochemical staining of slit2 in primary and metastatic prostatic adenocarcinoma. Transl Oncol 4:314–320

Biçaku E, Marchion DC, Schmitt ML, Münster PN (2008) Selective inhibition of histone deacetylase 2 silences progesterone receptor-mediated signaling. Cancer Res 68:1513–1519

Bojesen SE, Pooley KA, Johnatty SE, Beesley J, Michailidou K, Tyer JP et al (2013) Multiple independent variants at the TERT locus are associated with telomere length and risks of breast and ovarian cancer. Nat Genet 45:371–384

Bosviel R, Garcia S, Lavediaux G, Michard E, Dravers M, Kwiatkowski F et al (2012) BRCA1 promoter methylation in peripheral blood DNA was identified in sporadic breast cancer and controls. Cancer Epidemiol 36:e177–e182

Brennan K, Garcia-Closas M, Orr N, Fletcher O, Jones M, Ashworth A et al (2012) Intragenic ATM methylation in peripheral blood DNA as a biomarker of breast cancer risk. Cancer Res 72:2304–2313

Brose K, Bland KS, Wang KH, Arnott D, Henzel W, Goodman CS et al (1999) Slit proteins bind robo receptors and have an evolutionarily conserved role in repulsive axon guidance. Cell 96:795–806

Candelaria M, Gallardo-Rincón D, Arce C, Cetina L, Aguilar-Ponce JL, Arrieta O et al (2007) A phase II study of epigenetic therapy with hydralazine and magnesium valproate to overcome chemotherapy resistance in refractory solid tumors. Ann Oncol 18:1529–1538

Carone BR, Fauquier L, Habib N, Shea JM, Hart CE, Li R et al (2010) Paternally induced transgenerational environmental reprogramming of metabolic gene expression in mammals. Cell 143:1084–1096

Cheng JC, Yoo CB, Weisenberger DJ, Chuang J, Wozniak C, Liang G et al (2004) Preferential response of cancer cells to zebularine. Cancer Cell 6: 151–158

Cho YH, Yazici H, Wu HC, Terry MB, Gonzalez K, Qu M et al (2010) Aberrant promoter hypermethylation and genomic hypomethylation in tumor, adjacent normal tissues and blood from breast cancer patients. Anticancer Res 30:2489–2496

Choi JY, James SR, Link PA, McCann SE, Hong CC, Davis W et al (2009) Association between global DNA hypomethylation in leukocytes and risk of breast cancer. Carcinogenesis 30:1889–1897

Dallol A, Da Silva NF, Viacava P, Minna JD, Bieche I, Maher ER et al (2002) SLIT2, a human homologue of the Drosophila Slit2 gene, has tumor suppressor activity and is frequently inactivated in lung and breast cancers. Cancer Res 62:5874–5880

Dedeurwaerder S, Desmedt C, Calonne E, Singhal SK, Haibe-Kains B, Defrance M et al (2011a) DNA methylation profiling reveals a predominant immune component in breast cancers. EMBO Mol Med 3:726–741

Dedeurwaerder S, Fumagalli D, Fuks F (2011b) Unravelling the epigenomic dimension of breast cancers. Curr Opin Oncol 23:559–565

Dickinson RE, Dallol A, Bieche I, Krex D, Morton D, Maher ER et al (2004) Epigenetic inactivation of SLIT3 and SLIT1 genes in human cancers. Br J Cancer 91:2071–2078

Dickinson RE, Fegan KS, Ren X, Hillier SG, Duncan WC (2011) Glucocorticoid regulation of SLIT/ROBO tumour suppressor genes in the ovarian surface epithelium and ovarian cancer cells. Plos One 6:e27792

Dim DC, Jiang F, Qiu Q, Li T, Darwin P, Rodgers WH et al (2011) The usefulness of S100P, mesothelin, fascin, prostate stem cell antigen, and 14-3-3 sigma in diagnosing pancreatic adenocarcinoma in cytological specimens obtained by endoscopic ultrasound guided fine-needle aspiration. Diagn Cytopathol 42:193–199

Dobrovic A, Kristensen LS (2009) DNA methylation, epimutations and cancer predisposition. Int J Biochem Cell Biol 41:34–39

Dulaimi E, Hillinck J, Ibanez de Caceres I, Al-Saleem T, Cairns P (2004) Tumor suppressor gene promoter hypermethylation in serum of breast cancer patients. Clin Cancer Res 10:6189–6193

Dworkin AM, Huang THM, Toland AE (2009) Epigenetic alterations in the breast: Implications for breast cancer detection, prognosis and treatment. Semin Cancer Biol 19:165–171

Esteller M, Sparks A, Toyota M, Sanchez-Cespedes M, Capella G, Peinado MA et al (2000) Analysis of adenomatous polyposis coli promoter hypermethylation in human cancer. Cancer Res 60:4366–4371

Fabian CJ, Kimler BF, Mayo MS, Khan SA (2005) Breast tissue sampling for risk assessment and prevention. Endocr Relat Cancer 12:185–213

Fan J, Yin WJ, Lu JS, Wang L, Wu J, Wu FY et al (2008) ER alpha negative breast cancer cells restore response to endocrine therapy by combination treatment with both HDAC inhibitor and DNMT inhibitor. J Cancer Res Clin Oncol 134:883–890

Fang F, Turcan S, Rimner A, Kaufman A, Giri D, Morris LGT et al (2011) Breast cancer methylomes establish an epigenomicfoundation for metastasis. Sci Transl Med 3:75ra25
Fiskus W, Ren Y, Mohapatra A, Bali P, Mandawat A, Rao R et al (2007) Hydroxamic acid analogue histone deacetylase inhibitors attenuate estrogen receptor-alpha levels and transcriptional activity: a result of hyperacetylation and inhibition of chaperone function of heat shock protein 90. Clin Cancer Res 13:4882–4890
Flanagan JM, Munoz-Alegre M, Henderson S, Tang T, Sun P, Johnson N et al (2009) Gene-body hypermethylation of ATM in peripheral blood DNA of bilateral breast cancer patients. Hum Mol Genet 18:1332–1342
Graff JR, Herman JG, Lapidus RG, Chopra H, Xu R, Jarrard DF et al (1995) E-cadherin expression is silenced by DNA hypermethylation in human breast and prostate carcinomas. Cancer Res 55:5195–5199
Ha YS, Jeong P, Kim JS, Kwon WA, Kim IY, Yun SJ et al (2012) Tumorigenic and prognostic significance of RASSF1 A expression in low-grade (WHO grade 1 and grade 2) nonmuscle-invasive bladder cancer. Urology 79:1411.e1–1411.e6
Hanash SM, Baik CS, Kallioniemi O (2011) Emerging molecular biomarkers-blood-based strategies to detect and monitor cancer. Nat Rev Clin Oncol 8:142–150
Hansmann T, Pliushch G, Leubner M, Kroll P, Endt D, Gehrig A et al (2012) Constitutive promoter methylation of BRCA1 and RAD51 C in patients with familial ovarian cancer and early-onset sporadic breast cancer. Hum Mol Genet 21:4669–4679
Hatziapostolou M, Iliopoulos D (2011) Epigenetic aberrations during oncogenesis. Cell Mol Life Sci 68:1681–1702
Hayes DF, Isaacs C, Stearns V (2001) Prognostic factors in breast cancer: current and new predictors of metastasis. J Mammary Gland Biol Neoplasia 6:375–392
He L, Hannon GJ (2004) MicroRNAs: small RNAs with a big role in gene regulation. Nat Rev Genet 5:522–531
Hedenfalk I, Duggan D, Chen Y, Radmacher M, Bittner M, Simon R et al (2001) Gene-expression profiles in hereditary breast cancer. N Engl J Med 344:539–548
Heyn H, Carmona FJ, Gomez A, Ferreira HJ, Bell JT, Sayols S et al (2013) DNA methylation profiling in breast cancer discordant identical twins identifies DOK7 as novel epigenetic biomarker. Carcinogenesis 34:102–108
Hill VK, Ricketts C, Bieche I, Vacher S, Gentle D, Lewis C et al (2011) Genome-wide DNA methylation profiling of CpG islands in breast cancer identifies novel genes associated with tumorigenicity. Cancer Res 71:2988–2999
Hitchins MP, Wong JJL, Suthers G, Suter CM, Martin DIK, Hawkins NJ et al (2007) Inheritance of a cancer-associated MLH1 germ-line epimutation. N Engl J Med 356:697–705
Hoque MO, Prencipe M, Poeta ML, Barbano R, Valori VM, Copetti M et al (2009) Changes in CpG islands promoter methylation patterns during ductal breast carcinoma progression. Cancer Epidemiol Biomarkers Prev 18:2694–2700
Ignatiadis M, Reinholz M (2011) Minimal residual disease and circulating tumor cells in breast cancer. Breast Cancer Res 13:222
Iorio MV, Ferracin M, Liu CG, Veronese A, Spizzo R, Sabbioni S, Magri E et al (2005) MicroRNA gene expression deregulation in human breast cancer. Cancer Res 65:7065–7070
Iorio MV, Casalini P, Piovan C, Braccioli L, Tagliabue E (2011) Breast cancer and microRNAs: therapeutic impact. Breast 20:S63–70
Isaacs JS, Xu W, Neckers L (2003) Heat shock protein 90 as a molecular target for cancer therapeutics. Cancer Cell 3:213–217
Iwamoto T, Yamamoto N, Taguchi T, Tamaki Y, Noguchi S (2011) BRCA1 promoter methylation in peripheral blood cells is associated with increased risk of breast cancer with BRCA1 promoter methylation. Breast Cancer Res Treat 129:69–77
Izadi P, Noruzinia M, Fereidooni F, Nateghi MR (2012a) Association of poor prognosis subtypes of breast cancer with estrogen receptor alpha methylation in Iranian women. Asian Pac J Cancer Prev 13:4113–4117

Izadi P, Noruzinia M, Karimipoor M, Karbassian MH, Akbari MT (2012b) Promoter hypermethylation of estrogen receptor alpha gene is correlated to estrogen receptor negativity in Iranian patients with sporadic breast cancer. Cell J 14:102–109

Jacinto FV, Esteller M (2007) Mutator pathways unleashed by epigenetic silencing in human cancer. Mutagenesis 22:247–253

Jemal A, Bray F, Center MM, Ferlay J, Ward E, Forman D (2011) Global cancer statistics. CA Cancer J Clin 61:69–90

Jin J, You H, Yu B, Deng Y, Tang N, Yao G et al (2009) Epigenetic inactivation of SLIT2 in human hepatocellular carcinomas. Biochem Biophys Res Commun 379:86–91

Jing F, Yuping W, Yong C, Jie L, Jun L, Xuanbing T et al (2010) CpG island methylator phenotype of multigene in serum of sporadic breast carcinoma. Tumor Biol 31:321–331

Jirtle RL, Skinner MK (2007) Environmental epigenomics and disease susceptibility. Nat Rev Genet 8:253–262

Johnson SM, Grosshans H, Shingara J, Byrom M, Jarvis R, Cheng A et al (2005) RAS is regulated by the let-7 microRNA family. Cell 120:635–647

Jones PA, Baylin SB (2002) The fundamental role of epigenetic events in cancer. Nat Rev Genet 3:415–428

Jones PA, Liang G (2009) Rethinking how DNA methylation patterns are maintained. Nat Rev Genet 10:805–811

Khan SI, Aumsuwan P, Khan IA, Walker LA, Dasmahapatra AK (2012) Epigenetic events associated with breast cancer and their prevention by dietary components targeting the epigenome. Chem Res Toxicol 25:61–73

Kim JH, Shin MH, Kweon SS, Park MH, Yoon JH, Lee JS et al (2010) Evaluation of promoter hypermethylation detection in serum as a diagnostic tool for breast carcinoma in Korean women. Gynecol Oncol 118:176–181

Kim GE, Lee KH, Choi YD, Lee JS, Lee JH, Nam JH et al (2011) Detection of Slit2 promoter hypermethylation in tissue and serum samples from breast cancer patients. Virchows Arch 459:383–390

Kim JS, Chae Y, Ha YS, Kim IY, Byun SS, Yun SJ et al (2012) Ras association domain family 1 A: apromising prognostic marker in recurrent nonmuscle invasive bladder cancer. Clin Genitourin Cancer 10:114–120

Kohler C, Barekati Z, Radpour R, Zhong XY (2011) Cell-free DNA in the circulation as a potential cancer biomarker. Anticancer Res 31:2623–2628

Kristensen VN, Vaske CJ, Ursini-Siegel J, Van Loo P, Nordgard SH, Sachidanandam R et al (2012) Integrated molecular profiles of invasive breast tumors and ductal carcinoma in situ (DCIS) reveal differential vascular and interleukin signaling. Proc Natl Acad Sci U S A 109:2802–2807

Laird PW (2003) The power and the promise of DNA methylation markers. Nat Rev Cancer 3:253–266

Lee MN, Tseng RC, Hsu HS, Chen JY, Tzao C, Ho WL et al (2007) Epigenetic inactivation of the chromosomal stability control genes BRCA1, BRCA2, and XRCC5 in non-small cell lung cancer. Clin Cancer Res 13:832–838

Li Y, Meeran SM, Patel SN, Chen H, Hardy TM, Tollefsbol TO (2013) Epigenetic reactivation of estrogen receptor-α (ERα) by genistein enhances hormonal therapy sensitivity in ERα-negative breast cancer. Molcancer 12:9

Lustberg MB, Ramaswamy B (2011) Epigenetic therapy in breast cancer. Curr Breast Cancer Rep 3:34–43

Luu TH, Morgan RJ, Leong L, Lim D, McNamara M, Portnow J et al (2008) A phase II trial of vorinostat (suberoylanilide hydroxamic acid) in metastatic breast cancer: a California cancer consortium study. Clin Cancer Res 14:7138–7142

Marchion DC, Bicaku E, Daud AI, Richon V, Sullivan DM, Munster PN (2004) Sequence-specific potentiation of topoisomerase II inhibitors by the histone deacetylase inhibitor suberoylanilide hydroxamic acid. J Cell Biochem 92:223–237

Marlow R, Strickland P, Lee JS, Wu X, Pebenito M, Binnewies M et al (2008) Slits suppress tumor growth in vivo by silencing Sdf1/Cxcr4 within breast epithelium. Cancer Res 68:7819–7827

Martínez-Galán J, Torres B, del Moral R, Muñoz-Gámez JA, Martín-Oliva D, Villalobos M et al (2008) Quantitative detection of methylated ESR1 and 14-3-3-σ gene promoters in serum as candidate biomarkers for diagnosis of breast cancer and evaluation of treatment efficacy. Cancer Biol Ther 7:958–965

Matuschek C, Bölke E, Lammering G, Gerber PA, Peiper, M, Budach W et al (2010) Methylated APC and GSTP1 genes in serum DNA correlate with the presence of circulating blood tumor cells and are associated with a more aggressive and advanced breast cancer disease. Eur J Med Res 15:277–286

Mirza S, Sharma G, Prasad CP, Parshad R, Srivastava A, Gupta SD et al (2007) Promoter hypermethylation of TMS1, BRCA1, ERα and PRB in serum and tumor DNA of invasive ductal breast carcinoma patients. Life Sci 81:280–287

Mori T, Martinez SR, O'Day SJ, Morton DL, Umetani N, Kitago M et al (2006) Estrogen receptor-alpha methylation predicts melanoma progression. Cancer Res 66:6692–6698

Müller HM, Widschwendter A, Fiegl H, Ivarsson L, Goebel G, Perkmann E et al (2003) DNA methylation in serum of breast cancer patients: an independent prognostic marker. Cancer Res 63:7641–7645

Munster PN, Troso-Sandoval T, Rosen N, Rifkind R, Marks PA, Richon VM (2001) The histone deacetylase inhibitor suberoylanilide hydroxamic acid induces differentiation of human breast cancer cells. Cancer Rres 61:8492–8497

Munster PN, Thurn KT, Thomas S, Raha P, Lacevic M, Miller A et al (2011) A phase II study of the histone deacetylase inhibitor vorinostat combined with tamoxifen for the treatment of patients with hormone therapy-resistant breast cancer. Br J Cancer 104:1828–1835

Nass SJ, Ferguson AT, El-Ashry D, Nelson WG, Davidson NE (1999) Expression of DNA methyltransferase (DMT) and the cell cycle in human breast cancer cells. Oncogene 18:7453–7461

Ng SF, Lin RCY, Laybutt DR, Barres R, Owens JA, Morris MJ (2010) Chronic high-fat diet in fathers programs β-cell dysfunction in female rat offspring. Nature 467:963–966

Noruzinia M, Coupier I, Pujol P (2005) Is BRCA1/BRCA2-related breast carcinogenesis estrogen dependent? Cancer 104:1567–1574

Okumura H, Kita Y, Yokomakura N, Uchikado Y, Setoyama T, Sakurai H et al (2010) Nuclear expression of 14-3-3 sigma is related to prognosis in patients with esophageal squamous cell carcinoma. Anticancer Res 30:5175–5179

Ota D, Mimori K, Yokobori T, Iwatsuki M, Kataoka A, Masuda N et al (2011) Identification of recurrence-related microRNAs in the bone marrow of breast cancer patients. Int J Oncol 38:955–962

Parrella P (2010) Epigenetic signatures in breast cancer: clinical perspective. Breast Care(Basel) 5:66–73

Pei H, Ge H, Jiang R, Zhu H (2010) Expression and clinical significance of 14-3-3 sigma and heat shock protein 27 in colorectal cancer. Zhonghua Wei Chang Wai Ke Za Zhi 13:213–215

Pepe MS, Etzioni R, Feng Z, PotterJD, Thompson ML, Thornquist M et al (2001) Phases of biomarker development for early detection of cancer. J Natl Cancer Inst 93:1054–1061

Perou CM, Sørlie T, Eisen MB, van de Rijn M, Jeffrey SS, Rees CA et al (2000) Molecular portraits of human breast tumours. Nature 406:747–752

Piperi C, Themistocleous MS, Papavassiliou GA, Farmaki E, Levidou G, Korkolopoulou P et al (2010) High incidence of MGMT and RARbeta promoter methylation in primary glioblastomas: association with histopathological characteristics, inflammatory mediators and clinical outcome. Mol Med 16:1–9

Pu RT, Laitala LE, Alli PM, Fackler MJ, Sukumar S, Clark DP (2003) Methylation profiling of benign and malignant breast lesions and its application to cytopathology. Mod Pathol 16:1095–1101

Pujol P, This P, Noruzinia M, Stoppa-Lyonnet D, Maudelonde T (2004) Are the hereditary forms of BRCA1 and BRCA2 breast cancer sensitive to estrogens? Bull Cancer (Paris) 91:583–591

Qian B, Katsaros D, Lu L, Preti M, Durando A, Arisio R et al (2009) High miR-21 expression in breast cancer associated with poor disease-free survival in early stage disease and high TGF-beta1. Breast Cancer Res Treat 117:131–140

Qureshi SA, Bashir MU, Yaqinuddin A (2010) Utility of DNA methylation markers for diagnosing cancer. Int J Surg 8:194–198

Radpour R, Barekati Z, Kohler C, Lv Q, Bürki N, Diesch C et al (2011) Hypermethylation of tumor suppressor genes involved in critical regulatory pathways for developing a blood-based test in breast cancer. Plos One 6:e16080

Robert MF, Morin S, Beaulieu N, Gauthier F, Chute IC, Barsalou A et al (2003) DNMT1 is required to maintain CpG methylation and aberrant gene silencing in human cancer cells. Nat Genet 33:61–65.

Sharma G, Mirza S, Parshad R, Srivastava A, Gupta SD, Pandya P et al (2010) Clinical significance of promoter hypermethylation of DNA repair genes in tumor and serum DNA in invasive ductal breast carcinoma patients. Life Sci 87:83–91

Sharma VK, Vouros P, Glick J (2011) Mass spectrometric based analysis, characterization and applications of circulating cell free DNA isolated from human body fluids. Int J Mass Spectrom 304:172–183

Shinozaki M, Hoon DS, Giuliano AE, Hansen NM, Wang HJ, Turner R et al (2005) Distinct hypermethylation profile of primary breast cancer is associated with sentinel lymph node metastasis. Clin Cancer Res 11:2156–2162

Shukla S, Mirza S, Sharma G, Parshad R, Gupta SD, Ralhan R (2006) Detection of RASSF1 A and RAR? Hypermethylation in Serum DNA from Breast Cancer Patients. Epigenetics 1:88–93

Sidransky D (1997) Nucleic acid-based methods for the detection of cancer. Science 278:1054–1058

Sidransky D (2002) Emerging molecular markers of cancer. Nat Rev Cancer 2:210–219

Snell C, Krypuy M, Wong EM, kConFab investigators, Loughrey MB, Dobrovic A (2008) BRCA1 promoter methylation in peripheral blood DNA of mutation negative familial breast cancer patients with a BRCA1 tumour phenotype. Breast Cancer Res 10:R12

Stathis A, Hotte S, Hirte H, Chen EX, Webster S, Iacobucci A et al (2009) Phase I study of intravenous decitabine in combination with oral vorinostat in patients with advanced solid tumors and non-Hodgkin's lymphomas (NHL). J Clin Oncol 27:3528

Stearns V, Zhou Q, Davidson NE (2007) Epigenetic regulation as a new target for breast cancer therapy. Cancer Invest 25:659–665

Su C, Ren ZJ, Wang F, Liu M, Li X, Tang H (2012) PIWIL4 regulates cervical cancer cell line growth and is involved in down-regulating the expression of p14ARF and p53. FEBS Lett 586:1356–1362

Suijkerbuijk KPM, van Diest PJ, Van der Wall E (2010) Improving early breast cancer detection: focus on methylation. Ann Oncol 22:24–29

Terry MB, Delgado-Cruzata L, Vin-Raviv N, Wu HC, Santella RM (2011) DNA methylation in white blood cells: association with risk factors in epidemiologic studies. Epigenetics 6:828–837

Thomas S, Munster PN (2009) Histone deacetylase inhibitor induced modulation of anti-estrogen therapy. Cancer Lett 280:184–191

Umbricht CB, Evron E, Gabrielson E, Ferguson A, Marks J, Sukumar S (2001) Hypermethylation of 14-3-3 sigma (stratifin) is an early event in breast cancer. Oncogene 20:3348–3353

Van De Voorde L, Speeckaert R, Van Gestel D, Bracke M, De Neve W, Delanghe J et al (2012) DNA methylation-based biomarkers in serum of patients with breast cancer. Mutat Res 751:304–325

Veeck J, Esteller M (2010) Breast cancer epigenetics: from DNA methylation to microRNAs. J Mammary Gland Biol Neoplasia 15:5–17

Virmani AK, Rathi A, Sathyanarayana UG, Padar A, Huang CX, Cunnigham HT et al (2001) Aberrant methylation of the adenomatous polyposis coli (APC) gene promoter 1 A in breast and lung carcinomas. Clin Cancer Res 7:1998–2004

Walter BA, Gómez Macias G, Valera VA, Sobel M, Merino MJ (2011) miR-21 expression in pregnancy-associated breast cancer: a possible marker of poor prognosis. J Cancer 2:67–75

Wang R, Li LW, Wang RL, Fan QX, Zhao PR, Wang LX, Lu SH (2006) Demethylation of estrogen receptor gene and its re-expression in estrogen receptor-negative breast. Zhonghua Zhong Liu Za Zhi 28:894–897

Wang F, Zheng Z, Guo J, Ding X (2010) Correlation and quantitation of microRNA aberrant expression in tissues and sera from patients with breast tumor. Gynecol Oncol 119:586–593
Widschwendter M, Berger J, Hermann M, Muller HM, Amberger A, Zeschnigk M et al (2000) Methylation and silencing of the retinoic acid receptor-2 gene in breast cancer. Natl Cancer Inst 92:826–832
Widschwendter M, Apostolidou S, Raum E, Rothenbacher D, Fiegl H, Menon U et al (2008) Epigenotyping in peripheral blood cell DNA and breast cancer risk: a proof of principle study. PLoS One 16:e2656
Wojdacz TK, Thestrup BB, Overgaard J, Hansen LL (2011) Methylation of cancer related genes in tumor and peripheral blood DNA from the same breast cancer patient as two independent events. Diagn Pathol 6:116
Wong EM, Southey MC, Fox SB, Brown MA, Dowty JG, Jenkins MA et al (2010) Constitutional methylation of the BRCA1 promoter is specifically associated with BRCA1 mutation-associated pathology in early-onset breast cancer. Cancer Prev Res 4:23–33
Wu HC, John EM, Ferris JS, Keegan TH, Chung WK, Andrulis I et al (2011) Global DNA methylation levels in girls with and without a family history of breast cancer. Epigenetics 6:29–33
Wu HC, Delgado-Cruzata L, Flom JD, Perrin M, Liao Y, Ferris JS et al (2012) Repetitive element DNA methylation levels in white blood cell DNA from sisters discordant for breast cancer from the New York site of the Breast Cancer Family Registry. Carcinogenesis 33:1946–1952
XU X, Gammon MD, Hernandez-Vargas H, Herceg Z, Wetmur JG, Teitelbaum SL et al (2012) DNA methylation in peripheral blood measured by LUMA is associated with breast cancer in population-based study. FASEB J. 26:2657–2666
Yan PS, Perry MR, Laux DE, Asare AL, Caldwell CW, Huang TH (2000) CpG island arrays: an application toward deciphering epigenetic signatures of breast cancer. Clin Cancer Res 6:1432–1438
Yan PS, Venkataramu C, Ibrahim A, Liu JC, Shen RZ, Diaz NM et al (2006) Mapping geographic zones of cancer risk with epigenetic biomarkers in normal breast tissue. Clin Cancer Res 12:6626–6636
Yang X, Ferguson AT, Nass SJ, Phillips DL, Butash KA, Wang SM et al (2000) Transcriptional activation of estrogen receptor alpha in human breast cancer cells by histone deacetylase inhibition. Cancer Res 60:6890–6894
Yang X, Phillips DL, Ferguson AT, Nelson WG, Herman, JG, Davidson NE (2001a) Synergistic activation of functional estrogen receptor (ER)-alpha by DNA methyltransferase and histone deacetylase inhibition in human ER-alpha-negative breast cancer cells. Cancer Res 61:7025–7029
Yang X, Yan L, Davidson NE (2001b) DNA methylation in breast cancer. Endocr Relat Cancer 8:115–127
Yu F, Yao H, Zhu P, Zhang X, Pan Q, Gong C et al (2007) let-7 regulates self renewal and tumorigenicity of breast cancer cells. Cell 131:1109–1123
Zhu J, Yao X (2009) Use of DNA methylation for cancer detection: Promises and challenges. Int J Biochem Cell Biol 41:147–154
Zhu S, Wu H, Wu F, Nie D, Sheng S, Mo YY (2008) MicroRNA-21 targets tumor suppressor genes in invasion and metastasis. Cell Res 18:350–359

Chapter 11
Epigenetic of Retinoic Acid Receptor β2 Gene in Breast Cancer

Parvin Mehdipour

Contents

Abstract This chapter is aimed to focus on the multi-disciplinary nature of Retinoic acid receptor β2 (RARβ2) gene in breast cancer (BC) and highlighting the basic information as an evolutionay insight. The antiproliferative and proapoptotic capacities of Retinoids, derivatives of vitamin, play the crucial role in biological processes and chemopreventive agents against BC. Cause of the pyramid growth and progression in cancers has its roots in minor subpopulations of cancer stem cells. It is highlighted that in cancer stem cell model, the classical structure of tumorigenic and nontumorigenic cells is due to the native epigenetic diversity within the cancer cell populations.

P. Mehdipour (✉)
Department of Medical Genetics, School of Medicine, Tehran University of Medical Sciences, Poursina Street, Keshavarz Boulevard, P.O. Box 14176-13151, Tehran, Iran
e-mail: mehdipor@tums.ac.ir

P. Mehdipour (ed.), *Epigenetics Territory and Cancer*,
DOI 10.1007/978-94-017-9639-2_11

Altered expression of RARβ_2 could lead to tumorigenesis and retinoid resistance. Hypermethylation of RARβ2 interact with ERα/PR/HER2 as a triangle target genes in BC patients. Different environmental factors are considered as predisposing/ stimulator factors for methylation in ERα gene. Cancer family history as a preliminary risk factor, was inversely associated with the hypermethylated RARβ_2.

Hypermethylation of specific involved genes in BC may lead to scilecing of those genes which have influential impact on carcinogenic and progressive processes. Moreover, diagnostic and therapeutic paradigms rely on the a bridging system between epigenetic profiling and clinical characteristics of cancer patients. Performance of multi-target strategy by considering pedigree based analysis and molecular/cellular genetics, subsequently, bridging plan will be translated to the clinc. In this chapter, it was aimed to ladder the main facts in molecular and cell biological paradigm about *RARβ2* in BC which may lead to establish the more complementary prognostic based insights in direction of biomarker innovation, therapeutic strategy and more reliable clinical management for breast cancer patients.

Abbreviations

AML	Acute myelocytic leukemia
AOE	Axolotl oocyte extracts
ATRA	All-trans retinoic acid
AHPN or CD437	-[3-(1-adamantyl)-4-hydroxyphenyl]-2-naphthalene carboxylic acid
APC	Adenomatous polyposis coli
BC	Breast cancer
BRCA1	Breast cancer susceptibility gene 1
CTAG1 and CTAG2	Cancer testis antigen 1,2
CBS	Cell based strategy
CCND2	Cyclin D2
CCV	Complementary/confirmative/validitative
CDK2	Cyclin-dependent kinase 2
CDKN2A (p16INK4A)	Cyclin-dependent kinase inhibitor 2A, multiple tumor suppressor 1
CDH1	E-cadherin
CrbpI	Cellular retinol binding protein I
CSCs	Cancer stem cells
CST6	Cystatin-6 gene
DR5	Direct repeat five
DNMT3A	DNA methyltransferase gene
ES	Embryonic stem
EMT	Epithelial to mesenchymal transition
Er	Estrogen receptor
ES	Embryonic stem
EMT	Epithelial to mesenchymal transition
Er	Estrogen receptor
ERP	Epigenetic regulatory proteins

EZH2	Histone-lysine N-methyltransferase/enhancer of zeste
FISH	Fluorescence in situ hubridization
GSTP1	Glutathione S-transferase P1
4-HPR	*N*-(4-hydroxyphenyl) retinamide
HA	Histone acetylation
HAT	Histone acetyltransferases
HDAC	Histone deacetylases
Her2	Human epidermal growth factor receptor 2 (ERBB2 or CD340)
HMEC	Human mammary epithelial cell
HDAC	Histone deacetylase
HM	Histone methylation
HOXA PCDH BMI-1	HOXA gene polycomb protein group homolog Bmi-1
H3K27me3	Histone H3 lysine (K) 27
iPSCs	Induced pluripotent stem cells
IDH	Isocitrate dehydrogenase IDH2
IDH	Isocitrate dehydrogenase
IF	Immunofluoresence
IKK	IkappaB kinase
LSD2	Lysine-specific histone demethylases 1 (LSD1) and 2
LXN	Regulator latexin
MINT	Methylated-IN-tumor
MDS	Mmyelodysplastic Myelodysplastic syndromes
MET	Mesenchymal to epithelial transition EMT
MDS	Myelodysplastic syndromes
Pr (Pgr)	Progestron receptor
NES1	Normal epithelial cell-specific 1 or kallikrein 10
OncomiRs	Oncogenic miRNAs
PPP	Prognosis, Prediction and Preventive
PaC	Pancreatic and
PC	Prostate cancer
PRC	Polycomb repressive complex
RA	Retinoic acid receptors
RARRES1	Retinoic acid receptor responder 1
RASSF1A	RAS association domain family 1A
RARβ2	Retinoic acid receptor-β2
RARs	Retinoic acid receptors
RARg	Retinoic acid receptor g
TSG	Tumor suppressor gene
RXRs	Retinoid X receptors
RARα	Retinoic acid receptor Alfa
Rbp1	Retinol-binding protein, type 1
RIG-I/DDX58	Retinoic acid-inducible gene 1/DEAD (Asp-Glu-Ala-Asp) Box polypeptide 58
SHR	Steroid hormone receptors

SUZ12	Suppressor of zeste 12 homolog
TSGS	Transcriptional tumor suppressor genes silencing
TET2	Tet Methylcytosine dioxygenase 2
TWIST	Human basic helix-loop-helix DNA binding protein
TSG	Tumor-suppressor gene
TNBC	Triple negative breast cancer
TLR3	Toll-like receptor 3
TYRP1	Tyrosinase-related protein 1

11.1 Introduction

Incidence and mortality of cancer is rapidly growing and according to the previous report approximately 12.7 million individuals were diagnosed with cancer and almost 7.5 million were deceased (Chambon 1996) and these figures are rapidly increasing every year. Genetics and epigenetic play a complementary role in cancer and the reversibility of epigenetic alteration is remarkable with beneficial impact on the management of cancer patients through modification and its application as biomarker in cancer clinic.

Cancer is, basically, a cell cycle disease within the genetic paradigm Cancer development and progression are directed through a multi-highway journey and it is beyond the routine strategies which determine mutations, deletions, translocations and amplifications. Cancer has strong link to epigenetic inscription and is, fundamentally, known as the heritable system without capacity to modify the DNA sequence, but is a platform for cellular recognition of gene expression (Jones and Laird 1999).

Tumorigenesis is a multi-directed process in which genetic and epigenetic alterations at DNA, RNA, and protein levels play the key roles, leading to the evolutionary transformation of a single apparently normal cell on the way to s a pre-and/or malignant atatus (Geutjes et al. 2012; Jones and Baylin 2002). Cancer development is mainly related to many cellular and molecular imperfections. Furtheremore, epigenetic, micro- and macro-environmental factors play the key roles as the supportive avenue (Jones and Laird 1999; Hanahan and Weinberg 2011; Pirouzpanah et al. 2014a). As the matter of fact, epigenetic is aimed to study epigenome that direct to alter the manner of gene expression without affecting structural architecture of the genome. By highlighting the reversibility of epigenetic changes, genetic and epigenetic are as the interlink machinery and interact through the cancer transformation. Briefly, cancer development relies on different directive processes including genetics, epigenetic and environmental influences. But '*Cancer does not only deal with mutations, other complementary functions are required for the ongoing profile.*'

11.1.1 DNA Methylation

DNA domains of the human genome, range between 0.5 and 5 kb in which CpG islands, are rich of CG, and usually located in almost half of the promoters of genes. DNA methylation is characterized with an additional methyl group to carbon 5 of the cytosine within the dinucleotide of CpG islands (Lander et al. 2001). By methylation, the transcriptional silencing will be occurred, and DNA methyltransferases (DNMT) 3A and 3B is found to be the key element for formation of *de' novo* DNA methylation through conversion of cytosine residues of CpG dinucleotides into 5-methylcytosine (Tahiliani et al. 2009). Furthermore, 5-methylcytosine can be converted into 5-hydroxymethyl-2′-deoxycytidine by the Ten-Eleven-Translocation (TET) family enzymes as well. Specifically, DNA methylation in 5′ promoters is reported to suppress gene expression, and DNA methylation seems to act with a downstream manner of the promoters within the intra- and inter-genic domains (Maunakea et al. 2010).

However, a model was proposed in which the epigenetic instability of genomic domains that motivates methylation variability in cancer may be linked to tumor heterogeneity (Hansen et al. 2011).

11.1.2 Histone Modifications

Histones are proteins and by accompanying the coiled DNA around it, the nucleosomes are formed. A nucleosome consists of 147 base pairs of genomic DNA enfold twice around a conserved core histone octamer including H2A, H2B, H3, and H4. Besides, histones are capable to regulate gene expression. Histone tails may be involved in the posttranslational chemical modifications, including methylation, that constitute a code, known as the "histone code." Histone modifications could alter the chromatin structure, transcriptional repression, gene activation, and DNA repair (Kouzarides 2007. Moreover, the specialized machinery is capable for induction-, removal- and recognition- of histone modifications, transport of nucleosome, histone, or DNA-modifying enzymes.

11.1.3 Histone Acetylation

Histone acetylation (HA) is as a result of multi-events and the target of occurrence is found to be arginine-(R) and lysine-(K) residues. Two enzyme families including histone acetyltransferases (HAT) and histone deacetylases (HDAC) regulate HA. Besides, HA is also capable to promote transcription (Fig. 11.1). The final event would be formation of heterochromatin by acetylation of lysine 16 of histone 4 (H4K16) (Shahbazian and Grunstein 2007; Filippakopoulos et al. 2006).

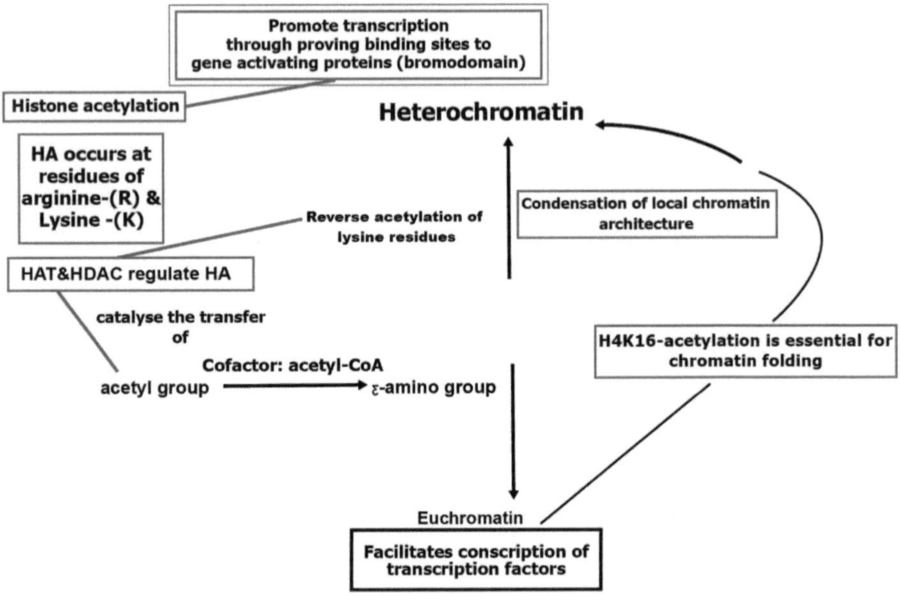

Fig. 11.1 Histone acetylation. *HA* Histone acetylation, *HAT* Histone acetyltransferases, *HDAC* Histone deacetylases

11.1.4 Histone Methylation

Histone Methylation (HM) also occurs at lysine-(K) and arginine-(R) residues without changing the chromatic architecture, but is active as binding sites for the alternative proteins that may be involved in chromatin condensation (Nielsen et al. 2001). Editing of methyl marks (MM)is performed by S-adenosylmethionine (SAM)- dependent methyltransferases(13- Tsukada et al. 2006) and erasing of MM by either the Jumonji family of demethylases or lysine-specific histone demethylases 1 (LSD1) and 2 (LSD2) (Shi et al. 2004). Arginine methylation of histone proteins is known to antagonize the alternative histone marks, followed by raising the histone code complexity (Guccione et al. 2007). A brief insight of HM-process is provided (Fig. 11.2).

Epigenetic modification play a key role in cancer development. In this platform, genomic hypomethylation is considered as a global event, and CPG methylation is responsible for silencing of tumor suppressor genes (Esteller 2008; Dobrovic and Simpfendorfer 1997; Deng et al. 1999; Hatzimichael et al. 2009; Hatzimichael et al. 2012; Esteller 2000, 2006) (Fig. 11.3). In addition, there are complementary events such as distraction of the histone modification territory, and histone modifiers include mutation of deacetylases, and amplification of methyltransferases and demethylases (Fraga et al. 2005; Ropero et al. 2006; Cloos et al. 2006). At a glance there are three major and interactive platforms in cancer and epigenetic modifications; (1) genomic hypomethylation, (2) CPG island, as a contrast territory against platform 1, and (3) Distraction of Histone modification territory (Fig. 11.3).

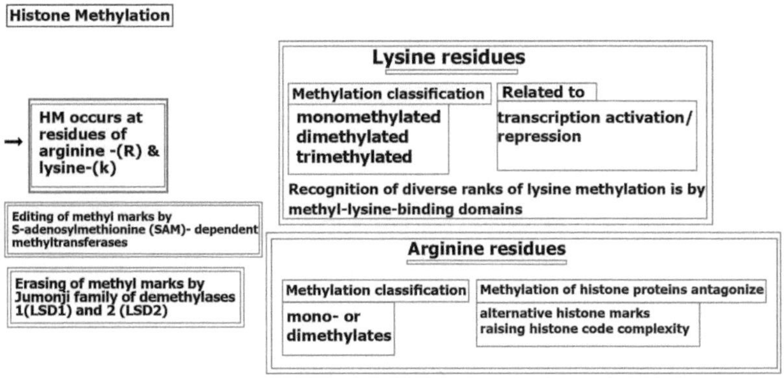

Fig. 11.2 Histone methylation. *HM* histone methylation, *(LSD1) and 2 (LSD2)* lysine-specific histone demethylases 1

Furthermore, epigenetic reservoir may be altered through protein activation or inactivation.

More specifically, the epigenomic stabilization and instabilization relies on enzymatic territory and could be predisposed to mutation in different malignancies (Hatzimichael et al. 2013; Yan et al. 2011; Simo-Riudalbas et al. 2011; Ley et al. 2010; Thol et al. 2011; Delhommeau et al. 2009; Weissmann et al. 2012) (Fig. 11.4). Besides, the metabolic genomic mutations are reported for isocitrate dehydrogenase (IDH) 1 and IDH2 genes in patients with myelodysplastic syndromes (MDS) and acute myelocytic leukaemia (AML) (Ward et al. 2010). In another study in AML,

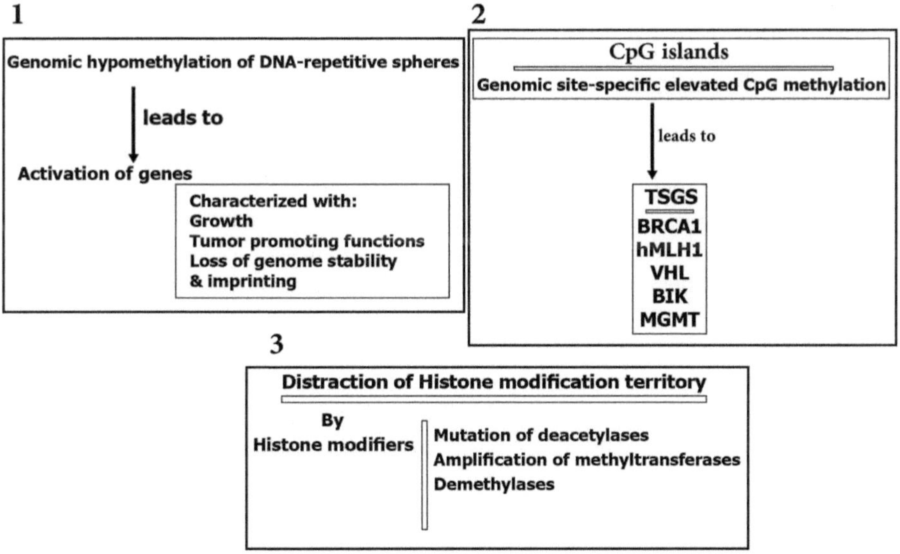

Fig. 11.3 Epigenetic modification in cancer

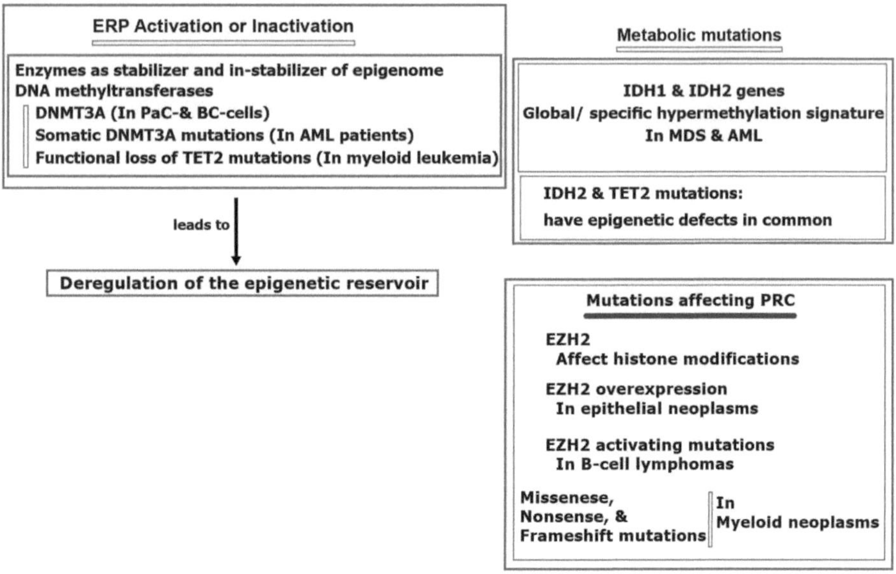

Fig. 11.4 Epigenomic regulation in cancer. *AML* acute myelocytic leukemia MDS myelodysplastic syndrome *BC* breast cancer, *DNMT3A* DNA methyltransferase gene, *ERP* epigenetic regulatory proteins, *EZH2* histone-lysine N-methyltransferase, *IDH* isocitrate dehydrogenase

IDH1 and IDH2 mutations revealed to have common epigenetic defects with TET2 mutations. However, specific hypermethylation signature has been found in AML patients with mutations of IDH1/2 (Figueroa et al. 2010). Interestingly, influential mutation on the Polycomb repressive complex (PRC), per se EZH2, are capable to modify histone elements. Moreover, EZH2 overexpression are found in the epithelial neoplasms and leukemia (Varambally et al. 2012; Benetatos et al. 2013; Simon and Lange 2008). Diversiy in the type of mutations include activating mutations of EZH2 in B-cell lymphomas, missense, nonsense, and frameshift mutations in myeloid malignancies (Sneeringe et al. 2010; Ernst et al. 2010; Nikolosk et al. 2010). Furthermore, silencing of tumor suppressor genes (TSGs) by promoter CpG methylation in breast tumorigenesis has been reported. They have highlighted that the TSG methylation in BCcould be used as a potential marker for cancer management. In a review article the epigenetic alterations in BC, biological and clinical insinuation are provided (Xian et al. 2013).

By considering the multistep carcinogenesis, divese capability of tumor cells through initiation and progression does not, solely, rely on the genetic alterations. But, epigenetic changes assist and enable the neoplastic cells to behave as the cancer cells in malignant phase with the possible reversible capacity (Fig. 11.5). However, still there are some unmasked facts in cancer epigenetics which require the harmonic and complementary approaches by considering the clinical follow-up data.

Details about epigenetic diversity in cancer is, previously, reviewd (Mehdipour et al. 2012) and the summary of crutial facts is provided (Fig. 11.6). In addition, epigenetic alterations play an essential role in malignant transformation.

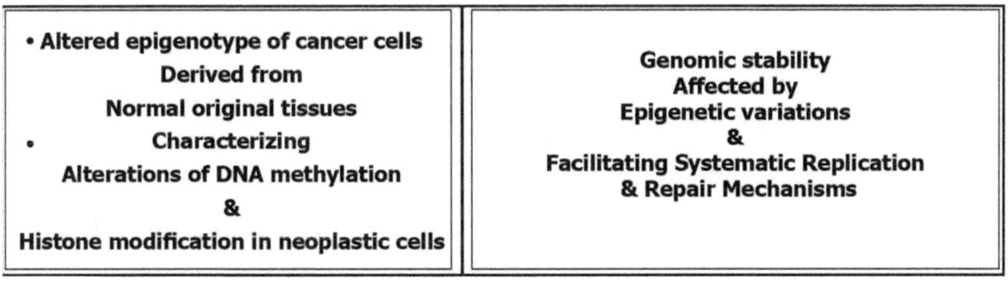

Fig. 11.5 Cancer epigenetic. *MDS* myelodysplastic syndromes, *PaC* pancreatic and, *PRC* polycomb repressive complex, *TSGS* transcriptional tumor suppressor genes silencing, *TET2* tet methylcytosine dioxygenase 2

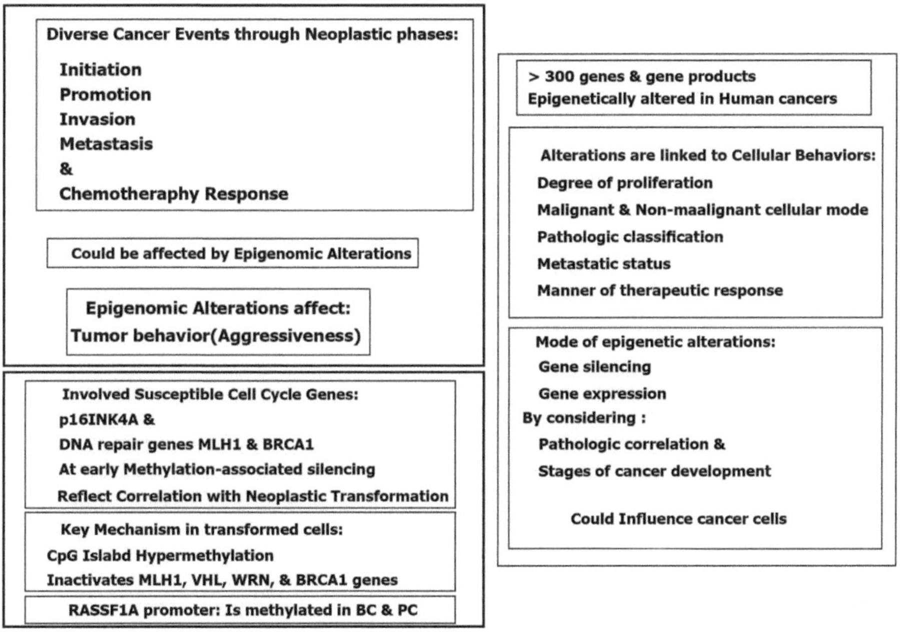

Fig. 11.6 Diversity and cancer epigenetic

11.2 Retinoids

Retinoids with capability of antiproliferative and proapoptotic impacts, are involved in the decisive biological processes and also act as the suppresser of carcinogenic development. Retinoic acid (RA), as a derivative of vitamin A or retinol, has a remarkable epigenetic regulatory role in gene transcription (Chambon 1996). It has also a crucial impact on the morphogenic progress and development of postpuberty mammary gland (Montesano et al. 2002; Wang et al. 2005). Retinoids and their products induce differentiation in various cell types as well.

Retinoids and all-trans retinoic acid, 13-cis retinoic acid, bexarotene as the Retinoids' natural metabolized and synthetic products have the key role in cell differentiation. Retinoids apply their actions by binding to the nuclear retinoic acid receptors (α, β, γ), in those the regulatory impact is due to transcriptional and homeostatic characteristics and their role in neoplastic transformation. By considering the retinoids' efficacy in cancer prevention and treatment, an achievement is the treatment of of leukemic subtypes characterized with chromosomal translocations. So far, the therapeutic limitation in the prevention and treatment of solid tumors may be due to the epigenetic silencing of retinoic acid receptor beta (RARβ) which could be developed by including assessment of RARβ and downstream genes in the solid tumor (Bistulfi et al. 2006; Connolly et al. 2013). Moreover, the "dualistic role" of the retinoic acid signaling pathway in cancer is reviewed (Coyle et al. 2013). They have emphasized on the quadro-radial factors including gene transcription, interactions with other transcription targets, apoptotic pathways, and the immune machinery. It was also referred to the therapeutic impact of retinoid on innovation of an appropriate cancer therapy. Then it could be stated that *retinoic acid could be considered as a hero in cancer*.

Retinoids are the promising targets in cancer prevention and treatment, especially in breast cancer. Partly, the therapeutic success in the prevention and treatment of solid tumors may may be due to the epigenetic silencing of RARβ (Connolly et al. 2013).

Retinoic acid receptors (RARs) belong to the nuclear hormone receptor superfamily, and are characterized with the heterodimeric partners, known as the retinoid X receptors (RXRs).

They are capable to regulate genes having a direct repeat five (DR5) retinoic acid response element (RARE) within the promoter regions. Regarding the pharmacologic levels of retinoid-derived ligands, all-trans retinoic acid (AT-RA) and 9-cis-retinoic acid, for RARs and RXRs are respectively aimed to transactive the heterodimeric partners. In addition, the reduction of RARβ2 or RARβ4 mRNA is reported in breast cancer cell lines (Hoffmann et al. 1990; Swisshelm et al. 1994). Besides, Reduction or lack of the RARβ2 mRNA expression in primary breast cancers have been also reported (Widschwendter et al. 1995; Zhang et al. 1994).

The biological effects of RAR are summarized (Evans and Kaye 1999; Chambon 1996; Altucci and Gronemeyer 2001; Balmer and Blomhoff 2002):

1. Retinoids, including retinoic acid (RA) regulate expression of genes and involved in cell proliferation, differentiation, and apoptosis. They play important role in normal development of embryon and health status of adult.
2. The multiple influential behaviors of retinoids are mediated by RARs and RXRs.
3. To guarantee the biological impact, the liganded RAR binding to RARE will lead to the expression or suppression of Retinoid target gene.
4. RA generates antiproliferative effects in tumor cells.
5. RARs directly mediate RA effects by regulating of gene expression.

Epigenetic silencing could repress the transcriptional function which lead to the reduction of RARβ2 expression (Xu et al. 1994; Houle et al. 1993; Liu et al. 1996).

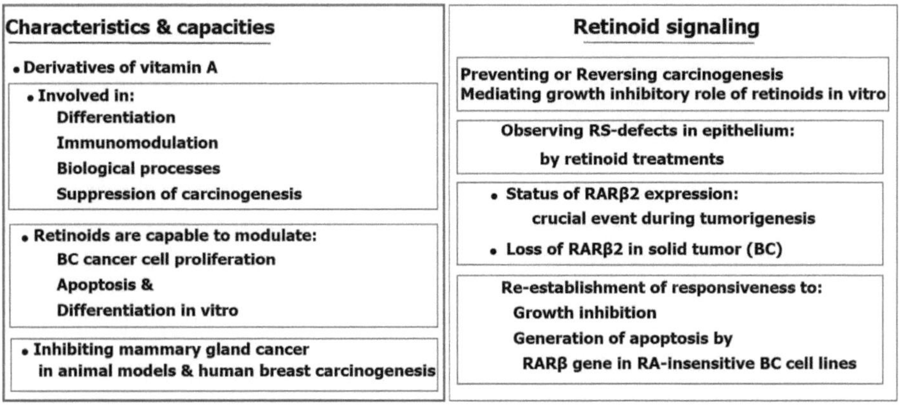

Fig. 11.7 Multi-influential capacities of retinoids

It is also reported that the RARβ-prime as a truncated, oncogenic RARβ protein is entirely expressed in breast cancer cell lines. Its worth to highlight that this isoform could block the functions of tumor suppressor of RARβ2 and RARβ4 protein isoforms as well (Lee et al. 1995; Sabichi et al. 1998). RARβ2 as a two edged sward, is capable to activate either tumor suppressor or antimetastatic programming. However, "the *RARβ2* could be defined as a reflective mirror with multi-potential territory which facilitates diverted interactions with variety of targets at biological and molecular levels" (Mehdipour et al. 2012).

Interstingly, an incorporated process is essential through the RA receptor α (RARα), RA signal elicit chromatin modifications in RA signal leads to the transcription of the RA receptor β2 which is located at chromosome 3p24 (Chambon 1996; Dilworth and Chambon 2001). Furthermore, RARβ2 maintain an independent transcription accompanied by only few downstream RA-responsive genes (Brand et al. 1988; Sucov et al. 1990; Husmann et al. 1991). Retinoids have also multi-influential capacities, however its informative characteristics is summarized (Fig. 11.7) (Mehdipour et al. 2012).

In RA-sensitive BC cell line, an inactive mode of transcription was found for RARβ2 in which notable value of repressive chromatin modifications including DNA hypermethylation could be characterized. Besides, simultaneously, with the conversion of RARβ2 alleles from a permissive transcriptional status into a non-permissive mode which is characterized by aberrant DNA hypermethylation, cells are converted to RA resistance (Ren et al. 2005). In addition, numerous proteins involved in retinol or RA metabolism found as unbalanced or down-regulated in breast cancer cells (Andreola et al. 2000; Guo et al. 2000; Mira-y-Lopez et al. 2000; Rexer et al. 2001), including retinol binding protein 1 (CRBP1) which is involved in the morphological differentiation of human breast epithelial cells (Farias et al. 2005). The CRBP1 is encoded by the *RBP1* gene, was characterized as a downstream RAR-regulated gene, as one of the RAR targets (Smith et al. 1991; Husmann et al. 1992). RARβ2 is also involved in transcription of RA-responsive gene and

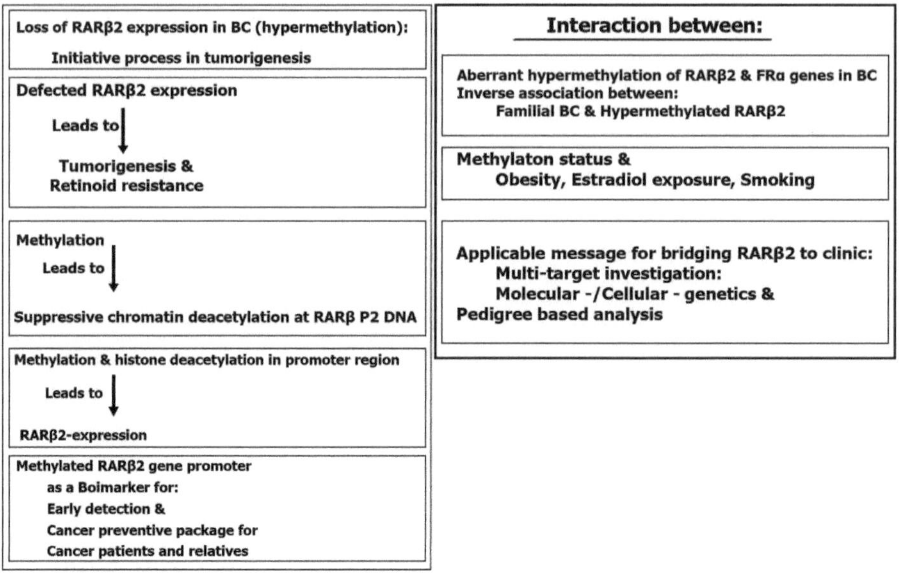

Fig. 11.8 The characteristics of RARβ2

retinoid signaling (Pozzi et al. 2006; Kawakami et al. 2005; Vermot et al. 2005a, b). RARβ has four isoforms with different biological functions of those RARβ2 play an influential capacities in carcinogenesis and therapy and could be listed as cancer preventive and therapeutic agents in BC. The characteristics of RARβ2 is summarized in Fig. 11.8.

Ectopic RARβ2 expression regulate downstream factors which are presentive of differential expression of transcription factors, signalling molecules and metabolic enzymes (Feinberg et al. 2006). RARβ2 is also capable to change the gene expression pattern of AT-RA in cell lines (Kurbel 2013). In addition, the ectopic RARβ2 expression, the type and origin of cell may have role in characterizing the expression profile. Besides, an interactive and regulatory manner may be involved between mode of RARβ2 expression and multi- pathways. So a global genetic programming may be involved for managing the status of gene regulation. In this regard, three upregulated genes by RARβ2 on chromosome Xq28 have been reported at cell line level including ARHGAP4 as a GTPase-activating prote, RPL10 as a 60S ribosomal proteand SSR4 as signal sequence receptor delta 4 (Tribioli et al. 1996; Oh et al. 2002; Mangelsdorf et al. 1994; OMIM: 300090). However, the importance of X chromosome may be due to the involvement of interactive targets in this chromosome with beta retinoic acid. Furthermore, the highlighted points of a review on RAR*β*2 gene in breast cancer is presented in Tables 11.1, 11.2, 11.3 and 11.4 (Mehdipour et al. 2012).

Cancer prevention is an optimal and end point hope in cancer world, in this regard the list of some environmental factors, their characteristics, and key mechanisms which are involved in the altered epigenetic and genetic make up is provided

Table 11.1 Distinctiveness and mechanisms of of RAR-β isoforms. (Adapted from Mehdipour et al. 2012)

Name of elements	Characteristic of element/mechanisms	Associated with/via/ related to/influenced by	Advantages/ disadvantages
RXR	Reduce the mucocutaneous toxicity	Associated with retinoid treatment	–
4-HPR	Inhibits proliferation of BC- cells without capacity to express RARs	Via retinoid receptor-independent mechanisms	Favorable preclinical outcome
	Induce its inhibitory effects	Related to its reduced toxicity compared with other retinoids	Limited therapeutic value in patients with advanced BC
	Induce RAR transcriptional activation and repression in breast cancer cells	–	–
	Preferentially accumulates in breast tissue	–	–
ATRA RARS	Inhibits the proliferation of BC cells is by inducing G1 cell cycle arrest	Sensitivity of ER-positive BC cells to the growth inhibitory effects of ATRA (by expressing RAR *a)*, but ER-negative cells are commonly resistant (due to lacking or small amount of RAR*a)*	
		RAR- β antagonist & expression of RAR-β antisense could inhibit ATRA sensitivity of ER-positive cells	

RXR retinoid X receptors; *4-HPR N*-(4-hydroxyphenyl) retinamide; *ATRA* all-trans retinoic acid; *RARs* retinoic acid receptors; *RAR-β* retinoic acid receptor

(Table 11.2). For instance, the dietary methionine, as a vital amino acid, plays a key role in epigenetic which could provide a normal firm for genome-wide DNA methylation patterns through cytosine methylation. In addition, the CpG methylation patterns is important during the early stages of embryon. Therefore, the nutritional behaviors could be translated to the gestations through ancestral line by influencing the epigenetic and genetic composition. So, the fate of individual health, not only, depends on each individual, but on the previous generations as well. Furthermore, research was conducted to investigate an association between folate intake, vitamins B(2), B(6), B(12) and methionine with methylation of promoter region in E-cadherin, p16, and RAR-β(2) genes in the primary BC tumors (Tao et al. 2011). They have stated that intake of these nutrient may be capable to alter promoter methylation in normal or dysplastic breast tissue. In recent publication the multidisciplinary role of epigenetics in cancer prevention as "Nutri- epigenetic" has been,

Table 11.2 The characteristics of some environmental/dietry factors in epigenetics. (Adapted from Mehdipour et al. 2012)

Factors	Characteristic of element	Involved mechanism	May lead to	Influenced by
Dietary methionine	Vital amino acid (dietary methionine)	Cytosine methylation/CpG methylation patterns	Normal genome-wide DNA methylation patterns	Dietary agents
Transposons	Transposable elements/mobile genetic	Heavily methylated/transcriptionally silent in somatic cells/ disturbed DNA methylation	Genetic mutations/ transcriptional defect/altered establishment and maintenance of epigenetic status	Cellular stress/ environmental and dietary agents
Dietary chemopreventive agents: butyrate, diallyl disulfide, and sulforaphane	–	HDAC inhibitory activity	–	Dietary agents
Resveratrol	A member of sirtuin family of NAD-dependent deacetylases	An inhibitor of SIRT1, a member of the sirtuin family of NAD-dependent deacetylases	Improves health and extends life span	Dietary agents
Green tea polyphenols and phenethyl isothiocyanate	Dual actions of DNMT and HDAC as inhibitors in cancer cells/on DNA and chromatin	Epigenetic modifiers	Cancer prevention	Dietary intervention

HDAC histone deacetylase, *SIRT1* silent information regulator type1

complementary, reviewed (Gerhauser 2013). They have highlighted the influential impact of natural chemopreventive agents on the expression or manner of action of histone modifying enzymes and DNA methyltransferases. The major deregulated events during carcinogenic process by epigenetic alterations include drug metabolism, cell cycle regulation, potential to repair DNA damage/ to induce apoptosis, response to inflammatory stimuli, cell signaling, cell growth control and differentiation. The provided scheme illustrates a complementary view of these machinery (Fig. 11.9). It was also emphasized that epigenetics has influential impact on gene regulation during developmental stages (Gerhauser 2013). It was also highlighted that epigenetic changes occur at early phase of cancer development. In addition they have stated that an interventions with chemopreventive agents, probably, starts early after birth, but I believe that the programming of epigenetic machinery has its root in two complementary territories including heritage and the life style of embryon,

Table 11.3 The functions of chemopreventive agents in epigenetic. (Adapted from Mehdipour et al. 2012)

Name of chemo-preventive agent	Alternative name/derivates	Function	Disadvantages	Advantages
Natural retinoids: retinoid 9-cis retinoic acid (9-cis RA, alitretinoin)	ATRA	Inhibit growth of BC- cells	Lacking clinical efficacy	
		Transactivates	Toxic side effects: Hyperlipidemia, mucocutaneous, liver toxicity	
		RARs and RXRs		
		Binds RAR		
		Does not bind RXR		
Synthetic N.R.-derivatives: The 9-cis RA, 4-HPR (fenretinide)	Synthetic derivative of ATRA	Inhibit growth of BC- cells		Higher potency
		Transactivate RAR responsive genes		Less toxicity
				Have significant RARs binding
Synthetic retinoids: LGD1069 (bexarotene, Targretin)	Synthetic derivative of 9-cis RA, 4-HPR	Inhibit growth of BC- cells	No significant binding RAR	Have selective binding of RXRs
		Transactivates	No transactivation of RAR responsive genes	Higher potency
		RXRs		Less toxicity
ATRA & Cell Cycle		Interaction between cell cycle regulators with antiproliferative effects of ATRA in BC cells		(1) decreased expression of cyclin D1 and D3, (2) activity of cdk2 and cdk4, and (3) expression and phosphorylation of pRb could be associated with growth inhibition induced by ATRA in BC cells
Retinoids/RAR- β gene		1-Retinoids alter gene expression in target cells		Important event in tumorigenesis may be loss of RAR- β 2 mRNA expression
		2- RAR β gene may be considered as a tumor suppressor		

Table 11.3 (continued)

Name of chemo-preventive agent	Alternative name/derivates	Function	Disadvantages	Advantages
RAR-β 2		1-Reduced RAR-β 2 mRNA expression has been observed in BC	1-RAR-β expression mediates the growth inhibitory effects of retinoids	RAR- β *b* was induced in 33 % of BC patient-streated with ATRA for 3 weeks
		2-RAR-β transcription is downregulated in BC cell lines and tumors, but upregulated in normal mammary epithelial cells	2- RAR-β2 mRNA induction is associated with growth inhibition in response to ATRA	
			3-resistance to ATRA could be associated with a collapse in RAR-β2 inducibility in BC cells *in vitro*	
ATRA& the protooncogenes jun and fos (AP-1)		1-AP-1 plays role in BC cell proliferation and transformation	AP-1 is associated with ATRA-mediated growth inhibition in BC cells	
		2- Its activity could be inhibited by ATRA		

N.R. natural retinoids, *ATRA* all-trans retinoic acid, *BC* breast cancer, *RAR* retinoic acid receptor, *RXR* retinoid X receptors, *RAR-β* retinoic acid receptor, *Rb* retinoblastoma, *4-HPR (Fenretinide) N*-(4-hydroxyphenyl retinamide)

i.e. maternal and also ancestral lines including paternal and maternal sides. In addition, a comprehensive insight has been, recently, provided in a book review (Gray SG 2014) in which the basic aspects of epigenetics have been explored.

11.2.1 Stratigical Approaches of Retinoic Acid Receptor β2 Gene

The key role of RAR-β2 gene is highlighted as a tumor suppressing gene which is epigenetically silenced through the course of carcinogenesis. It is shown that the combination of histone deacetylase and DNA methyltransferase are capable to inhibit and reverse the epigenetic silencing capacity of several growth regulatory genes in breast cancer cell lines (Mongan et al. 2005).

Table 11.4 Chemopreventive contribution of RARs and RXRs *in vivo*. (Adapted from Mehdipour et al. 2012)

RARs/RXRs contributions	Iso-forms of RARs	Generated by	Conserved modular structure consists	Selective ligand/or as combination	Mode of effect	Leading to
1-RARs	Alpha, beta, gamma	Alternative promoters and differential splicing	1-AF-1 or A/B (ATAF1) Domain	9-*cis*-RA	Suppression of carcinogenesis	Reduction in tumor burden
			2- Zinc-finger DBD or C (BD)1			
			3- CoR or D (HCB3) Domain			
			4- LBD or AF-2 or E (LBTA4) Domain			
			5- Variable F (CT) 5 Domain			
2-RXR				LGD1069 (targretin)	Suppression of carcinogenesis	Reduction in tumor burden
3- RXR				LGD1069 & tamoxifen	Increased efficacy	Increase in differentiation/ decrease in cellular proliferation
4* RXR				ligand LGD1069 & tamoxifen	Increased efficacy	Appropriate response to retinoid therapy

RA retinoic acid, *RARs* retinoic acid receptors, *RXRs* retinoid X receptor 1–3: in the *N*-nitroso-*N*-methylurea-induced rat mammary tumor model, *ATAF1* amino-terminal activating factor-1transcriptional activation domain; *BD 2* DNA-binding domain, *HCB 3* hinge/corepressor binding; *LBTA4* ligand-binding/transcriptional activation, *CT 5* carboxyl-terminal 5, 4*: in carcinogen-induced model characterize with no response to tamoxifen

Moreover, the chemopreventive impact of retinoids are related to RAR-β2 with its decreased expression in numerous malignant tissues. The first exon expressed in the RAR-β2 transcript is found to be methylated either in ZR-75-1 and SK-BR-3, or in six BC specimens (Widschwendter et al. 2001). They have emphasized on the grading and methylation mode of the lesions by considering the lack of expression of RAR-β2 in grade III and its remarkable carcinogenic impact of this gene on the breast.

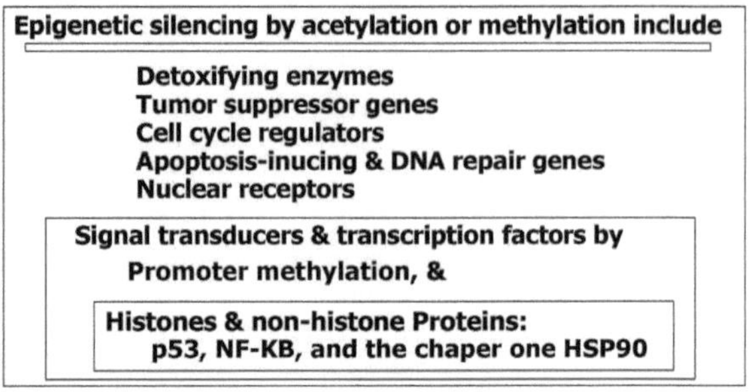

Fig. 11.9 Major deregulated events during carcinogenic process by epigenetic alterations

The strategic impact of RAR β2 is, partly, due to overexpression of RAR β2 gene, located at chromosome Xq28 and also the transcriptional regulatory mechanisms and immune response (Wallden et al. 2005). The antimetastatic potential of RAR β2 signalling was reported in human breast cancer cells in which following findings were reported:

1. Overexpression of tumor-cell antigens include cancer testis antigen (CTAG1 and CTAG2), involved in innate immune response i.e., retinoic acid-inducible gene 1/DEAD (Asp-Glu-Ala-Asp) box polypeptide 58 (RIG-I/DDX58), and tumor suppressor functions (i.e., Tyrosinase-Related Protein 1:TYRP1).
2. Reduced expression by RAR β2 includes CD164 as cell adhesion functions, and FABP6 is involved in metabolic or nutritional processes, and JUN as a transcription factor.
3. By considering expression profile, they have found diverse capability including "activated and repressed cellular activities in response to overexpression of RARβ2" that have emphasized interesting facts in the metastatic process.

Moreover, epigenetic alterations have key roles in tumorigenesis which has been reported in different types of malignancies, including breast cancer (Sadikovic et al. 2008). Misregulation of specific genes by either genetic or epigenetic alterations are considered as the key facts in cancer development (Sadikovic et al. 2008; Jones et al. 1999).

There are some pitfalls in tumor- formation, progression and management in breast tumors which could be due to epigenetic regulation, combat between epigenetic and genetic territories (which are summarized in Fig. 11.10 (Lo et al. 2008; Novak et al. 2009; Hinshelwood et al. 2008; Tommasi et al. 2009). However, the important points are reversibility of epigenetic chromatin modifications (Hendrix et al. 2007), and re-expression of specific targets (Clark et al. 2006). The questions are: (1) Does genomic milieu play key role in reprogramming of tumor suppressor genes? (2) How the required elements and transcriptional activation manage to induce re-expression of silenced territories.

Regarding the reprogramming of tumor suppressor gene, the occurance of silencing in the exogenous *RAR*β promoter of MCF-7 and HCC1954 cell lines was previously, created (Kondo et al. 2008). However, three questions are proposed (1)

How is the fate of cancer cell programming through the cell devisions? (2) How the micro- and macro- environmental factors manipulate this process? And (3) Is there any similarity between embryonic territory and post birth genomic and somatic statue?

One of the focal target in cancer prevention is diatry elements, amongst those folate intake is capable to decrease BC risk (Chen et al. 2014). Furthermore and as an example, a recent publication has focused on the essential and required elements by using a biomarker-based validity of food questionnaire (Pirouzpanah et al. 2014a). We have concluded that dietary folate and cobalamin were correlated with the BC-patients' fasting plasma concentrations. Besides, the dietary methyl group (DMG) have influential impact on the status of hypermethylation of some genes including *RARβ2*. In this regard, the lower dietary intake of folate or cobalamin, and higher intake of riboflavin or pyridoxine led to an increased incidence of breast tumors development by occuring the promoter methylation of *RARβ2 gene* (Pirouzpanah et al. 2014b).

However, the associations between DMG intake and the promoter hypermethylation mode of *RARβ2* and its expression level in BC patients was recently found by us (unpublished data). This report revealed an association between high dietary riboflavin and pyridoxine intakes with the hypermethylation status of *RARβ2*. By considering the cumulative nutrient-epigenetic interactions, we have emphasized that the individual who had high dietary vitamin B2 or vitamin B6, are more predisposed to breast tumors with *RARβ* hypermethylation. In addition, deficiencies of folate and cobalamin were shown to have simultaneous elavatory influence on frequency of tumors characterized with methylated *RARβ* gene.

11.2.2 Importance of Genetic and Epigenetic Role in Breast Cancer: Rar as a Focal Target

By considering the manner of epigenetic alteration, without change in DNA sequence, i.e., through an opposite direction of genetic changes, *the question is whether they cooperate or combat?* In this regard, the main fundamental strategies about the key role of genetic and epigenetic alterations is provided (Fig. 11.10). Moreover, in breast tumors there are the following facts at a glance (Allegrucci et al. 2011):

1. Uncharacteristic epigenetic regulation of cell cycle genes, apoptosis, DNA repair, cell adhesion and signalling leads to tumor formation, manage progression and drug resistance.
2. In initial developmental stages of breast tumor, epigenetic changes overcome genetic alterations (silencing of CDKN2A (p16INK4A), HOXA and PCDH gene clustering by DNA methylation with over-expression of polycomb proteins BMI-1, EZH2 & SUZ12 during Induced or spontaneously transformation) of human mammary epithelial cells. Methylation of homeobox genes in ductal carcinoma in situ and stage I in breast cancer is also highlighted.

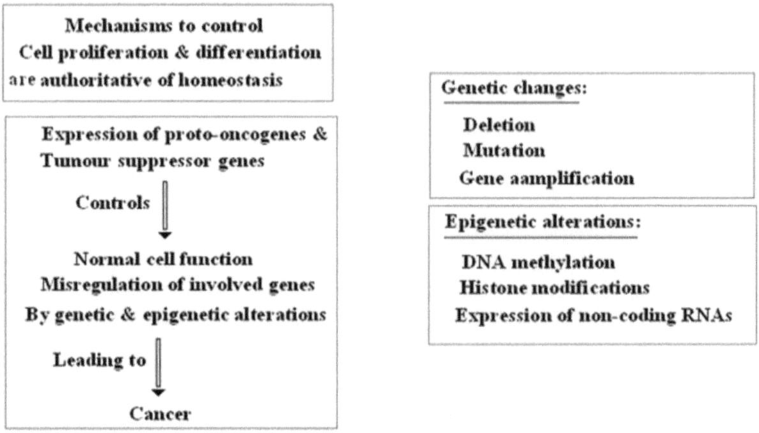

Fig. 11.10 Key role of genetic and epigenetic alterations

11.2.3 Characteristics Highlights

This item is rather directive and leading paradigm for the future of cancer strategies and managements:

1. Epigenetic chromatin modifications are reversible which is applicable for attenuation of cancer, like in emryo.
2. At breast cancer cell lines level: RARβ, CST6, CCND2 (Cyclin D2) were Re-expressed.
3. Epigenetic modifications of genomic regions mediate silencing of cancer-related genes.

The effect of cellular retinol binding protein I (CrbpI) loss on mammary at RA homeostasis using the Rbp1−/− mouse model was reported (Pierzchalski et al. 2013). It was laso highlighted that type I (CrbpI), is encoded by retinol-binding protein, type 1 (Rbp1). The Rbp assists vitamin A (retinol), acting as chaperone; and revealed to be, epigenetically, silenced in about 25 % of human breast cancers. Such harmonizing process have a key role in proliferation, apoptosis, differentiation, and migration.

Invstigation *in vitro* showed that the transcription factor RARβ2 is an effective inhibitor of breast cancer cells, proposing the loss of RARβ expression in primary breast cancer. It was reported that expression of the RARβ protein, as the translated product of the human RARβ4 transcript, is elevated in all of five breast tumor cell lines which was related to normal human mammary epithelial cells (Karen et al. 1999).

Loss of expression of RARβ2, is also observed in prostate cancers. Regarding the epigenetic mechanisms and its stable silencing, they have shown the manner of silencing of RARβ2 promoter by harboring two diverse tyrannical chromatin profiles at the same locus, highlighting "the polycomb-mediated epigenetic repression process in cell line prostate cancer (Moison et al. 2013). Morover, the RARβ4 protein is located in the cytoplasm and subnuclear section and its expression is remarkably raised in breast tumor cell lines. They have stated that "RARβ4 functions as a dominant-negative repressor of RAR-mediated growth suppression."

RARs regulate the genetic network in cancer including breast cancer. It was previously reported that RAR binding act simultaneously with estrogen receptor α (ERα) binding throughout the genome. Such cooperation is reflective of an extensive crosstalk between RA and estrogen signaling through which regulation of the breast cancer-associated genes would be possible (Hua et al. 2009). ERα- and RAR-binding crossroads of two critical nuclear hormone receptor signaling pathways creates a genomic global machinery to balance gene expression.

The characteristics of RA and RARβ2 and its interaction with RARα at cell line level is summarized (Ren et al. 2005):

1. Resistance to Growth-inhibitory action of Retinoic acid (RA).
2. Resistance of RA is associated with silencing and hypermethylation of RARβ2 gene.
3. RARβ2: RA-regulated tumor suppressor gene.
4. Epigenetically silent RARβ2 relates to lack of RA receptor α (RARα).
5. RARα regulates RARβ2 transcription by Arbitrating dynamic alteration of RARβ2 chromatin in the presence & absence of RA.
6. RARβ2 silencing can occur in absence of DNA methylation.
7. Restoration of RA signal at a silent RARβ2 through RARα leads to RARβ2 reactivation.
8. RARβ2 silencing and RA resistance lead to an unappropriate RA signal- incorporation at RARβ2 chromatin.

Now, does epigenetic change, such as RARβ2 methylation at diverse tumors' level is decisive sign to silence this tumor suppressor gene?

In addition, abnormal RARβ2 inactivity may cause repressive epigenetic altertaion at RARβ2, and consequently to RARβ2 silencing and RA resistance (Sirchia et al. 2000, 2002). Besides, normally, dynamic histone changes leads to RARβ2 transcription (in the presence and absence of RA) (Collingwood et al. 1999; Perissi et al. 2004; Dilworth et al. 2001; Xu et al. 1999).

By Bridging and messaging network between RA and RARβ2 chromatin, the following conclusion is provided:

1. RA resistance may be as the result of an aggravated and extended RARβ2 transcriptional repression; and on the basis of a substandard incorporation of RA signal at RARβ2, with a consequence of diverse factors including genetic, epigenetic, metabolic, micro- and macro- environment.
2. Abnormal RARβ2 function may be due to the lack of efficient RARα (as the upper regulator of RARβ2 transcription).
3. RARα has the role of keeping the chromatin of its direct target genes, such as RARβ2, poised for transcription yet inactive.

4. By RA-RARα binding, the chromatin-mode of the target genes will be renovateed from inactive into active due to the histone alteration, chromatin remodeling, and transcriptional activation.

Besides, in hormone-regulated genes, transcription is normally regulated by dynamic alterations of the chromatin condition in the presence and absence of the relevant hormone (s) (Collingwood et al. 1999; Perissi et al. 2004; Dilworth et al. 2001; Xu et al. 1999). This model may unmask the etiology of aberrant epigenetic silencing in cancer and aging. It was also stated that RARβ2 is commonly epigenetically silenced in RA-resistant cancer cells (Sirchia et al. 2001) and RA-resistant tumors (Sirchia et al. 2002). So, it was hypothesized that this fact is upon to an abnormal status of chromatin-repressive, subsequent lack of essential requirement including RA to facilitate the integration of RA signal at RARβ2. Furthermore, the accumulation of altered histone and DNA level is also presented (Bachman et al. 2003; Stirzaker et al. 2004). Such event in RARβ2, leads to CpG methylation.

Moreover, RA binding to RARα engage coactivator elements with histone acetyltransferase activity which is adequate to switch RARβ2 from a silent to a permissive mode (Perissi et al. 2004). Restoring RA-RARα signaling at RARβ2 is the self reactivation of the RARβ2 receptor (Chiba et al. 1997; Husmann et al. 1991; Sucov et al. 1990). However, Estrogen receptor alpha (ERα) is frequently epigenetically silenced in RARα-negative tumors (Ferguson et al. 1995). Besides, The aberrant hypermethylation of *RARβ2* and ERalpha at tumor level has been reported within the Iranian breast cancer patients (Pirouzpanah et al. 2010). As the matter of facts, obesity, duration of estradiol exposure, and smoking are considered as predisposing factors for development of methylation in ERalpha gene. Notably, familial BC was, inversely, correlated with the hypermethylated *RARβ2*. In addition, plasma folate and vitamin B12 levels were associated inversely with the hypermethylation status of ERalpha gene. These dat suggested that the mode of hypermethylation of specific genes is associated with environmental including lifestyle-related factors.

Interetingly, RARβ2 as a tumor suppressor gene, is involved in carcinogenesis of breast cancer and its silencing is linked to epigenetic chromatin alterations with influential capability on the promoter of RARβ P2. Suppression of chromatin deacetylation at RARβ P2 promoter is occurred due to the DNA methylation (Sirchia et al. 2002). In addition, they could maneuver on the level of histone reacetylation at RARβ P2 either *in vivo* or *in vitro.*

For more information, downregulation of the RARβ2 gene is reviewed (Widschwendter et al. 2001). The RARβ2gene as a tumor suppressor gene induces apoptosis and induction of RARβ2 leading to the chemopreventive and therapeutic effects of retinoids. It was highlighted that RARβ2 is reduced or lost, due to the involvement of 5′-region as a cause for loss of expression, through the progression of breat cancer. Interestingly, RARβ2 gene is revealed to be unmethylated either in benign breast tissue or in other normal tissues. In addition, loss of expression of RARβ2 gene was found in prostate cancer. In this regard the directive epigenetic mechanisms to the stable silencing was previously investigated in human prostate tumor cell lines (Moison et al. 2013).

Importantly, by considering gene expression and histopathological features, the methylation status of the RARβ2 gene, and the RARβ2 expression is found to be lower in malignant tissue than in fibroadenoma and normal tissue, but the methylation status was higher in malignant tissue than in normal (Sun et al. 2011). They have highlighted the hypermethylation as an initial event in carcinogenic process of breast.

11.3 About Cancer Stem Cell

Cancer stem cells (CSCs) as a cooperative network interact with many molecular and biological event. The characteristics of CSCs are previously highlighted (Clarke et al. 2006). CSC is a cell within a tumor property which instinctly acquires the capability to self-renew and to establish the heterogeneous lineages of cancer cells. CSC self renew capacity include two manners, (1) "Symmetrical selfrenewing cell division" in which the identical CSCs having self-renewal ability; and (2) "Asymmetrical self-renewing cell division" in which one stem cell and one more differentiated progenitor cell are characterized. CSCs are characterized by their capability to reiterate the generation of an incessantly growing neoplasm. The putative CSCs are also named as "tumorinitiating cell" and "tumorigenic cell". However, symmetrical division of stem cells may form two progenitor cells, leading to depletion of CSCs population or leading to cancer celll death which may be considered as an medication for cancer therapy.

The CSCs have been, initially, isolated in human breast carcinoma (Al-Hajj et al. 2003). The key related characteristics of ductal invasive carcinomas include; (1) Heterogenic intratumoral differentiation of this breast disease, (2) The epithelial to mesenchymal transition (EMT) status, (3) Expression of TWIST1, as a repressor of an EMT-inducing transcriptional factor in invasive lobular breast cancer, (4) Disuniting of invasive tumor cells in EMT-like condition, (5) different pathways are involved in formation of cancer stem cells, and (6) Distribution of associated tumor cell, are all correlated with the malignant progression leading to a poor clinical prognosis (Yang et al. 2004; Fujita et al. 2003; Xue et al. 2003; Blanco et al. 2002).

Due to these characteristics breast cancer is a focal territory in which two paradigms including cancer stem cells and EMT of cancer stem cell migration is remarkably highlighted.

Moreover, the intestinal gastric cancer, pancreatic cancer, and squamous cell carcinomas are also the examples for this category in which the intra-tumor heterogeneity and EMT could be detected (Nakajima et al. 2004; Rosivatz et al. 2002; McAlhany et al. 2004).

Three main fscts about stem cells in solid tumors are highlighted as followings:

1. Are less accessible.
2. Lack of the appropriate functional assay for tracing and quantifying normal stem cells from different organs.

3. Limitation in isolation of stem cells by the cell surface markers in human.

Regarding the genetic and epigenetic signatures of "Stemness" Clarke and his colleagues (Clarke et al. 2006) have stated very important considerations as:

1. "To identify true signatures, pure populations are necessary." This is especially true for cells expected to be rare, such as cancer stem cells, whose expression signature would be swamped by the majority of nonstem cells in a whole tumor sample.
2. "Even after a cancer stem cell signature from a particular type of tumor is identified, one cannot assume that a given signature is useful for identifying cancer stem cells in a different tumor type unless validated by a functional assay (such as an *in vivo* self-renewal assay as it is the most definitive at this point in time)."

However, as a complementary information the essential renewal assay *in vivo* is not adequate to be translated in human and more specific definition is required for evaluating the quantitative values of stemness in human cancers. Such attempts require the follow-up strategy during different stages of cancer patients. Although microarray and genomewide technologies are applied to unmask tendency in genetic and epigenetic for CSCs, but specific Cell Based Strategy (*CBS*) would, appropriately, solve the heterogeneity insight of diverse tumors.

Labelling a cell as stem cell and cancer stem cell is a rather difficult aim. There are some key questions; (1) Where is the stem cell originated from? Is it derived from other stem cell populations? Or (2) is it derived from an initially normal cell or (3) Or derived from cancer cell?

Now lets to characterize the capability of *in vitro* assay:

Reliability of an *in vitro* assay to define a cell as "stemness" and cancer stem cell:

1. Reflects the characteristic of an early stage.
2. The results require the complementary/confirmative/validitative (CCV) at vivo level.
3. The complementary functional assay (s) is required.
4. Gene activity profiles, gene expression signatures and/or cell surface marker are required.
5. There are some pitfalls in cell analyzing due to heterogenetic character of cancer cells and the techniques through which the results reflect a global insight and not individualized characteristics. To support this matter, an example is the expression assay by real time PCR which is considered as a global insight. Th spectrum for categorizing the cancer cell population is diverse and an exact grouping profile is essential to narrow and escape the subgrouping problem. Relative signature of cellular and molecular targets is required for providing more reliable definition for certain population of cancer cells.
6. There is a double edged sward including inactivation or activation of target gene (s) to reduce or produce stem cells respectively. This would support the therapeutic machinery in clinical application. Besides, what does matter would be resistance of cancer cells in target based therapy.

As the matter of fact and by considering the epithelial to mesenchymal transition/mesenchymal to epithelial transition (EMT/MET) processes, there are a triangle in cancer initiation including programming/reprogramming/cancer progression. Furthermore, stem cells are associated with embryonic stem (ES) cells or induced pluripotent stem cells (iPSCs), immortalized mammary epithelial cells that have undergone epithelial to EMT transition which is the reverse manner of the MET process (Mani et al. 2008; Morel et al. 2008). There is also a cooperation between two capacities including EMT and stemness through cancer progression, and separation of cells from the primry tumor that facilitate migration and metastasis. Interestingly, such evocative process of EMT is similar to the embryonic tissue development which is the state of art and fascinating journey of cells from far away departure station to diverse destinations (Thiery et al. 2009). Also, CSCs are capable to kernel new neoplasms to self-renew and lead to generate non-stem differentiated cells. CSCs is not apparently pluripotent which differentiate them from iPSCs; due to this point, cancer cells do not produce any cell type; instead may be considered as relapse of the original primary tumor cells occurs through metastatic process (Gupta et al. 2009; Polyak et al. 2009). However, the machinery of stem cell and invasion is rather a core item in cancer metastasis. In this regard, it worths to refer to the recent work which was focused on the "mouse haematopoietic stem cell regulator Latexin (LXN)" (Oldridge et al. 2013). This target was considerd as a unique "homologue of the retinoic acid receptor responder 1 (RARRES1) gene". The co-expression of either RARRES1 or LXN was suppressed by DNA methylation in prostate cancer (PC) cell lines and furthermore inhibition of RARRES1 and LXN led to increase the invasive capability of primary PC cell line.

Additionally, it is recently reported that phosphorylation of the RARγ2 play a key role for the neuronal differentiation embryonic stem cells in mouse (Al Tanoury et al. 2014).

11.3.1 Modelling Cancer Stem Cell

As a matter of fact the dialougue between genetics, epigenetics and the relevant diversities are the partial reason for heterogeneity of cancer cells.

Now, by considering the common and uncommon characteristics in variety of cancers, important points are adressed as; (1) Is there any cancer model for specific cancer? (2) However, it was stated that the cause of the pyramid growth and progression in cancers has its roots in "small subpopulations of cancer stem cells" (Reya et al. 2001; Dick 2008). Also, do cancer stem cells experience relatively irreversible epigenetic alteration (s) to form different population of nontumorigenic cancer cells? However, There is battle between diverse subpopulation of cells including tumorigenic and nontumorigenic which are differentiatable from each other by limited available expression profiling assays which was stated to be, morphologically, unclear (Al-Hajj et al. 2003).

Furthermore, the wonder is that 'Tumorigenic cancer stem cells may form diverse Nontumorigenic cells.' And *'How could we differentiate tumorigenic cells from nontumorigenic cells? Is it due the epigenetic event? Is it as a result of either epigenetic or genetic events?'*

There are few publication on cancer stem cell model in limited number of human cancers and in mouse (Kelly et al. 2007; Williams et al. 2007; Quintana et al. 2008; Quintana et al. 2008), but still the evolutionary/clonal model by Nowell seems to be more precisely (Nowell 1976). The focal points of this model include the heterogenic faetur of cancer stem cells (CSC), hetrogenic/homogenic feature of clones, diverse tumorigenic capability of CSC; and organization of tumor. In addition this model is characterized with diverse nature of tumorigenic and nontumorigenic cells which are affected by epigenetic in CSCs, and by genetic or epigenetic in clonal model. Furthermore, involvement of diversity in epigenetic and genetic is an explanation for therapeutic resistance in various cancers among tumorigenic cancer (Nowell 1976).

Diversity in epigenetic event between cancer stem cells and their progeny could verify the clinical manners of specific cancers, but clonal evolution has a key role in all neoplasms.

When the cancers are classically developed into epigenetically differentiable populations of tumorigenic and nontumorigenic cancer cells, the occurrence of clonal evolution in the CSCs, plays a key gate role (Barabe et al. 2007).

Some facts and questions in cancer cell heterogeneity include:

1. Tumorigenic capability due to epigenetic and/or genetic.
2. Does cancer cells differentiate to a nontumorigenic status?
3. Diverse behavior of tumorigenic cells from nontumorigenic cells depends on their epigenetic characteristics (Lapidot et al. 1994; Bonnet and Dick 1997; Al-Hajj et al. 2003; Singh et al. 2004; Ricci-Vitiani et al. 2007). Such explanations would be assumed by involvement of cancer stem cell model.
4. The cancer stem cells are not necessarily rare (Kelly et al. 2007).
5. By considering the clonal evolutionary model, genetic heterogeneity in cancer cells could lead to heterogenic feature, cellular function, and finally response to the therapeutic protocols.
6. Epigenetic diversity governs the ongoing events by leading to more complementary heterogenic behavior.
7. In cancer stem cell model, the classical structure of tumorigenic and nontumorigenic cells is due to the native epigenetic diversity within the cancer cell populations (Reya et al. 2001; Dick 2008). In this regard the provided image is indicative of such diversity in stem cells (Fig. 11.11).
8. Does clonal evolution leads to heterogeneity and forms a tumor consisting of a unique structure of tumor cells which have lots in common?
9. Does misregulated self-renewal pathways cause differentiable epigenetically tumorigenic and nontumorigenic cells in all cancers?

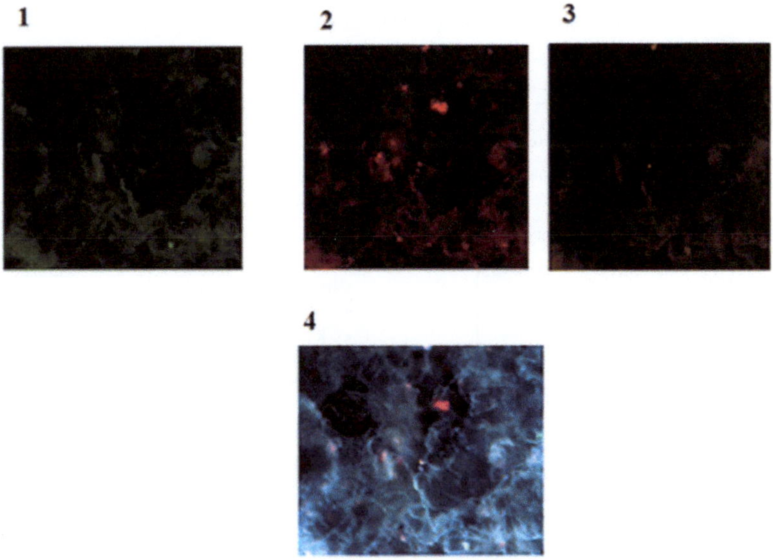

Fig. 11.11 Protein expression of p53, CD44 and CD24 in a patient affected with breast cancer. *1* Tumor of breast conjugated with FITC reflecting low expression of p53, *2* Same cells conjugated with Pe-Cy5 reflecting diverse expression of CD44 including low, medium and high (In limited cells), *3* Same cells conjugated with R-Pe reflecting lack of expression of CD24, accompanied by an isolated cell with low expression, *4* Co-expression of p53, CD44 and CD24 presenting the presence of limited cancer stem cell cooperating with CD44 Magnification:x200. From: P.Mehdipour's archive

11.4 An Interaction Insight in Epigenetic by Focusing on the ER, PR, HER2 Triangle Targets

One of the serious concern in breast cancer is the possible developmental capacity of malignancy in premalignant breast neoplasm. Such process by considering the status of DNA methylation including Methylated-IN-Tumor (MINT)17, MINT31, RARβ2 and RASSF1A genomic markers have been studied through benign, premalignant and malignant status of breast cancer (van Hoesel et al. 2013). They have found DNA hypermethylation at early stage of BC development with diverse degree of tumor during the cancer progression.

In a review article, the involved mechanisms in epigenetics and cell machinery were highlighted in which the altered methylation status was related to oncogenic behavior and cancer cell proliferation. In addition, DNA methylation and posttranslational histone modifications is shown to regulate gene expression without DNA sequence alteration. Besides, cellular functions including cell cycle, immunoresponses and signal transduction were highlighted (Itoh et al. 2013).

An epigenetic progenitor model was, previously, provided in human cancer (Feinberg et al. 2006) which was further used in invasive breast cancer (Kurbel 2013). This model was designed on the basis of expression status for three key molecules including estrogen receptors (ER), progesterone receptors (PgR), and

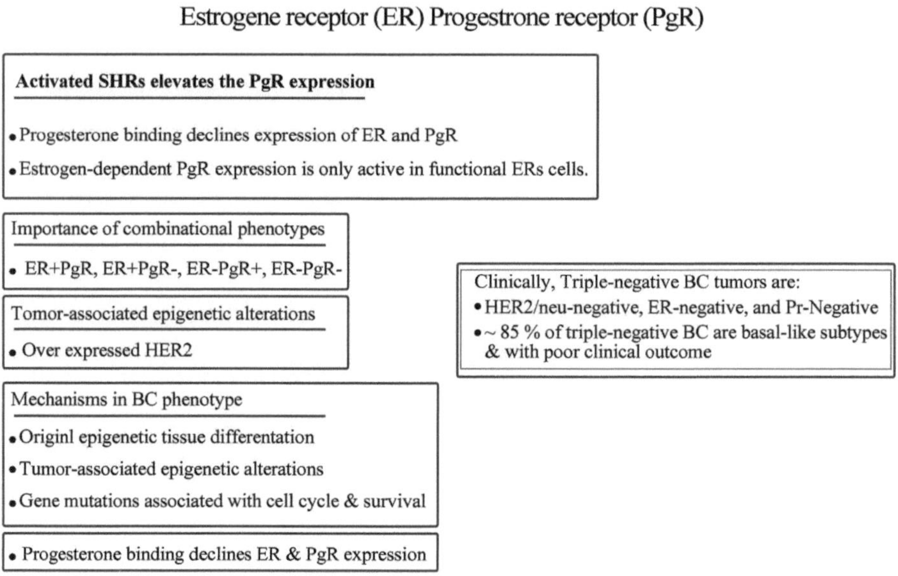

Fig. 11.12 ER, PgR- receptors- HER2 and epigenetic alterations as Triangle

Human epidermal growth factor receptor 2 (HER2) in breast tumors. Such effort is aimed to clarify an important question on the faith of these triangle target, whether they are residues of tissue stem cell or is related to the tumor epigenetic elements (Kurbel 2013). They have clarified the mode of expression in steroid hormone receptors (SHR) by defining "functional" SHRs which are derived from pretumoral tissue stem cells; and "dysfunctional" SHRs which are obtained from ER–PgR-negative cells during course of tumorigenesis. They have stressed on the combinational/functional strategy of SHRs including ER+PgR, ER+PgR−, ER−PgR+, and ER−PgR− in the tumor classification. They have emphasized that three separate mechanisms are involved in BC phenotype including "normal epigenetic tissue differentiation, tumor-associated epigenetic changes, and important gene mutations associated with cell division and survival."

These findings is summarized in a schematic overview (Fig. 11.12).

By considering the prognostic value, Er-positive breast tumors are more favorable than Er-negative cancers, but Her2/neu-positive breast cancers reveals to have a worse prognostic impact. In this regard, status of the promoter methylation was assayed within the promoter region CpG islands of breast tumor-related genes including RASSF1A, CCND2, GSPT1, TWIST, APC, NES1, RARβ2, and CDH1 (Sunami et al. 2008). It was, significantly, found that status of either Er or Her2/neu were related to the epigenetic changes of these genes.

It was also reported that diversity in epigenetic between ER-positive and ER-negative breast tumors occur at early duration of cancer development and continue through cancer progression (Sunami et al. 2008). They have also highlighted the diverse epigenetic characteristics as between; (1) HER2/neu-positive and HER2/

HER2/neu

- Key factor for BC management:
  Prognosis & Treatment
- Over-expressed in BC: 15% to 25%
- Associated with:
  Poor prognosis &
  Resistance to hormonal therapy
- Downregulated type is affected by ER
  Reverse correlation with ER expression

ER

- ER-negative tumors are:
  More malignant
  With poorer prognostic value than in
  ER-positive tumors
- Reduced or Lack of CDH1 expression &
  Lack of ER expression interact in BC-patients
- Correlations:
  Methylation of APC & ER positivity
  Status of methylation & ER alters with tumor progression
- Methylation status of RASSF1A, CCND2, GSTP1, TWIST, & APC:
  was higher in the ER-positive tumors

Fig. 11.13 Diverse epigenetic characteristics of Her2-neu and ER. *RASSF1A* RAS association domain family 1A, *CCND2* cyclin D2, *GSTP1* gluthatione S-transferase P1, *TWIST* human basic helix-loop-helix DNA binding protein, *ER* estrogene receptor, *APC* adenomatous polyposis coli, *CDH1* E- cadherin, *Her2-neu* human epidermal growth factor receptor 2

neu-negative breast tumors, (2) "double-negative" breast tumors, (3) HER2/neu -positive or ER- positive breast tumors (Fig. 11.13).

Epigenetic diversity in the relevant gene targets was observed between ER-positive and ER-negative breast cancers (Sunami et al. 2008). It was also highlighted that epigenetic caharacteristics vary in ER-positive breast tumors at early duration of cancer progression.

By considering the importance of functional insight, the signal copy number of Her2-neu plays crucial role for an appropriate clinical management in breast cancer patients. The provided image presents limited number of signal copy number and amplification of Her2-neu (Fig. 11.14).

As Table 11.5 presents, it could be concluded that the higher degree of hypermethylation is related to the positive status of ALN; and the less methylated

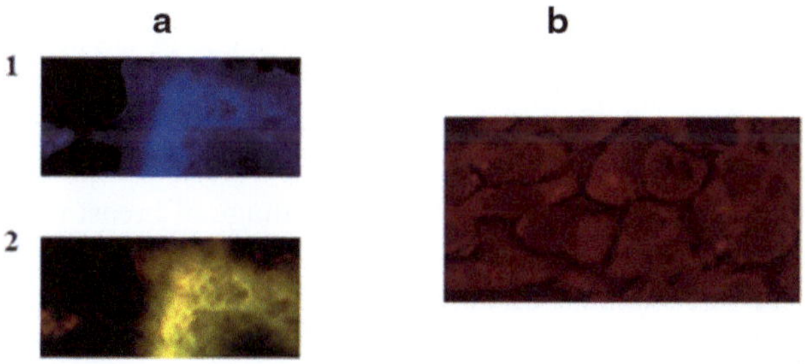

Fig. 11.14 Fluorescence in situ hybridization of Her2-neu in a patient affected with breast cancer. **a** *1* Breast tumor cells with dapi, *2* Same merged cells conjugated with FITC, R-pe and Pe-Cy5. This image is reflective of a harmonic co-expression fo Her2-neu, Er and RARβ2 in majority of cells, Magnification: x100. **b** Conjugated cells with Pe-Cy5 with limited signal copy number of Her2-neu and amplification Magnification: (**a**) x100, (**b**) x400. (*FISH* fluorescence in situ hybridization). (Adapted from: P. Mehdipour archive)

Table 11.5 Statue of methylation and prognostic factors

Status of methylation	*Status of prognostic factors*		
	ER+than Er-	Dual negative: ER-/HER2-	ALN + than ALN−
More methylated	RASSF1A &CCND2	–	–
More hypermethylatd	–	–	GSTP1
Less methylated	–	RASSF1A, GSTP1, &APC	–

RASSF1A RAS association domain family 1A, *CCND2* Cyclin D2, *GSTP1* Glutathione S-transferase P1, However, diverse clinical outcome between Er -positive and Er-negative is reflective of prognostic value and furthermore therapeutic factor in BC patients (Table 11.6)

Table 11.6 Diverse interactive profiles between Er and Her2/neu statue in breast cancer patiens

Status of gene	Age based incidence of Er- satus	Prognosis	Parity/birth timing
ER-positive	Increase after 50–54 years	Better	Inverse association
Er-negative	No increase	Worse	Non-inverse association
Her2/neu-positive/ overexpression	–	Poor/resistance to hormonal therapy (Tamoxifene)	–

catagory is related to the negative status of Her2/neu. In addition, it may be concluded that RASSF1A and GSTP1 has a two edged sward, i.e., the methylated status of RASSF1A would be higher when Er is positive, and it is lower when Er and Her2/neu are negative. But the higher and less degree of hypermethylation of GSTP1 relies on, (1) when ALN is positive and (2) when Er and Her2/neu are negative, respectively. Finally, Er and Her2/neu play a common influential prognostic role, in contrast ALN has its own territorial prognostic impact on the status of methylation; this statement refers to the importance of intra-somatic characteristics of auxiliary lymph node and more hypermethylation status of GSTP1 as a sole target gene. Besides, more unity of function is associated with dual negative status of Er/her2 and less methylation of RASSF1A, GSTP1, and APC as a cooperative/triangle genes with a more positive prognostic impact on breast cancer patients.

As a complementary presentation with IF, an image of breast tumor tissue of an affected patient with invasive ductal carcinoma of breast is provided (Fig. 11.15). Co-expression of three targets including RARβ2, CD44 and CD24 are assayed. RARβ2 is found to be hypermethylated in breast tumor.

Diverse expression in these targets are remarkable. The importance of stem cells have been explored in Sect. 11.3, but as the present image shows mode of interaction between stems cells and RARβ2, its methylation mode, seems to be crucial. So far, this fact has not been completely unmasked in breast cancer and possibly in other types of cancers and is under investigation in our ongoing project.

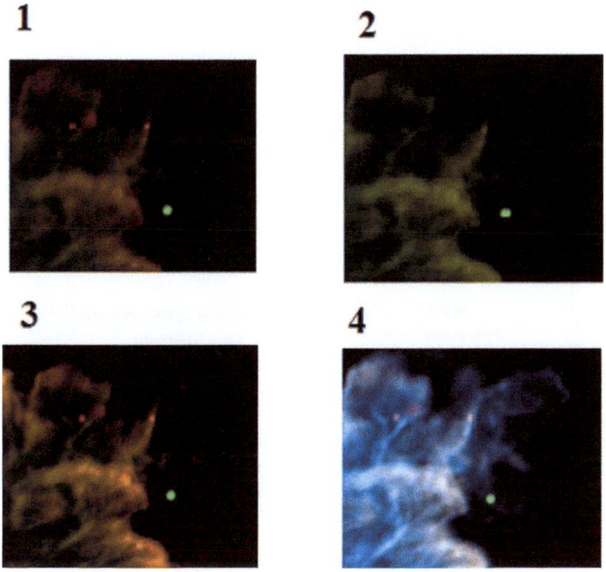

Fig. 11.15 Protein co-expression between CD44, CD24 and RARβ2 in a patient affected with breast cancer. *1* Co-expression of RARβ2/ Cd44, *2* Co-expression of RARβ2/ Cd24, *3* Co-expression of RARβ2/ Cd44/Cd24, *4* Co-expression of Dapi/ RARβ2/ CD44/CD24. (This image is retrieved from an unpublished data (P.Mehdipour's archive)

A publication was provided on triple negative breast cancer (TNBC) (Ordentlich et al. 2012) whis has highlighted the following Information:

1. TNBC lacks the required expression of its relevant involved genes which pave the ways towards an appropriate validation in targeted therapies.
2. It is known that chemotherapy, as a sole, may lead to development of resistance in breast
   cancer patients.
3. Genomic and molecular characteristic of TNBC has unmasked multiple roles in the
   epigenetic dysregulation of involved genes in cell differentiation and mormal growth.
4. Histone deacetylase (HDAC) inhibitors is capable to reverse the tumor epigenetic profile which
   leads to re-express the silenced genes encoding proteins including ERα, EGFR, and RARβ.
5. They have hypothesized that "combining epigenetic therapy using entinostat, with differentiation therapy using a retinoic acid receptor agonist- All Trans Retinoic Acid (ATRA) will provide an effective strategy for impeding the growth of TNBC and potentially sensitize tumors to commonly used chemotherapies (doxorubicin, carboplatin, paclitaxel)."

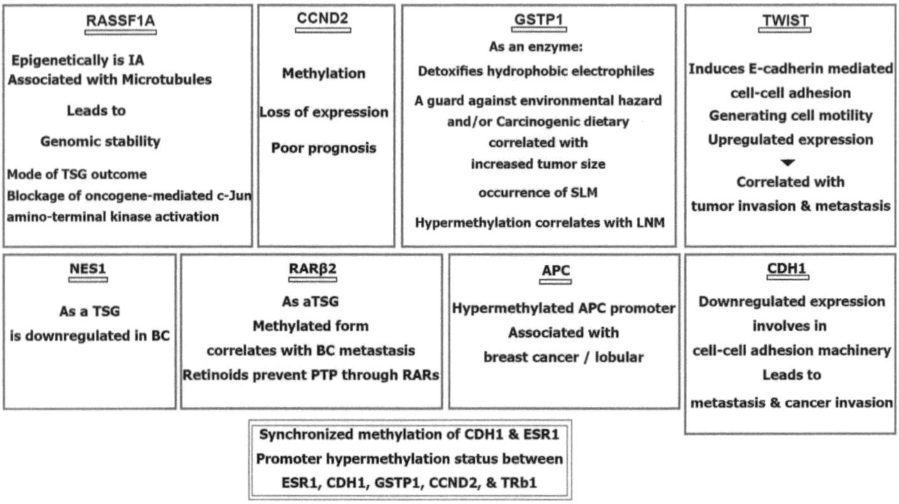

Fig. 11.16 Cascade of hypermethylation, gene silencing in breast cancer initiation and progression. *RASSF1A* RAS association domain family 1A; Location: 3p21.3; GenBank: AF132675, *CCND2* Cyclin D2; location: 12p13; GenBank: AF518005, *GSTP1* Glutathione S-transferase P1; Location: 11q13; GenBank: U12472, *TWIST* Human Basic Helix-Loop-Helix DNA Binding Protein; Location: 7p21.2; GenBank: U80998, *APC* adenomatous polyposis coli; Location: 5q21-q22; GenBank: M74088, *NES1* Normal Epithelial Cell-Specific 1 or Kallikrein 10; Location: 19q13.3-q13.4; GenBank AF024605, *RARβ2* Retinoic Acid Receptor-β2; Location: 3p24; GenBank: X07282, *CDH1* E-cadherin; GenBank: L08599, *TSG* Tumor-Suppressor Gene, *AC* Activated, *IA* Inactivated, *RAR* Retinoic Acid Receptor, *ER* Estrogen Receptor, *HER2/neu* Human Epidermal Growth Factor Receptor 2, *LN* Lymph Node, *NES1* Normal Epithelial Cell-Specific 1 or Kallikrein 10, *RAS* Association Domain Family 1A, *SLN* Sentinel Lymph Node, *RASSF1A* RAS Association Domain Family 1A, *CCND2* Cyclin D2, *GSTP1* Glutathione S-transferase P1, *TWIST* Human Basic Helix-Loop-Helix DNA Binding Protein

Furthermore, it was suggested that hypermethylation of the tumor suppressor gene *CDH1* may be considered as a key event in the metastasis of the axillary lymph node and the BC recurrence (Seung Pil Jung et al. 2013).

Hypermethylation as an epigenetic alteration is involved in the blockage of the target gene (s) at the prompter region which will lead to gene silencing (Sunami et al. 2008; Jin et al. 2001; Virmani et al. 2001; Sarrio et al. 2003; Mehrotra et al. 2004; Liu et al. 1996; Goyal et al. 1998; Zanardi et al. 2006; Parrella et al. 2004) (Fig. 11.16).

It is also reported that ERα methylation is correlated with ER negativity in Iranian primary/sporadic breast tumors. ERα is also associated with progesterone receptor negativity, and double receptor negative status (Izadi et al. 2012a). Furthermore, by considering the immunohistochemistry (IHC) marker based classification of ER, PR Her2, three main subtypes were highlighted as luminal A (ER+; PR+/−; HER-2−), luminal B (ER+;PR+/−; HER-2+), basal-like (ER−;PR-;HER2−) and Her2+(ER−; PR−; HER-2+) (Izadi et al. 2012b). They have found a correlation between ERα methylation and poor prognosis tumor subtypes (basal and Her2+) in

BC patients. According to this data, they have emphasized on the manner of aggressiveness in the pathogenesis of breast tumors.

Additionally, a model on the epigenetic alteration of HER2, ER, and PgR expression in BC is provided with the following basic elements and characteristics (Kurbel 2013):

1. HER2 overexpression is considered as the tumor-associated epigenetic alterations.
2. Capable to differentiate expression behavior as; a) ER and PgR expression, i.e., the "functional steroid hormone receptors inherited from pretumoral tissue stem cells" and acquired "dysfunctional" steroid hormone receptors through tumorigenesis which has derived from ER-PgR-negative cells.
3. The luminal A and luminal B tumor sub-types had heterogenic steroid receptor expression including functional and dysfunctional steroid receptors.
4. Future directions: Therapeutic impact in the premenopausal management is recommended.

Lets highlight the fact that cancer initiation and progression is due to the genetic and epigenetic changes which lead to inactivation of TSGs and activation of proto-oncogene. It was reported that the active Cyclin D2/Cyclin-dependent kinase 2 (CCND1/CDK2) complexes plays role in human mammary epithelial cell (HMEC) transformation (Junk et al. 2013). They have shown the common facts between HMEC transformation model and luminal breast cancer sub-types. They have finally suggested that "targeted inhibition of constitutive CCND1/CDK2 activity may enhance the effectiveness of current treatments for luminal breast cancer." As a complementary insight, the interaction between Ki67 either with Cyclin E or with CDC25A, as the negative regulators of cell cycle seems to be important (Fig. 11.17,). By focusing on the the nature of KI67-impact on the successful growth and proliferation of BC-tumors, the question is 'Does co-expression of Ki67/Cyclin E play a prominent key role or Ki67/ CDC25A?'. However as Fig. 11.17a, shows, more harmonic cooperation is notable between Ki67 with cyclin E. In addition very limited cells reveal to have high expression of the analysed targtes (Fig. 11.17a, b).

The next concern in cancer epigenetic is the mode of interaction between a gurdian of cells, p53, with two main gates of cancer cell cycle which dictate the fate of G1/S transition. This insight is under investigation in our ongoing project.

11.5 Impact of miRNA on Cancer Epigenetic

The small noncoding RNAs or nonprotein coding RNA, known as MicroRNAs (miRNAs or miRs) contains 21–23 nucleotides in length. The miRNAs regulate gene expression through the "sequence-selective targeting of mRNAs", which may lead to "translational repression or mRNA degradation ". By highlighting the critical impact of both DNA and RNA in epigenetic, the predicted number of miRNA in the human genome is previously, reported to be approximately 1000 which are

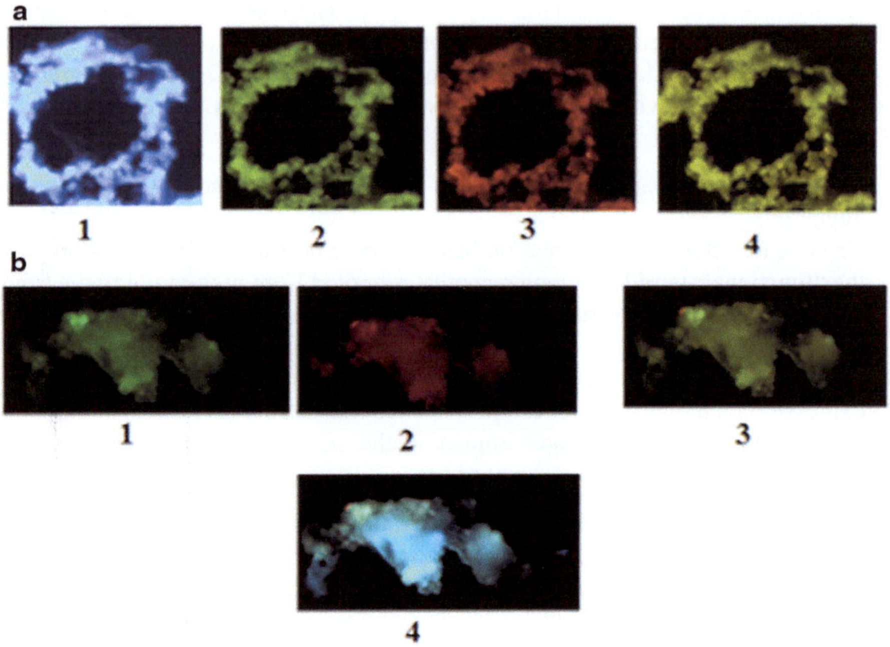

Fig. 11.17 Protein expression of KI67, Cyclin E and CDC25A in a patient affected with breast cancer. **a** Co-expression of Ki67 with cyclin E (Magnification:x100), *1* Breast tumor tissue of a patient with invasive ductal carcinoma, illustrating cells with dapi, *2* Same cells conjugated with FITC (*green*) reflective the expression of Ki67, *3* Same cells conjugated with R-Pe (*orange*) reflecting the mode of cyclin E expression, *4* Co-expression of Ki67 with Ki67 with cycin E, *5* The next concern in cancer epigenetic is the mode of interaction between a gurdian of cells, p53, with two main gates of cancer cell cycle which dictate the fate of G1/S transition. In this regard, images of co-expression between. **b** Co-expression of Ki67 with CDC25A (Magnification:x100), *1* Breast tumor tissue of a patient with invasive ductal carcinoma, illustrating cells with dapi, *2* Same cells conjugated with FITC (*green*) reflective the expression of Ki67, *3* Same cells conjugated with R-Pe (*orange*) reflecting the mode of CDC25A expression, 4 Co-expression of Ki67 with Ki67 with cycin E. (Modified from: Mehdipour et al. 2009)

supposed to target multiple protein coding transcripts (Rouhi et al. 2008). These elements are required in normal cells and their distorted expression could lead to carcinogenesis (Li and Faria 2004; Laird 2005; Calin et al. 2004a). It was also reported that the miRNA expression including *mir-125b, mir-145, mir-21*, and *mir-155* could be specified in the normal and malignant tissues (Iorio et al. 2005). They have characterized breast cancer histhopathologic nature on the basis of the expression profile of miRNAs, including tumor stage, proliferation index, ER, PR, and vascular invasion.

Some of the influential interaction between miRNAs and cancer epigenetic is summarized (Table 11.7).

Table 11.7 miRNA and cancer at a glance

Characteristics and findings	References
1. There is relationship between miRNAs and cancer	Yang et al. 2002; Li and Faria 2004; Cowland et al. 2007
2. Loss of miR-15 and miR-16 observed in 13q14-deleted chronic lymphocytic B-cell leukemia	
3. Significant association between miRNA expression & epigenetic machinery within a bilateral influential pattern in cancer	Laird 2005; Peter 2009; Garzon et al. 2006
4. miRNA could initiate catalytic functions leading to RNA splicing	Weston and Harris 1997; Garzon et al. 2006
5. miRNAs play a role in the epigenetic event of posttranscriptional gene modification	
6. There is a partial regulation of gene expression by miRNAs at the posttranscriptional level through a negative manner of messenger RNA regulation.	Calin et al. 2004b
7. miRNAs is an influential target in initiation and promotion of cancer	
8. Some of miRNA genes are located in cancer related genomic terittory or in fragile sites	
9. Specific bridging system is reported between the expression of specific miRNAs and cancer development	Cho 2007; Garzon et al. 2006; Yanaihara et al. 2006; Zhang et al. 2007
10. This machinery facilitate to distinguish dysplasia from cancer	
11. They might adjust specific tumor suppressor genes and/or oncogenes or probably upon the unknown predisposing factor (s) behaving as "tumor suppressor- or onco-miRNAs	Cho 2007; Garzon et al. 2006; Yanaihara et al. 2006; Zhang et al. 2007
12. This event may lead to cell differentiation, angiogenesis, proliferation, apoptosis or invasion in cancer, by affecting the relevant responsible genes	
13. Profiling of miRNA expression could clarify the role of other genes in	Cho 2007; Garzon et al. 2006; Yanaihara et al. 2006; Zhang et al. 2007
a) Histo-pathological features	
b) Tumorigenic process	
c) Prognostic and predictive values in different human cancers	
14. RASSF1A and RARβ promoter methylation and miR17, miR21, miR 124, and let-7a expression have highlighted differences of epigenetic regulation between male and female familial breast cancer (BC), also in comparison with sporadic BC	Pinto et al. 2013
15. miR17, miR21, and let-7a showed significant overexpression in familial compared to sporadic BC	
16. RASSF1A- and RAR β- overexpression were higher in BRCA1/2 carriers	
17. BRCA mutation carriers demonstrated significant overexpression of: miR17, let-7a, and of miR21	
18. RASSF1A is found to be involved in familial male BC, but miR17 and let-7a seemed to be implied in familial female BC	

Table 11.7 (continued)

Characteristics and findings	References
19. In breast tumors, correlation was found between	Piva et al. 2013
a) "Expression profile of oncogenic miRNAs (oncomiRs) or tumor suppressor miRNAs	
b) miR-221 is overexpressed in triple-negative primary BC.	
c) Oncosuppressor p27 ^{Kip1,} a substantiated miR-221 target, is downregulated in aggressive cancer cell lines	
d) Upregulation of, slug (as a key transcription factor) binds to the miR-221/miR-222 promoter, leading to over- expression of the miR-221/miR-222 cluster in BC cells.	
e) The Slug/miR-221 complex facilitates to link miR-221 activity to the Slug repressor downregulation, resulting to Slug/miR-221 upregulation and p27 ^{Kip1} downregulation.	
f) Applicability, is using "antisense miRNA (antagomiR) molecules targeting miR-221, inducing the down-regulation of Slug and the upregulation of p27 $^{Kip1.}$"	
20. Toll-like receptor 3 (TLR3) induces up-regulation of microRNA-29b, -29c, -148b, and -152 in tumor-derived cell lines and primary tumors	Galli et al. 2013
21. These microRNAs reverse expression of epigenetically silenced genes by targeting DNA methyltransferases.	
22. In cancer cells of DU145 and TRAMP-C1 prostate and MDA-MB-231 breast cancer	
a) Polyinosinic: polycytidylic acid-mediated activation of TLR3 was acapable to demethylate and reexpress the retinoic acid receptor beta (RARβ)	
b) Cancer cells, by becomimg sensitive to RAR βwill be led to apoptosis *in vitro* and *in vivo*	
c) Therapeutic achievement is due to the TLR3 agonist/retinoic acid cooperative strategy in prostate and breast cancer cells	

RAR retinoic acid receptors; *RASSF1A* ras association domain family 1 isoform A

11.6 Selected Therapeutic Aspects of RARβ2

The initial preventive and therapeutic impact of the retinoids has been successfully performed in the leukemias characterized with chromosomal translocations. However, in breast cancer further optimization is required for fenretinide prevention trials. More success is the phase III randomized trials of retinoids accompanied by chemotherapy in non–small cell lung cancer. The focal edge and restriction boundaries is found to be "epigenetic silencing of *RARβ*"which could be the key element for solid tumor management through an appropriate definitions and characterization of this gene (Roisin et al. 2013*)*. A key pitfall is the timing event of promoter methylation in tumor suppressor genes which is rather at early stage of tumor formation. It is reported that such events may be reversible by specific remedies. Upon these facts, in a review article, the epigenetic alterations of the selected TSGs together with their clinical impacts and their capabilities as the breast tumor markers and their therapeutic role have provided (Xiang et al. 2013). They have highlighted that epigenetic changes including promoter CpG methylation of TSGs as "the dual role of DNA methylation", have the influential impact on breast tumor development. This target was defined as a key marker for early detection, prognosis, prediction, and demethylation therapeutic based in breast cancer. However, the future hope for early detection through the appropriate prognosis, prediction and prevention (PPP) in direction of cancer therapy rely on the improved assessment of the epigenomic profile in solid tumors including BC. Fundamentally, down- regulation of RARβ2 is a frequent event during the breast carcinogenesis. Upon this fact, it was reported that the breast tumors with methylated RARβ P2 promoter failed to induce RARβ2 in primary breast tumors (Sirchia et al. 2002). They could manage to achieve remarkable growth inhibition by reactivating endogenous *RARβ2* transcription from either unmethylated or methylated *RARβ* P2 in breast cancer cell lines and Xenograft tumors.

Toll-like receptor 3 (TLR3) is characterized with its positive affect on the instinctive immune system to combat against viruses. It is reported that TLR3 activation is capable to up- miRNA-29b, -29c, -148b, and -152 in both BC cell line and primary tumors (Galli et al. 2013). An interaction between these miRNA and epigenetically silenced genes is found to be through demethylation and reexpression of the "oncosuppressor" RARβ. Finally sensitivity of cancer cells to RA could lead to apoptosis. Such strategy open the window towards the therapeutic management by "combined TLR3 agonist/retinoic acid treatment" in breast and prostate cancer (Galli et al. 2013).

An important challenging item *in vitro* assay is provided the diverse outcome in type of cell lines including early and late passages of primary breast cancer cells (Peng et al. 2011). They have investigated the effect of retinoids on BC cells. The "early passages of BC cells (EPBCCs)" are found to be sensitive to retinoids. By considering expression status of different sub-type or RAR including RARα, RARγ, RXRα, RXRβ proteins, RARβ5 and RARβ2, they have considered RARβ as the major target of retinoids in BC.

As a clinical benefit, exploration on drug discovery is the visualization of success in cancer world. A reliable target is found to be cancer stem cells (CSCs) of breast cancer, therefore the attempt was to revolutionize selected gene expression profile which is elite to CSCs. In this regard, All-trans retinoic acid (ATRA) was the best choice which was not associated with gene expression in CSCs. The mechanism of ATRA is through nuclear receptor and harmonize territory of cancer cells and is capable to induce apoptosis. By considering the previous work in cell lines, it was reported that, all cancer cell lines having CSC phenotype, do not respond to ATRA (Poornima Bhat-Nakshatri et al. 2013). They have emphasized on the role of interactive targets with CSC. However, they hoped for "developing ATRA based therapy for specific subtypes of breast cancer, which additionally considers biomarker driven patient selection and cancer genome-based combination therapies."

Furthermore, Xiao-Kun Zhang in 2014 has provided a research report "Retinoids and Their Receptors in Cancer" in breast cancer cells (www.sanfordburnham.org). The highlights include the following messages:

1. Their aim was to develop the new retinoids with anti-cancer potential. Retinoids are the appropriate target to prevent and treat different cancers.
2. The main pittfall of retinoids is the resistance of cancer cells to retinoids.
3. Regulation of anticancer capacity of retinoids.
4. The retinoids' anti cancer capacity is related to the nuclear receptors including the retinoic acid (RA) receptors (RARs) and the retinoid X receptors (RXRs).
5. Retinoids are capable to inhibit the growth of cancer cells.
6. Retinoids can promote apoptosis in BC cells.
7. Induction of apoptosis and growth inhibition is mainly governed by RARβ.
8. RARβ as a tumor suppressor gene has a key role in breast carcinogenesis.
9. Loss of RARβ may lead to therapeutic retinoid resistance of cancer cells.
10. Regarding the retinoid signaling, Trans-RA induces RARβ expression and could inhibit the growth of hormone-dependent BC cells. This is intercede by RAR/RXR heterodimer through binding to the RA response element (β RARE) in the promoter region of RARβ.
11. They have found a new pathway to induce RARβ in hormone-independent BC cells as well. In this regard, they have used "RXR-selective retinoids, such as 9-cis RA".
12. In this pathway, RXR/nur77 heterodimer binds to the same beta RARE.
13. Finally, they showed that upon the estimation of RAR, RXR and nur77, in both form including a RAR or a RXR signaling pathway, expression of RARβ and apoptosis in BC cells could be promoted.

Importantly, cell cycle transition and apoptosis play critical roles in experiencing therapeutic aspecpt in cancer which may lead to a reliable and trustable strategy. The 6-[3-(1-adamantyl)-4-hydroxyphenyl]-2-naphthalene carboxylic acid (AHPN or CD437) as a retinoic acid receptor g (RARg)-selective retinoid was investigated in human lung cancer cell lines (Marchetti et al. 1999). They have reported that AHPN/CD437 was capable due to the following findings:

1. Inhibiting lung cancer cell growth by induction of G0/G1 arrest and apoptosis.
2. Diverse expression of p53 and Bcl-2 which could be regulated by AHPN/CD437 in different cancer cell lines of lung.
3. Expression of nur77 plays a critical role in AHPN/CD437-induced apoptosis.
4. A novel pathway for retinoid-induced apoptosis is provided: "AHPN/CD437 or analogs" may provide more reliable therapeutic effectiveness in lung cancer.

Furthermore, transcription factor such as NF-kappaB was a reliable choice due to its overexpression in cancer cells by induction of antiapoptotic genes' expression which will lead to anticancer therapy resistence (Bayon et al. 2003). They have used retinoid antagonist MX781 as the inhibitor of NF-kappaB-dependent transcriptional activity at level of different tumors cell lines. The achieved results are summarized:

1. Complete inhibition capacity of MX781 on tumor necrosis factor alpha-mediated activation of IkappaB kinase (IKK), the upstream regulator of NF-kappaB.
2. Two more molecules including MX3350–1 and CD2325, as the retinoic acid receptor gamma-selective agonists, could also inhibiti IKK.
3. The other nonapoptotic retinoids including N-(4-hydroxyphenyl)-retinamide, and retinoic acid were not capable to inhibit IKK.
4. Inhibition capacity of IKK by application of retinoid-related composite and other small molecules is revealed to be correlated with:

    a. Reduction of cell proliferation.
    b. Increased apoptosis.
    c. Reduction of cell viability after overexpression of a molecularly altered IKK-beta kinase or the IkappaBalpha superrepressor.

5. The manner of the induction of apoptosis by the retinoid-related molecules was:

    a. Dependent on caspase activity.
    b. Independent on the retinoid receptors.
    c. All together, "retinoid receptor-independent mechanism of action" is found to be remarkable.

Reprogramming capacity of oocytes to cancer cells was aimed to study breast oncogenesis (Allegrucci et al. 2011). BC cells were directly reprogrammed by amphibian oocyte extracts and they have shown that epigenetic reprogramming in oocyte extracts led to reduce tumor growth in mouse xenografts. This article reflects "a new method to investigate tumor reversion by epigenetic reprogramming". The key points of this work include the followings:

1. Axolotl oocyte extracts (AOE) is capable to reverse epigenetic silencing of TSGs and tumorigenicity of BC cells in a mouse xenograft model.
2. Axolotl oocyte extracts have reprogramming capacity.
3. AOE reverses epigenetic silencing of TSGs and tumorigenicity of BC cells in a mouse xenograft model.
4. Such coordination, as a remarkable tool, could be applied for editing the TSGs.
5. Silencing and unmask the involved molecular events which are responsible for tumor growth.
6. Final aim is linking epigenetic changes to epigenetic therapies.

However, retinoic acid receptor β2 (*RARβ2*) gene as a TSG is capable to be silenced during tumorigenic process. By considering the inhibitory affect of Valproic acid (VPA) on histone deacetylase, combination of VPA with RA and the DNA methyltransferase inhibitor 5-aza-2′-deoxycytidine (Aza-dC), is reported to "defeat the epigenetic barriers to transcription of a prototypical silenced tumor suppressor gene, *RARβ2*, in human breast cancer cells." (Mongan et al. 2005). They have shown that these combination led to elevation of histone acetylation at the silenced *RARβ2* promoter of MCF-7 BC cells. The VPA was proposed as a combination therapeutic remedy in the human breast cancer.

The key elements of genetics and epigenetic profiles including the essential epigenetic changes through the course of carcinogenesis direct our knowledge towards an appropriate BC management. In this regard, a review on epigenetic mechanism involved in the progress and recurrence of breast cancer is published (Lo and Sukumar 2008). They emphasized on the impact of epigenetic changes in two direction including target based cancer therapy and predictive marker.

*RAR*β gene encodes RA through which development and differentiation of lung is facilitated. Besides, *RARβ2* and RARβ4 has repressor role in lung cancer, regarding these characteristics, it was shown that P2 promoter revealed to be methylated either in tumors of squamous cell lung carcinoma or in cell line (Virmani et al. 2000). They have stated that *RAR*β P2 promoter methylation is involved in silencing of *RAR*β2 and *RAR*β4 expression in squamous cell lung carcinoma. This finding was considered as a therapeutic move in lung cancer.

Expression of tumor suppressor Klotho, as a transmembrane protein, is found to be down-regulated and up regulated in normal breast samples and ductal hyperplasia respectively (Rubinek et al. 2012). They have reported that lack of klotho expression as an early developmental role in BC. This event plays a role in "DNA methylation and histone deacetylation in klotho silencing." Expression of Klotho and methylation, as a biomarkers in BC, was highlighted.

RARβ2, as a RA-regulated TSG could be hyermethylated and scilenced by a form of RA resistance. By considering the RA-resistant breast and prostate cancer cell lines, it was reported that interference of RA signal with RARα shapes aggravated form of the "repressed chromatin status of *RARβ2* including DNA methylation" followed by occurance of transcriptional silencing *of RARβ2* (Ren et al. 2005). *In addition*, the following g key points are highlighted:

1. Hypermethylation is not adequate to silence *RARβ2.*
2. Restorated RA signal by *RAR*α at an epigenetically silent *RARβ2* is decisive to restore a RARβ2 mode, acting as transcriptional factor.
3. *RARβ2* epigenetic silencing is correlated to the RA-resistant phenotype.
4. *RARβ2* epigenetic silencing leads to RA resistance.

However, these facts may pave the way towards an appropriate direction for cancer therapy. Furthermore, hypermethylation of *RAR*β promoter is recently reported at cell line level and in all prostate tumors with different Gleason score (Moison et al. 2014). They have suggested a remarkable harmony between DNA methylation and enhancer of zeste (EZH2)eZH2, the catalyzer of histone H3 lysine (K) 27 (H3K27me3), to silence *RAR*β through prostate tumorigenesis (Yu et al. 2014). By considering the therapeutic interaction and cancer management in this paper, it was reported that overexpression of DNA methyltransferases 1 and/or 3a could lead to the following outcomes:

1. Reduction in expression of estrogen receptor –alpha and breast cancer susceptibility gene 1 (BRCA1) in sporadic BC leading to a poor prognosis.
2. Promoter hypermethylation of ERα and BRCA1.

3. Shorter disease free survival or overall survival of patients characterized with age ≤50 years old, ERα-negative- or HER2-postive statue.
4. Poor prognosis in patients who have been given chemo–and endocrine–therapy.

11.7 Conclusions

Breast cancer (BC) is a malady characterized by alterations in genetic and epigenetic. Epigenetic silencing of tumor suppressor genes reveals to be an early event in carcinogenesis of breast. Epigenetic reprogramming through editing of gene silencing may pave the way to unmask the mechanisms involved in tumor initiation and progression and establish the therapeutic strategies in cancer.

According to the preclinical studies and clinical trials, retinoids, as structural and functional analogs of vitamin A could be considered as the chemopreventive against cancer development.

Retinoids play an influential role in differentiation and cell cycle arrest at G1 phase.

Deviant retinoid signaling occurs in different cancers including breast cancer.

Retinoic acid receptors *beta2 (RARβ2)* gene as a tumor suppressor gene, play an important role in the chemopreventive capacity of retinoids. However, the lack of *RARβ2* expression may occur in invasive breast carcinoma. Lack of *RARβ2* expression in BC may be partly due to loss of an allele of chromosome 3p24 and methylation of the other allele. Morever, such alteration will not necessarily facilitate the developmental process of resistance to retinoids.

In spite of expression loss of *RARβ2* in some BC patients, the responding capability to retinoid intervention may be conserved.

Interestingly, the methylation of *RARβ2* promoter region could be related to epigenetic gene silencing. The complete biallelic inactivation of the *RARβ2* gene may be due to the silencing mechanism, and could be reversible by demethylating agents.

Besides, environmental factors are considered as the catlytical elements in cancer development. The nutritional strategies could affect the epigenetic and genetic territories and facilitate the neoplasic process.

There is a great challenge regarding the demethylating agents with possible role in cancer prevention in individuals with a promoter methylated *RARβ2* and those who are predisposed to cancer development. This fact may be highlighted as an early tumor alteration, thereby, recognition of the methylated *RARβ2* in primary BC may be practical to differentiate tumors with a positive responsiveness capacity to RA therapy.

Practically, obesity, duration of estradiol exposure, and smoking are the predisposing factors for development of methylation in ERalpha gene. Besides, familial BC was, inversely, correlated with the hypermethylated *RARβ2*. In addition, plasma folate and vitamin B12 levels were inversly associated with the hypermethylation status of ERalpha gene. These dat suggested that the mode of hypermethylation of specific genes is associated with environmental including lifestyle-related factors.

Moreover, the miRNAs are an influential target in initiation and promotion of cancer and loss of different miRNA is found in varieties of malignancies including leukaemias. Different miRNAs play a role in the epigenetic event of posttranscriptional gene modification. There is a partial regulation of gene expression by miRNAs at the posttranscriptional level. Some of miRNA genes are located in cancer related genomic terittory or in fragile sites. Most importantly, this machinery facilitate to distinguish dysplasia from cancer. RASSF1A, RARβ promoter methylation and miR17, miR21, miR 124, and let-7a expression have highlighted differences of epigenetic regulation between male and female familial breast cancer (BC), also in comparison with sporadic BC. There is also cooperation between BRCA mutation and overexpression of: miR17, let-7a, and of miR21. At cell line level, cancer cells, by becomimg sensitive to RAR β will be led to apoptosis *in vitro* and *in vivo*. Besides, therapeutic achievement is due to the TLR3 agonist/retinoic acid cooperative strategy in prostate and breast cancer cells.

The initial preventive and therapeutic impact of the retinoids has been successfully performed in the leukemias characterized with chromosomal translocations.

The retinoids' anti cancer capacity is related to the nuclear receptors including the retinoic acid (RA) receptors (RARs) and the retinoid X receptors (RXRs). Retinoids are capable to inhibit the growth of cancer cells and promote apoptosis in BC cells. But, the main pittfall of retinoids is the resistance of cancer cells to retinoids. Finally, loss of RARβ may lead to therapeutic retinoid resistance of cancer cells.

References

Al-Hajj M, Wicha MS, Benito-Hernandez A, Morrison SJ, Clarke MF (2003) Prospective identification of tumorigenic breast cancer cells. Proc Natl Acad Sci U S A 100:3983–3988

Allegrucci C, Rushton MD, Dixon JE, Sottile V, Shah M, Kumari R et al (2011) Epigenetic reprogramming of breast cancer cells with oocyte extracts. Mol Cancer 10(1):7. doi:10.1186/1476-4598-10-7

Al Tanoury Z, Gaouar S, Piskunov A, Ye T, Urban S, Jost B et al (2014) Phosphorylation of the retinoic acid receptor RARγ2 is crucial for the neuronal differentiation of mouse embryonic stem cells. J Cell Sci 127:2095–105

Altucci L, Gronemeyer H (2001) The promise of retinoids to fight against cancer. Nat Rev Cancer 1:181–193

Andreola F, Giandomenico V, Spero R, De Luca LM (2000) Expression of a smaller lecithin:retinol acyl transferase transcript and reduced retinol esterification in MCF-7 cells. Biochem Biophys Res Commun 279:920–924

Bachman K, Park BH, Rhee I, Rajagopalan H, Herman JG, Baylin SB et al (2003) Histone modifications and silencing prior to DNA methylation of a tumor suppressor gene. Cancer Cell 3:89–95

Balmer JE, Blomhoff R (2002) Gene expression regulation by retinoic acid. J Lipid Res 43:1773–1808

Barabe F, Kennedy JA, Hope KJ, Dick JE (2007) Modeling the initiation and progression of human acute leukemia in mice. Science 316:600–604

Bauer KR, Brown M, Cress RD, Parise CA, Caggiano V (2007) Descriptive analysis of estrogen receptor (ER)-negative, progesterone receptor (PR)-negative, and HER2-negative invasive breast cancer, the so-called triple-negative phenotype: a population-based study from the California cancer Registry. Cancer 109:1721–1728

Bayon Y, Ortiz MA, Lopez-Hernandez FJ, Gao F, Karin M, Pfahl M et al (2003) Inhibition of IkappaB kinase by a new class of retinoid-related anticancer agents that induce apoptosis. Mol Cell Biol 23:1061–1074

Bean GR, Scott V, Yee L, Ratliff-Daniel B, Troch MM et al (2005) Retinoic acid receptor-beta2 promoter methylation in random periareolar fine needle aspiration. Cancer Epidemiol Biomark Prev 14:790–798

Benetatos L, Voulgaris E, Vartholomatos G, Hatzimichael E (2013) Non-coding RNAs and EZH2 interactions in cancer: long and short tales from the transcriptome. Int J Cancer 133:267–274

Bhat-Nakshatri P, Goswami CP, Badve S, Sledge GW Jr, Nakshatri H (2013) Identification of FDA-approved Drugs Targeting Breast Cancer Stem Cells Along With Biomarkers of Sensitivity. Sci Rep 3:2530

Bistulfi G, Pozzi S, Ming QR, Rossetti S, Sacchi N (2006) A Repressive epigenetic domino effect confers susceptibility to breast epithelial cell transformation: implications for predicting breast cancer risk. Cancer Res 66:10308–10314

Blanco MJ, Moreno-Bueno G, Sarrio D, Locascio A, Cano A, Palacios J et al (2002) Correlation of Snail expression with histological grade and lymph node status in breast carcinomas. Oncogene 21:3241–3246

Bonnet D, Dick JE (1997) Human acute myeloid leukemia is organized as a hierarchy that originates from a primitive hematopoietic cell. Nat Med 3:730–737

Brand N, Petkovich M, Krust A, Chambon P, de Thé H, Marchio A et al (1988) Identification of a second human retinoic acid receptor. Nat 332:850–853

Calin GA, Liu CG, Sevignani C, Ferracin M, Felli N, Dumitru CD et al (2004a) MicroRNA profiling reveals distinct signatures in B cell chronic lymphocytic leukemias. Proc Natl Acad Sci U S A 101:11755–11760

Calin GA, Sevignani C, Dumitru CD, Hyslop T, Noch E, Yendamuri S et al (2004b) Human microRNA genes are frequently located at fragile sites and genomic regions involved in cancers. Proc Natl Acad Sci U S A 101:2999–3004

Chambon P (1996) A decade of molecular biology of retinoic acid receptors. FASEB J 10:940–954

Chen P, Li C, Li X, Li J, Chu R, Wang H (2014) Higher dietary folate intake reduces the breast cancer risk: a systematic review and meta-analysis. Br J Cancer 110:2327–38

Chiba H, Clifford J, Metzger D, Chambon P (1997) Distinct retinoid X receptor-retinoic acid receptor heterodimers are differentially involved in the control of expression of retinoid target genes in F9 embryonal carcinoma cells. Mol Cell Biol 17:3013–3020

Cho WC (2007) OncomiRs: the discovery and progress of microRNAs in cancers. Mol Cancer 6:60

Clarke MF, Dick JE, Dirks PB, Eaves CJ, Jamieson CHM et al (2006) Cancer stem cells—perspectives on current status and future directions: AACR Workshop on cancer stem cells. Cancer Res 66(19)9339–9344

Cloos PA, Christensen J, Agger K, Maiolica A, Rappsilber J, Antal T (2006) The putative oncogene GASC1 demethylates tri- and dimethylated lysine 9 on histone H3. Nature 442:307–311

Collingwood TN, Urnov FD, Wolffe AP (1999) Nuclear receptors: coactivators, corepressors and chromatin remodeling in the control of transcription. J Mol Endocrinol 23:255–275

Connolly RM, Nguyen NK, Sukumar S (2013) Molecular pathways: current role and future directions of the retinoic Acid pathway in cancer prevention and treatment. Clin Cancer Res 19:1651–1659

Coyle KM, Sultan M, Thomas M, Kashani AV, Marcato P (2013) Retinoid signaling in cancer and its promise for therapy. J Carcinog Mutagen S7:006. doi: 10.4172/2157-2518.S7-006

Cowland JB, Hother C, Grønbaek K (2007) MicroRNAs and cancer. APMIS 115:1090–1106

Dobrovic A, Simpfendorfer D (1997) Methylation of the BRCA1 gene in sporadic breast cancer. Cancer Res 57:3347–3350

Delhommeau F, Dupont S, Della Valle V, James C, Trannoy S, Massé A Le Couedic JP, Robert F, Alberdi A, Lécluse Y et al (2009) Mutation in TET2 in myeloid cancers. N Engl J Med 360:2289–2301
Deng G, Chen A, Hong J, Chae HS, Kim YS (1999) Methylation of CpG in a small region of the hMLH1 promoter invariably correlates with the absence of gene expression. Cancer Res 59:2029–2033
Dick JE (2008) Stem cell concepts renew cancer research. Blood 112:4793–4807
Dilworth FJ, Chambon P (2001) Nuclear receptors coordinate the activities of chromatin remodeling complexes and coactivators to facilitate initiation of transcription. Oncogene 20:3047–3054
Ernst T, Chase AJ, Score J, Hidalgo-Curtis CE, Bryant C, Jones AV et al (2010) Inactivating mutations of the histone methyltransferase gene EZH2 in myeloid disorders. Nat Genet 42(722–726):2010
Esteller M (2006) Epigenetics provides a new generation of oncogenes and tumor-suppressor genes. Br J Cancer 94:179–183
Esteller M (2008) Molecular origins of cancer: epigenetics in cancer. N Engl J Med 358:1148–1096
Esteller M, Garcia-Foncillas J, Andion E, Goodman SN, Hidalgo OF, Vanaclocha V et al (2000) Inactivation of the DNA-repair gene MGMT and the clinical response of gliomas to alkylating agents. N Engl J Med 343:1350–1354
Evans TR, Kaye SB (1999) Retinoids: present role and future potential. Br J Cancer 80:1–8
Farias EF, Ong DE, Ghyselinck NB, Nakajo S, Kuppumbatti YS, Mira-y-Lopez R (2005) Cellular retinol-binding protein I, a regulator of breast epithelial retinoic acid receptor activity, cell differentiation, and tumorigenicity. J Natl Cancer Inst 97:21–29
Fraga MF, Ballestar E, Villar-Garea A, Boix-Chornet M, Espada J, Schotta G, Bonaldi T et al (2005) Villar-Garea loss of acetylation at Lys16 and trimethylation at Lys20 of histone H4 is a common hallmark of human cancer. Nature Genet 37:391–400
Feinberg AP, Ohlsson R, Henikoff S (2006) The epigenetic progenitor origin of human cancer. Rev Genet 7:21–33
Ferguson AT, Lapidus RG, Baylin SB, Davidson NE (1995) Demethylation of the estrogen receptor gene in estrogen receptor-negative breast cancer cells can reactivate estrogen receptor gene expression. Cancer Res 55:2279–2283
Filippakopoulos P, Picaud S, Mangos M, Keates T, Lambert JP, Barsyte-Lovejoy D et al (2006) Histone recognition and large-scale structural analysis of the human bromodomain family. Cell 149:214–231
Figueroa ME, Abdel-Wahab O, Lu C, Ward PS, Patel J, Shih A et al (2010) Leukemic IDH1 and IDH2 mutations result in a hypermethylation phenotype, disrupt TET2 function, and impair hematopoietic differentiation. Cancer Cell 18:553–567
Fujita N, Jaye DL, Kajita M, Geigerman C, Moreno CS, Wade PA (2003) MTA3, a Mi-2/NuRD complex subunit, regulates an invasive growth pathway in breast cancer. Cell 113:207–219
Galli R, Paone A, Fabbri M, Zanesi N, Calore F, Cascione L et al (2013) Toll-like receptor 3 (TLR3) activation induces microRNA-dependent reexpression of functional RARβ and tumor regression. Proc Natl Acad Sci U S A 110:9812–9817
Garzon R, Fabbri M, Cimmino A, Calin GA, Croce CM (2006) MicroRNA expression and function in cancer. Trends Mol Med 12:580–587
Gerhauser C (2013) Cancer chemoprevention and nutri-epigenetics: state of the art and future challenges. Top Curr Chem 329:73–132
Geutjes EJ, Bajpe PK, Bernards R (2012) Targeting the epigenome for treatment of cancer. Oncogene 31:3827–3844
Goyal J, Smith KM, Cowan JM, Wazer DE, Lee SW, B and V et al (1998) The role for NES1 serine protease as a novel tumor suppressor. Cancer Res 58:4782–4786
Gray SG (2014) Bookreview: "Epigenetics". Front.Genet. 5:104. doi:10.3389/fgene.2014.00104
Gururaj AE, Rayala SK, Vadlamudi RK, Kumar R (2006) Novel mechanisms of resistance to endocrine therapy: genomic and nongenomic considerations. Clin Cancer Res 12:1001–1007

Guccione E, Bassi C, Casadio F, Martinato F, Cesaroni M, Schuchlautz H et al (2007) Methylation of histone H3R2 by PRMT6 and H3K4 by an MLL complex are mutually exclusive. Nature 449:933–937

Guo X, Ruiz A, Rando RR, Bok D, Gudas LJ (2000) Esterification of all-trans-retinol in normal human epithelial cell strains and carcinoma lines from oral cavity, skin and breast: reduced expression of lecithin:retinol acyltransferase in carcinoma lines. Carcinogenesis 21:1925–1933

Gupta PB, Chaffer CL, Weinberg RA (2009) Cancer stem cells: mirage or reality? Nat Med 15:1010–1012

Houle B, Rochette-Egly C, Bradley WE (1993) Tumor-suppressive effect of the retinoic acid receptor β in human epidermoid lung cancer cells. Proc Natl Acad Sci U S A 90:985–989

Hatzimichael E, Dranitsaris G, Dasoula A, Benetatos L, Stebbing J, Crook T et al (2009) Von Hippel-Lindau methylation status in patients with multiple myeloma: a potential predictive factor for the development of bone Disease. Clinic Lymphoma Myeloma 9:239–242

Hatzimichael E, Dasoula A, Kounnis V, Benetatos L, Lo Nigro C, Lattanzio L et al (2012) Bcl2-interacting killer CpG methylation in multiple myeloma: a potential predictor of relapsed/refractory disease with therapeutic implications. Leuk Lymphoma 53:1709–1713

Hatzimichael E, Georgiou G, Benetatos L, Briasoulis E (2013) Gene mutations and molecularly targeted therapies in acute myeloid leukemia. Am J Blood Res 3:29–51

Hanahan D, Weinberg RA (2011) Hallmarks of cancer: the next generation. Cell 144:646–674

Hansen KD, Timp W, Bravo HC, Sabunciyan S, Langmead B, McDonald OG et al (2011) Increased methylation variation in epigenetic domains across cancer types. Nature Genet 43:768–775

Husmann M, Lehmann J, Hoffmann B, Hermann T, Tzukerman M, Pfahl M (1991) Antagonism between retinoic acid receptors. Mol Cell Biol 11:4097–4103

Husmann M, Hoffmann B, Stump DG, Chytil F, Pfahl M (1992) A retinoic acid response element from the rat CRBPI promoter is activated by an RAR/RXR heterodimer. Biochem Biophys Res Commun 187:1558–1564

Heard E (2005) Delving into the diversity of facultative heterochromatin: the epigenetics of the inactive X chromosome. Curr Opin Genet Dev 15:482–489

Heard E, Clerc P, Avner P (1997) X-chromosome inactivation in mammals. Annu Rev Genet 31:571–610

Hendrix MJ, Seftor EA, Seftor RE, Kasemeier-Kulesa J, Kulesa PM, Postovit LM (2007) Reprogramming metastatic tumor cells with embryonic microenvironments. Nat Rev Cancer 7:246–255

Hua S, KittlerR, White KP (2009) Genomic antagonism between retinoic acid and estrogen signaling in breast cancer. Cell 137(7):1259–1271

Hinshelwood RA, Clark SJ (2008) Breast cancer epigenetics: normal human mammary epithelial cells as a model system. J Mol Med 86:1315–1328

Husmann MJ, Lehmann B, Hoffmann T, Hermann M, Tzukerman M, Pfahl M (1991) Antagonism between retinoic acid receptors. Mol Cell Biol 11:4097–4103

Hoffmann B, Lehmann JM, Zhang XK, Hermann T, Husmann M, Graupner G et al (1990) A retinoic acid receptor-specific element controls the retinoic acid receptor-beta promoter. Mol Endocrinol 4:1727–1736

Hua S, Kittler R, White KP (2009) Genomic antagonism between retinoic acid and estrogen signaling in breast cancer. Cell 137:1259–1271

Iorio MV, Ferracin M, Liu C-G, Veronese A, Spizzo R, Silvia S et al (2005) MicroRNA gene expression deregulation in human breast cancer. Cancer Res 65:7065–7070

Itoh Y, Suzuki T, Miyata N (2013) Small-molecular modulators of cancer-associated epigenetic mechanisms. Mol Biosyst 9:873–896

Izadi P, Noruzinia M, Karimipoor M, Karbassian MH, Akbari MT (2012a) Promoter hypermethylation of estrogen receptor alpha gene is correlated to estrogen receptor negativity in iranian patients with sporadic breast cancer. Cell J 14:102–109

Izadi P, Mehrdad N, Foruzandeh F, Reza NM (2012b) Epigenetics and three main clinical aspects of breast cancer management. Asian Pacific J Cancer Prev 13:4113–4117

Junk DJ, Cipriano R, Stampfer M, Jackson MW (2013) Constitutive CCND1/CDK2 activity substitutes for p53 loss, or MYC or oncogenic RAS expression in the transformation of human mammary epithelial cells. PLoS ONE 8(2):e53776. doi:10.1371/journal.pone.0053776

Jung SP, Lee Y, Han KM, Lee SK, Kim S, Bae SY (2013) The Role of the *CDH1* promoter hypermethylation in the axillary lymph node metastasis and prognosis. J Breast Cancer 16:16–22

Jones PA, Laird PW (1999) Cancer epigenetics comes of age. Nature Genet 21:163–167

Jones PA, Baylin SB (2002) The fundamental role of epigenetic events in cancer. Nature Rev Genet 3:415–428

Jin Z, Tamura G, Tsuchiya T, Sakata K, Kashiwaba M, Osakabe M et al (2001) Adenomatous polyposis coli (APC) gene promoter hypermethylation in primary breast cancers. Br J Cancer 85:69–73

Karen J, Wang YB, Javaherian A, Vaccariello M, Fusenig NE (1999) 12-O-tetradecanoylphorbol-13-acetate induces clonal expansion of potentially malignant keratinocytes in a tissue model of early neoplastic progression. Cancer Res 59:474–481

Kurbel S (2013) Model of tumor-associated epigenetic changes of HER2, ER, and PgR expression in invasive breast cancer phenotypes. Tumor Biol 34:2011–2017

Kawakami Y, Raya A, Raya RM, Rodriguez-Esteban C, Belmonte JC (2005) Retinoic acid signalling links left-right asymmetric patterning and bilaterally symmetric somitogenesis in the zebrafish embryo. Nature 435:165–171

Kelly PN, Dakic A, Adams JM, Nutt SL, Strasser A (2007) Tumor growth need not be driven by rare cancer stem cells. Science 317:337

Kouzarides T (2007) Chromatin modifications and their function. Cell 128:693–705

Kondo Y, Shen L, Cheng AS, Ahmed S, Boumber Y, Charo C et al (2008) Gene silencing in cancer by histone H3 lysine 27 trimethylation independent of promoter DNA methylation. Nat Genet 40:741–750

Kurbel S (2013) Model of tumor-associated epigenetic changes of HER2, ER, and PgR expression in invasive breast cancer phenotypes. Tum Biol 34:2011–2017

Lander ES, Linton LM, Birren B, Nusbaum C, Zody MC, Baldwin J et al (2001) Initial sequencing and analysis of the human genome. Nature 409:860–921

Lee X, Si SP, Tsou HC, Peacocke M (1995) Cellular aging and transformation suppression: a role for retinoic acid receptor beta 2. Exp Cell Res 218:296–304

Leu YW, Yan PS, Fan M, Jin VX, Liu JC, Curran EM et al. (2004) Loss of estrogen receptor signaling triggers epigenetic silencing of downstream targets in breast cancer. Cancer Res 64:8184–8192

Ley TJ, Ding L, Walter MJ, McLellan MD, Lamprecht T, Larson DE et al (2010) DNMT3A mutations in acute myeloid leukemia. N Engl J Med 363(25):2424–2433

Liu Y, Lee MO, Wang HG, Li Y, Hashimoto Y, Klaus M et al (1996) Retinoic acid receptor β mediates the growth-inhibitory effect of retinoic acid by promoting apoptosis in human breast cancer cells. Mol Cell Biol 16:1138–1149

Liu XL, Wazer DE, Watanabe K, Band V (1996) Identification of a novel serine protease-like gene, the expression of which is down-regulated during breast cancer progression. Cancer Res 56:3371–3379

Li R, Faria TN, Boehm M, Nabel EG, Gudas LJ (2004) Retinoic acid causes cell growth arrest and an increase in p27 in F9 wild type but not in F9 retinoic acid receptor h2 knockout cells. Exp Cell Res 294:290–300

Li S, Rong M, Iacopetta B (2006) DNA hypermethylation in breast cancer and its association with clinicopathological features. Cancer Lett 237:272–280

Lapidot T, Sirard C, Vormoor J, Murdoch B, Hoang T, Cacere-Cortes J et al (1994) A cell initiating human acute myeloid leukaemia after transplantation into SCID mice. Nat 367:645–648

Laird PW (2005) Cancer epigenetics. Hum Mol Genet 14:R65–R76

Lo PK, Sukumar S (2008) Epigenomics and breast cancer. Pharmacogenomics 9:1879–1902

Mani SA, Guo W, Liao MJ, Eaton EN, Ayyanan A, Zhou AY, Brooks M, Reinhard F et al (2008) The epithelial-mesenchymal transition generates cells with properties of stem cells. Cell 133:704–715

Mangelsdorf DJ, Umesona K, Evans RM (1994) The retinoid receptors. In Sporn MB, Roberts AB, Goodman DS (eds) The retinoids: biology, chemistry, and medicine, 2nd edn. Raven Press, New York, pp. 319–349

Marchetti P, Zamzami N, Joseph B, Schraen-Maschke S, Méreau-Richard C, Costantini P et al (1999) The novel retinoid 6-[3-(1-adamantyl)-4-hydroxyphenyl]-2-naphtalene carboxylic acid can trigger apoptosis through a mitochondrial pathway independent of the nucleus. Cancer Res 59:6257–6266

Maunakea AK, Nagarajan RP, Bilenky M, Ballinger TJ, D'Souza C, Fouse SD (2010) Conserved role of intragenic DNA methylation in regulating alternative promoters. Nature 466:253–257

McAlhany SJ, Ayala GE, Frolov A, Ressler SJ, Wheeler TM, Watson JE et al (2004) Decreased stromal expression and increased epithelial expression of WFDC1/ps20 in prostate cancer is associated with reduced recurrencefree survival. Prostate 61:182–191

Mehdipour P, Pirouzpanah S, Sarafnejad A, Atri M, Shahrestani ST, Haidari M (2009) Prognostic implication of CDC25A and cyclin E expression on primary breast cancer patients. Cell Biol Int 33:1050–1056

Mehdipour P, Pirouzpanah S, Azari Yam A (2012) Retinoic acid receptorb2 gene in breast cancer. Eur J Clin Med Oncol 4:17–33

Mehrotra J, Vali M, McVeigh M, Kominsky SL, Fackler MJ, Lahti-Domenici J et al (2004) Very high frequency of hypermethylated genes in breast cancer metastasis to the bone, brain, and lung. Clin Cancer Res 10:104–3109

Mira-y-Lopez R, Zheng WL, Kuppumbatti YS, Rexer B, Jing Y, Ong DE (2000) Retinol conversion to retinoic acid is impaired in breast cancer cell lines relative to normal cells. J Cell Physiol 185:302–309

Moison C, Senamaud-Beaufort C, Fourrière L, Champion C, Ceccaldi A, Lacomme S et al (2013) DNA methylation associated with polycomb repression in retinoic acid receptor β silencing. FASEB J 27:1468–1478

Moison C, Assemat F, Daunay A, Tost J, Guieysse-peugeot AL, Arimondo PB (2014) Synergistic chromatin repression of the tumor suppressor gene *RARB* in human prostate cancers. Epigenetics 9:477–482

Mongan NP, Gudas LJ (2005) Valproic acid, in combination with all-trans retinoic acid and 5-aza-2′-deoxycytidine, restores expression of silenced RARβ2 in breast cancer cells. Mol Cancer Ther 4:477–486

Morel AP, Lievre M, Thomas C, Hinkal G, Ansieau S, Puisieux A (2008) Generation of breast cancer stem cells through epithelial- mesenchymal transition. PLoS One 3:e2888

Montesano R, Soulie P (2002) Retinoids induce lumen morphogenesis in mammary epithelial cells. J Cell Sci 115:4419–4431

Nass SJ, Herman JG, Gabrielson E, Iversen PW, Parl FF, Davidson NE et al (2000) Aberrant methylation of the estrogen receptor and E-cadherin 5′ CpG islands increases with malignant progression in human breast cancer. Cancer Res 60:4346–4348

Nielsen AL, Oulad-Abdelghani M, Ortiz JA, Remboutsika E, Chambon P, Losson R (2001) Heterochromatin formation in mammalian cells: interaction between histones and HP1 Proteins. Mol Cell 7:729–739

Nikoloski G, Langemeijer SM, Kuiper RP, Knops R, Massop M, Tönnissen ER et al (2010) Somatic mutations of the histone methyltransferase gene EZH2 in myelodysplastic syndromes. Nat Genet 42:665–667

Novak P, Jensen TJ, Garbe JC, Stampfer MR, Futscher BW (2009) Stepwise DNA methylation changes are linked to escape from defined proliferation barriers and mammary epithelial cell immortalization. Cancer Res 69:5251–5258

Nakajima S, Doi R, Toyoda E, Tsuji S, Wada M, Koizumi M et al (2004) N-cadherin expression and epithelial–mesenchymal transition in pancreatic carcinoma. Clin Cancer Res 10:4125–4133

Nowell PC (1976) The clonal evolution of Tumor cell population. Science 194:23–28

O'Brien CA, Pollett A, Gallinger S, Dick JE (2007) A human colon cancer cell capable of initiating tumor growth in immunodeficient mice. Nature 445:106–110

Oh HS, Kwon H, Sun SK, Yang CH (2002) QM, a putative tumor suppressor, regulates proto-oncogene c-yes. J Biol Chem 277:36489–36498

Oldridge EE, Walker HF, Stower MJ, Simms MS, Mann VM, Collins AT et al (2013) Retinoic acid represses invasion and stem cell phenotype by induction of the metastasis suppressors RARRES1 and LXN. Oncogenesis 2:e45. doi:10.1038/oncsis.2013.6

Ordentlich P, Nguyen N, Jin K, Sadik H, Han L, Sukumar S (2012) Reactivation of epigenetically silenced retinoic acid receptor-beta for therapy of breast cancer- from molecular mechanism to potential clinical applications. Cancer Res 72:Abstract nr P2-09-01

Parrella P, Poeta ML, Gallo AP, Prencipe M, Scintu M, Apicella A et al (2004) Nonrandom distribution of aberrant promoter methylation of cancer-related genes in sporadic breast tumors. Clin Cancer Res 10:5349–5354

Peng X, Green A, Shilkaitis A, Zhu Y, Bratescu L, Christov K (2011) Early *in vitro* passages of breast cancer cells are differentially susceptible to retinoids and differentially express RARβ isoforms. Int J Oncol 39:577–583

Perissi V, Aggarwal A, Glass CK, Rose DW, Rosenfeld MG (2004) A corepressor/coactivator exchange complex required for transcriptional activation by nuclear receptors and other regulated transcription factors. Cell 116:511–526

Peter ME (2009) Let-7 and miR-200 microRNAs: guardians against pluripotency and cancer progression. Cell Cycle 8:843–852

Pierzchalski K, Yu J, Norman V, Kane MA (2013) CrbpI regulates mammary retinoic acid homeostasis and the mammary microenvironment. FASEB J 27(5):1904–1916

Pinto R, Pilato B, Ottini L, Lambo R, Simone G, Paradiso A et al (2013) Different methylation and microRNA expression pattern in male and female familial breast cancer. J Cell Physiol 228:1264–1269

Pirouzpanah S, Taleban FA, Atri M, Abadi AR, Mehdipour P ((2010) The effect of modifiable potentials on hypermethylation status of retinoic acid receptor-beta2 and estrogen receptor-alpha genes in primary breast cancer. Cancer Causes Control 21:2101–2111

Pirouzpanah S, Taleban FA, Mehdipour P, Atri M, Hooshyareh-Rad A, Sabour S (2014a) The biomarker-based validity of a food frequency questionnaire to assess the intake status of folate, pyridoxine and cobalamin among Iranian primary breast cancer patients. Eur J Clin Nutr 68:316–323

Pirouzpanah S, Taleban FA, Mehdipour P, Atri M, Foroutan-Ghaznavi M (2014b) Plasma total homocysteine level in association with folate, pyridoxine, and cobalamin status among Iranian primary breast cancer patients. Nutr Cancer 66:1097–1108

Piva R, Piva R, Spandidos DA, Gambari R (2013) From microRNA functions to microRNA therapeutics: Novel targets and novel drugs in breast cancer research and treatment. Int J Oncol 43:985–994

Polyak K, Weinberg RA (2009) Transitions between epithelial and mesenchymal states: acquisition of malignant and stem cell traits. Nat Rev Cancer 9:265–273

Pozzi S, Rossetti S, Bistulfi G, Sacchi N (2006) RAR-mediated epigenetic control of the cytochrome P450 Cyp26a1 in embryocarcinoma cells. Oncogene 25:1400–1407

Quintana E, Shackleton M, Sabel MS, Fullen DR, Johnson TM, Morrison SJ (2008) Efficient tumor formation by single human melanoma cells. Nature 456:593–598

Ren M, Pozzi S, Bistulfi G, Somenzi G, Rossetti S, Sacchi N (2005) Impaired retinoic acid (RA) signal leads to RARβ2 epigenetic silencing and RA resistance. Mol Cell Biol 25:10591–10603

Ren MQ, Pozzi S, Bistulfi G, Somenzi G, Rossetti S, Sacchi N (2005) Impaired retinoic acid (RA) signal leads to RARβ2 epigenetic silencing and RA resistance. Mol Cell Biol 25:10591–10603

Rexer BN, Zheng WL, Ong DE (2001) Retinoic acid biosynthesis by normal human breast epithelium is via aldehyde dehydrogenase 6, absent in MCF-7 cells. Cancer Res 61:7065–7070

Reya T, Morrison SJ, Clark MF, Weissman IL (2001) Stem cells, cancer, and cancer stem cells. Nature 414:105–111

Ricci-Vitiani L, Lombardi DG, Pilozzi E, Biffoni M, Todaro M, Peschle C et al (2006) Identification and expansion of human colon-cancer-initiating cells. Nat 445:111–115

Roman SC, Ormandy J, Manning DL, Blamey RW, Nicholson RI, Sutherland RL et al (1993) Estradiol induction of retinoic acid receptors in human breast cancer cells. Cancer Res 53:5940–5945

Ricci-Vitiani L, Lombardi DG, Pilozzi E, Biffoni M, Todaro M, Peschle C et al (2007) Identification and expansion of human colon-cancer-initiating cells. Nature 445:111–115

Rishi AK, Shao ZM, Baumann RG, Li XS, Sheikh MS, Kimura S et al (1995) Estradiol regulation of the human retinoic acid receptor alpha gene in human breast carcinoma cells is mediated via an imperfect half-palindromic estrogen response element and Sp1 motifs. Cancer Res 55:4999–5006

Roman SD, Ormandy CJ, Manning DL, Blamey RW, Nicholson RI, Sutherland RL et al (2006) Cancer Stem cells—perspectives on current status and future directions: AACR Workshop on cancer stem cells. Cancer Res 66(19):9339–9344

Roisin M, Connolly RM, Nguyen NK, Sukumar S (2013) Molecular pathways: current role and future directions of the retinoic acid pathway in cancer prevention and treatment. Clin Cancer Res 19:1651–1659

Ropero S, Fraga MF, Ballestar E, Hamelin R, Yamamoto H, Boix-Chornet M et al (2006) A truncating mutation of HDAC2 in human cancers confers resistance to histone deacetylase inhibition. Nature Genet 38:56–569

Rosivatz E, Becker I, Specht K, Fricke E, Luber B, Busch R et al (2002) Differential expression of the epithelial–mesenchymal transition regulators Snail, SIP1, and Twist in gastric cancer. Am J Pathol 161:1881–1891

Rouhi A, Mager DL, Humphries RK, Kuchenbauer F (2008) MiRNAs, epigenetics, cancer. Mamm Genome 19:517–525

Rubinek T, Shulman M, Israeli S, Bose S, Avraham A, Zundelevich A et al (2012) Epigenetic silencing of the tumor suppressor klotho in human breast cancer. Breast Cancer Res Treat 133(2):649–657

Sadikovic B, Al-Romaih K, Squire JA, Zielenska M (2008) Cause and consequences of genetic and epigenetic alterations in human cancer. Curr Genomics 9:394–408

Sabichi AL, Hendricks DT, Bober MA, Birrer MJ (1998) Retinoic acid receptor β expression and growth inhibition of gynecologic cancer cells by the synthetic retinoid N-(4-hydroxyphenyl) retinamide. J Natl Cancer Inst 90:597–605

Sarrio D, Moreno-Bueno G, Hardisson D, Sanchez-Estevez C, Guo M, Herman JG, Gamallo C et al (2003) Epigenetic and genetic alterations of APC and CDH1 genes in lobular breast cancer: relationships with abnormal E-cadherin and catenin expression and microsatellite instability. Int J Cancer 106:208–215

Shahbazian MD, Grunstein M (2007) Functions of site-specific histone acetylation and deacetylation. Annu Rev Biochem 76:75–100

Shi Y, Lan F, Matson C, Mulligan P, Whetstine JR, Cole PA et al (2004) Histone demethylation mediated by the nuclear amine oxidase homolog LSD1. Cell 119:941–953

Shinozaki M, Hoon DS, Giuliano AE, Hansen NM, Wang HJ, Turner R et al (2005) Distinct hypermethylation profile of primary breast cancer is associated with sentinel lymph node metastasis. Clin Cancer Res 11:2156–2162

Simon JA, Lange CA (2008) Roles of the EZH2 histone methyltransferase in cancer epigenetics. Mutat Res 647:21–29

Simo-Riudalbas L, Melo SA (2011) Esteller M DNMT3B gene amplification predicts resistance to DNA demethylating drugs. Genes Chromosomes Cancer 50:527–534

Singh SK, Hawkins C, Clarke ID, Squire JA, Bayani J, Hide T et al (2004) Identification of human brain tumor initiating cells. Nature 432:396–401

Sirchia SM, Ren M, Pili R, Sironi E, Somenzi G, Ghidoni R et al (2002) Endogenous reactivation of the RARβ2 tumor suppressor gene epigenetically silenced in breast cancer. Cancer Res 62:2455–2461

Sirchia SM, Ferguson AT, Sironi E, Subramanyan S, Orlandi R, Sukumar S et al (2000) Evidence of epigenetic changes affecting the chromatin state of the retinoic acid receptor beta2 promoter in breast cancer cells. Oncogene 19:1556–1563

Sirchia SM, Ren M, Pili R, Sironi E, Somenzi G, Ghidoni R et al (2002) Endogenous reactivation of the RARβ2 tumor suppressor gene epigenetically silenced in breast cancer. Cancer Res 62:2455–2461

Smith WC, Nakshatri H, Leroy P, Rees J, Chambon P (1991) A retinoic acid response element is present in the mouse cellular retinol binding protein I (mCRBPI) promoter. EMBO J 10:2223–2230

Sommer KM, Chen LI, Treuting PM, Smith LT, Swisshelm K (1999) Elevated retinoic acid receptor β4 protein in human breast tumor cells with nuclear and cytoplasmic localization. Proc Natl Acad Sci U S A 96:8651–8656

Stirzaker C, Song JZ, Davidson B, Clark SJ (2004) Transcriptional gene silencing promotes DNA hypermethylation through a sequential change in chromatin modifications in cancer cells. Cancer Res 64:3871–3877

Sucov HM, Murakami KK, Evans RM (1990) Characterization of an autoregulated response element in the mouse retinoic acid receptor type β gene. Proc Natl Acad Sci U S A 87:5392–5396

Sun J, Xu X, Liu J, Liu H, Fu L, Gu L (2011) Epigenetic regulation of retinoic acid receptor β2 gene in the initiation of breast cancer. Med Oncol 28:1311–1318

Sunami E, Shinozaki M, Sim MS, Nguyen SL, Vu AT, Giuliano AE et al (2008) Estrogen receptor and HER2/neu status affect epigenetic differences of tumor-related genes in primary breast tumors. Breast Cancer Res 10(3):R46. doi:10.1186/bcr2098

Sneeringer CJ, Scott MP, Kuntz KW, Knutson SK, Pollock RM, Richon VM et al (2010) Coordinated activities of wild-type plus mutant EZH2 drive tumor-associated hypertrimethylation of lysine 27 on histone H3 (H3K27) in human B-cell lymphomas. Proc Natl Acad Sci U S A 107:20980–20985

Swisshelm K, Ryan K, Lee X, Tsou HC, Peacocke M, Sager R (1994) Down-regulation of retinoic acid receptor β in mammary carcinoma cell lines and its up-regulation in senescing normal mammary epithelial cells. Cell Growth Differ 5:133–141

Tahiliani M, Koh KP, Shen Y, Pastor WA, Bandukwala H, Brudno Y et al (2009) Conversion of 5-methylcytosine to 5-hydroxymethylcytosine in mammalian DNA by MLL partner TET1. Science 324:930–935

Tao MH, Mason JB, Marian C, McCann SE, Platek ME, Millen A et al (2011) Promoter methylation of E-cadherin, p16, and RAR-β(2) genes in breast tumors and dietary intake of nutrients important in one-carbon metabolism. Nutr Cancer 63(7):1143–1150

Thiery JP, Acloque H, Huang RY, Nieto MA (2009) Epithelial-mesenchymal transitions in development and disease. Cell 139:871–890

Thol F, Damm F, Lüdeking A, Winschel C, Wagner K, Morgan M et al (2011) Incidence and prognostic influence of DNMT3A mutations in acute myeloid leukemia. J Clin Oncol 29:2889–2896

Tsukada Y, Fang J, Erdjument-Bromage H, Warren ME, Borchers CH, Tempst P et al (2006) Histone demethylation by a family of JmjC domain-containing proteins. Nature 439:811–816

Tribioli C, Droetto S, Bione S, Cesareni G, Torrisi MR, Lotti LV et al (1996) An X chromosome-linked gene encoding a protein with characteristics of a rhoGAP predominantly expressed in hematopoietic cells. Proc Natl Acad Sci U S A 93:695–699

Tommasi S, Karm DL, Wu X, Yen Y, Pfeifer GP (2009) Methylation of homeobox genes is a frequent and early epigenetic event in breast cancer. Breast Cancer Res 11(1):R14. doi: 10.1186/bcr2233

Van Hoesel AQ, Sato Y, Elashoff DA, Turner RR, Giuliano AE, Shamonki JM, et al (2013) http://www.ncbi.nlm.nih.gov/pubmed/23652305. Assessment of DNA methylation status in early stages of breast cancer development. Br J Cancer. 108:2033-8. doi: 10.1038/bjc.2013.136

Varambally S, Dhanasekaran SM, Zhou M, Barrette TR, Kumar-Sinha C, Sanda MG et al (2012) Chinnaiyan AM. The polycomb group protein EZH2 is involved in progression of prostate cancer. Nature 419:624–629

Vermot J, Pourquie O (2005b) Retinoic acid coordinates somitogenesis and left-right patterning in vertebrate embryos. Nat 435:215–220

Vermot J, Gallego Llamas J, Fraulob V, Niederreither K, Chambon P, Dolle P (2005a) Retinoic acid controls the bilateral symmetry of somite formation in the mouse embryo. Science 308:563–566

Virmani AK, Rathi A, Zöchbauer-Müller S, Sacchi N, Fukuyama Y, Bryant D et al (2000) Promoter methylation and silencing of the retinoic acid receptor-β gene in lung carcinomas. J Natl Cancer Inst 92:1303–1307

Virmani AK, Rathi A, Sathyanarayana UG, Padar A, Huang CX, Cunnigham HT et al (2001) Aberrant methylation of the adenomatous polyposis coli (APC) gene promoter 1A in breast and lung carcinomas. Clin Cancer Res 7:1998–2004

Wallden B, Emond M, Swift ME, Disis ML, Swisshelm K (2005) Antimetastatic gene expression profiles mediated by retinoic acid receptor beta 2 in MDA-MB-435 breast cancer cells. BMC Cancer 5:140

Wang YA, Shen K, Wang Y, Brooks SC (2005) Retinoic acid signaling is required for proper morphogenesis of mammary gland. Dev Dyn 234:892–899

Ward PS, Patel J, Wise DR, Abdel-Wahab O, Bennett BD, Coller HA, Cross JR, Fantin VR, Hedvat CV, Perl AE et al (2010) The common feature of leukemia-associated IDH1 and IDH2 mutations is a neomorphic enzyme activity converting α-ketoglutarate to 2-hydroxyglutarate. Cancer Cell 17:225–234

Weissmann S, Alpermann T, Grossmann V, Kowarsch A, Nadarajah N, Eder C et al (2012) Landscape of TET2 mutations in acute myeloid leukemia. Leukemia 26:934–942

Weston A, Harris CC (1997) Chemical carcinogenesis. In: Holland JF, Frei E, Bast RC et al (eds) Cancer medicine, 4th edn, vol 1. Williams and Wikins, Baltimore, 261–276

Widschwendter M, Daxenbichler G, Dapunt O, Marth C (1995) Effects of retinoic acid and γ-interferon on expression of retinoic acid receptor and cellular retinoic acid-binding protein in breast cancer cells. Cancer Res 55:2135–2139

Widschwendter M, Berger J, Hermann M, Müller HM, Amberge A, Zeschnigk M et al (2005) Methylation and silencing of the retinoic acid receptor-β2 gene in breast cancer. JNCI J Natl Cancer Inst 92:826–832

Widschwendter M, Berger J, Hermann M, Müller HM, Amberger A, Zeschnigk M et al (2000) Methylation and silencing of the retinoic acid receptor-beta2 gene in breast cancer. J Natl Cancer Inst 92:826-832

Widschwendter M, Berger J, Müller HM, Zeimet AG, Marth C (2001) http://www.ncbi.nlm.nih.gov/pubmed/11501579. Epigenetic downregulation of the retinoic acid receptor-beta2 gene in breast cancer. J Mammary Gland Biol Neoplasia.6:193-201. Review.

Xiang TX, Yuan Y, Li LL, Wang H, Dan Y, Chen Y (2013) Aberrant promoter CpG methylation and its translational applications in breast cancer. Chin J Cancer 32:12–20

Xu XC, Ro JY, Lee JS, Shin DM, Hong WK, Lotan R (1994) Differential expression of nuclear retinoid receptors in normal, premalignant, and malignant head and neck tissues. Cancer Res 54:3580–3587

Xue C, Plieth D, Venkov C, Xu C (2003) Neilson EG. The gatekeeper effect of epithelial–mesenchymaltransition regulates the frequency of breast cancer metastasis. Cancer Res 63:3386–3394

Xu L, Glass CK, Rosenfeld MG (1999) Coactivator and corepressor complexes in nuclear receptor function. Curr Opin Genet Dev 9:140–147

Yan XJ, Xu J, Gu ZH, Pan CM, Lu G, Shen Y et al (2011) Exome sequencing identifies somatic mutations of DNA methyltransferase gene DNMT3A in acute monocytic leukemia. Nat Genet 43:309–317

Yanaihara N, Caplen N, Bowman E, Seike M, Kumamoto K, Yi M, et al (2006) http://www.ncbi.nlm.nih.gov/pubmed/16530703. Unique microRNA molecular profiles in lung cancer diagnosis and prognosis http://www.ncbi.nlm.nih.gov/pubmed/16530703. Cancer Cell. 9:189-98

Yang Q, Sakurai T, Kakudo K (2002) Retinoid, retinoic acid receptor β and breast cancer. Breast Cancer Res Treat 76:167–173

Yang J, Mani SA, Donaher JL, Ramaswamy S, Itzykson RA, Come C et al (2004) Twist, a master regulator of morphogenesis, plays an essential role in tumor metastasis. Cell 117:927–939

Yoo CB, Jones PA (2006) Epigenetic therapy of cancer: past, present and future. Nat Rev Drug Discov 5:37–50

Yu Z, Xiao Q, Zhao L, Ren J, Bai X, Sun M et al (2014) DNA methyltransferase 1/3a overexpression in sporadic breast cancer is associated with reduced expression of estrogen receptor-alpha/breast cancer susceptibility gene 1 and poor prognosis. Mol Carcinog. doi:10.1002/mc.22133

Zanardi S, Serrano D, Argusti A, Barile M, Puntoni M, Decensi A (2006) Clinical trials with retinoids for breast cancer chemoprevention. Endocr Relat Cancer 13:51–68

Zhang XK, Liu Y, Lee MO, Pfahl M (1994) A specific defect in the retinoic acid receptor associated with human lung cancer cell lines. Cancer Res 54:5663–5669

Zhang B, Pan X, Cobb GP, Anderson TA (2007) microRNAs as oncogenes and tumor suppressors. Dev Biol 302:1–12

Chapter 12
Retinoic Acid Receptor-β, From Gene to Clinic

Hassan Fazilaty and Parvin Mehdipour

Contents

Abstract Retinoic acid receptor-beta (RAR-β) is nuclear receptor which can be activated by corresponding ligands, including retinoic acid. It is encoded by *RARB* gene, which is conserved in several species, and code for a DNA binding protein that in complex to the ligand and other binding partners binds to specific sites and regulates the expression of several genes. RAR-β plays important roles in several developmental, physiologic, and pathogenic mechanisms in human. It is a tumor suppressor gene, which is not expressed in several cancers. Methylation status and micro-RNAs are key central regulators of the *RARB* expression, which are influenced by several factors, such as environment and diet. It is responsible for the induction of apoptosis, and for the chemo-preventive and therapeutic effects of anti-cancer drugs, therefor has been used a treatment against different kinds of cancer in several clinical trials.

P. Mehdipour (✉)
Department of Medical Genetics, School of Medicine,
Tehran University of Medical Sciences, Poursina Street,
P.O. Box: 14176-13151, Keshavarz Boulevard Tehran, Iran
e-mail: mehdipor@tums.ac.ir

H. Fazilaty
Instituto de Neurociencias CSIC-UMH, Avda. Ramon y Cajal s/n,
03550 San Juan de Alicante, Spain

P. Mehdipour (ed.), *Epigenetics Territory and Cancer*,
DOI 10.1007/978-94-017-9639-2_12

12.1 Introduction

Regulating a broad range of biological processes, retinoic acid receptor-beta (RAR-β) plays critical roles in homeostasis and its deficiencies is linked to a variety of disorders. RAR-β is a member of the thyroid-steroid hormone receptor superfamily of nuclear transcriptional regulators. This receptor is localized to the sub-nuclear compartmentsin cytoplasm, and binds to retinoic acid, which is the biological activated form of vitamin A, and governs key signaling pathways regulating embryonic morphogenesis, cell growth, and differentiation (Gudas et al. 1994).

12.2 Gene; DNA/RNA

RAR-β gene (RARB) is known to have conserved homologous members in different species including human, mouse, chimpanzee, dog, and chicken (Table 12.1), and more than 70 organisms have orthologs of this gene.

In human, *RARB* consists of 8 exons and 7 introns and spans about 423kb in short arm of chromosome 3. It is located in a chromosomal region which is the host of several genes including *CFL1P7*, *RNA5SP126*, *TOP2B*, *MIR4442*, *CRIP1P2*, *LOC101927874*, *RNA5SP125*, and *EIF3KP2*, placed in different directions, consonant oropposite, among which *CFL1P7* and *RNA5SP126* are located within *RARB*. Neighboring genes may have important associations to RAR-β functions, but little is known about their function until now, as they are mostly pseudogenes. (See Table12.2 for a short description and function of *RARB* neighboring genes).

Transcription from *RARB* produces 9 transcripts, 5 protein coding and 4 with no protein product, as a result of alternative splicing (Ensembl:ENSG00000077092), and differential promoter usage (Zelent et al. 1991).

Table 12.1 Retinoic acid receptor beta homologous genes. (Information is derived from National Center for Biotechnology Information (NCBI), Gene (http://www.ncbi.nlm.nih.gov/gene/5915))

Species	Chromosomal location	Gene symbol	Size	Other symbols
Human	3p24.2	*RARB*	423 kb	*HAP, RRB2, NR1B2, MCOPS12*
Mouse	14 A1-A3; 14 7.08 cM	*Rarb*	164 kb	*Hap, Nr1b2, RARbeta2, A830025K23*
Chimpanzee	3	*RARB*	172 kb	–
Chicken	2	*RARB*	316 kb	*RARBETA*
Dog	23	*RARB*	172 kb	–

Table 12.2 RARB neighboring genes. (Information is derived from National Center for Biotechnology Information (NCBI), Gene (http://www.ncbi.nlm.nih.gov/gene/5915))

Gene symbol	Gene full name	Direction	Gene type	Function
CFL1P7	cofilin 1 (non-muscle) pseudogene 7	Consonant	pseudo	Unknown
RNA5SP126	RNA, 5S ribosomal pseudogene 126	Opposite	pseudo	Unknown
TOP2B	topoisomerase (DNA) II beta 180kDa	Opposite	Protein coding	Regulatory roles in transcription and replication; increased proliferation in cancer
MIR4442	microRNA 4442	Opposite	miscRNA	Unknown
CRIP1P2	cysteine-rich protein 1 (intestinal) pseudogene 2	Consonant	pseudo	Unknown
LOC101927874	uncharacterized LOC101927874	Consonant	ncRNA	Unknown
RNA5SP125	RNA, 5S ribosomal pseudogene 125	Consonant	pseudo	Unknown
EIF3KP2	eukaryotic translation initiation factor 3, subunit K pseudogene 2	Consonant	pseudo	Unknown

12.3 Protein

RARB produces five protein products, as a result of alternative splicing, with the length of 455 amino acid (aa) being the biggest and 336 aa being the smallest one. Different isoforms differ in their N-terminal regions, which leads the generation of pleiotropic effects of retinoids. RAR-β has two conserved domains, including DNA binding domain of nuclear receptors (NR- DBD)-like superfamily, and ligand binding domain of retinoic acid receptor a member of nuclear receptor family (NR- LBD-RAR). NR-DBD is composed of two C4-type zinc fingers, and interacts with a specific DNA site in the upstream region of target genes and control the rate of transcription. NR- LBD-RAR, on the other hand, provides a binding site for retinoic acid. Upon binding of ligand RAR-β other RARs (α and γ) bind to specific RAR response elements (RAREs), composed of tandem 5'-AGGTCA-3', in the upstream of retinoid target genes. RAR complex recruit the corepressor proteins AMRT or NCoR when the ligand is absent, and therefor leads toward inactivation of corresponding genes (Marchler-Bauer et al. 2011; Marchler-Bauer et al. 2009; Marchler-Bauer and Bryant 2004). These two domains also exist in several other proteins as examples such as Protein NHR-128 isoform b, estrogen receptor (ER)-β, androgen receptor (AR) (Table 12.3).

Different isoforms of RAR-β are localized in specific subcellular location, including beta-1 and beta-2 in nucleus, and beta-4 in cytoplasm.

Table 12.3 Retinoic acid receptor beta homologous proteins*

Species	Protein symbol	Size	Other names
Human	RAR-β	335 aa	Retinoic acid receptor beta RAR-beta RAR-epsilon HBV-activated protein retinoic acid receptor beta 2 retinoic acid receptor beta 4 retinoic acid receptor beta 5 hepatitis B virus activated protein retinoic acid receptor beta variant 1 retinoic acid receptor beta variant 2 retinoic acid receptor, beta polypeptide nuclear receptor subfamily 1 group B member 2
Mouse	Rar-β	455 aa	Retinoic acid receptor beta RAR-beta nuclear receptor subfamily 1 group B member 2

(Information is derived from National Center for Biotechnology Information (NCBI), Protein (http://www.ncbi.nlm.nih.gov/protein/CAG38795.1))

12.4 Gene Regulation

RARB is tightly regulated, indicating its critical roles in vital cellular features. The mouse and human *RARB* promoters are highly homologous, containing conserved consensus TATA box, retinoic acid responsive element (RARE), TPA-responsive element (TRE), and cAMP-responsive element (CRE). It shows that in addition to the regulation by RARs, *RARB* can be regulated by several other transcription factors in a well-orchestrated manner (Shen et al. 1991; Dey et al. 1994).

It is shown that *RARB* is also regulated by microRNAs as well as epigenetic based mechanisms. *MIR-16* is located in chromosome 3, and encodes miR-16-2 which binds to three sites in *RARB* 3' UTR and inhibits its expression. It is expressed in a variety of cancers, and some normal cells, like different types of breast cancer, pancreatic cancer, prostate cancer, ependynoma primary tumors, renal cortex and medulla, and differentiated embryonic stem cells. miR-16-2 also inhibits nuclear casein kinase and cyclin-dependent kinase substrate 1(NUCKS1) (Hu et al. 2011) (miRTarBase). miR-1 and miR-206 are two other important microRNAs that inhibit *RARB* expression, and play pivotal roles in myogenesis (Goljanek-Whysall et al. 2012). miR-10a, which is a key mediator of pancreatic cancer metastasis, is shown to be effectively inhibited by RAR (Weiss et al. 2009). Epigenetics based regulation of RARB is evidently among crucial mechanisms that are involved in development and diseases (see below).

12.5 Function and Clinical Implication

RAR-β plays important roles in several developmental, physiologic, and pathogenic mechanisms in human. It takes parts in the development of embryonic digestive tract, myogenesis, eye, hindlimb, and nervous system differentiation, and has crucial negative effects on the regulation of cartilage, cell proliferation, transcription initiated by RNA polymerase II, and more importantly regulate apoptosis (Mark et al. 2009; Mendelsohn et al. 1994; Durston et al. 1989; Niederreither et al. 1999; Maden 2007; Zhu et al. 2009).

RAR-β is a tumor suppressor, and its expression is lost in several malignancies. It induces apoptosis, and is responsible for the chemo-preventive and therapeutic effects of anti-cancer drugs. The importance of RAR-β is notable, as its expression is reduced, or even lost, during different kinds of malignancies like breast cancer, and induction of its expression mediates growth arrest and apoptosis in breast tumor cells. It has been shown that RARs, including RAR-β, also controls cell adhesion by controlling the transcription of involved genes (Al Tanoury et al. 2014). Accordingly RAR-β is identified to actively impedes migration of RA associated breast cancer migration by inhibiting several key migratory proteins like moesin, c-Src, and focal adhesion kinase (FAK) (Inés et al. 2014).

12.5.1 Methylation Status and Diseases

It is demonstrated that *RARB*5'-region hypermethylation is responsible for suppression of *RARB* expression in different kinds of cancers (Seewaldt et al. 1995; Liu et al. 1996; Widschwendter et al. 2001; Youssef et al. 2004; Ivanova et al. 2002; Wang et al. 2003; Uray et al. 2009; Twelves et al. 2013; Lotan et al. 1995; Molinari et al. 2013; Cosialls et al. 2012; Chung et al. 2011). It presumably shed light to the important preventive and therapeutic role of retinoids in treatment of cancer. It is investigated that hypermethylated *RARB* is associated with diagnosis of breast cancer at younger age, with no family history (Pirouzpanah et al. 2010; Mehdipour et al.), high histological grade, high proliferation, increased tumor size, and metastasis (Marzese et al. 2012). In lung cancer, several alterations have been demonstrated in *RARB* (Gebert et al. 1991). Mutation in *RARB* is also shown to be associated to microphtalmia (Chitayat et al. 2007; Srour et al. 2013; Chassaing et al. 2013). Recessive and dominant mutations in *RARB* has shown to cause microphtalmia and diaphragmatic hernia, by loss or gain of function, respectively (Srour et al. 2013). In melanoma, it is shown that RAR-β interacts with $p14^{ARF}$ and plays roles toward the irreversible growth inhibition of malignant cells, therefore treatment of melanoma (Dahl et al. 2013). In Helicobacter Pylori derived gastric cancer, RARB is among genes with reduced expression levels, presumably regulated by epigenetic mechanisms (Cheng et al. 2013). Interestingly among risk factors, benzo(α) pyrene diol epoxide (BPDE) which is a carcinogen present in tobacco and environmental pollution in esophageal cancer, and bile acid which is an oncogene in gastrointestinal cancers, inhibit the expression of *RARB* (Song et al. 2005; Song and Xu 2001; Li et al. 2002). Methylation pattern of *RARB* in esophageal squamous cell carcinoma correlates with the development and severity of the disease (Li et al. 2014). Notably, BPDE induces miR-16-2, and as noted above, inhibits *RARB* (Hu et al. 2011). Unlike, the methylation pattern of *RARB* is lower in cancer stem cell rich populations of breast tumor (Park et al. 2012), which may play roles in lower proliferation rate of cancer stem cells.

Manipulation of methylation pattern of *RARB* is getting to be an interesting clinical procedure. Investigating the methylation pattern of *RARB* has been suggested to be a non-invasive biomarker for the prevention and diagnosis of prostate cancer (Gao et al. 2013). Genetic imbalance in the region containing *RARB*, on the other hand, has been shown to be a valuable target for personalized medicine (Ribeiro et al. 2014). Interestingly, epigenetic reprogramming seems to have considerable impact on tumor reversion, introducing epigenetic therapy. Axolotl extracts from oocyte reverses the epigenetic silencing of RARB, and therefore tumorigenicity of breast cancer cells, by arresting tumor growth (Allegrucci et al. 2011).

Considering the indispensable emerging role of immunotherapy, activation of *RARB* by immuno-modulation has shown to have antitumor effect. Toll-like receptor 3 upregulates a set of microRNAs, which induce re-expression of epigenetically silenced genes, including *RARB* in breast and prostate tumor cells, therefore induce apoptosis among cancer cells.

12.5.2 Clinical Trials

Retinoids have been the point of attention in clinical trials, giving their important possible participation in cancer treatment and prevention. Their roles have been well-established as regulators of cell proliferation, apoptosis, migration, and differentiation, and central participation in several key signaling pathways in preclinical studies. This led to launching several clinical trials, which revealed the using of RARs as remarkable treatment against cancer, as a chemo preventive agent (Dragnev et al. 2000). Fenretinide, which is the most widely studied retinoid in clinical trials, has been shown to be favorable on breast cancer chemoprevention. It showed a persistent manner in the reduction of breast cancer metastases in premenopausal women in a phase III trial, which assumingly was observed in accordance with the modulation IGF-1 (insulin-like growth factor-1), which have been shown as a prognosis marker for breast cancer risk in premenopausal women (Zanardi et al. 2006; Veronesi et al. 2006).

Retinoids have been FDA approved for being as a treatment in cutaneous T-cell lymphoma and acute promyelocytic leukemia (APL). Among those, Bexarotene, which is a rexinoid, has shown around 50 % response rate in patients (Duvic et al. 2001). The overall survival rate rises to 70 % in the case of ATRA (a synthetic retinoid) usage among APL patients concomitant with long term remissions (Tallman et al. 1997). Favorable results from treating this disease with retinoids come from the underlying chromosomal translocation and which generates the fusion protein PML/RARa, and also the ability of RAR-β (and other retinoids) in the induction of differentiation and apoptosis (Sirchia et al. 2000; Connolly et al. 2013). Application of ATRA is an HDAC (histone deacetylase) inhibitor and low-dose doxorubicin effectively target cancer stem cells (Connolly et al. 2013), thus likely inhibiting tumor progression and metastasis.

12.5.3 Environment and Lifestyle

Importantly, diet and environment regulate DNA methylation pattern and also *RARB*, taking important part in diseases like cancer. It is important to consider the pattern of lifestyle, and nutritional aspects in the development and prevention/treatment of cancer and other diseases. Some natural sources that have shown to have considerable impacts on gene regulation, for instance, changing the methylation pattern, may provide reliable treatments. Retinoic acid, folate, cobalamin, pyridoxine, riboflavin, EGCG (epigallacatechin-3 gallate), genistein, daidzein, curcumin, and selenium, are among factors that regulate epigenetic associated elements like DNA methyl transferases, S-adenosylmethionine transferase, and different microRNAs, thus, help to prevent cancer. Some of these elements directly target *RARB*, like EGCG, genistein, and daidzein (Pirouzpanah et al. 2013; Ross 2003; Stefanska et al. 2012).

Understanding different functions and implications of key players in development and disease associated processes, like *RARB*, may provide more reliable treatments for a variety of conditions like cancer. Signaling pathways, gene/protein interactions, regulatory factors, and environmental conditions, which are associated to *RARB*, need to be determined in all aspects in order to provide promising tools toward making better health condition for human.

References

Al Tanoury Z, Piskunov A, Andriamoratsiresy D, Gaouar S, Lutzing R, Ye T et al (2014) Genes involved in cell adhesion and signaling: a new repertoire of retinoic acid receptor target genes in mouse embryonic fibroblasts. J Cell Sci 127(3):521–533

Allegrucci C, Rushton MD, Dixon JE, Sottile V, Shah M, Kumari R et al (2011) Epigenetic reprogramming of breast cancer cells with oocyte extracts. Mol Cancer 10(1):7. doi:10.1186/1476-4598-10-7

Chassaing N, Ragge N, Kariminejad A, Buffet A, Ghaderi-Sohi S, Martinovic J et al (2013) Mutation analysis of the STRA6 gene in isolated and non-isolated anophthalmia/microphthalmia. Clin Genet 83(3):244–250. doi:10.1111/j.1399-0004.2012.01904.x

Cheng AS, Li MS, Kang W, Cheng VY, Chou JL, Lau SS et al (2013) Helicobacter pylori causes epigenetic dysregulation of FOXD3 to promote gastric carcinogenesis. Gastroenterology 144(1):122–133.e129. doi:10.1053/j.gastro.2012.10.002

Chitayat D, Sroka H, Keating S, Colby RS, Ryan G, Toi A et al (2007) The PDAC syndrome (pulmonary hypoplasia/agenesis, diaphragmatic hernia/eventration, anophthalmia/microphthalmia, and cardiac defect) (Spear syndrome, Matthew-Wood syndrome): report of eight cases including a living child and further evidence for autosomal recessive inheritance. Am J Med Genet Part A 143A(12):1268–1281. doi:10.1002/ajmg.a.31788

Chung JH, Lee HJ, Kim BH, Cho NY, Kang GH (2011) DNA methylation profile during multistage progression of pulmonary adenocarcinomas. Virchows Arc: Internatl J Pathol 459(2):201–211. doi:10.1007/s00428-011-1079-9

Connolly RM, Nguyen NK, Sukumar S (2013) Molecular pathways: current role and future directions of the retinoic acid pathway in cancer prevention and treatment. Clin Cancer Res 19(7):1651–1659

Cosialls AM, Santidrian AF, Coll-Mulet L, Iglesias-Serret D, Gonzalez-Girones DM, Perez-Perarnau A et al (2012) Epigenetic profile in chronic lymphocytic leukemia using methylation-specific multiplex ligation-dependent probe amplification. Epigenomics 4(5):491–501. doi:10.2217/epi.12.40

Dahl C, Christensen C, Jonsson G, Lorentzen A, Skjodt ML, Borg A et al (2013) Mutual exclusivity analysis of genetic and epigenetic drivers in melanoma identifies a link between p14 ARF and RARbeta signaling. Mol Cancer Res: MCR 11(10):1166–1178. doi:10.1158/1541-7786.MCR-13-0006

Dey A, Minucci S, Ozato K (1994) Ligand-dependent occupancy of the retinoic acid receptor beta 2 promoter in vivo. Mol Cell Biol 14(12):8191–8201

Dragnev KH, Rigas JR, Dmitrovsky E (2000) The retinoids and cancer prevention mechanisms. Oncologist 5(5):361–368

Durston A, Timmermans J, Hage W, Hendriks H, De Vries N, Heideveld M et al (1989) Retinoic acid causes an anteroposterior transformation in the developing central nervous system. Nature 340(6229):140–144

Duvic M, Hymes K, Heald P, Breneman D, Martin AG, Myskowski P et al (2001) Bexarotene is effective and safe for treatment of refractory advanced-stage cutaneous T-cell lymphoma: multinational phase II-III trial results. J Clin Oncol 19(9):2456–2471

Gao T, He B, Pan Y, Li R, Xu Y, Chen L et al (2013) The Association of Retinoic Acid Receptor Beta2 (RARβ2) Methylation Status and Prostate Cancer Risk: A Systematic Review and Meta-Analysis. PloS ONE 8(5):e62950

Gebert JF, Moghal N, Frangioni JV, Sugarbaker DJ, Neel BG (1991) High frequency of retinoic acid receptor beta abnormalities in human lung cancer. Oncogene 6(10):1859–1868

Goljanek-Whysall K, Pais H, Rathjen T, Sweetman D, Dalmay T, Munsterberg A (2012) Regulation of multiple target genes by miR-1 and miR-206 is pivotal for C2C12 myoblast differentiation. J Cell Sci 125(Pt 15):3590–3600. doi:10.1242/jcs.101758

Gudas LJ, Sporn MB, Roberts AB (1994) Cellular biology and biochemistry of the retinoids. Retin Biol Chem Med 1994:443–520

Hu Y, Correa AM, Hoque A, Guan B, Ye F, Huang J et al (2011) Prognostic significance of differentially expressed miRNAs in esophageal cancer. Internatl J Cancer J Internatl Du Cancer 128(1):132–143. doi:10.1002/ijc.25330

Inés FM, Gauna GV, Lis SM, Beatriz NS, Matias SA, María VRL (2014) Retinoic acid reduces migration of human breast cancer cells: role of retinoic acid receptor beta. J Cell Mol Med doi:10.1111/jcmm.12256.

Ivanova T, Petrenko A, Gritsko T, Vinokourova S, Eshilev E, Kobzeva V et al (2002) Methylation and silencing of the retinoic acid receptor-beta 2 gene in cervical cancer. BMC Cancer 2:4

Li M, Song S, Lippman SM, Zhang XK, Liu X, Lotan R et al (2002) Induction of retinoic acid receptor-beta suppresses cyclooxygenase-2 expression in esophageal cancer cells. Oncogene 21(3):411–418. doi:10.1038/sj.onc.1205106

Li RN, Yu FJ, Wu CC, Chen YK, Yu CC, Chou SH et al (2014) Methylation status of retinoic acid receptor beta2 promoter and global DNA in esophageal squamous cell carcinoma. J Surg Oncol 109(6):623–627

Liu Y, Lee MO, Wang HG, Li Y, Hashimoto Y, Klaus M et al (1996) Retinoic acid receptor beta mediates the growth-inhibitory effect of retinoic acid by promoting apoptosis in human breast cancer cells. Mol Cell Biol 16(3):1138–1149

Lotan R, Xu XC, Lippman SM, Ro JY, Lee JS, Lee JJ et al (1995) Suppression of retinoic acid receptor-beta in premalignant oral lesions and its up-regulation by isotretinoin. N Engl J Med 332(21):1405–1410. doi:10.1056/NEJM199505253322103

Maden M (2007) Retinoic acid in the development, regeneration and maintenance of the nervous system. Nat Rev Neurosci 8(10):755–765

Marchler-Bauer A, Bryant SH (2004) CD-Search: protein domain annotations on the fly. Nucl Acids Res 32(Web Server issue):W327–331. doi:10.1093/nar/gkh454

Marchler-Bauer A, Anderson JB, Chitsaz F, Derbyshire MK, DeWeese-Scott C, Fong JH et al (2009) CDD: specific functional annotation with the conserved domain database. Nucl Acids Res 37(Database issue):D205–210. doi:10.1093/nar/gkn845

Marchler-Bauer A, Lu S, Anderson JB, Chitsaz F, Derbyshire MK, DeWeese-Scott C et al (2011) CDD: a Conserved Domain Database for the functional annotation of proteins. Nucl Acids Res 39(Database issue):D225–229. doi:10.1093/nar/gkq1189

Mark M, Ghyselinck NB, Chambon P (2009) Function of retinoic acid receptors during embryonic development. Nucl Recept Signal 7:e002. doi:10.1621/nrs.07002.

Marzese DM, Hoon DS, Chong KK, Gago FE, Orozco JI, Tello OM et al (2012) DNA methylation index and methylation profile of invasive ductal breast tumors. J Mol Diagn 14(6):613–622. doi:10.1016/j.jmoldx.2012.07.001

Mehdipour P, Pirouzpanah S, Yam AA Retinoic Acid Receptor b2 Gene in Breast Cancer.

Mehdipour P, Pirouzpanah S, Azari Yam A (2012) Retinoic acid receptorb2 gene in breast cancer. Eur J Clin Med Oncol 4:17-33

Mendelsohn C, Lohnes D, Decimo D, Lufkin T, LeMeur M, Chambon P et al (1994) Function of the retinoic acid receptors (RARs) during development (II). Multiple abnormalities at various stages of organogenesis in RAR double mutants. Development 120(10):2749–2771

Molinari C, Casadio V, Foca F, Zingaretti C, Giannini M, Avanzolini A et al (2013) Gene methylation in rectal cancer: predictive marker of response to chemoradiotherapy? J Cell Physiol 228(12):2343–2349. doi:10.1002/jcp.24405

Niederreither K, Subbarayan V, Dollé P, Chambon P (1999) Embryonic retinoic acid synthesis is essential for early mouse post-implantation development. Nat Genet 21(4):444–448

Park SY, Kwon HJ, Choi Y, Lee HE, Kim SW, Kim JH et al (2012) Distinct patterns of promoter CpG island methylation of breast cancer subtypes are associated with stem cell phenotypes. Modern pathology: an official journal of the United States and Canadian academy of pathology. Inc 25(2):185–196. doi:10.1038/modpathol.2011.160

Pirouzpanah S, Taleban FA, Atri M, Abadi AR, Mehdipour P (2010) The effect of modifiable potentials on hypermethylation status of retinoic acid receptor-beta2 and estrogen receptor-alpha genes in primary breast cancer. Cancer Causes Control: CCC 21(12):2101–2111. doi:10.1007/s10552-010-9629-z

Pirouzpanah S, Taleban FA, Mehdipour P, Atri M, Hooshyareh-Rad A, Sabour S (2013) The biomarker-based validity of a food frequency questionnaire to assess the intake status of folate, pyridoxine and cobalamin among Iranian primary breast cancer patients. Eur J Clin Nutr. doi:10.1038/ejcn.2013.209

Ribeiro IP, Marques F, Caramelo F, Ferrão J, Prazeres H, Julião MJ et al (2014) Genetic imbalances detected by multiplex ligation-dependent probe amplification in a cohort of patients with oral squamous cell carcinoma-the first step towards clinical personalized medicine. Tumor Biology 35(5):4687–4695. doi: 10.1007/s13277-014-1614-9

Ross SA (2003) Diet and DNA methylation interactions in cancer prevention. Ann NY Acad Sci 983:197–207

Seewaldt VL, Johnson BS, Parker MB, Collins SJ, Swisshelm K (1995) Expression of retinoic acid receptor beta mediates retinoic acid-induced growth arrest and apoptosis in breast cancer cells. Cell growth & differentiation: the molecular biology journal of the American Association for. Cancer Res 6(9):1077–1088

Shen S, Kruyt FA, den Hertog J, van der Saag PT, Kruijer W (1991) Mouse and human retinoic acid receptor beta 2 promoters: sequence comparison and localization of retinoic acid responsiveness. DNA Seq: J DNA Seq Mapp 2(2):111–119

Sirchia SM, Ferguson AT, Sironi E, Subramanyan S, Orlandi R, Sukumar S et al (2000) Evidence of epigenetic changes affecting the chromatin state of the retinoic acid receptor beta2 promoter in breast cancer cells. Oncogene 19(12):1556–1563. doi:10.1038/sj.onc.1203456

Song S, Xu XC (2001) Effect of benzo[a]pyrene diol epoxide on expression of retinoic acid receptor-beta in immortalized esophageal epithelial cells and esophageal cancer cells. Biochem Biophys Res Commun 281(4):872–877. doi:10.1006/bbrc.2001.4433

Song S, Lippman SM, Zou Y, Ye X, Ajani JA, Xu XC (2005) Induction of cyclooxygenase-2 by benzo[a]pyrene diol epoxide through inhibition of retinoic acid receptor-beta 2 expression. Oncogene 24(56):8268–8276. doi:10.1038/sj.onc.1208992

Srour M, Chitayat D, Caron V, Chassaing N, Bitoun P, Patry L et al (2013) Recessive and dominant mutations in retinoic acid receptor beta in cases with microphthalmia and diaphragmatic hernia. Am J Hum Genet 93(4):765–772. doi:10.1016/j.ajhg.2013.08.014

Stefanska B, Karlic H, Varga F, Fabianowska-Majewska K, Haslberger A (2012) Epigenetic mechanisms in anti-cancer actions of bioactive food components-the implications in cancer prevention. Br J Pharmacol 167(2):279–297. doi:10.1111/j.1476-5381.2012.02002.x

Tallman MS, Andersen JW, Schiffer CA, Appelbaum FR, Feusner JH, Ogden A et al (1997) All-trans-retinoic acid in acute promyelocytic leukemia. N Engl J Med 337(15):1021–1028. doi:10.1056/NEJM199710093371501

Twelves D, Nerurkar A, Osin P, Dexter T, Ward A, Gui GP et al (2013) DNA promoter hypermethylation profiles in breast duct fluid. Breast Cancer Res Treat 139(2):341–350. doi: 10.1007/s10549-013-2544-8

Uray IP, Shen Q, Seo HS, Kim H, Lamph WW, Bissonnette RP et al (2009) Rexinoid-induced expression of IGFBP-6 requires RARbeta-dependent permissive cooperation of retinoid receptors and AP-1. J Biol Chem 284(1):345–353. doi:10.1074/jbc.M804721200

Veronesi U, Mariani L, Decensi A, Formelli F, Camerini T, Miceli R et al (2006) Fifteen-year results of a randomized phase III trial of fenretinide to prevent second breast cancer. Ann Oncol 17(7):1065–1071

Wang Y, Fang MZ, Liao J, Yang GY, Nie Y, Song Y et al (2003) Hypermethylation-associated inactivation of retinoic acid receptor beta in human esophageal squamous cell carcinoma. Clinical cancer research: an official journal of the American Association for. Cancer Res 9(14):5257–5263

Weiss FU, Marques IJ, Woltering JM, Vlecken DH, Aghdassi A, Partecke LI et al (2009) Retinoic acid receptor antagonists inhibit miR-10a expression and block metastatic behavior of pancreatic cancer. Gastroenterology 137(6):2136–2145. e2137

Widschwendter M, Berger J, Muller HM, Zeimet AG, Marth C (2001) Epigenetic down regulation of the retinoic acid receptor-beta2 gene in breast cancer. J Mammary Gland Biol Neoplasia 6(2):193–201

Youssef EM, Lotan D, Issa JP, Wakasa K, Fan YH, Mao L et al (2004) Hypermethylation of the retinoic acid receptor-beta(2) gene in head and neck carcinogenesis. Clin Cancer Res Off J Am Assoc Cancer Res 10(5):1733–1742

Zanardi S, Serrano D, Argusti A, Barile M, Puntoni M, Decensi A (2006) Clinical trials with retinoids for breast cancer chemoprevention. Endocr-Relat Cancer 13(1):51–68

Zelent A, Mendelsohn C, Kastner P, Krust A, Garnier JM, Ruffenach F et al (1991) Differentially expressed isoforms of the mouse retinoic acid receptor beta generated by usage of two promoters and alternative splicing. EMBO J 10(1):71–81

Zhu GH, Huang J, Bi Y, Su Y, Tang Y, He BC et al (2009) Activation of RXR and RAR signaling promotes myogenic differentiation of myoblastic C2C12 cells. Differ Res Biol divers 78(4):195–204. doi:10.1016/j.diff.2009.06.001

Chapter 13
Methylation in Colorectal Cancer

Pooneh Mokarram, Mehrdad Asghari Estiar and Hassan Ashktorab

Contents

H. Ashktorab (✉)
Department of Medicine and Cancer Center, Howard University College of Medicine
Washington DC, 20060, USA
e-mail: hashktorab@howard.edu

P. Mokarram
Department of Biochemistry, School of Medicine, Shiraz University of Medical Sciences, Shiraz, Iran

M. A. Estiar
Department of Medical Genetics, School of Medicine, Tehran University of Medical Sciences, Tehran, Iran

P. Mehdipour (ed.), *Epigenetics Territory and Cancer,*
DOI 10.1007/978-94-017-9639-2_13

Abstract Colorectal cancer (CRC) is the third most common cancer worldwide, but age-standardized incidence rates (ASRs) vary widely between different geographical regions. Distinct epidemiological and clinicopathological characteristics of CRCs based on their specific molecular profiles suggest different risk factors and pathways of transformation associated with colon carcinogenesis. Epigenetic events have been involved in the stepwise histological progression of CRC. Evidence for a mechanistic link between DNA methylation and histone deacetylation has also been demonstrated by treating cells with a combination of the DNA methyltransferase inhibitor and the histone deacetylase inhibitor. However, intrinsic and environmental factors that induce DNA methylation changes remain largely unknown. Therefore, in this chapter, our aim has been to define the molecular profiles including patterns of hypermethylation of the most important cancer candidate genes, polymorphism and mutation of specific genes in CRC in our studies, with relatively different environmental and genetic factors compared to Western countries. In addition, the study of DNA methylation in human disease represents an important frontier in medicine. Furthermore, hypermethylation of CpG islands is very common in cancer cells, coupled with the ability to detect methylation with a high degree of sensitivity, and has led to the development of several approaches for the detection of cancer in body fluids. Comparison of gene or protein expression patterns between several types of CRC should reveal fascinating insights into different mechanisms of CRC. Molecular profiling based on epigenetic alteration will eventually allow chemoresponsive patients to be identified with much greater accuracy.

13.1 Colorectal Cancer

Colorectal cancer (CRC) is the third most common cancer worldwide. However, age-standardized incidence rates (ASRs) differ extensively between different geographical regions, which may be indicative of variations in exposure rates to environmental carcinogens (Potter 1999; Kamangar et al. 2006). CRC is the leading cause of cancer-related mortality after lung and prostate cancers in men, and lung and breast cancer in women. It is also the second leading cause of cancer-related death in the United States (Potter et al. 1993; Jemal et al. 2005; Siegel et al. 2014).

Considering the recent advancements in screening and diagnostic methods, the number of new CRC cases and reported deaths has declined in worldwide. Colorectal cancer death rates declined by approximately 2% per year during the 1990s and by approximately 3% per year during the past decade. Progress in reducing colorectal cancer death rates can be accelerated by improving access to and use of screening and standard treatment in all populations (Siegel et al. 2014). However, it is estimated that approximately 783,000 new cases are diagnosed annually worldwide (Midgley and Kerr 1999; Siegel et al. 2014). However, Blacks have a higher incidence of colorectal cancer (CRC) and a younger age at diagnosis compared to Whites (Wallace et al. 2014). In 2010, it was estimated that over 140,000 Americans would be diagnosed with CRC, and that over 48,000 would die from the disease

(Jemal et al. 2010; Siegel et al. 2014). The CRC incidence rates vary with respect to race, ethnicity, and sex. Significantly different incidence rates have been reported in men and women. Black men have the highest rates; 62.0 out of every 100,000 black men were diagnosed with CRC in 2007. White men were second with an incidence rate of 51.5 per 100,000, followed by Hispanic men (44.8), Asian/Pacific Islander men (39.7), and American Indian/Alaska Native men (33.5) (Group USCW 2010; Siegel et al. 2014). In Spain, each year over 25,000 men and women are diagnosed with CRC, of which 13,000 will die (Quintero et al. 2009; Castells et al. 2014).

In 2014, an estimated 71,830 men and 65,000 women will be diagnosed with colorectal cancer and 26,270 men and 24,040 women will die of the disease according to incidence and mortality data were provided by the National Cancer Institute's Surveillance, Epidemiology, and End Results program and the North American Association of Central Cancer Registries and National Center for Health Statistics (Siegel et al. 2014).

CRC is also the third and fourth leading cause of cancer in Japan and China, respectively, with an annual mortality rate of over 22,000 men and 18,000 women (Division VAH 2006; Saif and Chu 2010). In Iran, over 51,000 people are diagnosed with cancer and 35,000 cancer-related deaths occur annually, which is the second highest incidence rate of cancer-related mortality in WHO's Eastern Mediterranean Region. In men, CRC is the third most common cancer with the age adjusted rate (ASR) of 8.3 per 100000, and the fourth most common cancer in women with an ASR of 6.5 per 105. It has been estimated that 3641 new cases of CRC are diagnosed in Iran each year, of which 2262 die, accounting for approximately 6.3 % of all cancer deaths in Iran (Parkin et al. 2005; Semnani et al. 2006; Sadjadi et al. 2003; Alireza et al. 2005).

Although CRC screening is potentially lifesaving, national guidelines indicate that many at risk individuals are not being screened. It has been estimated that as many as 60 % of deaths because of CRC could be prevented if all men and women aged $\geq$ 50 years were routinely screened (Group USCW 2014). Over 95 % of CRCs are adenocarcinomas, and about half of all patients with CRC develop local recurrence or distant metastasis during the course of their illness. The median survival time for CRC patients can vary from 4 to 22 months (Kim et al. 2010). The 5-year survival rate is highly dependent on the stage of tumor at the time of detection (Dashwood 1999). CRC develops either sporadically (SCRC, 85 %), or as part of hereditary cancer syndromes of hereditary non-polyposis colorectal cancer (HNPCC) and familial adenomatous polyposis (FAP) (15 %), or against an underlying inflammatory bowel disease (Gervaz et al. 2002, 2004; Haydon and Jass 2002; Jass et al. 2000). Of all patients with CRC, 11–16 % have at least one first-degree relative with CRC (Salovaara et al. 2000; Olsson and Lindblom 2003), but the prevalence would be much higher if second- or third-degree relatives are considered. In a previous study, among all probands of CRC, 53 % had a first-degree relative with cancer (Olsson and Lindblom 2003). Moreover, first-degree relatives of patients with CRC who do not have the criteria for FAP and HNPCC, have a more than 2-fold increased risk of developing tumors of the large intestine (Johns and Houlston 2001).

In developing countries, the epidemiology of SCRC is somewhat different from that of developed countries. Colorectal carcinomas in developing countries are generally characterized by a low prevalence, an onset at an early age, location on the left side, a weak development, and a low frequency of pre-adenomas (Abou-zeid et al. 2002; Adekunle and Ajao 1986; Chan et al. 2005; Parkin et al. 1992). Jordan and Egypt are two Muslim countries, with a common national border, similar culture, and eating habits. One study showed that these two countries also similar epidemiological characteristics with respect to SCRC (Abou-zeid et al. 2002; Adekunle and Ajao 1986; Chan et al. 2005; Parkin et al. 1992; Dajani et al. 1980). Most CRCs develop from adenomas, while few develop directly from epithelial cancer, mainly due to smoking and a Westernized diet, characterized by a high intake of meat and fat (Pehlivan et al. 2010). It is believed that the adenoma–carcinoma sequence is linked to the development of CRC in most patients, and three pathways have been identified—the microsatellite instability (MSI) and the chromosomal instability (CIN) pathways (Jass et al. 2002) and CIMP phenotype with epigenetic variation (Jass et al. 2002; Sugai et al. 2005). Discovery of these pathways has led to the assumption that CRC is a genetically heterogeneous disease (Jass et al. 2002; Sugai et al. 2005). CRC has a better prognosis with MSI than stage-matched microsatellite stable cancer (Gryfe et al. 2000; Kohonen-Corish et al. 2005; Popat et al. 2005; Sargent et al. 2010; Ribic et al. 2003; Colussi et al. 2013; Wright et al. 2005). Microsatellite unstable and stable CRCs have genomic and transcriptomic differences, some of which can be used as diagnostic, predictive, or prognostic markers (Popat et al. 2005, Anwar et al. 2004; Colussi et al. 2013; Koornstra et al. 2003; Munro et al. 2005; Popat et al. 2004). The development of CRC from an adenoma to a carcinoma may take several decades. Cancer is the result of an accumulation of genetic or epigenetic alterations that allow growth of neoplastic cells with phenotypic characteristics such as insensitivity to antigrowth signals, self-sufficiency in growth signals, evasion of apoptosis, limitless replicative potential, sustained angiogenesis, and the ability to invade tissues and metastasize (Hahn and Weinberg 2002; Hanahan and Weinberg 2011).

The CRC initiation and progression model, originally proposed by Fearon and Vogelstein (Fearon and Vogelstein 1990), initially identified the adenomatous polyposis coli (*APC*) gene, genes on 18q, and the *K-ras* and *p53* genes as those in which mutations contribute to the evolution of CRC. (Fearon and Vogelstein 1990; Huang et al. 1996; Kinzler and Vogelstein 1996; Lengauer et al. 1997; Liu et al. 1996; Parsons et al. 1993). However, they may represent alternative, multiple, and mutational pathways for colorectal cancerogenesis instead of representing a linear model of required accumulative mutations in the *APC, K-ras*, and *p53* genes (< 10 % of all CRCs have all mutations) (Smith et al. 2002), with specific associated chromosomal aberrations (Fearon and Vogelstein 1990) and specific clinical outcomes (Smith et al. 2002; Leslie et al. 2003; Conlin et al. 2005).

Significant differences have been observed in African-Americans compared with Caucasian with respect to CRC tumors in various epidemiological, clinical, and cytogenetic. Ashktorab et al identified genomic copy number irregularities in SCRC tumors from African-Americans, to find possible explanations for the observed disparities. They applied genome-wide array comparative genome hybridization (aCGH) using a 105k chip to identify copy number aberrations in samples from 15 African-Americans. A schematic form of aCGH has been seen in below and Fig. 13.1. Moreover, a comparative analysis was done in Caucasians using aCGH data as well as common colon cancer genes (CAN genes, 68 genes from Sjoblom et al. that are potentially involved in colon cancer). On average, 20 aberrations per patient were detected with more amplifications than deletions. Also, these deletions occurred primarily in chromosomes 4, 8, and 18. In >50 % of cases chromosomal duplications occurred on chromosomes 7, 8, 13, 20, and X. Chromosomes 3, 5, 7, 8, 20, and X had the highest rates of CGH irregularities. Some differences were also observed in the CIN profile compared with Caucasian alterations. Similar irregularities were seen with a few exceptions for the following genes; *THRB, RAF1, LPL, DCC, XIST, PCNT, STS,* and genes on the *20q12-q13* cytoband compared with Caucasians (Ashktorab et al. 2010).

Prominent African-Americans aberrations were chromosome X amplification in men and deletions in chromosomes 4, 8, and 18. In African-Americans several CAN genes were altered at high frequencies. *EXOC4, EPHB6, GNAS, MLL3,* and *TBX22* were the most frequently deleted genes and *HAPLN1, ADAM29, SMAD2,* and *SMAD4* underwent the most amplifications. One of the most deleted genes was *EPHB6* that is known to slow breast cancer cell lines invasiveness. Chromosome X amplification in men with CRC should be monitored. The observed CIN may have a key role in CRC in this population (Ashktorab et al. 2010; Brim et al. 2012). Among the ethnic groups, Ashktorab and colleagues found that African -Americans possessed the highest CRC the age-standardized incidence and mortality rates compared with other ethnic groups. Most CRCs originate from pre-existing adenoma. 30 % of the US adult population has adenomas (Nouraie et al. 2010). In African-Americans, CRC is more advanced and right-sided. African-Americans continue to have higher CRC death rates despite the reduced mortality rates as a result of screening (Nouraie et al. 2010; Lipkin and Higgins 1988; Kinzler and Vogelstein 1996; Jemal et al. 2007, 2014).

In a study, 1753 CRC cases were diagnosed from 1959 to 2006 in Howard University Hospital in America. The rate of in situ tumor reached a peak of 8.5 % in the 1990s ($p=0.0001$). Ashktorab and co-workers observed a decade-to-decade increasing rate of right-sided tumors, beginning with 36 % during 1959–1970 and reaching a peak of 60 % during 2001–2006 ($p=0.0001$). The recent increased rate of advanced and right-sided tumor were demonstrated which is consistent with SEER data and is highly important in developing strategies for CRC prevention and treatment for African-Americans (Nouraie et al. 2010). On the other hand, 5,013 colorectal polyps were diagnosed during 1959–2006, with tubular adenoma being the most frequent pathology (73 %). Right-sided polyps were mostly seen in the 1990s. Left-sided polyps were younger ($p<0.0001$), more hyperplasic (23 vs. 5 %; $p<0.0001$), and more frequent in women (56 vs. 52 %; $p=0.02$) compared with right-sided polyps. The frequency of right-sided adenoma significantly increased from 18 % in the 1960s to 51 % during 2001–2006 ($p<0.0001$). A higher neoplastic to hyperplastic polyps ratio (8:1) than what had been reported in Caucasians (7:1)

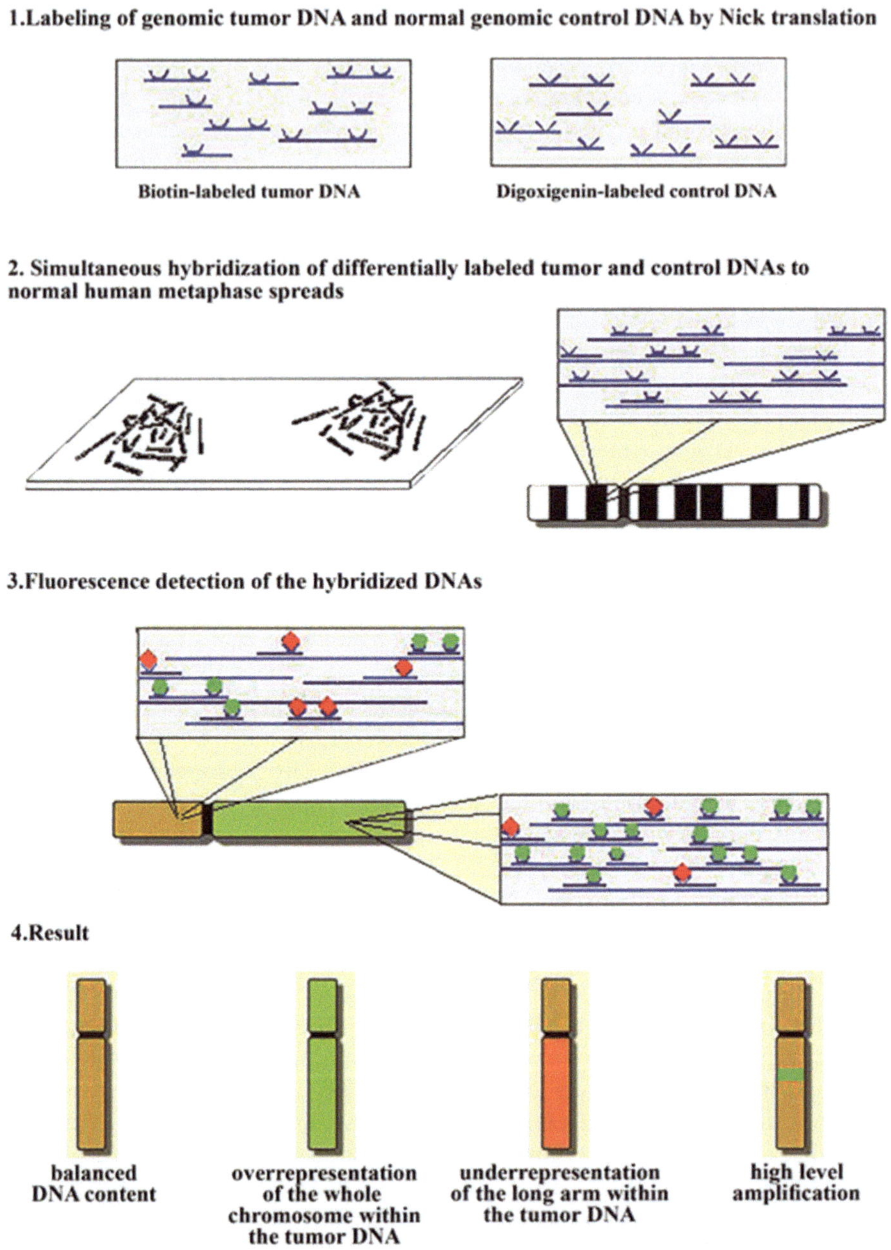

Fig. 13.1 Principle of array comparative genome hybridization (aCGH)

was observed. This implies a recent shift in polyps from the left side to the right side of the colon, which is in accordance with the increased colon cancer incidence in African-Americans. This, the researchers further emphasized on the key role of screening for reducing the incidence of colon cancer in African-Americans (Nouraie et al. 2010).

African-Americans microsatellite instability-high (MSI-H) cancers was high compare to microsatellite stable (MSS) and microsatellite instability-low (MSH-L) cancers. Moreover, most MSI-H tumors were well differentiated, proximal, and highly mucinous. MSI-H colorectal tumor was 2–3-folds more prevalent, while the defect in the percentage expression of mismatch repair (MMR) genes (*hMLH1* and *hMSH2*) was similar in African-American patients compared with American Caucasians. Here, the role of environmental or genetic factors that are more common in the development of CRC in African-Americans should not be underestimated (Ashktorab et al. 2005).

In this way, Brim et al showed the different DNA aberrations processes which can cause colorectal cancer (CRC). They conducted a comprehensive molecular characterization of 27 CRCs from Iranian patients. Array CGH was performed. The MSI phenotype and the methylation status of 15 genes were established using MSP. The CGH data was compared to two established lists of 41 and 68 cancer genes, respectively, and to CGH data from African Americans. A maximum parsimony cladogram based on global aberrations was established. The numbers of aberrations seem to depend on the MSI status. MSI-H tumors displayed the lowest number of aberrations. MSP revealed that most markers were methylated, except *RNF182* gene. *P16* and *MLH1* genes were primarily methylated in MSI-H tumors. Seven markers with moderate to high frequency of methylation (*SYNE1, MMP2, CD109, EVL, RET, LGR* and *PTPRD*) had very low levels of chromosomal aberrations. All chromosomes were targeted by aberrations with deletions more frequent than amplifications. The most amplified markers were *CD248, ERCC6, ERGIC3, GNAS, MMP2, NF1, P2RX7, SFRS6, SLC29A1* and *TBX22*. Most deletions were noted for *ADAM29, CHL1, CSMD3, FBXW7, GALNS, MMP2, NF1, PRKD1, SMAD4* and T*P53*. Aberrations targeting chromosome X were primarily amplifications in male patients and deletions in female patients. A finding similar to what were reported for African American CRC patients. Therefore, this first comprehensive analysis of CRC Iranian tumors reveals a high MSI rate. The MSI tumors displayed the lowest level of chromosomal aberrations but high frequency of methylation (Brim et al. 2014). The MSI-L was predominantly targeted with chromosomal instability in a way similar to the MSS tumors. Chromosome 20 consistently in 20q13 region and its corresponding genes were amplified as shown in Fig. 13.2. The global chromosomal aberration profiles showed many similarities with other populations but also differences that might allow a better understanding of CRC's clinico-pathological specifics in this population.

In spite of the characteristics of CRC ethnicity, differences in clinical presentation and surgical management of right- and left-sided large bowel cancer are well known (Fig. 13.2). Tumors that are on the right side usually present at a more advanced stage and are accompanied by symptoms such as weight loss and anemia. On the contrary, left-sided tumors often present with rectal bleeding, change in bowel habit, and tenesmus. However, the molecular pathology of carcinomas might differ with respect to the side they are one in the large bowel. These variations and

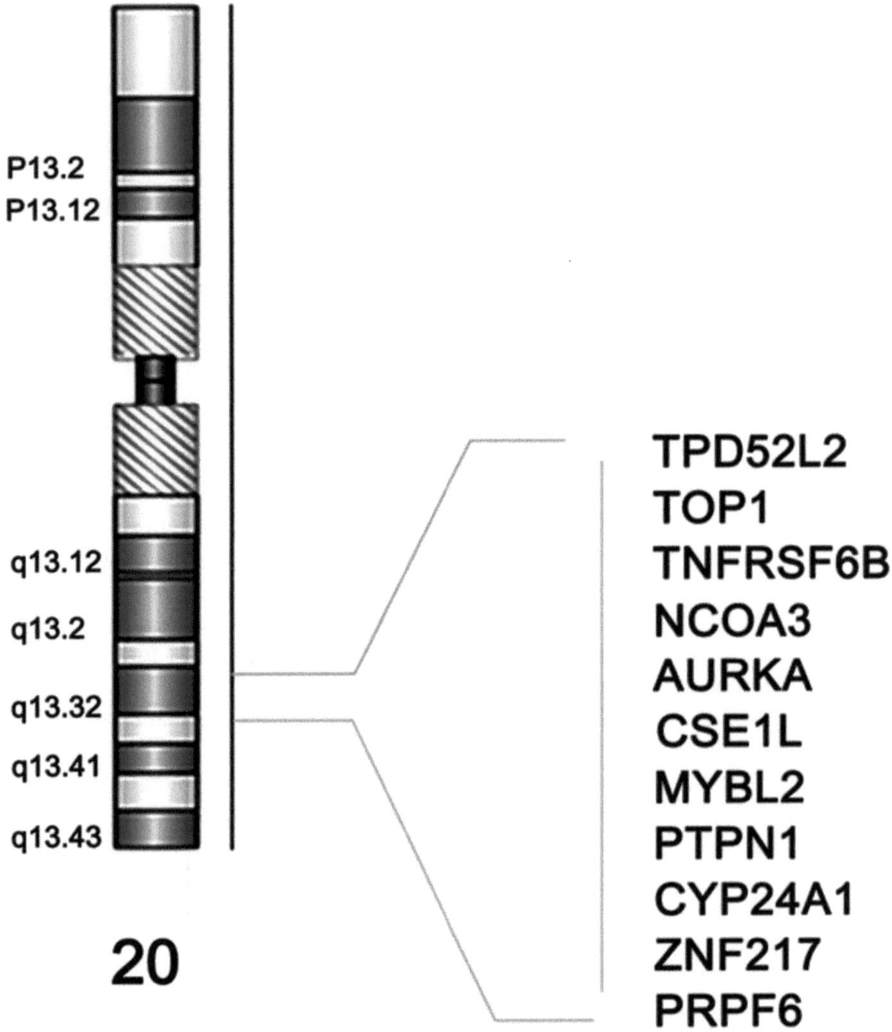

Fig. 13.2 Chromosome 20's consistently amplified 20q13 region and its corresponding genes

differences will become more significant with advancements in systemic treatments (Richman and Adlard 2002; Fig. 13.3).

In a study conducted by Ellidokuz and colleagues, polyps and cancers were more prevalent in the left colon. Right-sided SCRCs are more fatal than left-colon cancers (Ellidokuz et al. 2003). Most SCRCs are located proximally in women and distally in men (Potter et al. 1993). Researchers have found that age is an important factor in the distribution of colorectal polyps and cancers, and a proximal trend is observed with increasing age (Samowitz et al. 2007).

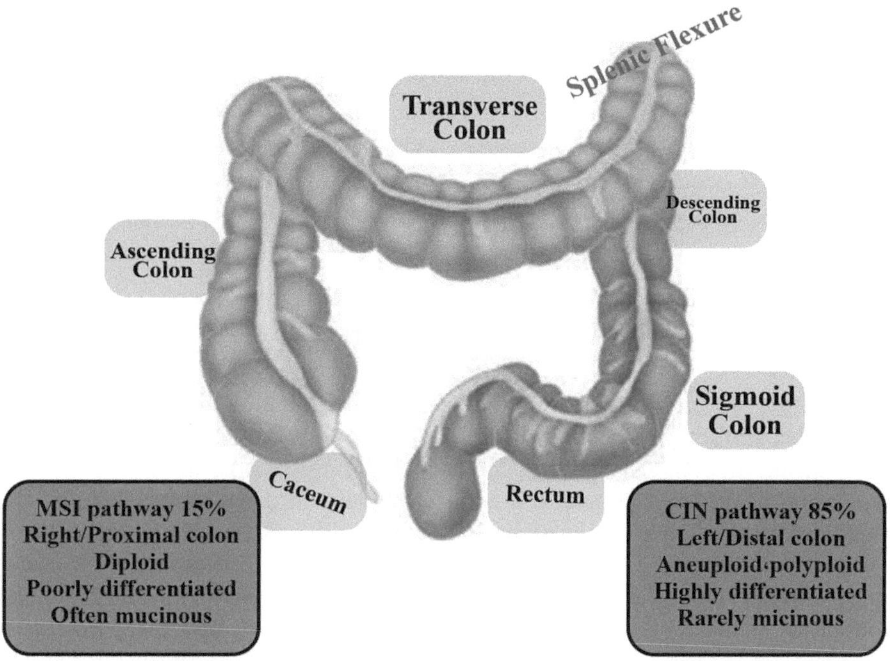

Fig. 13.3 Right-sided tumors are classified as splenic flexure (*caecum, ascending colon and transverse colon*), whereas left-sided tumors arise distantly to this site (*descending colon, sigmoid colon, rectum*). Two separate mutational pathways (*MSI, CIN*) have been shown

A different embryological origin has been assumed for proximal and distal colons suggesting that tumors originating at any of these sites could develop along different pathways. The midgut eventually develops into distal duodenum, jejunum, ileum, caecum, appendix, ascending colon, and proximal two thirds of the transverse colon. The hindgut develops into the distal third of the transverse colon, the sigmoid colon, rectum, and upper two thirds of the anal canal. The different origins of the proximal and distal colons also lead to different vascular supplies. The proximal colon is served by the superior mesenteric artery and the distal colon by the inferior mesenteric artery. Moreover, proximal and distal colons differ in expression of several antigens, metabolism of glucose, polyamines and butyric acid, as well as in bile acid consumption, and composition and density of the bacterial population (Bufill 1990; Pocard et al. 1995; Distler and Holt 1997). Bufilland co-workers found different gene expressions in 87 genes in the ascending and descending fetal colon, indicating the occurrence of additional changes in gene expression in postnatal development (Bufill 1990). However, significant anatomic differences have already been found in the embryonic colon (Glebov et al. 2003). Gene expression was studied in fetal (17–24 weeks gestation) proximal and distal colon. More than 1000 genes were expressed differentially in adult ascending versus descending colon, with 165 genes showing >2-fold and 49 genes showing >3-fold differences in expression (Table 13.1).

Table 13.1 Representative gene that show more than three-fold differences in expression in ascending versus descending adult colon

Gene	Fold changes
Home box DB	8.15
Retinoic acid receptor responder	11.83
Alpha actin 2 associated LIM protein	14.46
acyl coenzyme A oxidase 1 Palmitoyl	−13.1
Megrin A, B	−15
Ethanol amine kinase	−19.74

Therefore, developmental and biologic differences in proximal and distal colon may reflect differing susceptibilities to neoplastic transformation. Insights on the mechanisms of CRC can be provided by comparing patterns of gene or protein expression in different populations or in proximal and distal tumors. One of the most important clinical applications is related to the better selection of patients needing chemotherapy. Chemoresponsive patients will be identified more accurately through molecular profiling. For example: Patients with MSI-H tumors had a modestly better prognosis than those with microsatellite-stable (MSS) or MSI-low (MSI-L) cancers, yet also did not seem to benefit from adjuvant fluorouracil (FU)-based chemotherapy. This resistance to FU is presumably due to incorporation of FU metabolites into DNA rather than inhibition of its effective target, thymidylate synthase (Kimmie and Schrag 2010).

In one study, sporadic CRCs were analyzed for microsatellite instability, expression status of mismatched repair genes (*hMLH1, hMSH2*), and presence of the *BRAF* (*V600E*) in Omani, Iranian, African-Americans patients to analyze the difference between molecular genetics of CRC in different populations. Tumors with *BRAF* mutations were on the left side in Omanis while 88 % of such mutations were found on the right side of the colon in African-American patients. The highest and lowest rates of microsatellite instability tumors at two or more markers (MSI-H) were seen in African-Americans and Omanis (31 vs. 13 %), most of which were located in the proximal colon in African-American and Iranian patients. *hMLH1* gene expression defects were seen approximately two times more in African-Americans and Iranians compared with Omanis (77 vs 38 %). Tumors in all patients had *BRAF* mutations and mains had the most mutations (19 %). These findings suggest that Omani and Iranian patients experience CRC at a younger age. The incidence of MSI-H was lower in these groups compared with the African-American patients. This emphasizes on the key role of *hMLH1* expression and *BRAF* mutation in MSI-H CRC in these populations compared with other populations (Brim et al. 2008).

On the other hand, 36 (23.8 %) tumors out of 151 were MSI+ in Iranians, especially in those with proximal tumors (OR = 10.4; 95 %CI = 3.9–27.8) and in smokers (OR = 2.9; 95 %CI = 1.3–6.7). *MTHFR* 677CT+ TT genotype was strongly associated with MSI (OR = 2.6; 95 %CI = 1.3–5.3) and a positive relationship was found between the hypermethylation of mismatch repair genes and the incidence of MSI

($p=0.001$) Therefore, the researchers suggested the *MTHFR* 677CT+ TT variant genotype as a risk factor for MSI+ cancer (Naghibalhossaini et al. 2010). A study on the Iranian population provided insights into the mechanisms of colorectal tumor development. *K-ras, p53* mutations were significantly more common in left-sided than in right-sided tumors, indicating differences in the carcinogenesis pathways in these tissues. This finding is consistent with previous reports (Russo et al. 2005; Calistri et al. 2005; Toribara and Sleisenger 1995). This could be explained by the fact that the left-sided bowel lumen exposed to ingested carcinogens and mutagens more than the right-sided bowel lumen. Moreover, simultaneous *p53* and *K-ras* mutations were rarely seen in the same tumor. In the mentioned study, the researchers evaluated the hot spot genetic changes leading to the development of CRC in three key genes (*APC, K-ras,* and *p53*) in a CRC series from southern Iran for the first time. Mutations in exons 5 and 7 of *p53*, and exons 1 or 2 of *K-ras* genes were investigated in 151 sporadic CRC tumors by PCR-SSCP, and *K-ras* results were confirmed by pyrosequencing. *p53* was the most frequently mutated gene, with a frequency of 62/151 (41.1 %). This finding is consistent with previous studies on CRC, with reported frequencies ranging from 40 to 50 % of *p53* alterations in CRC (Soong et al. 2000; Soussi et al. 2000; Leroy et al. 2014). *K-ras* gene mutations were identified in 46 (30.5 %) of 151 cases which is in line with the reported frequency of 27.4 % (Smith et al. 2002). Mutations of *K-ras* result in specific amino acid substitutions that lead to permanent activation of the encoded p21 ras protein (Barbacid, 1987). It is well known that the activating *K-ras* mutations cluster at codons 12/13 (GTP-binding domain). Similar with previous reports, most samples of *K-ras* mutations were found in codon 12, with a smaller number of nucleotide substitutions in codon 13. No *K-ras* mutations were detected in codon 61 in Iranian patients. This confirms that codons 12 and 13 are also the hot spots of mutations in the Iranian population. *K-ras* gene mutations were localized more in distal (41.8 %) than proximal (13.3 %) tumors ($p=0.001$). Most *K-ras* mutations were base pair transitions which mostly occurred at the second bases of codons 12 and 13. All mutations in codon 12 and 13 were G→A transitions (data not published). Ras proteins are key components of signal transduction pathways leading from cell surface receptors to the control of cell proliferation, differentiation or death. Active Ras as well as mutated Ras stimulates the RAS-RAF-MEK-ERK-MAP kinase signaling pathway through tyrosine kinase receptors (RTKs) and *EGF* ligand. Therefore, anti-*EGFR* therapies via monoclonal antibody or *EGFR*-tyrosine kinase inhibitors are able to block cell proliferation, survival, invasion and metastasis. When the *EGFR* is blocked wild type *K-ras* doesnot signal and the tumor cells do not proliferate. However, mutated *K-ras* is permanently turned on, tumor cells proliferate.

On the other hand, Mokarram and colleagues found that one of the critical factors for cancer progression is the hypermethylation of the specific locus near the *MGMT* start codon. In patients with CRC, the assessment of *MGMT-B* which is associated with *K-ras* mutation has high prognostic value (Mokarram et al. 2012). Moreover, 41.1 % of Iranian cases carried a pathogenic *p53* mutation, which is similar to those reported for CRC in other populations (Mcdermott et al. 2002; Iacopetta 2003; Russo et al. 2005). The type of *p53* mutations was not determined in the aforementioned

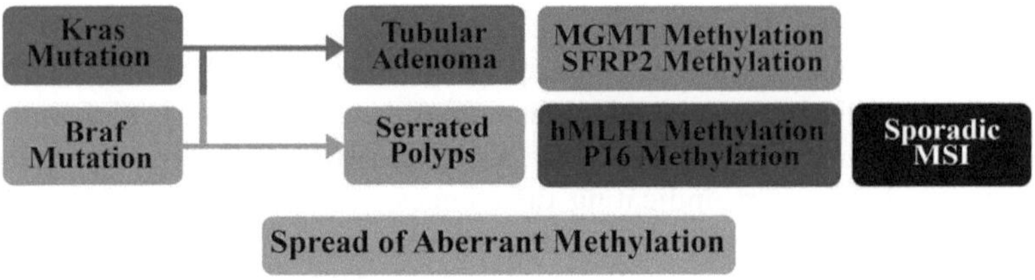

Fig. 13.4 Methylation pattern of polyp to cancer

study. However, in other populations (Russo et al. 2005; Béroud and Soussi 2003; Olivier et al. 2003; Soussi and Beroud 2003), most detected *p53* mutations were G→A transitions at CpGs (Pfeifer 2006). This type of mutation could be the result of spontaneous deamination or preferential adduct for mutation at methylated CpGs (Soussi and Beroud 2003; Pfeifer 2006; Greenblatt et al. 1994), followed by G→A transitions at non-CpGs that might be associated with hypemethylation of *MGMT* in our population (Mokarram et al. 2012). However, since simultaneous mutations in *K-ras* and *p53* were very rare ($p=0.007$) and the frequency of *p53* mutations in tumors differed in the left and right colon, it seems that *MGMT* does not follow the same mechanism in two genes.

Other study showed when neoplasia gains *K-ras* mutations, the neoplasia would transform into tubular adenoma with villous architecture and accumulate dense methylation in the *SFRP2/MGMT* promoter, consequently becoming non-MSI cancer (Takeda 2011; Fig. 13.4).

21/62 (33.8%) mutations in left-sided tumors were in exon 5, and 12/62 (19.4%) were in exon 7. In the right colon, only 6/62 (9.6%) mutations occurred in exon 5, while 12/62 (19.4%) of mutations occurred in exon 7. 11/62 (17.7%) of the simultaneous mutations in both exon 5, 7 were localized in the left colon. Mokarram and co-workers did not find alterations in *p53* (exons 5 or 7) or *K-ras* genes in 35.7% of Iranian patients with cancer. Therefore, these mutations possibly lie on alternate pathways in the development of CRC which depends on epigenetic mechanisms (data not published).

The researchers found the MSI-H phenotype in 23.8% of the patients which is higher than reports from Western countries but not in African-Americans (Gryfe and Gallinger 2001; Peltomaki 2001). Their finding was consistent with the results reported from the North of Iran (Bishehsari et al. 2006). The high prevalence of MSI-H could be related to the relatively high number of right-sided tumors found in the studied group (43.3%). Most MSI-H tumors in the mentioned study were right sided (72.2%) which is consistent with results from developed countries indicating a relatively high rate of MSI-H in proximal colon cancer (Ashktorab et al. 2005; Urso et al. 2008). A higher percentage of *K-ras* gene mutations were detected in non-MSI-H tumors (40/46, 87%) compared with MSI-H tumors (6/46, 13%, $p=0.04$). Similar results were obtained for total mutations (individuals with mutation in any of two genes, *p53* or *K-ras*) and MSI status ($p=0.04$). This finding is consistent with another report showing that *K-ras* mutation is less frequent in patients with sporadic MSI-H with *hMLH1* hypermethylation compared with non-

MSI-H CRC without *hMLH1* methylation and MSS CRC (Oliveira et al. 2004). There is an inverse relationship between microsatellite instability and *K-ras*, *p53* mutations regardless of the type of microsatellite. However, the existence of such a relationship is controversial among different studies (Fujiwara et al. 1998).

A previous report from Iran showed that sporadic MSS and sporadic MSI-H CRCs shared similar *K-ras* mutation frequencies. However, the type of *K-ras* mutations differed between MSI-H and MSS subgroups, suggesting a link between this specific type of mutation and MMR defect in MSI-H CRCs. Defects in other DNA repair systems, such as O6-methylguanine-DNA methyltransferase (*MGMT*) activity might be related to inability to protect from G→A transition in *K-ras* induced by alkylating agents (Esteller et al. 2000; Lees et al. 2004; Qi et al. 2005), which was shown in a recent study (Mokarram et al. 2012).

The relatively low frequency of tumors with mutations in both genes (*K-ras* and *p53*) as well as the fact that 57 % of tumors had mutations in only one gene (*p53*, or *K-ras*), implies that the progressive accumulation of multiple mutations in these genes is not a prerequisite for tumor development and does not represent a synergistic evolutionary pathway. Therefore, the simultaneous alterations in *K-ras* and *p53* are not frequent, suggesting that the sequential *K-ras-p53* module is not obligatory in the progression of CRC in Iranian population. While some studies have shown that *K-ras* mutations may be lost through selection in the progression of tumors from adenoma to carcinoma, (Pretlow et al. 1993; Ines et al. 2014) we found no statistically significant difference in *K-ras* mutations between stage I and II. However, a significant difference was found in the frequency of *p53* mutations between stage I and II, suggesting that *p53* inactivation may be an important determinant of tumor progression. In contrast to other reports suggesting that MSS and MSI-L tumors have a common molecular background (Bouzourene et al. 2000; Laiho et al. 2002). Mokarram and co-workers observed more *p53* mutations in the MSI-L tumors in lower stages (I, II). It seems that MSI-L CRCs were distinct from both MSI-H and MSS cancers in our study population. Correlations between *p53* mutations and tumor stage correspond with the notion that *p53* mutations accumulate during CRC progression (Bouzourene et al. 2000; Akkiprik et al. 2007). An important aspect of this study is that two major gene mutations were analyzed in Iranian patients with CRC in relation to prognosis for the first time. Although we and other researchers showed that there are genetic variations between population and proximal and distal CRCs, the issue of whether proximal and distal CRC should be considered as a single entity or two distinct entities is still debated. Further studies addressing the heterogeneity of the pathogenetic pathways leading to sporadic CRC in different populations as well identifying biological and/or molecular differences between proximal and distal CRCs should be done in this regard. With respect to the obtained data, future therapies must be targeted on individual patients, based on the detailed understanding of their genetic background and the nature of the mutation present in individual tumors.

13.2 Epigenetic and CRC

Epigenetics include variations in chromatin structure that modulate the use of the genome by histone modification, chromatin remodeling, histone variant composition, DNA methylation, and noncoding RNAs. Most of these modifications and chromatin changes are reversible and, therefore, are unlikely to be propagated through the germ line. In today's modern terms, epigenetics can be molecularly defined as "the sum of the alterations to the chromatin template that collectively establish and propagate different patterns of gene expression (transcription) and silencing from the same genome that is associated with alteration in cancers (Roloff and Nuber 2005; Zoratto et al. 2014; Wolffe and Matzke 1999). Multiple, cumulative epigenetic and DNA sequence changes in the cells' genome would lead to tumor. Sequence changes include deletions in chromosome regions accompanied with gene loss that may be associated with negative cell- cycle regulation (tumor suppressor genes), mutations that may activate or inactivate a number of proteins, gene amplifications entailing an over expression of specific genes, and even loss or gain of entire chromosomes.

The etiology of human cancer changing from a benign neoplasm to a malignant tumor has been explained by pathways involving the accumulation of genetic and/or epigenetic alterations (Kinzler and Vogelstein 1996; Vogelstein et al. 2013). Mutations in oncogenes and tumor suppressor genes comprise classic genetic alterations. The "classic" genotype comprises only about 10% of CRCs (Smith et al. 2002, Rennie and Nelson, 1998). Besides these genetic alterations, the initiation and promotion of cancer could occur by epigenetic mechanisms. These mechanisms can be defined as heritable changes in gene function that are not due to changes in DNA sequence (Jones and Baylin 2002). Three widely accepted mechanisms for the "epigenetic regulation" of genes are (a) changing patterns of DNA methylation, (b) histone acetylations/deacetylations, and (c) alterations in stable circuits and regulatory feedback loops for growth and transcription factors (Rennie and Nelson 1998).

Ever since Fearon and Vogelstein defined their colorectal carcinogenesis model in 1990, a deeper understanding of this process is constantly developing (Fearon and Vogelstein 1990; Alberto Morán et al. 2010). Currently, only 60% of CRCs emerge from adenomas via the suppressor route, also called chromosome instability (CI), which is initiated by a mutation in gene *APC*—this is the classical description of colorectal carcinogenesis corresponding to the adenoma- carcinoma sequence (Fearon and Vogelstein 1990; Alberto Morán et al. 2010). Moreover, two additional alternative, non–excluding routes, are now considered; the microsatellite instability (MSI) pathway, associated with Lynch syndrome and a small proportion of sporadic cases, and the methylator phenotype pathway, most recently identified and referred to as CIMP (CpG Island Methylator Phenotype) in the English-language literature. Therefore, CRC is being increasingly classified into various phenotypes according to its molecular profile (Fearon and Vogelstein 1990; Vilar et al. 2014). Therefore, from a molecular viewpoint it is classified based on the predominant cell event (CIN, MSI, CIMP) or, equivalently, according to the event-initiating factor (suppressor pathway for CIN; mutation pathway for MSI; methylator pathway for CIMP). Mutual exclusion can be seen in some colorectal carcinogenesis pathways such as MSI and CIN, whereas in CIMP there may be some overlap. Since MSI arises through a process of methylation of the *hMLH1* promoter, it is not surprising that a significant overlap between CIMP and sporadic MSI has been reported (Toyota et al. 1999). With respect to the CIMP phenotype and mutation overlapping, the methylation status of three cancer-related genes (*APC2, p14ARF*, and *ECAD*) were evaluated in colorectal carcinogenesis and their association with the mutational status of *BRAF* and *K-ras* among Iranian patients with CRC (Naghibalhossaini et al. 2011). The frequencies of *APC2, E-CAD,* and *p14* methylation were 92.6, 40.4, and 16.7 %, respectively. However, V600E mutation was not found in the *BRAF* gene in any of the samples and gene methylation was not related to *K-ras* mutations. These findings implied a distinct molecular pathway for *APC2, p14,* and *ECAD* methylation different from those previously described for CRCs with *BRAF* or *K-ras* mutations (Naghibalhossaini et al. 2011).

In another study, the *MTHFR* CT genotype was associated with an increased risk of tumor methylation (OR = 2.5; 95 % CI, 1.1–5.6). Methylated tumors were more frequent in the high methyl donor, especially in those who had the *MTHFR* CT and CT/TT genotypes ($p = 0.01$, $p = 0.002$, respectively). Hence, the risk of promoter methylation in tumor-specific genes was higher when serum folate/vitamin B(12) levels were higher. Such a relationship is modified by *MTHFR* C677T genotypes (Mokarram et al. 2008).

The aberrant methylation profile of four genes (*APC, Axin1, Axin2,* and *GSK3β*) was studied in an unselected series of 112 sporadic CRCs using methylation specific PCR to better understand the methylation silencing of the WNT pathway during colorectal carcinogenesis. The *Axin2* C148T genotype was assessed in patients with CRC as well as healthy controls with PCR-RFLP. Among 18.75 % of CRCs at least one had methylated gene and 7.1 and 11.9 % of tumors had experienced promoter methylation in *Axin2* and *APC* genes, respectively (Naghibalhossaini et al. 2012). *Gsk3β* and *Axin1* gene were not methylated in these tumor series. Promoter methylation of *Axin2* was seen in women more than men implying that this type of methylation is sex-related ($p = 0.002$). Patients with distal tumors had a lower risk of developing CRC compared with proximal ones (OR = 0.3; 95 % CI 0.1–0.9, $p = 0.04$). These findings depict the minor role of *Axin1* and *GSK3β* methylation in carcinogenesis of CRC (Naghibalhossaini et al. 2012). In general, there is a strong correlation between polymorphism in specific genes and CIMP profile (Curtin et al. 2009; Slattery et al. 2009).

The Wnt signaling pathways are a group of signal transduction pathways made of proteins that pass signals from outside of a cell through cell surface receptors to the inside of the cell. Dysregulation of WNT signaling pathway was first identified for its role in carcinogenesis.

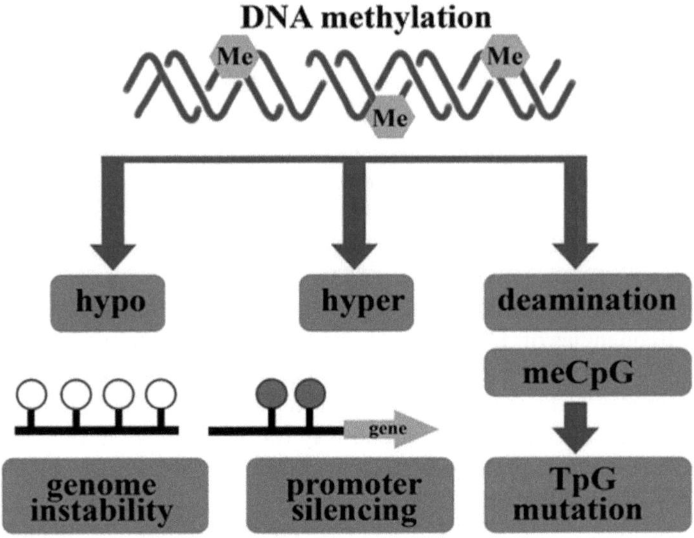

Fig. 13.5 Epigenetic alteration involving DNA methylation can lead to cancer by various mechanisms

13.2.1 The Role of DNA Methylation and Chromatin Modification in CRC

Deviant changes in DNA methylation and histone modification are two of the most common types of epigenetic alterations in cancer. Such alterations direct gene expression through maintaining restricted and permissive chromatin states and occur at multiple layers of regulation. They can also be controlled by cancer cells for oncogenic gain (Eden et al. 2003; Ashktorab et al. 2009). Hypo or hyper- DNA methylation and loss of cytosine methylation results in genome instability (Fig. 13.5).

DNA methylation (an enzyme-induced chemical modification to the DNA structure) comprises the major form of epigenetic information in mammalian cells. A methyl (CH3) group is covalently bonded to the 5-carbon on the cytosine base of CpG island in promoter. The methyl group is provided by S-adenosyl methionine (SAM), and this is converted to S-adenosyl homocysteine (SAH) in the process. In a pathway dependent on folate and cobalamin this is recycled back to SAM (Wajed et al. 2001).

CpG methylation occurs in vast areas in the genome in which the sequences are repeated such as centromeres and transposon elements (Wang and Leung 2004) CpG island shores (Irizarry et al. 2009), noncoding regions (ie, enhancer regions and miRNAs; Varambally et al. 2008), and gene bodies (silencing alternative transcription start sites; Guenther et al. 2007). CpG sites are seen in about 60 % of gene promoters (Bird 2002).

In mammals, the interplay of three independently encoded DNMT'S (*DNMT1*, *DNMT3A,* and *DNMT3B*) is influential in global cytosine methylation patterns. The role of fourth DNMT (DNMT2) which has been cloned needs to be proven in vitro or in vivo (Okano et al. 1998; Robertson 2001; Yoder and Bestor 1998; Subramaniam et al. 2014). The enzyme Dnmtl is specific for CpG and has significant activity against non-methylated DNA. As a maintenance enzyme, *DNMT1* preserves existing methylation patterns after cell replication, and its deletion leads to apoptosis (Bird 2002) as well as death in mice if lost during embryonic development (Endres et al. 2001). In contrast, *DNMT3A* and -3B are de novo methyltransferases that methylate previously unmethylated DNA. Although their enzymic class is the same and their catalytic domains are similar to some extent, they might play different roles in tumorigenesis; cancer progression may be promote by *DNMT3A* deletion (Gao et al. 2011), while deletion of *DNMT3B* might inhibit oncogenesis (Nosho et al. 2009; Linhart et al. 2007). Transition between active and non- methylated gene promoter due to DNA methylation is mediated by Methyl-CpG binding proteins (MeCP) and Sin3A.

Ashktorab and colleagues studied the effect of gene methylation in the development of CRC (Wang et al. 2013). Their results showed that *EDN2* and *EDN3* genes were hypermethylated in primary human colon cancers and in a panel of human colon cancer cell lines. Therefore, epigenetic inactivation of ET-2 and ET-3 occurs frequently in both rat and human colon cancers (Wang et al. 2013). Sojblom et al and Schuebel et al also introduced CAN genes (including 13 genes, *GPNMB, SYNE1, APC2, EVL, MMP-2, ICAM-5 PTPRD, CDH5, LGR6, STARD8, CD109, RNF182* and showed that genetic and epigenetic alterations of these markers are important in the progression of CRC (Schuebel et al. 2007; Sjoblom et al. 2006). They described a microarray of whole human transcriptome screen for identifying the genes that are silenced by promoter hypermethylation in human CRC. Therefore, candidate cancer genes are identified in single tumors with high validation efficiency. Moreover, as shown in Fig. 13.6, the important relationship the altered tumor genome and the gene hypermethylome was identified by comparing candidate hypermethylated genes with recently identified mutated genes in CRC (Schuebel et al. 2007; Sjoblom et al. 2006).

Mokarram and co-workers studied CAN gene methylation and *CHD5* protein expression CRC tissue microarrays (TMA) using immunohistochemical staining. Methylation-specific PCR was used to study the status of promoter methylation in CAN genes in Iranians and African-Americans (Mokarram et al. 2009). The researchers also studied microsatellite instability (Mokarram et al. 2009). More than 65 % of *SYNE1, MMP2, APC2, GPNMB, EVL, PTPRD,* and *STARD8* genes were methylated, whereas the methylation was <50 % for *LGR6, RET, CD109,* and *RNF* in both populations (Mokarram et al. 2009). However, a significant difference was observed for chromodomain-helicase-DNA-binding protein 5 (*CHD5*), *ICAM5,* and *GPNMB* methylation among the two studied populations (Mokarram et al. 2009). MSI-H was respectively found in 31 and 28 % of tumors in African-Americans and Iranians. African-Americans had a significantly higher methylation rate with respect to *GPNMB, ICAM5,* and *CHD5* genes (Mokarram et al. 2009). The low expression of *CHD5* in CRC was correlated with *CHD5* promoter CpG island hypermethylation. *CHD5* is a tumor suppressor which is frequently downregulated in a variety of human cancers.

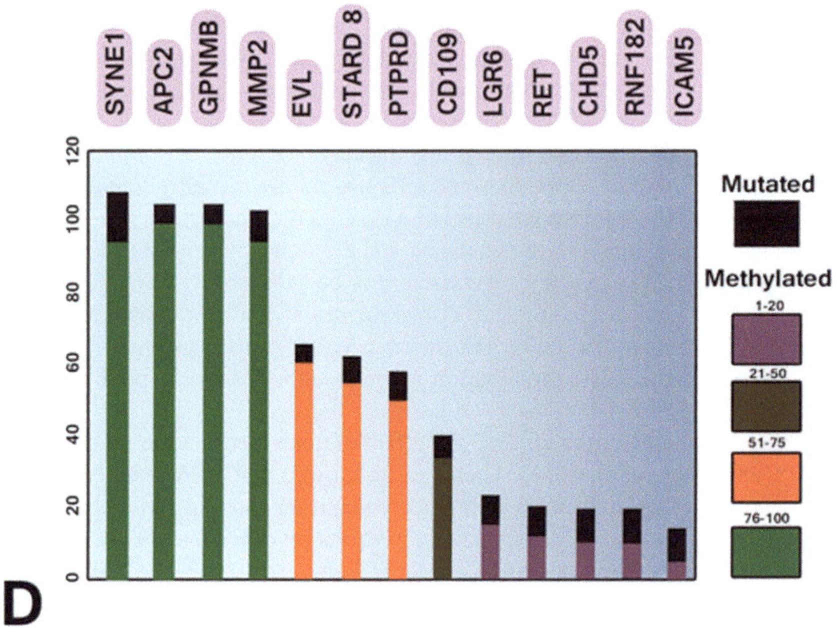

Fig. 13.6 Comparing methylation and mutation frequency of cancer genes in CRC tumor samples. (Schuebel et al. 2007)

The effect of microRNA-211 (miR-211)-regulated *CHD5* expression was investigated on colorectal tumorigenesis. MiR-211 was predicted to target *CHD5* using Target Scan software analysis. An exogenous miR-211 CRC cell line (HCT-116[miR-211]) was generated using lentiviral transduction and used as a model for in vitro and in vivo studies. The expression level of miR-211 in HCT-116(miR-211) cells had a 16-fold upregulation rate compared with vector control cells (HCT-116[vector]). The *CHD5* protein level in HCT-116(miR-211) cells was decreased by 50 % as a result of the direct binding of exogenous miR-211 to the 3'-untranslated region (3'-UTR) of *CHD5* mRNA. HCT-116 (miR-211) cells had a higher cell proliferation level, tumor growth, and cell migration compared with HCT-116 (vector) cells under in vitro and in vivo conditions, using MTT, colony formation, flow cytometry, scratch assay, and tumor xenografts, respectively. Moreover, *p53* pathway-associated regulatory proteins (MDM2, Bcl-2, Bcl-xL, and Bax) were induced changes as a result of the enforced expression of miR-211 in HCT-116 cells. Ashktorab's results show that *CHD5* is a direct target of miR-211 regulation (Cai et al. 2012). Tumor cell growth is promoted by this enforced expression, partially by downregulating the expression level of the *CHD5* tumor suppressor. Therefore, a better understanding of the association of between miR-211-regulated *CHD5* expression and *CHD5* function in colorectal tumorigenesis is provided (Cai et al. 2012).

Epigenetic alterations occur early in the progression process and often precede malignancy (Wang and Tang 2008). The adenoma to carcinoma progression sequence and the hyperplastic polyp-serrated adenoma to carcinoma sequence are concurrent molecular changes that are influential in the development of colon cancer (Petko et al. 2005). These sequences result from the progressive accumulation of genetic and epigenetic alterations that transform normal colonic epithelium to colon adenocarcinoma (Grady et al. 2001). Two types of hyperplastic polyp associated with CRC can be defined through methylation patterns (Wynter et al. 2004).

According to the classic model for genetic alteration of colon cancer, the altered DNA methylation is shown to occur very early. One alternative mechanism includes proteins that bind selectively to methylated DNA. MeCP2 is one of these proteins which were initially identified in 1992. Structurally, it has a methyl-CpG binding domain (MBD) which recognizes a symmetrically methylated CpG dinucleotide via contacts in the major groove of the double helix, and a transcriptional repression domain (TRD) that interacts with other regulatory proteins.

Another mechanism by which DNA methylation could inhibit transcription is through blocking the access of transcription factors through MBD binding. It was found that MeCP2 could recruit histone deacetylase (HDAC). Therefore, DNA methylation was able to repress transcription and result in a chromatin structural change by recruiting MBD's and their associated HDAC's to methylated DNA. As a result, core histone tails would be locally deacetylated, in turn leading to tighter DNA packaging. This would ultimately reduce the transcription factors' access to their binding sites. Recent studies have shown a link between four of the MBD-containing proteins (MeCP2, MBD1, MBD2, and MBD3) and aspects of the chromatin remodeling machinery in addition to HDAC. An understanding of the mechanism of repression came from the realization that MeCP2 associates with the Sin3a co-repressor complex and depends on histone deacetylation for its action. This finding showed that DNA methylation could be read by MeCP2 and provides a signal to alter chromatin structure. (Wynter et al. 2004; Laird 2005; Hall et al. 2013; Pancione et al. 2010; Cowley et al. 2005; Curradi et al. 2002).

By treating cells with a combination of the DNA methyltransferase inhibitor *5-azaCdR* and the histone deacetylase inhibitor trichostatin A (TSA) a link between DNA methylation and histone deacetylation can be determined. When the doses of 5-azaCdR were low, re-expression was low and the demethylation of hypermethylated CpG-island-associated genes were minimal. However, the same genes were robustly activated when a combination of 5-azaCdR and TSA were used, while TSA per se had no effect. This showed that DNA methylation and histone deacetylation work together to silence transcription. Moreover, DNA methylation was dominant over the histone acetylation status (Robertson 2001; Laird 2005; Cowley et al. 2005; Curradi et al. 2002; Toiyama et al. 2014).

A considerable amount of DNA methylation in embryonic stem cells seems to occur independent from CpG sites (Lister et al. 2009). The Ten-Eleven-Translocation (TET) oxidase family converts 5-methylcytosine to 5-hydroxymethylcytosine which is a step towards demethylation. This process does not seem to be limited to CpG islands (Ito et al. 2011). The exact roles of these phenomena in epigenetic regulation are still not fully understood.

Core histone modifications (ie, two copies of H2A, H2B, H3, and H4 proteins) are also crucial in epigenetic regulation apart from direct DNA manipulation. Epigenetic silencing of gene transcription does not occur only by promoter methylation, but is mediated by a complex series of molecular events that remodel chromatin configuration (Gronbek et al. 2007; Goossens et al. 2014). Recent studies show chromatin changes occurring during tumorigenesis. Cancer cells display palmate changes in histone methylation and histone acetylation patterns (Sharma et al. 2010). Histone methylation and acetylation for gene activation or deactivation depend on the modified residue of the histone (Vo and Millis 2012).

To moderate transcription, histones bind selectively and release DNA as nucleosomes. DNA is regulated by adding acetyl, methyl, phosphoryl, ubiquityl, or sumoyl groups, producing a dynamic epigenetic histone code (Rodríguez-Paredes and Esteller 2011). These histone marks are deposited or removed by a plethora of proteins and clinical targets, including histone acetyltransferases (HATs), histone deacetylases (HDACs), histone methyltransferases (HMTs), histone demethylases (HDMs), kinases, phosphatases, and others (Campos and Reinberg 2009; Beck et al. 2011). The positive histone charge is contradicted by the acetylation of lysines at specific sites (such as on H3K4) by HATs, and permits negatively charged DNA to form a configuration ready for transcription. However, acetyl groups are removed from histones by HDACs, and consequently DNA is shielded from expression because the oppositely charged histone is bind to the DNA. Although there are some exceptions, gene expression is usually silenced by DNA methylation. Moreover, histone marks can activate or silence genes, depending on the target residues, the targeted histone, and extent of the alteration (Sawan and Herceg 2010). With respect to histone modification and cancer incidence, the status of global histone acetylation (by measuring H3, H4 acetylation of lysine residues, which also occur over large regions of chromatin including coding regions and non-promoter sequences) and the expression of histone deacetylase 2 (HDAC2) was assessed in CRC tissues. Global histone H4 acetylation and HDAC2 expression in colon adenoma and carcinoma (Ashktorab et al. 2009). High levels of HDAC2 nuclear expression were detected in 81.9, 62.1, and 53.1 % of CRC, adenoma, and normal tissues, respectively ($p=0.002$). In moderate to well differentiated tumors increased global expression levels were seen for H4K12 and H3K18 acetylation as opposed to poorly differentiated tumors ($p=0.02$). A significant correlation was found between HDAC2 expression and progression of adenoma to carcinoma, when comparing cancer and non-cancer tissues. Thus, a significant association between HDAC2 expression and CRC progression is implied (Ashktorab et al. 2009).

According to Fig. 13.7 many combinations of histone modifications are known to establish specific states for activating or inhibiting expression. For example, histone H3 marks such as trimethylation of lysine (K) 4 over the promoter and K36 over the gene body (H3K4me3 and H3K36me3, respectively) establish permissive chromatin states (Barski et al. 2007; Rosenfeld et al. 2009). For example, histone H3-K9 methylation induces gene silencing and histone H3- K4 induces gene activation.

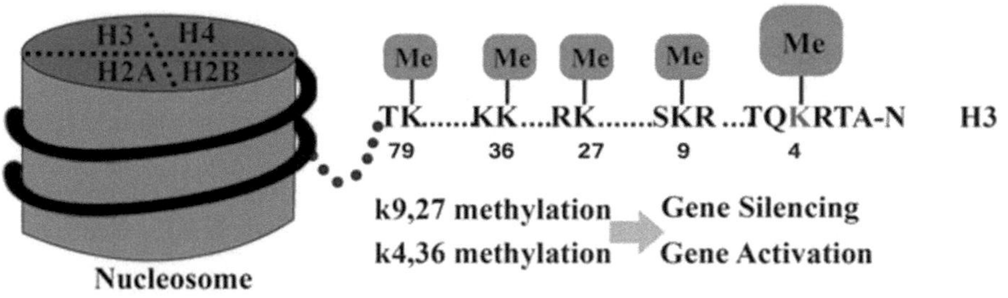

Fig. 13.7 Histone Methylation can lead to gene expression activation or inhibition

Monoacetylation of H3K9 and H3K14 (H3K9ac1 and H3K14ac1) (Li et al. 2007), the presence of variant histones such as H2A.Z (Hu et al. 2012), and methylation of enhancer elements downstream (H3K4me1) are among other enabling marks (Heintzman et al. 2009). An additional contributory layer is represented by nucleosome positioning. This is determined by the intrinsic binding affinity of the DNA sequence, competitive binding of surrounding proteins, and translocation activity by adenosine triphosphate-dependent remodeling complexes (Gaffney et al. 2012).

However, classically H3K27me3, H3K9me3, and H3K9me2 indicate the heterochromatin configurations that compact nucleosomes (Baylin and Jones 2011). They also indicate the presence of inhibitory proteins such as CTCF and HP1 (Wen et al. 2012; Keller et al. 2012) in addition to dense, localized DNA methylation. Polycomb group (Polycomb repression complex [PRC]) proteins can initiate repression, which methylate the histones in promoter regions. Such PRC regulators, including zeste 2 (EZH2) enhancer (Asangani et al. 2012), are themselves regulated by noncoding RNAs such as miRNA-101 to inhibit expression (Banerjee et al. 2011; Sakurai et al. 2012). EZH2 induces methylation of H3K27 which leads to the transcriptional repression.

Conclusively, the importance of the epigenome in coordinating transcription and downstream biological events have been underestimated considering the interplay between systems that previously thought to be independent. The complexity of such a system enables flexible conditions for guiding development and physiology, as well as guiding states such as tumorigenesis.

13.2.2 Hypermethylated Gene Promoters in CRC

Mammalian DNA contains 5-methylcytosine, the genomic distribution of which is specific for each cell type and is largely established during embryonic development. In normal tissues, DNA methylation patterns partially depend on the relative levels and activities of DNMT'S and DNA demethylases whose expression is regulated at both the transcriptional and posttranscriptional level (Turker and Bestor 1997; Jost and Bruhat 1997). Mammalian DNMT'S recognize CpG dinucleotides at specific sites within the genome called "CpG Island," a CpG-rich region often encompassing the promoter and transcription start site of the associated gene possessing de novo methylation activity (Bock et al. 2007; Jones and Takai 2001). Approximately half of all human genes contain CpG islands in the 5′ area of their gene promoter (Bock et al. 2007; Jones and Takai 2001). Tumor suppressor or oncogenes are regulated through epigenetic marks. It is widely accepted that the activation of oncogenes or the hypermethylation of tumor suppressors cause pre-neoplastic lesions that in turn trigger SCRCs (Tanaka et al. 2006).

In nearly 35 % of CRCs the mechanism by which methylation occurs in the promoter regions of various genes plays is highly important (Snover 2011). CRCs can be differentiated from other tumors by features such as proximal colon preference, occurring more in women and in older age, and poor differentiation (Van Rijnsoever et al. 2002; Samowitz et al. 2005; Ogino et al. 2006). A number of studies have clearly indicated that the pattern of DNA methylation in tumors is strongly influenced by age, sex, and anatomic site (Wiencke et al. 1999; McMichael and Potter 1985; Vaiserman et al. 2014; Yamashita et al. 2014; Mcmichael and Potter 1983).

So far, three types of altered DNA methylation patterns have been identified in human cancer:hypomethylation, hyper-methylation, and loss of imprinting (LOI) (Schulz 1998; Bressan et al. 2014). Loss of the differential expression of parental allele is mostly seen in embryonal tumors (Reik and Surani 1989; Rainier et al. 1993). Many tumors, especially those in advanced stages experience DNA hypomethylation, which is widely known to be a genome wide event (Rainier et al. 1993; Bedford and Van Helden 1987). Promoter hypomethylation with an associated increase in gene transcription has also been defined. The hypermethylation of DNA takes place at special regulatory sites in the promoter regions or repetitive sequences (Sakai et al. 1991; Okada et al. 2010; Herman et al. 1994; Graff et al. 1995) and has tumor specificity (Baylin et al. 1997, 2012; Tanaka et al. 2006; Fan and Beck 2004; Kwabi-Addo et al. 2007). A heavy density of cytosine methylation in the CpG islands of the tumor suppressor gene promoters can lead to a complete block of transcription (Fan and Beck 2004), and many types of cancer use this mechanism to inactivate tumor suppressor genes and initiate cancer formation and progression (Sakai et al. 1991; Graff et al. 1995; Baylin et al. 1997, 2012).

Studies show that the earliest steps in tumorigenesis are depicated as abnormal clonal expansion which evolves during the stress of cell renewal. Factors such as aging, chronic injury, inflammation and epigenetic alteration could drive tumor progression via abnormal clonal expansion. Also, there is strong association between DNA methylation and histone modification to regulate gene expression. CpG island methylation affects a number of genes in colon cancer, and many studies have reported the significance of the epigenetic alterations in the pathogenesis of colon cancer.

13.2.3 Classification of Cancer Candidate Genes

Mokarram et al divided Cancer candidate genes into 5 classes which are classified in Fig. 13.8 (Mokarram et al. 2009).

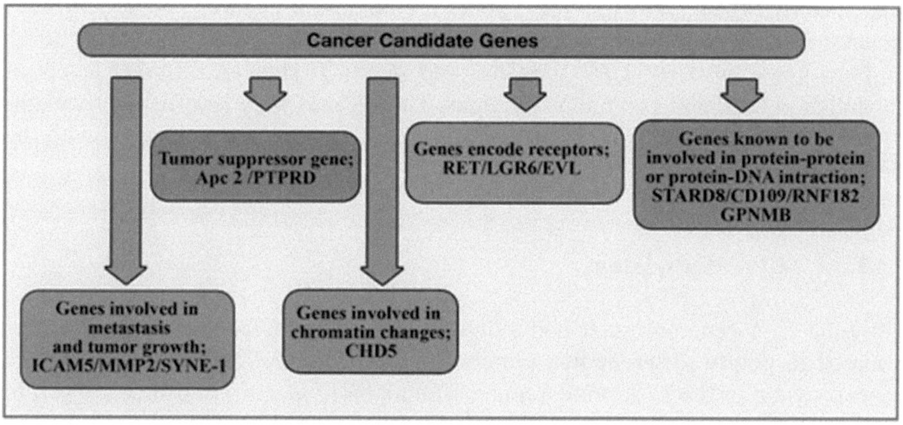

Fig. 13.8 Cancer candidate genes in CRC

The methylation profile of 13 genes in African-Americans and Iranian CRC tissue was investigated. Fresh ($n = 51$) sporadic primary CRC tumor samples were collected from surgical patients at several hospitals affiliated to Shiraz University of Medical Sciences, Shiraz, Southern Iran from July 2003 to September 2005. Also, 51 African-Americans CRC samples (19 fresh and 32 formalin fixed and paraffin embedded, from Howard University) sex and age-matched patients were also collected (Mokarram et al. 2009). The two groups of samples were analyzed using MSP analysis of 13 genes. It was hypothesize that in different populations CAN genes might show a different profile. It should also be mentioned that these genes could be inactivated through epigenetic silencing. In both populations, most analyzed genes were highly methylated. However, their levels of methylation differed in the two study populations, and from one gene to another. The synaptic nuclear envelope encoding protein (*SYNE-1* gene) was methylated in all analyzed samples; while the *RNF182* promoter, that encodes a ring finger protein, was not methylated in any of the samples (Mokarram et al. 2009). Function has been known for *RNF182, SYNE-1* protein is involved in the process of cytokinesis (Fan and Beck 2004; Liu et al. 2008). The latter protein and KIF3B protein facilitate the accumulation of membrane vesicles at the spindle midbody. How this gene's methylation is involved in a carcinogenic process is currently unknown. More than 65 % methylation was achieved for *SYNE1, MMP2, APC2, GPNMB, EVL, PTPRD,* and *STARD8,* while it was less than 50 % for *LGR6, RET, CD109,* and *RNF.*

13.2.3.1 Methylation and Age

Although there is a global methylating process that is age dependent in other studies, (Kwabi-Addo et al. 2007), the methylation pattern of 13 genes was independent from the patients' age. The results obtained from the 13 CAN genes in the mentioned study reflect show that these genes are important in carcinogenic processes that are dependent on methylation.

Four genes (*CD109, CHD5, LGR6,* and *ICAM-5*) showed different levels of methylation depending on tumor locations. These genes were less frequently methylated in the left colon than in the right-side, which could be associated with the descending methylation gradient from the proximal to the distal colon.

13.2.3.2 *EVL* Methylation

Only the *EVL* gene promoter had a higher level of methylation from well differentiated to poorly differentiated tumors. Bournier and colleagues found that the co-expression of the *EVL* protein along with alpha-II spectrin reinforce the cell to cell interaction (Bournier et al. 2006). The down regulation of the *EVL* gene by hypermethylation in all poorly differentiated tumors and more than 75 % of well and moderately differentiated tumors possibly make the cells contact loose and ready to become invasive. This finding could be justified by the fact that this gene's methylation increases by tumor stage. While *EVL* methylation was found only in 56 % of stage I tumors, this gene was methylated in 86 % of stage IV tumors.

13.2.3.3 *RET* Methylation

A similar stage dependent increase in methylation was observed for *RET* gene where 44 % of tumors were methylated at stage I, while 71 % were methylated at stage III and IV. No explanation for this result can be currently suggested, knowing that *RET* gene is a proto-oncogene (Herman et al. 1998) activated by gene translocations and mutations (Hedayati et al. 2006). However, the low methylation of *RET*, may be consistent with its tumorgenetic function in colon cancer progression. The *LGR6* gene (encoding a Leucin-rich repeat-containing G-protein-coupled receptor) seemingly displayed a stage-dependent methylation process. Therefore, it is thus involved in cell proliferation via signal transduction (Shu et al. 2000). However, this gene's function could not explain the methylation decrease from stage I to IV (78 to 29 %). The methylation profile was similar for MSI-H as well as non-MSI-H tumors for all genes, except the *PTPRD* gene which encodes a protein that is a member of the protein tyrosine phosphatase (PTP) family. This finding confirms dissociation between the microsatellite instability (MIN) phenotype and the CIMP in colon cancer tumors which has been previously established.

13.2.3.4 Different Methylation Profile was Found Between Iranians and African-Americans

Cell growth, differentiation, mitotic cycle, and oncogenic transformation are among the various cellular processes that are regulated by PTPs which are signaling molecules. Mori et al. have shown that a similar gene PTPRO is highly methylated in MSI-H tumors supporting this gene's increased methylation status in the MSI tumors (Mori et al. 2004). Different methylation profile was found between Iranians and African-Americans for only four genes, *GPNMB* (89 vs 100 %), *CHD5* (47 vs 78 %), *LGR6* (31 vs 49 %), and *ICAM-5* (7.5 vs 40 %).

13.2.3.5 *GPNMB* Gene Could Delay Growth and Reduce Metastatic Potential

GPNMB, a type I transmembrane glycoprotein, shows expression in the low metastatic human melanoma cell lines. However, it does not show such expression in high metastatic cell lines (Kuan et al. 2006). Therefore, since this gene is methylated more in African-Americans, it could account for the high aggressiveness and fast progression of colon tumors in this population.

13.2.3.6 *ICAM-5* Gene Encodes a Type I Transmembrane Glycoproteins that is a Member of the Intercellular Adhesion Molecule (ICAM) Family

This finding is also reinforced by the fact that another gene involved in metastasis, *ICAM-5,* is highly methylated in African-Americans compared with Iranians (40 vs 7.5 %). *ICAM-5* gene encodes a type I transmembrane glycoproteins that is a member of the intercellular adhesion molecule (ICAM) family. The high methylation rate of *ICAM-5* leads to reduced cell-to-cell adhesion in corresponding tumor cells and in turn increased their invasive potential. This finding is consistent with the results of the *GPNMB* product, is similar to the *ICAM-5* gene product, and leads to cumulative effects that all increase the invasiveness and metastatic potential of the colon tumor cells in African-Americans (Mokarram et al. 2009).

13.2.3.7 *CHD5* Gene Methylation in African-American CRC Tumors Could Explain the High Incidence and Aggressiveness of CRC

Unlike *GPNMB* and *ICAM-5, CHD-5* is involved in early tumorigenic processes at the chromatin remodeling and controls events such as that proliferation, apoptosis, and senescence via the p19Arf/*p53* pathway (Bagchi et al. 2007). The higher level of methylation of this gene in African-Americans compared with Iranians (78 vs 47 %) might indicate the higher level of colon cancer in African-Americans (Mokarram et al. 2009).

The expression profiles of many genes are affected by chromatin modification. Chromatin modification affects tumor progression rate and aggressiveness. Colon tumors from African-Americans showed increased global Histone H4 acetylation and HDAC2 expression. Race might be the driving force for the high profile gene methylation since MSI in both African-Americans and Iranians had little difference. Environmental factors including smoking and drinking may also play a role in the distinct MSI and methylation level in these two populations.

The mentioned methylation study confirmed the biomarker status of many of the CAN genes that were shown to be highly methylated in the African-American and Iranian populations despite similarities in the MSI level.

A thorough analysis of both populations might need to be performed on the patients' established cell lines using agents targeting both whole genome methylation (5-azacytidine) and/or the chromatin modification inhibitor (TSA) followed by differential microarray expression experiments to gain a global insight of all epigenetic processes taking place in the tumorigenesis within these two groups of patients.

13.2.4 Inactivation of MMR Genes in CRC

The most common cause of the inactivation of MMR genes (especially *MLH1* and *MSH2*) in CRC is epigenetic silencing due to methylation in the promoter region. MMR deficiency via mutation or epigenetic silencing display MSI phenotype in hereditary or sporadic CRC and *hMLH1* imperfection is more common in CRC.

In one study, 77 % of MSI-H tumors in both African-Americans and Iranians and 38 % of tumors in Omanis were found to have defects in *hMLH1* gene expression (Brim et al. 2008). The aberrant methylation of *MLH1* occurs in > 80 % of sporadic microsatellite instability (MSI) CRC, and the restoration of *MLH1* expression and function by demethylating the *MLH1* promoter in MSI CRC cell lines suggests that such aberrant methylation is a cause rather than a consequence of colorectal carcinogenesis (Herman et al. 1998; Kane et al. 1997).

On the other hand, Promoter methylation of *p14ARF* could be a significant alteration leading to CRC with MSI-L. Target CpG sites in GABRA1 and LAMA2 experienced aberrant with high frequency in tumor tissues compared with matched tumor-adjacent normal tissues in another study (Lee et al. 2012).

UNC5C and *DCC* share the ability to transmit cell death signals in the absence of their ligand (Ackerman et al. 1997; Hong et al. 1999; Llambi et al. 2001; Sanchez-Cespedes et al. 2000) and thus act as tumor suppressors in CRC (Ackerman et al. 1997; Mazelin et al. 2004). The netrin-1 receptors are aberrantly methylated in primary CRC (Kim et al. 2009; Shin et al. 2007) and are significantly correlated with Dukes' stage C (Hibi et al. 2009b; Hibi et al. 2009a). UNC5C inactivation occurs early, whereas *DCC* loss is seen in advanced CRC more often (Chen et al. 2005 and Han et al. 2012). Therefore, it is implied that epigenetic alterations in the netrin-1 receptors do not occur randomly in CRC and may be related to its malignant potential. In 82 and 69 % of primary colon cancers, aberrant methylation was observed in netrin-1 receptor and UNC5C genes, respectively (Hibi et al. 2009a; Hibi et al.2009b).

A study showed that methylation at the *p16INK4A* promoter is higher in colon cancer tumors when compared with the normal tissue of the same individuals

(Yoruker et al. 2012). *CDKN2A* promoter hypermethylation has been described in 12–51 % of CRCs and is often included in the panel of markers used to assess the CIMP phenotype (Shima et al. 2011).

Frequent *APC* hypermethylation is also found in CRC tissue but a promoter that is repressed in normal gastric tissue has been shown to be the site of hypermethylation in gastric cancers. Hence, frequent *APC* hypermethylation at major promoter sites in these tissues is important (Sproul et al. 2012). The following genes displayed alterations in methylation pattern from that of the mucosa of the non-cancer tissues compared with neoplasmic mucosa: *MGMT, hMLH1, p16 INK4a, MINT1, MINT31* and with a great level of changes: *COX2, cyclin A1* and *CDX1, RAR, MYOD1, p15 INK4b, CDH13, CXX1, p73* and *WT1* (Xu et al. 2004). One study investigated the methylation of 12 genes (*APC, COX-2, DAP-kinase, E-cadherin, GSTP1, hMLH1, MGMT, p14, p16, RASSF1A, THBS1,* and *TIMP3*) in normal colon mucosa, colon adenoma, and CRC. All the mentioned 12 genes were methylated in CRC and colon adenoma, except for *GSTP1* gene. Also, normal colon mucosa was methylated for *APC* only (Lee et al. 2004).

Some studies also indicate an association between *MGMT* and CDKN2A methylation and *K-ras* mutations (Hawkins et al. 2009). Higher methylation levels were observed in tumor samples from patients with multiple lesions compared with those with solitary tumors regarding the following genes: *MGMT, CDKN2A, SERF1, TMEFF2, HS3ST2, RASSF1A,* and *GATA4* (Gonzalo et al. 2010).

The frequencies of *APC2, E-CAD*, and *p14* methylation were reported to be 92.6, 40.4 and 16.7 %, respectively, in another study (Gonzalo et al. 2010). NDRG2, THBS4, and Desmocollin 3 (DSC3) were also downregulated in CRC compared to benign colorectal tissues. Other studies showed that the vimentin gene (VIM), usually activated in mesenchymal cells, was highly methylated in colorectal carcinoma (Shirahata et al. 2010) and all connexins including GJA1, GJA9, GJB1, GJB2, GJC1 and GJD3 were hypermethylated in colon cancer cell lines (Sirnes et al. 2011).

Other researchers investigated the methylation level of several genes using tissue samples that included normal mucosa, adenomas, and carcinomas of the colorectum. The eleven studied genes were *ADAMTS1, CDKN2A, CRABP1, HOXA9, MAL, MGMT, MLH1, NR3C1, PTEN, RUNX3,* and *SCGB3A1. ADAMTS1, CDKN2A, CRABP1, MLH1, NR3C1, RUNX3,* and *SCGB3A1* genes showed increased methylation levels from adenomas to carcinomas; while *HOXA9, MAL*, and *MGMT* were similarly methylated in all tumor stages. *PTEN* was not methylated in carcinomas, and therefore the researchers did investigate this gene in any of the tumor stages (Ahlquist et al. 2008).

A new hypermethylated gene called Phosphatase and Actin Regulator 3 (PHACTR3) was found that showed increased DNA methylation levels by more than 70-fold in advanced neoplasia (Bosch et al. 2012). Researchers evaluated the methylation level of two different regions from the RASSF2 and *SFRP2* promoters in DNA from various tissue samples. Normal colonic mucosa specimen was mostly unmethylated, colorectal adenoma samples were partially methylated, and a CRC specimen was extensively methylated (Nagasaka et al. 2009). In primary CRC tissues, oncostatin M receptor-β (OSMR) and β-1,4-galactosyltransferase-1 (B4GALT) are highly methylated. Such methylation levels are rarely seen in corresponding normal adjacent mucosa or in non-malignant normal colon tissues. *PAPSS2, TUBG2, NTRK2, B4GALT1* and *SFRP4* are methylated in colon cancer tissues samples (Kim and Deng 2007). Methylation of *BMP3, EYA2, ALX4, WIF-1, EGFR, SFRP-1, OST2* and vimentin was detected in tissue specimens of patients with colon cancer (Shirahata et al. 2010; Zou et al. 2007; Tokuyama et al. 2010). The hypermethylation of the KEAP1 promoter region leads to increased nuclear Nrf2 and downstream ARE gene expression in CRC cell and tissues (Hanada et al. 2012).

Conclusively, one of the major epigenetic modifications influential in the physiological control of genome expression is DNA methylation. As shown in Table 13.2 methylation patterns are extensively changed in cancer cells, thus leading to better differentiation of cancer cells from normal tissues. Measuring the extent of DNA methylation can help cancer diagnosis by identifying methylated genes that are commonly expressed.

13.2.5 Epigenetic Gene Silencing Role in the Evolution of CRC—Importance for Early Tumor Progression Stages (ie; IBD, Polyp)

Inflammatory bowel disease (IBD) is a heterogeneous disease strongly associated with chronic uncontrolled inflammation of the intestines, and has two distinct disease categories, including Crohn's disease (CD) and ulcerative colitis (UC) (Hartnett and Egan 2012). Each year 30,000 people worldwide are diagnosed with IBD (Hanauer 2006). Chronic Inflammation is thought to be the root cause of tumor development and associated with about 20% of all human cancers (Hartnett and Egan 2012; Hanauer 2006; Ullman and Itzkowitz 2011). CRC is a serious complication in patients with IBD. These patients are 60% more likely to develop CRC than the general population (Herrinton et al. 2012; Goel et al. 2011). IBD-associated CRC (IBD-CRC) accounts for about 10–15% of mortality rates in patients with IBD (Munkholm 2003). Screening of such patients using colonoscopy and intestinal biopsy might be difficult and inefficient as a cancer surveillance method (Chambers et al. 2005; Mattar et al. 2011). In order to overcome such difficulties and improve diagnosis and surveillance, noninvasive diagnostic tool and well-validated molecular markers for the early detection, prognosis and monitoring of IBD patients seems necessary. Epigenetic alterations, particularly alteration in DNA methylation are believed to play an early etiopathogenic role in IBD (Hartnett and Egan 2012; Petronis and Petroniene 2000; Tahara et al. 2009; Lin et al. 2011; Olaru et al. 2012; Cooke et al. 2012). Table 13.3 shows the association of some epigenetic alterations in IBD.

Chronic inflammation is associated with infections or autoimmune disease precedes tumor development and can contribute to it through induction of oncogenic mutations, genomic instability, early tumor promotion, and enhanced angiogenesis.

Inflammation acts at all stages of tumorigenesis. It may contribute to tumor initiation through different mechanisms such as mutation, genomic instability and epigenetic modifications. Inflammation activates tissue repair responses, induces proliferation of premalignant cells, and enhances their survival and metastasis.

Table 13.2 Methylation of the genes in CRC

Gene	Name	Histological type (% methylation, cases)	References
3OST2	3-O-Sulfotransferase 2	CRC (71.4%)	(Tokuyama et al. 2010)
ALX4	Aristaless-like homeobox-4	CRC (64%, 30/47), adenoma (85%, 11/13), normal colon mucosa (0%, 0/21)	(Ebert et al. 2006)
APC	Adenomatosis polyposis coli	MSS CRC (28%, 7/25), MSI CRC (36%, 10/28)	(Miotto et al. 2004)
APC	Adenomatosis polyposis coli	Normal tissue (0%, 0/21), CRC (21,%, 10/47), liver metastasis (42%, 10/24)	(Miotto et al. 2004)
AXIN 2	AXIN 2 (CONDUCTIN)	MSI CRC (50%, 5/10), MSS CRC (0%, 0/10)	(Miotto et al. 2004)
BNC1	Basonucleolin	Colon tumor (92%, 22/24), normal colon tissue (54%, 13/24)	(Shames et al. 2006)
BNIP3	BCL2/adenovirus E1B interacting protein 3	CRC (66%, 40/61)	(Tan et al. 2007)
CD133	Prominin 1	CRC (62%, 10/16), normal colon control (0%, 0/19)	(Lenhard et al. 2005)
CDH1	E-cadherin	MSS CRC (42%, 10/24), MSI CRC (39%, 11/28)	(Miotto et al. 2004)
CDH13	H-cadherin	CRC (38%, 23/61)	(Jensen et al. 2008)
CDH13	H-cadherin	CRC (32%, 27/84)	(Hibi et al. 2004)
CDH13	H-cadherin	CRC (65%, 42/65)	(Xu et al. 2004)
CDH4	R-cadherin	CRC (78%, 38/49), normal tissue (29%, 5/17), CRC or adenoma (100%, 10/10), normal tissue (0%, 0/10)	(Miotto et al. 2004)
CDKN2A/p14	Cyclin-dependent kinase inhibitor 2A, alternated reading frame	Adenoma from FAP-Pts (41%, 13/32), multiple adenoma Pts (69%, 20/29), MSI-H CRC Pts (86%, 12/14), MSS/MSI-L CRC Pts (88%, 14/16)	(Wynter et al. 2006)
CDKN2A/p14	Cyclin-dependent kinase inhibitor 2A, alternated reading frame	MSS CRC (12%, 3/24), MSI CRC (39%, 17/28)	(Miotto et al. 2004)

Table 13.2 (continued)

Gene	Name	Histological type (% methylation, cases)	References
CDKN2A/p14	Cyclin-dependent kinase inhibitor 2A, alternated reading frame	CRC (32%, 61/188)	(Kang et al. 2008)
CDKN2A/p16	Cyclin-dependent kinase inhibitor 2A	Adenomas (34%, 14/41)	(Petko et al. 2005)
CDNK2A/p16	Cyclin-dependent kinase inhibitor 2A	Adenoma from FAP-Pts (52%, 17/33), multiple adenoma Pts (45%, 13/29), MSI-H CRC Pts (64%, 9/14), MSS/MSI-L CRC Pts adenoma (63%, 10/16)	(Wynter et al. 2006)
CDNK2A/p16	Cyclin-dependent kinase inhibitor 2A	MSS CRC (28%, 7/25), MSI CRC (36%, 10/28)	(Miotto et al. 2004)
CDX1	Audal type homeobox transcription factor 1	CRC (100%, 65/65)	(Xu et al. 2004)
COX2	Prostaglandin-endoperoxide synthase 2	CRC(72%,47/65)	(Xu et al. 2004)
CXCL12	Chemokine (C-X-C motif) ligand 12	CRC (76.2%, 16/21), AP matched normal mucosa (0%, 0/19)	(Wendt et al. 2006)
Cyclin A1	Cyclin A1	CRC (100%, 65/65)	(Xu et al. 2004)
DCC	Deleted in colorectal carcinoma	CRC (56%, 28/50)	(Hibi et al. 2009a)
EphA7	Eph receptor A7	CRC (49%, 37/75)	(Wang et al. 2005)
ESR1	Estrogen receptor-α	Lymph node from stage I & II CRC Pts (UICC stage) (31%, 15/49), disease-free (86%, 13/15), local recurrence (14%, 2/15)	(Harder et al. 2009)
ESR1	Estrogen receptor-α	UC with neoplasia UC without neoplasia CRC (86%, 36/42), Adenoma (67%, 6/9), matched normal colon (56%, 9/16), noncancerous normal colon (12%, 2/16)	(Tominaga et al. 2005)
GATA4	GATA binding protein 4	CRC (70%, 63/90), noncancerous colon mucosa (6%, 5/88)	(Hellebrekers et al. 2009)
GATA5	GATA binding protein 5	CRC (79%, 61/77), noncancerous colon mucosa(13%, 13/100)	(Hellebrekers et al. 2009)

Table 13.2 (continued)

Gene	Name	Histological type (% methylation, cases)	References
GJA4	Gap junctions A4	CRC (60%, 29/48)	(Sirnes et al. 2011)
GJB6	Gap junctions B6	CRC (25%, 12/48)	(Sirnes et al. 2011)
GJD2	Gap junctions D2	CRC (96%, 46/48)	(Sirnes et al. 2011)
ITGA4	Integrin, alpha 4	Adenoma (75%, 27/36), adenocarcinoma (92%, 69/75), colon mucosa (6%, 2/32)	(Ausch et al. 2009)
LAMA2	Laminin α2	CRC (80.6%), normal tissue (13.4%)	(Lee et al. 2012)
MAL	T cell differentiation protein	CRC (80%, 49/61), adenomas (71%, 45/63), normal mucosa (4%, 1/23)	(Lind et al. 2008)
MGMT	O-6-methylguanine-DNA methyltransferase	Adenomas (49% 19/39)	(Petko et al. 2005)
MGMT	O-6-methylguanine-DNA methyltransferase	MSS CRC (40%, 10/25), MSI CRC(39%, 11/28)	(Miotto et al. 2004)
MGMT	O-6-methylguanine-DNA methyltransferase	Adenoma from FAP-Pts (66%, 22/33), multiple adenoma Pts (41%, 12/29), MSI-H CRC Pts (43%, 6/14), MSS/MSI-L CRC Pts (53%, 8/15)	(Wynter et al. 2006)
MLH1	MutL homolog 1, colon cancer, nonpolyposis type 2	MSS (0%, 0/25) CRC, MSI CRC (39%, 11/28)	(Miotto et al. 2004)
MSX1	Msh homeobox 1	Colon tumor (87%, 21/24), normal colon tissue (42%, 10/24)	(Shames et al. 2006)
MYOD	Myogenic factor 3	CRC (69%, 45/65)	(Xu et al. 2004)
NDRG2	N-myc downstream-regulated gene 2	CRC (27%, 8/30), matched normal colon (0%, 0/30)	(Piepoli et al. 2009)
NDRG4	N-myc downstream-regulated gene 4	CRC (1st, 86%, 71/83), CRC (2nd, 70%, 84/128), noncancerous colon mucosa (4%, 2/48)	(Melotte et al. 2009)
NMDAR2	N-methyl-D- aspartate receptor-2A	CRC (82%, 82/100), matched normal colon mucosa (15%, 15/100), noncancerous normal colon (9%, 1/11)	(Kim et al. 2007b)

Table 13.2 (continued)

Gene	Name	Histological type (% methylation, cases)	References
NTRK2	Neurotrophin tyrosine kinase receptor type 2	CRC (83%, 25/30), noncancerous normal colon (8%, 1/13)	(Kim et al. 2009)
OSMR	Oncostatin M receptor-β	CRC (80%, 80/100), matched normal colon mucosa (4%, 4/100), noncancerous normal colon (0%, 0/13)	(Kim et al. 2009)
p15INK4	Cyclin-dependent kinase inhibitor 2B	CRC (68%, 44/65)	(Xu et al. 2004)
p73	Tumor protein p73	CRC (63%, 41/65)	(Xu et al. 2004)
PAPSS2	3′-phosphoadenosine 5′-phosphosulfate synthase 2	CRC (70%, 21/30), noncancerous normal colon (8%, 1/13)	(Kim et al. 2009)
PTGIS	Prostaglandin I2 (prostacyclin) synthase	Adenoma (30, 3/10), CRC (43%, 43/100)	(Frigola et al. 2005)
RAR-β	Retinoic acid receptor-β	CRC (85%, 55/65)	(Xu et al. 2004)
RASSF1 A	Ras association (RalGDS/AF-6) domain family 1	MSI+ CRC (52%, 16/31), HNPCC (30%, 6/20)	(Frigola et al. 2005)
RASSF1 A	Ras association (RalGDS/AF-6) domain family 1	CRC (81%, 39/48), normal colon mucosa (49%, 19/39)	(Sakamoto et al. 2004)
RASSF1 A	Ras association (RalGDS/AF-6) domain family 1	CRC (21%, 14/64)	(Brandes et al. 2005)
RASSF2A	Ras association (RalGDS/AF-6) domain family 2	CRC (73%, 106/146), matched normal Colon (12%, 2/17)	(Park et al. 2007)
RASSF2A	Ras association (RalGDS/AF-6) domain family 2	CRC (42%, 51/122)	(Akino et al. 2005)
RASSF2A	Ras association (RalGDS/AF-6) domain family 2	CRC (70%, 121/30), matched normal colon (0%, 0/30)	(Hesson et al. 2005)
RASSF2A	Ras association (RalGDS/AF-6) domain family 2	Adenoma (43%, 21/49), CRC (42%, 51/122)	(Akino et al. 2005)

Table 13.2 (continued)

Gene	Name	Histological type (% methylation, cases)	References
RECK	Reversion-inducing-cysteine-rich protein with kazal motifs	CRC (44%, 11/25)	(Cho et al. 2007)
RIL	A LIM domain gene mapping to 5q31	CRC (70%, 30/43), adenoma (85%, 11/13), normal colon (5%, 1/22)	(Boumber et al. 2007)
RUNX3	Runt-related transcription factor 3	CRC (34%, 31/92)	(IMAMURA et al. 2005)
RUNX3	Runt-related transcription factor 3	Polyps (74%, 64/87), normal colon (16%, 2/12), HP (89%, 17/19), SA (86%, 12/14), sTAs (44%, 7/17), FAP polyps (74%, 28/38)	(Subramaniam et al. 2009)
SPARC	Osteonectin	CRC (100%, 20/20), normal colon mucosa (15%, 3/20)	(Yang et al. 2007)
TFPI2	Tissue factor pathway inhibitor-2	(62%) primary colon colon carcinomas	(Hibi et al. 2010)
TPEF/HPP1	Transmembrane protein with EGF-like and two follistatin-like domains 2	Normal tissue (5%, 1/21), CRC (77%, 36/47), liver metastasis (79%, 19/24)	(Ebert et al. 2005)
TPEF/HPP1	Transmembrane protein with EGF-like and two follistatin-like domains 2	Adenoma from FAP-Pts (31%, 10/32), multiple adenoma Pts (65%, 17/26), MSI-H CRC Pts (64%, 9/14), MSS/MSI-L CRC Pts (54%, 7/13)	(Wynter et al. 2006)
TUBG2	γ-Tubulin gene 2	CRC (97%, 29/30)	(Kim et al. 2009)
UNC5A	unc-5 homolog A	CRC (68%, 34/50)	
UNC5C	unc-5 homolog C	CRC (62%, 112/147), AP (63.5%, 33/52), matched normal colon mucosa (6% 9/147)	(Shin et al. 2007)
Vimentin	Vimentin	CRC (65%, 31/45)	(Shirahata, 2010)
Vimentin	Vimentin	CRC Pts (62%, 95/153), HD (2%, 1/46)	(Chen et al. 2005)
WNT5A	Wingless-type MMTV integration site family, member 5A	CRC (48%, 14/29), matched normal colon mucosa (13%, 2/15)	(Ying et al. 2008)

Table 13.2 (continued)

Gene	Name	Histological type (% methylation, cases)	References
DFNA5	Autosomal dominant 5	CRC (65%, 65/100), matched normal colon mucosa (3%, 3/100), noncancerous normal colon (9%, 1/11)	(Kim et al. 2009)
DKK-3	Dickkopf homolog 3	CRC (52.3%, 67/128), adjacent nontumor tissues (5%, 1/20)	(Yu et al. 2009)
EphA1	Eph receptor A1	CRC (49%, 26/53)	(Herath et al. 2009)

Table 13.3 Gene Methylation in IBD. (Lin et al. 2011; Cooke et al. 2012)

Gene ID	Gene ID
BCL6	*HIF1A*
TJP1	*APC*
TNFSF	*BGN*
EYA4	*GAS1*
PTPN6	*LAT*
PURA	*EGF*
GATA6	*RIPK4*
TCOF1	*FANCC*
THRAP2	*GFPT1*
GBGT1	*ANKRD9*
PAQR6	*TNFSF12–13*
HFE	*EVI1*
IFNGR2	*THRAP2*
GATA4	*ABCC2*
GATA5	*EFNB3*
HACE1	*IL1B*
MAS1	*FLJ20712*
IL18BP	*GABRA5*
LMTK2	*HOXB2*
MMP3	*MAPK10*
RHOH	*MAS1*
ITGA4	*PITPNM2*
LRRC3B	*MS4A4A*
DOK2	*SERPINA5*
CHML	*FABP3*
VWF	*NEU1*
MGMT	*S100A4*
FUT7	*GPR116*
TMEM116	*GPX3*
TNFSF4	*MAGEL2*
FCGBP	*SLC22A18*
SPATA22	*STAT5A*
TNFRSF1A	*ASCL2*
NOTCH1	*IL16*
PDE1B	*RUNX1T1*
PDGFRB	*SP11*
MLH1	*ICAM3*
MSX1	*MMRN2*
MYOD	*GCET2*

Table 13.3 (continued)

Gene ID	Gene ID
CYP2D6	*RBM13*
COG8	*KCNK4*
FOLR1	*AKAP2*
HTR2A	*CD28*

Genetic susceptibility is influenced by luminal microbial; stimulate immune responses in pathogenic or protective way. Environmental triggers are necessary to initiate or reactivate disease expression. As shown in Fig. 13.9 long lasting intestinal inflammation stimulates cell proliferation in the mucosal and could eventually lead to the low grade of dysplasia, high grade dysplasia and ultimately carcinoma.

Reactive oxygen species (ROS) and reactive nitrogen intermediate (RNI) produced by inflammatory cells may cause mutations in epithelial cells. Also, cytokines produced by inflammatory cells can elevate intracellular ROS and RNI in premalignant cells. In addition, as seen in Fig. 13.9 inflammation can result in epigenetic changes that favor tumor initiation. Tumor associated inflammation contribute further Ros, RNI, and cytokine production.

13.2.5.1 DNA Methylation May Play an Important Role in IBD Susceptibility

Currently more than 32 susceptibility loci have been identified for IBD (Yamazaki et al. 2005; Duerr et al. 2006; Hampe et al. 2006; Rioux et al. 2007; Burton et al. 2007; Parkes et al. 2007). However, all these genetic risk factors can only account for approximately 20 % of the genetic risk (Xavier and Rioux 2008; Barrett et al. 2008), suggesting that other factors, including possible epigenetic factors, are involved in the pathogenesis of IBD (Petronis and Petroniene 2000). Heritable and reversible epigenetic alteration has been recognized as a factor affecting disease pathogenesis (Moss and Wallrath 2007; Shames et al. 2007). DNA methylation and histone modification are the most studied epigenetic events. DNA methylation and its crucial role has been extensively investigated in various cancers (Wilson et al. 2007; Robertson 2005) and several other human diseases including IBD. Abnormal DNA methylation has been observed in UC patients in the estrogen receptor (ER), *P14ARF, E-cadherin* gene and other genes (Maeda et al. 2006). Moreover, there is evidence that epigenetic factors are involved in the regulation of gene activity as a factor in the etiopathogenesis of IBD (Tahara et al. 2009).

DNA hypermethylation is associated with gene silencing and is often observed in CpG islands of cancer-related genes in IBD. Aberrant cancer specific methylation of genes such as *p16, E-cadherin, hMLH1,* and *p14* has been reported in IBD associated neoplasia (Dhir et al. 2008). These epigenetic changes may accelerate the development of IBD-CRC (Levin, 1992; Petronis and Petroniene 2000; Kellermayer 2012). CpG island methylation phenotype and global DNA methylation

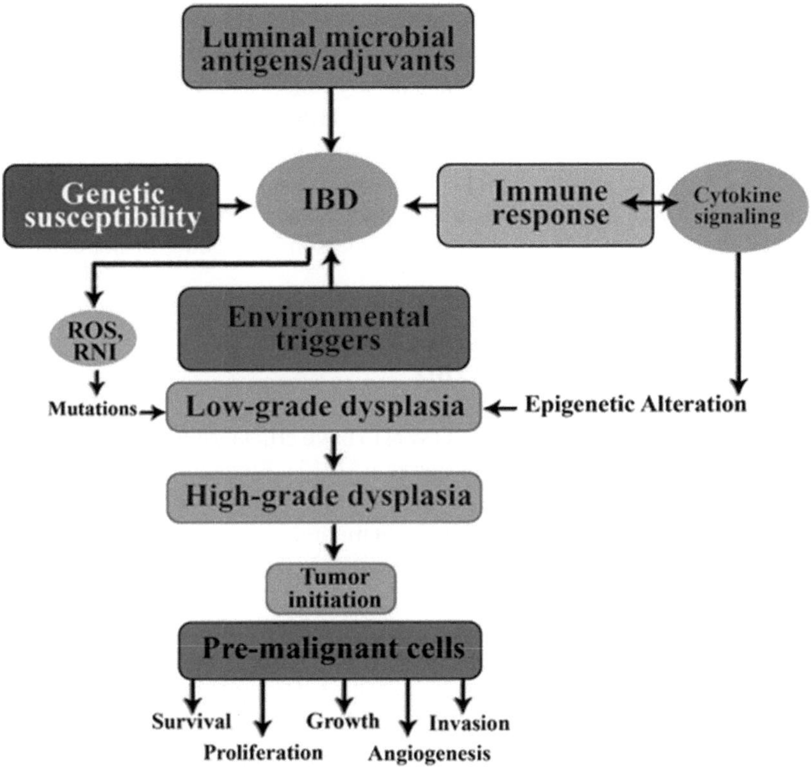

Fig. 13.9 Role of inflammation in tumor initiation

with LINE-1 assay were also observed in 17 and 58 % of UC-related cancers, respectively (Konishi et al. 2007). Loss of *hMLH1* protein expression associated with gene hypermethylation is common in Microsatellite instability (MSI) positive UC cancers (Fleisher et al. 2000). Methylation of hyperplastic polyposis gene 1 (*HPP1*) was observed in 50 % UC adenocarcinoma and 40 % of dysplasias but not in normal mucosa (Sato et al. 2002). Also, methylation of *E-cadherin* (CDH1) was detected in 93 % of the patients with dysplasias (Azarschab et al. 2002). Moreover, the increased level of methylation is widespread in the colon affected by inflammation and occurs early in the process of carcinogenesis (Issa et al. 2001).

DNA methylation in CRC has been previously studied leading to the identification of several tumor-associated DNA methylation patterns (Ahmed 2007; Wong et al. 2007). Previous IBD-related DNA methylation studies have focused on the development of CRC in patients with UC. Initial studies reported that four genes were highly methylated in high-grade dysplastic epithelium from patients with UC (Toyota et al. 2002). Subsequent observations of disease-related methylation status were made for the ER, *HPP1, MLH1,* RAB32, *MGMT,* and *P14ARF* (Issa et al. 2001; Fujii et al. 2005; Schuebel et al. 2007) and *E-cadherin* genes from patients with UC (Azarschab et al. 2002).

Inflammatory rectal mucosa from patients with UC showed increased methylation level and frequency with respect to the MDR1 gene promoter, especially in patients with shorter disease duration with a younger onset (Tahara et al. 2009). Recently, the expression of the MD-2, a critical TLR4 coreceptor that is upregulated in IBD but not in normal or diverticulitis patients, is upregulated by demethylation of the MD-2 promoter region (Vamadevan et al. 2010). Multiple genetic risk loci are common in both CD and UC, indicating a significant shared genetic component in the etiopathophysiology of these diseases (Anderson et al. 2009; Fisher et al. 2008).

13.2.5.2 Aberrant Wingless Signaling Pathways (WNT) in Patients with IBD

Aberrant wingless signaling pathways (WNT) have crucial roles for cancer progression as early progression events (Anastas et al. 2012). Epigenetic alteration of WNT signaling genes has been reported in up to 80% of patients with CRC (You et al. 2008; Barker and Clevers 2006). Differential methylations were identified in tissues of patients with IBD, suggesting that IBD-associated changes in DNA methylation may be disease subtype specific.

WNT Ligands interact with the Fz receptor and LRP co-receptor and thus initiate the WNT signaling cascade (Anastas et al. 2012). These interactions subsequently lead to the formation of a complex composed of *APC,* AXIN, and Glycogen synthase kinase-3β (GSK-3β). Together, *APC*/AXIN/GSK-3β makes up the β-catenin destruction complex. In the absence of WNTs stimulation, the *APC*/AXIN/GSK-3β complex phoshphorylate β-catenin. Phosphorylated β-catenin is degraded by proteosome system (Baylin and Ohm 2006; Fyang et al. 2014) β-catenin is the cytoplasmic protein and the key component of the WNT signaling pathway. In the presence of WNTs stimulation, or loss of *APC* due to mutation, the destruction complex is destabilized, resulting in excessive amounts of β-catenin. β-catenin then translocates to the nucleus, where it associates with the Tcf family of transcription factors, and promotes the transcription of many target genes including *c-MYC* and *cyclin D1* whose products increase cell proliferation (Noah et al. 2013). Secreted Frizzled-related proteins (*SFRPs*) are tumor suppressor genes that function as antagonists of WNT signaling. *SFRP2* bind to WNTs and prevent their interaction with Fz and LRP (Baylin and Ohm 2006). Dysregulation of WNT signaling pathway may be involved in IBD-CRC owing to epigenetic silencing of key tumor suppressor genes such as *APC* and *SFRP2* (Dhir et al. 2008).

Although, there are few studies addressing the effects of WNT gene methylation in IBD, little is known about the role of WNT related genes in IBD initiation and progression in this region. Also, since the mutation of WNT signaling genes is rare in neoplasia associated with IBD (Dhir et al. 2008),

Mokarram and et al showed the association of promoter methylation *APC1A, SFRP2* genes from WNT pathway occur during the IBD-associated carcinogenesis (personal communication). Our study related to IBD patients demonstrates for the first time hypermethylation of the *SFRP2* not *APC1A* gene promoter region was involved in IBD developing lesions.

13.2.6 The Role of Kras and MGMT Methylation in CRC

On the other hand, the activation of the Raf/ MEK/ ERK (MAPK) kinase pathway through either *K-ras* or *BRAF* mutation was detected in 30 % of UC related cancers. Non-dysplastic UC mucosa of patients with UC cancers show *K-ras*, but not *BRAF* mutation, indicating that *K-ras* mutations are initiating events in UC carcinogenesis (Aust et al. 2005) Several studies have also reported that reduced expression and function of DNA repair enzyme induced by oxidative stress during inflammation disease (Switzeny et al. 2012). DNA Mismatch repair elements and O6-methylguanine-DNA methyltransferase (O6-*MGMT*) are DNA repair proteins. *MGMT* loss of activity associated with creating a permissive state for G→A mutations in protooncogenes such as *K-ras* and *p53* (Mokarram et al. 2012). *K-ras* mutations occur in about 60 % of CRCs and are an early event in UC carcinogenesis (Aust et al. 2005). *MGMT* promoter has two loci (Mokarram et al. 2012) described as *MGMT*-A and *MGMT*-B. Recently, it has been shown that methylation silencing of *MGMT*-B was significantly associated with the *K-ras* gene mutation rather than *MGMT*-A in patients with CRC (Mokarram et al. 2012) and analyzing the promoter profile methylation of *MGMT*-B could be of high prognostic value for patients with CRC. Because of this correlation, it is hypothesized that methylation of *MGMT*-B as a caretaker gene may be responsible for dysregulation of signaling through *K-ras* mutation in CRC lesions. On the other hand, our study related to IBD patients demonstrates for the first time hypermethylation of the *MGMT* gene promoter region was involved in IBD developing lesions (personal communication). Overall, methylation of *MGMT* and *SFRP2* in IBD patients may provide a method for early detection of IBD-associated neoplasia which will be independent of the pathologist.

13.2.7 DNA Methylation in Polyp

CRCs develop as a result of the transformation of normal colonic mucosal epithelium to cancer through a series of precursor lesions with genetic and epigenetic changes termed the adenoma to carcinoma sequence and the hyperplastic polyp-serrated adenoma to carcinoma sequence. Epigenetic events have been involved in the stepwise histological progression such as adenoma-carcinoma and hyperplastic polyp/serrated adenoma-carcinoma sequences in the development of CRC (Ashktorab et al. 2014; Dhir et al. 2011). These precursor lesions experience CGI methylation of tumor suppressor genes as well as genetic changes (*APC* and *K-ras* mutations). Aberrant crypt foci (ACF), known as the earliest precursor lesions, experienced aberrant CGI methylation. Such methylation was also seen in adenomas; although it was less frequent than in adenocarcinoma. However, CGI methylation increased as early adenomas progressed towards advanced adenomas. In different stages of the adenoma-carcinoma progression sequence the methylation frequency gene-specific. The *MGMT, p16INK4a,* and *HLTF* genes had the highest methylation rate during this sequence whereas adenocarcinoma had the highest methylated

MLH1 and *TIMP3*. Therefore, depending on the gene, the tumor initiation or progression process is affected by the inactivation of tumor suppressor genes caused by aberrant DNA methylation.

Recently, colorectal polyps have been classified into two major groups; serrated polyps and conventional adenomas (Spring et al. 2006; Pereyra et al. 2014; Spring et al. 2006). More CGI methylation and *BRAF* mutations were seen in serrated polyps (including hyperplastic polyps), sessile serrated adenomas, sessile adenomas, and mixed polyps.

BRAF mutations occur rarely in tubular adenomas. Hyperplastic polyps are mostly diminutive lesions are less likely to develop into cancer. However, sessile serrated adenomas, serrated adenomas and mixed polyps have a higher risk of progressing into the cancer. Sessile serrated adenomas were prevalent in about 9% of patients who were undergoing colonoscopy compared with sessile adenomas (0.7%) and mixed polyps (1.7%). In the aforementioned study, sessile serrated adenomas were linked to *BRAF* mutations, proximal location, female sex, and presence of multiple polyps. In other studies, associations were found between serrated adenomas and CIMP and MSI-H cancers, tending to show near diploid DNA indices, more frequent allelic imbalance at 18q, and less frequent allelic imbalance at 5q or *K-ras* mutations. Serrated polyps and tubular adenomas also differed with respect to the proportions of specific methylated genes (Spring et al. 2006; Kim et al. 2006; Kim et al. 2014).

The neoplastic potential of ACF is still unknown. However, some ACF lesions harbor *K-ras* or *APC* mutation and methylation of specific gene loci such as *RASSF1A, CRBP1, MINT31, CDH13, MINT1, SLC5A8,* and *MGMT*. It is a precancerous condition of the colorectum such as hyperplastic polyposis (HPP), the polyps are morphologically distinguished from typical small hyperplastic polyps. Similar to adenomas, the polyps in HPP can evolve through a histological progression sequence that culminates in colon adenocarcinomas. A serrated adenoma intermediate could also give rise to hyperplastic polyps in some CRCs. The HP-serrated adenoma–adenocarcinoma sequence is linked with DNA methylation of *HPP1, CDKN2A/p14, CDKN2A/p16,* and *MLH1* and mutations in *BRAF* and is more common in the proximal colon. Moreover, detection of DNA methylation in normal colorectal mucosa may serve as a useful diagnostic biomarker for high cancer risk in HPP.

Figure 13.10. represents the morphologic and genetic alterations along a putative serrated adenoma-carcinoma pathway in comparison to the classic form of adenomatous polyp progression. Similar differences are seen in proximal and distal pathway of tumorigenesis.

When SSA polyps and HP polyps are analyzed to look at mutations in key cell cycle regulatory genes, it found that HPs have a higher mutation rate in *K-ras* but SSAs frequently have mutation in *BRAF* (Pereyra et al. 2014). SSAs tend to display CpG island methylation to a greater degree than sporadic HPs. Minoo and co-workers investigated a series of serrated polyps and matched normal mucosa to find distinguishing molecular features in patients with and without HPP. They found higher DNA methylation in sessile serrated adenomas and normal colorectal mucosa in HPP. A high level of CpG island methylator phenotype (CIMP, a subset of colon tumors with accumulation of type C methylation) in the normal mucosa of

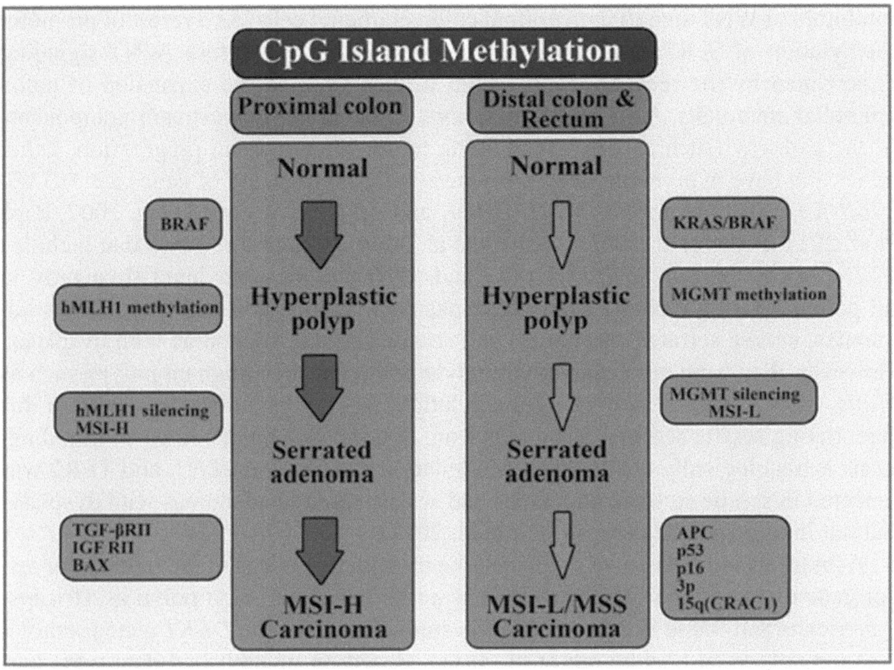

Fig. 13.10 Pattern of genetic and epigenetic alterations in polyp and CRC based on location of tumorigenesis

patients with HPP was also observed. In some forms of HPP, the earliest manifestation of transformation might be hyper-methylation of multiple gene promoters in normal colorectal mucosa (Kim et al. 2010; Minoo et al. 2006).

13.2.7.1 Aberrant CpG Island Hypermethylation in Polyp

Since aberrant CpG Island DNA promoter hypermethylation can be detected not only in colorectal polyps and cancers, but also in sera and stool, epigenetic changes are considered as a diagnostic hallmark during the initiation and progression process of tumors. This pathway implicates the inactivation of DNA mismatch repair genes tracked by mutations in the microsatellite repeat sequences in the genes that are important in tumor progression as TGF RII and BAX, which leads to uncontrolled cellular growth and decreased apoptosis. Loss of *APC* functions with activation of WNT pathway is observed in most familial adenomatosis polyposis-associated and sporadic CRC. Studies have reported that promoter regions of genes that are responsible for encoding secreted frizzled related proteins (*SFRPs*) undergo abnormal methylation in colorectal neoplasms (Kim and Deng 2007; Papa et al. 2011; Petko et al. 2005; Belshaw et al. 2004). *SFRPs* are antagonists of WNT signaling in normal colon epithelial cells, by competing with WNT proteins for binding to their receptors frizzled (FRZ) on the cell surface. *SFRPs* act as constitutive inhibitors of WNT signaling in normal colon epithelial cells. As a result of promoter methylation of SFRP genes, SFRP expression is lost. Therefore, WNT signaling is activated by the receptor FRZ, which in turn leads to the expansion of colon epithelial stem cells. As a result, mutations occur in the downstream components of the pathway (such as *APC*) facilitating tumor initiation and progression. Other genes that have hypermethylated promoters with the silencing of genes are *MGMT, p16INK4a, HIC1, RASSF1, ESR1, HPP1,* and *MSH2* (Kim and Deng, 2007, Papa et al. 2011, Petko et al. 2005, Belshaw et al. 2004). A subset of genes that included *EVL, GATAs, HIN-1, SFRPs, SOX17* and *SYNE1* were methylated frequently in all premalignant gastrointestinal adenomas such as tubular adenomas, villous adenomas, sessile serrated adenomas, and sessile serrated adenomas with dysplasia. However, they were not frequently methylated in non-premalignant polyps such as HPPs. CDX2, *hMLH1* and TLR2 methylation may be diagnostically useful in differentiating sessile serrated adenomas from hyperplastic polyps in cases that diagnosis is histologically challenging. Methylation of CDX2, *hMLH1*, and TLR2 was detected in sessile serrated adenomas and sessile serrated adenomas with dysplasia but not in hyperplastic polyps (Dhir et al. 2011).

Ashktorab and colleagues evaluated the methylation status of the tumor suppressor gene dickkopf homolog 1 (*DKK1*) as a risk factor for colon polyp in Africans-Americans and found that in 96% of the studied samples the *DKK1* gene promoter was unmethylated (Ashktorab et al. 2011). Based on unpublished data were performed by Mokarram et al in patients with polyp, at least one of two genes (*SFRP2* and *MGMT*) was methylated in 93.8% of the patients. Both genes were methylated in 45.8% of the patients, implying that half of the patients had both methylated genes. These data revealed the potential diagnostic value of these two genes in polyps and adenomas. By comparing *MGMT* and *SFRP2* genes and considering that the extensive methylation in *MGMT* is associated with the development of polyp and adenoma. Therefore, *MGMT* has a key role in the pathologies of patients with polyp and adenoma.

13.2.8 Detection of CRC by Microarray or DNA Methylation Assays

Currently, molecular biology has turned medical oncology into a more interesting topic of study since molecular changes influential in progression can be studied in detail using molecular biology. Therefore, the carcinogenesis process is better understood and novel prognostic markers and therapeutic targets can be discovered. Although clinical and pathological parameters are at hand for the classification and prognostic stratification of cancer, inadequacy might still exist in daily practice due to the vast biological and genetic heterogeneity of cancer.

Also, since patient-tailored therapies are becoming increasingly common, it is necessary to gain a more comprehensive knowledge of downstream signaling pathways in order to discover new tumor targets and as a result develop novel biological drugs. Here, gene expression profiling analysis with microarray technologies is a newly emerging field that allows the simultaneous mapping of thousands of gene expression in as little as one tumor sample. Fig. 13.11 shows how microarray performs the gene expression evaluation.

DNA microarrays can survey virtually the entire expressed genome. Studies showd that a small amount of high quality RNA from tumor or non-tumor tissues is labeled and hybridized on the surface of chip which is composed of spotted cDNA clones or probes spotted or synthesized on the surface of the chip. DNA microarrays analysis is most useful when it can be integrated with clinical, imaging and histological data. DNA microarray accompanied with clinical data provides complementary types of information, so most microarray studies use combination of these approaches.

Gene microarray analysis is becoming an increasingly valued tool in studying human cancers theoretically and practically. Researchers are now able to investigate the expression of ≤50,000 genes using complementary DNA or oligonucleotide

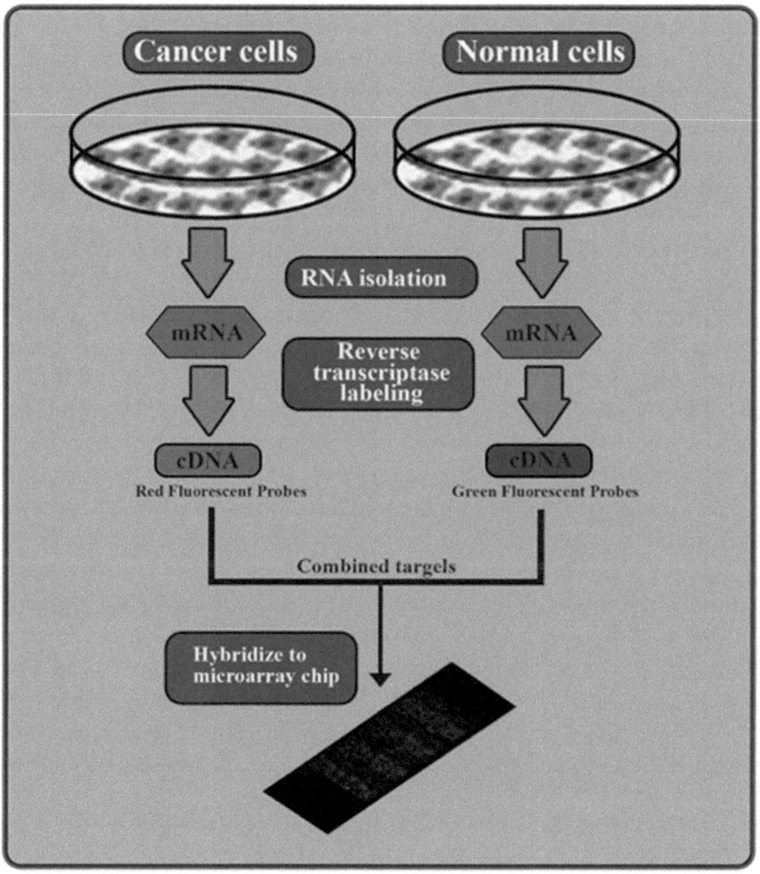

Fig. 13.11 DNA microarray to compare gene expression in cancer and normal samples

arrays, which facilitates an improved understanding of the complex relationships between genes. Moreover, this approach may provide insight as to how cellular gene expression might yield important prognostic and/or predictive information.

Understanding the molecular signature of tumors could provide physicians with a highly sensitive, quantitative, reproducible, and individualized approach for making treatment decisions as well as provide a more accurate assessment of a patient's tumor burden when compared with current methods that rely on conventional clinical criteria.

The various clinical applications of microarray technologies could be divided into three categories in CRC; [1] studies on carcinogenesis process, [2] studies on prognosis prediction, [3] studies on treatment response prediction.

One of the earliest applications for microarray in preclinical settings is examining tumor specimens versus normal tissues to differentiate the gene expression of normal and neoplastic tissue. The hypothesis was that these genes could provide insight into the changes in the subcellular machinery that eventually gives rise to neoplasia, directly or indirectly. Several studies have identified differential gene profiles in normal versus cancerous tissues (Li et al. 2004; Bandres et al. 2004; Abajo et al. 2012). Gene microarray has proven valuable in prognostic studies on hormonal and cytotoxic treatment in breast cancer. CRC is also an interesting field in microarrays research because it is a biological tumorigenesis model with distinct molecular alterations (Vogelstein et al. 1988). Moreover, for distinguishing high risk from low risk CRC, traditional clinical and pathological parameters are not always adequate due to the unavailability of validated molecular markers with prognostic value. And finally, although there is currently controversy regarding data on molecular markers that have predictive tumor response value, some cytotoxic and biological medications are extensively used in routine clinical practice (Spano et al. 2008).

In addition, CpG Island Methylator Phenotype (CIMP) is one of the underlying mechanisms in colorectal cancer (CRC). Study aimed to define a methylome signature in CRC through a methylation microarray analysis and a compilation of promising CIMP markers from the literature (Ashktorab et al. 2014). Illumina HumanMethylation27 (IHM27) array data was generated and analyzed based on statistical differences in methylation data (1st approach) or based on overall differences in methylation percentages using lower 95 % CI (2nd approach). Pyrosequencing was performed for the validation of nine genes. A meta-analysis was used to identify CIMP and non-CIMP markers that were hypermethylated in CRC but did not yet make it to the CIMP genes' list. Our 1st approach for array data analysis demonstrated the limitations in selecting genes for further validation, highlighting the need for the 2nd bioinformatics approach to adequately select genes with differential aberrant methylation. A more comprehensive list, which included non-CIMP genes, such as *APC, EVL, CD109, PTEN, TWIST1, DCC PTPRD, SFRP1, ICAM5, RASSF1A, EYA4, 30ST2, LAMA1, KCNQ5, ADHEF1* and *TFPI2* was established. Array data are useful to categorize and cluster colonic lesions based on their global methylation profiles; however, its usefulness in identifying robust methylation markers is limited and rely on the data analysis method. Ashktorab et al. have identified 16 non-CIMP-panel genes for which we provide rationale for inclusion in a more comprehensive characterization of CIMP+ CRCs (Ashktorab et al. 2014). The

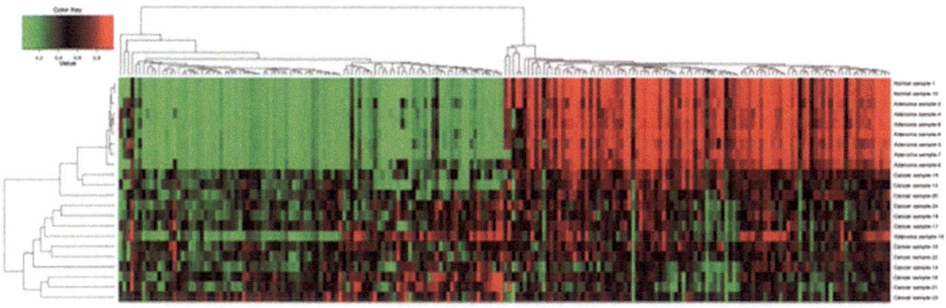

Fig. 13.12 Cluster profile analysis of colonic samples

identification of a definitive list for methylome specific genes in CRC will contribute to better clinical management of CRC patients.

The methylation profiles of 12 cancers, 8 adenomas and 2 normal colonic tissues were generated by Cluster profile analysis of colonic samples in Fig. 13.12 (Using the IHM27). A clustering of the different samples based on their methylome profiles led to a clear resolution between the cancer samples and the adenomas (Ashktorab et al. 2013).

Bioinformatics approach helped us for validation representative CpG sites. Four CpG sites (*SAP130*, *RAD54L*, *PFDN5*, and *PTPN12*) were selected for further validation because they showed the largest differences between cancer and adenoma; all of these sites were hypermethylated in cancer. Additionally, these four CpG sites were located within the promoter region of the genes and were statistically significant based on the FDR adjusted *p* values. Based on the criteria set forth in "Bioinformatics Approach #2," in Fig. 13.13, these markers were considered to be eligible for further validation of the array data. Validation of *RAD54L* indicated that only 1 out of 40 patients showed hypermethylation (data not shown), suggesting that our 2nd bioinformatics approach was more stringent than the 1st approach and potentially better suited to selecting genes for validation (Ashktorab et al. 2013).

Furthermore, DNA methylation patterns in CRC, especially in AAs, have not been systematically explored and remain poorly understood. Here, Ashktorab et al performed DNA methylome profiling to identify the methylation status of CpG islands within candidategenes involved in critical pathways important in the initiation and development of CRC. They used reduced representation bisulfite sequencing (RRBS) in colorectal cancer and adenoma tissues that were compared with DNA methylome from a healthy AA subject's colon tissue and peripheral blood DNA. The identified methylation markers were validated in fresh frozen CRC tissues and corresponding normal tissues from AA patients diagnosed with CRC at Howard University Hospital. They identified and validated the methylation status of 355 CpG sites located within 16 gene promoter regions associated with CpG islands. Fifty CpG sites located within *CpG* islands in genes *ATXN7L1*, *BMP3*, *EID3*, *GAS7*, *GPR75*, and *TNFAIP2* were significantly hypermethylated in tumor vs. normal tissues ($p<0.05$). Figure 13.14 shows the average methylation of promoters in tumor

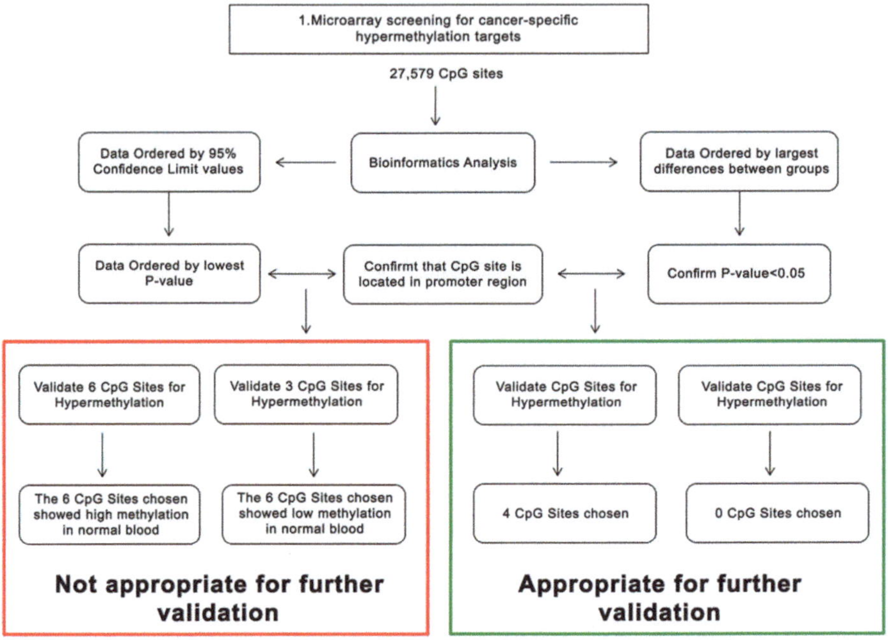

Fig. 13.13 Schematic representations of approaches used to identify appropriate genes for validation of CpG island methylation resulting from Illumina Human Methylation 27 Array data

and normal based on RefSeq genes annotations. The methylation status of *BMP3, EID3, GAS7,* and *GPR75* was confirmed in an independent, validation cohort. Ingenuity pathway analysis mapped three of these markers (*GAS7, BMP3,* and *GPR*) in the insulin and TGF-β1 network the two key pathways in CRC. In addition to hypermethylated genes, LINE-1 repeat elements were progressively hypomethylated in the normal-adenoma-cancer sequence. Therefore, DNA methylome profiling based on RRBS is an effective method for screening aberrantly methylated genes in CRC. Novel hypermethylated genes, as well as hypomethylated LINE-1 sequences, may serve as potential biomarkers for CRC in African Americans. Discovered biomarkers were intimately linked to the insulin/TGF-B1 pathway, further strengthening the association of diabetic disorders with colon oncogenic transformation (Ashktorab et al 2014).

13.2.8.1 Microarray and Prediction of Cancer Status or Treatment

It may be relevant to identify specific signatures of metastases in order to personalize patient treatment. In particular, because the liver is the most frequent site of metastases in CRC, liver metastases represent a research field of great interest that was less investigated and thus less known biologically. In fact, most studies have analyzed metastatic tissue in comparison with primary tumor to better understand

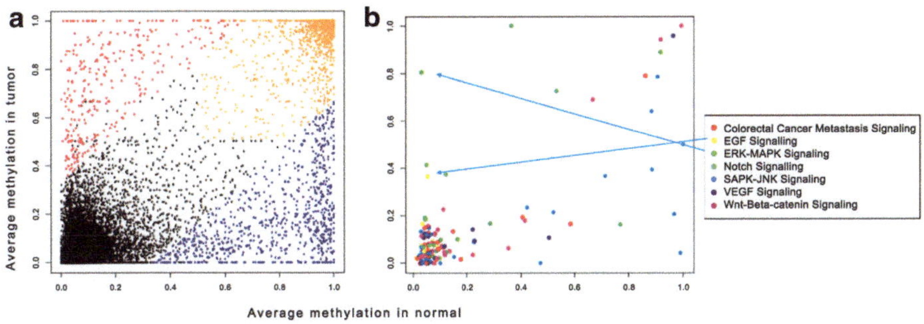

Fig. 13.14 Average methylation of promoters in tumor and normal. Promoters that were hypermethylated only in tumor are shown in *red*; those hypermethylated only in normal in *blue* and promoters' hypermethylated in both are shaded *orange*. (B) Promoters from 3A filtered to include genes associated with various Gene Ontology (*GO*) terms. The *Y-axis* shows the average methylation in tumor and the *X-axis* shows the average in normal colon

the development process of the tumor as well as to identify gene expression patterns predictive of metastatic potential. However little information is available about its molecular background. Many genes codifying for proteins involved in cell adhesion, migration, angiogenesis and proliferation have been linked to the development of colorectal liver metastases. However, a real genetic signature for metastatic tissue has still not been defined (Friederichs et al. 2005). Burness and colleagues identified 23 pathways that differed in expression in a site-specific segregation of genes between liver and peritoneal metastases (Burness et al. 2007). The gene expression profile of synchronous and metachronous liver metastases was studied using Affymetrix platform that lead to the identification of *EGFr* pathway and the pathway mainly related to angiogenesis which were upregulated in metachronous and synchronous lesions, respectively (Pantaleo et al. 2007). These results suggest that liver metastases might have a genetic signature that can be used as a basis for selecting the right treatment. Moreover, based on the hypothesis that advanced CRC might be a multiform disease, different molecular features could yield different medical treatments.

Also, discovering new molecular targets in order to produce new biological drugs is significantly challenging. The wide list of genes identified in all previously described studies could represent a source of potential therapeutic target because most of these genes are involved in key mechanisms of tumor development; from cell proliferation and differentiation to cell survival. Weir and co-workers state that the systematic understanding of the molecular basis of all types of cancer requires at least three steps:"comprehensive characterization of recurrent genomic aberrations, elucidation of their biological role in cancer pathogenesis, and evaluation of their utility for diagnostics, prognostics and therapeutics" (Weir et al. 2007; De Divitiis et al. 2014; Coghlin et al. 2014). Therefore, a comprehensive whole proteomic and genomic characterization, including newer technologies such as microarrays (gene, Methyl), comparative genomic hybridization, SNPs arrays, and miRNA, is strictly necessary to discover novel therapeutic targets and increase treatment options for CRC.

As in breast cancer, surgery can be curative for patients with CRC who are diagnosed with early-stage disease or liver-limited metastatic disease. Moreover, for patients with CRC, there are several treatment options. Understanding the gene-expression pattern of the tumor would be invaluable in determining the true prognosis of the patient, deciding if adjuvant treatment is needed, and choosing the best treatment regimen to achieve the most optimal benefit that is the subject of pharmacogenomics studies. Approximately 75 % of patients with CRC present with respectable disease and for those with stage I/II disease (N0 M0), the 5-year survival rate is 90 % (Pazdur et al. 2003). In patients with stage III disease, the risk is significantly higher, with approximately 30–40 % of patients eventually relapsing and dying from metastatic disease, implying that a relatively large fraction of patients might have occult disease at the time of initial presentation, which should warrant a more aggressive adjuvant therapy after surgical resection.

Understanding the underlying basic biology of neoplasia and identifying potential, new therapeutic targets and predicting the likelihood of metastasis based on having the "metastatic signature" in the primary tumor specimen is among the applications of gene profile identification in a metastatic tumor. The hypothesis is that finding the expression of such genes in the primary tumor would indicate that metastasis has likely already occurred.

Croner and colleagues studied the feasibility of using genomic analysis to predict the likelihood of lymph node involvement (Croner et al. 2005, 2010, 2014). In that study, the assessment of the primary tumor as a means of determining prognosis was compared with the standard means of assessing lymph node status postoperatively via conventional histopathology. The goal was to establish a gene profile of tumors with lymphatic metastasis and to determine whether these data could improve conventional clinicopathologic variables. Analysis of conventional parameters resulted in a positive prediction rate of 53–61 % with a sensitivity of 42 % and specificity of 72 %. However, microarray prediction rates, specificity, and sensitivity ranged between 62–67 %, 76–83 %, and 38–48 %, respectively. The accuracy of predictions made using standard clinical means were improved $\leq$12 % with the inclusion of information from microarray analysis, suggesting that the incorporation of gene array data with standard staging and evaluation procedures after surgery could more accurately predict which patients are at a higher risk of relapse. Taken together, the results of these studies suggest that gene profiling of the primary tumor at the time of surgery could be useful in predicting a patient's risk of relapse beyond other clinical variables (Eschrich et al. 2005).

Censored survival data were matched against tumor microarrays using significance analysis of the Microarray program (Tusher et al. 2001). This method yielded a 43-gene model that appeared to correlate strongly with patient survival outcome (Arango et al. 2005) The model was then validated using a previously studied group of 95 Danish patients (Dukes B and Dukes C). Although the Dukes staging system failed to discriminate between the survival outcomes of these patients, the outcome predicted by this model significantly correlated with the real outcome. This finding is clinically significant because it is estimated that as many as half of patients with Dukes C disease who are cured by surgery receive unnecessary adjuvant treatment with chemotherapy because of inaccuracies in the current staging system. Additional studies have also confirmed the value of genome profiling of patients with Dukes C disease, showing that tumors might be able to more accurately identify patients who need additional treatment compared with those who do not (Arango et al. 2005).

Another clinically useful application of microarray technologies is predicting which treatment regimens would be more successful in treating CRC as well as other malignancies; thus individualizing patient care. Before microarray analysis, 1 or 2 gene polymorphism analyses were used to predict response to and/or potential for adverse reactions. For example, the homozygotes for 6 TA repeats within the uridine diphosphate glucuronosyl transferase gene are linked to the increased incidence of diarrhea, emesis, and fatigue (Massacesi et al. 2006). Single-base polymorphisms in the orotate phosphoribosyl transferase seem to predict the response to 5-fluorouracil (5-FU) (Kitajima et al. 2006). However, data from these studies have been contradictory and/or inconclusive, implying that response and resistance are influenced by a series of genes and/or gene families that could best be studied using microarray technology.

More recently, a number of studies are validating predictive microarrays that can be used for decision-making on treatment. In 2006, Matsuyama and co-workers reported the results of a gene microarray analysis that influenced the sensitivity of hepatic metastases from CRC to treatment with 5-FU (Matsuyama et al. 2006). Among 81 genes that were identified in preliminary analyses, 4 of which were identified as differentially expressed in responding versus non-responding patients ($P<0.05$). This finding led to the development of a response index based on the expression of these genes. In a validation set, 9 of 11 patients with a positive response index responded to 5-FU-based therapy compared with only 1 of 11 patients with a negative response index.

13.2.8.2 Microarray and Response to Treatment/Resistance to Chemotherapy

Arango and colleagues reported the use of a microarray of 30 genes in the prediction of response to oxaliplatin (Arango et al. 2004). In this assessment of human colon cancer cell lines, evaluation of the expression of single genes, such as *p53*, *TS*, *TP*, and mismatch repair complex, was found inadequate in predicting the ability of the cell to undergo programmed cell death in response to treatment with oxaliplatin. However, assessment of 30 gene-expression profiles using microarray analysis was predictive of response. Validation of these results in patients treated with oxaliplatin-based regimens is still awaited.

Resistance to chemotherapy continues to confound treatment of patients with cancer. One of the challenges in understanding how cells become resistant is that tumors might become refractory through numerous alterations in multiple pathways.

Gene microarrays might provide clarity in these multiple changes, thus allowing for the identification and development of new agents to overcome resistance.

Recent studies have evaluated the feasibility of gene microarray analysis as a tool for understanding drug resistance mechanisms. At the 2006 American Society of Clinical Oncology Gastrointestinal Cancers Symposium, Martinez-Cardus and colleagues reported the results of a study assessing the molecular aberrations associated with acquired resistance to oxaliplatin (Martinez-Cardus et al. 2006). Continuous drug exposure generated four distinct oxaliplatin resistant cell lines. RNA assessment of these lines identified 32 genes, 15 of which were upregulated and 17 of which were down-modulated. Several of the gene products were identified as part of the P13 kinase and JNK/p38 pathway.

Boyer and co-workers developed human CRC cell lines from the drug-sensitive HCT116 via continuous exposure to oxaliplatin and 5-FU, leading to resistant subclones (Boyer et al. 2006). Gene microarray profiling was used to dynamically examine genetic changes that occurred in these cell lines during drug exposure, and the correlation of these gene sets with real-time reverse-transcriptase PCR. As a result of this dynamic evaluation and in contrast to previous studies, changes in resistant versus sensitive cell lines were identified. However, researchers were unable to determine whether these changes occurred as a response to treatment or were present before therapy and were outgrown via selective pressure. A panel of genes was identified that had a strong concordance to those expressed differentially using reverse-transcriptase PCR. Three genes were subanalyzed; prostate-derived factor, calretinin, and spermidine/spermine N1-acctyl transferase. Functionally, the roles of these genes suggest that they might be good targets for therapeutic inhibition to overcome drug resistance. Prostate-derived factor appeared to be able to make the cells resistant to death in response to treatment and was secreted, suggesting a role in tumor progression. Similarly, calretinin was induced during treatment and appeared to confer resistance to apoptosis. Downregulation of calretinin expression in HCT116 parental cells using small interfering RNA resulted in a dramatic decrease in 5_FU and oxaliplatin-induced apoptosis. These data suggest that microarray analysis might be an important tool in the identification of future therapeutic targets to overcome drug resistance.

Naghibalhossaini et al demonstrated a significantly elevated 5-FU resistance in 3D culture of the CEA-expressing CHO transfectants in comparison to the 3D spheroids of parental CHO. These findings suggest that the CEA level may be a suitable biomarker for predicting tumor response to 5-FU-based chemotherapy in CRC (Eftekhar and Naghibalhossaini 2014).

Mokarram and colleagues studied the potential diagnostic value of *MGMT* in CRC in early stages such as those observed in patients with IBD or polyps (Mokarram et al. 2012). They found that *MGMT* hypermethylation seemingly preceded the *K-ras* mutation in the adenoma-carcinoma sequence. This finding is important since the *MGMT* promoter methylation occurs early in carcinogenesis process, implying that examining the primary tumor for this sensitivity marker to temozolamide will suffice; because CRCs that lose *MGMT* expression may respond impressively to temozolomide. When *MGMT* expression is lost DNA repair in cells is compromised and may lead to cancer formation. Temozolomide acts through DNA methylation at the O6–guanine site, inducing base pair mismatch. In a one-step methyl transfer reaction, the methyl group at O6-site is removed by the DNA repair enzyme *MGMT*. *MGMT* is irreversibly inactivated because of the alkyl transfer group and leading to its ubiquitination and proteasome degradation. Hence, DNA is protected from methylation damage. At lower *MGMT* (i.e. as a result of methylated *MGMT* promoter), the methyl group is not removed from O6-guanine. So, methylated guanine binds thymine instead of cytosine, which is to be replaced by adenine in the next cell division (G→A mutation).

Alkylating agents have not been effective in the treatment of CRC. Alkylating agents whose effects were not exerted on the O6-guanine, unlike temozolamide, were used and thus their effect did not depend on *MGMT*. However, it has been clearly demonstrated that cancer cells with inactive *MGMT* (because of its promoter methylation), are definitely chemosensitive; 27–40 % of the CRCs had this mutation. Therefore, such patients might benefit from temozolamide (or dacarbazine). Also, CRCs with *K-ras* G→A mutation experienced *MGMT* methylation more (71 %), implying that the inactivation of *MGMT* through its promoter methylation could lead to somatic mutations in *K-ras*. Furthermore, a G→A mutation in *K-ras* could be caused by this epigenetic effect (Shacham-Shmueli et al. 2011).

NF-κB is activated in response to TMZ in a MMR- and AKT-dependent manner and confers protection against drug-induced cell growth inhibition. These findings suggest that a clinical benefit could be obtained by combining TMZ with NF-κB inhibitors (Pietrantonio et al. 2014; Caporali et al. 2012).

Therefore, the ability to rapidly and quantitatively analyze multiple genes has made it possible to not only investigate single genes, but evaluating entire gene families and or signaling pathways, providing a more complete molecular fingerprint of the tumor phenotype. This approach has led to the use of genetic analysis in many areas beneficial to patients with CRC, including the potential to predict whether metastasis is likely to have occurred, additional treatment is warranted, and/or response to a certain chemotherapy regimen is optimal. Finally, these analyses are yielding important clues into the pathogenesis nature of drug resistance with the potential to therapeutically alter the refractory nature of the tumor. Although this technique is currently not a replacement for standard clinicopathological evaluations, data gathered in this manner are being used to complement standard evaluation approaches.

The application of microarray technologies on carcinogenesis studies aims to identify specific alterations on gene expression according to tumor development and diagnose and classify tumors based on molecular features. Studies of class comparison between normal mucosa, adenoma and carcinoma or between primary tumor and metastases, as well as between left-side and right-side tumors are performed to discover distinctive genetic signatures belonging to each class (Table 13.4).

Table 13.4 Example of some studies 2005–200 using microarray chip. (Croner et al. 2005; Kwong et al. 2005; Friedrich et al. 2005; Komuro et al. 2005; Birkenkamp-Demtroder et al. 2005; Kita et al. 2005, 2006; Groene et al. 2006; Lin et al. 2007; Ki et al. 2007; Kleivi et al. 2007; Grade et al. 2007; Bianchini et al. 2007; Kim et al. 2008; D'Arrigo et al. 2005; Eschrich et al. 2005; Arango et al. 2005; Barrier et al. 2005, 2006, 2007; Cavalieri et al. 2007; Yamasaki et al. 2007; Shimizu et al. 2005; Ghadimi et al. 2005; Kim et al. 2007; Del Rio and Chalbos 2007; Rimkus et al. 2008)

Years	Authors	Samples	Aim
2005	Croner et al.	30 Carcinomas and 30 normal mucosa	Carcinogenesis process
2005	Kwong et al.	28 primary tumors, 10 normal mucosa, 10 liver metastases	
2005	Friederichs et al.	25 primary tumors and 6 normal mucosa	
2005	Kumuro et al.	89 primary tumors	
2005	Birkenkamp-Demtroder	25 primary tumors	
2006	Kita et al.	12 flat adenomas and 12 normal mucosa	
2006	Groene et al.	36 primary tumors	
2006	Bianchini et al	25 primary tumors, 13 normal mucosa	
2007	Lin et al.	48 primary tumors and 28 liver metastases	
2007	Ki et al.	27 primary tumors, 27 liver metastases	
2007	Kleivi et al.	18 primary tumors and 4 carcinomatoses	
2007	Grade et al.	73 primary tumors and 30 normal mucosa	
2008	Kim et al.	5 serrated adenomas and 5 normal mucosa Adenomas (49% 19/39)	
2005	D'Arringo et al.	10 primary metastatic tumors, 10 primary non-metastatic tumors	Prognosis prediction
2005	Eschrich et al.	75 primary tumors	
2005	Arango et al.	281 primary tumors	
2005	Barrier et al.	12 primary tumors and 12 normal mucosa	
2005	Barrier et al.	18 primary tumors and 18 normal mucosa	
2006	Barrier et al.	50 primary tumors	
2007	Cavalieri et al.	19 primary tumors	
2007	Yamasaki et al.	58 primary tumors 34 liver metastases	
2005	Shimizu et al.	In vitro	Treatment response prediction
2005	Ghadimi et al.	30 advanced rectal cancer	
2007	Kim et al.	31 advanced rectal cancer	
2007	Del Rio et al.	21 primary advanced tumors	
2008	Rimkus et al.	43 advanced rectal cancer	

13.3 Transcriptome-Wide Approach

In a study evaluating 13,023 genes in 11 breast and 11 CRCs the researcher found that approximately 90 mutant genes are accumulated in individual tumors on average (Sjoblom et al. 2006). However, not all these genes contribute to neoplastic processes. By benefiting from certain criteria the researchers identified 189 genes which were mutated frequently, an average of 11 per tumor. It seems that these genes affect many cellular functions such as transcription, adhesion, and invasion. Moreover, most of these genes were not known to be genetically altered in tumors. Therefore, the genetic landscape of breast and CRC is outlined and new targets are provided for diagnostic and therapeutic intervention as well as new insights to tumor biology (Sjoblom et al. 2006). A transcriptome-wide approach was developed by Schuebel and colleagues to recognize genes affected by transcriptional silencing and promoter CpG island DNA hypermethylation in CRC. Cell lines were screened and then tumor-specific hypermethylation were assessed in primary human CRC samples. They found that almost 5% of known genes may be promoter methylated in an individual tumor. Individual tumors experienced larger numbers of gene hypermethylations as well as a hypermethylation was more frequent in individual genes that had genetic or epigenetic changes as shown in Fig. 13.6 (Schuebel et al. 2007).

13.3.1 Epigenetic Alteration is More Important in Cancer Development

The main considerations emerging from these studies are as follows:firstly, genetic modifications paralleling clinical progression develop into the carcinogenesis process. Secondly, most genes involved during the carcinogenesis process are implicated in cell proliferation, migration and adhesion however, there are some genes that have unknown functions and should undergo further genomic and proteomic investigation. Thirdly, although each tumor stage might have specific genetic signatures, more homogeneous studies should be done to ensure a reliable molecular classification useful in clinical practice. And finally, the identification of genes differently expressed during tumor progression may lead to a stage-specific tumor treatment as well as to the discovery of novel therapeutic targets.

Gene expression profiling using microarray technologies has several potential clinical applications in CRC, ranging from the study on mechanisms involved in tumor development, to the identification of gene signatures with prognostic and predictive value, and to the discovery of novel tumor targets.

Standardized methodological and analytical protocols are required for achieving data as homogeneously as possible to be easily compared. Also, novel approaches for isolating integer and non-contaminated tumor tissue, such as laser capture microdissection (LCM) should be investigated. Furthermore, all data should be stored in sharing databases available to every investigator. Also, tissue banks, possibly containing both primary tumor and metastatic specimens, in association with suitable clinical information, may be a solution for the lack of fresh tissue samples available which limit large population studies. Moreover, since biological material are now much more available, serial and different genomic and proteomic analyses are done on the same sample, enabling the global molecular analysis of tissue. Furthermore it could be challenging to investigate the feasibility of microarray analysis from paraffin-embedded tissue in large series because so far few data are available on the sensitivity of this approach (Shacham-Shmueli et al. 2011). For extracting real informative data, small sets of the most significant genes should be selected among the pool of all differentially expressed genes and then retrospectively or prospectively validated by conventional quantitative assays, such as reverse-transcriptase PCR.

13.3.2 Epigenetic Therapy in CRC

Although the body of literature on DNA methylation and its biological functions in mammals is expanding, it is far from complete. For instance, unlike genetic mutations, we know very little about the rate of changes in CpG methylation in mammals and the intrinsic and environmental factors that induce changes in DNA methylation patterns. Accumulating evidence has indicated that changes in DNA methylation and histone modification may contribute to the pathogenesis of many complex diseases. Thus, the modulation of epigenetic states of the genome has the potential to evolve into a new therapeutic approach for the treatment of these diseases.

It was previously shown that the rate of cell death is affected by the DNA fragmentation factor by generating PARP-1-activating DNA breaks. This factor is composed of a caspase-3-activated DNase (CAD) and its inhibitor (ICAD). The researchers investigated whether ICAD-deficient colon epithelial cells accumulate additional genetic modifications making, them resistant to death and ultimately leading to a tumorigenic phenotype. The researchers found a possible association between ICAD deficiency and colon malignancy in humans. ICAD expression levels were considerably compromised in colon cancer tissues compared with normal tissues; and ICAD cells would have tumorigenic phenotype when DNA is damaged by a low dose of irradiation. When death was induced by dimethylhydrazine, a significant resistance was observed in colon epithelial cells derived from ICAD in mice, due to a decrease in PARP-1 activation. ICAD (-/-) mice had significantly more tumors with larger sizes compared with normal wild mice. Since the phenotype of ICAD (-/-) mice was not significantly associated with elevated precancerous aberrant crypt foci, it is suggested that it is linked to tumor progression not tumor initiation. In fact, array comparative genomic hybridization showed an association between ICAD deficiency and severe genomic instability consisting of amplifications as well as sizable deletions that affected several cancer-related genes such as *RAF-1, GSN, LMO3,* and *Fzd6* independently of *p53*. The mentioned results confirm the involvement of ICAD deficiency in colon carcinogenesis. Moreover, this deficiency increases susceptibility to carcinogen-induced tumorigenesis through apoptosis and genomic instability (Errami et al. 2013). Since some types of cancer are resistant to treatment, it might be possible to work on their epigenetic.

13.3.3 Epigenetic Targeting

Epigenetic transcriptional repression has been demonstrated in a wide variety of tumor types and occurs in tumor suppressor genes, Wang et al. 2014, cell-cycle genes, DNA repair genes, and genes involved in invasion and metastasis. For many of these genes, it has been shown that their re-expression in tumor cells can lead to suppression of cell growth or altered sensitivity to existing anticancer therapies. Since compounds have been identified that can readily reverse epigenetic silencing, there is increasing interest in epigenetic regulation of gene expression as a basis for new approaches to cancer treatment (Strathdee and Brown 2002; Marks et al. 2001). Many of these compounds are small molecules that have pharmacological properties that enable easy delivery to tumors. This is inconsistent with the challenge of delivering gene therapy to reverse genetic silencing caused by gene mutation in tumors (Brown and Strathdee 2002).

DNA methylation could lead to gene silencing through MBD proteins that recruit histone methylases (HMTs). When Lys9 residue of histone H3 is methylated and the chromatin silencer HP1 is subsequently recruited, gene repression occurs. HP1 is a family of heterchromatic adaptor molecules implicated in both gene silencing and supra-nucleosomal chromatin structure (Bannister et al. 2001; Lachner et al. 2001; Ashktorab et al. 2009, 2014). Although these observations argue that DNA methylation is a key signal leading to histone modifications, chromatin remodeling and gene silencing, this signaling can also operate in the opposite direction. The disruption of histone methylation can be caused by the dim-5 gene mutation, which encodes a protein homologous to the chromatin-associated protein Suv39h found in mammalian cells. Similarly, increased histone acetylation in cells treated with HDAC inhibitors can also lead to DNA demethylation (Cervoni and Szyf 2001).

Taken together, studies demonstrate the emerging concept that crosstalk between these different mechanisms of epigenetic regulation (DNA methylation, histone methylation, etc.) is essential for appropriate gene transcription control. In order to produce effective epigenetic drugs, strategies should be focused on distinguishing important targets, considering the different epigenetic layers engaging in this complex crosstalk. For example, DNMT inhibitors, such as 2-deoxy-5-azacytidine, appear to be the most active compounds for inducing re-expression of epigenetically silenced genes in tumor cell models. However, HDAC inhibitors can increase levels of gene expression and work together with DNMT inhibitors to induce gene re-expression (Cameron et al. 1999; Huang et al. 2014; Venkateswaran 2014).

13.3.4 Epigenetic Chemotherapeutic Targets

Two important classes of chemical compounds, inhibitors of epigenetic enzymes, including:1) DNMTIs and 2) HDACIs have undergone major preclinical investigation and clinical development to tackle mechanisms of tumor progression and resistance.

Nucleoside DNMTIs comprise 5-aza-cytidine, 5-aza-2-deoxycytidine, 5-fluoro-2- deoxycytidine and zebularine. Non-nucleoside DNMTI comprises small molecule inhibitors such as RG108. Third classes of DNTMIs are oligonucleotides such as MG98 (Crea et al. 2011). Other type of epigenetic chemotherapeutic targets is antisense therapy. Nucleoside DNMTIs are incorporated into DNA and prevent the resolution of a covalent reaction intermediate which leads to DNMT being trapped and inactivated in the form of a covalent protein-DNA adduct. Thus, cellular DNMT is depleted and genomic DNA is demethylated because of continued DNA replication (Lyko and Brown 2005). DNMT activity is blocked by non-nucleoside DNMTIs through binding to the catalytic region of DNMTs. MG98, a specific oligonucleotide compound, can suppress DNMT expression by antisense mechanisms that are being currently studied (Crea et al. 2011).

Subclass of HDAC:Seven classes of HDACIs have been developed so far. Four of them are currently investigated in the clinic:short-chain fatty acids, cyclic peptides, hydroxamic acids, and benzamides. Despite their structural diversity, they all act by inhibiting HDACs identified in humans. When these enzymes are inhibited acetylation in histones occurs and accumulates. This event will be then followed by changes in cellular processes that are defective in cancer (Crea et al. 2011).

The increased methylation patterns of CpG islands observed in tumor cells are only rarely detected in normal cells. Methylation of genes in tumor cells could provide a tumor-specific target for new therapies. Most studies have focused on therapies that reverse methylation as a means of switching on genes that will suppress tumor growth or modify sensitivity to existing therapies (Strathdee and Brown 2002; Shiovitz et al. 2014; Kisiel et al. 2014; Costello and Plass 2001). Proof-of-principle experiments have shown that when these genes are re-expressed by gene re-introduction, tumor growth is suppressed or sensitized to existing therapies. The lack of methylation of such genes in normal cells provides the potential for tumor specificity. Whether inhibiting DNA methylation would be toxic for normal cells is question with respect to developing DNMT inhibitors to be used in patients. DNA methylation is clearly important during development. Homozygous loss of any of the three known mammalian DNA methyltransferases (*DNMT1,* 3a and 3b) is lethal in mice, and DNMT activities during embryogenesis are probably vital for establishing the correct gene expression pattern (Li et al. 1992; Okano et al. 1999). However, adult tissue might require much lower DNA methylation, with its primary role being maintenance of the bulk, non-coding portion of the genome in a transcriptionally inactive state, effectively increasing the specificity of transcription factors for their target sites within genes. Indeed, combined genetic and pharmacological reduction of DNMT in adult mice had no significant toxicity, and yet reduced the levels of ApcMin- induced intestinal neoplasia (Laird et al. 1995). It

Table 13.5 Nucleoside analog inhibitors of DNA methylation, and inhibitors of histone deacethylation

Compound	Cancer type
DNA methylation inhibitors	
5-Azacytidine 5-Aza-CR Vidaza	MDS; Hematologic malignancies
5-Aza-2-deoxytidine 5-Aza-CdR Dacogen	MDS; Hematologic malignancies
Zebularine 1-b-D-ribofuranosyl-2 (1H)-pyrimidinone	N/A
HISTONE deacetylase inhibitors	
4-Phenylbutyrate (PBA)	Refractory solid tumors
Suberoylanilide hydroxamic acid (SAHA)	Solid tumors and hematologic malignancies
NVP-LAQ824	N/A
Depsipeptide FK-228 FR901228	Advanced neoplasms, CLL, AML, and T-cell lymphoma
MS-275	Solid tumors and lymphoma

has been argued that this is because of the reduced levels of 5-methylcytosine in the DNA of these mice, leading to a reduced frequency of gene mutation because of the disproportionately high mutation rate of 5-methylcytosine residues. Therefore, the inhibition of DNMT activity in somatic cells might be chemo-preventative instead of toxic or carcinogenic. Table 13.5 shows the summary of clinical trials according to epigenetic targeting.

13.3.5 Clinical Implications and Applications

Few effective inhibitors of DNMTs are currently known. 5-azacytidine and 2-deoxy-5-azacytidine (also known as Decitabine) are two closely related drugs that have been used for inhibiting DNA methylation in tissue culture and reactivating various genes that were silenced by methylation (Jones 1985; Rajaii et al. 2014). Decitabine induces cell differentiation and has been used in clinical trials for the treatment of several haematopoietic disorders (Pinto and Zagonel 1993). Although its use in the activation of genes silenced by methylation is limited by its toxicity prolonged low-dose schedules (Pinto and Zagonel 1993) or combinations with other drugs could overcome these limitations (Plumb et al. 2000). For example, in mouse xenograft models, treatment with relatively low doses of decitabine can reactivate *MLH1*; a methylation-silenced gene (Plumb et al. 2000). The *MLH1* protein is important in determining the sensitivity to several important chemotherapeutic agents.

Increased sensitivity to carboplatin, temozolomide, and epirubicin was observed in treated xenografts. Therefore, combining epigenetic drugs with existing therapies is clinically promising. In cells that have been re-expressed as a result of treatment with DNMT inhibitors (i.e. 5-azacytidine), the occurrence of epigenetic silencing is possible. Therefore, over time tumor cells would die due to the epigenetic reversal of silencing of tumor suppressor genes. This demethylation can be used for appropriate scheduling of a cytotoxic or cytokine whose anti-tumor effects are being resisted due to gene methylation.

HDAC activity is important in the transcriptional repression of methylated sequences. Cameron et al. found that the combination of decitabine and an inhibitor of HDAC, trichostatin A, caused a synergistic reactivation of *MLH1* and *TIMP3* gene expression in the CRC cell line RKO (Cameron et al. 1999). A phase I clinical trial aimed at assessing the clinical potential of this synergistic interaction has been initiated using the combination of 5-azacytidine and another HDAC inhibitor, phenyl butyrate (Crea et al. 2011).

The use of antisense oligonucleotides also inhibits DNA methylation. Antisense oligonucleotides directed against the *DNMT1* mRNA reduce *DNMT1* protein levels and induce demethylation and expression of the *p16* tumor suppressor gene in human tumor cells. It also inhibits tumor growth in mouse models. This *DNMT1* antisense molecule has also been used in phase I and II clinical trials.

Several HDAC inhibitors suppress growth of tumors in animal models (Marks et al. 2001; Yan et al. 2000; Wagner et al. 2014). No toxicity has been observed at doses that inhibit tumor growth. At least four HDAC inhibitors (phenylbutyrate, hydroxamic acid-based HDAC inhibitors, SAHA and pyroxamide, and the cyclic tetrapeptide FR901228) have been mentioned in clinical trials as cancer therapeutics. CI-994 (N-acetyldinaline), which is also in phase I clinical trial, inhibits histone deacetylation, but apparently not through the direct inhibition of HDACs.

Epigenetic therapies have shown relevant activity in the treatment of hematological malignancies leading to the approval of some drugs. Minor efficacy has been reported in solid tumors, despite the high number of clinical trials performed in the past 25 years. Currently, it is believed that the reason for the poor success in solid tumors is mainly because of the use of high doses and short term administration of epigenetic drugs. In fact, DNMTIs exert their epigenetic clinical activity in myelodysplastic syndromes when administered at low doses, for several successive days and for multiple cycles (Issa 2005; Oki et al. 2007). These conditions allow the survival of the cells but determine changes in their gene expression profile, thus favoring cell differentiation, decrease in cell proliferation and increased apoptosis (Jones and Taylor 1980). Instead, at high doses, DNMTIs induce cytotoxic effects. Under these circumstances, suspension of DNA synthesis through antimetabolic activity rather than epigenetic effects inhibits cell growth. Most of the trials performed with epigenetic drugs as single- agents in solid tumors were aimed at investigating the maximum tolerated dose (MTD) in previously treated patients with advanced disease rather than investigating their epigenetic effects. Results of such trials usually showed high grade toxicity associated with low grade or lack of anticancer activity both for DNMTIs (van Groeningen et al. 1986; Abele et al. 1987; Newman

Table 13.6 Epigenetic drugs (Crea et al. 2011)

Agent	Class	Combined drugs
5-azacytidine 5-aza-2-deoxycycytidine 5-fluoro-2-deoxycytidine	Nucleoside DNMTIs Nucleoside DNMTIs Nucleoside DNMTIs	None or Valproic acid or Sodium Phenylbutyrate None or Carboplatin Tetrahydrouridine
Zebularine RG108 MG98 Romidepsin Valproic acid	Nucleoside DNMTIs Non-Nucleoside DNMTIs Antisense oligonucleotide Cyclic peptides HDACIs Aliphatic peptides HDACIs	Oxalipalatin None None None None or Epirubicin
Vorinistat	Aliphatic peptides HDACIs	Doxorubicin or 5-FU or 5-FU and Oxaliplatin
Sodium phenylbutyrate Etinostat	Aliphatic peptides HDACIs Benzamides HDACIs	None or 5-FU None
Mocetinostat Belinostat Panobinostat	Benzamides HDACIs Hydroxamic acids HDACIs Hydroxamic acids HDACIs	None None or 5-FU None or Epirubicin

et al. 2002) and HDACIs (Vansteenkiste et al. 2008) On the other hand, it has been shown that, the administration of low doses of DNMTIs may lead to the reactivation of methylated tumor suppressor genes in solid tumors as well (Schrump et al. 2006). However, this effect is transient and may be reversed when the administration of the drug(s) is suspended. This may hamper the use of these drugs as single agents in solid tumors. Similar observations have been reported for HDACIs (Wu et al. 2001; Kelly et al. 2005; Prince et al. 2009). Table 13.6 shows some types of epigenetic drugs (Crea et al. 2011).

13.4 CRC Detection in Serum or Stool

CRC screening is the most efficient strategy for reducing the mortality rate of this disease. As the gold standard, colonoscopy is highly sensitive in detecting and removing early lesions; however, it is also invasive and expensive (Frazier et al. 2000; Rabeneck et al. 2008). Simple and noninvasive methods such as stool testing are better for population-wide screening (De Visser et al. 2005; Ahlquist et al. 2014). In follow-up to the guaiac-based Faecal Occult Blood Test (FOBT), the more sensitive immunochemical fecal occult blood test (Fecal Immunochemical Test or FIT) (Oort et al. 2010; Hol et al. 2009; Van Rossum et al. 2008) is now widely used in screening programs in Europe and Japan. It is expected to reduce the mortality rate due to CRC by around 30% (van Veen and Mali 2009; Zeller et al. 2014).

FOBT and FIT do not detect most advanced adenomas (Imperiale et al. 2004). This testing method still needs to improve probably by molecular stool tests like those testing done for tumor DNA in stool. For this purpose, multiple assays have been developed, although having suboptimal sensitivity (Ahlquist et al. 2000; Tagore et al. 2003; Whitney et al. 2004; Kutzner et al. 2005; Ahlquist et al. 2008; Bosch et al. 2011). Improved testing has been achieved through the combination of mutation markers with DNA methylation (Ahlquist et al. 2008; Itzkowitz et al. 2008; Itzkowitz et al. 2007; Lidgard et al. 2013).

With the American Cancer Society goal of screening 75 % of guideline-eligible people by 2015, it is important to develop an accurate, noninvasive, early detection method (Levin et al. 2008). A noninvasive procedure would increase adherence with CRC screening guidelines and reduce the number of patients reluctant to be screened.

It is necessary to develop early detection and preventive strategies that benefit from biomarkers assigning patients to suitable screening or surveillance programs. Discovery of novel markers that are highly specific and sensitive will also improve strategies for the management of cancer by facilitating the rapid determination of tumor responses to novel therapies. A few molecular tumor markers are already being used in clinical settings (Sidransky 2002). However, limited sensitivity and specificity for diagnosis or prognosis have restricted the wide application of conventional and/or newly developed markers in clinical practice (Smith et al. 1999).

As mentioned, in order to design both diagnostic and therapeutic strategies, understanding the molecular basis of CRC is of utmost importance. In this regard, CIMP screening is most relevant, as well as the most recently identified carcinogenesis pathway, as is the screening for specific CpG island methylations in the genes associated with colorectal carcinogenesis (Veganzones-de-Castro et al. 2012; Ashktorab et al. 2014).

13.4.1 Methylated Marker in Serum for Cancer Detection

The fact that focal hypermethylation of CpG islands is very common in cancer cells, coupled with the ability to detect methylation with a high degree of sensitivity, has led to the development of several approaches for the detection of cancer in body fluids. Acquired changes in CpG island methylation can be detected in a background of normal cells following conversion of cytosines to uracil yet leaving 5-methylcytosine intact in DNA treated with sodium bisulfate. PCR approaches such as methylation-specific PCR, in which primers are designed to amplify only methylated regions, are very sensitive. Other methods include techniques based on real-time PCR such as "MethyLight," where a fluorescent probe that can only bind to methylated DNA is used to detect methylation patterns. These techniques can detect one methylated allele in a background of about 1000–10,000 alleles. Thus, the acquisition of an abnormal methylation pattern can be easily detected; these approaches are applicable to mixtures of cells or even various biological fluids such as plasma, urine, or sputum. Cancer detection through the identification of altered cytosine methylation is quite robust because of the inherent stability of DNA compared to RNA or proteins. Also, since altered methylation patterns are often cancer specific, these approaches may be able to distinguish between different types of cancer.

Methylated DNA is found in several body fluids such as stool and blood. Therefore, using DNA methylation as a marker for CRC detection and screening is a very attractive strategy. Methylation markers, alone or combined, have also yielded promising results (Glockner et al. 2009; Hellebrekers et al. 2009; Huang et al. 2007; Kim et al. 2009; Leung et al. 2007; Melotte et al. 2009; Wang and Tang 2008; Zhang et al. 2014). In CRC, since DNA methylation occurs in early disease stages, methylation markers are important (Derks et al. 2006).

Recently, researchers have focused on cell-free methylated DNA based biomarkers in serum or stool (Philipp et al. 2012; Bosch et al. 2012; Hong and Ahuja 2013; Summers et al. 2013; Zhai et al. 2012). DNA shed into serum or stool provides a more accurate sampling of abnormal cells than random punch biopsies among patients with IBD-associated cancer. Stool DNA testing for *SFRP-1* promoter hypermethylation is a sensitive and specific screening tool for sporadic CRC (Zhang et al. 2007; Guo et al. 2011; Saito et al. 2014).

Several studies also suggest that methylated DNA released in the circulation could be used as a prognostic marker for early tumor detection. The presence of aberrantly methylated genes such as *SEPT9, HLTF,* and *HPP1* DNA in plasma is highly correlated with the occurrence of CRC and tumor size, stage, grade, and metastatic disease, respectively (Tetzner et al. 2009; Lange et al. 2012; Herbst et al. 2009). *APC* gene-promoter sequences were unmethylated in 88 % in patients after obtaining peripheral blood DNA, of which 66 % had polyps and 33 % were polyp-free using their blood DNA. This sequence may be an indicator of risk for polyp formation and an important screening tool (Ashktorab et al. 2007 and 2014).

Mokarram et al showed that if the findings on the methylation of *MGMT*-B and *SFRP2* are validated in future studies, they can be used as serum or stool-based DNA methylation tools for early detection of patients with IBD before the occurrence of cancer (data not published). Studies on molecular markers that detect genetic and epigenetic modifications in human cancers have provided new insights in cancer detection approaches (Ashktorab et al. 2014). Novel epigenetic markers, alongside user friendly and sensitive assay methods, will consequently improve the detection, treatment, and overall prognosis of cancer malignancy. Differences in methylation patterns among tumors may be correlated with clinical features of patients and can serve as markers in cancer classification.

References

Abajo A, Bitarte N, Zarate R, Boni V, Lopez I, Rodriguez J, Bandres E, Garcia-Foncillas J (2012) Identification of colorectal cancer metastasis markers by an angiogenesis-related cytokine-antibody array. World J Gastroenterol 18(7):637–645

Abele R, Clavel M, Dodion P, Bruntsch U, Gundersen S, Smyth J et al (1987) The EORTC early clinical trials cooperative group experience with 5-aza-2′-deoxycytidine (NSC 127716) in patients with colorectal, head and neck, renal carcinomas and malignant melanomas. Eur J Cancer Clin Oncol 23:1921–1924

Abou-zeid AA, Khafagy W, Marzouk DM, Ahmed Alaa M (2002) Colorectal cancer in Egypt. Dis Col Rectum 45:1255–1260

Ackerman SL, Kozak LP, Przyborski SA, Rund LA, Boyer BB, Knowles BB (1997) The mouse rostral cerebellar malformation gene encodes an UNC-5-like protein. Nature 24:838–842

Adekunle O, Ajao G (1986) Colorectal cancer in adolescent Nigerians. Scand J Gastroenterol 21:183–186

Ahlquist DA (2014) Multi-target stool DNA test: a new high bar for noninvasive screening. Dig Dis Sci. Dec 10

Ahlquist DA, Skoletsky JE, Boynton KA, Harrington JJ, Mahoney DW, Pierceall WE et al (2000) Colorectal cancer screening by detection of altered human DNA in stool: feasibility of a multi-target assay panel. Gastroenterology 119:1219–1227

Ahlquist DA, Sargent DJ, Loprinzi CL, Levin TR, Rex DK, Ahnen DJ et al (2008) Stool DNA and occult blood testing for screen detection of colorectal neoplasia. Ann Intern Med 149:441–450

Ahlquist T, Lind GE, Costa VL, Meling GI, Vatn M, Hoff GS et al (2008) Gene methylation profiles of normal mucosa, and benign and malignant colorectal tumors identify early onset markers. Mol Cancer 7:94

Ahmed FE (2007) Colorectal cancer epigenetics: the role of environmental factors and the search for molecular biomarkers. J Env Sci Health C 25:101–154

Akino K, Toyota M, Suzuki H, Mita H, Sasaki Y, Ohe-Toyota M et al (2005) The Ras effector RASSF2 is a novel tumor-suppressor gene in human colorectal cancer. Gastroenterology 129:156–169

Akkiprik M, Ataizi-Chelikel C, Ducunceli F, Sonmez O, Gulluodlu BM, Sav A et al (2007) Clinical significance of *p53*, *K-ras* and *DCC* gene alterations in the stage I-II colorectal cancers. J Gastrointestin Liver Dis 16:11

Alireza S, Mehdi N, Ali M (2005) Cancer occurrence in Iran in 2002, an international perspective. Asian Pac J Cancer Prev 6:359

Anastas JN, Nastas JN, Moon RT (2012) WNT signaling pathways as therapeutic targets in cancer. Nat Rev Cancer 13:11–26

Anderson CA, Massey DC, Barrett JC, Prescott NJ, Tremelling M, Fisher S et al (2009) Investigation of Crohn's disease risk loci in ulcerative colitis further defines their molecular relationship. Gastroenterology 136:523–529

Anwar S, Frayling I, Scott N, Carlson G (2004) Systematic review of genetic influences on the prognosis of colorectal cancer. Br J Surg 91:1275–1291

Arango D, Wilson A, Shi Q, Corner G, Aranes M, Nicholas C et al (2004) Molecular mechanisms of action and prediction of response to oxaliplatin in colorectal cancer cells. Br J Cancer 91:1931–1946

Arango D, Laiho P, Kokko A, Alhopuro P, Sammalkorpi H, Salovaara R et al (2005) Gene-expression profiling predicts recurrence in Dukes' C colorectal cancer. Gastroenterology 129:874–884

Asangani IA, Ateeq B, Cao Q, Dodson L, Pandhi M, Kunju LP et al (2012) Characterization of the EZH2-MMSET histone methyltransferase regulatory axis in cancer. Mol Cell 49:80–93

Ashktorab H, Smoot DT, Farzanmehr H, Fidelia-Lambert M, Momen B, Hylind L et al (2005) Clinicopathological features and microsatellite instability (MSI) in colorectal cancers from African Americans. Int J Cancer 116:914–919

Ashktorab H, Begum R, Akhgar A, Smoot D, Elbedawi M, Daremipouran M et al (2007) Folate status and risk of colorectal polyps in African Americans. Dig Dis Sci 52:1462–1470

Ashktorab H, Belgrave K, Hosseinkhah F, Brim H, Nouraie M, Takkikto M et al (2009) Global histone H4 acetylation and HDAC2 expression in colon adenoma and carcinoma. Dig Dis Sci 54:2109–2117

Ashktorab H, Schaffer AA, Daremipouran M, Smoot DT, Lee E, Brim H (2010) Distinct genetic alterations in colorectal cancer. PLoS ONE 5:e8879

Ashktorab H, Nguza B, Fatemi M, Nouraie M, Smoot DT, Schaffer AA et al (2011) Case-control study of vitamin D, dickkopf homolog 1 (*DKK1*) gene methylation, VDR gene polymorphism and the risk of colon adenoma in African Americans. PloS ONE 6:e25314

Ashktorab H, Rahi H, Wansley D, Varma S, Shokrani B, Lee E et al (2013) Toward a comprehensive and systematic methylome signature in colorectal cancers. Epigenetics 8:807–815
Ashktorab H, Daremipouran M, Goel A, Varma S, Leavitt R, Sun X et al (2014) DNA methylome profiling identifies novel methylated genes in African American patients with colorectal neoplasia. Epigenetics 9:1–10
Ausch C, Kim YH, Tsuchiya KD, Dzieciatkowski S, Washington MK, Paraskeva C et al (2009) Comparative analysis of PCR-based biomarker assay methods for colorectal polyp detection from fecal DNA. Clin Chem 55:1559–1563
Aust DE, Haase M, Dobryden L, Markwarth A, Lohrs U, Wittekind C et al (2005) Mutations of the *BRAF* gene in ulcerative colitis-related colorectal carcinoma. Int J Cancer 115:673–677
Azarschab P, Porschen R, Gregor M, Blin N, Holzmann K (2002) Epigenetic control of the E-cadherin gene (CDH1) by CpG methylation in colectomy samples of patients with ulcerative colitis. Genes Chromosomes Cancer 35:121–126
Bagchi A, Papazoglu C, Wu Y, Capurso D, Brodt M, Francis D et al (2007) *CHD5* Is a tumor suppressor at human 1p36. Cell 128:459–475
Bandres E, Catalan V, Sola I, Honorato B, Cubedo E, Cordeo E et al (2004) Dysregulation of apoptosis is a major mechanism in the lymph node involvement in colorectal carcinoma. Oncol Rep 12:287–292
Banerjee R, Mani RS, Russo N, Scanlon CS, Tsodikov A, Jing X et al (2011) The tumor suppressor gene rap1GAP is silenced by miR-101-mediated EZH2 overexpression in invasive squamous cell carcinoma. Oncogene 30:4339–4349
Bannister AJ, Zegerman P, Partridge JF, Miska EA, Thomas JO, Allshire R et al (2001) Selective recognition of methylated lysine 9 on histone H3 by the HP1 chromo domain. Nature 410:120–124
Barbacid M (1987) Ras genes. Ann Rev Biochem 56:779–827
Barker N, Clevers H (2006) Mining the Wnt pathway for cancer therapeutics. Nat Rev Drug Discov 5:997–1014
Barrett JC, Hansoul S, Nicolae DL, Cho JH, Duerr RH, Rioux JD et al (2008) Genome-wide association defines more than 30 distinct susceptibility loci for Crohn's disease. Nat Genet 40:955–962
Barrier A, Boelle PY, Lemoine A, Tse C, Brault D, Chiappini F et al (2005) Gene expression profiling of nonneoplastic mucosa may predict clinical outcome of colon cancer patients. Dis Colon Rectum 48:2238–2248
Barrier A, Lemoine A, Boelle PY, Tse C, Brault D, Chiappini F et al (2005) Colon cancer prognosis prediction by gene expression profiling. Oncogene 24:6155–6164
Barrier A, Boelle PY, Roser F, Gregg J, TSE C, Brault D et al (2006a) Stage II colon cancer prognosis prediction by tumor gene expression profiling. J Clin Oncol 24:4685–4691
Barrier A, Roser F, Boelle P, Franc B, Tse C, Brault D et al (2006) Prognosis of stage II colon cancer by non-neoplastic mucosa gene expression profiling. Oncogene 26:2642–2648
Barski A, Cuddapah S, Cui K, Roh TY, Schones DE, Wang Z et al (2007) High-resolution profiling of histone methylations in the human genome. Cell 129:823–837
Baylin SB (2012) The cancer epigenome: its origins, contributions to tumorigenesis, and translational implications. Proc Am Thorac Soc 9(2):64–65
Baylin SB, Jones PA (2011) A decade of exploring the cancer epigenome-biological and translational implications. Nat Rev Cancer 11:726–734
Baylin SB, Ohm JE (2006) Epigenetic gene silencing in cancer-a mechanism for early oncogenic pathway addiction? Nat Rev Cancer 6:107–116
Baylin SB, Herman JG, Graff JR, Vertino PM, Issa JP (1997) Alterations in DNA methylation: a fundamental aspect of neoplasia. Adv Cancer Res 72:141–196
Beck D, Bonasio R, Kaneko S, Li G, Margueron R, Oda H et al (2011) Chromatin in the nuclear landscape. In: Cold Spring Harbor Symposia on Quantitative Biology, 2011. Cold Spring Harbor Laboratory Press
Bedford MT, Van Helden PD (1987) Hypomethylation of DNA in pathological conditions of the human prostate. Cancer Res 47:5274–5276

Belshaw NJ, Elliott GO, Williams EA, Bradburn DM, Mills SJ et al (2004) Use of DNA from human stools to detect aberrant CpG island methylation of genes implicated in colorectal cancer. Cancer Epidemiol Biomarkers Prev 13:1495–1501

Beroud C, Soussi T (2003) The UMD-*p53* database: new mutations and analysis tools. Hum Mutat 21:176–181

Bianchini M, Levy E, Zucchhini C, Pinski V, Macagno C, De Sanctis P et al (2006) Comparative study of gene expression by cDNA microarray in human colorectal cancer tissues and normal mucosa. Int J Oncol 29:83–94

Bird A (2002) DNA methylation patterns and epigenetic memory. Genes Dev 16:6–21

Birkenkamp-Demtroder K, Olesen SH, Sorensen FB, Laurberg S, Laiho P, Aaltonen LA et al (2005) Differential gene expression in colon cancer of the caecum versus the sigmoid and rectosigmoid. Gut 54:374–384

Bishehsari F, Mahdavinia M, Malekzadeh R, Verginelli F, Catalano T, Sotoudeh M et al (2006) Patterns of *K-ras* mutation in colorectal carcinomas from Iran and Italy (a Gruppo Oncologico dell'Italia Meridionale study): influence of microsatellite instability status and country of origin. Ann Oncol 17:91–96

Bock C, Walter J, Paulsen M, Lengauer T (2007) CpG island mapping by epigenome prediction. PLoS Comput Biol 3:e110

Bosch LJ, Carvalho B, Fijneman RJ, Jimenez CR, Pinedo HM, Van Engeland M et al (2011) Molecular tests for colorectal cancer screening. Clin Colorectal Cancer 10:8–23

Bosch LJ, Mongera S, Sive Droste JST, Oort FA, Van Turenhout ST, Penning MT et al (2012) Analytical sensitivity and stability of DNA methylation testing in stool samples for colorectal cancer detection. Cell Oncol 35:309–315

Boumber YA, Kondo Y, Chen X, Shen L, Gharibyan V, Konishi K et al (2007) RIL, a LIM gene on 5q31, is silenced by methylation in cancer and sensitizes cancer cells to apoptosis. Cancer Res 67:1997–2005

Bournier O, Kroviarski Y, Rotter B, Nicolas G, Lecomte MC, Dhermy D (2006) Spectrin interacts with *EVL* (Enabled/vasodilator-stimulated phosphoprotein-like protein), a protein involved in actin polymerization. Biol Cell 98:279–293

Bouzourene H, Gervaz P, Cerottini JP, Benhattar J, Chaubert P, Saraga E et al (2000) *p53* and Ki- ras as prognostic factors for Dukes' stage B colorectal cancer. Eur J Cancer 36:1008–1015

Boyer J, Allen WL, Mclean EG, Wilson PM, Mcculla A, Moore S et al (2006) Pharmacogenomic identification of novel determinants of response to chemotherapy in colon cancer. Cancer Res 66:2765–2777

Brandes JC, Van Engeland M, Wouter KA, Weijenberg MP, Herman JG (2005) CHFR promoter hypermethylation in colon cancer correlates with the microsatellite instability phenotype. Carcinogenesis 26:1152–1156

Bressan FF, Therrien J, Filion F, Perecin F, Smith LC, Meirelles FV (2014) 331 abnormal DNA methylation patterns and allele-specific expression of imprinted genes in bovine-induced pluripotent stem cells. Reprod Fertil Dev 27(1):254

Brim H, Mokarram P, Naghibalhossaini F, Saberi-Firoozi M, Al-Mandhari M, Al-Mawaly K et al (2008) Impact of *BRAF, MLH1* on the incidence of microsatellite instability high colorectal cancer in populations based study. Mol Cancer 7:68

Brim H, Lee E, Abu-Asab MS, Chaouchi M, Razjouyan H, Namin H et al (2012) Genomic aberrations in an African American colorectal cancer cohort reveals a MSI-specific profile and Chromosome X amplification in male patients. PloS ONE 7:e40392

Brim H, Abu-Asab MS, Nouraie M, Salazar J, Deleo J, Razjouyan H et al (2014) An integrative CGH, MSI and candidate genes methylation analysis of colorectal tumors. PloS ONE 9:e82185

Brown R, Strathdee G (2002) Epigenomics and epigenetic therapy of cancer. Trends Mol Med 8:43–48

Bufill JA (1990) Colorectal cancer: evidence for distinct genetic categories based on proximal or distal tumor location. Ann Intern Med 113:779–788

Burness M, Xu H, Beresnev T, Pingpank J (2007) Site-specific gene expression profiles and novel molecular prognostic factors in patients with lower gastrointestinal adenocarcinoma diffusely metastatic to liver or peritoneum. Ann Surgical Oncol 14:3460–3471

Burton PR, Clayton DG, Cardon LR, Craddock N, Deloukas P, Duncanson A et al (2007) Genome-wide association study of 14,000 cases of seven common diseases and 3000 shared controls. Nature 447:661–678

Cai C, Ashktorab H, Pang X, Zhao Y, Sha W et al (2012) MicroRNA-211 expression promotes colorectal cancer cell growth in vitro and in vivo by targeting tumor suppressor *CHD5*. PLoS ONE 7:e29750

Calistri D, Rengucci C, Seymour I, Lattuneddu A, Monti F, Saragoni L et al (2005) Mutation analysis of *p53*, *K-ras*, and *BRAF* genes in colorectal cancer progression. J Cell Physiol 204:484–488

Cameron EE, Bachman KE, Myohanen S, Herman JG (1999) Synergy of demethylation and histone deacetylase inhibition in the re-expression of genes silenced in cancer. Nat Genet 21:103–107

Campos EI, Reinberg D (2009) Histones: annotating chromatin. Annu Rev Genet 43:559–599

Caporali S, Levati L, Graziani G, Muzi A, Grazia AM et al (2012) NF-κB is activated in response to temozolomide in an AKT-dependent manner and confers protection against the growth suppressive effect of the drug. J Transl Med 10:252

Castells A, Quintero E (2014) Programmatic screening for colorectal cancer: the COLONPREV study. Dig Dis Sci. doi:10.1007/s10620-014-3446-2

Cavalieri D, Dolara P, Mini E, Luceri C, Castagnini C, Toti S et al (2007) Analysis of gene expression profiles reveals novel correlations with the clinical course of colorectal cancer. Oncol Res 16:535–548

Cervoni N, Szyf M (2001) Demethylase activity is directed by histone acetylation. J Biol Chem 276:40778–40787

Chambers W, Warren B, Jewell D (2005) Cancer surveillance in ulcerative colitis. Br J Surg 92:928–936

Chan AO, Soliman AS, Zhang Q, Rashid A, Bedeir A, Houlihan PS et al (2005) Differing DNA methylation patterns and gene mutation frequencies in colorectal carcinomas from Middle Eastern countries. Clin Cancer Res 11:8281–8287

Chen WD, Han ZJ, Skoletsky J, Olson J, Sah J, Myeroff L et al (2005) Detection in fecal DNA of colon cancer-specific methylation of the nonexpressed vimentin gene. J Natl Cancer Inst 97:1124–1132

Cho CY, Wang JH, Chang HC, Chang CK, Hung WC (2007) Epigenetic inactivation of the metastasis suppressor RECK enhances invasion of human colon cancer cells. J Cell Physiol 213:65–69

Coghlin C, Murray G (2014) Biomarkers of colorectal cancer: recent advances and future challenges. Proteomics Clin Appl

Colussi D, Brandi G, Bazzoli F, Ricciardiello L (2013) Molecular pathways involved in colorectal cancer: implications for disease behavior and prevention. Int J Mol Sci 14(8):16365–16385

Conlin A, Smith G, Carey FA, Wolf CR, Steele RJ (2005) The prognostic significance of *K-ras*, *p53*, and *APC* mutations in colorectal carcinoma. Gut 54:1283–1286

Cooke J, Zhang H, Greger L, Silva AL, Massey D, Dawson C et al (2012) Mucosal genome-wide methylation changes in inflammatory bowel disease. Inflamm Bowel Dis 18:2128–2137

Costello JF, Plass C (2001) Methylation matters. J Med Genet 38:285–303

Cowley SM, Iritani BM, Mendrysa SM, Xu T, Cheng PF et al (2005) The mSin3A chromatin-modifying complex is essential for embryogenesis and T-cell development. Mol Cell Biol 25:6990–7004

Crea F, Nobili S, Paolicchi E, Perrone G, Napoli C, Landini I et al (2011) Epigenetics and chemoresistance in colorectal cancer: an opportunity for treatment tailoring and novel therapeutic strategies. Drug Resist Updat 14:280–296

Croner RS, Peters A, Brueckl WM, Matzel KE, Klein-Hitpass L, Brabletz T et al (2005) Microarray versus conventional prediction of lymph node metastasis in colorectal carcinoma. Cancer 104:395–404

Croner RS, Schellerer V, Demund H, Schildberg C, Papadopulos T, Naschberger E, Stürzl M, Matzel KE, Hohenberger W, Schlabrakowski A (2010) One step nucleic acid amplification (OSNA) - a new method for lymph node staging in colorectal carcinomas. J Transl Med 8:83

Curradi M, Izzo A, Badaracco G, Landsberger N (2002) Molecular mechanisms of gene silencing mediated by DNA methylation. Mol Cell Biol 22:3157–3173

Curtin K, Samowitz WS, Wolff RK, Caan BJ, Ulrich CM, Potter JD et al (2009) MSH6 G39E polymorphism and CpG island methylator phenotype in colon cancer. Mol Carcinog 48:989–994

D'Arrigo A, Belluco C, Ambrosi A, Digito M, Esposito G, Bertola A et al (2005) Metastatic transcriptional pattern revealed by gene expression profiling in primary colorectal carcinoma. Int J Cancer 115:256–262

Dajani YF, Zayid I, Malatjalian DA, Kamal MF (1980) Colorectal cancer in Jordan and Nova Scotia a comparative epidemiologic and histopathologic study. Cancer 46:420–426

Dashwood R (1999) Early detection and protection of colorectal cancer. Oncol Rep 6:227–281

De Divitiis C, Nasti G, Montano M, Fisichella R, Iaffaioli RV, Berretta M (2014) Prognostic and predictive response factors in colorectal cancer patients: between hope and reality. World J Gastroenterol 20(41):15049–15059

De Visser M, Van Ballegoojien M, Bloemers S, Van Deventer, Jansen J, Jespersen J, Kluft C et al (2005) Report on the Dutch consensus development meeting for implementation and further development of population screening for colorectal cancer based on FOBT. Anal Cell Pathol 27:17–29

Del Rio MM, Chalbos P et al (2007) Gene expression signature in advanced colorectal cancer patients select drugs and response for the use of leucovorin, fluorouracil, and irinotecan. J Clin Oncol 25:773–780

Derks S, Postma C, Moerker PT, Van Den Bosch SM, Carvalho B, Hermsen MA et al (2006) Promoter methylation precedes chromosomal alterations in colorectal cancer development. Anal Cell Pathol 28:247–257

Dhir M, Montgomery EA, Glockner SC, Schuebel KE, Hooker CM, Herman JG et al (2008) Epigenetic regulation of WNT signaling pathway genes in inflammatory bowel disease (IBD) associated neoplasia. J Gastrointestin Surg 12:1745–1753

Dhir M, Yachida S, Neste LV, Glockner SC, Jeschke J, Pappou EP et al (2011) Sessile serrated adenomas and classical adenomas: an epigenetic perspective on premalignant neoplastic lesions of the gastrointestinal tract. Int J Cancer 129:1889–1898

Distler P, Holt P (1997) Are right-and left-sided colon neoplasms distinct tumors? Dig Dis 15:302–311

Division VAH (2006) Vital Statistics 2006. In: Statstics and information department, MSS Ministry of Health Labour and Welfare (ed)

Duerr RH, Taylor KD, Brant SR, Rioux JD, Silverberg MS, Daly MJ et al (2006) A genome-wide association study identifies IL23R as an inflammatory bowel disease gene. Science 314:1461–1463

Ebert MP, Mooney SH, Tonnes-Priddy L, Lograsso J, Hoffmann J, Chen J et al (2005) Hypermethylation of the TPEF/*HPP1* gene in primary and metastatic colorectal cancers. Neoplasia (New York) 7:771

Ebert M, Model F, Mooney S, Hale K, Lograsso J, Tonnes-Priddy L et al (2006) Aristaless-like Homeobox-4 Gene Methylation is a potential marker for colorectal adenocarcinomas. Gastroenterology 131:1418–1430

Eden A, Gaudet F, Waghmare A, Jaenisch R (2003) Chromosomal instability and tumors promoted by DNA hypomethylation. Science 300:455–455

Eftekhar E, Naghibalhossaini F (2014) Carcinoembryonic antigen expression level as a predictive factor for response to 5-fluorouracil in colorectal cancer. Mol Biol Rep 41:459–466

Ellidokuz E, Kundak I, Akpinar H, Okan A, Bektaser C, Fuzun M (2003) Kolorektap polip ve kanser Iokalizasyoni arasindaki iliski. Kocaatepe Tip Derg 4:49–53

Endres M, Fan G, Meisel A, Dirnagl U, Jaenisch R (2001) Effects of cerebral ischemia in mice lacking DNA methyltransferase 1 in post-mitotic neurons. Neuroreport 12:3763–3766

Errami Y, Brim H, Oumouna-Benachour K, Oumouna M, Naura AS, Kim H et al (2013) ICAD deficiency in human colon cancer and predisposition to colon tumorigenesis: linkage to apoptosis resistance and genomic instability. PloS ONE 8:e57871

Eschrich S, Yang I, Bloom G, Kwong KY, Boulware D, Cantor A et al (2005) Molecular staging for survival prediction of colorectal cancer patients. J Clin Oncol 23:3526–3535

Esteller M, Garcia-Foncillas J, Andion E, Goodman SN, Hidalgo OF, Vanaclocha V et al (2000) Inactivation of the DNA-repair gene *MGMT* and the clinical response of gliomas to alkylating agents. N Engl J Med 343:1350–1354

Fan J, Beck KA (2004) A role for the spectrin superfamily member *Syne-1* and kinesin II in cytokinesis. J Cell Sci 117:619–629

Fearon E, Vogelstein B (1990) A genetic model for colorectal tumorigenesis. Cell 61:759–767

Fleisher AS, Esteller M, Harpaz N, Leytin A, Rashid A, Xu Y et al (2000) Microsatellite instability in inflammatory bowel disease-associated neoplastic lesions is associated with hypermethylation and diminished expression of the DNA mismatch repair gene, *hMLH1*. Cancer Res 60:4864–4868

Fisher SA, Tremelling M, Anderson CA, Gwilliam R, Bumpstead S, Prescott NJ et al (2008) Genetic determinants of ulcerative colitis include the ECM1 locus and five loci implicated in Crohn's disease. Nat Genet 40:710–712

Frazier AL, Colditz GA, Fuchs CS, Kuntz KM (2000) Cost-effectiveness of screening for colorectal cancer in the general population. JAMA 284:1954–1961

Friederichs J, Rosenberg R, Mages J, Janssen KP, Maeckl C, Nekarda H et al (2005) Gene expression profiles of different clinical stages of colorectal carcinoma: toward a molecular genetic understanding of tumor progression. Int J Colorectal Dis 20:391–402

Frigola J, Munoz M, Clark SJ, Moreno V, Capella G, Peinado MA (2005) Hypermethylation of the prostacyclin synthase (PTGIS) promoter is a frequent event in colorectal cancer and associated with aneuploidy. Oncogene 24:7320–7326

Fujii S, Tominaga K, Kitajima K, Takeda J, Kusaka T, Fujita M et al (2005) Methylation of the oestrogen receptor gene in non-neoplastic epithelium as a marker of colorectal neoplasia risk in longstanding and extensive ulcerative colitis. Gut 54:1287–1292

Fujiwara T, Stolker JM, Watanabe T, Rashid A, Longo P, Eshleman JR et al (1998) Accumulated clonal genetic alterations in familial and sporadic colorectal carcinomas with widespread instability in microsatellite sequences. Am J Pathol 153:1063–1078

FYang Y, Yang JJ, Tao H, Jin WS (2014) New perspectives on β-catenin control of cell fate and proliferation in colon cance. ood Chem Toxicol. 74C:14–19

Gaffney DJ, Mcvicker G, Pai AA, Fondufe-Mittendorf YN, Lewellen N, Michelini K et al (2012) Controls of nucleosome positioning in the human genome. PLoS Genet 8:e1003036

Gao Q, Steine EJ, Barrasa MI, Hockemeyer D, Pawlak M, Fu D et al (2011) Deletion of the de novo DNA methyltransferase *Dnmt3a* promotes lung tumor progression. Proc Natl Acad Sci U S A 108:18061–18066

Gervaz P, Cerottini JP, Bouzourene H, Hahnloser D, Doan CL, Benhattar J et al (2002) Comparison of microsatellite instability and chromosomal instability in predicting survival of patients with T3N0 colorectal cancer. Surgery 131:190–197

Gervaz P, Bucher P, Morel P (2004) Two colons-two cancers: paradigm shift and clinical implications. J Surg Oncol 88:261–266

Ghadimi BM, Grade M, Difilippantonio MJ, Varma S, Simon R, Monatgna C et al (2005) Effectiveness of gene expression profiling for response prediction of rectal adenocarcinomas to preoperative chemoradiotherapy. J Clin Oncol 23:1826–1838

Glebov OK, Rodriguez LM, Nakahara K, Jenkins J, Cliatt J, Humbyrd CJ et al (2003) Distinguishing right from left colon by the pattern of gene expression. Cancer Epidemiol Biomarkers Prev 12:755–762

Glockner SC, Dhir M, YI JM, Mcgarvey KE, Van Neste L, Louwagie J et al (2009) Methylation of *TFPI2* in stool DNA: a potential novel biomarker for the detection of colorectal cancer. Cancer Res 69:4691–4699

Goel GA, Kandiel A, Achkar JP, Lashner B (2011) Molecular pathways underlying IBD-associated colorectal neoplasia: therapeutic implications. Am J Gastroenterol 106:719–730

Gonzalo V, Lozano JJ, Munoz J, Balaguer F, Pellise M, De Miguel CR et al (2010) Aberrant gene promoter methylation associated with sporadic multiple colorectal cancer. PLoS ONE 5:e8777

Goossens-Beumer IJ, Benard A, van Hoesel AQ, Zeestraten EC, Putter H, Böhringer S, Liefers GJ, Morreau H, van de Velde CJ, Kuppen PJ (2014) Age-dependent clinical prognostic value of histone modifications in colorectal cancer. Transl Res. pii: S1931-5244(14)00406-X

Grade M, Hormann P, Becker S, Hummonn AB, Wangsa D, Varma S et al (2007) Gene expression profiling reveals a massive, aneuploidy-dependent transcriptional deregulation and distinct differences between lymph node-negative and lymph node-positive colon carcinomas. Cancer Res 67:41–56

Grady WM, Rajput A, Lutterbaugh JD, Markowitz SD (2001) Detection of aberrantly methylated *hMLH1* promoter DNA in the serum of patients with microsatellite unstable colon cancer. Cancer Res 61:900–902

Graff JR, Herman JG, Lapidus RG, Chopra H, Xu R, Jarrard DF et al (1995) *E-cadherin* expression is silenced by DNA hypermethylation in human breast and prostate carcinomas. Cancer Res 55:5195–5199

Greenblatt M, Bennett W, Hollstein M, Harris C (1994) Mutations in the *p53* tumor suppressor gene: clues to cancer etiology and molecular pathogenesis. Cancer Res 54:4855–4878

Groene J, Mansmann U, Meister R, Staub E, Roepcke S, Heinze M et al (2006) Transcriptional census of 36 microdissected colorectal cancers yields a gene signature to distinguish UICC II and III. Int J Cancer 119:1829–1836

Gronbek K, Hother C, Jones PA (2007) Epigenetic changes in cancer. Apmis 115:1039–1059

Group UCSW (2014) United States cancer statistics: 1999–2006 incidence and mortality web-based report. Atlanta

Gryfe R, Gallinger S (2001) Microsatellite instability, mismatchrepair deficiency, and colorectal cancer. Surgery 130:17–20

Gryfe R, Kim H, Hsieh ET, Aronson MD, Holowaty EJ, Bull SB et al (2000) Tumor microsatellite instability and clinical outcome in young patients with colorectal cancer. N Engl J Med 342:69–77

Guenther MG, Levine SS, Boyer LA, Jaenisch R, Young RA (2007) A chromatin landmark and transcription initiation at most promoters in human cells. Cell 130:77–88

Guo Y, Guo W, Chen Z, Kuang G, Yang Z, Dong Z (2011) Hypermethylation and aberrant expression of Wnt-antagonist family genes in gastric cardia adenocarcinoma. Neoplasma 58(2):110–117

Hall E, Dayeh T, Kirkpatrick CL, Wollheim CB, Dekker Nitert M, Ling C (2013) DNA methylation of the glucagon-like peptide 1 receptor (GLP1R) in human pancreatic islets. BMC Med Genet 14:76

Hampe J, Franke A, Rosenstiel P, Till A, Teuber M, Huse K et al (2006) A genome-wide association scan of nonsynonymous SNPs identifies a susceptibility variant for Crohn disease in ATG16L1. Nat Genet 39:207–211

Han Y, Shao Y, Lin Z, Qu YL, Wang H, Zhou Y, Chen W, Chen Y, Chen WL, Hu FR, Li W, Liu Z (2012) Netrin-1 simultaneously suppresses corneal inflammation and neovascularization. Invest Ophthalmol Vis Sci 53(3):1285–1295

Hanh WC, Weinberg RA (2002) Rules for making human tumor cells. N Engl J Med 347:1593–1603

Hanada N, Takahata T, Zhou Q, Ye X, Sun R, Itoh J et al (2012) Methylation of the KEAP1 gene promoter region in human colorectal cancer. BMC Cancer 12:66

Hanahan D, Weinberg RA (2011) Hallmarks of cancer: the next generation. Cell 144:646–674

Hanauer SB (2006) Inflammatory bowel disease: epidemiology, pathogenesis, and therapeutic opportunities. Inflamm Bowel Dis 12:S3–9

Harder J, Engelstaedter V, Usadel H, Lassmann S, Werner M, Baier P et al (2009) CpG-island methylation of the ER promoter in colorectal cancer: analysis of micrometastases in lymph nodes from UICC stage I and II patients. Br J Cancer 100:360–365

Hartmentt L, Egan LJ (2012) Inflammation, DNA methylation and colitis-associated cancer. Carcinogenesis 33:723–731

Hawkins NJ, Lee JH, Wong JJ, Kwok C, Ward RL, Hitchins MP (2009) *MGMT* methylation is associated primarily with the germline C > T SNP (rs16906252) in colorectal cancer and normal colonic mucosa. Mod Pathol 22:1588–1599

Haydon AM, Jass JR (2002) Emerging pathways in colorectal-cancer development. Lancet Oncol 3:83–88

Hedayati M, Nabipour I, Rezaei-Ghaleh N, Azizi F (2006) Germline *RET* mutations in exons 10 and 11: an Iranian survey of 57 medullary thyroid carcinoma cases. Med J Malaysia 61:564

Heintzman ND, Hon GC, Hawkins RD, Kheradpour P, Stark A Harp LF et al (2009) Histone modifications at human enhancers reflect global cell-type-specific gene expression. Nature 459:108–112

Hellebrekers DM, Lentjes MH, Van Den Bosch SM, Melotte V, Wouters KA, Daenen KM et al (2009) *GATA4* and GATA5 are potential tumor suppressors and biomarkers in colorectal cancer. Clin Cancer Res 15:3990–3997

Herath N, Doecke J, Spanevello M, Leggett B, Boyd A (2009) Epigenetic silencing of EphA1 expression in colorectal cancer is correlated with poor survival. Br J Cancer 100:1095–1102

Herbst A, Wallner M, Rahmig K, Stieber P, Crispin A, Lamerz R et al (2009) Methylation of helicase-like transcription factor in serum of patients with colorectal cancer is an independent predictor of disease recurrence. Eur J Gastroenterol Hepatol 21:565–569

Herman JG, Latif F, Weng Y, Lerman MI, Zbar B, Liu S et al (1994) Silencing of the VHL tumor-suppressor gene by DNA methylation in renal carcinoma. Proc Natl Acad Sci U S A 91:9700–9704

Herman JG, Umar A, POlyak K, Graff JR, Ahuja N, Issa JPJ et al (1998) Incidence and functional consequences of *hMLH1* promoter hypermethylation in colorectal carcinoma. Proc Natl Acad Sci U S A 95:6870–6875

Herrinton LJ, Liu L, Levin TR, Allison JE, Lewis JD, Velayos F (2012) Incidence and mortality of colorectal adenocarcinoma in persons with inflammatory bowel disease from 1998 to 2010. Gastroenterology 143:382–389

Hesson LB, Wilson R, Morton D, Adams C, Walker M, Maher ER et al (2005) CpG island promoter hypermethylation of a novel Ras-effector gene RASSF2A is an early event in colon carcinogenesis and correlates inversely with *K-ras* mutations. Oncogene 24:3987–3994

Hibi K, Nakayama H, Kodera Y, Ito K, Akiyama S, Nakao A (2004) *CDH13* promoter region is specifically methylated in poorly differentiated colorectal cancer. Br J Cancer 90:1030–1033

Hibi K, Mizukami H, Shirahata A, Goto T, Sakata M, Saito M et al (2009a) Aberrant methylation of the UNC5C gene is frequently detected in advanced colorectal cancer. Anticancer Res 29:271–273

Hibi K, Mizukami H, Shirahata A, Goto T, Sakata M, Sanada Y (2009b) Aberrant methylation of the netrin-1 receptor genes UNC5C and *DCC* detected in advanced colorectal cancer. World J Surg 33:1053–1057

Hibi K, Goto T, Kitamura YH, Yokomizo K, Sakuraba K, Shirahata A et al (2010) Methylation of *TFPI2* gene is frequently detected in advanced well-differentiated colorectal cancer. Anticancer Res 30:1205–1207

Hol L, Wilschut J, Van Ballegooijen M, Van Vuuren A, Van Der Valk H, Reijerink J, Van Der Togt A et al (2009) Screening for colorectal cancer: random comparison of guaiac and immunochemical faecal occult blood testing at different cut-off levels. Br J Cancer 100:1103–1110

Hong K, Hinck L, Nishiyama M, Poo MM, Tessier-Lavigne M, Stein E (1999) A ligand-gated association between cytoplasmic domains of UNC5 and *DCC* family receptors converts netrin-induced growth cone attraction to repulsion. Cell 97:927–941

Hong L, Ahuja N (2013) DNA Methylation biomarkers of stool and blood for early detection of colon cancer. Genet Test Mol Biomarkers 17:401–406

Hu G, Cui K, Northrup D, Liu C, Wang C, Tang Q et al (2012) H2A. Z facilitates access of active and repressive complexes to chromatin in embryonic stem cell self-renewal and differentiation. Cell Stem Cell 12:180–192

Huang J, Papadopoulos N, Mckinley AJ, Farrington SM, Curtis LJ, Wyllie AH et al (1996) *APC* mutations in colorectal tumors with mismatch repair deficiency. Proc Natl Acad Sci U S A 93:9049–9054

Huang ZH, Li LH, Yang F, Wang JF (2007) Detection of aberrant methylation in fecal DNA as a molecular screening tool for colorectal cancer and precancerous lesions. World J Gastroenterol 13:950

Huang J, Schriefer AE, Yang W, Cliften PF, Rudnick DA (2014)Identification of an epigenetic signature of early mouse liver regeneration that is disrupted by Zn-HDAC inhibition. Epigenetics 9(11):1521–1531

Iacopetta B (2003) T*P53* mutation in colorectal cancer. Hum Mut 21:271–276

Imamura Y, Hibi K, Koike M, Fujiwara M, Kodera Y, Ito K et al (2005) *RUNX3* promoter region is specifically methylated in poorly-differentiated colorectal cancer. Anticancer Res 25:2627–2630

Imperiale TF, Ransohoff DF, Itzkowitz SH, Turnbull BA, Ross ME (2004) Fecal DNA versus fecal occult blood for colorectal-cancer screening in an average-risk population. N Engl J Med 351:2704–2714

Ines C, Donia O, Rahma B, Ben Ammar A, Sameh A, Khalfallah T, Abdelmajid BH, Sabeh M, Saadia B (2014) Implication of K-ras and p53 in colorectal cancer carcinogenesis in Tunisian population cohort. Tumour Biol 35(7):7163–7175. doi:10.1007/s13277-014-1874-4

Irizarry RA, Ladd-Acosta C, Wen B, Wu Z, Montano C, Onyango P et al (2009) The human colon cancer methylome shows similar hypo-and hypermethylation at conserved tissue-specific CpG island shores. Nat Genet 41:178–186

Issa JP (2005) Optimizing therapy with methylation inhibitors in myelodysplastic syndromes: dose, duration, and patient selection. Nat Clin Pract Oncol 2:24–29

Issa JP, Ahuja N, Toyota M, Bronner MP, Brentnall TA (2001) Accelerated age-related CpG island methylation in ulcerative colitis. Cancer Res 61:3573–3577

Ito S, Shen L, Dai Q, Wu SC, Collins LB, Swenberg JA et al (2011) Tet proteins can convert 5-methylcytosine to 5-formylcytosine and 5-carboxylcytosine. Science 333:1300–1303

Itzkowitz SH, Jandorf L, Brand R, Rabeneck L, Schroy PC III, Sontag S et al (2007) Improved fecal DNA test for colorectal cancer screening. Clin Gastroenterol Hepatol 5:111–117

Itzkowitz S, Brand R, Jandorf L, Durkee K, Millholland J, Rabeneck L et al (2008) A simplified, noninvasive stool DNA test for colorectal cancer detection. Am J Gastroenterol 103:2862–2870

Jass JR, Whitehall VL, Young J, Leggett BA (2000) Emerging concepts in colorectal neoplasia. Gastroenterology 123:862–876

Jass JR, Young J, Leggett BA (2002) Evolution of colorectal cancer: change of pace and change of direction. J Gastroenterol Hepatol 17:17–26

Jemal A, Murray T, Ward E, Samuels A, TiwarI RC, Ghafoor A et al (2005) Cancer statistics, 2005. CA Cancer J Clin 55:10–30

Jemal A, Siegel R, WardE, Murray T, Xu J, Thun MJ (2007) Cancer statistics, 2007. CA Cancer J Clin 57:43–66

Jemal A, Siegel R, Xu J, Ward E (2010) Cancer statistics, 2010. CA Cancer J Clin 60:277–300

Jemal A, Siegel RL, Ma J, Islami F, DeSantis C, Sauer AG, Simard EP, Ward EM (2014) Inequalities in premature death from colorectal cancer by state. J Clin Oncol. pii: JCO.2014.58.7519

Jensen L, Lindebjerg J, Byriel L, Kolvraa S, Cruger DG (2008) Strategy in clinical practice for classification of unselected colorectal tumours based on mismatch repair deficiency. Colorectal Dis 10:490–497

Johns LE, Houlston RS (2001) A systematic review and meta-analysis of familial colorectal cancer risk. Am J Gastroenterol 96:2992–3003

Jones PA (1985) Effects of 5-azacytidine and its 2′-deoxyderivative on cell differentiation and DNA methylation. Pharmacol Ther 28:17–27

Jones PA, Baylin SB (2002) The fundamental role of epigenetic events in cancer. Nat Rev Genet 3:415–428

Jones PA, Takai D (2001) The role of DNA methylation in mammalian epigenetics. Science 293:1068–1070

Jones PA, Taylor SM (1980) Cellular differentiation, cytidine analogs and DNA methylation. Cell 20:85–93

Jost JP, Bruhat A (1997) The formation of DNA methylation patterns and the silencing of genes. Progress Nucleic Acid Res Mol Biol 57:217–248

Kamangar F, Dores GM, Anderson WF (2006) Patterns of cancer incidence, mortality, and prevalence across five continents: defining priorities to reduce cancer disparities in different geographic regions of the world. J Clin Oncol 24:2137–2150

Kane MF, Loda M, Gaida GM, Lipman J, Mishra R, Goldman H et al (1997) Methylation of the *hMLH1* promoter correlates with lack of expression of *hMLH1* in sporadic colon tumors and mismatch repair-defective human tumor cell lines. Cancer Res 57:808–811

Kang MY, Lee BB, Ji YI, Jung EH, Chun HK, Song SY et al (2008) Association of interindividual differences in *p14ARF* promoter methylation with single nucleotide polymorphism in primary colorectal cancer. Cancer 112:1699–1707

Keller C, Adaixo R, Stunnenberg R, Woolcock KJ, Hiller S, Buhler M (2012) HP1(Swi6) Mediates the recognition and destruction of heterochromatic RNA transcripts. Mol Cell 47:215–227

Kellermayer R (2012) Epigenetics and the developmental origins of inflammatory bowel diseases. Can J Gastroenterol 26:909–915

Kelly WK, O'Connor OA, Krug LM, Chiao JH, Heaney M, Curley T et al (2005) Phase I study of an oral histone deacetylase inhibitor, suberoylanilide hydroxamic acid, in patients with advanced cancer. J Clin Oncol 23:3923–3931

Ki DH, Jeung HC, Park CH, Kang SH, Lee GY, Lee WS et al (2007) Whole genome analysis for liver metastasis gene signatures in colorectal cancer. Int J Cancer 121:2005–2012

Kim YS, Deng G (2007) Epigenetic changes (aberrant DNA methylation) in colorectal neoplasia. Gut Liver 1:1–11

Kim YH, Petko Z, Dzieciatkowski S, Lin L, Ghiassi M, Stain S et al (2006) CpG island methylation of genes accumulates during the adenoma progression step of the multistep pathogenesis of colorectal cancer. Genes Chromosomes Cancer 45:781–789

Kim IJ, Lim SB, Kang HC, Chang HJ, Ahn SA, Park HW et al (2007) Microarray gene expression profiling for predicting complete response to preoperative chemoradiotherapy in patients with advanced rectal cancer. Dis Colon Rectum 50:1342–1353

Kim M, Chang X, Nagpal J, Yamashita K, Baek J, Dasgupta S et al (2007) The N-methyl-D-aspartate receptor type 2A is frequently methylated in human colorectal carcinoma and suppresses cell growth. Oncogene 27:2045–2054

Kim K, Park U, Wang J, Lee J, Park S, Kim S et al (2008) Gene profiling of colonic serrated adenomas by using oligonucleotide microarray. Int J Colorectal Dis 23:569–580

Kim MS, Louwagie J, Carvalho B, Sive Droste JST, Park HL, Chae YK et al (2009) Promoter DNA methylation of Oncostatin M receptor-β as a novel diagnostic and therapeutic marker in colon cancer. PLoS ONE 4:e6555

Kim MS, Lee J, Sidransky D (2010) DNA methylation markers in colorectal cancer. Cancer Metastasis Rev 29:181–206

Kim MJ, Lee EJ, Chun SM, Jang SJ, Kim D, Lee D, Youk E (2014) The significance of ectopic crypt formation in the differential diagnosis of colorectal polyps. Diagn Pathol 9(1):212

Kimmie N, Schrag D (2010) Microsatellite instability and adjuvant fluorouracil chemotherapy: a mismatch? J Clin Oncol 28:3207–3210

Kinzler K, Vogelstein B (1996) Lessons from hereditary colorectal cancer. Cell 87:159–170

Kisiel JB, Yab TC, Taylor WR, Mahoney DW, Ahlquist DA (2014) Stool methylated DNA markers decrease following colorectal cancer resection-implications for surveillance. Dig Dis Sci 59(8):1764–1767

Kita H, Hikichi Y, Hikami K, Tsuneyama K, Cui ZG, Osawa H et al (2006) Differential gene expression between flat adenoma and normal mucosa in the colon in a microarray analysis. J Gastroenterol 41:1053–1063

Kitajima M, Takita N, Hata M, Maeda T, Sakamoto K, Kamano T et al (2006) The relationship between 5-fluorouracil sensitivity and single nucleotide polymorphisms of the orotate phosphoribosyl transferase gene in colorectal cancer. Oncol Rep 15:161–165

Kleivi K, Lind GE, Diep CB, Meling GI, Brandal LT, Nesland JM et al (2007) Gene expression profiles of primary colorectal carcinomas, liver metastases, and carcinomatoses. Mol Cancer 6:2

Kohonen-Corish MR, Daniel JJ, Chan C, Lin BP, Kwun SY, Dent OF et al (2005) Low microsatellite instability is associated with poor prognosis in stage C colon cancer. J Clin Oncol 23:2318–2324

Komuro K, Tada M, Tamoto E, Kawakami A, Matsunaga A, Teramoto KI et al (2005) Right-and left-sided colorectal cancers display distinct expression profiles and the anatomical stratification allows a high accuracy prediction of lymph node metastasis. J Surg Res 124:216–224

Konishi K, Shen L, Wang S, Meltzer SJ, Harpaz N, Issa JPJ (2007) Rare CpG island methylator phenotype in ulcerative colitis-associated neoplasias. Gastroenterology 132:1254–1260

Koornstra J, De Jong S, Hollema H, De Vries E, Kleibeuker J (2003) Changes in apoptosis during the development of colorectal cancer: a systematic review of the literature. Crit Rev Oncol/Hematol 45:37–53

Kuan CT, Wakiya K, Dowell JM, Herndon JE, Reardon DA, Graner MW et al (2006) Glycoprotein nonmetastatic melanoma protein B, a potential molecular therapeutic target in patients with glioblastoma multiforme. Clin Cancer Res 12:1970–1982

Kutzner N, Hoffmann I, Linke C, Thienel T, Grzegorczyk M, Urfer W et al (2005) Non-invasive detection of colorectal tumours by the combined application of molecular diagnosis and the faecal occult blood test. Cancer Lett 229:33–41

Kwabi-Addo B, Chung W, Shen L, Ittmann M, Wheeler T, Jelinek J et al (2007) Age-related DNA methylation changes in normal human prostate tissues. Clin Cancer Res 13:3796–3802

Kwong KY, Bloom GC, Yang I, Boulware D, Coppola D, Haseman J et al (2005) Synchronous global assessment of gene and protein expression in colorectal cancer progression. Genomics 86:142–158

Lachner M, O'Carroll D, Rea S, Mechtler K, Jenuwein T (2001) Methylation of histone H3 lysine 9 creates a binding site for HP1 proteins. Nature 410:116–120

Laiho P, Launonen V, Lahermo P, Esteller M, Guo M, Herman JG et al (2002) Low-level microsatellite instability in most colorectal carcinomas. Cancer Res 62:1166–1170

Laird PW (2005) Cancer epigenetics. Hum Mol Genet 14:R65–R76

Laird PW, Jackson-Grusby L, Fazeli A, Dickinson SL, Edward Jung W, Li E et al (1995) Suppression of intestinal neoplasia by DNA hypomethylation. Cell 81:197–205

Lange CP, Campan M, Hinoue T, Schmitz RF, Van Der Meulen-De AE, Slingerland H et al (2012) Genome-scale discovery of DNA-methylation biomarkers for blood-based detection of colorectal cancer. PloS ONE 7:e50266

Lee S, Hwang KS, Lee HJ, Kim JS, Kang GH (2004) Aberrant CpG island hypermethylation of multiple genes in colorectal neoplasia. Lab Invest 84:884–893

Lees NP, Harrison KL, Hall CN, Margison GP, Povey AC (2004) Reduced *MGMT* activity in human colorectal adenomas is associated with *K-ras* GC→ AT transition mutations in a population exposed to methylating agents. Carcinogenesis 25:1243–1247

Lee S, OH T, Chung H, Rha S, Kim C, Moon Y et al (2012) Identification of GABRA1 and LAMA2 as new DNA methylation markers in colorectal cancer. Int J Oncol 40:889–898

Lengauer C, Kinzler KW, Vogelstein B (1997) DNA methylation and genetic instability in colorectal cancer cells. Proc Natl Acad Sci U S A 94:2545–2550

Lenhard K, Bommer GT, Asutay S, Schauer R, Brabletz T, Goke B et al (2005) Analysis of promoter methylation in stool: a novel method for the detection of colorectal cancer. Clin Gastroenterol Hepatol 3:142–149

Leslie A, Pratt NR, Gillespie K, Sales M, Kernohan NM, Smith G et al (2003) Mutations of *APC*, *K-ras*, and *p53* are associated with specific chromosomal aberrations in colorectal adenocarcinomas. Cancer Res 63:4656–4661

Leung WK, To KF, Man EP, Chan MW, Hui AJ, Ng SS et al (2007) Detection of hypermethylated DNA or cyclooxygenase-2 messenger RNA in fecal samples of patients with colorectal cancer or polyps. Am J Gastroenterol 102:1070–1076

Levin B (1992) Inflammatory bowel disease and colon cancer. Cancer 70:1313–1316

Levin B, Lieberman DA, Mcfarland B, Smith RA, Brooks D, Andrews KS et al (2008) Screening and surveillance for the early detection of colorectal cancer and adenomatous polyps, 2008: a joint guideline from the American Cancer Society, the US Multi-Society Task Force on Colorectal Cancer, and the American College of Radiology. CA Cancer J Clin 58:130–160

Leroy B, Anderson M, Soussi T (2014) TP53 mutations in human cancer: database reassessment and prospects for the next decade. Hum Mutat 35(6):672–688

Li E, Bestor TH, Jaenisch R (1992) Targeted mutation of the DNA methyltransferase gene results in embryonic lethality. Cell 69:915–926

Li M, Lin YM, Hasegawa S, Shimokawa T, Murata K, Kameyama M et al (2004) Genes associated with liver metastasis of colon cancer, identified by genome-wide cDNA microarray. Int J Oncol 24:305

Li B, Carey M, Workman JL (2007) The role of chromatin during transcription. Cell 128:707–719

Lidgard GP, Domanico MJ, Bruinsma JJ, Light J, Gagrat ZD, Oldham-Haltom RL, Fourrier KD, Allawi H, Yab TC, Taylor WR, Simonson JA, Devens M, Heigh RI, Ahlquist DA, Berger BM (2013) Clinical performance of an automated stool DNA assay for detection of colorectal neoplasia. Clin Gastroenterol Hepatol 11(10):1313–1318

Lin HM, Chatterjee A, Lin YH, Anjomshoaa A, Fukuzawa R, Mccall J et al (2007) Genes associated with liver metastasis of colon cancer, identified by genome-wide cDNA microarray. Oncol Rep 17:1541–1549

Lin Z, Hegarty J, Cappel J, Yu W, Chen X, Faber P et al (2011) Identification of disease-associated DNA methylation in intestinal tissues from patients with inflammatory bowel disease. Clin Genet 80:59–67

Lind GE, Ahlquist T, Kolberg M, Berg M, Eknes M, Alonso MA et al (2008) Hypermethylated *MAL* gene-a silent marker of early colon tumorigenesis. J Transl Med 6:13

Linhart HG, Lin H, Yamada Y, Moran E, Steine EJ, Gokhale S et al (2007) *Dnmt3b* promotes tumorigenesis in vivo by gene-specific de novo methylation and transcriptional silencing. Genes Dev 21:3110–3122

Lipkin M, Higgins P (1988) Biological markers of cell proliferation and differentiation in human gastrointestinal diseases. Adv Cancer Res 50:1–24

Lister R, Pellizzola M, Dowen RH, Hawkins RD, Hon G, Tonti-Filippini et al (2009) Human DNA methylomes at base resolution show widespread epigenomic differences. Nature 462:315–322

Liu B, Parsons R, Papapoulos N, Nicolaides NC, Lynch HT, Wtson P et al (1996) Analysis of mismatch repair genes in hereditary non-polyposis colorectal cancer patients. Nat Med 2:169–174

Liu QY, Lei JX, Sikorska M, Liu R (2008) A novel brain-enriched E3 ubiquitin ligase RNF182 is up regulated in the brains of Alzheimer's patients and targets ATP6V0C for degradation. Mol Neurodegener 3:4. doi:10.1186/1750-1326-3-4

Llambi F, Causeret F, Bloch-Gallego E, Mehlen P (2001) Netrin-1 acts as a survival factor via its receptors UNC5H and *DCC*. EMBO J 20:2715–2722

Lyko F, Brown R (2005) DNA methyltransferase inhibitors and the development of epigenetic cancer therapies. J Natl Cancer Inst 97:1498–1506

Maeda O, Ando T, Watanabe O, Ishiguro K, Ohmiya N, Niwa Y et al (2006) DNA hypermethylation in colorectal neoplasms and inflammatory bowel disease: a mini review. Inflammopharmacology 14:204–206

Marks PA, Richon VM, Breslow R, Rifkind RA (2001) Histone deacetylase inhibitors as new cancer drugs. Curr Opin Oncol 13:477–483

Martinez-Cardus A, Martinez-Balibrea E, Bandres E (2006) Gene expression profile related to oxaliplatin resistance in a panel of sensitive human colorectal cancer (CRC) cell lines with acquired resistance to the drug. The 2006 American Society of Clinical Oncology Gastrointestional Cancers Symposium. San Francisco

Massacesi C, Terrazzino S, Marcucci F, Rocchi MB, Lippe P, Bisonni R et al (2006) Uridine diphosphate glucuronosyl transferase 1A1 promoter polymorphism predicts the risk of gastrointestinal toxicity and fatigue induced by irinotecan-based chemotherapy. Cancer 106:1007–1016

Matsuyama R, Togo S, Shimizu D, Momiyama N, Ishikawa T, Ichikawa Y et al (2006) Predicting 5-fluorouracil chemosensitivity of liver metastases from colorectal cancer using primary tumor specimens: three-gene expression model predicts clinical response. Int J Cancer 119:406–413

Mattar MC, Lough D, Pishvaian MJ, Charabaty A (2011) Current management of inflammatory bowel disease and colorectal cancer. Gastrointestin. Cancer Res 4:53

Mazelin L, Bernet A, Bonod-Bidaud C, Pays L, Arnaud S, Gespach C et al (2004) Netrin-1 controls colorectal tumorigenesis by regulating apoptosis. Nature 431:80–84

Mcdermott U, Longley D, Johnston P (2002) Molecular and biochemical markers in colorectal cancer. Ann Oncol 13:235–245

Mcmichael AJ, Potter JD (1983) Do intrinsic sex differences in lower alimentary tract physiology influence the sex-specific risks of bowel cancer and other biliary and intestinal diseases? Am J Epidemiol 118:620–627

Mcmichael AJ, Potter JD (1985) Host factors in carcinogenesis: certain bile-acid metabolic profiles that selectively increase the risk of proximal colon cancer. J Natl Cancer Inst 75:185–191

Melotte V, Lentjes MH, Van Den Bosch SM, Hellebrekers DM, De Hoon JP, Wouters KA et al (2009) N-Myc downstream-regulated gene 4 (NDRG4): a candidate tumor suppressor gene and potential biomarker for colorectal cancer. J Natl Cancer Inst 101:916–927

Midgley R, Kerr D (1999) Colorectal cancer. The Lancet 353:391–399

Minoo P, Baker K, Goswami R, Chong G, Foulkes WD, Ruszkiewicz AR et al (2006) Extensive DNA methylation in normal colorectal mucosa in hyperplastic polyposis. Gut 55:1467–1474

Miotto E, Sabbioni S, Veronese A, Calin GA, Gullini S, Liboni A et al (2004) Frequent aberrant methylation of the CDH4 gene promoter in human colorectal and gastric cancer. Cancer Res 64:8156–8159

Mokarram P, Naghibalhossaini F, Firoozi MS, Hosseini SV, Izadpanah A, Salahi H et al (2008) Methylenetetrahydrofolate reductase C677T genotype affects promoter methylation of tumor-specific genes in sporadic colorectal cancer through an interaction with folate/vitamin B12 status. World J Gastroenterol 14:3662

Mokarram P, Kumar K, Brim H, Naghibalhossaini F, Saberi-Firoozi M, Nouraie M et al (2009) Distinct high-profile methylated genes in colorectal cancer. PLoS ONE 4:e7012

Mokarram P, Zamani M, Kavousipour S, Naghibalhossaini F, Irajie C, Sarabi MM et al (2012) Different patterns of DNA methylation of the two distinct O6-methylguanine-DNA methyltransferase (O6-*MGMT*) promoter regions in colorectal cancer. Mol Biol Rep 40:3851–3857

Morán A, Ortega P, de Juan C, Fernández-Marcelo T, Frías C, Sánchez-Pernaute A, Torres AJ, Díaz-Rubio E, Iniesta P, Benito M (2010) Differential colorectal carcinogenesis: molecular basis and clinical relevance. World J Gastrointest Oncol 2(3):151–158

Mori Y, Yin J, Sato F, Sterian A, Simms LA, Selaru FM et al (2004) Identification of genes uniquely involved in frequent microsatellite instability colon carcinogenesis by expression profiling combined with epigenetic scanning. Cancer Res 64:2434–2438

Moss TJ, Wallrath LL (2007) Connections between epigenetic gene silencing and human disease. Mutat Res 618:163–174

Munkholm P (2003) Review article: the incidence and prevalence of colorectal cancer in inflammatory bowel disease. Aliment Pharmacol Ther 18:1–5

Munro A, Lain S, Lane D (2005) *P53* abnormalities and outcomes in colorectal cancer: a systematic review. Br J Cancer 92:434–444

Nagasaka T, Tanaka N, Cullings HM, Sun DS, Sasamoto H, Uchida H et al (2009) Analysis of fecal DNA methylation to detect gastrointestinal neoplasia. J Natl Cancer Inst 101:1244–1258

Naghibalhossaini F, Mokarram P, Khalili I, Vasei M, Hosseini SV, Shktorab H et al (2010) *MTHFR* C677T and A1298C variant genotypes and the risk of microsatellite instability among Iranian colorectal cancer patients. Cancer Genet Cytogenet 197:142–151

Naghibalhossaini F, Hosseini HM, Mokarram P, Zamani M (2011) High frequency of genes' promoter methylation, but lack of *BRAF* V600E mutation among Iranian colorectal cancer patients. Pathol Oncol Res 17:819–825

Naghibalhossaini F, Zamani M, Mokarram P, Khalili I, Rasti M, Mostafavi-Pour Z (2012) Epigenetic and genetic analysis of WNT signaling pathway in sporadic colorectal cancer patients from Iran. Mol Biol Rep 39:6171–6178

Newman E, Longmate J, Lenz H, Carroll M, Stalter S, Lim D et al (2002) Phase I and clinical pharmacokinetic evaluation of the DNA methyltransferase inhibitor 5-fluoro-2′-deoxycytidine: a California Cancer Consortium study. In Proceedings of the American Society of Clinical Oncology, 2002. 108a

Noah TK, Lo YH, Price A, Chen G, King E, Washington MK et al (2013) SPDEF Functions as a colorectal tumor suppressor by inhibiting β-Catenin activity. Gastroenterology 5:1012–1023

Nosho K, Shima K, Irahara N, Kure S, Baba Y, Kirkner GJ et al (2009) *DNMT3B* expression might contribute to CpG island methylator phenotype in colorectal cancer. Clin Cancer Res 15:3663–3671

Nouraie M, Hosseinkhah F, Brim H, Zamanifekri B, Smoot DT, Shktorab H (2010) Clinicopathological features of colon polyps from African-Americans. Dig Dis Sci 55:1442–1449

Ogino S, Cantor M, Kawasaki T, Brahmandam M, Kirkner GJ, Weisenberger DJ et al (2006) CpG island methylator phenotype (CIMP) of colorectal cancer is best characterised by quantitative DNA methylation analysis and prospective cohort studies. Gut 55:1000–1006

Okada T, Suehiro Y, Ueno K, Mitomori S, Kaneko S, Nishioka M, Okayama N, Sakai K, Higaki S, Hazama S, Hirata H, Sakaida I, Oka M, Hinoda Y (2010) TWIST1 hypermethylation is observed frequently in colorectal tumors and its overexpression is associated with unfavorable outcomes in patients with colorectal cancer. Genes Chromosomes Cancer 49(5):452–462

Okano M, Xie S, Li E (1998) Dnmt2 is not required for de novo and maintenance methylation of viral DNA in embryonic stem cells. Nucleic Acids Res 26:2536–2540

Okano M, Bell DW, Haber DA, Li E (1999) DNA methyltransferases *Dnmt3a* and *Dnmt3b* are essential for de novo methylation and mammalian development. Cell 99:247–257

Oki Y, Aoki E, Issa JPJ (2007) Decitabine-bedside to bench. Crit Rev Oncol/Hematol 61:140–152

Olaru AV, Cheng Y, Agarwal R, Yang J, David S, Abraham JM et al (2012) Unique patterns of CpG island methylation in inflammatory bowel disease-associated colorectal cancers. Inflamm Bowel Dis 18:641–648

Olivier M, Hussain SP, Caron DFromentelC, Hainaut P, Harris CC (2003) T*P53* mutation spectra and load: a tool for generating hypotheses on the etiology of cancer. IARC Scientific Publications, Lyon, pp 247–270

Oliveira C, Westra JL, Arango D, Ollikainen M, Domingo E, Ferreira A et al (2004) Distinct patterns of KRAS mutations in colorectal carcinomas according to germline mismatch repair defects and *hMLH1* methylation status. Hum Mol Genet 13:2303–2311

Olsson L, Lindblom A (2003) Family history of colorectal cancer in a Sweden county. Fam Cancer 2:87–93

Oort F, Terhaar SDrosteJ, Van Der Hulst R, Van Heukelem H, Loffeld R, Wesdorp I et al (2010) Colonoscopy-controlled intra-individual comparisons to screen relevant neoplasia: faecal immunochemical test vs. guaiac-based faecal occult blood test. Aliment Pharmacol Ther 31:432–439

Pancione M, Sabatino L, Fucci A, Carafa V, Nebbioso A, Forte N, Febbraro A, Parente D, Ambrosino C, Normanno N, Altucci L, Colantuoni V (2010) Epigenetic silencing of peroxisome proliferator-activated receptor γ is a biomarker for colorectal cancer progression and adverse patients' outcome. PLoS One 5(12):e14229

Pantaleo M, Astolfi A, Nannini M (2007) Gene expression profiling differences between synchronous and metachronous liver metastases of colorectal cancer. AACR-NCI-EORTC conference on molecular targets and cancer therapeutics. San Francisco 26:15062

Papa Z, Pavai Z, Denes L, Brinzaniuk K, Jung I (2011) Hyperplastic polyps and serrated adenomas: precancerous lesions with mixed immunophenotype. Rom J Morphol Embryol 52:797–802

Park HW, Kang HC, Kim IJ, Jang SG, Kim K, Yoon H et al (2007) Correlation between hypermethylation of the RASSF2A promoter and *K-ras/BRAF* mutations in microsatellite-stable colorectal cancers. Int J Cancer 120:7–12

Parkes M, Barrett JC, Prescott NJ, Tremelling M, Anderson CA, Fisher SA et al (2007) Sequence variants in the autophagy gene IRGM and multiple other replicating loci contribute to Crohn's disease susceptibility. Nat Genet 39:830–832

Parkin DM, Muir CS, Whelan S, Gao Y, Ferlay J, Powell J (1992) Cancer incidence in five continents, vol 6. IARC Scientific Publications, Lyon

Parkin DM, Bray F, Ferlay J, Pisani P (2005) Global cancer statistics, 2002. CA Cancer J Clin 55:74–108

Parsons R, Li GM, Longley MJ, Fang WH, Papadopoulos N, Jen J et al (1993) Hypermutability and mismatch repair deficiency in RER(+) tumor cells. Cell 75:1227–1236

Pazdur R, Coia LR, Hoskins WJ, Wagman LD (2003) Cancer management: a multidisciplinary approach. F. A. Davis Company, Philadelphia

Pehlivan S, Artac M, Sever T, Bozcuk H, Kilincarslan C, Pehlivan M (2010) Gene methylation of *SFRP2, P16,* DAPK1, *HIC1,* and *MGMT* and KRAS mutations in sporadic colorectal cancer. Cancer Genet Cytogenet 201:128–132

Peltomaki P (2001) Deficient DNA mismatch repair: a common etiologic factor for colon cancer. Hum Mol Genet 10:735–740

Pereyra L, Gómez EJ, González R, Fischer C, Eraña GB, Torres AG, Correa L, Mella JM, Panigadi GN, Luna P, Pedreira SC, Cimmino DG, Boerr LA (2014)

Pereyra L, Gómez EJ, González R, Fischer C, Eraña GB, Torres AG, Correa L, Mella JM, Panigadi GN, Luna P, Pedreira SC, Cimmino DG, Boerr LA (2014) Finding sessile serrated adenomas: is it possible to identify them during conventional colonoscopy? Dig Dis Sci 59(12):3021–3026

Pietrantonio F, Perrone F, De Braud F, Castano A, Maggi C, Bossi I et al (2014) Activity of temozolomide in patients with advanced chemorefractory colorectal cancer and *MGMT* promoter methylation. Ann Oncol 25:404–408

Petko Z, Ghiassi M, Shuber A, Gorham J, Smalley W, Washington MK et al (2005) Aberrantly methylated *CDKN2A, MGMT,* and *MLH1* in colon polyps and in fecal DNA from patients with colorectal polyps. Clin Cancer Res 11:1203–1209

Petronis A, Petroniene R (2000) Epigenetics of inflammatory bowel disease. Gut 47:302–306

Pfeifer G (2006) Mutagenesis at methylated CpG sequences. DNA methylation: basic mechanisms. CTMI 301:259–281

Philipp AB, Stieber P, Nagel D, Neumann J, Spelsberg F, Jung A et al (2012) Prognostic role of methylated free circulating DNA in colorectal cancer. Int J Cancer 131:2308–2319

Piepoli A, Cotugno R, Merla G, Gentile A, Augello B, Quitadamo M et al (2009) Promoter methylation correlates with reduced NDRG2 expression in advanced colon tumour. BMC Med Genomics 2:11

Pinto A, Zagonel V (1993) 5-Aza-2'-deoxycytidine (Decitabine) and 5-azacytidine in the treatment of acute myeloid leukemias and myelodysplastic syndromes: past, present and future trends. Leukemia 7:51–60

Plumb JA, Strathdee G, Sludden J, Kaye SB, Brown R (2000) Reversal of drug resistance in human tumor xenografts by 2′-deoxy-5-azacytidine-induced demethylation of the *hMLH1* gene promoter. Cancer Res 60:6039–6044

Pocard M, Salmon R, Muleris M, Remvikos Y, Bara J Dutrillaux B et al (1995) Two colons-two cancers? Proximal or distal adenocarcinoma: arguments for a different carcinogenesis. Bull Cancer 82:10

Popat S, Matakidou A, Houlston RS (2004) Thymidylate synthase expression and prognosis in colorectal cancer: a systematic review and meta-analysis. J Clin Oncol 22:529–536

Popat S, Hubner R, Houlston R (2005) Systematic review of microsatellite instability and colorectal cancer prognosis. J Clin Oncol 23:609–618

Potter JD (1999) Colorectal cancer: molecules and populations. J Natl Cancer Inst 91:916–932

Potter JD, Slattery ML, Bostick RM, Gapstur SM (1993) Colon cancer: a review of the epidemiology. Epidemiol Rev 15:499–545

Pretlow TP, Brasitus *TA,* Fulton NC, Cheyer C, Kaplan EL (1993) *K-ras* mutations in putative preneoplastic lesions in human colon. J Natl Cancer Inst 85:2004–2007

Prince HM, Bishton MJ, Harrison SJ (2009) Clinical studies of histone deacetylase inhibitors. Clin Cancer Res 15:3958–3969

Qi J, Zhu YQ, Huang MF, Yang D (2005) Hypermethylation of CpG island in O6-methylguanine-DNA methyltransferase gene was associated with *K-ras* G to A mutation in colorectal tumor. World J Gastroenterol 11:2022–2025

Quintero E, Andreu M, Lanas A, Pique JM (2009) Estrategias para la detección precoz del cáncer colorrectal. La Prevención del Cáncer Colorrectal en Espana, 21 need complete details

Rabeneck L, Paszat LF, Hilsden RJ, Saskin R, Leddin D, Grunfeld E et al (2008) Bleeding and perforation after outpatient colonoscopy and their risk factors in usual clinical practice. Gastroenterology 135:1899–1906

Rainier S, Johnson LA, Dobry CJ, Ping AJ, Grundy PE, Feinberg AP (1993) Relaxation of imprinted genes in human cancer. Nature 22:747–749

Rajaii F, Asnaghi L, Enke R, Merbs SL, Handa JT, Eberhart CG (2014) The demethylating agent 5-Aza reduces the growth, invasiveness, and clonogenicity of uveal and cutaneous melanoma. Invest Ophthalmol Vis Sci 55(10):6178–6186

Reik W, Surani MA (1989) Genomic imprinting and embryonal tumours. Nature 338:112–113

Rennie PS, Nelson CC (1998) Epigenetic mechanisms for progression of prostate cancer. Cancer Metastasis Rev 17:401–409

Ribic CM, Sargent DJ, Moore MJ, Thibodeau SN, French AJ, Goldberg RM et al (2003) Tumor microsatellite-instability status as a predictor of benefit from fluorouracil-based adjuvant chemotherapy for colon cancer. N Engl J Med 349:247–257

Richman S, Adlard J (2002) Left and right sided large bowel cancer: have significant genetic differences in addition to well known clinical differences. Br Med J 324:931

Rimkus C, Friederichs J, Boulesteix AL, Theisen J, Mages J, Becker K et al (2008) Microarray-based prediction of tumor response to neoadjuvant radiochemotherapy of patients with locally advanced rectal cancer. Clin Gastroenterol Hepatol 6:53–61

Rioux JD, Xavier RJ, Taylor KD, Silverberg MS, Goyette P, Huett A et al (2007) Genome-wide association study identifies new susceptibility loci for Crohn disease and implicates autophagy in disease pathogenesis. Nat Genet 39:596–604

Robertson KD (2001) DNA methylation, methyltransferases, and cancer. Oncogene 20:3139–3155

Robertson KD (2005) DNA methylation and human disease. Nat Rev Genet 6:597–610

Rodriguez-Paredes M, Esteller M (2011) Cancer epigenetics reaches mainstream oncology. Nat Med 17:330–339

Roloff TC, Nuber UA (2005) Chromatin, epigenetics and stem cells. Eur J Cell Biol 84:123–135

Rosenfeld JA, Wang Z, Schones DE, Zhao K, Desalle R, Zhang MQ (2009) Determination of enriched histone modifications in non-genic portions of the human genome. BMC Genomics 10:143

Russo A, Bazan V, Iacopetta B, Kerr D, Soussi T, Gebbia N (2005) The T*P53* colorectal cancer international collaborative study on the prognostic and predictive significance of *p53* mutation: influence of tumor site, type of mutation, and adjuvant treatment. J Clin. the International Society for Cellular 23:7518–7528

Sadjadi A, Malekzadeh R, Derakhshan MH, Sepehr A, Nouraie M, Sotoudeh M et al (2003) Cancer occurrence in Ardabil: results of a population-based cancer registry from Iran. Int J Cancer 107:113–118

Saif MW, Chu E (2010) Biology of colorectal cancer. Cancer J 3:196–201

Sakai T, Toguchida J, Ohtani N, Yandell DW, Rapaport JM, Dryja *TP* (1991) Allele-specific hypermethylation of the retinoblastoma tumor-suppressor gene. Am J Hum Genet 48:880

Sakamoto N, Terai T, Ajioka Y, Abe S, Kobayasi O, Hirai S et al (2004) Frequent hypermethylation of *RASSF1A* in early flat-type colorectal tumors. Oncogene 23:8900–8907

Sakurai T, Bilim VN, Ugolkov AV, Yuuki K, Tsukigi M, Motoyama T et al (2012) The enhancer of zeste homolog 2 (EZH2), a potential therapeutic target, is regulated by miR-101 in renal cancer cells. Biochem Biophysical Res Commun 422:607–614

Salovaara R, Loukola AA, Kristo P, Kaariainen H, Ahtola H, Eskelinen M et al (2000) Population-based molecular detection of hereditary nonpolyposis colorectal cancer. J Clin Oncol 18:2193–2200

Samowitz WS, AAlbertsen H, Herrick J, Levin TR, Sweeeney C, Murtaugh MA et al (2005) Evaluation of a large, population-based sample supports a CpG island methylator phenotype in colon cancer. Gastroenterology 129:837–845

Samowitz WS, Slattery ML, Sweeney C, Herrick J, Wolff RK, Albertsen H (2007) *APC* mutations and other genetic and epigenetic changes in colon cancer. Mol Cancer Res 5:165–170

Sanchez-Cespedes M, Esteller M, Wu L, Nawroz-Danish H, Yoo GH, Koch WM, Jen J et al (2000) Gene promoter hypermethylation in tumors and serum of head and neck cancer patients. Cancer Res 60:892–895

Saito T, Mitomi H, Imamhasan A, Hayashi T, Mitani K, Takahashi M, Kajiyama Y, Yao T (2014) Downregulation of sFRP-2 by epigenetic silencing activates the β-catenin/Wnt signaling pathway in esophageal basaloid squamous cell carcinoma. Virchows Arch 464(2):135–143

Sargent DJ, Marsoni S, Monges G, Thibodeau SN, Labianca R, Hamilton SR, French AJ, Kabat B, Foster NR, Torri V, Ribic C, Grothey A, Moore M, Zaniboni A, Seitz JF, Sinicrope F, Gallinger S (2010) Defective mismatch repair as a predictive marker for lack of efficacy of fluorouracil-based adjuvant therapy in colon cancer. J Clin Oncol 28(20):3219–3226

Sato F, Shibata D, Harpaz N, Xu Y, Yin J, Mori Y et al (2002) Aberrant methylation of the *HPP1* gene in ulcerative colitis-associated colorectal carcinoma. Cancer Res 62:6820–6822

Sawan C, Herceg Z (2010) Histone modifications and cancer. Adv Genet 70:57–85

Schrump DS, Fischette MR, Nguyen DM, Zhao M, Li X, Kunst TF et al (2006) Phase I study of decitabine-mediated gene expression in patients with cancers involving the lungs, esophagus, or pleura. Clin Cancer Res 12:5777–5785

Schuebel KE, Chen W, Cope L, Glockner SC, Suzuki H, Yi JM et al (2007) Comparing the DNA hypermethylome with gene mutations in human colorectal cancer. PLoS Genet 3:e157

Schulz W (1998) DNA methylation in urological malignancies (review). IntJ Oncol 13:151–218

Semnani S, Sadjadi A, Fahimi S, Nouraie M, Naeimi M, Kabir J et al (2006) Declining incidence of esophageal cancer in the Turkmen Plain, eastern part of the Caspian Littoral of Iran: a retrospective cancer surveillance. Cancer Detect Prev 30:14–19

Shacham-Shmueli E, Beny A, Geva R, Blachar A, Figer A, Aderka D (2011) Response to temozolomide in patients with metastatic colorectal cancer with loss of *MGMT* expression: a new approach in the era of personalized medicine? J Clin Oncol 29:e262–e265

Shames DS, Girard L, Gao B, Sato M, Lewis CM, Shivapurkar N et al (2006) A genome-wide screen for promoter methylation in lung cancer identifies novel methylation markers for multiple malignancies. PLoS Med 3:e486

Shames DS, Minna JD, Gazdar AF (2007) DNA methylation in health, disease, and cancer. Cur mol med 7:85–102

Sharma S, Kelly TK, Jones PA (2010) Epigenetics in cancer. Carcinogenesis 31:27–36

Shima K, Nosho K, Baba Y, Cantor M, Meyerhardt JA, Giovannucci EL et al (2011) Prognostic significance of *CDKN2A* (*p16*) promoter methylation and loss of expression in 902 colorectal cancers: cohort study and literature review. Int J Cancer 128:1080–1094

Shimizu D, Ishikawa T, Ichikawa Y, Togo S, Hayasizaki Y, Okazaki Y et al (2005) Prediction of chemosensitivity of colorectal cancer to 5-fluorouracil by gene expression profiling with cDNA microarrays. Int J Oncol 27:371–376

Shin SK, Nagasaka T, Jung BH, Matsubara N, Kim WH, Carethers JM et al (2007) Epigenetic and genetic alterations in Netrin-1 Receptors UNC5C and *DCC* in human colon cancer. Gastroenterology 133:1849–1857

Shiovitz S, Bertagnolli MM, Renfro LA, Nam E, Foster NR, Dzieciatkowski S, Luo Y, Lao VV, Monnat RJ Jr, Emond MJ, Maizels N, Niedzwiecki D, Goldberg RM, Saltz LB, Venook A, Warren RS, Grady WM (2014) Alliance for clinical trials in oncology. CpG island methylator phenotype is associated with response to adjuvant irinotecan-based therapy for stage III colon cancer. Gastroenterology 147(3):637–645

Shirahata A, Sakuraba K, K, Goto T, Saito M, Ishibashi K, Kigawa G et al (2010) Detection of vimentin (VIM) methylation in the serum of colorectal cancer patients. Anticancer Res 30:5015–5018

Shu SY, Kudo M, Chen T, Nakabayashi K, Bhalla A, Van Der Spek PJ et al (2000) The three subfamilies of leucine-rich repeat-containing G protein-coupled receptors (*LGR*): identification of *LGR6* and LGR7 and the signaling mechanism for LGR7. Mol Endocrinol 14:1257–1271

Siegel R, Desantis C, Jemal A (2014) Colorectal cancer statistics. CA Cancer J Clin 64(2):104–117

Sirnes S, Honne H, Ahmed D, Danielsen SA, Rognum TO, Meling GI et al (2011) DNA methylation analyses of the connexin gene family reveal silencing of GJC1 (Connexin45) by promoter hypermethylation in colorectal cancer. Epigenetics 6:602–609

Sjoblom T, Jones S, Wood LD, Parsons DW, Lin J, Barber TD et al (2006) The consensus coding sequences of human breast and colorectal cancers. Science 314:268–274

Slattery ML, Wolff RK, Curtin K, Fitzpatrick F, Herrick J, Potter JD et al (2009) Colon tumor mutations and epigenetic changes associated with genetic polymorphism: insight into disease pathways. Mutat Res 660:12–21

Smith TJ, Davidson NE, Schapira DV, Grunfeld E, Muss HB, Vogel VG et al (1999) American Society of Clinical Oncology 1998 update of recommended breast cancer surveillance guidelines. J Clin Oncol 17:1080–1080

Smith G, Carey FA, Beattie J, Wilkie MJ, Lightfoot TJ, Coxhead J et al (2002) Mutations in *APC*, Kirsten-ras, and *p53*-alternative genetic pathways to colorectal cancer. Proc Natl Acad Sci U S A 99:9433–9438

Snover DC (2011) Update on the serrated pathway to colorectal carcinoma. Hum Pathol 42:1–10

Soong R, Powell B, Elsaleh H, Gnanasampanthan G, Smith D, Goh H et al (2000) Prognostic significance of T*P53* gene mutation in 995 cases of colorectal carcinoma: influence of tumour site, stage, adjuvant chemotherapy and type of mutation. Eur J Cancer 36:2053–2060

Soussi T, Beroud C (2003) Significance of T*P53* mutations in human cancer: a critical analysis of mutations at CpG dinucleotides. Hum Mutat 21:192–200

Soussi T, Dehouche K, Beroud C (2000) *p53* website and analysis of *p53* gene mutations in human cancer: forging a link between epidemiology and carcinogenesis. Hum Mutat 15:105–113

Spano JP, Milano G, Vignot S, Khayat D (2008) Potential predictive markers of response to EGFR-targeted therapies in colorectal cancer. Crit Rev Oncol/Hematol 66:21–30

Spring KJ, Zhao ZZ, Karamatic R, Walsh MD, Whitehall VL, Pike T et al (2006) High prevalence of sessile serrated adenomas with *BRAF* mutations: a prospective study of patients undergoing colonoscopy. Gastroenterology 131:1400–1407

Sproul D, Kitchen RR, Nestor CE, Dixon JM, Sims AH, Harrison DJ et al (2012) Tissue of origin determines cancer-associated CpG island promoter hypermethylation patterns. Genome Biol 13:84

Strathdee G, Brown R (2002) Aberrant DNA methylation in cancer: potential clinical interventions. Expert Rev Mol Med 2:1–17

Subramaniam MM, Chan JY, Soong R, Ito K, Yeoh KG, Wong R (2009) *RUNX3* inactivation in colorectal polyps arising through different pathways of colonic carcinogenesis. Am J Gastroenterol 104:426–436

Subramaniam D, Thombre R, Dhar A, Anant S (2014) DNA methyltransferases: a novel target for prevention and therapy. Front Oncol 4:80

Sugai T, Habano W, Jiao YF, Suzuki M, Takagi R, Otsuka K et al (2005) Analysis of allelic imbalances at multiple cancer-related chromosomal loci and microsatellite instability within the same tumor using a single tumor gland from colorectal carcinomas. Int J Cancer 114:337–345

Summers T, Langan RC, Nissan A, Brucher BL, Bilchik AJ, Protic M et al (2013) Serum-based DNA methylation biomarkers in colorectal cancer: potential for screening and early detection. J Cancer 4:210

Switzeny OJ, Mullner E, Wagner KH, Brath H, Aumuller E, Haslberger AG (2012) Vitamin and antioxidant rich diet increases *MLH1* promoter DNA methylation in DMT2 subjects. Clin Epigenetics 4:19

Tagore KS, Lawson MJ, Yucaitis JA, Gage R, Orr T, Shuber AP et al (2003) Sensitivity and specificity of a stool DNA multitarget assay panel for the detection of advanced colorectal neoplasia. Clin Colorectal Cancer 3:47–53

Tahara T, Shibata T, Nakamura M, Yamashita H, Yoshioka D, Okubo M et al (2009) Effect of MDR1 gene promoter methylation in patients with ulcerative colitis. Int J Mol Med 23:521

Tan SH, Ida H, Lau QC, Goh BC, Chieng WS, Loh M et al (2007) Detection of promoter hypermethylation in serum samples of cancer patients by methylation-specific polymerase chain reaction for tumour suppressor genes including *RUNX3*. Oncol Rep 18:1225–1230

Tanaka H, Deng G, Matsuzaki K, Kakar S, Kim GE, Miura S et al (2006) *BRAF* mutation, CpG island methylator phenotype and microsatellite instability occur more frequently and concordantly in mucinous than non-mucinous colorectal cancer. Int J Cancer 118:2765–2771

Takeda M, Nagasaka T, Dong-Sheng S (2011) Expansion of CpG methylation in the *SFRP2* promoter region during colorectal tumorigenesis. Acta Med Okayama 65:169–177

Tetzner R, Model F, Weiss G, Schuster M, Distler J, Steiger KV et al (2009) Circulating methylated *SEPT9* DNA in plasma is a biomarker for colorectal cancer. Clin Chem 55:1337–1346

Toiyama Y, Okugawa Y, Goel A (2014) DNA methylation and microRNA biomarkers for noninvasive detection of gastric and colorectal cancer. Biochem Biophys Res Commun 455(1-2):43–57

Tokuyama Y, Takahashi T, Okumura N, Nonaka K, Kawaguchi Y, Yamaguchi K et al (2010) Aberrant methylation of heparan sulfate glucosamine 3-O-sulfotransferase 2 genes as a biomarker in colorectal cancer. Anticancer Res 30:4811–4818

Tominaga K, Fujii S, Mukawa K, Fujita M, Ichikawa K, Tomita S et al (2005) Prediction of colorectal neoplasia by quantitative methylation analysis of estrogen receptor gene in nonneoplastic epithelium from patients with ulcerative colitis. Clin Cancer Res 11:8880–8885

Toribara NW, Sleisenger MH (1995) Screening for colorectal cancer. N Engl J Med 332:861–867

Toyota M, Ahuja N, Ohe-Toyota M, Herman JG, Baylin SB, Issa JPJ (1999) CpG island methylator phenotype in colorectal cancer. Proc Natl Acad Sci U S A 96:8681–8686

Toyota M, Itoh F, Kikuchi T, Satoh A, Obata T, Suzuki H et al (2002) DNA methylation changes in gastrointestinal disease. J Gastroenterol 37:97–101

Turker MS, Bestor TH (1997) Formation of methylation patterns in the mammalian genome. Mutat Res 386:119–130

Tusher VG, Tibshirani R, Chu G (2001) Significance analysis of microarrays applied to the ionizing radiation response. Proc Natl Acad Sci U S A 98:5116–5121

Ullman TA, Itzkowitz SH (2011) Intestinal inflammation and cancer. Gastroenterology 140:1807–1816

Urso E, Pucciarelli S, Agostini M, Maretto I, Mescoli C, Bertorelle R et al (2008) Proximal colon cancer in patients aged 51-60 years of age should be tested for microsatellites instability. A comment on the Revised Bethesda Guidelines. Int J Colorectal Dis 23:801–806

Vaiserman A (2014) Early-life exposure to endocrine disrupting chemicals and later-life health outcomes: an epigenetic bridge? Aging Dis 5(6):419–429

Vamadevan AS, Fukata M, Arnold ET, Thomas LS, Hsu D, Abreu MT (2010) Regulation of Toll-like receptor 4-associated MD-2 in intestinal epithelial cells: a comprehensive analysis. Innate Immun 16:93–103

Van Groeningen CJ, Leyva A, Obrien AM, Gall HE, Pinedo HM (1986) Phase I and pharmacokinetic study of 5-aza-2′-deoxycytidine (NSC 127716) in cancer patients. Cancer Res 46:4831–4836

Van Rijnosever M, Grieu F, Elsaleh H, Joseph D, Iacopetta B (2002) Characterisation of colorectal cancers showing hypermethylation at multiple CpG islands. Gut 51:797–802

Van Rossum LG, Van Rijn AF, Laheij RJ, Van Oijen MG, Fockens P, Van Krieken HH et al (2008) Random comparison of guaiac and immunochemical fecal occult blood tests for colorectal cancer in a screening population. Gastroenterology 135:82–90

Van Veen W, Mali W (2009) Colorectal cancer screening: advice from the Health Council of the Netherlands. Nederlands tijdschrift voor geneeskunde 153:A1441

Vansteenkiste J, Van Cutsem E, Dumez H, Chen C, Ricker JL, Randolph SS et al (2008) Early phase II trial of oral vorinostat in relapsed or refractory breast, colorectal, or non-small cell lung cancer. Invest New Drugs 26:483–488

Varambally S, Cao Q, Mani RS, Shankar S, Wang X, Ateeq B et al (2008) Genomic loss of microRNA-101 leads to overexpression of histone methyltransferase EZH2 in cancer. Science 322:1695–1699

Veganzones-De-Castro S, Rafael-Fernandez S, Vidaurreta-Lazaro M, De-La-Orden V, Mediero-Valeros B, Fernandez C et al (2012) *p16* gene methylation in colorectal cancer patients with long-term follow-up. Rev Esp Enferm Dig 104:111

Venkateswaran K, Verma A, Bhatt AN, Agrawala P, Raj HG, Malhotra S, Prasad AK, Wever OD, Bracke ME, Saso L, Parmar VS, Shrivastava A, Dwarakanath BS (2014) Modifications of cell signalling and redox balance by targeting protein acetylation using natural and engineered molecules: implications in cancer therapy. Curr Top Med Chem 14(22):2495–2507

Vilar E, Mork ME, Cuddy A, Borras E, Bannon SA, Taggart MW, Ying J, Broaddus RR, Luthra R, Rodriguez-Bigas MA, Lynch PM, You YQ (2014) Role of microsatellite instability-low as a diagnostic biomarker of Lynch syndrome in colorectal cancer. Cancer Genet. 207(10-12):495–502. pii: S2210-7762(14)00228-2

Vo AT, Millis RM (2012) Epigenetics and breast cancers. Obstet Gynecol Int 2012:1–10

Vogelstein B, Fearon ER, Hamilton SR, Kern SE, Preisinger AC, Leppert M et al (1988) Genetic alterations during colorectal-tumor development. N Engl J Med 319:525–532

Vogelstein B, Papadopoulos N, Velculescu VE, Zhou S, Diaz LA Jr, Kinzler KW (2013) Cancer genome landscapes. Science 339(6127):1546–1558

Wagner T, Brand P, Heinzel T, Krämer OH (2014) Histone deacetylase 2 controls p53 and is a critical factor in tumorigenesis. Biochim Biophys Acta 1846(2):524–538

Wajed SA, Laird PW, Demeester TR (2001) DNA methylation: an alternative pathway to cancer. Ann Surg 234:10

Wallace K, Burke CA, Ahnen DJ, Barry EL, Bresalier RS, Saibil F, Baron JA (2014) The association of age and race and the risk of large bowel polyps. Cancer Epidemiol Biomarkers Prev. pii: cebp.1076.2014

Wang Y, Leung FC (2004) An evaluation of new criteria for CpG islands in the human genome as gene markers. Bioinformatics 20:1170–1177

Wang DR, Tang D (2008) Hypermethylated *SFRP2* gene in fecal DNA is a high potential biomarker for colorectal cancer noninvasive screening. World J Gastroenterol 14:524

Wang J, Kataoka H, Suzuki M, Sato N, Nakamura R, Tao H et al (2005) Downregulation of EphA7 by hypermethylation in colorectal cancer. Oncogene 24:5637–5647

Wang R, Lohr CV, Fischer K, Dashwood WM, Greenwood JA, Ho E et al (2013) Epigenetic inactivation of endothelin-2 and endothelin-3 in colon cancer. Int J Cancer 132:1004–1012

Wang X, Kuang YY, Hu XT (2014) Advances in epigenetic biomarker research in colorectal cancer. World J Gastroenterol 20(15):4276–4287

Weir BA, Woo MS, Getz G, Perner S, Ding L, Beroukhim R et al (2007) Characterizing the cancer genome in lung adenocarcinoma. Nature 450:893–898

Wen B, Wu H, Loh YH, Briem E, Daley G, Feinberg A (2012) Euchromatin islands in large heterochromatin domains are enriched for CTCF binding and differentially DNA-methylated regions. BMC Genomics 13:566

Wendt M, Johanesen P, Kang-Decker N, Binion D, Shah V, Dwinell M (2006) Silencing of epithelial CXCL12 expression by DNA hypermethylation promotes colonic carcinoma metastasis. Oncogene 25:4986–4997

Whitney D, Skoletskky J, Moore K, Boynton K, Kann L, Brand R et al (2004) Enhanced retrieval of DNA from human fecal samples results in improved performance of colorectal cancer screening test. J Mol Diagn 6:386–395

Wiencke JK, Zheng S, Lafuente A, Lafuente MJ, Grudzen C, Wrensch MR et al (1999) Aberrant methylation of *p16INK4a* in anatomic and gender-specific subtypes of sporadic colorectal cancer. Cancer Epidemiol Biomarkers Prev 8:501–506

Wilson AS, Power BE, Molloy PL (2007) DNA hypomethylation and human diseases. Rev Cancer 1775:138–162

Wolffe AP, Matzke MA (1999) Epigenetics: regulation through repression. Science 286:481–486

Wong JJL, Hawkins NJ, Ward RL (2007) Colorectal cancer: a model for epigenetic tumorigenesis. Gut 56:140–148

Wright C, Dent O, Newland R, Barker M, Chapuis P, Bokey E et al (2005) Low level microsatellite instability may be associated with reduced cancer specific survival in sporadic stage C colorectal carcinoma. Gut 54:103–108

Wu JT, Archer SY, Hinnebusch B, Meng S, Hodin RA et al (2001) Transient vs. prolonged histone hyperacetylation: effects on colon cancer cell growth, differentiation, and apoptosis. Am J Physiol-Gastrointestin Liver Physiol 280:G482–490

Wynter C, Walsh M, Higuchi T, Leggett B, Young J, Jass J (2004) Methylation patterns define two types of hyperplastic polyp associated with colorectal cancer. Gut 53:573–580

Wynter CV, Kambara T, Walsh MD, Leggett BA, Young J, Jass JR et al (2006) DNA methylation patterns in adenomas from FAP, multiple adenoma and sporadic colorectal carcinoma patients. Int J Cancer 118:907–915

Xavier RJ, Rioux JD (2008) Genome-wide association studies: a new window into immune-mediated diseases. Nat Rev Immunol 8:631–643

Xu XL, Yu J, Zhang HY, Sun MH, Gu J, Du X et al (2004) Methylation profile of the promoter CpG islands of 31 genes that may contribute to colorectal carcinogenesis. World J Gastroenterol 10:3441–3454

Yamasaki M, Takemasa I, Komori T, Watanabe S, Sekimoto M, Doki Y et al (2007) The gene expression profile represents the molecular nature of liver metastasis in colorectal cancer. Int J Oncol 30:129–138

Yamashita K, Waraya M, Kim MS, Sidransky D, Katada N, Sato T, Nakamura T, Watanabe M (2014) Detection of methylated CDO1 in plasma of colorectal cancer; a PCR study. PLoS One 9(12):e113546

Yamazaki K, Mcgovern D, Ragoussis J, Paolucci M, Butler H, Jewell D et al (2005) Single nucleotide polymorphisms in TNFSF15 confer susceptibility to Crohn's disease. Hum Mol Genet 14:3499–3506

Yan PS, Perry MR, Laux DE, Asare AL, Caldwell CW, Huang THM (2000) CpG island arrays: an application toward deciphering epigenetic signatures of breast cancer. Clin Cancer Res 6:1432–1438

Yang E, Kang HJ, Koh KH, Rhee H, Kim NK, Kim H (2007) Frequent inactivation of SPARC by promoter hypermethylation in colon cancers. Int J Cancer 121:567–575

Yang Y, Yang JJ, Tao H, Jin WS (2014) New perspectives on β-catenin control of cell fate and proliferation in colon cancer. Food Chem Toxicol 74C:14–19

Ying J, Li H, Yu J, Ng KM, Poon FF, Wong SCC et al (2008) WNT5A exhibits tumor-suppressive activity through antagonizing the Wnt/β-catenin signaling, and is frequently methylated in colorectal cancer. Clin Cancer Res 14:55–61

Yoder JA, Bestor TH (1998) A candidate mammalian DNA methyltransferase related to pmt1p of fission yeast. Hum Mol Genet 7:279–284

Yoruker EE, Mert U, Bugra D, Yamaner S, Dalay N (2012) Promoter and histone methylation and *p16ink4a* gene expression in colon cancer. Exp Ther Med 4:865–870

You J, Nguyen AV, Albers CG, Lin F, Holcombe RF (2008) Wnt pathway-related gene expression in inflammatory bowel disease. Dig Dis Sci 53:1013–1019

Yu J, Tao Q, Cheng YY, Lee KY, Ng SS, Cheung KF, Tian L et al (2009) Promoter methylation of the Wnt/β-catenin signaling antagonist Dkk-3 is associated with poor survival in gastric cancer. Cancer 115:49–60

Zeller G, Tap J, Voigt AY, Sunagawa S, Kultima JR, Costea PI, Amiot A, Böhm J, Brunetti F, Habermann N, Hercog R, Koch M, Luciani A, Mende DR, Schneider MA, Schrotz-King P, Tournigand C, Tran Van Nhieu J, Yamada T, Zimmermann J, Benes V, Kloor M, Ulrich CM, von Knebel Doeberitz M, Sobhani I, Bork P (2014) Potential of fecal microbiota for early-stage detection of colorectal cancer. Mol Syst Biol 10:766. doi:10.15252/msb.20145645

Zhai R, Zhao Y, Su L, Cassidy L, Liu G, Christiani DC (2012) Genome-wide DNA methylation profiling of cell-free serum DNA in esophageal adenocarcinoma and Barrett esophagus. Neoplasia (New York) 14:29

Zhang W, Bauer M, Croner RS, Pelz JO, Lodygin D, Hermeking H et al (2007) DNA stool test for colorectal cancer: hypermethylation of the secreted frizzled-related protein-1 gene. Dis Colon Rectum 50:1618–1627

Zhang H, Qi J, Wu YQ, Zhang P, Jiang J, Wang QX, Zhu YQ (2014) Accuracy of early detection of colorectal tumours by stool methylation markers: a meta-analysis. World J Gastroenterol. 20(38):14040–14050

Zoratto F, Rossi L, Verrico M, Papa A, Basso E, Zullo A, Tomao L, Lo Russo G, Tomao S (2014) Focus on genetic and epigenetic events of colorectal cancer pathogenesis: implications for molecular diagnosis. Tumour Biol 35(7):6195–6206

Zou H, Harrington J, Rego RL, Ahlquist DA (2007) A novel method to capture methylated human DNA from stool: implications for colorectal cancer screening. Clin Chem 53:1646–1651

Chapter 14
Malignant Rhabdoid Tumor: Epigenetic Mechanism of Tumorogenesis

Sima Kheradmand Kia

Contents

Abstract This chapter explains the epigenetic mechanism of tumor formation in an aggressive and rare childhood malignancy, malignant rhabdoid tumors (MRT). In the majority of these tumors hSNF5/INI1 is inactivated via deletion or mutation. hSNF5/INI1 is a member of the ATP-dependent hSWI-SNF chromatin remodeling complex. This gene is a tumor suppressor gene. hSNF5 can function in both transcription activation and repression of genome. Re-expression of hSNF5 in MRT cells causes an accumulation in G0/G1, cellular senescence and apoptosis. Cellular senescence is largely the result of direct transcriptional activation of the tumor-suppressor $p16^{INK4a}$ by hSNF5. Whole genome expression profiling of hSNF5 cells revealed expression change of many E2F targets, including mitotic control genes and pre-replication complex. The balance between SWI/SNF activation and Polycomb group (PcG) silencing affects epigenetic control of the *INK4b-ARF-INK4a*

S. Kheradmand Kia (✉)
Department of Biochemistry and Centre for Biomedical Genetics, Erasmus University Medical Centre, P. O. Box 2040, 3000 CA, Rotterdam, The Netherlands
e-mail: Kheradmandkia@erasmusmc.nl

P. Mehdipour (ed.), *Epigenetics Territory and Cancer*,
DOI 10.1007/978-94-017-9639-2_14

locus in MRT cells. PcG proteins regulate higher order chromatin structure dynamically, to balance cell proliferation and differentiation. SWI/SNF mediates eviction of the Polycomb group repressive complex1 (PRC1) and Polycomb group repressive complex2 (PRC2) and extensive chromatin reprogramming.

14.1 Introduction

Chromatin is a dynamic structure that modulates access of regulatory factors to the genetic material of eukaryotes and hence controls DNA replication machinery, transcription and repair at higher level. The regulated alteration of chromatin structure, termed remodeling, can be accomplished by covalent modification of histones or by the action of ATP-dependent remodeling complexes. Remodelers are nucleosome-translocating motors that utilize the energy of ATP to disrupt histone-DNA contacts (Fyodorov and Kadonaga 2002; Lia et al. 2006; Saha et al. 2005; Whitehouse et al. 2003; Zhang et al. 2006).

The multi-subunit SWI/SNF ATP-dependent chromatin remodeling complexes are highly conserved molecular motors that play crucial roles in diverse cellular processes, including gene expression and genome duplication during cell cycle. Human SNF5/INI1 is a universal SWI/SNF subunit and a tumor suppressor gene that is frequently lost in malignant rhabdoid tumors (MRTs), rare but highly aggressive pediatric cancers (Biegel et al. 1999; Sevenet et al. 1999a; Sevenet et al. 1999b; Versteege et al. 1998).

Re-expression of hSNF5 in MRT cells caused G0/G1 stall, cellular senescence and apoptosis (Oruetxebarria et al. 2004). Cellular senescence is largely the result of direct transcriptional activation of the tumor-suppressor $p16^{INK4a}$ by hSNF5 (Oruetxebarria et al. 2004). hSNF5 acts as a transcriptional coactivator, which is required for the recruitment of the BRG1 containing SWI/SNF chromatin remodeling complex to the *p16*^{INK4a} promoter. The increased *p16*^{INK4a} levels result in inhibition of the cyclin D1-CDK4 complex, thus retaining pRb in its hypo phosphorylated anti proliferative state (Oruetxebarria et al. 2004). Loss of hSNF5 function in MRT cells promotes chromosomal instability by compromised mitosis (Vries et al. 2005). Since cancers resulted from loss of hSNF5 is extremely aggressive, insight in the involved pathways might be relevant for understanding other forms of cancer.

14.2 Malignant Rhabdoid Tumor, Cause and Disease Mechanism

14.2.1 Malignant Rhabdoid Tumor and Atypical Teratoid Rhabdoid Tumor

The term malignant rhabdoid tumor (MRT) has been used to describe a heterogeneous group of neoplasms, having distinct so-called “rhabdoid” cytologic features in common. The rhabdoid cell is a medium-sized, round-to-oval cell with distinct borders, an eccentric nucleus, and a prominent nucleolus. MRT may arise either de novo from non-neoplastic cells or through tumor progression from other types of neoplasms. Rhabdoid tumors were reported in many tissues including kidney, liver, soft tissue, and central nervous system (Biegel et al. 1999; Rorke et al. 1995; Sevenet et al. 1999b; Versteege et al. 1998). The cerebellum is the most common location for primary intracerebral MRT. MRT of the CNS is known as atypical teratoid/rhabdoid tumor (AT/RT). ATRT/MRTs are very rare but highly aggressive cancer of early childhood and despite intensive therapies 80–90 % of children die within 1 year of diagnosis. About 50 % of atypical teratoid/rhabdoid tumors arise in the posterior fossa, 40 % are supratentorial, and the rest are pineal, spinal, or multifocal. Both ATRTs and MRTs are characterized by the presence of rhabdoid cells carrying vacuolated nuclei and Periodic acid-Shiff (PAS) cytoplasmic inclusions, however the histological diagnosis can be difficult (Haas et al. 1981). Biallelic inactivating mutations and deletions of the SWI/SNF core subunit SNF5/INI1 have been identified in the majority of kidney malignant rhabdoid tumors and brain atypical teratoid/rhabdoid tumors (ATRT). However, despite loss of immunostaining for the SNF5 protein at least 20 % of cases do not have genomic alterations of *hSNF5* (Sevenet et al. 1999b; Sevenet et al. 1999a; Versteege et al. 1998; Biegel et al. 1999). In several cases, germline mutations in *SNF5/INI1* gene accompanied by somatic loss or mutation of the remaining allele were documented in patients with ATRT/MRTs indicating that *SNF5/INI1* is a classical tumor suppressor gene (Sevenet et al. 1999b; Biegel et al. 1999). Loss of *SNF5/INI1* has been also detected in a number of tumors histologically distinct from ATRT/MRTs such as pediatric choroid plexus carcinoma, meningioma, medulloblastoma (Roberts and Orkin 2004). Recurrent hemizygous- homozygous deletions of 7q35–q36.1, involving Contactin Associated Protein-Like 2 (CNTNAP2) locus, hypermethylation of CNTNAP2 and a novel R157C missense mutation have been reported in MRT specimens, cell lines and in a primary case (Takita et al. 2014).

14.2.2 The SWI/SNF Complex and Cancer

ATP-dependent chromatin-remodeling factors (remodelers) are critical for the transmission, maintenance and expression of the eukaryotic genome. They function by mobilizing nucleosomes at the sites of DNA replication/repair and transcription activation/repression thus opening or closing chromatin for DNA-binding proteins. ATP-dependent chromatin remodeling complex can affect gene expression, cell cycle progression, and cell differentiation (Becker and Horz 2002). The multi subunit SWI/SNF complex is evolutionary highly conserved and present in all eukaryotes (Mohrmann and Verrijzer 2005). Thus far there is strong evidence supporting a role for SWI/SNF complexes in cancer development, as several subunits possess intrinsic tumor suppressor activity or are required for the activity of other tumor suppressor genes (Versteege et al. 1998; Sevenet et al. 1999a; Sevenet et al. 1999b; Biegel et al. 1999; Decristofaro et al. 2001; Judkins et al. 2004; Wong et al. 2000; Reisman et al. 2003). Inactivating mutations or aberrant expression of the genes encoding SWI/SNF core subunits have been found in different human tumor cell lines and primary tumors (Decristofaro et al. 2001). In humans, the SWI/SNF family includes both the BAF and PBAF subclasses. BRG1 and BRM mutations are frequently observed in various tumor cell lines including pancreatic, breast, lung and prostate cancer cells. Lack of expression of both SWI/SNF ATPase subunits BRG1 and hBRM seem to correlate with poorer prognosis in patients with non-small-cell lung cancer (Reisman et al. 2003). Human *SNF5* (hSNF5, INI1 or SMARCB1) located in the chromosomal region 22q11.2, is a core component of the hSWI/SNF and it has been implicated in gene regulation, cell division and tumorogenesis (Johnson et al. 2005). Mice heterozygous for SNF5 are predisposed to tumors with features similar to the human MRTs, which are frequently metastatic to the lung and/or lymph node. SNF5-deficient tumors undergo loss of heterozygosity (LOH), which results in SNF5 depletion (Bultman et al. 2000; Klochendler-Yeivin et al. 2000; Guidi et al. 2001; Roberts et al. 2000). Unlike SNF5-deficient tumors, BRG1-deficient tumors occur at different locations such as the neck or inguinal regions, display different features of epitheloid origin, and appear not to undergo LOH (Wong et al. 2000).

The SWI/SNF complex associates directly with cancer-related molecules such as BRCA1 and c-Myc and beta-catenin (Bochar et al. 2000; Takayama et al. 2000; Cheng et al. 1999; Barker et al. 2001). BRCA1 activates p53-dependent transcription, which is abrogated by a dominant-negative mutant form of BRG1 (Bochar et al. 2000). SNF5 also interacts directly with c-Myc and this interaction is important for trans-activating the function of c-Myc (Cheng et al. 1999). SWI/SNF complex has been shown to interact directly with MLL, the human homolog of Drosophila Trithorax (Trx), through SNF5 subunit. Collectively these observations point to the function of SWI/SNF complexes as tumor suppressor gene.

14.2.3 SNF5 is Mutated in Malignant Rhabdoid Tumor

A significant majority of malignant rhabdoid tumors (MRTs) carry specific, biallelic, inactivating mutations in *hSNF5*. Loss of *SNF5/INI1* has been also detected in pediatric choroid plexus carcinoma, meningioma, medulloblastoma and primitive neuroectodermal tumors which are histologically distinct from ATRT/MRTs (Roberts and Orkin 2004). Deletion of *SNF5* is reported in epithelioid sarcoma with high frequency (Sullivan et al. 2013).

Inherited mutation of *SNF5* lead to rhabdoid predisposition syndrome (Biegel et al. 1999; Sevenet et al. 1999b).

As the majority of these extremely aggressive cancers have an entirely normal karyotype with only loss of *SNF5* which may result in extensive epigenetic changes (Douglass et al. 1990; Sansam and Roberts 2006). Although most of the biallelic mutations in *SNF5* are deletions, truncating nonsense mutations, or frame shifts, a number of point mutations resulting in single amino acid substitutions (Fig. 14.1a),

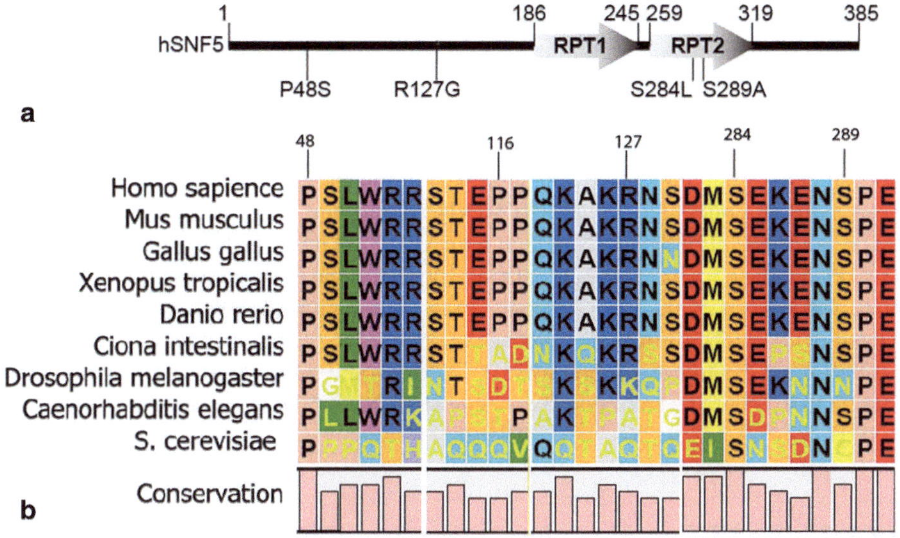

Fig. 14.1 Cancer associated amino acid substitution of hSNF5 (**a**) Schematic representation of hSNF5 depicting the two repeats (RPT1 and RPT2) and cancer-associated amino acid substitution mutations. (**b**) Conservation of cancer associated amino acid substitution mutations from Human to Yeast

have been identified in tumors as well (Sevenet et al. 1999b; Cho et al. 2006). These include proline 48 to serine (P48S), arginine 127 to glycine (R127G), and serine 284 to leucine (S284L) and proline 116 to threonine. S284 is located within one of the most highly conserved regions of SNF5, which forms part of direct repeat 2 (RPT2) and it is conserved during the evolution (Fig. 14.1b). Loss of hSNF5 function in MRT cells promotes chromosomal instability by compromised mitosis (Vries et al. 2005).

14.3 Epigenetic Mechanism of SNF5-Tumorogenesis

14.3.1 SNF5 is a Tumor Suppressor Gene

The SWI/SNF complex is involved in various cellular processes that are potentially associated with tumor formation including DNA synthesis, virus integration, DNA repair, and mitotic gene regulation. Numerous studies to dissect the connection between these activities and tumor formation are currently in progress. The molecular mechanisms underlying tumor development in mice with inactivation of BRG1 or SNF5 are still unclear.

The ability of SNF5 to function as a tumor suppressor has been confirmed in studies utilizing SNF5-deficient mice. Mice strain carrying reversibly inactivating

SNF5/INI1 allele by applying LoxP-Cre recombination system has been generated to find the mechanisms of tumorogenesis caused by loss of SNF5/INI1 (Roberts et al. 2002). All of the resultant mice develop short latency highly aggressive tumors such as $CD8^+$ T cell lymphomas and rare rhabdoid tumors (Roberts et al. 2002).

Although the molecular mechanisms for SNF5/INI1 function in cell survival in normal cells is not yet known for mammals, in recent years significant progress has been made in understanding of SNF5/INI1 role in tumor suppression. It appeared, that re-expression of SNF5/INI1 in human MRT cell lines leads to an accumulation in G0/G1 phase, cellular senescence and in some cases apoptosis (Oruetxebarria et al. 2004; Vries et al. 2005; Ae et al. 2002; Versteege et al. 2002; Betz et al. 2002; Zhang et al. 2002). Cellular senescence is largely the result of direct transcriptional activation of the tumor-suppressor $p16^{INK4a}$ by hSNF5 (Ae et al. 2002; Betz et al. 2002; Oruetxebarria et al. 2004; Versteege et al. 2002; Vries et al. 2005; Zhang et al. 2002).

14.3.2 SNF5 Can Function in Transcription Activation and Repression of the Genome

To study the function of hSNF5 in MRT cells, two distinct strategies to re-express hSNF5 in G401 and Mon, two different MRT cell lines, has been established. Induction of the hSNF5 gene in G401 cell line was under control of the Lac repressor-operator system; in parallel hSNF5 or GFP has been transduced to Mon cells using lentiviral transfection. The expression level of SNF5 in both system were comparable.

Moreover, the levels of induced exogenous hSNF5 expression were comparable to the endogenous levels in a variety of cell lines (Oruetxebarria et al. 2004; Doan et al. 2004; Moshkin et al. 2007).

To elucidate the pathways controlled by hSNF5 in rhabdoid tumor cells, we performed cDNA microarray analysis. Our gene expression profiling results suggested that majority of the up-regulated genes encode proteins with functions in, extra cellular matrix remodeling, adhesion or cell migration (*SERPINE2*, *ITGB5*, *MAP1B*), apoptosis (*DR6,FAS*, *CASP4*, *GAS6*, *ADAM19*), and cancer related pathway or other specialized functions (*CDKN2A*, *ETS2* and *TRIM22*). The majority of the genes down-regulated by hSNF5 encoded proteins that play key roles during cell cycle, like, *CENPE*, *POLD3*,*CENPA*, *CDC25A*, *CCNF* and *MAD2* but remain unchanged upon induction of mutant hSNF5. These results suggested that mis-expression of mitotic checkpoint components might cause the abnormal ploidy of MRT cells. For example, over-expression of MAD2 and its regulator E2F1 was implicated in mitotic defects leading to aneuploidy (Hernando et al. 2004). Interestingly, in our microarray experiments, both genes were down-regulated following hSNF5 induction. We used QRT-PCR to corroborate our microarray results

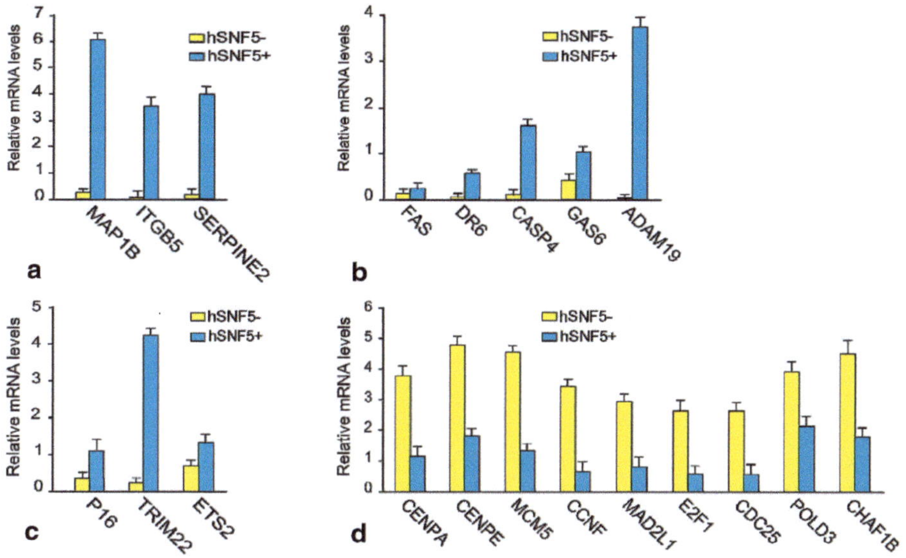

Fig. 14.2 QRT-PCR analysis of gene regulated by hSNF5 identified by whole-genome expression profiling. RT-qPCR analysis of gene expression in MRT cells reveals hSNF5-dependent induction of genes involved in (**a**) apoptosis pathway (**b**) Cell migration and invasion (**c**) Other pathway and (**d**) depletion of E2F and E2F target genes affected. Cells were collected 48 h following transduction with lentiviruses expressing either GFP (*yellow bars*) or hSNF5 (*blue bars*). mRNA levels were plotted as percentage of *GAPDH* mRNA. The bar graphs represent the mean of three independent biological replicates, each analyzed by three separate qPCR reactions. Standard deviations are indicated

(Fig. 14.2). We found that *MAD2*, *E2F1* and E2F1 target genes are highly expressed in MRT cells, but are strongly down-regulated following hSNF5 induction.

Further we transduced GFP or SNF5 in MRT cells (G401 and MON), Hela cell and U2OS cell with lentivirus (Fig. 14.3a). Western immunoblot analysis of extracts from cells transduced by GFP (lanes 1, 3 and 5) or cells transduced with lentiviruses expressing SNF5 (lanes 2, 6 and 8) revealed expression of hSNF5 in G401 and Mon cells (lane 2,4) and over-expression of hSNF5 in Hela and U2OS cells (lane 6 and 8). Antibodies directed against histone H3 were used as a loading control. QRT-PCR revealed induction of *FAS*, *ETS2*, *TRIM22*, *GAS6*, *ADAM19* and *SERPINE2* in MRT cells upon induction of SNF5 while they remain unchanged or down-regulated in either Hela or U2OS cells (Fig. 14.3b–14.3g).

To identify hSNF5 direct target genes of hSNF5 we performed chromatin immuneprecipitation on selected set of genes identified as potential hSNF5 targets upon hSNF5 transduced MRT cells. MRT cells were transduced with lentivirus (or were induced with IPTG). Chromatin was prepared from cells at 48 h post-stimulation and subjected to chromatin immunoprecipitation with an antibody specific for

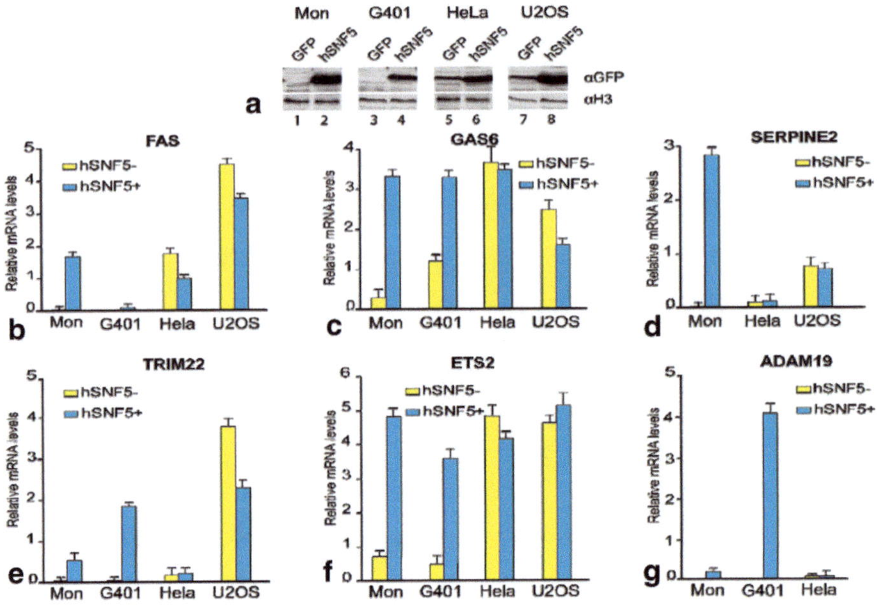

Fig. 14.3 RTQ-PCR analysis of gene regulated by hSNF5 in different cells (**a**) Western immunoblotting analysis of hSNF5 expression in MON. G401, Hela and U2OS cells transduced with lentiviruses expressing either GFP (*lanes 1, 3, 5, 7*) or hSNF5 (*lanes 2, 4, 6, 8*). Cell lysates were resolved by SDS-PAGE and analyzed by Western immunoblotting using antibodies directed against hSNF5. Histone H3 serves as a loading control. (**b–e**) RTQ-PCR analysis of genes upon induction or over expression of hSNF5 in two MRT cells (*Mon, G401*) Hela and U2OS cells. PCR revealed induction of (**b**) FAS, (**c**) GAS6, (**d**) SERPINE2, (**e**) TRIM22, (**f**) ETS2 and (**g**) ADAM19 in MRT cells upon induction of SNF5 while they remain unchanged or down-regulated in either Hela or U2OS cells. Procedures were as described in the legend to Fig. 14.2

SMARCB1 (hSNF5). As shown in Fig. 14.4a our results indicate that, while initially absent from the *TRIM22*, *ETS2, ANGPT2* and *MAD2L* genes, SNF5 is specifically recruited to these promoters in response to *SNF5* expression. *hSNF5* expression does not result in recruitment of SNF5 to the adjacent *P14*, *ADAM19* and *INTGB5* promoters. SNF5 is also recruiting BRG1 to the promoter of *TRIM22, ETS2, ANGPT2* and *MAD2L* but not *ADAM19* and *INTGB5* (Fig. 14.4). Taken together, these results demonstrate that re-expression of hSNF5 in MRT cells results in induction and specific recruitment of SNF5 and BRG1 to the *TRIM22*, *ETS2, ANGPT2* and *MAD2L* promoters but *not P14, ADAM19* and *INTGB5* promoters.

All ChIP data presented in this study are the result of at least three independent experiments. The abundance of specific DNA sequences in the immuneprecipitates was determined by qPCR and corrected for the independently determined amplification curves for each primer set. Background levels were determined by ChIP using species and isotype-matched immunoglobins directed against an unrelated (GST) protein. ChIPs with antibodies directed against SNF5 and BRG1 were analyzed by qPCR using primer sets corresponding to *MAD2, TRIM22, ETS2, ANGPT2,* $p14^{Arf}$, *INTGB5* and *ADAM19* promoters.

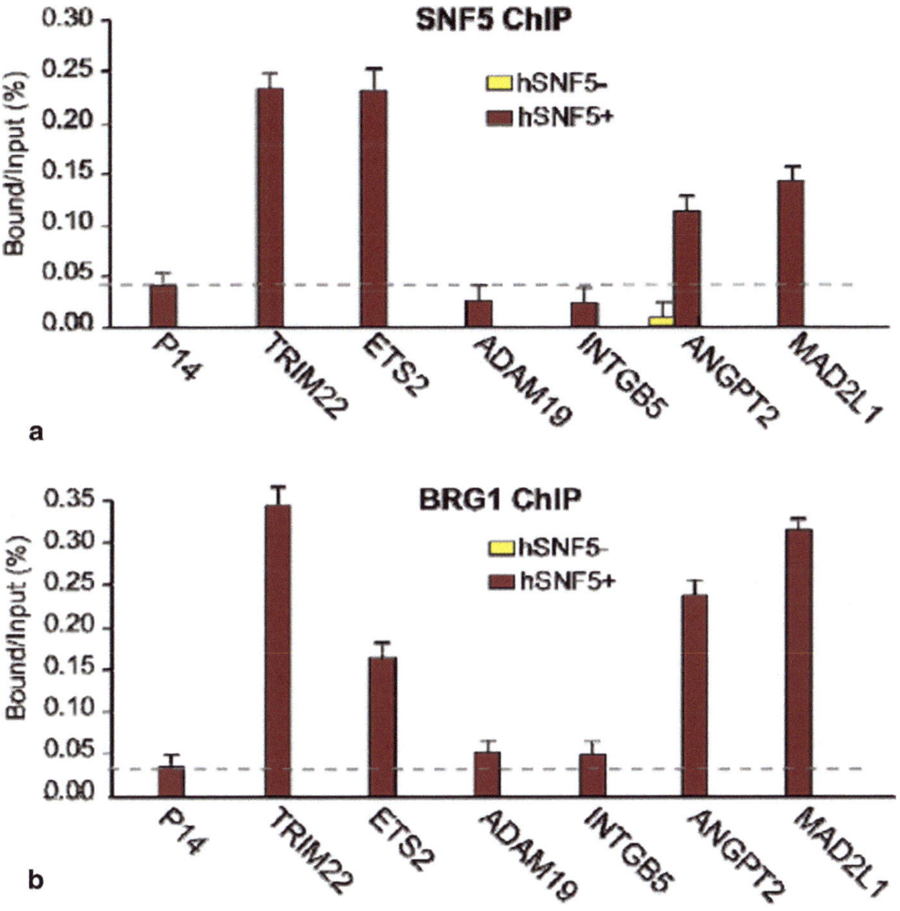

Fig. 14.4 hSNF5 Mediates BRG1 Recruitment to the MAD2, TRIM22, ETS2 and ANGPT2 promoters (**a**) ChIP-qPCR analysis of hSNF5 binding to the MAD2,TRIM22, ETS2 and ANGPT2, p14Arf, INTGB5 and ADAM19 revealed that hSNF5 binds directly to the MAD2, TRIM22, ETS2 and ANGPT2 promoters, but not to p14Arf. Cross-linked chromatin was isolated from MRT cells that either lack- (*light bars*) or express hSNF5 (*dark bars*). (**b**) RG-1 binding to the MAD2, TRIM22, ETS2 and ANGPT2 promoters is hSNF5-dependent, as revealed by ChIP-qPCR using antibodies directed against BRG-1

ChIP signal levels for each region are presented as percentage of input chromatin.

14.3.3 SNF5 is an Activator of P16 Tumor Suppressor Gene

SNF5/INI1 binds to the promoter of $p16^{ink4a}$ tumor suppressor gene and recruits BRG1-containing SWI/SNF complex resulting in transcription activation (Oruetxebarria et al. 2004; Kia et al. 2008). $p16^{ink4a}$ gene encodes a specific cyclin-

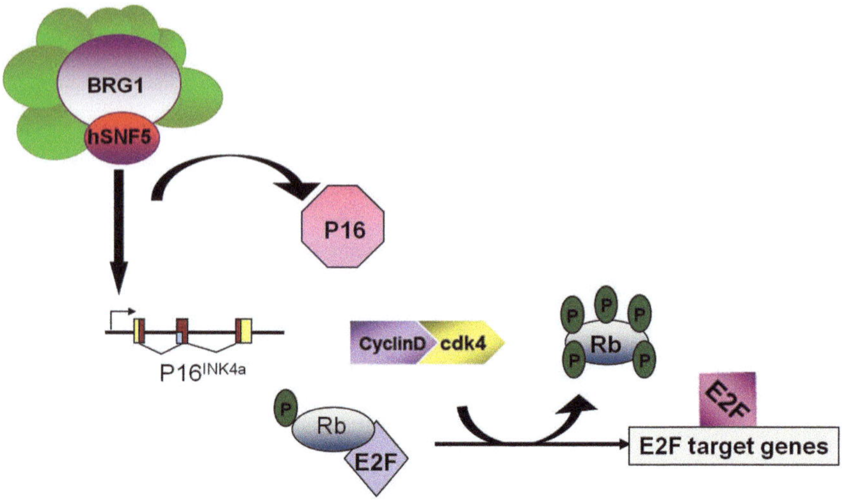

Fig. 14.5 Mammalian ATRT/MRT tumor cells lack SNF5/INI1 function resulting in increased CDK4/cyclin D1 activity

dependent kinase (CDK)4/CDK6 inhibitors, which targets retinoblastoma (pRb) protein for phosphorylation (Lowe and Sherr 2003). Hyperphosphorylated pRb dissociates from E2F transcription factor allowing S phase specific genes expression and promoting cell proliferation. Therefore, increased $p16^{ink4a}$ expression upon SNF5/INI1 induction in MRT cells results in pRb hypophosphorylation and inhibits cell cycle progression (Fig. 14.5). Interestingly, MRT cell lines lacking $p16^{ink4a}$ activity or expressing $p16^{ink4a}$-insensitive mutant of CDK4 continue to grow after SNF5/INI1 re-expression supporting a role of SNF5/INI1 in $\mathrm{p16^{ink4a}}$-CDK4/Cyclin D1-pRB/E2F pathway (Oruetxebarria et al. 2004; Vries et al. 2005). Several groups have also shown that SNF5/INI1 can repress Cyclin D1 expression in ATRT/MRTs by recruiting HDACs to the promoter, or it can function in the repression of E2F target genes via direct association with pRb, thereby causing cell-cycle arrest in G1 phase (Zhang et al. 2000; Zhang et al. 2002). In agreement with a role of SWI/SNF ATP-dependent chromatin remodelers in the regulation of pRb/E2F transcriptional circuitry, genome-wide expression profiling revealed up-regulation of some of the E2F-target genes in MRT cell lines, including mitotic checkpoint gene *MAD2* (Vries et al. 2005). Over-expression of Mad2 leads to chromosomal instability and tumorogenesis (Sotillo et al. 2007; van Deursen 2007).

The SWI/SNF complexes lack sequence specific DNA binding, and are therefore thought to be recruited to specific promoters via interactions with DNA binding proteins. It is still of interest to identify other DNA binding regulators that activate $p16^{INK4a}$ through association with the hSNF5 chromatin-remodeling factor. ETS2 is a transcriptional activator of $p16^{INK4a}$ which has been shown to bind to the $p16^{INK4a}$ promoter (Ohtani et al. 2001). ChIP data revealed that ETS2 and SNF5 are co-recruited to the P16 promoter in MRT cells (Fig. 14.6). shRNA knock down

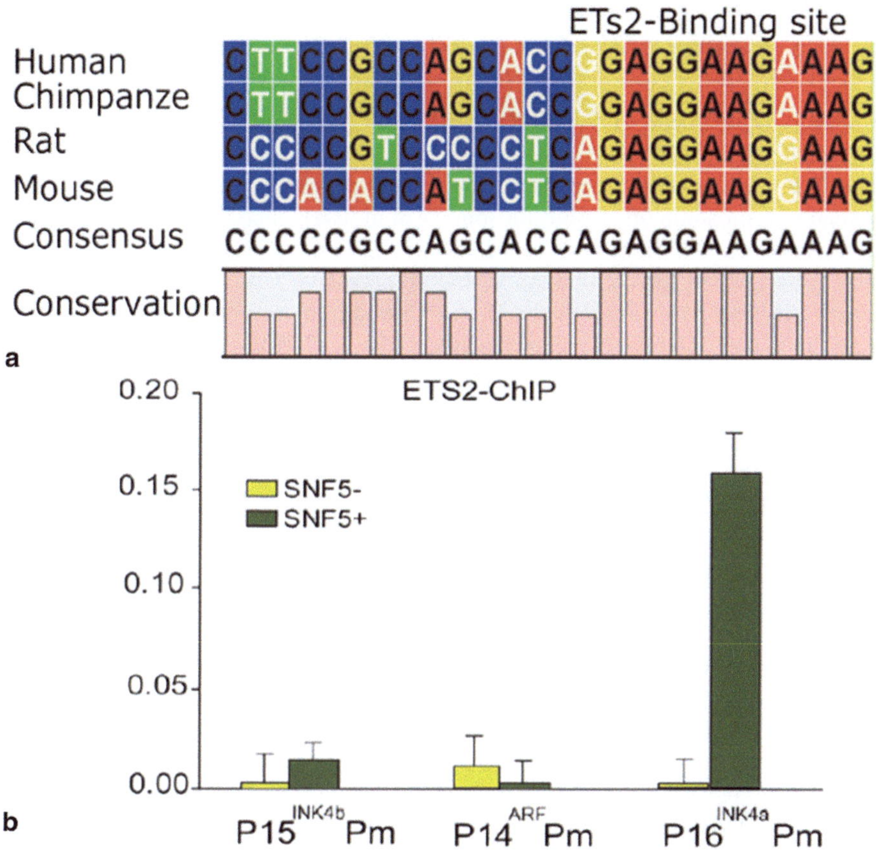

Fig. 14.6 hSNF5 and ETS2 are co-recruited to the p16INK4a Promoter. (**a**) Conservation of ETS2 binding site at p16INK4a promoter during evolution (**b**) ETS2 binding to the p16INK4a promoter is hSNF5-dependent, as revealed by ChIP-qPCR using antibodies directed against ETS2. ChIPs using antibodies directed against ETS2. Cross-linked chromatin was prepared from MRT cells lacking hSNF5, but expressing GFP (*light green bars*), or cells expressing hSNF5 (*dark green bars*). ChIPs were analyzed by qPCR using primer sets specific for the Ink4-Arf locus, revealing that ETS2 binding peaks at the p16INK4a promoter. Upon induction of hSNF5, ETS2 is co-recruited

analysis of ETS2, and studies of $p16^{INK4a}$ activation upon induction of hSNF5 will be informative useful to address the recruitment to and role of hSNF5 in activation of p16INK4a.

This triggers pRb hyperphosphorylation and unleashes transcription of E2F-dependent genes causing increased proliferation, chromosomal instabilities and cancer. Re-expression of SNF5/INI1 reverses cell proliferation and leads to G0/G1 arrest, senescence and apoptosis primarily due to activation of $P16^{INK4a}$ tumor suppressor gene expression and repression of *Cyclin D1* gene expression.

14.3.4 The Mechanism of SNF5- Dependent Transcriptional Control of the INK4b-ARF-INK4a Locus in MRT Cells

14.3.4.1 The INK4-ARF locus

The INK4B-ARF-INK4A locus, spanning an approximately 40 kb stretch of the human chromosome 9p21 (chromosome 4 in mouse), encodes three distinct tumor suppressors, *P15iINK4B*, *P16INK4A* and *P14ARF* (*p19AR*F in mice) whose expression enhances the growth-induction and -suppressive functions of the retinoblastoma (RB) and p53, respectively. Both p15 and p16 are able to induce cell cycle arrest in G1 by inhibiting cyclin dependent kinase CDK4 and CDK6 to inactivate retinoblastoma (RB) (Ortega et al. 2002). Since p16 can only form stable, binary complex with both CDK4 and CDK6, p16 is likely the most effective inhibitor of CDK4 and CDK6 (Parry et al. 1999). The unrelated p14 protein acts via MDM2 to activate the key check point protein TRP53, thereby inducing either cell cycle arrest (both in G1 and G2) or apoptosis (Pomerantz et al. 1998). The locus $p\mathit{14}^{ARF}$ (named ARF because it uses the second exon of INK4A in an alternative translational reading frame) is implicated in various types of cancer (reviewed in (Gil and Peters 2006; Sharpless 2005)). In mammalian cells, products of the INK4A-ARF locus play major roles in senescence and tumor suppression in different contexts, whereas the adjacent *INK4B* gene is generally associated with transforming growth factor beta (TGF-beta)-mediated growth arrest (Lowe and Sherr 2003). Mouse models indicate that both $p16^{INK4a}$ and $p14^{ARF}$ are tumor suppressors while mice deficient for $p15^{Ink4b}$ show only a very subtle tumor predisposition (Latres et al. 2000). Mice specifically defective for $p14^{ARF}$ are highly tumor-prone but mice deficient for all Ink4b-Arf-Ink4a locus genes have are more tumor prone and develop a wider spectrum of tumor than $p16^{Ink4a}$ mutant mice, with skin tumor and soft tissue sarcomas (Kamijo et al. 1997; Sharpless et al. 2004). It is reported that $\mathrm{p15^{Ink4b}}$ serves as a backup of $\mathrm{p16^{Ink4a}}$ (Krimpenfort et al. 2007). In chicken cells, which lack $p16^{Ink4a}$, $p15^{Ink4b}$ has the major role rather than $p14^{Arf}$ in senescence (Kim et al. 2006). In a variety of tumors, $P16^{INK4a}$ is inactivated through epigenetic silencing, involving PcG (Polycomb groups) proteins and DNA methylation (Gil and Peters 2006; Jones and Baylin 2007; Sparmann and van Lohuizen 2006). Significantly, the PcG protein BMI1 promotes oncogenesis in mice through silencing of the *Ink4a-Arf* locus (Jacobs et al. 1999). Both the Polycomb-repressive Complex1 (PRC1) and Polycomb-repressive Complex2 (PRC2) directly bind and silence the *Ink4a-Arf* locus (Bracken et al. 2007; Kotake et al. 2007; Kia et al. 2008). In MRT cells, human embryonic fibroblast (TIG3) and Human neonatal fibroblast, both PRC1 and PRC2 bind to INK4A and INK4B (Kheradmand Kia et al. 2009; Kia et al. 2008). It has been shown that depletion of EZH2 subunit of the Polycomb-repressive Complex2 (PRC2) in response to stress causes the loss of H3K27me3, displacement of BMI1 subunit of the Polycomb-repressive Complex1 (PRC1) and transcription activation of INK4A (Bracken et al. 2007). Depletion of EZH2 during aging and differentiation causes displacement of BMI1 and activation of INK4A and INK4B

(Kheradmand Kia et al. 2009; Kia et al. 2008). Numerous studies showed that promoter hypermethylation of CpG island (CGI) sequences is the most frequent pathway for inactivation of *P 16*^{INK4A} in human carcinomas,including those that arise in the lung, oropharynx, bladder,cervix, liver, colon, pancreas, and other sites (Baylin et al. 1998). The H3K36 demethylase JmjC domain-containig histone demethylase 1b in primary MEFs regulates cell proliferation and senescence through Ink4b (He et al. 2008). The *Ink4/Arf* locus is normally expressed at very low levels in most tissues of young organisms (Krishnamurthy et al. 2004). It is well established that the *Ink4/Arf* locus is activated during organismal ageing in both rodents and humans, and the levels of *p16*^{INK4a} constitute an impressively good overall biomarker of ageing (Kim and Sharpless 2006).These observations point to P16INK4A both serving as a brake for the proliferation of cancer cells, and also limiting the long-term renewal of stem cells. The Ink4a-Arf locus responds to stress signals, limiting cell proliferation and modulating oncogene-induced apoptosis (Lowe and Sherr 2003). A challenging issue which remains is to understand the interplay between signaling and PcG control of INK4B-ARF-INK4A locus.

14.3.4.2 SWI/SNF and Polycomb Group Proteins Act Antagonistically on INK4-ARF Locus

Polycomb Group Complex

Polycomb group proteins were first identified in Drosophila Melanogaster as mutants deregulating Hox gene expression pattern during fly early development. The PcG are required to maintain chromatin in a repressed state while the trithorax-group (trxG) proteins (including the hSWI/SNF complex) are necessary for the maintenance of transcriptional activity of several developmental genes. The PcG is a diverse group of proteins that form at least three different complexes: Polycomb repressive complex 1 and 2 (PRC1 and PRC2) and pleiohomeotic (Pho) repressive complex (PhoRC) (Schwartz and Pirrotta 2007).

The main function of the PRC1 proteins is to inhibit chromatin remodeling and maintain the repressed state of chromatin by out-competing the TrxG protein complexes such as SWI/SNF chromatin-remodeling complex. PRC2 on the other hand is known to be the initiator of the suppression process in which chromatin and or DNA are marked for repression. The key component of PRC2 is the SET domain H3 methyltransferase protein enhancer of Zeste (E (Z)). E (Z), when assembled in the complex, methylates H3K27. In human cells, PRC2 can physically associate with Histone deacetylases (HDACs 1 and 2). HDACs can deacetylate H3K27 to make it available for methylation by PRC2 (Kuzmichev et al. 2002). In vivo, tri-methylation of H3K27 is characteristic of PcG target genes (Schwartz et al. 2006). The H3K27me3 mark is thought to act as a docking site for chromo-domain of the CBX family proteins, which recruit other members of the PRC1 complex (Bernstein et al. 2006). Genome wide mapping of polycomb target genes revealed that EZH2 depletion alone causes INK4b induction (Bracken et al. 2006).

Mammalian Polycomb group (PcG) proteins are essential transcription silencers that control multiple development processes, including stem cell self-renewal, cell differentiation and have been implicated in several types of cancers (Boyer et al. 2006; Lee et al. 2006; Lund and van Lohuizen 2004; Valk-Lingbeek et al. 2004).

Polycomb Group Complex and INK4-ARF Locus

A number of reports have demonstrated a role for Polycomb silencers, including PRC1 and PRC2 upstream of *P15INK4B*, *P14ARF* and *P16INK4A* (Core et al. 2004; Gil et al. 2004; Kheradmand Kia et al. 2009; Bracken et al. 2007; Martin et al. 2013; Jacobs et al. 1999). The SWI/SNF and PcG proteins act antagonistically on INK4-ARF locus. It is clearly established that PRC1 and PRC2 mainly act upstream of *P16INK4A* and specifically localize within a region spanning approximately 800 bp. Interestingly, restoration of SWI/SNF in MRT cells causes removal of Polycomb silencers (PRC1 and PRC2) from the *P16INK4A* promoter (Kia et al. 2008). However it is not consistent with a model in which PRC1 blocks SWI/SNF mediated chromatin remodeling as has been previously reported from experiments conducted in vitro (Shao et al. 1999). Instead, the silencing by PcG complexes is a less rigid more dynamic process subject to removal in response to SWI/SNF expression (Kia et al. 2008).

Concomitant with the decrease in H3K27me3, the active H3K4me3 mark is strongly induced at *P16INK4A* and *P15INK4B* (Kia et al. 2008). The prominent H3K4me3 methyltransferase MLL1 is the human homologue of *Drosophila* TRX, the founding member of the trxG (Canaani et al. 2004). During hSNF5-mediated $p16^{INK4a}$ activation, the trxG activators SWI/SNF and MLL1 replace the PcG silencers PRC1 and PRC2.

Here we show that YY1 and PcG silencers concomitantly bind to the INK4A/B locus. Strikingly, induction of hSNF5 leads to the removal of YY1 and PcG proteins (Fig. 14.7). YY1 is the human homolog of Drosophila Pleiohomeotic (PHO)—a key sequence specific DNA binding recruiter of PcG complex.

It is not clear how PcG proteins and specifically those bound to downstream regions exert their repression over target genes. Long-range chromatin interactions by chromatin looping are thought to be one of the potential mechanisms to explain PcG action over broad distances in *cis* (Cleard et al. 2006; Comet et al. 2006; Kadauke and Blobel 2009; Schwartz and Pirrotta 2007). The first direct evidence comes from work in *Drosophila* showing that all major PcG-bound elements at the BX-C multi-gene locus—including PREs and core promoters—physically associate by chromatin long-range interactions in the repressed state (Lanzuolo et al. 2007). The PcGs and histone marks help to silence target genes via the looping process in human cells (Schwartz and Pirrotta 2007; Tiwari et al. 2008b). The PcGs also bind to *P16INK4A* locus and 3 kb upstream of *P15INK4B* in the repress state and is removed upon induction of hSNF5 during aging and differentiation (Kia et al. 2008).

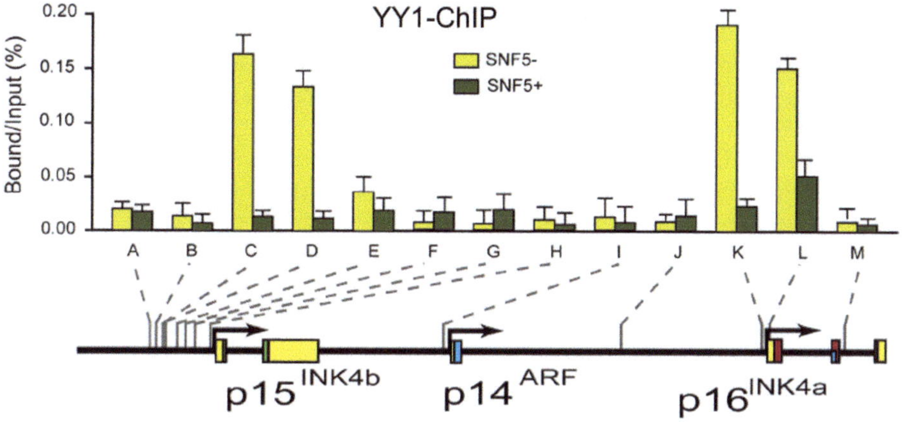

Fig. 14.7 Restoration of SWI/SNF Causes Removal of YY1 from P15INK4B and P16INK4A Promoters. (**a**) Restoration of SWI/SNF Causes Removal of YY1 from INK4a/b locus. ChIPs using antibodies directed against YY1. Cross-linked chromatin was prepared from MRT cells lacking hSNF5, but expressing GFP (*light green bars*), or cells expressing hSNF5 (*dark green bars*). ChIPs were analyzed by qPCR using primer sets specific for the regions indicated by A–M along the INK4B-ARF-INK4A locus, revealing that YY1 binding peaks at the p16INK4a promoter and 3 kb upstream of p15INK4b. Upon induction of hSNF5 YY1 is removed or strongly reduced. QPCR primer sets correspond to the upstream of p15INK4b promoter (A–H), the p14ARF promoter (I), an intergenic control region (J), and various regions of the p16INK4a locus (sets K–M). Primer sets K and L cover the p16INK4A promoter. The positions of the amplified regions on the INK4B-ARF-INK4A locus are indicated at the bottom

Methylation of INK4-ARF Locus

Hypermethylation of CpG island (CGI) sequences are a nearly universal somatic genome alteration in cancer. DNA methylation, occurring on cytosine bases in CpG dinucleotides, is an important epigenetic mechanism of gene regulation in eukaryotic cells. CpG Island of the $p16^{INK4A}$ promoter is hypermethylated in various cancers (Merlo et al. 1995). Alterations of SWI/SNF remodeling complex activity in various mammalian cells and organs have been implicated in transcriptional silencing through site-specific genomic methylation (Banine et al. 2005; Dennis et al. 2001; Gibbons et al. 2000). hSNF5, a core subunit of SWI/SNF induces demethylation of $p16^{INK4A}$ to promote transcriptional activation. DNMT3b is involved in de-novo methylation during development. Induction of hSNF5 causes removal of DNMT3b from the $p16^{INK4A}$ promoter that can explain alterations of methylation in this region (Kia et al. 2008). Higher expression of DNMT3b has shown in human oesophageal squamous cell carcinoma in correlation with low $p16^{INK4A}$ expression (Simao Tde et al. 2006). Disruption of DNMT1 and DNMT3b resulted in demethylation of the $p16^{INK4A}$ in other human cancer cells (Rhee et al. 2002).

EZH2 a Key for Differentiation, Aging and Cancer

EZH2 is the catalytic subunit of PRC2, which is a highly conserved histone methyltransferase that targets lysine-27 of histone H3. This methylated H3K27 chromatin mark is commonly associated with silencing of differentiation genes in organisms ranging from flies to human. EZH2 is frequently overexpressed in wide variety of cancerous tissue types including prostate, breast, lymphoma, myeloma, bladder, colon, skin, liver, endometrial, lung, gastric (Simon and Lange 2008). EZH2 and EED co-immuneprecipitate with all three human DNMTs and silencing of certain target genes requires both EZH2 and DNMTs (Vire et al. 2006). PRC2 is required for the expression of proliferative genes. EZH2 and EED are targets of the pRB-E2F pathway, and deregulation of the pathway, as is frequently observed in human cancer, would result in higher levels of EZH2 and EED. The *EZH2* and *EED* promoters are direct targets of the E2F transcription factors *in vivo*. Overexpression of EZH2, like MYC shortens the G_1 phase of the cell cycle, results in accumulation of cells in the S phase of the cell cycle, and confers a proliferative advantage by suppression of *Macrophage Stimulating 1* (*MST1*) (Kuser-Abali et al. 2014). The role of E2F and pRB in the control of embryonic development is likely related to their ability to regulate the abundance of the PRC2 complex (Bracken et al. 2003). EZH2 is accumulated in undifferentiated progenitor cell population, such as hematopoitic cells (Su et al. 2003). Genome wide mapping of PRC2 target genes in different cells in human, mouse and Drosophila by ChIP-on-ChIP method have been performed by different groups. These studies revealed that PRC2 target genes are highly enriched for transcription factors and signalling components that control cell differentiation (Simon and Lange 2008). Correspondence between silenced genes bound by Oct4, Sox2 and Nanog, which promote expression of proliferation genes and silencing of differentiation genes, with PRC2 target genes suggests that PRC2 is a key co-repressor in ES cells (Lee et al. 2006; Pasini et al. 2007). It has been shown that epidermal basal cells like ES cells are rich in Ezh2 and other PcG proteins, but as they differentiate, Ezh2 expression is turned off and the cells exit the cell cycle concomitant with induction of *INK4b/a*. How PcG genes themselves are regulated still remains unresolved. It is not mechanistically clear how PcG proteins repress their targets, PcG and histone marks may silence target genes through the looping process. It has been shown that PcG-occupied region can form chromatin loops and physically interact in *cis* around a single gene in mammalian cells. Ezh2 knock down was shown to affect this long-range interaction (Tiwari et al. 2008a). Analysis of the spatial organization of the INK4/ARF locus *in vitro* by 3C (Chromosome Conformation Capture) technology revealed that at least in Mon cells, Neonatal fibroblasts and Megacaryocyte-Erythrocyte progenitor cells, there is a physical and spatial interaction between $p15^{INK4B}$ and $p16^{INK4A}$ but not $p14^{ARF}$ (Dekker et al. 2002). Importantly, $p15^{INK4B}$ loses its physical interaction with $p16^{INK4A}$ upon induction of hSNF5, aging and differentiation. It is shown that inhibition of EZH2 in atypical rhabdoid teratoid tumor cells suppresses the self-renewal (Alimova et al. 2013). There is strong evidence that PcGs play a role in generating the repressive loop at the INK4/ARF locus (Dekker et al. 2002).

14.4 Conclusion

Human SNF5/INI1 is a subunit of SWI/SNF and tumor-suppressor lost in malignant rhabdoid tumors (MRTs), rare but highly aggressive pediatric cancers.

We performed genome-wide gene expression profiling to address the role of hSNF5 in tumorigenesis. This investigation helped us to gain insight into the corruption of cellular pathways resulting from loss of hSNF5 tumor suppressor in MRTs. Our genome wide expression suggests that hSNF5 can function in both transcription activation and repression of the genome.

This study identified hSNF5/INI1 target genes and provided evidence that hSNF5/INI1 may modulate cell cycle control through the regulation of the p16 INK4A-cyclinD/CDK4-pRb-E2F pathway. We have shown that the majority of the up-regulated genes encoded proteins with functions in extracellular matrix remodeling, adhesion or cell migration (*SERPINE2, ITGB5, MAP1B*), apoptosis (*DR6, FAS, CASP4, GAS6, ADAM19*), and cancer related pathway or other specialized functions (CDKN2A, *ETS2* and *TRIM22*). Those genes did not change or were down-regulated upon induction of mutant SNF5. shRNA knock down analysis of candidate effectors together with Chromatin immunoprecipitation studies (as we have shown for ETS2, MAD2L1, TRIM22 and CDKN2A) to determine the direct targets of SNF5 will be useful to address the role of hSNF5 in tumorigenesis. hSNF5 is critical for the recruitment of the SWI/SNF complex to the activated P16INK4A promoter. Restoration of SWI/SNF functionality through hSNF5 re-expression overcomes epigenetic silencing and mediates *P16INK4A* transcriptional activation in MRT cells. The coordinate induction of $p15^{INK4B}$ and $p16^{INK4A}$ but not $p14^{ARF}$ during differentiation or aging is accompanied by down-regulation of EZH2 and reduced locus occupancy of PcG repressors. EZH2 is required for coordinate silencing of $p15^{INK4B}$ and $p16^{INK4A}$. EZH2 depletion (KD) causes loss of PcG repressors on the INK4B & INK4A loci and resolution of the repressive loop (Kheradmand Kia et al. 2009). Therefore PcG proteins dynamically regulate higher order chromatin structure to balance proliferation and differentiation of human cells.

The main conclusion from our work is a model for reactivation of the p16INK4A–Rb pathway by the SWI/SNF complex in MRT cells, which emphasizes the close interconnectivity of epigenetic pathways; i.e. polycomb silencing, histone methylation, DNA methylation and ATP dependent chromatin remodeling.

References

Ae K, Kobayashi N, Sakuma R, Ogata T, Kuroda H, Kawaguchi N, Shinomiya K, Kitamura Y (2002) Chromatin remodeling factor encoded by ini1 induces G1 arrest and apoptosis in ini1-deficient cells. Oncogene 21(20):3112–3120

Alimova I, Birks DK, Harris PS, Knipstein JA, Venkataraman S, Marquez VE, Foreman NK, Vibhakar R (2013) Inhibition of EZH2 suppresses self-renewal and induces radiation sensitivity in atypical rhabdoid teratoid tumor cells. Neuro Oncol 15(2):149–160

Banine F, Bartlett C, Gunawardena R, Muchardt C, Yaniv M, Knudsen ES, Weissman BE, Sherman LS (2005) SWI/SNF chromatin-remodeling factors induce changes in DNA methylation to promote transcriptional activation. Cancer Res 65(9):3542–3547

Barker N, Hurlstone A, Musisi H, Miles A, Bienz M, Clevers H (2001) The chromatin remodelling factor Brg-1 interacts with beta-catenin to promote target gene activation. EMBO J 20(17):4935–4943

Baylin SB, Herman JG, Graff JR, Vertino PM, Issa JP (1998) Alterations in DNA methylation: a fundamental aspect of neoplasia. Adv Cancer Res 72:141–196

Becker PB, Horz W (2002) ATP-dependent nucleosome remodeling. Annu Rev Biochem 71:247–273

Bernstein E, Duncan EM, Masui O, Gil J, Heard E, Allis CD (2006) Mouse polycomb Proteins bind differentially to methylated histone H3 and RNA and are enriched in facultative heterochromatin. Mol Cell Biol 26(7):2560–2569

Betz BL, Strobeck MW, Reisman DN, Knudsen ES, Weissman BE (2002) Re-expression of hSNF5/INI1/BAF47 in pediatric tumor cells leads to G1 arrest associated with induction of p16ink4a and activation of RB. Oncogene 21(34):5193–5203

Biegel JA, Zhou JY, Rorke LB, Stenstrom C, Wainwright LM, Fogelgren B (1999) Germ-line and acquired mutations of INI1 in atypical teratoid and rhabdoid tumors. Cancer Res 59(1):74–79

Bochar DA, Wang L, Beniya H, Kinev A, Xue Y, Lane WS, Wang W, Kashanchi F, Shiekhattar R (2000) BRCA1 is associated with a human SWI/SNF-related complex: linking chromatin remodeling to breast cancer. Cell 102(2):257–265

Boyer LA, Plath K, Zeitlinger J, Brambrink T, Medeiros LA, Lee TI, Levine SS, Wernig M, Tajonar A, Ray MK, Bell GW, Otte AP, Vidal M, Gifford DK, Young RA, Jaenisch R (2006) Polycomb complexes repress developmental regulators in murine embryonic stem cells. Nature 441(7091):349–353

Bracken AP, Pasini D, Capra M, Prosperini E, Colli E, Helin K (2003) EZH2 is downstream of the pRB-E2F pathway, essential for proliferation and amplified in cancer. EMBO J 22(20):5323–5335

Bracken AP, Dietrich N, Pasini D, Hansen KH, Helin K (2006) Genome-wide mapping of polycomb target genes unravels their roles in cell fate transitions. Genes Dev 20(9):1123–1136

Bracken AP, Kleine-Kohlbrecher D, Dietrich N, Pasini D, Gargiulo G, Beekman C, Theilgaard-Monch K, Minucci S, Porse BT, Marine J-C, Hansen KH, Helin K (2007) The polycomb group proteins bind throughout the INK4A-ARF locus and are disassociated in senescent cells. Genes Dev 21(5):525–530

Bultman S, Gebuhr T, Yee D, La Mantia C, Nicholson J, Gilliam A, Randazzo F, Metzger D, Chambon P, Crabtree G, Magnuson T (2000) A Brg1 null mutation in the mouse reveals functional differences among mammalian SWI/SNF complexes. Mol Cell 6(6):1287–1295

Canaani E, Nakamura T, Rozovskaia T, Smith ST, Mori T, Croce CM, Mazo A (2004) ALL-1/MLL1, a homologue of Drosophila TRITHORAX, modifies chromatin and is directly involved in infant acute leukaemia. Br J Cancer 90(4):756–760

Cheng SW, Davies KP, Yung E, Beltran RJ, Yu J, Kalpana GV (1999) c-MYC interacts with INI1/hSNF5 and requires the SWI/SNF complex for transactivation function. Nat Genet 22(1):102–105

Cho YM, Choi J, Lee OJ, Lee HI, Han DJ, Ro JY (2006) SMARCB1/INI1 missense mutation in mucinous carcinoma with rhabdoid features. Pathol Int 56(11):702–706

Cleard F, Moshkin Y, Karch F, Maeda RK (2006) Probing long-distance regulatory interactions in the *Drosophila melanogaster* bithorax complex using dam identification. Nat Genet 38(8):931–935

Comet I, Savitskaya E, Schuettengruber B, Negre N, Lavrov S, Parshikov A, Juge F, Gracheva E, Georgiev P, Cavalli G (2006) PRE-mediated bypass of two Su(Hw) insulators targets PcG proteins to a downstream promoter. Dev Cell 11(1):117–124

Core N, Joly F, Boned A, Djabali M (2004) Disruption of E2F signaling suppresses the INK4a-induced proliferative defect in M33-deficient mice. Oncogene 23(46):7660–7668

Decristofaro MF, Betz BL, Rorie CJ, Reisman DN, Wang W, Weissman BE (2001) Characterization of SWI/SNF protein expression in human breast cancer cell lines and other malignancies. J Cell Physiol 186(1):136–145

Dekker J, Rippe K, Dekker M, Kleckner N (2002) Capturing chromosome conformation. Science 295(5558):1306–1311

Dennis K, Fan T, Geiman T, Yan Q, Muegge K (2001) Lsh, a member of the SNF2 family, is required for genome-wide methylation. Genes Dev 15(22):2940–2944

Doan DN, Veal TM, Yan Z, Wang W, Jones SN, Imbalzano AN (2004) Loss of the INI1 tumor suppressor does not impair the expression of multiple BRG1-dependent genes or the assembly of SWI/SNF enzymes. Oncogene 23(19):3462–3473

Douglass EC, Rowe ST, Valentine M, Parham D, Meyer WH, Thompson EI (1990) A second nonrandom translocation, der(16)t(1;16)(q21;q13), in Ewing sarcoma and peripheral neuroectodermal tumor. Cytogenet Cell Genet 53(2–3):87–90

Fyodorov DV, Kadonaga JT (2002) Dynamics of ATP-dependent chromatin assembly by ACF. Nature 418(6900):897–900

Gibbons RJ, McDowell TL, Raman S, O'Rourke DM, Garrick D, Ayyub H, Higgs DR (2000) Mutations in ATRX, encoding a SWI/SNF-like protein, cause diverse changes in the pattern of DNA methylation. Nat Genet 24(4):368–371

Gil J, Peters G (2006) Regulation of the INK4b-ARF-INK4a tumour suppressor locus: all for one or one for all. Nat Rev Mol Cell Biol 7(9):667–677

Gil J, Bernard D, Martinez D, Beach D (2004) Polycomb CBX7 has a unifying role in cellular lifespan. Nat Cell Biol 6(1):67–72

Guidi CJ, Sands AT, Zambrowicz BP, Turner TK, Demers DA, Webster W, Smith TW, Imbalzano AN, Jones SN (2001) Disruption of Ini1 leads to peri-implantation lethality and tumorigenesis in mice. Mol Cell Biol 21(10):3598–3603

Haas JE, Palmer NF, Weinberg AG, Beckwith JB (1981) Ultrastructure of malignant rhabdoid tumor of the kidney. A distinctive renal tumor of children. Hum Pathol 12(7):646–657

He J, Kallin EM, Tsukada Y-i, Zhang Y (2008) The H3K36 demethylase Jhdm1b/Kdm2b regulates cell proliferation and senescence through p15Ink4b. Nat Struct Mol Biol 15(11):1169–1175

Hernando E, Nahle Z, Juan G, Diaz-Rodriguez E, Alaminos M, Hemann M, Michel L, Mittal V, Gerald W, Benezra R, Lowe SW, Cordon-Cardo C (2004) Rb inactivation promotes genomic instability by uncoupling cell cycle progression from mitotic control. Nature 430(7001):797–802

Jacobs JJ, Kieboom K, Marino S, DePinho RA, van Lohuizen M (1999) The oncogene and polycomb-group gene bmi-1 regulates cell proliferation and senescence through the ink4a locus. Nature 397(6715):164–168

Johnson CN, Adkins NL, Georgel P (2005) Chromatin remodeling complexes: ATP-dependent machines in action. Biochem Cell Biol 83(4):405–417

Jones PA, Baylin SB (2007) The Epigenomics of Cancer. Cell 128(4):683–692

Judkins AR, Mauger J, Ht A, Rorke LB, Biegel JA (2004) Immunohistochemical analysis of hSNF5/INI1 in pediatric CNS neoplasms. Am J Surg Pathol 28(5):644–650

Kadauke S, Blobel GA (2009) Chromatin loops in gene regulation. Biochim Biophys Acta 1789(1):17–25

Kamijo T, Zindy F, Roussel MF, Quelle DE, Downing JR, Ashmun RA, Grosveld G, Sherr CJ (1997) Tumor suppression at the mouse ink4a locus mediated by the alternative reading frame product p19 ARF. Cell 91(5):649–659

Kheradmand Kia S, Solaimani Kartalaei P, Farahbakhshian E, Pourfarzad F, von Lindern M, Verrijzer CP (2009) EZH2-dependent chromatin looping controls INK4a and INK4b, but not ARF, during human progenitor cell differentiation and cellular senescence. Epigenetics Chromatin 2(1):16

Kia SK, Gorski MM, Giannakopoulos S, Verrijzer CP (2008) SWI/SNF mediates polycomb eviction and epigenetic reprogramming of the INK4b-ARF-INK4a locus. Mol Cell Biol 28(10):3457–3464

Kim WY, Sharpless NE (2006) The Regulation of INK4/ARF in Cancer and aging. Cell 127(2):265–275

Kim SH, Rowe J, Fujii H, Jones R, Schmierer B, Kong BW, Kuchler K, Foster D, Ish-Horowicz D, Peters G (2006) Upregulation of chicken p15INK4b at senescence and in the developing brain. J Cell Sci 119(12):2435–2443

Klochendler-Yeivin A, Fiette L, Barra J, Muchardt C, Babinet C, Yaniv M (2000) The murine SNF5/INI1 chromatin remodeling factor is essential for embryonic development and tumor suppression. EMBO Rep 1(6):500–506

Kotake Y, Cao R, Viatour P, Sage J, Zhang Y, Xiong Y (2007) pRB family proteins are required for H3K27 trimethylation and polycomb repression complexes binding to and silencing p16INK4a tumor suppressor gene. Genes Dev 21(1):49–54

Krimpenfort P, Ijpenberg A, Song J-Y, van der Valk M, Nawijn M, Zevenhoven J, Berns A (2007) p15Ink4b is a critical tumour suppressor in the absence of p16Ink4a. Nature 448(7156):943–946

Krishnamurthy J, Torrice C, Ramsey MR, Kovalev GI, Al-Regaiey K, Su L, Sharpless NE (2004) Ink4a/Arf expression is a biomarker of aging. J Clin Invest 114(9):1299–1307

Kuser-Abali G, Alptekin A, Cinar B (2014) Overexpression of MYC and EZH2 cooperates to epigenetically silence MST1 expression. Epigenetics 9(4):634–643

Kuzmichev A, Nishioka K, Erdjument-Bromage H, Tempst P, Reinberg D (2002) Histone methyltransferase activity associated with a human multiprotein complex containing the Enhancer of Zeste protein. Genes Dev 16(22):2893–2905

Lanzuolo C, Roure V, Dekker J, Bantignies F, Orlando V (2007) Polycomb response elements mediate the formation of chromosome higher-order structures in the bithorax complex. Nat Cell Biol 9(10):1167–1174

Latres E, Malumbres M, Sotillo R, Martin J, Ortega S, Martin-Caballero J, Flores JM, Cordon-Cardo C, Barbacid M (2000) Limited overlapping roles of P15(INK4b) and P18(INK4c) cell cycle inhibitors in proliferation and tumorigenesis. EMBO J 19(13):3496–3506

Lee TI, Jenner RG, Boyer LA, Guenther MG, Levine SS, Kumar RM, Chevalier B, Johnstone SE, Cole MF, Isono K-i, Koseki H, Fuchikami T, Abe K, Murray HL, Zucker JP, Yuan B, Bell GW, Herbolsheimer E, Hannett NM, Sun K, Odom DT, Otte AP, Volkert TL, Bartel DP, Melton DA, Gifford DK, Jaenisch R, Young RA (2006) Control of developmental regulators by polycomb in human embryonic stem cells. Cell 125(2):301–313

Lia G, Praly E, Ferreira H, Stockdale C, Tse-Dinh YC, Dunlap D, Croquette V, Bensimon D, Owen-Hughes T (2006) Direct observation of DNA distortion by the RSC complex. Mol Cell 21(3):417–425

Lowe SW, Sherr CJ (2003) Tumor suppression by Ink4a-Arf: progress and puzzles. Curr Opin Genet Dev 13(1):77–83

Lund AH, van Lohuizen M (2004) Epigenetics and cancer. Genes Dev 1(18):2315–2335

Martin N, Popov N, Aguilo F, O'Loghlen A, Raguz S, Snijders AP, Dharmalingam G, Li S, Thymiakou E, Carroll T, Zeisig BB, So CW, Peters G, Episkopou V, Walsh MJ, Gil J (2013) Interplay between homeobox proteins and polycomb repressive complexes in p16INK(4)a regulation. EMBO J 32(7):982–995

Merlo A, Herman JG, Mao L, Lee DJ, Gabrielson E, Burger PC, Baylin SB, Sidransky D (1995) 5' CpG island methylation is associated with transcriptional silencing of the tumour suppressor p16/CDKN2/MTS1 in human cancers. Nat Med 1(7):686–692

Mohrmann L, Verrijzer CP (2005) Composition and functional specificity of SWI2/SNF2 class chromatin remodeling complexes. Biochim Biophys Acta 1681(2–3):59–73

Moshkin YM, Mohrmann L, van Ijcken WFJ, Verrijzer CP (2007) Functional Differentiation of SWI/SNF Remodelers in Transcription and Cell Cycle Control. Mol Cell Biol 27(2):651–661

Ohtani N, Zebedee Z, Huot TJ, Stinson JA, Sugimoto M, Ohashi Y, Sharrocks AD, Peters G, Hara E (2001) Opposing effects of Ets and Id proteins on p16INK4a expression during cellular senescence. Nature 409(6823):1067–1070

Ortega S, Malumbres M, Barbacid M (2002) Cyclin D-dependent kinases, INK4 inhibitors and cancer. Biochim Biophys Acta 1602(1):73–87

Oruetxebarria I, Venturini F, Kekarainen T, Houweling A, Zuijderduijn LM, Mohd-Sarip A, Vries RG, Hoeben RC, Verrijzer CP (2004) P16INK4a is required for hSNF5 chromatin remodeler-induced cellular senescence in malignant rhabdoid tumor cells. J Biol Chem 279(5):3807–3816

Parry D, Mahony D, Wills K, Lees E (1999) Cyclin D-CDK Subunit arrangement is dependent on the availability of competing INK4 and p21 class inhibitors. Mol Cell Biol 19(3):1775–1783

Pasini D, Bracken AP, Hansen JB, Capillo M, Helin K (2007) The polycomb group protein Suz12 Is required for embryonic stem cell differentiation. Mol Cell Biol 27(10):3769–3779

Pomerantz J, Schreiber-Agus N, Liégeois NJ, Silverman A, Alland L, Chin L, Potes J, Chen K, Orlow I, Lee H-W, Cordon-Cardo C, DePinho RA (1998) The Ink4a tumor suppressor gene product, p19Arf, interacts with MDM2 and neutralizes MDM2's inhibition of p53. Cell 92(6):713–723

Reisman DN, Sciarrotta J, Wang W, Funkhouser WK, Weissman BE (2003) Loss of BRG1/BRM in human lung cancer cell lines and primary lung cancers: correlation with poor prognosis. Cancer Res 63(3):560–566

Rhee I, Bachman KE, Park BH, Jair KW, Yen RW, Schuebel KE, Cui H, Feinberg AP, Lengauer C, Kinzler KW, Baylin SB, Vogelstein B (2002) DNMT1 and DNMT3b cooperate to silence genes in human cancer cells. Nature 416(6880):552–556

Roberts CW, Orkin SH (2004) The SWI/SNF complex-chromatin and cancer. Nat Rev Cancer 4(2):133–142

Roberts CW, Galusha SA, McMenamin ME, Fletcher CD, Orkin SH (2000) Haploinsufficiency of Snf5 (integrase interactor 1) predisposes to malignant rhabdoid tumors in mice. Proc Natl Acad Sci U S A 97(25):13796–13800

Roberts CW, Leroux MM, Fleming MD, Orkin SH (2002) Highly penetrant, rapid tumorigenesis through conditional inversion of the tumor suppressor gene Snf5. Cancer Cell 2(5):415–425

Rorke LB, Packer R, Biegel J (1995) Central nervous system atypical teratoid/rhabdoid tumors of infancy and childhood. J Neurooncol 24(1):21–28

Saha A, Wittmeyer J, Cairns BR (2005) Chromatin remodeling through directional DNA translocation from an internal nucleosomal site. Nat Struct Mol Biol 12(9):747–755

Sansam CG, Roberts CW (2006) Epigenetics and cancer: altered chromatin remodeling via snf5 loss leads to aberrant cell cycle regulation. Cell Cycle 5(6):621–624

Schwartz YB, Pirrotta V (2007) Polycomb silencing mechanisms and the management of genomic programmes. Nat Rev Genet 8(1):9–22

Schwartz YB, Kahn TG, Nix DA, Li XY, Bourgon R, Biggin M, Pirrotta V (2006) Genome-wide analysis of polycomb targets in *Drosophila melanogaster*. Nat Genet 38(6):700–705

Sevenet N, Lellouch-Tubiana A, Schofield D, Hoang-Xuan K, Gessler M, Birnbaum D, Jeanpierre C, Jouvet A, Delattre O (1999a) Spectrum of hSNF5/INI1 somatic mutations in human cancer and genotype-phenotype correlations. Hum Mol Genet 8(13):2359–2368

Sevenet N, Sheridan E, Amram D, Schneider P, Handgretinger R, Delattre O (1999b) Constitutional mutations of the hSNF5/INI1 gene predispose to a variety of cancers. Am J Hum Genet 65(5):1342–1348

Shao Z, Raible F, Mollaaghababa R, Guyon JR, Wu CT, Bender W, Kingston RE (1999) Stabilization of chromatin structure by PRC1, a polycomb complex. Cell 98(1):37–46

Sharpless NE (2005) INK4a/ARF: A multifunctional tumor suppressor locus. Mutat Res 576(1–2):22–38

Sharpless NE, Ramsey MR, Balasubramanian P, Castrillon DH, DePinho RA (2004) The differential impact of p16INK4a or p19ARF deficiency on cell growth and tumorigenesis. Oncogene 23(2):379–385

Simao Tde A, Simoes GL, Ribeiro FS, Cidade DA, Andreollo NA, Lopes LR, Macedo JM, Acatauassu R, Teixeira AM, Felzenszwalb I, Pinto LF, Albano RM (2006) Lower expression of p14ARF and p16INK4a correlates with higher DNMT3B expression in human oesophageal squamous cell carcinomas. Hum Exp Toxicol 25(9):515–522

Simon JA, Lange CA (2008) Roles of the EZH2 histone methyltransferase in cancer epigenetics. Mutat Res 647(1–2):21–29

Sotillo R, Hernando E, Diaz-Rodriguez E, Teruya-Feldstein J, Cordon-Cardo C, Lowe SW, Benezra R (2007) Mad2 overexpression promotes aneuploidy and tumorigenesis in mice. Cancer Cell 11(1):9–23

Sparmann A, van Lohuizen M (2006) Polycomb silencers control cell fate, development and cancer. Nat Rev Cancer 6(11):846–856

Su IH, Basavaraj A, Krutchinsky AN, Hobert O, Ullrich A, Chait BT, Tarakhovsky A (2003) Ezh2 controls B cell development through histone H3 methylation and Igh rearrangement. Nat Immunol 4(2):124–131

Sullivan LM, Folpe AL, Pawel BR, Judkins AR, Biegel JA (2013) Epithelioid sarcoma is associated with a high percentage of SMARCB1 deletions. Mod Pathol 26(3):385–392

Takayama MA, Taira T, Tamai K, Iguchi-Ariga SM, Ariga H (2000) ORC1 interacts with c-Myc to inhibit E-box-dependent transcription by abrogating c-Myc-SNF5/INI1 interaction. Genes Cells 5(6):481–490

Takita J, Chen Y, Kato M, Ohki K, Sato Y, Ohta S, Sugita K, Nishimura R, Hoshino N, Seki M, Sanada M, Oka A, Hayashi Y, Ogaw S (2014) Genome-wide approach to identify second gene targets for malignant rhabdoid tumors using high-density oligonucleotide microarrays. Cancer Sci 105(3):258–264

Tiwari VK, Cope L, McGarvey KM, Ohm JE, Baylin SB (2008a) A novel 6 C assay uncovers polycomb-mediated higher order chromatin conformations. Genome Res 18(7):1171–1179

Tiwari VK, McGarvey KM, Licchesi JD, Ohm JE, Herman JG, Schubeler D, Baylin SB (2008b) PcG proteins, DNA methylation, and gene repression by chromatin looping. PLoS Biol 6(12):2911–2927

Valk-Lingbeek ME, Bruggeman SWM, van Lohuizen M (2004) Stem cells and cancer: the polycomb connection. Cell 118(4):409–418

van Deursen JM (2007) Rb loss causes cancer by driving mitosis mad. Cancer Cell 11(1):1–3

Versteege I, Sevenet N, Lange J, Rousseau-Merck MF, Ambros P, Handgretinger R, Aurias A, Delattre O (1998) Truncating mutations of hSNF5/INI1 in aggressive paediatric cancer. Nature 394(6689):203–206

Versteege I, Medjkane S, Rouillard D, Delattre O (2002) A key role of the hSNF5/INI1 tumour suppressor in the control of the G1-S transition of the cell cycle. Oncogene 21(42):6403–6412

Vire E, Brenner C, Deplus R, Blanchon L, Fraga M, Didelot C, Morey L, Van Eynde A, Bernard D, Vanderwinden J-M, Bollen M, Esteller M, Di Croce L, de Launoit Y, Fuks F (2006) The polycomb group protein EZH2 directly controls DNA methylation. Nature 439(7078):871–874

Vries RG, Bezrookove V, Zuijderduijn LM, Kia SK, Houweling A, Oruetxebarria I, Raap AK, Verrijzer CP (2005) Cancer-associated mutations in chromatin remodeler hSNF5 promote chromosomal instability by compromising the mitotic checkpoint. Genes Dev 19(6):665–670

Whitehouse I, Stockdale C, Flaus A, Szczelkun MD, Owen-Hughes T (2003) Evidence for DNA translocation by the ISWI chromatin-remodeling enzyme. Mol Cell Biol 23(6):1935–1945

Wong AK, Shanahan F, Chen Y, Lian L, Ha P, Hendricks K, Ghaffari S, Iliev D, Penn B, Woodland AM, Smith R, Salada G, Carillo A, Laity K, Gupte J, Swedlund B, Tavtigian SV, Teng DH, Lees E (2000) BRG1, a component of the SWI-SNF complex, is mutated in multiple human tumor cell lines. Cancer Res 60(21):6171–6177

Zhang HS, Gavin M, Dahiya A, Postigo AA, Ma D, Luo RX, Harbour JW, Dean DC (2000) Exit from G1 and S phase of the cell cycle is regulated by repressor complexes containing HDAC-Rb-hSWI/SNF and Rb-hSWI/SNF. Cell 101(1):79–89

Zhang ZK, Davies KP, Allen J, Zhu L, Pestell RG, Zagzag D, Kalpana GV (2002) Cell cycle arrest and repression of cyclin D1 transcription by INI1/hSNF5. Mol Cell Biol 22(16):5975–5988

Zhang Y, Smith CL, Saha A, Grill SW, Mihardja S, Smith SB, Cairns BR, Peterson CL, Bustamante C (2006) DNA translocation and loop formation mechanism of chromatin remodeling by SWI/SNF and RSC. Mol Cell 24(4):559–568

Chapter 15
Epigenetics of Thyroid Cancer

Javad Mohammadi-Asl

Contents

Abstract With the development of tumor research, it is discovered that epigenetic modifications such as methylation of CpG islands in promoter region or histone modification, as well as genetic alterations including mutation, deletion, and DNA amplification, are closely related with the tumorigenesis and development of cancer. Epigenetics of human malignancy has become an emerging research area due to growing understanding of specific epigenetic pathways, markers, as well as rapid development of diagnostic technologies. Epigenetic alternation of the DNA has been widely studied in the case of thyroid malignancies. Some of thyroid related genes as well as tumor suppressing genes have seen to be epigenetically regulated during malignancy. This chapter tends to shed light on this epigenetic alteration in different kinds of thyroid cancer.

J. Mohammadi-Asl (✉)
Department Of Medical Genetics, Faculty of Medicine, Ahvaz Jundishapur University of medical sciences, Ahvaz, Iran
email: mohammadiasl@gmail.com

P. Mehdipour (ed.), *Epigenetics Territory and Cancer*,
DOI 10.1007/978-94-017-9639-2_15

Abbreviations

FTC:	Follicular thyroid carcinoma
PTC:	Papillary thyroid carcinoma
DTC:	Differentiated thyroid carcinoma
UTC:	Undifferentiated thyroid carcinoma
ATC:	Anaplastic thyroid carcinoma
MTC:	Medullary thyroid carcinoma
BTA:	Benign thyroid adenoma
RASSF1A:	Ras association domain family 1A gene
RASSF1:	Ras association domain family 1
TSG:	Tumor suppressor gene
ATT:	Adjacent thyroid tissues
BRAF:	v-raf murine sarcoma viral oncogene homolog B
qMSP:	Quantitative methylation-specific PCR
PCR:	Polymerase chain reaction
PTEN:	Phosphatase and tensin homolog
INK4a:	Inhibitor of cyclin-dependent kinase type 4
MTS1:	Multiple tumor suppressive gene 1
PI3K:	Phosphatidylinositol 3-kinase
Akt:	Activation of the prosurvival kinase, protein kinase
PIP3:	Phosphatidylinositide-3,4,5-trisphosphate
RXR:	Retinoid X receptors
RAR:	Retinoic acid receptor
RARβ2:	Retinoic acid nuclear receptor beta 2
DI:	Deiodinase
RA:	Retinoid acid
TIG1:	Tazarotene-induced gene 1
Ni3S2:	Nickel subsulfide
HME1:	Human mammary epithelium-specific marker 1
DNA:	Deoxyriboncleic acid
TSHR:	Thyroid-stimulating Hormone Receptor
CHNG1:	Hypothyroidism, Congenital, Nongoitrous, 1
hTSHR-I:	Thyrotropin (TSH) Receptor-1
MT1G:	Metallothionein 1G
CRABP1:	Cellular Retinoic Acid Binding Protein 1
HBB:	Beta globi
HNF3β:	Hepatocyte nuclear factor 3β
FoxA2:	Forkhead box protein A2
PKG:	cGMP-dependent protein kinase
TSH-α:	Thyroid-stimulating-hormone-alpha
SLC26A4:	Solute carrier family 26, member 4
FHIT:	Fragile histidine triad
TTF-1:	Thyroid transcription factor 1
SERPINA5:	Serpin peptidase inhibitor clade A member5
RIZ1:	Retinoblastoma-interacting zinc-finger protein 1
PRDM2:	PR domain containing 2

15.1 Introduction

Four types of thyroid cancer have been introduced do far, comprising 98% of all thyroid cancer, and 1% of all malignancies: follicular thyroid carcinoma (FTC), papillary thyroid carcinoma (PTC), both also classified as differentiated thyroid carcinoma (DTC), undifferentiated or anaplastic thyroid carcinoma (UTC) and medullary thyroid carcinoma (MTC) (Gimm 2001). It is generally accepted that different forms of thyroid cancer can develop from each other. For instance, ATC can develop from DTC, and DTC, predominantly FTC, can develop from benign thyroid adenoma (BTA) (Hou et al. 2008). A number of genes have been identified as being implicated in the process of oncogenesis including initiation, development, and malignancy steps.

Here, some important genes, which are epigenetically regulated in thyroid malignancy, are listed.

15.1.1 RASSF1

Ras association domain family 1A gene (RASSF1A), the longest isoform of Ras association domain family 1 (RASSF1) (Oliveira et al. 2005), is a tumor suppressor gene located at 3p21.3, which belongs to a six member family with tumor suppressing potentials. It has been widely discover that epigenetic alteration in RASSF1A promoter plays an important role in various kinds of cancer including thyroid, lung, ovarian, colorectal, breast, hepatocellular pituitary, parathyroid, cervical carcinoma and endometria cancer (Schagdarsurengin et al. 2003; Yeo et al. 2005; Choi et al. 2006; Pallarés et al. 2008; Qian et al. 2005; Oliveira et al. 2005; Juhlin et al. 2010; CUI et al. 2011). Animal studies showed that knockout mice lacking *RASSF1* exon 1α is prone to spontaneous tumorigenesis (Tommasi et al. 2005). Interestingly, allelic loss and abnormal promoter methylation but not mutations are the main mechanisms of *RASSF1A* inactivation, however, tumors harboring a mutation in other tumor suppressor genes (TSGs) had a larger number of hypermethylation events, showing a link between genetic and epigenetic control of this gene (Dammann et al. 2003) (Fig. 15.1).

Buffy coat sample analysis also displayed an increased promoter hypermethylation of *RASSF1A* in children with thyroid carcinoma (Wong et al. 2004). Genome-wide DNA methylation profiling showed that follicular tumors of thyroid anchorage higher levels of methylation in comparison with normal thyroid tissues (Mancikova et al. 2014). Similarly, a 4.2 times hypermethylation have been observed in PTC

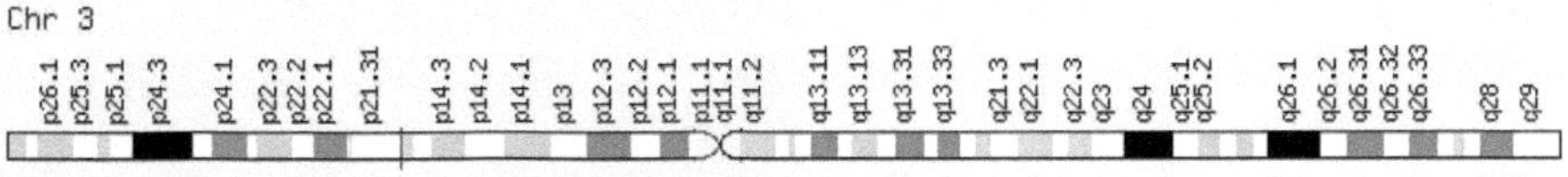

Fig. 15.1 RASSF1 chromosome location (The red line on the p arm)

compared to normal tissues of thyroid (Kunstman et al. 2013). These epigenetic changes were also confirmed in PTC patients when compared to adjacent thyroid tissues (ATT) (TANG and SU 2010). Comparing different forms of thyroid malignancy, this gene was detected to be hypermethylated in 71 % of thyroid carcinomas and it is more frequent in aggressive forms, which reach 80 % in undifferentiated thyroid Carcinoma (UTC) and medullary thyroid carcinoma (MTC) (Schagdarsurengin et al. 2002).

15.1.2 BRAF

V-raf sarcoma viral oncogene homolog B1 (BRAF) is a member of Raf kinase family function as an intracellular mediator of MAPK signaling pathway which is mutated in numerous cancers (Greco et al. 2009). The point mutation T1799A exchanging valine to glutamate exists in 40–70 % of PTC patients (Lee et al. 2011). This mutation causes BRAF to be consecutively and oncogenically active (Knauf and Fagin 2009). An association also found between this activating mutation and hypermethylation status and shed light on its role in the pathogenesis of thyroid cancer. It is demonstrated that tumors with *BRAF* mutations display a 3.6 fold incensement in methylation sites compared to normal tumors (Ellis et al. 2013). Interestingly, quantitative methylation-specific PCR (qMSP) analysis discovered an adverse correlation between *BRAF* mutation and *RASSF1A* methylation (Brait et al. 2012; Hoque et al. 2005).

15.1.3 P16

Genetic and epigenetic alterations in p16^{INK4a} (*MTS1*) encodedat 9p21 region has been reported in diverse cancers. This tumor suppressor gene prevents cell progression through the G1 phase and its involvement in tumor suppression pathways makes it a significant factor in development of cancer (Bartoletti et al. 2007; Jones et al. 1996; Wang et al. 2002; Serrano et al. 2000). The two most common mechanisms of *p16* inactivation are homozygous deletion or hypermethylation of the gene; it is reported that respectively 44 and 50 % of PTC and FTC patients show promoter hypermethylation in this gene (Boltze et al. 2003) (Fig. 15.2).

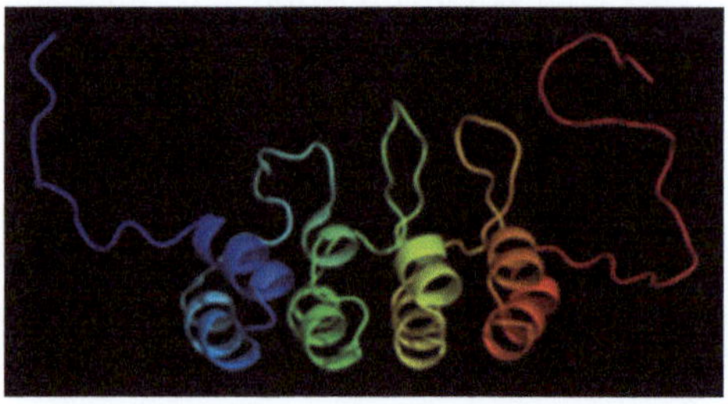

Fig. 15.2 P16 ribbon structure

In addition, 90 % of thyroid tumors with *p16* inactivation harbor also *RASSF1A* inactivation (Schagdarsurengin et al. 2002). Interestingly, its expression statue changes during differentiated thyroid tumorigenesis (Ferru et al. 2006) so that it is induced in differentiated thyroid cancer and blocked during progression toward the undifferentiated status. 89 % of ATC patients lack the expression of p16 that is expressed in 46 % of malignant follicular-derived lesions of the Thyroid (Lee et al. 2008; Barroeta et al. 2006). Because of *p16* over expression in follicular thyroid cells, it is believed this gene is not necessary for tumor initiation but it may be involved in progression and metastasis of thyroid cancer (Ishida et al. 2007; Ball et al. 2007; Ferenc et al. 2004; Wang et al. 2013). Although some mentioned it as a prognostic indicator (Mohammadi-asl et al. 2011; LUO et al. 2004; Gerdes et al. 2002), others observed its under or lack of expression in malignant diseases (Kiss et al. 2008; Lam et al. 2007; Huang et al. 2001) (Fig. 15.3).

15.1.4 *PTEN*

The tumor suppressor gene *PTEN* mapping on 10q 23.3 acts as a phosphatase in downstream of PI3K/Akt pathway and terminate the signaling process via dephosphorylation of PIP3. The germline mutations within this TSG cause variety of syndromes and can been seen in melanoma and thyroid cancers (Zhou et al. 2000; Xing 2008). *PTEN* promoter hypermethylation reported in several malignancies such as melanoma (Zhou et al. 2000), endometrial carcinoma (Salvesen et al. 2001), and thyroid cancer (Wei et al. 2013) not in gastric cancer (Sato et al. 2002). It is stated that 45.7 % of PTC patients harbor methylated *PTEN* promoters (Alvarez-Nuñez et al. 2006) (Fig. 15.4).

15.1.5 *RARβ*

Retinoic acid receptors consist of two family as retinoid X receptors (RXRs) and retinoic acid receptors (RARs), which are necessary for transduction of retinoid signals. Each of these nuclear hormone receptor families contains α, β and γ isoforms involving in suppression of PI3K/Akt pathway through either PI3K or phosphatase PTEN targeting (Bastien and Rochette-Egly 2004). Tumor suppressor gene retinoic acid receptor-beta2 (*RARβ2*) encoded at 3p24 (Wu et al. 2006) exists in a hypermethylated condition in several cancers including 22 % of PTC patients (Cras et al. 2007; Li et al. 2014; Russo et al. 2011). Since no mutation reported in its relation to cancer, epigenetic control by methylation, and loss of heterozygocity have been considered as inactivation processes. Evaluation of type I iodothyronine 5'-deiodinase (5' DI) documented that respond to retinoid therapy using retinoid acid (RA) may correlate with disregulation of expression in RA receptors including RARβ (Schmutzler et al. 2004). In addition, there is a close relationship between methylation of tazarotene-induced gene 1 (TIG1), also known as RAR-responsive

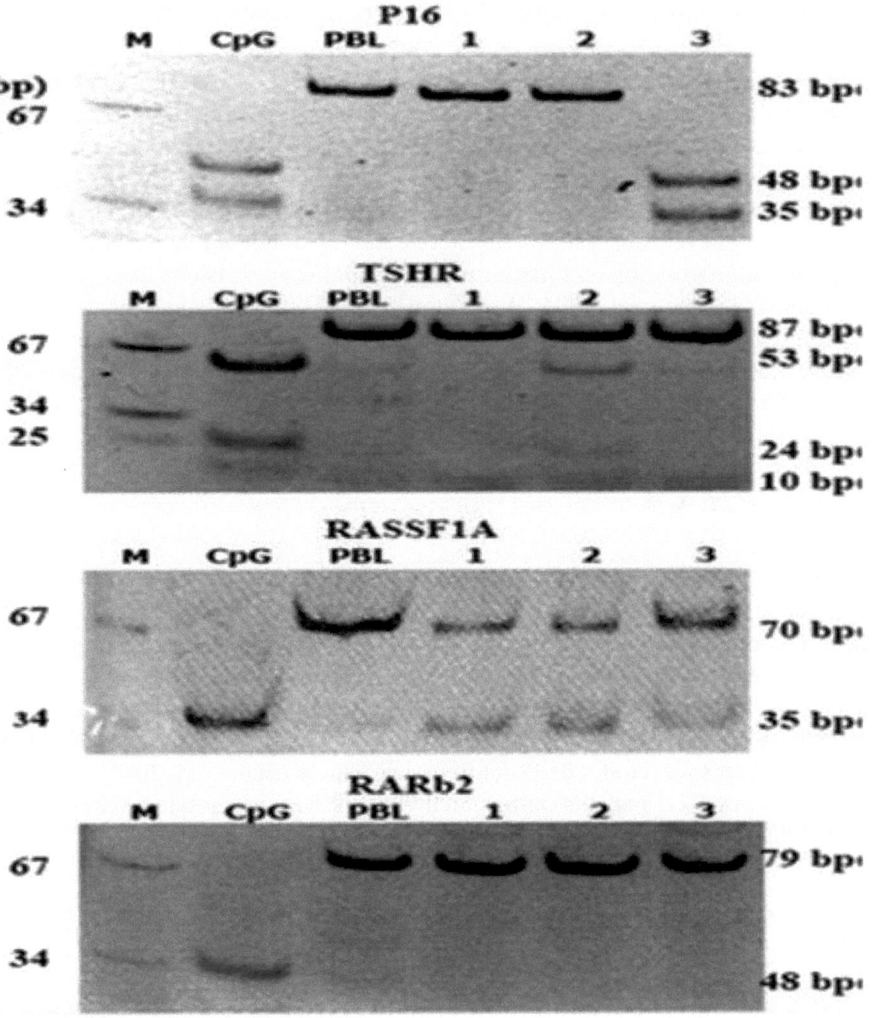

Fig. 15.3 Digestion results of PCR products of P16 (by TaqI) TSHR (by TaqI), RASSF1A (by RsaI) and RARb2 (by Bst UI) MARKER (pUC18 mSPI), CpG universal methylated DNA, PBL peripheral lymphocyte DNA. Lanes 1–3, patient's DNA. (Mohammadi-asl et al. 2011)

1 gene, and *RARβ* methylation confirming that induction of TIG1 occurs in a RAR-specific manner (Zhang et al. 2004). Interestingly, retinoic acid-resistant thyroid cancer cells called FTC238 harboring the same methylation pattern in RARβ, compared with retinoic acid-sensitive thyroid cancer cells (FTC133), which bear hypermethylated histones on RARβ promoter (Cras et al. 2007). Furthermore, the expression of RARs suggested to be considered as an indicator for distinguishing adenoma and well-differentiated thyroid carcinoma (Gauchotte et al. 2013). Methylation of RARβ has been shown to be chemically modified; epigenetic modifiers such as nickel subsulfide (Ni3S2) (ZHANG et al. 2011) and PD153035 (Grunt et al.

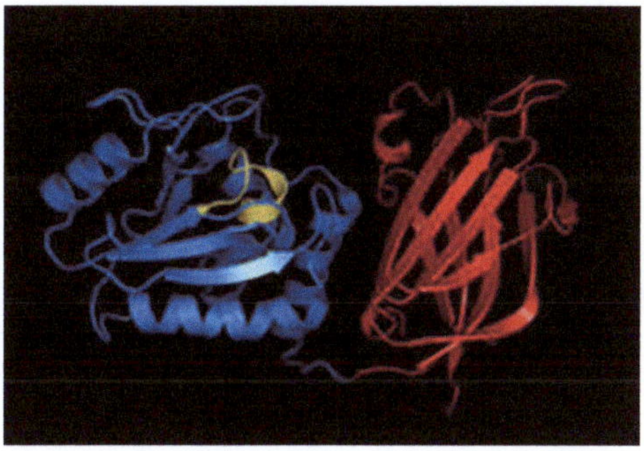

Fig. 15.4 PTEN ribbon structure

2005) cause hyper- and hypomethylation of RARβ, respectively. Interestingly, re-expression of *RARβ2* using demethylating agent 5-aza-2-deoxycytidine causes a significant inhibition of thyroid cancer cell growth (Miasaki et al. 2008). A study showed that smoking, which is a well-known risk factor in malignancies, induces an aberrant hypermethylation in RARβ and cause thyroid tumorigenesis (Kiseljak-Vassiliades and Xing 2011) (Fig. 15.5).

15.1.6 The 14-3-3σ

14-3-3σ also named human mammary epithelium-specific marker 1 (HME1), is a cell cycle regulator which acts as a G2/M arresting molecule expressing in malignant thyroid tissues (Lal et al. 2008). Interestingly, 14-3-3σ observed to be expressed in all papillary carcinomas but not follicular carcinomas and adenomas, and it proposed that this protein might not be required for development of thyroid follicular tumors (Ito et al. 2003). In addition, it has been postulated that this molecule that

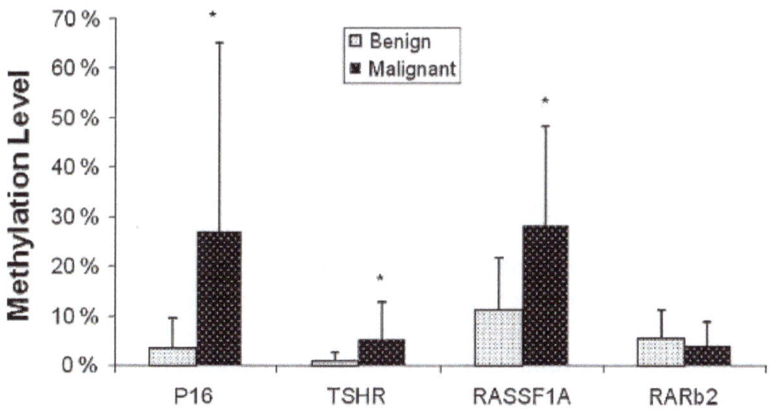

Fig. 15.5 Quantitative methylation level of four genes in benign and malignant thyroid tumors. (Mohammadi-asl et al. 2011)

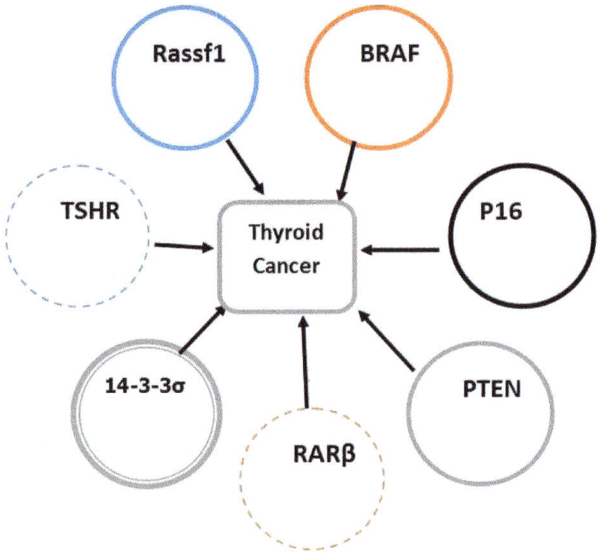

Fig. 15.6 Thyroid cancer and candidate genes

expresses under direct activation of p53, is induced after DNA damaging and plays a significant roles during several kind of malignancies including breast, prostate, skin and colon cancers (LODYGIN and HERMEKING 2005). Epigenetic analysis suggested that 14-3-3σ expression in thyroid cancer cell lines is epigenetically controlled by an alteration in CpG methylation (Lal et al. 2008). More studies are required to elucidate its function in thyroid malignancies.

15.1.7 TSHR

Thyroid-stimulating Hormone Receptor (TSHR) also known as LGR3, CHNG1 or hTSHR-I located on 14q31 is a membrane protein and a major regulator of thyroid cell metabolism. TSHR protein is a receptor for thyrothropin and thyrostimulin, which its activity is mediated by adenylate cyclase. (http://www.ncbi.nlm.nih.gov/gene/7253). Methylation of CpG islands in 5" flanking areas of the gene is known to be associated with carcinogenesis, development, progression and decreased tumor recurrence in PTC (Xu et al. 2010; Dai et al. 2010; Smith et al. 2007) and introduced as a diagnostic marker of malignancy to distinguish FTC from benign adenoma (Xing et al. 2003b). In addition, this methylation pattern occurred more preferentially in undifferentiated thyroid carcinoma (UTC) compared with PTC and FTC (Schagdarsurengin et al. 2006). Like previous genes, there is an association between T1799A *BRAF* mutation and *TSHR* methylation in PTC (SHI et al. 2009). It is also shown that transcription of *TSHR* is regulated through a methylation-dependent mechanism in Rat (Yokomori et al. 1998) (Fig. 15.6).

Table 15.1 Comparison of the methylation status in candidate genes

Methylated genes in PTC patients (%)				Reference
P16	TSHR	RASSF1A	RARβ2	
60	72	24	48	(Mohammadi-asl et al. 2011)
44	–	–	–	(Boltze et al. 2003)
41	–	–	–	(Lam et al. 2007)
27	–	–	–	(Wang et al. 2013)
25	–	62	–	(Schagdarsurengin et al. 2002)
–	41.1	–	–	(Xu et al. 2010)
–	59	–	–	(Smith et al. 2007)
–	68	–	–	(Dai et al. 2010)
–	43	–	–	(XiaoGuang et al. 2009)
–	–	20	–	(Xing et al. 2004)
–	–	32	–	(Nakamura et al. 2005)
–	–	62	–	(Schagdarsurengin et al. 2010)

PTC: papillary thyroid carcinoma

15.1.8 Other Genes

Growing epigenetic studies have updated the list of genes, which are epigenetically controlled during thyroid carcinogenesis (Table 15.1). Further researches are needed to clarify the role of epigenetic regulation in thyroid malignancy (Table 15.2).

15.2 Conclusion

Methylation-induced gene silencing appears to affect multiple genes in thyroid tissue and increases with cancer progression. Many tumor suppressor and tumor related genes are aberrantly methylated in thyroid cancer, suggesting a role of this epigenetic event in early and late thyroid tumorigenesis. The mechanisms that might mediate methylation and demethylation in carcinogenesis remain obscure, and there are questions as to whether the methylation changes are a cause or consequence of cellular transformation and clonal expansion. So, identification of all gene promoters methylated in cancer cells, known as the cancer methylome, would greatly advance our understanding of gene regulatory networks in tumorigenesis. Future studies need to emphasize the mechanistic aspects of these two types of epigenetic alterations to uncover new molecular mechanisms in thyroid tumorigenesis and to provide novel therapeutic targets for thyroid cancer. Until the results of these trials become available, research on epigenetic alterations in thyroid cancer must continue with the ultimate objective of developing more effective treatments for these tumors.

Table 15.2 Other genes in thyroid cancer

Gene	Position	Methylation Site	Type of Thyroid Cancer	Expression status	Reference
MT1G CRABP1	16q13 15q24	5-UTR	PTC	Down-regulated	(Huang et al. 2003)
HBB	11p15.5	5-UTR	ATC	Down-regulated	(Onda et al. 2005)
HNF3β/FoxA2	20p11	5-UTR	ATC PTC	Down-regulated	(Akagi et al. 2008)
PKG	17q21	5-UTR	PTC	Down-regulated	(Han and Kim 2009)
TSH-α	6q12	5-UTR	PTC	Up-regulated	(Han et al. 2009)
SLC26A4	7q31	5-UTR	various	Down-regulated	(Xing et al. 2003a)
FHIT	3p14.2	5-UTR	DTC	Down-regulated	(Yin et al. 2010)
TTF-1	9q34.13	5-UTR	UDC(ATC)	Down-regulated	(Kondo et al. 2009)
SERPINA5	14q32.1	5-UTR	PTC	Down-regulated	(Lee et al. 2013)
RASSF2	3p21.3	5-UTR	Various cell lines	Down-regulated	(Schagdarsuren-gin et al. 2010)
RIZ1(PRDM2)	1p36	5-UTR	Various cell lines	Down-regulated	(Lal et al. 2006)

15.3 Summary

Abnormalities of genomic methylation patterns have been attributed a role in carcinogenesis since large-scale demethylation of the genome was thought be an early event in multistep carcinogenesis. Up to now, several important gene have been introduced, whose methylation status is changed in thyroid cancers. There is a special focus on tumor suppressor genes whose local methylations were held to be involved in their silencing state. It have been demonstrated that aberrant TSHR gene methylation in human epithelial thyroid cancers is a molecular pathway underlying the silencing of this gene in these cancers. In addition, DNA methylation may directly affect the binding of the gene with transcription factors. Also, study on classical tumor suppressor genes such as RARβ2, whose methylation was associated with tumor aggressiveness showed that methylation-mediated gene silencing is an important mechanism in thyroid tumorigenesis. There is some hypothesis regarding the relationship between epigenetic and genetic modifications and the signaling pathways such as MAP kinase and PI3K pathways. Researches show that RASSR1A methylation was inversely associated with BRAF mutation, suggesting that epigenetic disruption of this tumor suppressor gene may play a role in thyroid tumorigenesis through signaling pathways other than the MAP kinase pathway. Interestingly, genetic alterations that could activate both the MAP kinase and PI3K pathways were found. Recently, it is also demonstrated that aberrant methylation of the PTEN gene was associated with activating genetic alterations in the PI3K/Akt pathway in thyroid tumors, suggesting a self-enhancing mechanism for the PI3K/Akt signaling through the epigenetic silencing of PTEN gene as a consequence of activation of this pathway. It is concluded that a cross talk between genetic and epigenetic alterations may occur in thyroid cancer through aberrant signaling of major molecular pathways. More studies are needed in order to clarify the exact role(s) of other candidate genes such as P16, 14-3-3σ with an emphasis on understanding of signaling pathways involved in their functions during thyroid cancers. In recent years, we are at the verge of the development of novel diagnostic, prognostic and therapeutic strategies for this common endocrine malignancy.

References

Akagi T, Luong Q, Gui D, Said J, Selektar J, Yung A, Bunce C, Braunstein G, Koeffler H (2008) Induction of sodium iodide symporter gene and molecular characterisation of HNF3β/FoxA2, TTF-1 and C/EBPβ in thyroid carcinoma cells. Br J Cancer 99(5):781–788

Alvarez-Nuñez F, Bussaglia E, Mauricio D, Ybarra J, Vilar M, Lerma E, Leiva Ad, Matias-Guiu X (2006) PTEN promoter methylation in sporadic thyroid carcinomas. Thyroid 16(1):17–23

Ball E, Bond J, Franc B, DeMicco C, Wynford-Thomas D (2007) An immunohistochemical study of p16 INK4a expression in multistep thyroid tumourigenesis. Eur J Cancer 43(1):194–201

Barroeta JE, Baloch Z, Lal P, Pasha TL, Zhang PJ, LiVolsi V (2006) Diagnostic value of differential expression of CK19, Galectin-3, HBME-1, ERK, RET, and p16 in benign and malignant follicular-derived lesions of the thyroid: an immunohistochemical tissue microarray analysis. Endocr Pathol 17:225–234

Bartoletti R, Cai T, Nesi G, Roberta Girardi L, Baroni G, Dal Canto M (2007) Loss of P16 expression and chromosome 9p21 LOH in predicting outcome of patients affected by superficial bladder cancer. J Surg Res 143(2):422–427

Bastien J, Rochette-Egly C (2004) Nuclear retinoid receptors and the transcription of retinoid-target genes. Gene 328:1–16

Boltze C, Zack S, Quednow C, Bettge S, Roessner A, Schneider-Stock R (2003) Hypermethylation of the CDKN2/p16 INK4A promotor in thyroid carcinogenesis. Pathol Res Pract 199(6):399–404

Brait M, Loyo M, Rosenbaum E, Ostrow KL, Markova A, Papagerakis S, Zahurak M, Goodman SM, Zeiger M, Sidransky D (2012) Correlation between BRAF mutation and promoter methylation of TIMP3, RARβ2 and RASSF1A in thyroid cancer. Epigenetics 7(7):710

Choi Y-L, Kang SY, Choi JS, Shin YK, Kim SH, Lee S-J, Bae D-S, Ahn G (2006) Aberrant hypermethylation of RASSF1A promoter in ovarian borderline tumors and carcinomas. Virchows Arch 448(3):331–336

Cras A, Darsin-Bettinger D, Balitrand N, Cassinat B, Soulie A, Toubert M, Delva L, Chomienne C (2007) Epigenetic patterns of the retinoic acid receptor β2 promoter in retinoic acid-resistant thyroid cancer cells. Oncogene 26 (27):4018–4024

Cui H-y, Zhou X-r, Han Q (2011) Methylation status of RASSF1A gene promoter in cervical carcinoma and its clinical significance. Chin Gen Pract 14:007

Dai Y, Cai D, Chen H, Zhang Y, Li J (2010) Transcription and promoter hypermethylation of thyroid stimulating hormone receptor gene in human papillary thyroid carcinoma. Nan Fang Yi Ke Da Xue Xue Bao. J Southern Med Univ 30(1):114–117

Dammann R, Schagdarsurengin U, Liu L, Otto N, Gimm O, Dralle H, Boehm BO, Pfeifer GP, Hoang-Vu C (2003) Frequent RASSF1A promoter hypermethylation and K-ras mutations in pancreatic carcinoma. Oncogene 22(24):3806–3812

Ellis RJ, Wang Y, Stevenson HS, Boufraqech M, Patel D, Nilubol N, Davis S, Edelman DC, Merino MJ, He M (2013) Genome-wide methylation patterns in papillary thyroid cancer are distinct based on histological subtype and tumor genotype. J Clin Endocrinol Metab 99(2):329–37
Ferenc T, Lewinski A, Lange D, Niewiadomska H, Sygut J, Sporny S, Jarzab B, Salacinska-Los E, Kulig A, Wloch J (2004) Analysis of P161NK4A protein expression in follicular thyroid tumors. Pol J Pathol 55(4):143–148
Ferru A, Denis S, Guilhot J, Gibelin H, Tourani J, Kraimps J, Larsen C, Karayan-Tapon L (2006) Expression of TAp73 and ΔNp73 isoform transcripts in thyroid tumours. Eur J Surg Oncol 32(2):228–230
Gauchotte G, Lacomme S, Brochin L, Tournier B, Cahn V, Monhoven N, Piard F, Klein M, Martinet N, Rochette-Egly C (2013) Retinoid acid receptor expression is helpful to distinguish between adenoma and well-differentiated carcinoma in the thyroid. Virchows Arch 462(6):619–632
Gerdes B, Ramaswamy A, Ziegler A, Lang SA, Kersting M, Baumann R, Wild A, Moll R, Rothmund M, Bartsch DK (2002) p16INK4a is a prognostic marker in resected ductal pancreatic cancer: an analysis of p16INK4a, p53, MDM2, an Rb. Ann Surg 235(1):51
Gimm O (2001) Thyroid cancer. Cancer letters 163(2):143–156
Greco A, Borrello MG, Miranda C, Degl'Innocenti D, Pierotti MA (2009) Molecular pathology of differentiated thyroid cancer. Q J Nucl Med Mol Imaging 53(5):440–453
Grunt TW, Puckmair K, Tomek K, Kainz B, Gaiger A (2005) An EGF receptor inhibitor induces RAR-β expression in breast and ovarian cancer cells. Biochem biophys Res Commun 329(4):1253–1259
Han K-H, Kim T-J (2009) Effects of promoter methylation on the expression levels of plakoglobin gene in both the aro thyroid cancer cell line and cancer tissues. Korean J Clin Lab Sci 41(4):180–188
Han KH, Son KS, Hong JE, Kim SJ (2009) Promoter hypermethylation and up-regulation of thyroid-stimulating-hormone-alpha (TSH-α) in thyroid cancer. Genes Genomics 31(5):341–347
Hoque M, Rosenbaum E, Westra W, Xing M, Ladenson P, Zeiger M, Sidransky D, Umbricht C (2005) Quantitative assessment of promoter methylation profiles in thyroid neoplasms. J Clin Endocrinol Metab 90(7):4011–4018
Hou P, Ji M, Xing M (2008) Association of PTEN gene methylation with genetic alterations in the phosphatidylinositol 3-kinase/AKT signaling pathway in thyroid tumors. Cancer 113(9):2440–2447
Huang GW, Mo WN, Kuang GQ, Nong HT, Wei MY, Sunagawa M, Kosugi T (2001) Expression of p16, nm23-H1, E-cadherin, and CD44 gene products and their significance in Nasopharyngeal Carcinoma. Laryngoscope 111(8):1465–1471
Huang Y, de la Chapelle A, Pellegata NS (2003) Hypermethylation, but not LOH, is associated with the low expression of MT1G and CRABP1 in papillary thyroid carcinoma. Int J Cancer 104(6):735–744
Ishida E, Nakamura M, Shimada K, Higuchi T, Takatsu K, Yane K, Konishi N (2007) DNA hypermethylation status of multiple genes in papillary thyroid carcinomas. Pathobiology 74(6):344–352
Ito Y, Miyoshi E, Uda E, Yoshida H, Uruno T, Takamura Y, Miya A, Kobayashi K, Matsuzuka F, Matsuura N (2003) 14-3-3σ possibly plays a constitutive role in papillary carcinoma, but not in follicular tumor of the thyroid. Cancer Lett 200(2):161–166
Jones C, Shaw J, Wyllie F, Gaillard N, Schlumberger M, Wynford-Thomas D (1996) High frequency deletion of the tumour suppressor gene P16 INK4a(MTS1) in human thyroid cancer cell lines. Mol Cell Endocrinol 116(1):115–119
Juhlin CC, Kiss NB, Villablanca A, Haglund F, Nordenström J, Höög A, Larsson C (2010) Frequent promoter hypermethylation of the APC and RASSF1A tumour suppressors in parathyroid tumours. PLoS ONE 5(3):9472
Kiseljak-Vassiliades K, Xing M (2011) Association of cigarette smoking with aberrant methylation of the tumor suppressor gene RARβ2 in papillary thyroid cancer. Front Endocrinol 2:99
Kiss N, Geli J, Lundberg F, Avci C, Velazquez-Fernandez D, Hashemi J, Weber G, Höög A, Ekström T, Bäckdahl M (2008) Methylation of the p16INK4A promoter is associated with

malignant behavior in abdominal extra-adrenal paragangliomas but not pheochromocytomas. Endocr-Relat Cancer 15(2):609–621

Knauf JA, Fagin JA (2009) Role of MAPK pathway oncoproteins in thyroid cancer pathogenesis and as drug targets. Curr Opin Cell Biol 21(2):296–303

Kondo T, Nakazawa T, Ma D, Niu D, Mochizuki K, Kawasaki T, Nakamura N, Yamane T, Kobayashi M, Katoh R (2009) Epigenetic silencing of TTF-1/NKX2-1 through DNA hypermethylation and histone H3 modulation in thyroid carcinomas. Lab Invest 89(7):791–799

Kunstman JW, Korah R, Healy JM, Prasad M, Carling T (2013) Quantitative assessment of RASSF1A methylation as a putative molecular marker in papillary thyroid carcinoma. Surgery 154(6):1255–1262

Lal G, Padmanabha L, Smith BJ, Nicholson RM, Howe JR, O'Dorisio MS, Domann FE (2006) RIZ1 is epigenetically inactivated by promoter hypermethylation in thyroid carcinoma. Cancer 107(12):2752–2759

Lal G, Padmanabha L, Provenzano M, Fitzgerald M, Weydert J, Domann FE (2008) Regulation of 14-3-3σ expression in human thyroid carcinoma is epigenetically regulated by aberrant cytosine methylation. Cancer Lett 267(1):165–174

Lam AKY, Lo CY, Leung P, Lang BHH, Chan WF, Luk JM (2007) Clinicopathological roles of alterations of tumor suppressor gene p16 in papillary thyroid carcinoma. Ann Surg Oncol 14(5):1772–1779

Lee J-J, Au AY, Foukakis T, Barbaro M, Kiss N, Clifton-Bligh R, Staaf J, Borg Å, Delbridge L, Robinson BG (2008) Array-CGH identifies cyclin D1 and UBCH10 amplicons in anaplastic thyroid carcinoma. Endoc-Relt Cancer 15(3):801–815

Lee SJ, Lee MH, Kim DW, Lee S, Huang S, Ryu MJ, Kim YK, Kim SJ, Kim SJ, Hwang JH (2011) Cross-regulation between oncogenic BRAFV600E kinase and the MST1 pathway in papillary thyroid Carcinoma. PloS ONE 6(1):e16180

Lee EK, Chung K-W, Yang SK, Park MJ, Min HS, Kim SW, Kang HS (2013) DNA methylation of MAPK signal-inhibiting genes in papillary thyroid Carcinoma. Anticancer Res 33(11):4833–4839

Li RN, Yu FJ, Wu CC, Chen YK, Yu CC, Chou SH, Lee JY, Cheng YJ, Wu MT, Wu I (2014) Methylation status of retinoic acid receptor beta2 promoter and global DNA in esophageal squamous cell carcinoma. J Surg Oncol 109(6):623–7

LODYGIN D, HERMEKING H (2005) The role of epigenetic inactivation of 14-3-3σ in human cancer. Cell Res 15(4):237–246

LUO Y, CHEN J, LV Y (2004) Methylation of p16 promotor's CpG region and its association with clinic pathological characteristics of gliomas. J Shanghai Medica (University) 5:009

Mancikova V, Buj R, Castelblanco E, Inglada-Pérez L, Diez A, Cubas AA, Curras-Freixes M, Maravall FX, Mauricio D, Matias-Guiu X (2014) DNA methylation profiling of well-differentiated thyroid cancer uncovers markers of recurrence free survival. Int J Cancer 135(3):598–610

Miasaki F, Vivaldi A, Ciampi R, Agate L, Collecchi P, Capodanno A, Pinchera A, Elisei R (2008) Retinoic acid receptor beta2 re-expression and growth inhibition in thyroid carcinoma cell lines after 5-aza-2′-deoxycytidine treatment. J Endocrinol Invest 31(8):724–730

Mohammadi-asl J, Larijani B, Khorgami Z, Tavangar SM, Haghpanah V, Kheirollahi M, Mehdipour P (2011) Qualitative and quantitative promoter hypermethylation patterns of the P16, TSHR, RASSF1A and RARβ2 genes in papillary thyroid carcinoma. Med Oncol 28(4):1123–1128

Nakamura N, Carney JA, Jin L, Kajita S, Pallares J, Zhang H, Qian X, Sebo TJ, Erickson LA, Lloyd RV (2005) RASSF1A and NORE1A methylation and BRAFV600E mutations in thyroid tumors. Lab Investigation 85(9):1065–1075

Oliveira C, Velho S, Domingo E, Preto A, Hofstra RM, Hamelin R, Yamamoto H, Seruca R, Schwartz S (2005) Concomitant RASSF1A hypermethylation and KRAS/BRAF mutations occur preferentially in MSI sporadic colorectal cancer. Oncogene 24(51):7630–7634

Onda M, Akaishi J, Asaka S, Okamoto J, Miyamoto S, Mizutani K, Yoshida A, Ito K, Emi M (2005) Decreased expression of haemoglobin beta (HBB) gene in anaplastic thyroid cancer and recovery of its expression inhibits cell growth. British J Cancer 92(12):2216–2224

Pallarés J, Velasco A, Eritja N, Santacana M, Dolcet X, Cuatrecasas M, Palomar-Asenjo V, Catasús L, Prat J, Matias-Guiu X (2008) Promoter hypermethylation and reduced expression of RASS-F1A are frequent molecular alterations of endometrial carcinoma. Mod Pathol 21(6):691–699

Qian ZR, Sano T, Yoshimoto K, Yamada S, Ishizuka A, Mizusawa N, Horiguchi H, Hirokawa M, Asa SL (2005) Inactivation of RASSF1A tumor suppressor gene by aberrant promoter hypermethylation in human pituitary adenomas. Lab Investigation 85(4):464–473

Russo D, Damante G, Puxeddu E, Durante C, Filetti S (2011) Epigenetics of thyroid cancer and novel therapeutic targets. J Mol Endocrinol 46(3):R73–R81

Salvesen HB, MacDonald N, Ryan A, Jacobs IJ, Lynch ED, Akslen LA, Das S (2001) PTEN methylation is associated with advanced stage and microsatellite instability in endometrial carcinoma. Int J Cancer 91(1):22–26

Sato K, Tamura G, Tsuchiya T, Endoh Y, Sakata K, Motoyama T, Usuba O, Kimura W, Terashima M, Nishizuka S (2002) Analysis of genetic and epigenetic alterations of the PTEN gene in gastric cancer. Virchows Arch 440(2):160–165

Schagdarsurengin U, Gimm O, Hoang-Vu C, Dralle H, Pfeifer GP, Dammann R (2002) Frequent epigenetic silencing of the CpG island promoter of RASSF1A in thyroid carcinoma. Cancer Res 62(13):3698–3701

Schagdarsurengin U, Wilkens L, Steinemann D, Flemming P, Kreipe HH, Pfeifer GP, Schlegelberger B, Dammann R (2003) Frequent epigenetic inactivation of the RASSF1A gene in hepatocellular carcinoma. Oncogene 22(12):1866–1871

Schagdarsurengin U, Gimm O, Dralle H, Hoang-Vu C, Dammann R (2006) CpG island methylation of tumor-related promoters occurs preferentially in undifferentiated carcinoma. Thyroid 16(7):633–642

Schagdarsurengin U, Richter AM, Hornung J, Lange C, Steinmann K, Dammann RH (2010) Frequent epigenetic inactivation of RASSF2 in thyroid cancer and functional consequences. Mol Cancer 9(1):264

Schmutzler C, Hoang-Vu C, Ruger B, Kohrle J (2004) Human thyroid carcinoma cell lines show different retinoic acid receptor repertoires and retinoid responses. Eur J Endocrinol 150(4):547–556

Serrano J, Goebel SU, Peghini PL, Lubensky IA, Gibril F, Jensen RT (2000) Alterations in the p16INK4a/CDKN2A tumor suppressor gene in gastrinomas. J Clin Endocrinol Metab 85(11):4146–4156

SHI X-g, Cheng J-x, Guan H-x, Teng W-p (2009) Hypermethylation and Protein Expression of TSHR and NIS Genes in Papillary Thyroid Cancer and Their Association with BRAF Mutation. J China Med University 6:002

Smith JA, Fan C-Y, Zou C, Bodenner D, Kokoska MS (2007) Methylation status of genes in papillary thyroid carcinoma. Arch Otolaryngol-Head Neck Surg 133(10):1006–1011

TANG J-d, SU X-l (2010) Research of CpG island methylation status of NIS and RASSF1A gene promoters in papillary thyroid carcinomas. China J Mod Med 21:021

Tommasi S, Dammann R, Zhang Z, Wang Y, Liu L, Tsark WM, Wilczynski SP, Li J, You M, Pfeifer GP (2005) Tumor susceptibility of Rassf1a knockout mice. Cancer Res 65(1):92–98

Wang S, Du J, Liu Y (2002) P16 gene deletion and 5'CpG island methylation in endometrial carcinoma. Tumor 22 (3):213–214

Wang P, Pei R, Lu Z, Rao X, Liu B (2013) Methylation of p16 CpG islands correlated with metastasis and aggressiveness in papillary thyroid carcinoma. J Chin Med Asso 76(3):135–139

Wei F, Wang Z, Wu Y (2013) The clinical significance of tumor suppressor gene methylation, expression in nodular thyroid disease. Endocr Abstr 32:572

Wong IH, Chan J, Wong J, Tam PK (2004) Ubiquitous aberrant RASSF1A promoter methylation in childhood Neoplasia1. Clin Cancer Res 10(3):994–1002

Wu D-L, Sui F-Y, Jiang X-M, Jiang X-H (2006) Methylation in esophageal carcinogenesis. World J Gastroenterol 12(43):6933

XiaoGuang S, JianXin C, HaiXia G, WeiPing T (2009) Hypermethylation and protein expression of TSHR and NIS genes in papillary thyroid cancer and their association with BRAF mutation. J China Med University 38(6):401–404

Xing M (2008) Recent advances in molecular biology of thyroid cancer and their clinical implications. Otolaryngol Clin North Ame 41(6):1135–1146

Xing M, Tokumaru Y, Wu G, Westra WB, Ladenson PW, Sidransky D (2003a) Hypermethylation of the pendred syndrome gene SLC26A4 is an early event in thyroid tumorigenesis. Cancer Res 63(9):2312–2315

Xing M, Usadel H, Cohen Y, Tokumaru Y, Guo Z, Westra WB, Tong BC, Tallini G, Udelsman R, Califano JA (2003b) Methylation of the thyroid-stimulating hormone receptor gene in epithelial thyroid tumors a marker of malignancy and a cause of gene silencing. Cancer Res 63(9):2316–2321

Xing M, Cohen Y, Mambo E, Tallini G, Udelsman R, Ladenson PW, Sidransky D (2004) Early occurrence of RASSF1A hypermethylation and its mutual exclusion with BRAF mutation in thyroid tumorigenesis. Cancer Res 64(5):1664–1668

Xu J, Ge M, Ling Z, CHENG L, XU J-j (2010) Studies on methylation of 5′-CpG island in the promoter region of TSHR gene in papil-lary thyroid carcinomas. Tumor 30(8):696–699

Yeo W, Wong N, Wong WL, Lai P, Zhong S, Johnson PJ (2005) High frequency of promoter hypermethylation of RASSF1A in tumor and plasma of patients with hepatocellular carcinoma. Liver Int 25(2):266–272

Yin D-T, Wang L, Sun J, Yin F, Yan Q, Shen R, He G, Gao J-X (2010) Association of the promoter methylation and protein expression of Fragile Histidine Triad (FHIT) gene with the progression of differentiated thyroid carcinoma. Int J Clin Exp Pathol 3(5):482

Yokomori N, Tawata M, Saito T, Shimura H, Onaya T (1998) Regulation of the rat thyrotropin receptor gene by the methylation-sensitive transcription factor GA-binding protein. Mol Endocrinol 12(8):1241–1249

Zhang J, Liu L, Pfeifer GP (2004) Methylation of the retinoid response gene TIG1 in prostate cancer correlates with methylation of the retinoic acid receptor beta gene. Oncogene 23(12):2241–2249

Zhang J, Zhang J, Li M, Wu Y, Fan Y, Zhou Y, Tan L, Shao Z, Shi H (2011) Methylation of RAR-β2, RASSF1A, and CDKN2A genes induced by nickel subsulfide and nickel-carcinogenesis in rats. Biomed Environ Sci 24(2):163–171

Zhou X-P, Gimm O, Hampel H, Niemann T, Walker MJ, Eng C (2000) Epigenetic PTEN silencing in malignant melanomas without PTEN mutation. Am J Pathol 157(4):1123–1128

Chapter 16
An Introduction to Impact of Bio-Resonance Technology in Genetics and Epigenetics

Mohammad Ebrahimi, Sabokhi Sharifov, Maryam Salili and Larysia Chernosova

Contents

Abstract According to the WHO, chronic diseases have major economic and social impacts. Despite the increasing scientific efforts to identify the etiology and mechanisms of chronic diseases and to treat them, the prevalence of these diseases in the world is expanding. One concept describing the etiology and mechanisms of chronic diseases is based on "Epigenetic Changes". Epigenetic changes are permanent changes in gene expression due to Chromatin conformation changes that do not involve any change in DNA sequence. Depending on the time-scale these changes can be persistent through DNA replication. In the eukaryotic nucleus, the nuclear chromatin cluster has electric oscillation capacity. The natural frequency of an oscillating chromatin region is determined by the physical properties of DNA-protein complexes in that region, which can be changed by its epigenetic state and associated protein factors. These changes can be detected using Bio-resonances method and therefore be used to early detection of chronic diseases. It works on the

M. Ebrahimi (✉)
The Bio-signal and Immunculus Scientific Group, No 4,
The Research Center for New Technologies in Life Science Engineering,
Shahid Oroji all, 16th Azar St, Keshaverz, Tehran, Iran
e-mail: dr.mohamadebrahimi@gmail.com

S. Sharifov · L. Chernosova
Center of Intellectual Medical Systems IMEDIS, Moscow, Russia

M. Salili
Firoozgar Hospital, Iran University of Medical Sciences, Tehran, Iran

P. Mehdipour (ed.), *Epigenetics Territory and Cancer*,
DOI 10.1007/978-94-017-9639-2_16

basis of spectral analysis of magnetic fields of living organisms which enables therapist to differentiate normal from abnormal conditions. It is proposed that the electromagnetic waves as epigenetic factors could effect on chromatin dynamic changes and activate or suppress biochemical processes in organism and play a critical role in development or treatment of chronic diseases. This chapter has attempted to demonstrate the opinions of the authors on this issue and its relationship with genetic, epigenetic and also its application in medicine.

Keywords Bioresonance therapy · Biophoton · Epigenetics · Genetic

16.1 Introduction

If one searches the Internet about Bioresonance, he or she may find the definition of Bioresonace as a pseudo-scientific medical concept. Pseudoscience describe as a claim or practice which is presented as scientific, but does not adhere to a valid scientific method and lacks supporting evidence or acceptability. In contrast, science is "a set of methods designed to understand, describe and interpret and aimed at building a testable body of knowledge open to rejection or confirmation". Does the Bioresonance really have the Pseudoscience or it has the scientific character? Medical literature provides the following differences between science and pseudoscience:

1. The primary goal of science is to achieve a more complete and more integrated understanding of the physical world. But, Pseudoscience is more likely to be determined by ideological, cultural, or even commercial goals.
2. As a rule, most of the scientific areas are the subjects of intense study and research which result in the continual expansion of knowledge in the discipline. The field of Pseudoscience has evolved very little since it was first established. The small amount of research and experimentation that has been carried out is generally done more to justify the belief than to extend it. (Nearly every new finding raises new questions that beg exploration. There is little evidence of this in the pseudoscience.)
3. Scientific explanations must be stated in clear, unambiguous terms. But, pseudoscientific explanations tend to be unclear and ambiguous, often invoking scientific terms in uncertain contexts.
4. Scientific ideas and concepts must stand or fall on their own facts, based on existing knowledge and on scientific evidence. Pseudoscientific concepts tend to be made by individual egos and personalities, almost always by individuals who are not in contact with mainstream science.
5. Science is a process in which each principle must be tested in the crucible of experience and remains subject to being questioned or rejected at any time. But for pseudoscience, the major beliefs and principles of the field are often not falsifiable, and are unlikely ever to be altered or shown to be wrong. (Allchin 2004; Martin 1994; Phelan 2008).

In relation to the above-mentioned, in general, in the basic and clinical level, numerous positive studies have been derived from Bioresonace method that has been conducted by international and scientific workgroups (Gernert 2008; Grass and Kasper 2008; Imaizumi et al. 1984; Kobayashi et al. 1999a; Mansfield 2005; Popp et al. 1984; Quickenden and Que Hee 1974; Tilbury and Cluickenden 1988; OJu and Gogoleva 2000; Gogoleva 2001; Islamov et al. 2002; Huang et al. 2005; Nienhaus and Galle 2006; Rahlfs and Rozehnal 2008; Schuller and Galle 2007; Adamo et al. 1989; Herrmanna and Galleb 2011; Pihtili et al. 2009; Chen et al. 2010; Prelević 2011). However, few researchers did not confirm the therapeutic effectiveness of the bioresonance method (Schöni et al. 1997). Therefore, the continuous controversial debates in this field are going on. In this review article, we discuss the scientific aspects of Bioresonance and Biophoton technology in relation to Genetic and Epigenetic Science.

16.1.1 History of Electrophysiology, Bioresonance and Biophoton

Most people are now familiar with ECG, EEG and MRI Scans. None of these diagnostic apparatuses would work if we were not energetic organisms.

Carlo Matteucci was a physicist and neurophysiologist who was a pioneer in the study of Bioelectricity. *Carlos Matteucci*, in the 1830's, proved that an electrical current is generated by injured tissues.

*Emil du Bois-Reymond*was a physician and physiologist, he is known as the father of experimental Electrophysiology because of the discovery of "Nerve reaction potential". In1843, *Dubois-Reymondin* constructed a galvanometer for detecting electrical current and used the terms "Muscular current" and "negative variation" for first time.

Nikola Tesla in 1920 developed the Tesla coil during his experimentations with high frequency phenomena. A Tesla coil is an electrical resonant transformer circuit. It is used to produce high-voltage, low-current, high frequency alternating-current electricity. Tesla coil is used in the production of the Multi-Wave Oscillator apparatus (MWO) (Carlson 2005; Roland Hans Penner 1995).

Tesla collaborated with French engineer, *Georges Lakhovsky* to complete The Multi-Wave Oscillator. *Tesla* and *Lakhovsky* thought the nucleus of the cell with its "filament strands" is similar to an electronic oscillating circuit, capable of sending and receiving vibratory information. *Lakhovsky* believed that every cell in the body has its own rate of internal vibration. He postulated that all living cells (plants, people, bacteria, parasites, etc.) possess resonance. *Lakhovsky* proposed that not only do all living cells produce and radiate oscillations of very high frequencies, but also they receive and respond to oscillations imposed upon them from outside sources.

According to *Lakhovsky*, the approach to stand microbial vibrational disturbance in body is to produce harmonic broad spectrum radio frequency electromagnetic waves and send them into the system and then, through the principle of sympathetic vibration, each cell in body responds to external vibrations to which it has a harmonic likeness. Therefore, the healthy cells would be more resistant to vibrational attack from virus and bacteria.

This method resulted in the invention of the Multi-Wave Oscillator (MWO) apparatus. The MWO and other similar devices continued to be used in clinics throughout Europe, but the technology seems to have been almost forgotten in America. MWO's have been documented to be of value in treating cancer, arthritis, and other illnesses. *Lakhovsky's* article and patents can be found online at:http://www.rexresearch.com/lakhov/lakhusps.htm.

In 1920, *RR Rife* who was an American inventor had finished building the world's first universal microscope. *Rife* was an optical engineer and technician with great skills. With this unbelievable microscope, He could see a live virus for the first time. *Rife* carefully identified the individual spectroscopic characteristic (reflected or absorbed) of each microbe, using a split spectroscope attachment. In his study, he gradually rotated block quartz prisms in order to focus a single wavelength light upon the examined micro-organism. In this way, he established that every molecule oscillates at its own distinct frequency. *Rife* claimed to have documented a "Mortal Oscillatory Rate" for various pathogenic organisms, and to be able to destroy the organisms by vibrating them at this particular rate. (Rife 2013; Rosenow 1965; Montgomery 2003; Bird 1976).

In 1937, *Harold Burr* a Professor of Anatomy at the Yale University began a series of experiments to find characteristics of the bio-magnetic field of living organisms. *Dr. Burr* discovered that all living things—from men to animals and plants—have electro-dynamic fields, which can be measured and mapped with standard voltmeters. *Dr. Burr* was able to demonstrate a specific technique for measuring the microvolt levels in living organisms. (Burr et al. 1936).

In 1939, *Semyon Valentina Kirlian*, a Russian inventor and researcher, discovered an approach for visualizing bio-fields in living organisms. It is known as *Kirlian* photography. In this method if an object on a photographic plate is connected to a high-voltage source, an image is produced on the photographic plate. The technique has been variously known as "electrography", "electrophotography", "corona discharge photography".

Kirlian photography involves emitting a high frequency, high voltage, ultra-low current to the object being photographed. It travels through and reacts with the complex systems of living organisms. This influx of electrical energy amplifies and makes the organisms biologically visible. The subject and the film interact to produce a corona of multi-frequency energy waves, which are captured by the camera (Andrew et al. 1979). Although acupuncture therapy began in China in the seventeenth century, it has been under investigation since the 1900s in the West. In 1951, a Russian researcher *Jean Niboyet* found out that acupuncture points have a lower skin resistance than other points of the body (Helene et al. 2002).

In 1953, *Dr Reinhold Voll*, a German medical doctor, developed an electronic testing device for finding acupuncture points electrically. He was successful in finding acupuncture points and demonstrating that these points have different resistance from the adjacent tissues when facing an electrical current. *Dr. Voll* made up a diagnostic system based on electro-conductivity of acupuncture points. He also introduced a special scale to interpret the results efficiently.

He found out that, for example, patients with lung cancer have abnormal readings on the acupuncture points referred to as lung points.

He also was successful in combining the ancient acupuncture knowledge with western medicine in order to introduce electro-acupuncture as a novel method. According to *Voll*, the resistance of the acupuncture point is the measuring scale of energy in a particular organ and an indicator of its ability to function. This method makes measuring and registering of the condition and function of the body organs possible.

The overall function of a person can be recorded in this way and the source of the cause can be located. The system that has been developed on the basis of these findings is called "Electro Acupuncture according to *Voll*" (Voll 1974a, b; Peter 1984).

In 1941, *Albert Szent-Gyorgyi*, who won the Nobel Prize in Physiology in 1937 published an article entitled, "Towards a New biochemistry," Which suggested that energy, in living systems, may be transmitted by conduction bands.

He suggested that the double bonds in the protein backbone provide free or mobile electrons and these electrons (energy) can move through proteins. He proposed that these electrons belong to the whole system and not to one or two atoms. A great number of molecules can join together to form an energy continuum, along which, energy may travel. This is a "whole-system" perspective on energy transfer, and offers a basis for a variety of bio-energy diagnostics and therapies (Szent-Gyorgyi 1894; Szent-Gyorgyi 1960).

Professor *Kim Bong Han* was a North Korean medical surgeon at Pyongyang Medical University. He discussed "the primo-vascular system" in reports that were published during the early 1960s. *Kim* was able to show the existence of neuro-anatomical basis of acupuncture meridians by injecting radioactive phosphorous (P32) into acupuncture points on a rabbit's abdomen and tracing its flow.

He traced the uptake of the substance into the nearby tissue and discovered that the isotope was actively taken up along a fine duct-like tubule system (approximately 0.5–1.5 microns in diameter). The energy conduit followed the path of the classical acupuncture meridians. Later, researchers in South Korea replicated Han's work. They discovered novel threadlike structures in the cerebral ventricles of rabbits that are proposed as sites of quantum communication. (Soh et al. 2013; Avijgan and Avijgan 2013).

Dr. Helmut Schimmel designed a simplified form of *Dr Voll's* device, which is known as the Vegatest or the "Photon Resonance Test". The original technique started in 1953 by *DrVoll*, was a complex procedure involving measuring hundreds of acupuncture points. But, with the Vegatest, all measurements are carried out using one single acupuncture point instead of hundreds, as the system is based on measuring against test ampoules rather than against the organ-linked points themselves (Schimmel and Penzer 1997; Katelaris et al. 1991; Voll 1974a, b).

In the 1950's, *Dr. George Goodheart*, discovered that the muscles of the body, in the presence of certain substances, would become either weaker or stronger. This finding is part of a diagnostic system called "Applied Kinesiology". Its basic idea is that every organ dysfunction is accompanied by a specific muscle weakness, which enables diseases to be diagnosed through muscle-testing procedures. Based on this phenomenon, a simple arm or leg check can monitor the body's response to any given substance (Haas et al. 1994).

Dr. Hunt is a retired Professor in the UCLA Department of Physiological Sciences. She was the first to discover the relationship between variations in bio-energy patterns and human behavior. Dr. Hunt began to quantify human bio-energy, and found that it contains information related to physiological and conscientious levels of human body. In 1970's, she recorded the electrical energy from the body's surface (Hunt 1996).

In 1977, *Dr. Franz Morell* and an electronics engineer *Mr. Erich Rasche* developed the "MORA-Therapie", (for MOrell and RAsche), which is a medical device for bioresonance therapy. The MORA system, which is one of the bioresonance modification devices can analyze the healthy oscillations, amplify them and returns them to the patient's body. Abnormal oscillations are omitted and changed via a process of filtration and wave inversion (Herrmanna and Galleb 2011; Chen et al. 2010; Schöni et al. 1997).

Scientific observations showed that salamanders are able to regenerate limbs, while frogs, that are only one evolutionary stage before salamanders, have lost this potential. In 1980, Dr. *Robert O. Becker* tried to find the reason for these differences. He measured the electrical differences between the two animals at the end of a limb and found that both showed a positive potential. However, the salamander's limb stump soon reversed in polarity to a negative potential, which gradually returned to zero over the days that the limb re-grew. When *Becker* artificially used a negative potential on the frog's healing limb stump, the frog grew a new limb. *Dr. Becker* also predicted that living organisms could be influenced by external electromagnetic fields as the fields interacted with the direct currents that flow within the organism (Becker 1963, 1972; Becker et al. 1962, 1974).

During the last years of the Russian Soviet Union, the country's space medical program concentrated on sending men into space for long periods of time. These cosmonauts were in space with no access to medical services. This led to the Russian Government having to develop electronic devices to treat their cosmonauts' health issues in space. Using principles of Bioresonance therapy they developed a device called the Skenar. This is a small, computerized electro-therapy device that sends an electric impulse into the body, reads the impulse coming back from the body and then alters the next impulse it sends out to the body. This is repeated until the body reaches a state of electrical normality. The Skenar is certified by the European Common Market for pain control. In the US the Skenar is also registered with the FDA as a biofeedback device for muscular disorders (Dunwell 2011; Grinberg 1996; Nozdrachev 1996; Zavitaev 1996).

Dr. Bruce Lipton is an American developmental biologist, is best known for promoting the idea that genes and DNA can be manipulated by the person's beliefs. Dr. *Br. Lipton* began examining the principles of quantum physics and how to integrate them into the understanding of the cell's information processing systems and internal bio-signaling. He's spent his life studying human biology and behavior. He produced breakthrough findings on the cell membrane, which revealed this outer layer of the cell was an organic homologue of a computer chip, the cell's equivalent of brain. His research at Stanford University's School of Medicine, between 1987 and 1992, showed that the environment, co-operating though the membrane, controlled the behavior and physiology of the cell, turning genes on and off. His discoveries, which ran counter to the established scientific view that life is controlled by the genes, presaged one of today's most important fields of study, the science of Epigenetics. Results derived from these studies defined the molecular pathways connecting the mind and body. According to *Dr. Lipton*, gene activity can change on a daily basis. If the perception in your mind is reflected in the chemistry of your body, and if your nervous system reads and interprets the environment and then controls the blood's chemistry, then you can literally change the fate of your cells by altering your thoughts. Many subsequent papers by other researchers have since validated his concepts and ideas (Lipton and Konigsberg 1972; Lipton and Jacobson 1974; Konigsberg et al. 1975; Lipton 1977, 1988, 1998, 2001, 2005a, b; Lipton and Schultz 1979; Lipton et al. 1991).

It is now recognized that the environment, can control the activity of our genes. Environment controls gene activity through a process known as epigenetic control.

Today many medical centers use electro-diagnostic devices to improve diagnostic and select their recommended treatments. Bioresonance is named in different terminology. The diagnostic procedure is most commonly referred to as Electroacupuncture according to *Voll* (EAV) or Electro Dermal Screening (EDS), but some practitioners call it bioelectric functions diagnosis (BFD), bio resonance therapy (BRT), bio-energy regulatory technique (BER), Biocybernetic Medicine (BM), computerized electro dermal screening (CEDS), computerized electro dermal stress analysis (CDCSA), electro dermal testng (EDT), limbic stress assessment (LSA), meridian energy analysis (MEA), or point testing.

Recently, the term INFORMATIVE MEDICINE has established itself as a very appropriate description of the complementary medical therapy. This means, that in contrast to the classical medicine, healing is achieved here through INFORMATION FROM BODY rather than material substances.

16.2 What is the Logic of Bio-Energy Methods?

Some of the complementary and alternative medical device and approaches are based in part on energetic aspects of life. Bioresonance testing is based on the science of biophysics (see below and Fig. 16.1).

Bioresonance therapy (BRT) is based upon the knowledge that the entire body is held together at the subatomic level by waves and photons.

Every material is made up of atoms, whether it is a virus, bacteria or a human being. Atoms themselves are made up of subatomic particles—protons, neutrons and electrons. All subatomic particles share a fundamental property: They have "in-

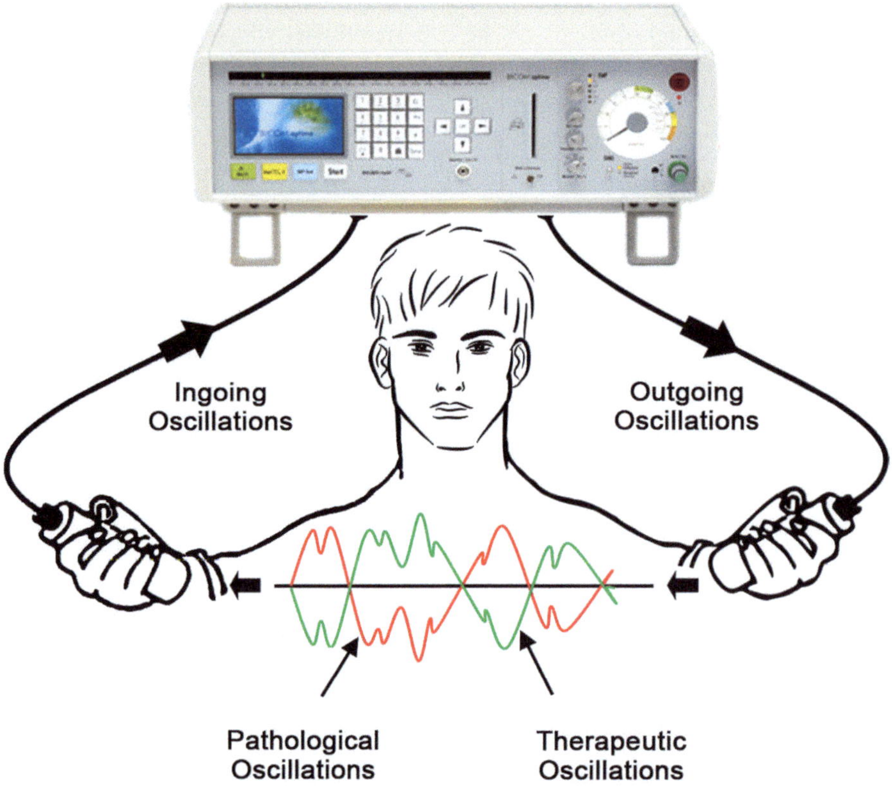

Fig. 16.1 Schematic diagrams of Bioresonance therapy

trinsic angular momentum," or spin. This means they rotate in one direction, just like a planet. Physicists discovered that subatomic particles behave like energy and radiate energy into their surroundings in specific patterns, called waves.

Subatomic particles have dual characteristics as both particles and wave forms. Subatomic particles vibrate at different rates or frequencies based in part on changes in temperature and thermodynamics. In their waveform state, quantum particles emit a frequency vibration that extends indefinitely. In this state, subatomic particles are present in all space in what is known as superposition. In the superposition state, they are also in contact with every other subatomic particle in the universe. This interconnection provides a huge amount of information transfer between all of the building blocks of our universe, including our own body. Each bacterium, each virus, organic substance has its own specific resonant frequency. (Cottingham and Greenwood 2007).

One kind of biological resonance is sunlight. If light, as the electro-magnetic oscillation of a defined frequency touches skin, it triggers regulatory reactions, such as pigmentation or the formation of vitamin D. Light's effect on the circadian rhythms of all or most animals has been well documented. Clearly, that huge number of other frequencies encountered during life also has some kind of effect on the organism. (Baehr et al. 1999; Holick 2004).

Researchers have been able to study the distinct wave patterns of normally-functioning body systems and organs as well as the oscillations of allergens, viruses, bacteria, and toxins.

Dr. Franz Morell is the father of bioresonance therapy. At the beginning of 1953 Dr. *Morell* was a member of the group investigating electro-acupuncture testing under the direction of *Dr. Voll. Voll* discovered that by making measurements of skin resistance at acupuncture points, diagnoses about the condition of the meridian energies could be made. He also revealed that this technique could be used to test allergic reactions to allergens. This is a way of testing the effects of harmful substances, allergens as well as drugs on the body. This test and therapy method is known as electro-acupuncture.

Morell developed electro-acupuncture further by discovering that the reversal of polarity in a body or material oscillations using the appropriate type of device led to "obliteration phenomena" in the body. This resulted, for example, in a form of allergy therapy which was practiced as "allergy obliteration". This rotation of the information on an allergy or a body's own oscillation is known as inversion (Herrmanna and Galleb 2011; Chen et al. 2010).

Biophotons were discovered in 1992, when the Russian embryologist *Alexander G.Gurwitsch* (1874–1954) performed an experiment with onion roots. He found that some effect from the dividing cells at the tip of one root stimulated the division of cells in the other root and called it "mitogenetic radiation" (Beloussov 1997).

Gurwitsch was persuaded that this radiation is an expression of morphogenetic fields within the organism that structure and organize the life processes in the cell and the organism. In developmental biology, a morphogenetic field is a group of cells able to respond to discrete, localized biochemical signals leading to the development of specific morphological structures or organs. Later, many other researchers, included *Popp* and his colleagues all over the world have not only demonstrated the existence and ubiquity of biophoton emission beyond any reasonable doubt, but also established its properties, developed and tested a number of hypotheses about its possible biological functions, done a lot of theoretical work towards explanation of biophoton theory and started to develop a number of practical applications for the use of biophoton measurements of microorganisms, plants, animals and humans.

Popp noted that a healthy cell stores light the longest. A healthy cell radiates coherent light, while a diseased cell radiates chaotic light. A large increase in biophoton flux during mitosis arises from the generation of a large amount of information, while an increase at the time of death is due to the usual thermodynamic cooling that occurs at the sudden destruction of a large amount of information. (Cohen and Popp 1997; Popp et al. 2002).

A biophoton is a photon of non-thermal origin in the visible and ultraviolet spectrum emitted from a biological system. The term biophoton used in this narrow sense should not be confused with the broader field of biophotonics, which studies the general interaction of light with biological systems.

Biochemical reaction via biotransformation phases and oxidative stress by reactive oxygen and nitrogen species and/or catalysis by enzymes is a common event in the biomolecular microenvironment. Such reactions can lead to the formation of triplet excited species, which release photons upon returning to a lower energy level in a process analogous to phosphorescence (Giuseppe and Waldemar 1995).

The study done by *Ankush Prasad* and *Pave Pospısil* revealed that the oxidation of linoleic acid by hydroxyl radical and intrinsic lipoxygenase results in the ultra-weak photon emission (Prasad and Pospısil 2011).

Actually, the human body emits biophotons, also known as ultra-weak photon emissions (UPE), with a visibility 1000 times lower than the sensitivity of our naked eye. While not visible to us, these particles of light or waves are part of the visible electromagnetic spectrum (380–780 nm) and are detectable via sophisticated modern instrumentation (Schwabl and Klima 2005; Niggli et al. 2005; Artem'ey et al. 1967).

Seemingly biophotons are used by the cells of many living organisms to communicate, that facilitates energy/information transfer which is several orders of magnitude faster than chemical diffusion. According to Yan Sun and his collogues, "Cell to cell communication by biophotons have been demonstrated in plants, bacteria, animals, neutrophil granulocytes and kidney cells (Sun et al. 2010).

Researchers were able to demonstrate different spectral light stimulation at one end of the spinal sensory or motor nerve roots resulted in a significant increase in the biophotonic activity at the other end". Researchers interpreted their finding to suggest that light stimulation can generate biophotons that conduct along the neural fibers, probably as neural communication signals. The change of biophotonic activity is noticeable under physiological and pathological conditions. For example, mechanical, thermal and chemical stresses, mitochondrial respiration, the cell cycle and cancer growth lead to these biophotonic activities (Sun et al. 2010; Tilbury 1992; Slawinski et al. 1992; Niggli 1993; Amano et al. 1995; Kataoka et al. 2001; Nakano 1989; Yoon 2005).

While Reactive oxygen species (ROS) and radical theory of biophoton origin is relatively simple and easily understandable due to more or less common biochemical approach, DNA theory of biophoton origin is much more complex (Fig. 16.2).

Popp discovered that photons provided the vehicle for which information was transmitted. They transmit information within a cell and between cells. Popp demonstrated that DNA of living cells is the major source of biophoton storages and emissions. In this theory the DNA helix in cell nucleus is considered to be quantum electrodynamic cavity that is constantly excited by metabolic activity of cell.

According to the biophoton theory developed on the base of these discoveries, the biophoton light is stored in the cells of organism-more precisely, in the DNA molecules of their nuclei—and a dynamic web of light constantly released and absorbed by the DNA may connect cell organelles, cells, tissues and organs within the body, and serve as the organisms main communication network and as the principal regulating instance for all life process. Popp believed that cancer cells can be detected by the biophoton emission of the cancerous cells and these cells can potentially

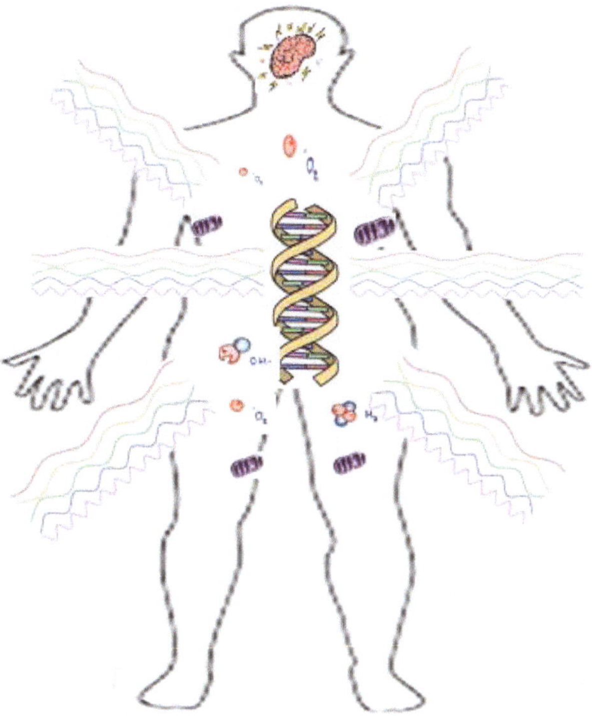

Fig. 16.2 Sources of electromagnetic waves. Biophotons and electromagnetic waves are emitted by the human body and can be released through oxidative reaction, DNA configuration changing, mental intention, and may modulate fundamental processes within cell-to-cell communication and DNA

be destroyed by biophotons. *Popp* discovered cells of an organism communicate by chemical-massager molecule or by light (Gisel 2009; Popp et al. 1984).

A Bonghan duct, also known as a primo vessel, was identified by *Bonghan Kim* in the 1960s, is a thread-like structure found on the surface of mammalian organs, blood vessels, lymphatic vessels and under the skin (Stefanov and Kim 2012). *Bonghan* ducts renamed as Primo vascular system (PVS) by the Seoul National University (SNU) research group in 2002 (Soh et al. 2011). More recently, the vessels were isolated and observed using confocal laser scanning microscopy (CLSM) and transmission electron microscopy (TEM), showing they were movable on the endocardium of the bovine atrium and ventricle (Lee and Bae 2011). The liquid carried within the PVS consists of various microparticles, such as DNA, proteins, and hormones. It is proposed that the PVS is a circulatory system in which microparticles, such as extracellular DNA (eDNA) and microvesicles, are floating and interacting (Lee and Lee 2013).

Experiment conducted by *Bonghan Kim* and *Sang-Hyun Park* showed that PVS has electrical signals similar to those from smooth-muscle-like cells.

In 1791 *Galvan* observed that injured tissue would generate electrical currents which was steady state or DC (direct current) in character (Piccolino 1998). *Burr* (1972) established, with the aid of voltmeters and electrodes, that every living organism possesses what he has termed as L-field (life-field)—a voltage difference between two points on, or close to, the surface of the living form. A complete listing of *Burr's* articles can be found in the Yale Journal of Biology and Medicine (Burr 1936).

Emission of photons in the visible range by animal cells and tissues has been described for a variety of organs and by many researchers. With the use of photomultiplier tubes, emissions of photons in the visible range have been already detected from the liver, heart, lung, nerves, skin and muscles (Kim et al. 2003; Cadenas 1980; Blokha 1968; Cohen and Popp 1997).

Modern research has confirmed the observations of *Burr*. Not only does every event in the body, either normal or pathological, produce electrical changes, it also produces alterations of the magnetic fields in the spaces around the body. This can guide to possible diagnostic applications in connection with bioresonance.

16.3 Epigenetics and Bioresonance

Epigenetic changes are continual changes in gene expression that do not involve any change in DNA sequence. They may last for varying times-within a long-lived cell, from cell to cell during development, or sometimes from parents to offspring. *Arthur Riggs* and colleagues defined Epigenetics as "the study of mitotically and/or meiotically heritable changes in gene function that cannot be explained by changes in DNA sequence". (Russo et al. 1996). In parallel to the term "genome" that defines the complete set of genetic information contained in the DNA of an organism, "epigenome" generally refers to the complete set of characteristics of epigenetic pathways in an organism. Researchers have identified four types of epigenetic pathways: DNA methylation, histone modification, nucleosome remodeling, and non-coding RNA-mediated pathways.

These epigenetic pathways intertwine with each other to regulate expression of genes and it is likely that other pathways beyond these four known ones be discovered in the future. (Van Vliet et al. 2007). Normal and abnormal physiological responses to environmental stimuli may be mediated by epigenetic mechanisms. Epigenetic states are reversible and can be modified by environmental factors.

The three-dimensional conformation of chromosomes in the nucleus is important for many cellular processes, including the regulation of gene expression, DNA replication, and chromatin structure (Cremer and Cremer 2001).

Oscillation is the repetitive variation, typically in time, of a central value (often a point of equilibrium) or between two or more different states. Familiar examples include a swinging pendulum and AC power. Oscillations occur not only in physical systems but also in biological systems, from human society to the brain Oscillations occur when a system is disturbed from a position of stable equilibrium. This displacement from equilibrium changes periodically over time. Thus, Oscillations are said to be periodic, and display periodic motions in human and animal cells and organs that connect with neighboring organs and environment. The harmonic oscillator has a single degree of freedom. More complicated systems have more degrees of freedom, for example two masses and three springs (each mass being attached to fixed points and to each other). In such cases, the behavior of each variable influences of the others. This leads to a coupling of the oscillations of the individual degrees of freedom. For example, two pendulum clocks (of identical frequency) mounted on a common wall will tend to synchronize. Coupled oscillators are oscillators connected in a way that energy can be transferred between them. As the number of degrees of freedom becomes arbitrarily large, a system approaches continuity; examples include a string or the surface of a body of water. Such systems have an infinite number of normal modes and their oscillations occur in the form of waves that can characteristically propagate. In the eukaryotic nucleus, DNA is packed into a periodic nucleoprotein complex, known as chromatin. The nuclear chromatin organized as clustered and has electric oscillation capacity. The coupling strengths of chromatin regions are determined by physical interactions among chromatin-associated proteins, the electromagnetic fields around the oscillating chromosomal regions, and the hydrogen and other bonds linking different chromatin regions within the same chromosome. The natural frequency of an oscillating chromatin region is determined by the physical properties of DNA-protein complexes in that region, which can be changed by its epigenetic state and the protein factors associated with it (Zhao and Zhan 2012). On the other hand, experiments confirmed that Biophotons can be absorbed by natural chromophores such as porphyrin rings, flavi1nic, pyridinic rings, lipid chromophores and caromatic amino acids, etc. (Gao and Xing 2009; Mazhul' and Shcherbin 1999).

We now know that the photon can exchange between the bio-systems. It also was shown that the excision exchange supposedly constitutes the effective system of signaling and regulation of the bio-system development. It seems that such signaling to the large extent regulates the homogeneity of bio-system growth, preventing the large fluctuations of its global form and defines its morphogenesis.

Experimental results show that under the different stress conditions the photon rates from bio-system can rise in short time significantly, probably, as the consequence of intensive internal signaling (Mayburov 2009).

Experiment conducted by *Peter P. Gariaev* and co-workers in Moscow confirmed that the chromosomes and DNA produce "laser radiations". They suggested (1) that there are genetic "texts", similar to natural context-dependent texts in human language; (2) that the chromosome apparatus acts simultaneously both as a sender and receiver of these genetic texts, respectively decoding and encoding them; (3) the chromosome continuum acts like a dynamical holographic gate, which displays weak laser light and electro-acoustic fields. The distribution of the character frequency in genetic texts is fractal, so the nucleotides of DNA molecules are able to form holographic pre-images of biostructures (Gariaev 2001). He supposed that genetic information, except for the coding form, exists in a quantum (wave) form. This model enables a fundamentally different way to cure people who suffered from cancer, viral diseases, bacterial infections, and degenerative processes in organs and tissues. The disruption of the electromagnetic energy system can disrupt DNA transcription; suppress T-cell and NK-cell activity all leading to chronic degenerative diseases, depression, and other problems. Electrons absorb and emit photons, which is why the DNA electrons are storage houses for biophotons. It is believed that the specific vibratory rate of each biophoton is what activates specific gene sequencing via resonance. It has been documented that DNA repair can be activated by using a frequency of *528 hz*. At that precise frequency the clustered water molecules that surround the DNA structure form a perfect six-sided hexagon.

16.4 Conclusion

All cells have small electrically powered pumps inside of them. Healthy cells, according to Nobel Prize winner *Otto Warburg*, have cell voltages of 70–90 millivolts. Bioelectric signals are generated by specific ion channels and pumps within cell membranes. The segregation of charges achieved by ion fluxes through such transporter proteins gives rise to a trans-membrane voltage potential (McCaig and Rajnicek 2005).

Meanwhile, all living cells of plants, animals and humans constantly emit ultra-weak biophotons in the optical range of the spectrum, which is associated with their physiological states. The intensity of biophotons is in direct correlation with, organ energy metabolism, organ activity, organ blood flow, organ health status and oxidative processes (Kobayashi et al. 1999b).

The biophoton light is stored in the cells, almost exclusively inside the DNA molecules, managing processes, alike a dynamic web of light, which is constantly released and absorbed. *Frohlich* argued that as organisms are made up of strong bipolar molecules packed rather densely together, electric and elastic forces can constantly interact. Cells and organisms display their own rhythms of activity that are partly internally regulated, but they also respond to external energy (Fröhlich 1980).

Bio-mechanical resonance is created when a small periodic stimulus of the same natural vibration period of a cell, tissue, or even a molecule, is used to produce a large amplitude vibration of the cell, tissue, or molecule.

Biophysicists view the body as an interconnected bio-energetic organism. The key to understanding bioresonance lies in understanding the fact that all vital processes in the organism are influenced and controlled by electromagnetic oscillations. These electromagnetic oscillations are super-ordinate to the biochemical processes and control them. Cell associations and organs oscillate in particular frequency ranges. Thus, an oscillation spectrum arises in the organism.

Electrons also absorb and emit photons, which is why the electron rich DNA is storage house for biophotons. It is now thought that the unique vibratory rate of each biophoton is what activates specific gene sequencing through what is known as resonance. The vibratory energy of biophotons are able to induce responses in other biophotons—within the same cell and neighboring cells—in fact, throughout the entire organism.

DNA, RNA, ribosomes, and mitochondria are all proton, electron and photon apparatuses. Photons have the ability to knock electrons out of their atomic and molecular orbits. They are able to direct electrons to where they are needed to run metabolic processes. Enzymes capture and transfer electrons and protons along a path to various protein molecules in order to activate each protein's specific function.

The nuclear chromatin has electric oscillation capacity and biophotons can be absorbed and emitted by chromosomes. It is known that cells receive, store, and emit quantum packets of light-photons. From a biological standpoint, the term "biophoton" is more appropriate. Electrons also absorb and emit photons, which is why the electron rich DNA is storage house for biophotons. Calculations show that the helix form of the DNA molecule exhibits the ideal geometric form of a hollow resonator that allows it to store light very effectively. *Blank M* Supposed that DNA seems to possess the two structural characteristics of fractal antennas, electronic conduction and self-symmetry. (Blank and Goodman 2011).

The DNA is directly attached to the nucleus, specifically at the Telomeres—which is one of the reasons telomeres are so important, they receive and amplify the initial electric current received at the nuclear membrane—and at heterochromatin (highly condensed areas of DNA).

Molecular rearrangements in DNA are affected through epigenetic modifications. Direct methylation of CpG residues as well as many different modifications modifiable to histones produces molecular rearrangements of nucleotide segments that will produce differential electron orbital configurations. A very important feature of the molecular encoding of electromagnetic information within the atomic structure of DNA is the role played by Transposons. It is the Transposons that direct RNA-mediated DNA epigenetic regulation (Fedoroff 2012).

It is showed that weak electromagnetic (EM) fields interact with gene promoter in DNA can lead to the stimulation of protein synthesis. Scientific evidence confirmed that weak electromagnetic fields have effect on electron transfer on DNA molecule that may change the transcription and translation process in cells (Blank and Goodman 2008).

It is now thought that the unique vibratory rate of each biophoton is what activates specific gene sequencing through what is known as resonance. The vibratory energy of biophotons is able to induce responses in other biophotons—within the same cell and without to neighboring cells—in fact, throughout the entire organism.

Changes in environmental factors can lead to variation in electric oscillation in chromosome which in turn may result to the fluctuations in epigenetic pattern of organism.

Disease can be considered as the disturbance of biochemical sequences and electromagnetic oscillations order in the body, which is triggered by exogenous and endogenous stimuli. It is at the energetic and vibrational level that the physical processes shape the transfer of energy and the flow of bio-energetic information in the living system.

References

Adamo AM, Llesuy SF, Pasquini JM, Boveris A (1989) Brain chemiluminescence and oxidative stress in hyperthyroid rats. Biochem J 263:273–277

Allchin D (2004) Pseudohistory and pseudoscience. Sci Educ 13:179–195

Amano T, Kobayashi M, Devaraj B, Usa M, Inaba H (1995) Ultraweak biophoton emission imaging of transplanted bladder cancer. Urol Res 23:315–318

Andrew AM, Becker R, Ullrich B (1979) Kirlian photography: potential for use in diagnosis. Psychoenergetic Syst 3:47–54

Artem'ey VV, Goldobin AS, Gus'kov LN (1967) Recording the optical emission of a nerve. Biophysics 12:1278–1280

Avijgan M, Avijgan M (2013) Can the primo vascular system (Bong Han duct system) be a basic concept for qi production. Int J Integr Med 1:20

Baehr EK, Fogg LF, Eastman CI (1999) Intermittent bright light and exercise to entrain human circadian rhythms to night work. Am J Physiol-Regul Integr Comp Physiol 277:R1598–R1604

Becker RO (1963) Electron paramagnetic resonance in non-irradiated bone. Nature 28:1304–1305

Becker RO (1972) Stimulation of partial limb regeneration in rats. Nature 235:109–111

Becker RO, Bachman CH, Slaughater WH (1962) Longitudinal direct-current gradients of spinal nerves. Nature 196:675–676

Becker RO, Chapin S, Sherry R (1974) Regeneration of the ventricular myocardium in amphibians. Nature 248:145–147

Beloussov LV (1997) Life of Alexander G Gurwitsch and his relevant contribution to the theory of morphogenetics field. Int J Dev Biol 41:771

Bird C (1976) What has become of the Rife Microscope. New Age J, 41–47

Blank M, Goodman R (2008) A mechanism for stimulation of biosynthesis by electromagnetic fields: charge transfer in DNA and base pair separation. J Cell Physiol 214:20–26

Blank M, Goodman R (2011) DNA is a fractal antenna in electromagnetic fields. Int J Radiat Biol 87:409–415

Blokha VV et al (1968) The ultraweak glow of muscles on stimulation. Biophysics 13:1084–1085

Burr HS, Lane CT, Nims LF (1936) A vacuum tube micro-voltmeter for the measurement of bioelectric phenomena. http://www.ncbi.nlm.nih.gov/pubmed/21433705. Yale J Biol Med 9(1):65–76

Cadenas E (1980) Spectral analysis of the hydroperoxideinduced chemiluminescence of the perfused lung. FEBS Lett 111:413–418

Carlson B (2005) Inventor of dreams. Sci Am 292(3):66

Chen T, Guo ZP, Zhang YH, Gao Y (2010) Effect of MORA bioresonance therapy in the treatment of Henoch-Schonlein purpura and influence on serum antioxidant enzymes. J Clin Dermato 139:283–285

Cohen S, Popp FA (1997) Biophoton emission of the human body. J Photochem Photobiol B Biol 40:187–189

Cottingham WN, Greenwood DA (2007) An introduction to the standard model of particle physics. Cambridge University Press, Cambridge, pp 1–18

Cremer T, Cremer C (2001) Chromosome territories, nuclear architecture and gene regulation in mammalian cells. Nat Rev Genet 2:292–301

Dunwell R (2011) SCENAR technology. NZ J Nat Med 3:67–69

Fedoroff NV (2012) Transposable elements, epigenetics, and genome evolution. Science 338:758–767

Fröhlich H (1980) The biological effects of microwaves and related questions. Adv Electronics Electron Phys 53:85–152

Gao X, Xing D (2009) Molecular mechanisms of cell proliferation induced by low power laser irradiation. J Biomed Sci 16:4

Gariaev PP (2001) The DNA-wave Biocomputer. http://www.rialian.com/rnboyd/dna-wave.doc.2001

Gernert D (2008) How to reject any scientific manuscript. J Sci Explor 22:233–243
Gisel HR (2009) In foodture we trust. Xulon, Tallahassee, p. 264 (ISBN 1624199690)
Giuseppe C, Waldemar A (1995) From free radicals to electronically excited species. Free Radic Biol Med 19:103–114
Gogoleva EF (2001) New approaches to diagnosis and treatment of fibromyalgia in spinal osteochondrosis. Ter Arkh 73:40–45
Grass F, Kasper S (2008) Humoral phototransduction: light transportation in the blood, and possible biological effects. Med Hypotheses 71:314–317
Grinberg YA (1996) SCENAR therapy: the effectiveness from the point of view of methods of electrotherapy. SCENAR therapy and SCENAR expertise. Compilation Art 2:18–33
Haas M, Peterson D, Hoyer D, Ross G (1994) Muscle testing response to provocative vertebral challenge and spinal manipulation: a randomized controlled trial of construct validity. J Manip Physiol Ther 17:141–148
Helene M, Langevin, Jason A (2002) Relationship of acupuncture points and meridians to connective tissue planes. Anat Rec (NEW ANAT) 269:257–265
Herrmanna E, Galleb M (2011) Retrospective surgery study of the therapeutic effectiveness of MORA bioresonance therapy with conventional therapy resistant patients suffering from allergies, pain and infection diseases. Eur J Integr Med 3:e237–e244
Holick MF (2004) Sunlight and vitamin D for bone health and prevention of autoimmune diseases, cancers, and cardiovascular disease. Am J Clin Nutr 80:1678S–1688
Huang S, Sun Z, Fang Y (2005) Klinische Behandlung vom allergischen Schnupfen und Bronchialasthma der Kinder mit dem Bioresonanztherapiegerät. Zhejiang Med J 27:457–458
Hunt VV (1996) Infinite mind: science of the human vibrations of consciousness. Malibu Publishing, Malibu, p 364
Imaizumi S, Kayama T, Suzuki J (1984) Chemiluminescence in hypoxic brain-the first report. Correlation between energy metabolism and free radical reaction. Stroke 15:1061–1065
Islamov BI, Balabanova RM, Funtikov VA (2002) Effect of bio-resonance therapy on antioxidant system in lymphocytes in patients with rheumatoid arthritis. Bull Exp Biol Med 134:248–250
Kataoka Y, Cui Y, Yamagata A, Niigaki M, Hirohata T, Oishi N, Watanabe Y (2001) Activity-dependent neural tissue oxidation emits intrinsic ultraweak photons. Biochem Biophys Res Commun 285:1007–1011
Katelaris CH, Weiner JM, Heddle RJ, Stuckey MS, Yan KW (1991) Vega testing in the diagnosis of allergic conditions. Med J Aust 155:113–114
Kim JD, Choi C, Lim JK (2003) Biophoton emission from rat liver. J Korean Phys 42:427–430
Kobayashi M, Takeda M, Ito K, Kato H, Inaba H (1999a) Two-dimensional photon counting imaging and spatiotemporal characterization of ultraweak photon emission from a rat's brain in vivo. J Neurosci Methods 93:163–168
Kobayashi M, Takeda M, Sato T (1999b) In vivo imaging of spontaneous ultraweak photon emission from a rat's brain correlated with cerebral energy metabolism and oxidative stress. Neurosci Res 34:103–113
Konigsberg UR, Lipton BH, Konigsberg IR (1975) The regenerative response of single mature muscle fibers isolated in vitro. Dev Biol 45:260–275
Lee B-C, Bae KH (2011) Network of endocardial vessels. Cardiology J 118:1–7
Lee BC, Lee HS (2013) Evidence for the fusion of extracellular vesicles with/without DNA to form specific structures in fertilized chicken eggs, mice and rats. Micron 44:468–474
Lipton BH (1977) A fine structural analysis of normal and modulated cells in myogenic culture. Dev Biol 60:26–47
Lipton BH (1988) The evolving science of chiropractic philosophy. Today's Chiropr 27:16–19
Lipton BH (1998) Nature, nurture and the power of love. J Prenat Perinat Psychol Health 13:3–10
Lipton BH (2001) Nature, nurture and human development. J Prenat Perinat Psychol Health 16:167–180
Lipton BH (2005a) Insight into cellular consciousness. Bridges ISSEEEM Org 12:5–9
Lipton BH (2005b) The biology of belief: unleashing the power of consciousness, matter and miracles. Mountain of Love Productions, Inc and Elite Books, San Rafael

Lipton BH, Jacobson AG (1974) Analysis of normal somite development. Dev Biol 38:73–90

Lipton BH, Konigsberg IR (1972) A fine structural analysis of the fusion of myogenic cells. J Cell Biol 53:348–363

Lipton BH, Schultz E (1979) Developmental fate of skeletal muscle satellite cells. Science 205:1292–1924

Lipton BH, Bensch KG, Karasek MA (1991) Endothelial cell transdifferentiation: phenotypic characterization. Differentiation 46:117–133

Mansfield JW (2005) Biophoton distress flares signal the onset of the hypersensitive reaction. Trends Plant Sci 10:307–309

Martin M (1994) Pseudoscience, the paranormal, and science education. Sci Educ 3:357–371

Mayburov S (2009) Biophoton production and communications. Nanotechnology and nanomaterials. MGOU, Moscow, pp 351–358

Mazhul' VM, Shcherbin DG (1999) Phosphorescent analysis of lipid peroxidation products in liposomes. Biofizika 44:676–681

McCaig CD, Rajnicek AM (2005) Controlling cell behavior electrically: current views and future potential. Physiol Rev 85:943–978

Montgomery S (2003) The rise and fall of a scientific genius: the forgotten story of Royal Raymond Rife. NZ Med J 116:1177

Nakano M (1989) Low-level chemiluminescence during lipid peroxidations and enzymatic reactions. J Biolumin Chemilumin 4:231–240

Nienhaus J, Galle M (2006) Placebokontrollierte Studie zur Wirkung einer standardisierten MORA Bioresonanztherapie auf funktionelle Magen-Darm-Beschwerden. Forsch Komplementarmed 13:28–34

Niggli HJ (1993) Artificial sunlight irradiation induces ultra-weak photon emission in human skin fibroblasts. J Photochem Photobiol 18:281–285

Niggli HJ, Salvatore T, Lee AA, Scordino A, Musumeci F, Giuseppe P (2005) Laser-ultraviolet-A-induced ultraweak photon emission in mammalian cells. J Biomed Opt 10:24–26

Nozdrachev AD (1996) Chemical structure of the peripheral autonomic (visceral) reflex. Usp Physiol Sci 27:28–60

Oju M, Gogoleva EF (2000) Outpatient bioresonance treatment of gonarthrosis. Ter Arkh 72:50–53

Peter M (1984) The uses and limitation of acupuncturepoint measurement, German electroacupuncture or electroacupuncture according to Voll. Am J Acupunct 12:33–42

Phelan SE (2008) What is complexity science, really. Emergence 3:120–136

Piccolino M (1998) Animal electricity and the birth of electrophysiology: the legacy of Luigi Galvani. Brain Res Bull 46:381–407

Pihtili A, Cuhhadraroglu C, Kilicaslan Z, Issever H, Erkan F (2009) The effectiveness of bioresonance method in quitting smoking. Clinical report 2009 University Istanbul, Turkey: Department of Medicine

Popp FA, Nagl W, Li KH, Scholz W, Weingärtner O, Wolf R (1984) Biophoton emission. New evidence for coherence and DNA as source. Cell Biophys 6:33–52

Popp FA, Chang JJ, Herzog A, Yan Z, Yan Y (2002) Evidence of non-classical (squeezed) light in biological systems. Phys Lett A 293:98–102

Prasad A, Pospısil P (2011) Linoleic acid-induced ultra-weak photon emission from *Chlamydomonas reinhardtii* as a tool for monitoring of lipid peroxidation in the cell membranes. Plos ONE 6(7):e22345

Prelević R (2011) Quantum-informational medicine and bioresonance technology. Symposium of quantum-informational medicine QIM 2011: acupuncture-based & consciousness-based holistic approaches & techniques, Belgrade, 23-25 Sep 2011

Quickenden TI, Que Hee SS (1974) Weak luminescence from the yeast *Sachharomyces-Cervisiae*. Biochem Biophys Res Commun 60:764–770

Rahlfs VW, Rozehnal A (2008) Wirksamkeit und Verträglichkeit der Bioresonanzbehandlung. Erfahrungsheilkunde 57:462–468

Rife R (2013) From Wikipedia. http://en.wikipedia.org/wiki/Royal_Rife

Roland Hans Penner J (1995) The strange life of Nikola Tesla. Kolmogorov-Smirnov Publishing, Basingstoke
Rosenow E (1965) Observations with the Rife microscope of filter-passing forms of microorganisms. Science 76:192–193
Russo VA, Martienssen RA, Riggs AD (1996) Epigenetic mechanisms of gene regulation. Cold Spring Harbor Laboratory Press, Woodbury, pp 5–27
Schimmel HW, Penzer V (1997) Functional medicine: the origin and treatment of chronic diseases, 2nd edn. Alden, Oxford, p 356 (Title No 2639. ISBN 3-7760-1639-6)
Schöni MH, Nikolaizik WH, Schöni-Affolter F (1997) Efficacy trial of bioresonance in children with atopic dermatitis. Int Arch Allergy Immunol 112:238–246
Schuller J, Galle M (2007) Untersuchung zur Prüfung der klinischen Wirksamkeit elektronisch abgespeicherter Zahn- und Gelenksnosoden bei Erkrankungen des Rheumatischen Formenkreises. Forsch Komplemented 14:289–296
Schwabl H, Klima H (2005) Spontaneous ultraweak photon emission from biological systems and the endogenous light field. Forsch Komplementarmed Klass Naturheilkd 12:84–89
Slawinski J, Ezzahir A, Godleweski M (1992) Stress-induced photon emission from perturbed organisms. Experientia 48:1041–1058
Soh KS (2011) Current state of research on the primo vascular system. In: Soh KS, Kang KA, Harrison DK (eds) The primo vascular system, its role in cancer and regeneration. Springer, New York, pp 25–39
Soh K-S, Kang KA, Ryu YH (2013) 50 years of Bong-Han theory and 10 years of primo vascular system. Evid-Based Complementary Altern Med. doi:dx.doi.org/10.1155/2013/587827
Stefanov M, Kim J (2012) Primo vascular system as a new morphofunctional integrated system. J Acupunct Meridian Stud 5(5):193–200. doi:10.1016/j.jams.2012.07.001
Sun Y, Wang C, Dai J (2010) Biophotons as neural communication signals demonstrated by in situ biophoton autography. Photochem Photobiol Sci 9:315–322
Szent-Gyorgyi A (1960) Introduction to a submolecular biology. Academic, New York, pp 91–103
Szent-Gyorgyi AP (1894) Woods hole marine biological laboratory, Massachusetts. Papers from 1894 to 1995, including photographs, oral histories, published articles, video recordings and lectures. profiles.nlm.nih.gov/WG/Views/
Tilbury RN (1992) The effect of stress factors on the spontaneous photon emission from microorganisms. Experientia 48:1030–1104
Tilbury RN, Cluickenden TI (1988) Spectral and time dependence studies of the ultraweak bioluminescence emitted by the bacterium *Escherichia coli*. Photobiochem Photobiophys 47:145–150
Van Vliet J, Oates NA, Whitelaw E (2007) Epigenetic mechanisms in the context of complex diseases. Cell Mol Life Sci 64:1531–1538
Voll R (1974a) Twenty years of electroacupuncture therapy using low-frequency current pulses. Am J Acupunct 3:291–314
Voll R (1974b) Verification of acupuncture by mans of electroacupuncture according to Voll. Am J Acupunct 6:5–15
Yoon YZ (2005) Changes in ultraweak photon emission and heart rate variability of epinephrine-injected rats. Gen Physiol Biophys 24:147–159
Zavitaev YA (1996) SCENAR examples of single SCENAR application. SCENAR therapy and SCENAR expertise. Compilation Art 2:81–82
Zhao Y, Zhan Q (2012) Electric oscillation and coupling of chromatin regulate chromosome packaging and transcription in eukaryotic cells. Theor Biol Med Modelling 9:27

Chapter 17
Essence of Cancer Epigenetic: A Harmonic Art for the Future

Parvin Mehdipour

Contents

Abstract This chapter provides an evolutionary mode in the field of Cancer epigenetic and it was aimed to bridge between different chapters. Epigenetic as a ***Reservoir*** in cancer open diverse avenues including fundamental, diagnostics, preventive and therapeutic aspects. Three new insights as the multi-potential and multi-directional models are provided including '*a map of Cancer Development*', '*Epigenetic Programming*', and '*Diverse- methylation process, target and function at different ages*'. These models are indicative of capability and cooperative manner of epigenetics in Science and Medicine. In addition, different aspects of 16 provided chapters have been presented at a glance. Finally, I tried to provide a new insight in the field of *Evolutionary Epigenetic* as two new terms including '*Medical Epi-anthropology*' and '*Cancer Epi-anthropology*'**.** Through these directions, exploring about our ancestral lines and origin of the malignant and non- cancerous would be possible. By considering the importance of translational paradigm and pharmacogenetics, epigenetic is a reliable choice in cancer drug innovation. Besides, the impact of other genetic and cellular alterations as a challenging item in cancer therapy is highlighted. The final message is 'considering epigenetic alteration by application of the complementary maneuvering through the cellular/molecular platforms to achieve the most effective therapeutic mélange'.

P. Mehdipour (✉)
Department of Medical Genetics, School of Medicine, Tehran University of Medical Sciences, Poursina Street, P.O. Box: 14176-13151 Keshavarz Boulevard, Tehran, Iran
e-mail: mehdipor@tums.ac.ir

P. Mehdipour (ed.), *Epigenetics Territory and Cancer*,
DOI 10.1007/978-94-017-9639-2_17

Abbreviation

ATM:	Ataxia-telangiectasia mutated gene
ATO:	Arsenic trioxide
BC:	Breast cancer
BPP:	Beneficial preventive programming
CTCs:	Circulating TUMOR cells
DI:	Diverse impact
DF:	Diverse function
DIn:	Diverse interaction
DC:	Diverse cell territories
DA:	Different ages
DP:	Developmental process
EP:	Embryonic programming
FP:	Fetal programming
EP:	Environmental programming
EF:	Environmental factors
ED:	Embryonic development
FD:	Fetal development
FTC:	Follicular thyroid cancer
GBEF:	Genetic based environmental factors
14-3-3σ:	Human mammary epithelium-specific marker 1 (HME1)
HEF:	Hazard environmental factors
PNP:	Pre-natal programming
P/N-PP:	Positive/negative preventive programming
PND:	Pre-natal development
Pos-ND:	Post-natal development
RASSF1A:	Ras association domain family 1A gene
SCZ:	Schizophrenia

17.1 Final Words

Mini-chapter 17 is reflective of an evolutionary manner in cancer epigenetic and a bridging insight of all provided chapters and direction for the future.

The interactive regulatory system characterizes and differentiates epigenetic as an extraordinary molecular biological territory. By laddering through Epigenetic in cancer, the bridging conclusion would lead to the new imminent in this field. The Epigenetic as a ***Reservoir*** in cancer, hopefully, opens a window to achieve the essence of discovery.

As far as genetic makeup concerns, cells have lots in common but with diverse functional characteristics. The requirements in the epigenetic mechanisms include gene expression and epigenetic modification. Epigenetic has its own historical background which, continuously, alters through the life. However, in cancer development, the command platform of epigenetic is beyond DNA methylation, RNA interference and histone modification (Fig. 17.1).

Epigenetic as the result of genes/environmental interaction, through embryonic at the right time and precise direction and well defined target, at either somatic or germ cells, Pre-natal and post-natal periods, may be reversible. By considering the mosaicism and clonal diversity and expansion, it is worth to emphasize on ***inheritable capacity*** through somatic cells. Epigenetic interact with gene expression, genome stability, specific expression of imprinted genes, cell cycle, X-chromosome inactivation, and chromosomes. It has also key impact on gene expression, cell differentiation, tumor- initiation, development and progression. Epigenetic processes which lead to a change of genetic activity and inactivity have meanwhile been reported for all phyla of animals (Lyon 1974):

At a glance, the key topics in cancer Epigenetic consist of the followings aspects (Chap. 1):

1. Aberrant identification of methylation patterns.
2. Hypermethylation of CGIs in tumor suppressor genes that leads to genes' silencing.
3. Hypomethylation leads to activation of oncogenes.
4. Balance between "stability and susceptibility to developmental and environmental stimuli".
5. Prevention.
6. Personalized clinical management within the pedigree of proband.

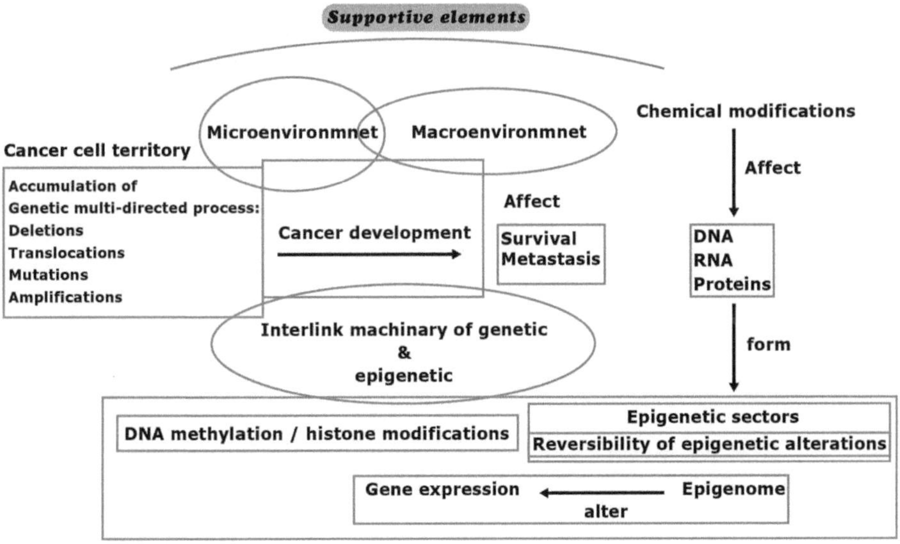

Fig. 17.1 The map of cancer development. (Adopted from P. Mehdipour's archive)

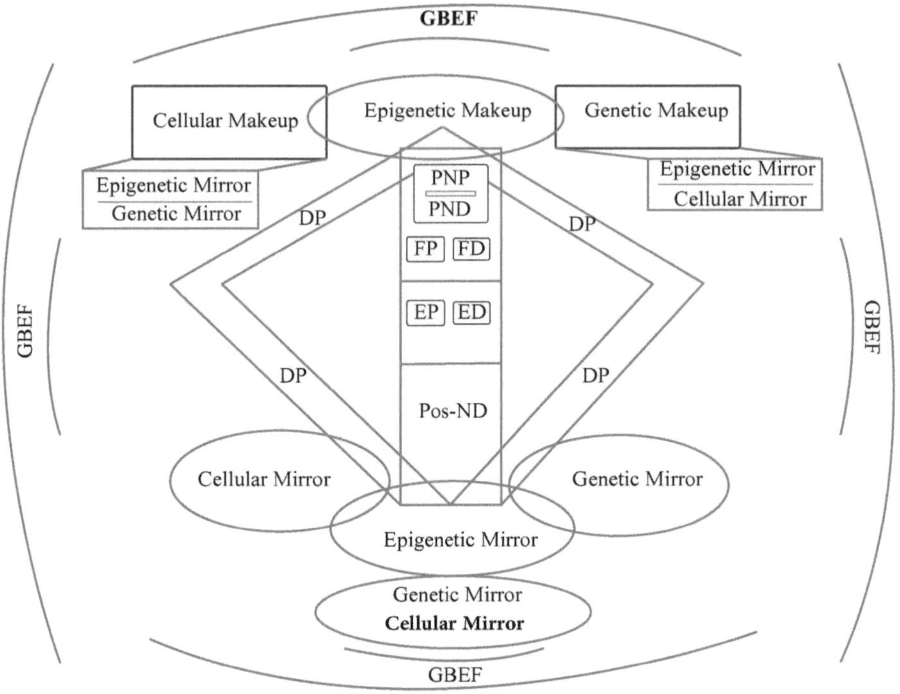

Fig. 17.2 Epigenetic programming. (This scheme illustrate the interactive programming and the process of multi-events through the life). (*BPP* beneficial preventive programming (as positive PP), *DP* developmental process, *ED* embryonic development, *EF* environmental factors, *EP* environmental programming, *EP* embryonic programming, *FP* fetal programming, *FD* fetal development, *GBEF* genetic based environmental factors, *HEF* hazard environmental factors (as negative PP), *PNP* pre-natal programming, *PND* pre-natal development, *Pos-ND* post-natal development, *P/N-PP* positive/negative preventive programming). (Adopted from P. Mehdipour's archive)

However, the key question is 'How is the epigenetic machinery programmed?' (Fig. 17.2)

Manner of developmental process is considered as an event through the whole period of life which is reflective of the inhibitory and stimulatory elements. The outcome is the matter of cooperation or inhibition.

The tower of programming includes the embryonic -, fetal-, pre-natal-, and post natal- development, as a non-stop process. *Is it an infinitive event*?

Moreover, Epigenetic is capable to translate its influential potentials at different stages of life including pre-zygotic, post-zygotic, through embryonic developmental period and post- natal duration, but with diverse impacts. In fact, we face the diverse impact (DI), diverse function (DF), diverse interaction (DIn), in diverse cell territories (DC), and at different ages (DA) (Fig. 17.3).

The rapid evolution in the field of Epigenetics seems to be linked to developmental process.

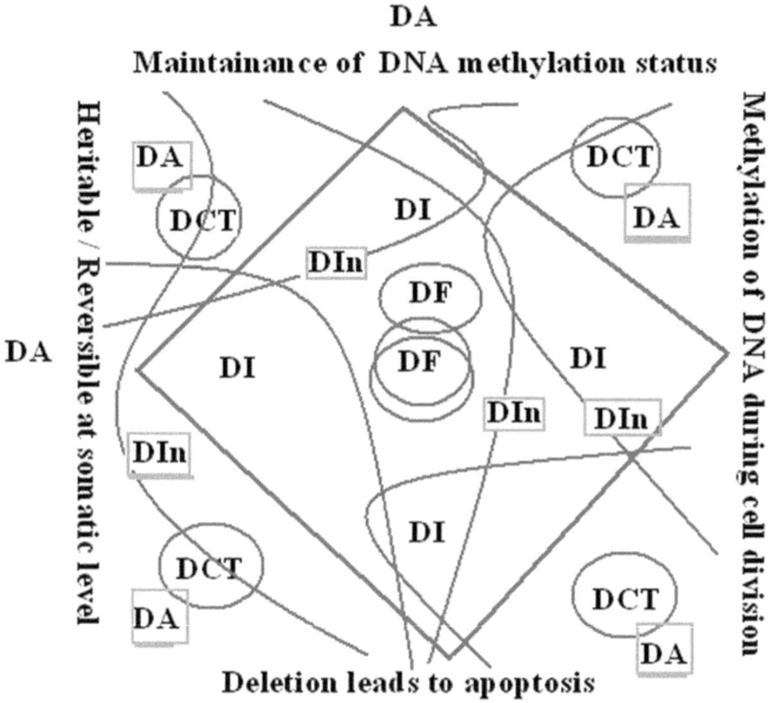

Fig. 17.3 Diverse- methylation process, target and function at different ages. (*DA* different ages, *DCT* diverse cell territories, *DI* diverse impact, *DF* diverse function, *DIn* diverse interaction). (Adopted from P. Mehdipour's archive)

Developmental Epigenetic leads to alteration of genetic motion and indolence which have been recognized in the animals with bilateral symmetry, also known as phyla animals (Lyon 1974).

17.2 Book Chapters at a Glance

The present book is divided to 17 chapters under 4 major sections including fundamental (Chaps. 1–5), brain (Chaps. 6–9), breast (Chaps. 10–12), sporadic section (Chaps. 13–16), and final mini-chapter 17.

Chapter 1 reflects a deep exploring insight of fundamental epigenetic facts in cancer.

In spite of the fact that Epigenetics is a rather new paradigm in cancer, but its fundamental roots are mainly linked to 1949, 1961 and 1974. In this chapter a triangle as "the cellular differentiation, gene expression, and epigenetic regulation" are the highlighted mechanisms.

I believe that cancer is a Cellular/Genetic disease and not only "Genetic disease". However Chap. 1 reflects a fact that cell cycle phases are characterized with the dual individualized and/or complimentary behaviour.

The authors of this chapter have provided the basic aspects, essential methods and guidelines in the epigenetic research and diagnosis. They have clarified imprinting, epigenetic marks and mosaic formation during the ontogenesis. In addition, they have highlighted the importance of new field as "combination of constitutional chromosome aberrations with epigenetic alterations". Finally, the developmental and epigenetics based therapeutic strategies are provide.

Chapters 2–5 focus on the ongoing fundamental paradigms which may lead to the applicable strategies in cancer patients.

In **Chap. 2**, the focal point is circulating tumor cells (CTCs) and their characteristics through migration. The Epigenetic alterations, silencing of tumor suppressor, metastatic suppressors genes, functional events and micro -environmental factors have been discussed. The aim of authors was to bridge between CTCs and personalized therapy. The interactive machinery between epigenetic regulations, miRNA through EMT process led to highlight an interesting conclusion. This has highlighted the dual cooperation of the CpG island neighborhood of "miRNA-200c and miRNA-141 transcription". Importantly, the diverse role of different miRNA in different carcinogenic and metastatic processes including epigenetic EMT regulation was discussed with remarkable note on the crucial role of "miR-9 as a pro-metastatic miRNA and a negative regulator of the key metastasis suppressor, E-cadherin".

Finally, they have emphasized on the role of miRNA in CTCs which may lead to therapeutic innovation. However, my explanation is the presence of cross-talk within miRNA platform, microenvironment, cellular/molecular territories and epigenetic behavior.

Chapter 3 provides an interesting paradigm, i.e., Retrotransposons, especially LINE-1 elements in cell biology and cancer cell biology. However, the focal points include diversity, variation and evolution which create the pyramids contributed by many cellular and molecular targets including Retrotransposons.

Chapter 4 is reflective of the crucial role s of miRNA within the width spectrum including biological atmosphere and mechanisms in different cancers. This platform is constructed through direction towards the innovation of "epigenetic drugs".

In this chapter, the dual machinery between epigenome and miRNome has been presented. Besides, the role of miRNAs and epigenetic in varieties of cancers were explored. Finally, a desirable aim through the epigenetic based drug innovation, by deregulation of miRNAs, has been highlighted. They have hypothesized that by associating the unique arsenic metabolism with global DNA hypomethylation, a new avenue may be opened to the upregulation of the epigenetically regulated miRNA in the Arsenic trioxide (ATO) -treated cancer cells. Interestingly, the dual function of these miRNAs as a tumor and/or metastatic suppressor in a cascade of biological targets including cell cycle regulation, apoptosis, angiogenesis, invasion, and metastasis are highlighted. Finally, an interactive therapeutic insight between specific drug and the epigenetically regulated miRNAs is a key point within the paradigm of "epigenome and miRNome".

Chapter 5 provides an extraordinary parallel and cooperative archetype in which soul and body meet each other in an unexpected moment of the individuals' life. Cancer is a cellular/molecular/psychosomatic disease with full of complexity. Diversity, variation, heterogeneity and evolution create an unpredicted condition in which the basic and clinical management seem to be thorny. This chapter, by considering the epigenetic alterations, provides an interactive network which bridges Schizophrenia to different cancers in human, at cell line level and in animal models through profiling of molecular/pathways functional targets and environmental aspects. Furthermore, the immunological aspects in cancer and SCZ are also explored. However, the opposite manner of behavior in cancer and SCZ could be considered as a 'Natural Gift' for achieving more appropriate and effective strategies to combat both diseases.

So, by unmasking the key facts in both diseases, the hope for therapeutic reliability may be emerged.

After a brief focus on Chaps. 1–5, the Chaps. 6–15 will present Epigenetic in specific cancers.

Chapter 6–9, are included in brain section. The methylation area is the focal point of Chaps. 6–8 followed by Chap. 9 in which a therapeutic achievement in brain tumor is highlighted:

In **Chap. 6**, the Ataxia-telangiectasia mutated (ATM) gene as a master molecular target and its product as a multi-functional protein are explored. However, the fate of ATM signaling depends on type of tissue and cell, stage of development; DNA damage and hazard environmental factors. The multi-platform characteristics of ATM also include cooperation with MDM2, p53 and *p21* genes. Another triangle correlation exists between ATM, chromosomal abnormalities/breakage and telomere length. In AT-patients, ATM governs an interaction between nuclear matrix and telomere which could lead to telomere -and chromosomal stability. Such manner also is true for neuroblats of Drosophila, so these examples indicates that the ATM function is highly conserved and stabled within different species.

ATM is not just a master, but it is a magic- gene as well and is widely present within different territories in the body, for instance, it is detectable either in nucleus or cytoplasm in neural cells of cerebellum. Other examples include its crucial role in insulin metabolism pathway, and in mitochondrial diseases. However, ATM is, influentially, involved in different malignant and non-malignant diseases.

At a glance, ATM, with its multi-influential protein, is characterized with multi-directive, multi- potential and multi-interactive manners. The focal points in our research reveal to be the assessment of correlation between *ATM* promoter methylation, its protein expression and telomere length. The question is how does *ATM* promoter methylation interact with expression of other genes with the specific methylation status? Then we may clarify the determining factors in survival of cancer patients.

Chapter 7 In this chapter, the fundamental aspects of MCPH1 gene, its protein and its role in cancer and selected non-cancerous diseases have been explored. *Microcephalin gene* (*MCPH1*) has been initially found in microcephaly, but due to its interactive manner with cell cycle checkpoint molecules, DNA repair proteins, response to DNA damage and repair, its additional role, as a tumor suppressor gene, alteration in its promoter methylation was linked to the down-regulated MCPH1 protein expression in different neoplastic diseases including brain tumors.

The diverse molecular and functional aspects of MPCH1 in various neoplastic disorders including brain tumors have provided the peculiarities of this gene in diseases. Now the question is how mutation in this gene has preventive impact against cancer? And what is the relevant scenario?

Chapter 8 *P53* gene has a remarkable Antitumorigenesis role in various types of cancers. An interaction between *p53* promoter methylation and its protein expression in brain tumors is required to be well defined. In this chapter the fundamental aspects of *p53* gene and the status of methylation are explored. By considering the controversy in diverse methylation status in brain tumors, interestingly, all of our patients revealed to harbor promoter unmethylated p53 and low protein expression of this gene. That is the remarkable point, and I would like to emphasize on the reverse correlation between an unmethylated p53 gene and the manner of function of this gene which is reflected to be in an opposite direction. This matter may be related to the methylated status of p53 protein, so unmethylated p53 gene versa methylated protein is an outstanding cross-talk. Now, I would like to propose some questions; (1) in spite of the suppressive nature of p53 gene, is there any resistant strategy in this gene? This is notable by unmethylated status of it. (2) What about low protein expression in 93 % of tumors? (3) Is p53 protein function more harmonic with other cellular and molecular targets? (4) Is p53 protein more sensitive in response to the mocro- and micro- environmental situation? and (5) Does p53 protein act more harmonic with some neighboring cellular/molecular partners by sympathizing behavior and accepting the methylation process? This is a great challenge and will be, hopefully, clarified through our ongoing project by aiming towards a novel therapeutic insight.

As a final conclusion on Chaps. 6–8, there is a crosstalk between ATM, p53, and MCPH1 within the cell cycle sphere in brain tumors.

In **Chap. 9**, the predictive role of O^6-methylguanine DNA methyltransferase (MGMT), as a therapeutic tool in brain tumors, especially in malignant gliomas; and the methodological requirements are presented. As the matter of fact the influential benefit of MGMT is due to its component as alkylating cytotoxic agent which characterized it as a possible remedy in the therapy of central nervous system tumors. However, diverse therapeutic strategies are linked to DNA methylation or expression of MGMT, but in one exception case in which combinational protocol including radiotherapy plus temozolomide was applied, it was related to RNA expression. This matter highlights the methodological limitation as well. The clinical managements are rather crucial strategies in brain tumors, therefore some questions in this chapter include; (1) Regardless of positive or negative status of MGMT-expression, why surgical option has a unique influential impact on the outcome in varieties of brain tumors?; (2) why in patients who have not received chemotherapy, the mode of MGMT does not affect survival? And (3) how the interactive barriers in therapy of brain tumors could be classified?

Moving forward, the breast Section include Chaps. 10–12 in which the predictive, preventive and therapeutic approaches are provided.

Chapter 10 is focused on the Epigenetics and three crucial clinical insights in breast cancer; and by emphasizing on Epigenetic alterations, the triangle including diagnosis, tumor classification/prognosis and treatment were discussed. A profile of involved genes per se, BRCA1, RASSF1A, APC, RAR-β, MGMT, stratifin, E-cadherin have been highlighted which are targets for epigenetic modifications. Furthermore three remarkable receptors including estrogen receptor (ER), PR, and HER2/neu with diverse functions and impacts on breast cancer progression were discussed.

However, it is shown that a profile of complementary molecular and cellular targets cooperate within the epigenetics territory. In this chapter the insight of "*predict and prevent*" is also highlighted, in this regard tracing the DNA methylation patterns in lymphocytes and serum is shown to be remarkable. And finally, the role of miRNA revealed to be important in diagnosis and prognosis. Finally, the Epigenetic based therapeutic aspects include Inhibitor-, HDACi-, Combination-, nutrigenomic based- and miRNA-therapy.

In **Chap. 11,** by laddering through retinoic acid receptor gene *beta* 2 (RARβ2), manner of its cooperation will be partially unmasked and maneuvering in breast cancer management will be achieved. However, Retinoic acid could be considered as a hero in cancer and Retinoic acid receptor gene is just a cozy corner of the cellular territory with dual actions directs other systems, and is governed by an intelligent and interactive biological system. The basic information has been provided at a glance, followed by the characteristics, definitions and mechanisms of retinoids at a glance. Interactions of epigenetics with environmental factors, including nutrition, the role of chemopreventive agents in epigenetic, cancer stem cells, the main target receptors and genes, protein expression, miRNA and the therapeutic insight of *RARβ* are presented. Finally, cancer world faces an enormous challenge in the demethylating agents in two group of individuals including; (1) Those with a promoter methylated *RARβ2* and (2) Those who are predisposed to cancer development. Therefore, detection of the methylated *RARβ2* in primary BC may be useful to consider tumors with a positive responsiveness capacity to RA therapy.

Chapter 12 In this chapter the feasibility of translational approach in Retinoic acid receptor-β is explored at a glance. By considering the methylation status, the importance of genetic factors at a triangle level including DNA, RNA and protein; and the bridging system between functional and clinical insights are challenged. The final step of this journey is accomplished by providing complementary information on environmental factors and a required system as clinical trial. RAR-β as a key gene induces apoptosis, and has influential impact on chemo-prevention that pave the way towards cancer therapy. As it is presented in this chapter, RAR-β will found their way towards cancer prevention and therapy. In this regard examples are provided and by considering tumor progression and metastasis, it seems that histone deacetylase and doxorubicin are capable to target cancer stem cells.

Final message is that life style including diet which has strong link to DNA methylation of RAR-β. So by improving the nutritional guide, the health status would be relatively stabilized.

Chapter 13 deals with Methylation in the Colorectal Cancer. The strong link between the hazard environmental factors and colorectal cancer (CRC) is well known. In this chapter the histological progression, as well as the step by step evolutionary pattern is highlighted. The molecular characteristics and the mode of hypermethylation of involved genes are presented. The authors have emphasized on the key role of epigenetic on our life. They have also noted that molecular/epigenetic based pattern is capable to shape the accurateness of chemotherapy in CRC patients. Importance of CPG methylation, environmental based of DNA methylation was also discussed within the paradigm of CRC therapy related to the facts in the genomic epigenetic modulation and regulation. Furthermore, epigenetic targeting and silencing through channel of cell cycle genes and re-expression were considered as the strength element in anticancer therapeutic innovation. However, differences between the methylation characteristics in normal cells and cancer cells would pave the way toward the novel therapeutic strategies. The final focus in this chapter is the translational paradigm. However, the experimental work in mouse models resulted in the cellular death "due to the epigenetic reversal of silencing of tumor suppressor genes". Moreover, the end of the epigenetic based therapy is more promising in the hematopoietic diseases rather than in solid tumors. It's worth to stress on the fact that CRC like other cancers is a complex disease with full of peculiarities in which the single therapy, even target based would not be fruitful.

Chapter 14 focuses on the epigenetic mechanism of tumorigenesis in Malignant Rhabdoid Tumor (MRT). A fascinating genetic/cellular/and epigenetic network forms a platform for cellular senescence, apoptosis, mitotic control genes, pre-replication complex, chromatin behavior, cell proliferation and differentiation. The author has provided a quadratic model in which the reactivation of the p16INK4a –Rb pathway by the SWI/SNF complex in MRT cells has been considered. This model is reflective of the '***Communicative Epigenetics***'.

Chapter 15 presents Epigenetics of Thyroid Cancer in which the thyroid related genes have been explored. Different tumor suppressor genes are capable of diverse functions. In this regard, different characteristics and challenging items for the specific involved gene in thyroid cancers have been discussed for the following genes:

1. In **RASSF1,** promoter hypermethylation as a negative epigenetic regulator, and *global* methylation.
2. **p16INK4a (*MTS1*)** has preventive behavior on cell progression through the G1 phase and is also involved in cancer development.
3. **PTEN** a master gene is involved in varieties of cancers.
4. **14-3-3σ,** expression of this target, as a G2/M arresting molecule is observed in malignant thyroid tissues including in papillary carcinomas but not in follicular carcinomas and adenomas. These facts reflects the specific pathological based of 14-3-3σ- function.

5. **TSHR** with its multi-functional characteristics is known as a diagnostic marker of malignancy to distinguish FTC from benign adenoma.

Chapter 16 provides an introduction to Bio-Energy and Bio-Resonance technology by linking this paradigm to Genetic and Epigenetic. There is a great challenges about bio-resonance, therefore, this paradigm could be initially explained as the pool of electrical power within the cells and production of biophotons which are stored inside the DNA. This chapter is reflective of new insight on Bio-mechanical resonance within the cells, tissues and molecules. Bio-resonance relies on a triangle pyramid including Biophysicists, electromagnetic oscillations, and body as an interconnected bio-energetic organism. The scenario is partially due to either a combat between Photons and electrons, or their cooperation at different levels including DNA, RNA, ribosomes and mitochondria.

Specifically, this chapter deals with epigenetic modifications and transposans which are connected to "different electron orbital configurations". It was stated that "Transposons direct RNA-mediated DNA epigenetic regulation." The backbone of r-Transposons has been deeply explored in Chap. 3. In addition, an interesting link between environmental hazards, diverse electric oscillation in chromosome and epigenetic pattern of organism has been also highlighted. As the final statement, ***all is about the pool of bio-energic system within the living territories; and the hope depends on 'how we could create an avenue through which diagnosis, prediction and prevention of diseases including cancer would be achievable'.***

Furthermore, at a glance, it was aimed to apply the bridging and complementary manner in order to present diverse insights in this book. The selected target organs in different chapters of this book are reflective of scenarios in which the multi-manner behaviors including diversity and harmony are remarkable; besides each organ is characterized with its own background and developmental history. They are different targets, but they have some characters in common. In fact the human body is reflective of a diverse community; the populations which pave the way, co-operate, support, combat, and play roles in fating and shaping our future health and malady. The organs' platforms provide a discovery opportunity to unmask the facts by composing and harmonizing different cellular and molecular events.

Here we arrive at the end of our journey, our group was capable to feel and translate the scientific and clinical challenges through the great pleasure channel of cooperation, understanding, happiness and patience.

Now, the query is '*Reversibility*' of neoplastic process, this is an '*ART of NATURE*' proposing some Questions; (1) I*s it possible, to treat cancer through Epigenetics* as a sole? And (2) what is the interactive spectrum between the epi-based therapy and micro-/macro- environmental factors? (3) How long this therapeutic strategy will last? (4) What about another round of reversibility of epigenetic alteration? And (5) how often does reversibility occur during the patients' life?

A recent book review entitles Epigenetics consists of fundamental and clinical aspects. Different insights including basic and clinical implication in different diseases were explored. In addition the role of epigenetic in disease development was also discussed. However, information on Cancer Epigenetic was found to be incomplete which included limited discussion on the roles of "aberrant DNA CpG Methylation and histone post-translational modification patterns, plus the role of miRNAs in regulating DNA methyltransferases" (Gray 2014).

Epigenetics, epigenomics and epigenome are meaningfull versions of an extraordinary territory within different species, but there are unmasked facts about ***Evolutionary Epigenetic.*** An interesting report is recently published which has provided a new insight on diversity of methylation map between ancient hominids and present humans (Gokhman et al. 2014). They have also found a significant association between differentially methylated regions and disease. This report provides a new avenue through which ***Medical Epi-anthropology*** and ***Cancer Epi-anthropology*** could be explored.

As a matter of applicability and translatability, the final aim in Cancer Epigenetic reveals to be the appropriate therapy and the cancer world is witness of continuous efforts performed by scientists, clinicians and different health sectors. However, no one is fully satisfied with the outcome. The Epi-based therapeutic aspects are explored within different chapters of this book. In addition, the key facts about E pi-based drugs are recently published (Heerboth et al. 2014).The matter of fact for therapeutic innovation mainly is highlighted to be inhibition of epigenetic alteration. They have discussed different currently epigenetic drugs in use.

Now a crucial question may be proposed as 'what about the impact of other genetic and cellular alterations?' which is a challenging item in cancer therapy. In addition the characteristics of Pharmacogenetic are rather unique in different patients who harbor diverse cellular and molecular alteration. The final message is to consider epigenetic alteration accompanied by the complementary maneuvering through the cellular/molecular platforms to achieve the therapeutic cocktail.

References

Gokhman D, Lavi E, Prüfer K, Fraga MF, Riancho JA, Kelso J, Pääbo S, Meshorer E, Carmel L (2014) Reconstructing the DNA methylation maps of the Neandertal and the Denisovan. Science 344:523–527

Gray SG (2014) Book review: "Epigenetics". Front Genet 5:104. doi:10.3389/fgene.2014.00104 (Lyle A (Ed) Epigenetics. Garland Sci. Taylor & Francis Group, LLC)

Heerboth S, Lapinska K, Snyder N, Leary M, Rollinson S, Sarkar S (2014) Use of epigenetic drugs in disease. An overview. Genet Epigenetics 6:9–19. doi:10.4137/GEG.S12270

Lyon MF (1974) Mechanisms and evolutionary origins of variable X-chromosome activity in mammals. Proc R Soc Lond B 187:243–268

Index

P. Mehdipour (ed.), *Epigenetics Territory and Cancer*,
DOI 10.1007/978-94-017-9639-2

MIX
Papier aus verantwortungsvollen Quellen
Paper from responsible sources
FSC® C105338
FSC
www.fsc.org

If you have any concerns about our products,
you can contact us on
ProductSafety@springernature.com

In case Publisher is established outside the EU,
the EU authorized representative is:
Springer Nature Customer Service Center GmbH
Europaplatz 3, 69115 Heidelberg, Germany

Printed by Libri Plureos GmbH
in Hamburg, Germany